Interhemispheric
Climate Linkages

Interhemispheric Climate Linkages

Edited by

Vera Markgraf

Institute of Arctic and Alpine Research
University of Colorado
Boulder, Colorado

ACADEMIC PRESS

A Harcourt Science and Technology Company

San Diego   San Francisco   New York   Boston   London   Sydney   Tokyo

This book is printed on acid-free paper.

Copyright © 2001 by ACADEMIC PRESS

All Rights Reserved.
No part of this publication may be reproduced or transmitted in any form or by any
means, electronic or mechanical, including photocopy, recording, or any information
storage and retrieval system, without permission in writing from the publisher.

Requests for permission to make copies of any part of the work should be mailed to:
Permissions Department, Harcourt Inc., 6277 Sea Harbor Drive,
Orlando, Florida 32887-6777

Academic Press
A Harcourt Science and Technology Company
525 B Street, Suite 1900, San Diego, California 92101-4495, USA
http://www.academicpress.com

Academic Press
Harcourt Place, 32 Jamestown Road, London NW1 7BY, UK
http://www.academicpress.com

Library of Congress Catalog Card Number: 00-104375

International Standard Book Number: 0-12-472670-4

PRINTED IN THE UNITED STATES OF AMERICA
00 01 02 03 04 05 MM 9 8 7 6 5 4 3 2

Contents

PART

I

PRESENT DAY CLIMATES

CHAPTER

1

Interhemispheric Effects of Interannual and Decadal ENSO-Like Climate Variations on the Americas

MICHAEL D. DETTINGER, DAVID S. BATTISTI,
RENE D. GARREAUD, G. J. MCCABE, JR.,
AND CECILIA M. BITZ

CHAPTER

2

Interannual to Multidecadal Climate Variability and Its Relationship to Global Sea Surface Temperatures

DAVID B. ENFIELD AND ALBERTO M. MESTAS-NUÑEZ

CHAPTER

3

Polar Air Outbreaks in the Americas: Assessments and Impacts During Modern and Past Climates

JOSE A. MARENGO AND JEFFREY C. ROGERS

CHAPTER

4

Globality and Optimality in Climate Field Reconstructions from Proxy Data

M. N. EVANS, A. KAPLAN, M. A. CANE, AND R. VILLALBA

PART

II

HUMAN DIMENSIONS, THE LAST MILLENIUM OF CLIMATE CHANGE

CHAPTER

5

The Effects of Explosive Volcanism on Simple to Complex Societies in Ancient Middle America

PAYSON SHEETS

CHAPTER

10

Decadal-Scale Climatic Variability Along the Extratropical Western Coast of the Americas: Evidence from Tree-Ring Records

R. VILLALBA, R. D. D'ARRIGO, E. R. COOK,
G. C. JACOBY, AND G. WILES

PART

III

LONG-TERM CLIMATE VARIABILITY: Geomorphic Evidence

CHAPTER

11

Glaciation During Marine Isotope Stage 2 in the American Cordillera

CHALMERS CLAPPERTON AND GEOFFREY SELTZER

CHAPTER

12

Late Quaternary Eolian Records of the Americas and Their Paleoclimatic Significance

DANIEL R. MUHS AND MARCELO ZÁRATE

CHAPTER

13

Periods of Wet Climate in Cuba: Evaluation of Expression in Karst of Sierra de San Carlos

JESÚS M. PAJÓN, ISMAEL HERNÁNDEZ,
FERNANDO ORTEGA, AND JORGE MACLE

Lacustrine Evidence

CHAPTER

14

Identifying Paleoenvironmental Change Across South and North America Using High-Resolution Seismic Stratigraphy in Lakes

D. ARIZTEGUI, F. S. ANSELMETTI, G. O. SELTZER,
K. D'AGOSTINO, AND K. KELTS

CHAPTER

15

Holocene Climate Patterns in the Americas Inferred from Paleolimnological Records

SHERILYN C. FRITZ, SARAH E. METCALFE,
AND WALTER DEAN

Vegetation Evidence

CHAPTER

18

Neotropical Savanna Environments in Space and Time: Late Quaternary Interhemispheric Comparisons

HERMANN BEHLING AND HENRY HOOGHIEMSTRA

CHAPTER

19

Holocene Vegetation and Climate Variability in the Americas

ERIC C. GRIMM, SOCORRO LOZANO-GARCÍA,
HERMANN BEHLING, AND VERA MARKGRAF

CHAPTER

20

Late Glacial Vegetation Records in the Americas and Climatic Implications

MARIE-PIERRE LEDRU AND PHILIPPE MOURGUIART

CHAPTER

21

The Midlatitudes of North and South America During the Last Glacial Maximum and Early Holocene: Similar Paleoclimatic Sequences Despite Differing Large-Scale Controls

CATHY WHITLOCK, PATRICK J. BARTLEIN, VERA MARKGRAF, AND ALLAN C. ASHWORTH

CHAPTER

22

Late Glacial Climate Variability and General Circulation Model (GCM) Experiments: An Overview

DOROTHY M. PETEET

CHAPTER

23

Pole-Equator-Pole Paleoclimates of the Americas Integration: Toward the Big Picture

VERA MARKGRAF AND GEOFFREY O. SELTZER

Contributors

Numbers in parentheses indicate the pages on which the authors' contribution begin.

Mark B. Abbott (87) Department of Geosciences, Morrill Science Center, University of Massachusetts, Amherst, Massachusetts 01003-5520

F. S. Anselmetti (227) Department of Geology, Eidgenössische Technische Hochschule, ETH Zentrum, CH-8092 Zürich, Switzerland

D. Ariztegui (227) Department of Geology, Eidgenössische Technische Hochschule, ETH Zentrum, CH-8092 Zürich, Switzerland

Allan C. Ashworth (391) Department of Geosciences, North Dakota State University, Fargo, North Dakota 58105-5517

Patrick Bartlein (391) Department of Geography, University of Oregon, Eugene, Oregon 97403-1251

David S. Battisti (1) Joint Institute for Study of the Atmosphere and the Oceans, University of Washington, Seattle, Washington 98195-1640

Hermann Behling (293, 307, 325) Center for Tropical Marine Ecology, Fahrenheitstrasse 1, D-28359 Bremen, Germany

Michael W. Binford (87) Department of Geography, University of Florida, Gainesville, Florida 32611

Cecilia M. Bitz (1) Quarternary Research Center, University of Washington, Seattle, Washington 98195

Jose A. Boninsegna (141) Laboratorio de Dendrocronología, Instituto Argentino de Nivología y Glaciología—Centro Regional de Investigaciones Cientificos y Tecnicos (IANIGLA-CRICYT), 5500 Mendoza, Argentina

J. Platt Bradbury (265) Institute of Arctic and Alpine Research, University of Colorado, Boulder, Colorado 80309-0450

Mark Brenner (87) Department of Geological Sciences, University of Florida, Gainesville, Florida 32611-2120

M. B. Bush (293) Department of Biological Sciences, Florida Institute of Technology, Melbourne, Florida 32901

M. A. Cane (53) Lamont-Doherty Earth Observatory of Columbia University, Palisades, New York 10964

Isabel Cartajena (105) Departamento de Antropología, Universidad de Chile, Ignacio Carera Pinto 1045, Santiago, Chile

Chalmers Clapperton (173) University of Aberdeen, Aberdeen, AB9 2UF, Scotland

P. A. Colinvaux (293) Marine Biological Laboratory, Woods Hole, Massachusetts 02453

E. R. Cook (155) Tree-Ring Laboratory, Lamont-Doherty Earth Observatory of Columbia University, Palisades, New York 10964

Jason H. Curtis (87) Department of Geological Sciences, University of Florida, Gainesville, Florida 32611

K. D'Agostino (227) Department of Earth Sciences, Syracuse University, Syracuse, New York 13244

R. D. D'Arrigo (155) Tree-Ring Laboratory, Lamont-Doherty Earth Observatory, Columbia University, Palisades, New York 10964

P. E. De Oliveira (293) Instituto de Geosciencias, Universidad de São Paulo, São Paulo, Brazil

Walter Dean (241) U.S. Geological Survey, MS 980, Denver Federal Center, Denver, Colorado 80225

Michael D. Dettinger (1) U.S. Geological Survey, Scripps Institute of Oceanography, La Jolla, California 92093

David B. Enfield (17) National Oceanic and Atmospheric Administration, NOAA/AOML-PHOD, Miami, Florida 33149

M. N. Evans (53) Lamont-Doherty Earth Observatory of Columbia University, Palisades, New York 10964

Sherilyn C. Fritz (241) Department of Geosciences, University of Nebraska, Lincoln, Nebraska 68588

Rene D. Garreaud (1) Joint Institute for Study of the Atmosphere and the Oceans, University of Washington, Seattle, Washington 98195 and Departmento Geofísica, Universidad de Chile, Santiago, Chile

Eric C. Grimm (295, 325) Illinois State Museum, Research and Collections Center, Springfield, Illinois 62703

Martin Grosjean (105, 265) Department of Physical Geography, University of Bern, CH-3012 Bern, Switzerland

S. Haberle (293) Department of Geography, Monash University, Clayton, Victoria 3168, Australia

Ismael Hernández (217) Instituto de Geofisíca y Astronomía, Ministerio de Ciencia, Tecnología y Medio Ambiente (CITMA), La Habana, Cuba

David A. Hodell (87) Department of Geological Sciences, University of Florida, Gainesville, Florida 32611-2120

H. Hooghiemstra (293, 307) Hugo de Vries Laboratory Department of Palynology and Paleo/Actuo-Ecology, University of Amsterdam, NL-1098 SM Amsterdam, The Netherlands

Malcolm K. Hughes (141) Laboratory of Tree-Ring Research, University of Arizona, Tucson, Arizona 85721

G. C. Jacoby (155) Tree-Ring Laboratory, Lamont-Doherty Earth Observatory of Columbia University, Palisades, New York 10964

A. Kaplan (53) Lamont-Doherty Earth Observatory of Columbia University, Palisades, New York 10964

K. Kelts (227) Department of Geology, Eidgenössische Technische Hochschule, ETH Zentrum, CH-8092 Zürich, Switzerland and Limnological Research Center, University of Minnesota, Minneapolis, Minnesota 55455-0219

Marie-Pierre Ledru (293, 371) Institut de Recherche pour le Développement—Universidade de São Paulo, Instituto de Geociências, Dptp Geologia Sedimentar e Ambiental, São Paulo, SP, Brazil

B. W. Leyden (293) Department of Geology, University of South Florida, Tampa, Florida 33620

Socorro Lozano-García (325) Universidad Nacional Autónoma de México, Instituto de Geología, México D.F. 04510 Mexico

B. H. Luckman (119) Department of Geography, The University of Western Ontario, London, Ontario N6A 5C2, Canada

Jorge Macle (217) Archivo Nacional de Cuba, Ministerio de Ciencia, Tecnología y Medio Ambiente (CITMA), La Habana, Cuba

Jose A. Marengo (31) Centro de Previsao de Tempo e Estudos de Clima (CPTEC)/Instituto Nacional de Pesquisas Espaciais (INPE), 12630-000 Cachoira Paulista, São Paulo, Brazil

Vera Markgraf (325, 391, 433) Institute of Arctic and Alpine Research, University of Colorado, Boulder, Colorado 80309-0450

G. J. McCabe, Jr. (1) U.S. Geological Survey, MS972, Denver Federal Center, Lakewood, Colorado 80225

Alberto M. Mestas-Nuñez (17) Cooperative Institute for Marine and Atmospheric Studies, Miami, Florida 33149

Sarah E. Metcalfe (241) Department of Geography, University of Edinburgh, Edinburgh EH8 9XP, United Kingdom

Philippe Mourguiart (371) Institut de Recherche pour le Développement—Université de Pau et des Pays de l'Adour, Dpt d'Ecologie, Campus Universitaire d'Anglet, Unité de Formation et de Recherche (UFR) Sciences et Technologies "Côte Basque", F-64600 Anglet, France

Daniel R. Muhs (183) United States Geological Survey, Denver Federal Center, Denver, Colorado 80225

Lautaro Núñez (105) Instituto de Investigaciones Arqueológicas y Museo, Universidad Católica del Norte, San Pedro de Atacama, Chile

Fernando Ortega (217) Centro de Antropología, Ministerio de Ciencia, Tecnología y Medio Ambiente (CITMA), La Habana, Cuba

Jesús M. Pajón (217) Instituto de Geofisica y Astronomía, Ministerio de Ciencia, Tecnología y Medio Ambiente (CITMA), La Habana, Cuba

Dorothy M. Peteet (417) National Aeronautics and Space Administration/Goddard Space Flight Center, Institute for Space Studies (NASA/GISS), New York, New York 10025 and Lamont-Doherty Earth Observatory of Columbia University, Palisades, New York 10964

Jeffrey C. Rogers (31) Department of Geography, Ohio State University, Columbus, Ohio 43210

Michael F. Rosenmeier (87) Department of Geological Sciences, University of Florida, Gainesville, Florida 32611-2120

M. L. Salgado-Labouriau (293) Departmento de Geosciencias, Universidade de Brasilia, Campus Universitario, Asa Norte, 70910-900 Brasilia, D.F., Brazil

Geoffrey O. Seltzer (173, 277, 433) Department of Earth Sciences, Syracuse University, Syracuse, New York 13244

Payson Sheets (73) Department of Anthropology, University of Colorado, Boulder, Colorado 80309-0233

Scott Stine (265) Department of Geology, California State University, Hayward, California 94542

M. Stute (293) Lamont-Doherty Earth Observatory of Columbia University, Palisades, New York 10964

Florence Sylvestre (265) Laboratoire de Geologie, Université d'Angers, F-49045 Angers, France

R. Villalba (53, 119, 155) Departamiento de Dendrocronología e Historia Ambiental, Instituto Nacional de Glaciología y Nivologia, Centro Regional de Investigaciones Cientificas (IANIGLA-CRICYT), 5500 Mendoza, Argentina

R. Webb (293) National Oceanographic and Atmospheric Administration—Climate Diagnostic Center (NOAA/CDC), Boulder, Colorado 80303

Cathy Whitlock (391) Department of Geography, University of Oregon, Eugene, Oregon 97403

G. Wiles (155) Department of Geology, The College of Wooster, Wooster, Ohio 44691

Marcelo Zárate (183) Instituto Nacional de Glaciología y Nivologia,Centro Regional de Investigaciones Cientificas (IANIGLA-CRICYT), 5500 Mendoza, Argentina

Foreword

This is an exciting time for those studying global change and the paleosciences. After almost a decade of planning and operation, the vision set forth in the early planning documents for the International Geosphere Biosphere Program—Past Global Change Project (IGBP—PAGES) is now bearing fruit. We recognize that global science issues require global-scale science projects and these projects need clear intellectual and administrative frameworks within which the scientific community can organize and focus its efforts on. To this end, the PAGES Project initiated and sponsored a series of Pole-Equator-Pole (PEP) transects as an organizational framework directed towards the ultimate establishment of a global observational network for paleoenvironmental information. The Americas transect, PEP 1, represents one important building block of the global network, the inter-American paleoenvironmental research program designed to address questions about the dynamics of transequatorial atmospheric linkages and to determine the hierarchy of climate control over the last glacial–interglacial transition. PEP 1 is also especially designed to link the marine and terrestrial records along the eastern Pacific coast and to understand the importance of the equatorial trans-Pacific and Amazonian basin in climatic linkages.

It is also the task of the PEP 1 Project to provide the guidance for the paleoscience effort in the Americas, clarifying the most promising research directions, fostering the essential international cooperation, and planning for future research. The community needs to navigate a path through a changing landscape of scientific objectives, community interests, and funding agency requirements. It seems as if the scientific questions grow in complexity as the resources become increasingly limited. However, through the clarity and force of its arguments, and because of the urgency of the global change issue, the PEP 1 Project has succeeded in steering a course through these difficulties. The project is now *mature,* and it is evident that PEP 1 is a solid contributor to understanding paleoenvironmental processes and the temporal framework within which environmental change occurs.

The international paleoscience community has made remarkable strides in advancing our understanding of global change in the Americas. This publication is the latest example of the close collaboration of paleoscientists from countries throughout the Western Hemisphere contributing to the common goal of solving some of the most vexing global environmental issues. In this volume, PEP 1 scientists present the results of their recent efforts. It is the nature of science, however, that this morning's accomplishments provide this afternoon's questions. Therefore, one might consider this volume a *progress report* with the expectation of further advances as we come to better appreciate and clarify the paleo-perspective.

A final word—all the effort, planning, and good intentions would not amount to much without consistent leadership. For PEP 1, that unselfish leadership is provided by Vera Markgraf. Implementation requires leadership, and implementation at the global level requires an extraordinary level of leadership. With good cheer and great energy, Vera has led the community and, when required, has pushed the community. She has fashioned an international scientific contribution in which we may all take pride.

Herman Zimmerman
September 24, 1999
Washington, D.C.

Preface

The Earth's climate has never been stable. Climate has varied on all time scales and will continue to vary in the future, irrespective of the extent to which human activities will affect it. The length of the instrumental climate record, however, is insufficient to fully understand natural climate variability, and the causes of climate change and paleoclimate records need to be assessed. Because no single paleoclimate record adequately represents past global climate variability, regional paleoclimate evidence needs to be assembled and integrated to create the understanding of the larger-scale picture.

The chapters assembled in this volume present for the first time a systematic approach documenting how regional climate and paleoclimate records can be linked and integrated between the northern and southern hemispheres. These chapters are the outcome of a meeting in Merida, Venezuela, March 1998, that brought together the scientific community from the Americas under the Pole-Equator-Pole (PEP 1) initiative, a program of the International Geosphere Biosphere Program—Past Global Change Project (IGBP-PAGES). The chapters review climate and paleoclimatic evidence from the Western Hemisphere from a range of different disciplines and for different time intervals during the late Pleistocene and Holocene and compare the responses of different civilizations to environmental and climate change. In taking an interhemispheric view of past climate variability the behavior and interaction of the larger-scale climate systems are assessed, such as the El Niño/Southern Oscillation, the Inter Tropical Convergence Zone, the monsoons, the westerlies, etc.

The integrated climate and paleoclimate evidence is based on a wide range of different climate proxy records. Depending on the specific type of climate indicator, there are differences in the temporal and spatial resolution as well as in the climate signal interpreted from the data. The temporal resolution ranges from annual, represented by meteorological and tree ring records, to decadal and millennial resolution, represented by vegetation, lake, glacier, eolian, and speleothem records. The spatial resolution of the records is primarily a reflection of the density of the data network, which for some disciplines and specific areas, such as the tropics, continues to hamper the development of the broader picture of climate change. The spatial aspect combined with the fact that every climate proxy senses only a very specific portion of the climate spectrum, present one of the major challenges in the overall integration of present and past climates.

Apart from the scientific results spawned by this effort, the most rewarding aspect of this interhemispheric paleoclimate correlation project was to develop an international community collaborating for a common goal. A number of new collaborative projects have resulted from the increased communication between the members of the community, and the recognition of gaps in data networks has led to enhanced efforts in specific disciplines and regions. One aspect I personally feel especially rewarding is the potential of integrating the paleoclimate efforts with those developed in the realm of archaeology and sociology, addressing the effect as well as the response of civilizations to environmental and climate change. It is hoped that following this attempt many other such integrative studies will be undertaken so that ultimately we will understand the Earth system as a whole.

Vera Markgraf
Boulder, Colorado

Acknowledgments

I wish to thank all the authors contributing to this volume for their dedication in assembling and discussing the wealth of present and past environmental and climate data from the Americas in context of enhancing our understanding of interhemispheric linkages. Every chapter represents a major effort of interdisciplinary and international collaboration and I hope the authors are as pleased as I am with the outcome.

I wish to also thank the reviewers for giving their time and expertise in the process of editing this book. I especially appreciated the painstaking efforts of Ms. Diana Miller, technical editor, for her assistance during all stages of production.

Finally I acknowledge the financial support from the U.S. National Science Foundation through grants ATM-95-26139 and ATM-97-29145. These grants facilitated several national and international meetings that ultimately lead to the completion of this volume.

PRESENT DAY CLIMATES

CHAPTER

1

Interhemispheric Effects of Interannual and Decadal ENSO-Like Climate Variations on the Americas

MICHAEL D. DETTINGER, DAVID S. BATTISTI, RENE D. GARREAUD,
G.J. McCABE, Jr., and CECILIA M. BITZ

Abstract

Interannual El Niño/Southern Oscillation (ENSO) and decadal ENSO-like climate variations of the Pacific basin are important contributors to the year-to-year (and longer) variations of the climate of North and South America. Analysis of historical observations of global sea surface temperatures (SSTs), global 500 mbar pressure surfaces, and Western Hemisphere hydroclimatic variations that are linearly associated with the ENSO-like climate variations yields striking cross-equatorial symmetries as well as qualitative similarities between the climatic expressions of the interannual and decadal processes. These similarities are impressive because, at present, the mechanisms that are believed to drive the two timescales of ENSO-like variability include several candidates that are quite dissimilar in terms of physical processes and locations. Despite potentially different source mechanisms, both interannual and decadal ENSO-like climate variations yield wetter subtropics (when the ENSO-like indices are in positive, El Niño–like phases) and drier midlatitudes and tropics (overall) over the Americas in response to equatorward shifts in westerly winds and storm tracks in both hemispheres. The similarities of their continental surface climate expressions may impede separation of the two ENSO-like processes in paleoclimatic reconstructions. Copyright © 2001 by Academic Press.

Resumen

Variaciones asociadas con El Niño/Oscilación Sur (ENSO) a escala de tiempo de caracter interanual y decadal en el Océano Pacífico, resultan contribudores importantes del clima de Norte y Sur America. Analisis de los registros historicos de la temperatura de superficie del mar (SSTs), del campo global de nivel de la superficie a 500 mbar, y de variaciones hidroclimáticas en el Hemisferio Occidental asociadas en forma lineal con variaciones de origen ENSO, exhiben simetrias marcadas en ambos hemisferios en la mayoria de sus expresiones de variabilidad climatica, en ambas escalas—interanual y decadal.

Estas similaridades son notables, porque, al presente, los mecanismos que se creen responsables por las variaciones climaticas en estas dos escalas de variación temporal, vinculadas con fluctuaciones generadas por el fenomeno ENSO, incluyen varios candidatos, los cuales son asociados con distintos procesos físicos y zonas geográficas de acción. No obstante estos mecanismos potencialmente distintos, variaciones climaticas de escala interanual y decadal resultan en el desarrollo de condiciones mas lluviosas que lo normal en las regiones subtropicales (durante la fase calida del ENSO), y condiciones mas seca que lo normal en las latitudes medias y en los tropicos en general sobre los continentes Americanos. Esto ocurre en respuesta a los cambios de vientos de superficie en la zona ecuatorial del Pacifico, y cambios de las trayectorias de tormentas en ambos hemisferios. La similaridad de las expresiones climaticas en superficies en ambos hemisferios y escalas temporales, si acaso hará mas difícil la separación de las señales de indole paleoclimaticas de estas dos escalas de variación del sistema ENSO.

1.1. INTRODUCTION

A climatologist who specializes in modern climate variations was immediately struck upon attending a workshop of the Pole-Equator-Pole (PEP 1) Paleoclimate of the Americas Program by the need to reconcile the various past-climate reconstructions with climatic processes and conditions that we can recognize from instrumental climate variations. Some past climate variations are recognizable; many are not. Indeed, on interannual and decadal timescales, the number of *typical* patterns of historical climate variation is limited. In the present climate, three climatic phenomena dominate climate variations in the Americas on interannual to decadal timescales. The best known is the El Niño/Southern Oscillation (ENSO) phenomenon, which dominates global climate variations on interannual timescales ranging (mostly) from 3 to 6 years; many climate studies have described this quasi-regular phenomenon of tropical air–sea interactions centered in the equatorial Pacific and its teleconnections to many parts of the globe (see reviews in Diaz and Markgraf, 1992; Allan et al., 1996; Wallace et al., 1997). The decadal (10- to 50-year timescales) climate processes centered in the Pacific and Atlantic basins are less well known. On timescales longer than interannual, the dominant climate phenomenon in the Pacific Ocean overlying atmosphere has an ENSO-like spatial distribution (Zhang et al., 1997; hereafter ZWB) of surface temperatures and atmospheric circulations and, as will be shown here, interhemispheric climate effects on the Americas. The physical processes responsible for this decadal ENSO-like variability remain uncertain, but are tied to well-documented Pan-Pacific changes in the atmosphere and ocean in the 1950s and mid-1970s (Ebbesmeyer et al., 1991). Finally, decadal variability in the climate of the Atlantic basin also has a basinwide signature (Deser and Blackmon, 1993; Hurrell, 1995; Enfield and Mestas-Nuñez, 2000), but it is somewhat distant from the cordillera of the Americas and thus will not be addressed here.

In this chapter, modern interannual and decadal climate variations of the Pacific basin will be outlined, and their interhemispheric climatic effects on the Pacific Ocean and the Americas will be compared. Several

indices may be used to describe the historical variations of the ENSO and decadal ENSO-like phenomena, but it is often difficult to be sure that the two phenomena have been separated so that none of the climatic effects are attributed twice. Thus, a new set of uncorrelated indices for the two processes will be developed and used, alongside a pair of indices that have been used in the literature, to characterize oceanic, atmospheric, and land surface climate variations associated with the ENSO-like phenomena. The construction of indices, and data sets that will be compared to those indices, will be described in Section 1.2. In subsequent sections, oceanic, atmospheric, and continental conditions that are linearly related to the interannual and decadal phenomena will be delineated by simple correlation and regression analyses. In the final two sections, the results will be discussed and conclusions will be summarized.

1.2. DATA AND DEFINITIONS

Interannual and decadal climate variations in the Pacific basin can be traced through literally dozens of physical and biological indices (e.g., Ebbesmeyer et al., 1991). In this chapter, two sets of indices based on sea surface temperature (SST) patterns of global scale will be used to depict the North Pacific climate and its effects on the Americas. Two indices developed by ZWB—the Cool Tongue (CT) index and the Global Residual (GR) index—will be discussed for the most part. However, for completeness, a second pair of indices based on a linear combination of some more commonly available time series—the Southern Oscillation Index (SOI), the CT index, and the Pacific (inter)Decadal Oscillation (PDO) index—also will be considered. The origins and significance of these various indices are discussed later.

Global SSTs were analyzed by ZWB using linear regressions and principal component (PC) analyses to separate interannual ENSO-related SST variations from other interannual and decadal variations and to describe those other SST variations. Interannual ENSO-related SST variations were characterized by the CT index, which is the average of SST anomalies from 6°N to 6°S, 180° to 90°W, an index previously used by Deser and Wallace (1990). This region is the locus of cool SSTs in the eastern equatorial Pacific, where the El Niño process has its clearest surface expression. Monthly mean SST anomalies on a 5° × 5° grid, from an updated version of the global United Kingdom Meteorological Office Historical Sea Surface Temperature Dataset (HSSTD; Folland and Parker, 1990, 1995), were used to form this series (Fig. 1a), and the CT series was high-pass filtered to remove variations with periods longer

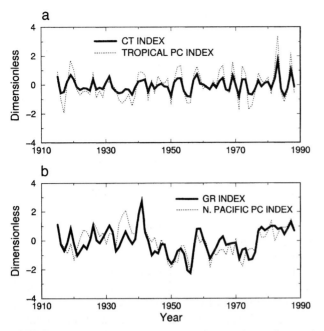

FIGURE 1 (a) Unfiltered Cool Tongue (CT) index and rotated "tropical" Principal Component (PC) index of dominantly interannual El Niño/Southern Oscillation (ENSO) variability in the tropical Pacific, and (b) Global Residual (GR) index and rotated "North Pacific" PC index of dominantly decadal ENSO-like variability in the Pacific basin. The indices are described in the text. The CT and GR indices are identical to those of Zhang et al. (1997).

than 6 years, forming a strictly interannual series called CT*. Then, the entire SST data set was regressed against the CT* time series, and the best linear fits at each grid point were subtracted from the grid-point SSTs to arrive at the time series of residuals. By construction, these residuals are uncorrelated with the CT* series. A PC analysis of these residual series yielded a leading mode of "non-ENSO" SST variation that was, surprisingly, very ENSO-like in its spatial patterns. This leading mode is called the GR index by ZWB and represents the decadal ENSO-like variations of the global and, especially, the Pacific SSTs (Fig. 1b). Although CT* and GR are not correlated, CT and GR are ($r = -0.55$). In this chapter, the interannual and decadal climate variations in the Pacific basin will be characterized in terms of these same CT and GR series; CT, rather than CT*, is used because it is more similar to such commonly used ENSO indices as the unfiltered SST indices along the equatorial Pacific in so-called Niño-3 and Niño-4 regions (e.g., Allan et al. 1996) than the band-pass-filtered CT*.

The differing timescales of the CT and GR series can be demonstrated by spectral analysis. Shown in Fig. 2 are multitaper power spectra (Percival and Walden, 1993; Lees and Park, 1995) of the two series. The CT

spectrum (solid curve) indicates variance spanning a broad range of frequencies centered near 0.2 cycles/year, corresponding to a period of 5 years. In contrast, GR has a spectrum that increases rather steadily from high to low frequencies, with a local maximum near 0.16 cycles/year. Thus, GR is dominated by low-frequency variations with timescales between about 6 years and several decades; CT is dominated by interannual variations in the 3- to 7-year range. Recall that the CT series used here has not been temporally filtered; therefore, the interannual dominance of CT shown reflects the true mix of frequencies in the tropical Pacific ENSO variations. The GR index is derived as a residual from an interannually filtered CT* series (and its reflection in global SSTs), and thus, as a measurement of North Pacific climate variation, it is biased by design toward the more decadal parts of Pacific climate variability.

Alternative approaches and indices for characterizing the interannual and decadal modes of Pacific climate variation have been developed in the literature. Enfield and Mestas-Nuñez (Chapter 2) used complex empirical orthogonal function (EOF) analyses of global SSTs to arrive at similar modes. Mantua et al. (1997) used the CT index and a Pacific Decadal Oscillation (PDO) index to describe essentially the same variability. The PDO index is defined by Mantua et al. (1997) as the first PC of monthly SSTs, from the HSSTD, poleward of 20°N in the Pacific basin, initially calculated from anomalies for the period 1900–93. This index has

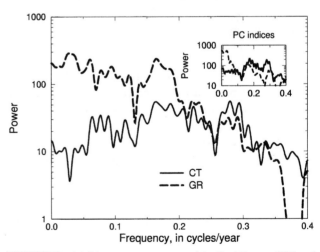

FIGURE 2 Multitaper power spectra of the Cool Tongue (CT) and Global Residual (GR) climate indices. The inset is a plot of the corresponding power spectra of rotated Principal Component (PC) series: "tropical" PCs (solid curve) and "North Pacific" PCs (dashed curve). In the CT and GR spectra, at the lowest frequencies, powers > 80–200 exceed 95% confidence levels—not shown here—in comparison to red noise (Mann and Lees 1996); at the highest frequencies, powers > 30 reach 95% confidence levels.

a strong decadal character, much like the GR index (and is correlated with GR at $r = +0.74$), but has not had any ENSO variability explicitly removed; indeed, its correlation with the CT series is $r = +0.38$. The PDO was intended to reflect the decadal variability of the North Pacific in a simple index, but it has a substantial expression throughout the tropical and South Pacific (see Fig. 3d in this chapter or Fig. 2a of Mantua et al., 1997).

Tropical ENSO variability historically—e.g., Troup (1965) and most more recent references—has been indexed by the SOI and SST anomalies in selected equatorial regions. The SOI is usually defined as the difference between standardized monthly departures of sea level pressure (SLP) from their normal seasonal cycles at Tahiti and Darwin, Australia. When SLPs are low in the eastern tropical Pacific near Tahiti and high in the western tropical Pacific near Darwin (when SOI is negative), the tropical easterly winds become weak and the resulting slackening of equatorial upwelling of cool deep water in the eastern equatorial Pacific, together with a relaxation of warm waters from the western Pacific into the usually cool eastern Pacific, leads to warmer than normal SSTs in the eastern equatorial Pacific. These episodes of warm equatorial SSTs in the central and eastern Pacific have far-reaching climatic consequences and have traditionally been called El Niños. When the SLP anomalies are reversed (high at Tahiti and low at Darwin) and the SOI has a positive value, the tropical easterlies strengthen and the eastern equatorial Pacific cools anomalously in episodes that recently have been called La Niñas, to emphasize their contrasts with El Niños. Thus, the SOI index is an atmospheric measure of ENSO processes that complements, but largely reflects, the CT index, with which it is highly anticorrelated ($r = -0.73$). The PDO index is, in part, due to ENSO, as is indicated by the correlation between the PDO and SOI indices ($r = -0.51$).

Although the CT and GR indices of ZWB will be the focus of much of the remaining discussion, it was also useful to develop yet another pair of indices for interannual and decadal ENSO-like variations, a pair of indices that—by design—are entirely uncorrelated with each other but which are not derived from any explicit temporal filtering. In order to develop these new indices, three commonly used climate indices—SOI, CT, and PDO—were joined in a simple trivariate PC analysis. The two most influential of the resulting components were weighted heavily on (1) SOI and CT, the tropical indices, and (2) PDO, the extratropical index, respectively. The benefit of this PC analysis was that, by construction, it yielded statistically uncorrelated series that described the tropical and extratropical variations without arbitrary filtering of either. The two components were then rotated by the Varimax methodology

(Richman 1986), although this served only to reduce the amount of mixing between PDO or the tropical indices in the PCs (nudging the loadings toward even more one-sided weighting of one or the other set in each case). The two resulting, still uncorrelated, rotated PC series (dotted curves, Figs. 1a and 1b) will be called the tropical PC and the North Pacific PC, respectively, although it is clear from later results that the *North Pacific* PC reflects pan-Pacific climate variations that may yet have their roots in the tropics. The weights of SOI, CT, and PDO in each PC are given in Table 1. When the more commonly used Niño-3 SST average, from 5°S to 5°N, 150° to °W, is substituted for CT in this calculation, the amount of mixing between indices that is required to arrive at uncorrelated tropical and North Pacific climate indices is even smaller.

The power spectra of the resulting PC series are shown in Fig. 2 (inset) and are similar in character to the spectra of CT and GR, with the tropical PC naturally echoing CT's interannual character and the North Pacific PC following GR's strong decadal dominance. Although no temporal filtering was applied when these PCs were derived, the North Pacific PC has narrower spectral peaks than does the GR series, concentrated near the periods > 25 years and near 7 years, and the tropical PC has power concentrated in a more clearly defined interannual frequency range—0.16–0.3 cycles/year (3–7 years)—than does CT. Indeed, the tropical PC is virtually identical to ZWB's CT*, the CT index filtered to remove variability longer than 6 years.

The oceanic and atmospheric patterns that accompany the interannual ENSO and the ENSO-like climate variations indexed by the CT and GR series and by the tropical and North Pacific PCs will be described in sections 1.3 and 1.4 by using regressions of various atmospheric fields, and correlations of SST fields, with the series. In Section 1.5, the climatic consequences of this variability at the surface of the Americas will be described by similar methods. SST variations are depicted from the HSSTD set used in developing the CT and GR indices. Atmospheric forms of the variations are depicted in terms of the recently released reanalyzed global atmospheric fields of 500-mbar height

TABLE 1   **Principal Component (PC) Loadings for Varimax-Rotated Factors of October–September, 1915–88, Average Indices of Pacific Climate Variability**

Index	Tropical PC	North Pacific PC
SOI	−0.85	−0.38
CT	+0.95	+0.10
PDO	+0.21	+0.97

anomalies from 1958–1996 on a 2.5° × 2.5° grid (Kalnay et al., 1996) from the National Centers for Environmental Prediction and the National Center for Atmospheric Research (NCAR). The reanalyzed fields provide global coverage, but are of questionable reliability at high latitudes where observations are particularly sparse and can be problematic along major coastlines where there are particularly steep gradients in the density of observations; overall, however, the fields provide the best available global depiction of atmospheric conditions in the last 40 years. Also considered is a transformation of the 500-mbar fields that represents monthly variations in high-frequency storminess. As is described by Bitz and Battisti (1999), daily 500-mbar height anomalies at each grid point were band-pass filtered to isolate variability with timescales of between 2 and 8 days; the monthly root mean squares of these high-frequency fluctuations were then used to map changes in monthly mean storminess. For the Northern Hemisphere, these analyses will be extended back to the first half of the twentieth century by applying the same analysis of high-frequency variations to 5° × 5°-gridded daily Northern Hemisphere SLP fields from NCAR (Trenberth and Paolino, 1980).

The continental effects of the ENSO-like climate variations are depicted in terms of surface air temperatures, precipitation, and streamflow in North and South America. Temperatures from an updated version of the monthly, 5° × 5°-gridded temperature anomaly set of Jones et al. (1986a,b; Carbon Dioxide Information Analysis [CDIAC] at Oak Ridge National Laboratory; Product NDP020), from 1904–90, and land-precipitation anomalies on a similar grid from Eischeid et al. (1991, 1995), for the same time period, are analyzed. Attendant streamflow variations are analyzed in the monthly records cataloged in a Western Hemisphere subset of the global streamflow set compiled by Dettinger and Diaz (2000); this subset includes long-term monthly streamflow records for 248 sites in the Americas. The series were drawn from among the U.S. Geological Survey's Hydroclimatic Data Network for United States streamflow (Slack and Landwehr 1992; updated subsequently to 1995 by L. Riddle, Scripps Institution of Oceanography, La Jolla, CA) and from several public domain international sources; as much as possible, they are free from overwhelming human influences. Also included in the present analyses are 13 proprietary Brazilian flow series provided by Jose Marengo, Instituto Nacional de Pesquisas Espacias, courtesy of Eletrobras and Eletronorte in Brazil.

The results presented here are comparisons between climate indices, SSTs, atmospheric circulation patterns, precipitation, surface air temperatures, and streamflow anomalies averaged over "water years" that begin Oc-

tober 1 and end September 30. The decision to analyze the (northern water year) October–September averages is arbitrary because seasons reverse across the equator and thus are not strictly comparable. No single definition of annual averages will be best suited everywhere; however, Dettinger et al. (2000) have shown that persistence of interannual teleconnections across the Americas is sufficiently long to allow useful interpretation of climatic and hydrologic responses within such a water year. Decadal variations also are expected to yield slowly varying teleconnections that can be studied within this water year.

1.3. SEA SURFACE TEMPERATURE VARIATIONS

Both pairs of climate indices used to characterize interannual and decadal ENSO-like variability—CT/GR and the tropical/North Pacific PCs—are derived from SSTs. Thus, initially, the global SST variations associated with the indices will be outlined. The patterns of SST associated with the CT and GR series are indicated by correlations mapped in Figs. 3a and 3b; positive correlations indicate regions where the ocean is, on average, warmer when the index is more positive and cooler when the index is more negative. As can be expected from the preceding descriptions of the indices, the CT index is more (positively) correlated than the GR index with SSTs in the tropical Pacific. Conversely, the GR index is more (negatively) correlated with SSTs of the North Pacific. Extratropical SSTs in the southern Pacific are about equally correlated to the two indices. Both indices are modestly correlated with SSTs in the tropical Indian Ocean. Despite the differences in emphasis between the CT and GR correlations, the two most remarkable aspects of the SST patterns associated with the two indices are (1) the symmetry of each correlation pattern about the equator in the Pacific basin and (2) the overall similarity between the CT and GR patterns. This similarity led ZWB to describe the decadal variations that dominate the GR index as *decadal ENSO-like variability* and motivated this study. The strong cross-equatorial symmetries of the SST patterns are reflected in strong symmetries of atmospheric circulation patterns and hydroclimatic responses analyzed in subsequent sections of this chapter.

The rotated tropical and North Pacific PC indices are correlated to SSTs in patterns (Figs. 3c and 3d) that are similar to CT and GR, respectively. However, SST correlations with the tropical PCs emphasize the tropical Pacific SSTs (relative to extratropical SSTs) even more than the CT SST correlations do. SST correlations with the North Pacific PCs emphasize the North Pacif-

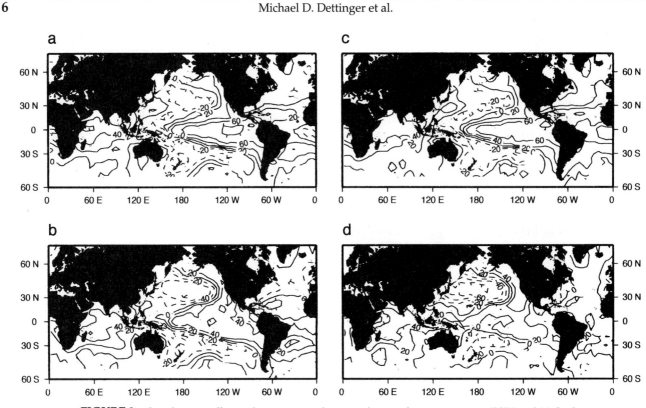

FIGURE 3 Correlation coefficients between annual-averaged sea surface temperatures (SSTs) and (a) Cool Tongue (CT) index (1903–90), (b) Global Residual (GR) index (1903–90), (c) rotated tropical Principal Component (PC) index (1914–90), and (d) North Pacific PC index (1914–90). The contour interval is 0.2, dashed where negative. Correlations $> +0.2$ or < -0.2 pass a two-tail Student t-test of being different from zero at 95% significance levels.

ic temperatures even more than the GRs do. Once again, southern SSTs are about equally correlated with the two PC series.

Thus, the two pairs of Pacific climate indices correspond to similar patterns of SST variation. The CT corresponds largely to ENSO variations by its very definition; the tropical PC emphasizes those variations as an outgrowth of the parsimony provided by all PC analyses. As representations of ENSO variations in the tropical Pacific, the CT and tropical PC series have dominantly interannual timescales (Fig. 2). The GR and, even more so, the North Pacific PC series capture variations that are not correlated with CT variations, are dominantly decadal (although neither has been limited to that timescale; Fig. 2), and are similar to SST patterns in the world oceans associated with ENSO. How the interannual and decadal climate indices can be related to such similar SST patterns, while having such different (and, in the case of the PCs, entirely uncorrelated) temporal variations, is something of a mystery, but these similarities will be echoed in the atmospheric expressions of the time series as well as in their hydroclimatic expressions over the Americas.

Miller and Schneider (1998) recently listed six cate-

gories of mechanisms that may contribute to the decadal timescales of SST variations in the North Pacific:

• Stochastic atmospheric forcing with a low-frequency SST response (e.g., Barsugli and Battisti 1998)
• Decadal tropical forcing of midlatitude SSTs (e.g., Trenberth, 1990; Graham, 1994)
• Decadal midlatitude ocean–atmosphere interactions of ocean gyre strength and wind stress curls (e.g., Latif and Barnett, 1994; and as a North Atlantic analog, Deser and Blackmon, 1993)
• Tropical–extratropical interactions through subduction of midlatitude (atmospherically forced) SST anomalies into the subsurface ocean to provide source waters of upwelling in the El Niño region of the tropical Pacific (e.g., Gu and Philander, 1997)
• Slow oceanic wave teleconnections from the tropics to the extratropical oceans (e.g., Jacobs et al., 1994)
• Intrinsic decadal vacillation of midlatitude ocean currents (e.g., Jiang et al., 1995)

To this list, we would add the slow vestiges of irregular interannual ocean atmosphere climate variations. Aperiodic variations on interannual scales will yield decadal components upon averaging to decadal time-scales, and those decadal components will necessarily be similar in appearance to the interannual climate processes. Some combination of these mechanisms, then, presumably generates decadal Pacific SST variations that, in turn, influence and even drive parts of the global climate system with clear, if irregular, decadal timescales.

Notably, the present analysis demonstrates remarkable interhemispheric symmetries in expressions of the decadal climate variations associated with the North Pacific modes. This interhemispheric symmetry may provide clues as to which of the processes listed previously is most likely to be dominant, e.g., those that would be expected to yield strong symmetries about the equator. Resolution of the question of which mechanisms dominate is, however, beyond the scope of the present analysis and will require a combination of ocean–atmosphere modeling with careful analyses of the details of global atmospheric and oceanic observations over the last 50 years.

1.4. ATMOSPHERIC ASSOCIATIONS

Superposed on these SST patterns, and interacting with them, are the atmospheric counterparts (and consequences) of the CT and GR SST variations. Regression coefficients relating 500-mbar height anomalies (deviations of the heights of 500 mbar pressure surfaces in the atmosphere from their long-term seasonal averages) to the CT and GR series are shown in Figs. 4a and 4c. Note that regression coefficients are related to correlation coefficients by

$$\beta = r\sigma_{500} / \sigma_{clim},$$

where β is a regression coefficient, r is the corresponding correlation coefficient, σ_{500} is the local standard deviation of 500-mbar height anomalies, and σ_{clim} is the standard deviation of the climate index. Hence, these maps can be interpreted as indicating the typical atmospheric anomaly that accompanies a modest excursion (1 σ) of the climate index. Negative coefficients relate the 500-mbar height anomalies and CT over vast

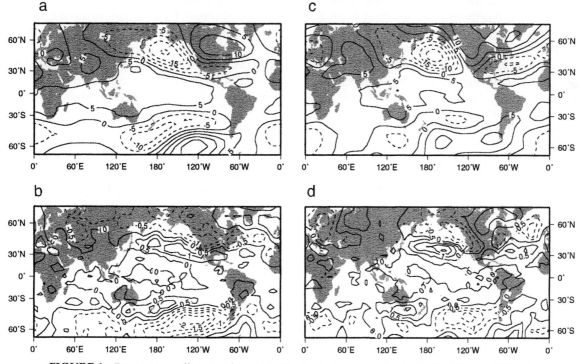

FIGURE 4 Regression coefficients relating (a) 500-mbar height anomalies to the Cool Tongue (CT) index, (b) 500-mbar storminess to the CT index, (c) 500-mbar height anomalies to the Global Residual (GR) index, and (d) 500-mbar storminess to the GR index. The contour intervals in (a) and (c) are 5 m per climate-index standard deviation, and in (b) and (d) they are 0.5 m per climate-index standard deviation. Storminess in panels (b) and (d) is measured by the monthly standard deviation of band-pass-filtered (3–10 days) 500-mbar height anomalies.

Siberian and North Pacific regions, whereas a north-west-southeast slanted dipole of positive and negative coefficients is indicated over the South Pacific (Fig. 4a). The negative coefficients correspond to lower than normal 500-mbar heights when CT is positive (El Niños). The broad region of negative coefficients over the North Pacific corresponds to an average 500-mbar height decline of as much as 20 m for a 1 σ rise in CT. Positive CTs (El Niños) are associated with deeper than normal Aleutian Lows, and the resulting steep north-south coefficient gradients near 35°N, 160°W in Fig. 4a correspond to anomalously strong 500-mbar height gradients, and anomalously strong westerly winds, when CT is positive. Thus, anomalies in atmospheric circulation patterns associated with positive CT values (El Niños) bring storms and precipitation to the southwestern and southeastern parts of North America. The circulation pattern associated with positive CTs also results in enhanced midlatitude blocking and the diversion of storms away from northwestern North America. The effects of these circulation changes on North American climate and hydrology will be discussed in Section 1.5.

The dipole of positive and negative coefficients over the southeast Pacific in Fig. 4b corresponds to diversions of low-pressure systems and attendant storms toward the subtropical belt of South America when CT is positive (El Niños), in a rough parallel to the changes in circulation in the Northern Hemisphere. Observations of 500-mbar heights in the far south are rare enough, however, that the coefficients may reflect dynamic consistency requirements of the reanalysis process (Kalnay et al., 1996) more than an observed feature of the atmosphere. Anomalously strong westerly winds are associated with positive CT along a belt extending from the Southern Ocean near the international date line to the southern Archipiélago de los Chonos of Chile. These patterns have been described as the Pacific South American pattern, a primary mode of Southern Hemisphere atmospheric variability (Mo and Ghil, 1987; Kidson, 1988; Mo and Higgins, 1998; Garreaud and Battisti, 1999).

Together, the low-pressure bands (of negative coefficients) associated with positive CT in the Northern and Southern Hemispheres reflect the extratropical extensions of the SOI seesaw of pressures. The Southern Oscillation pattern is not well represented at 500 mbars in the tropics (Fig. 4a), but along 30°N and 30°S it appears as the equatorward increase in coefficients (toward positive values in the tropics). In the tropical eastern Pacific, the increase in height at 500 mbars is mirrored as a decrease in pressure in the lower troposphere and a slackening of easterly trade winds during El Niños.

The relations between CT and midlevel winds and low-pressure systems are inferred from regression coefficients mapped in Fig. 4a. Corresponding relations between CT and storminess can be inferred from regression coefficients relating monthly standard deviations of high-frequency 500-mbar height anomalies to CT, as in Fig. 4b. Positive values in Fig. 4b correspond to regions in which 500-mbar heights are more variable, on 3- to 10-day timescales, than normal when CT is positive; regions with negative values are less variable and less stormy. Positive CTs (El Niños) are associated with less than normal storminess across Siberia, Alaska, northwestern Canada, and the northwestern United States. By this measure, the southwestern United States is more stormy than normal during El Niños, but the eastern Pacific storm track, near 35°N, 160°W, is clearly enhanced, is extended farther eastward over the southern United States, and is displaced farther south than normal. Storminess may also be enhanced during El Niños over northeastern Canada. A similar analysis (not shown here) of monthly standard deviations of high-frequency SLP variability in the Northern Hemisphere (where daily SLP records span the entire twentieth century) indicates that the associations of SLP-based storminess with CT closely parallel those in the northern half of Fig. 4b both before and after the 1950s (the period encompassing the bulk of available 500-mbar height observations).

In the Southern Hemisphere, positive CTs (El Niños) are associated with greater storminess over the central Pacific along the band from 30°–50°S. Between South America and the date line, storminess appears to be diverted northward from the midlatitudes (40°–60°S) toward about 30°S. Less storminess is indicated over southern Argentina.

The atmospheric expressions of GR are broadly similar to CT's influences, but—as with SSTs—with different spatial emphases. In Fig. 4c, notice that regression coefficients relating 500-mbar height anomalies to GR are dominated by negative coefficients over the North Pacific and a weaker north-south dipole of negative and positive coefficients over the South Pacific. The negative coefficients over the North Pacific are very similar to coefficients associated with CT. However, the less similar circulation anomalies over North America indicate that the circulations associated with CT are more zonal than those associated with GR. In the Southern Hemisphere, the dipole of coefficients is much less intense, implying relatively less expression of GR than CT in Southern Hemisphere 500-mbar heights and winds.

Storminess patterns associated with GR loosely parallel those associated with CT, but variations of GR (Fig. 4d) seem to yield storminess anomalies that are more focused over the North Pacific and the Pacific Northwest, with a much weaker expression over the southern United States than is associated with CT. In the midlatitude Pacific of both hemispheres, the relationship between storminess anomalies and the average 500-mbar height anomalies during ENSO variations is similar to the relationship between storminess anomalies and the average 500-mbar heights during the decadal ENSO-like variability: a decrease (increase) in storminess is found on the poleward (equatorward) flank of the negative midlatitude height anomaly in the central Pacific (e.g., on the flanks of the Aleutian Low) when the indices are positive. The Northern Hemisphere storminess patterns associated with GR (Fig. 4d) also are reflected in correlations of pre-1950s GR values with SLP-based storminess indices from that earlier period (not shown).

Taken together, the panels of Fig. 4 suggest qualitative similarities in the atmospheric expression of interannual ENSO variability (CT) and decadal ENSO-like variability (GR). These atmospheric similarities, and the SST similarities discussed previously, will be related to their similar hydroclimatic effects in the Americas in Section 1.5. Overall, the atmospheric expressions of both CT and GR (as well as tropical and North Pacific PC indices—not shown) are remarkably symmetric about the equator, especially over the Pacific Ocean: positive CT and positive GR variations are associated with equatorward diversions of westerlies, low-pressure systems, and storms from the midlatitude Pacific basin toward subtropical latitudes. These cross-equatorial symmetries may reflect tropical roots for both CT and GR (and for the tropical and North Pacific PCs). These strong interhemispheric symmetries of ENSO-like variations of Pacific climate are likely to produce strong interhemispheric climate variations along the entire length of the North and South American cordillera on both interannual and decadal timescales.

1.5. CLIMATIC EFFECTS IN THE AMERICAS

Most of the paleoclimatic records of climate variability studied as parts of PEP 1 are proxies for either near-surface temperature variations or precipitation/soil moisture variations over the Americas. The interannual and decadal ENSO-like climate variations depicted previously play important roles in determining those variations—roles to be delineated in this section.

Annual-scale temperature, precipitation, and streamflow variations associated with CT and GR are illustrated in Fig. 5, which shows regression coefficients relating deviations of October–September averages of precipitation and temperatures from normal to CT

(Figs. 5a and 5c) and GR (Figs. 5d and 5f), respectively. Also shown are correlation coefficients between streamflow and CT (Fig. 5b) and GR (Fig. 5e). Correlations are indicated by circles with radii proportional to the coefficients. Sizing each circle with a radius proportional to the correlation (and regression) coefficient makes the area of the circle proportional to the fraction of variance that is associated with the climate index at each site.

1.5.1. Precipitation and Streamflow

Strong spatial similarities between CT- and GR-forced patterns of precipitation are immediately obvious from a comparison of Figs. 5a and 5d. Streamflow variations with CT and GR also are similar (Figs. 5b and 5e) and mostly reflect, and even amplify (Dettinger et al. 2000), basin-scale precipitation variations. Both CT and GR are positively correlated with precipitation and streamflow in the southwestern United States, Central America, Paraguay, Uruguay, and northern Argentina. Both CT and GR are negatively correlated with precipitation and streamflow in much of northwestern North America, in the tropical parts of South America, east of the Andes, and possibly—although data are particularly sparse for this area—in southernmost South America. Composite averages (not shown here) of precipitation and streamflow during negative and positive phases of CT and GR indicate that drier than normal conditions in tropical South America and in the northwestern United States are more extreme during positive phases of both CT and GR and contribute more to the regression relations shown in Fig. 5 than do wet conditions during negative CTs and GRs. Positive values of both CT and GR also contribute more (wetness) to the regression relations for the southwestern United States and Paraguay–Uruguay–Argentina than do negative phases. Negative phases may be more important contributors to the regression relations in the northeastern United States. Dettinger et al. (1998) and Cayan et al. (1998) also found clear similarities between ENSO and decadal precipitation patterns over western North America.

In the midlatitudes of the Northern Hemisphere, the precipitation and streamflow patterns shown in Fig. 5 reflect the equatorward shifts in midlatitude westerlies and storm tracks associated with positive values of CT and GR that were indicated in Fig. 4 and discussed previously. When CT is positive, more storms, precipitation, and streamflow occur in the southern United States and, across the equator, in Paraguay–Uruguay; the northwestern United States and much of western Canada experience less stormy, drier, and warmer conditions.

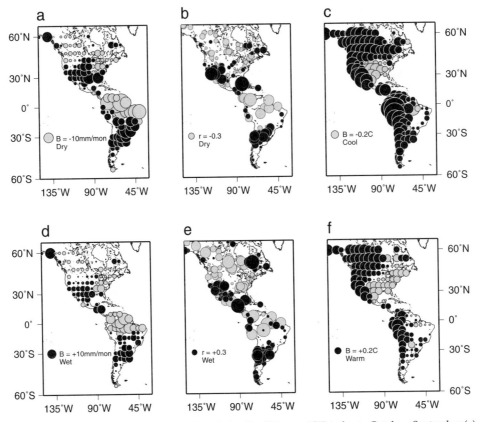

FIGURE 5 Regression coefficients—B—relating Cool Tongue (CT) index to October–September (a) precipitation, 1904–90, and (c) surface air temperatures, 1904–90; and correlation coefficients—r—between CT and (b) streamflows, with periods of record ranging as long as 1904–90 but commonly on the order of 40 years. Panels (d)–(f) are the same as (a)–(c) except for Global Residual (GR) index. Radii of circles are proportional to the magnitude of regression and correlation coefficients: black for positive relations, white for negative. Circles inset near 30°S, 135°W indicate scale of influences: B is the regression coefficient and r is the correlation coefficient (r = 0.3 is significantly different from zero at the 95% confidence level for the average length of streamflow records).

The patterns of precipitation, streamflow, SST, and circulation anomalies over the Americas associated with CT and GR are remarkably similar. However, there are notable differences in spatial emphasis. Precipitation and streamflow anomalies in the tropical parts of South America and in the subtropical bands across North and South America are larger and more consistent (spatially) in the CT responses than in the GR responses (Figs. 5a and 5b and Figs. 5d and 5e). Precipitation and, especially, streamflow responses to GR are larger in northwestern North America, México, and Central America than are the corresponding CT responses. Again, we are struck by the remarkable interhemispheric symmetries of ENSO-like climate variations in the western Americas.

As was expected from their close ties to CT and GR, regression coefficients relating the rotated PC series (tropical PC and North Pacific PC) to precipitation and

streamflow (Figs. 6a and 6b and 6d and 6e) are similar to the corresponding patterns for CT and GR. The precipitation and streamflow patterns associated with the two PC series are also remarkably similar to each other, even though the two PCs are uncorrelated time series; the pattern correlation between the precipitation coefficients in Figs. 6a and 6d is +0.41, and the pattern correlation between streamflow correlations in Figs. 6b and 6e is +0.42. Furthermore, three-fourths of the correlations have the same signs in Figs. 6b and 6e. As with CT and GR, positive tropical PC and North Pacific PC excursions are associated with wetter than normal southwestern United States and Paraguay–Uruguay–Patagonia and drier than normal tropical South America, northwestern North America, and southernmost South America. As with CT and GR, the decadal North Pacific PC is more strongly expressed in northwestern North American precipitation and streamflow varia-

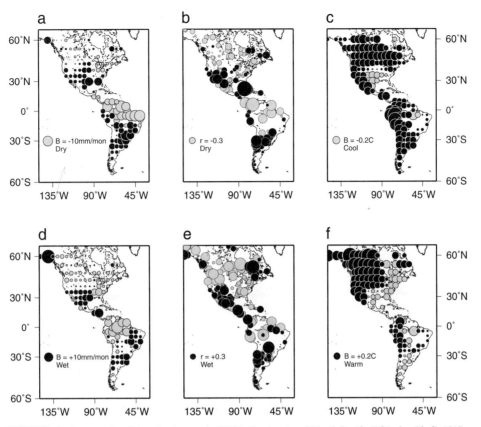

FIGURE 6   Same as Fig. 5, but for the tropical PC index (a–c) and North Pacific PC index (d–f), 1915–88, respectively.

tions (and perhaps in southernmost South America) than is the tropical PC, whereas precipitation and streamflow in most other regions more strongly express the tropical PC series. These parallels and connections can also be identified by computing the leading PCs of streamflow and precipitation in the Americas and then correlating the variations of those leading modes with global SSTs (Dettinger et al., 2000) to obtain SST correlation patterns that closely resemble the combined ENSO and decadal ENSO-like SST patterns studied here.

1.5.2. Surface Air Temperature

Positive CT and GR indices are associated with warmer than normal surface conditions in much of the Americas. The patterns indicated for CT temperatures (Fig. 5c) and GR temperatures (Fig. 5f), however, are not as similar as the precipitation patterns discussed previously. Positive CTs (El Niños) are associated with warm temperatures in Canada, the western United States, Central America, much of tropical South America (on both sides of the Andes), Paraguay–Uruguay–

Patagonia, and southeasternmost Brazil (see also Diaz and Kiladis 1992). El Niños bring cool temperatures to the southeastern United States and to the eastern Amazon basin. Positive GRs are associated with warmer than normal conditions in northwestern North America and cooler than normal conditions across southeastern and eastern North America. Temperature conditions elsewhere are not as closely related to the GR index as they are to CT. These temperature relations are largely echoed in regression coefficients for the tropical and North Pacific PCs (Figs. 6c and 6f). As was the case with CT and GR, the spatial patterns of temperature relations with the two PC series are not as similar as the relations for precipitation and streamflow; the pattern correlation between Figs. 6c and 6f is only +0.16.

These temperature responses to CT and GR are as expected from the atmospheric circulation changes associated with the two indices. When positive, both CT and GR tend to route anomalously southerly winds over western North America as circulation patterns become more zonal over much of the North Pacific. The winds then tend to be anomalously northward near the west coast of North America from northern México to

Alaska (Fig. 4a). Warmer than normal American tropics derive from warmer SSTs and drier conditions in tropical South America (Diaz and Kiladis 1992). In subtropical South America, poleward deviations of the westerlies also result in warmer temperatures overall. Cooler temperatures in the southeastern United States with positive CTs and along the entire eastern United States are consistent with more storminess and anomalous northerly winds, respectively, as indicated by the details of the 500-mbar height regression coefficients shown in Figs. 4a and 4c.

1.6. DISCUSSION

The interannual and decadal ENSO-like variations of the Pacific climate system described here make important contributions to modern climatic and hydrologic variations of the Americas. For example, the numbers of temperature, precipitation, and streamflow series in which various amounts of variance are attributable to a combination of the tropical and North Pacific PCs are summarized in Table 2. Because the PC series are uncorrelated, the percentage of a series variance attributable to them is simply the sum of the squares of their respective correlations with the series. More than 30% of the variance is explained by various combinations of the two PC series in over 25% of all the weather series considered here; almost a third of the temperature series and a fifth of the streamflow series owe half or more of their year-to-year variance to the interannual and decadal ENSO-like climate variations.

Perhaps the most important implication of these strong dependences upon the interannual and decadal ENSO-like climate variations is that strong interhemispheric symmetries are to be expected in, at least, temperature, precipitation, and streamflow variations. Other forms of climate variation (e.g., North Atlantic climate oscillations) may be distinguishable from the ENSO-like variations simply by their equatorial asymmetric influences on the Americas.

Given the important role of these ENSO-like climate variations, a capacity to recognize their variations in paleoclimatic proxies is desirable. Reconstructions of these ENSO-like climate variations could form a groundwork for assessing the stability and long-term contributions of the Pacific climate system to the modern climate of the Americas. Given the possibly quite different mechanisms that underlie the interannual and decadal ENSO-like variations, it also would be desirable to reconstruct their separate contributions in paleoclimatic proxies. Separate reconstructions would help us to understand how stable and reliable these important contributors are to modern climate within their differing timescales in the face of natural and human-induced climate changes.

The interannual and decadal ENSO-like climate variations are influential over large areas of the PEP 1 region, and both affect temperatures, precipitation, streamflow, and (presumably) soil moisture—all of which may be amenable to a variety of paleoclimatic reconstructions. At the same time, however, the interannual and decadal ENSO-like variations may yield spatial patterns of variation of proxies that are difficult to separate from each other. If the analyses of instrumental records presented previously are any indication, it will be difficult to use paleoclimate proxies to distinguish the two forms of Pacific climate variability from patterns of temperature and, especially, precipitation proxies alone. Indeed, PC analyses of precipitation and streamflow in the Americas (Dettinger et al., 2000) similarly found that interannual and decadal hydroclimatic variations associated with ENSO-like atmospheric forcings could not be separated orthogonally, whereas other, less influential combinations of ENSO-like and North Atlantic climate variability could be. Perhaps, whether ENSO related or decadal in nature, the similar SST patterns associated with the ENSO-like variations cause such rapid and similar responses from the atmosphere that the particular processes that determine those SST variations matter little to their atmospheric and continental outcomes (e.g., Deser and Blackmon, 1993). Therefore, it may be exceedingly difficult to be certain which of the ENSO-like climate forces are providing the interannual or decadal fluctuations in North and South American proxies. We have limited this analysis to instrumental records, which span roughly 100 years; careful scrutiny of multiple-series, spatial networks of paleoclimate proxies to determine how often the interhemispheric ENSO-like pattern recurs and over what timescales it is present clearly is warranted.

The changes in atmospheric circulation and conti-

TABLE 2   Percentages of Temperature, Precipitation, and Streamflow Series from Sites Indicated in Fig. 5, with More Than Threshold Fractions of Variance Explained by the Combination of the Tropical and North Pacific PC Series

Variance threshold fraction	Fraction of variance			
	>0.2	>0.3	>0.4	>0.5
Temperature	58	42	38	30
Precipitation	35	27	25	14
Streamflow	39	35	30	21

nental climate associated with the ENSO-like climate variations represent temporary reorganizations of distributions of heat and water on global scales. Because of this large scale and the strong similarities between the influences of ENSO and decadal ENSO-like climate variations on the Americas, these climate variations also have important effects on the geochemical makeup of the atmosphere. Overall, the positive (El Niño–like) CT and GR forcings are associated with drier tropical conditions in the Americas and mostly wetter extratropical conditions (Figs. 5a and 5b and 5d and 5e), along with mostly warmer temperatures (Figs. 5c and 5f); these responses are part of a global pattern of drier tropical lands during El Niño–like periods (Dettinger et al., 2000). On interannual timescales, these influences affect vegetation and surface fluxes of CO_2 in ways that are observable on global scales (Siegenthaler, 1990; Keeling et al., 1995; Dettinger and Ghil, 1998). The strong similarities of effects of interannual ENSO and decadal ENSO-like climate influences in the Americas (and elsewhere) suggest that similar ENSO-like variations of global CO_2 concentrations may occur on decadal timescales.

Finally, the similarities between the influences of ENSO and decadal ENSO-like climate variations on the Americas, together with their differing timescales, ensure that sometimes CT and GR—and their respective precipitation, streamflow, and temperature patterns—share the same signs and sometimes they are in opposition. When the signs of the indices agree, the precipitation, streamflow, and temperature patterns in Figs. 5 and 6 can be expected to reinforce each other; when the indices disagree, the patterns may interfere and reduce the overall precipitation, streamflow, and temperature responses. These constructive and destructive influences have been documented by Gershunov and Barnett (1998) and Gershunov et al. (1999) over the United States, and results presented here indicate that similar influences can be expected in many other regions of North and South America.

1.7. CONCLUSIONS

During the twentieth century, three phenomena have dominated climate variations in the Americas on interannual and decadal timescales. The ENSO phenomenon dominates global climate variations on interannual timescales, and ENSO-like variations over the Pacific basin and North Atlantic climate variations dominate climate variations on decadal timescales. In this chapter, historical oceanic and atmospheric patterns of climate variability associated with ENSO and ENSO-like variations of the Pacific basin are analyzed,

and their influences on the temperatures, precipitation, and hydrology of the Americas are compared.

Several climate indices that reflect the ENSO interannual and ENSO-like decadal climate variations have been developed in the literature, including the SOI, Niño-3 temperatures, and the CT index for the tropical ENSO variations and the PDO index and so-called GR index for decadal ENSO-like variability. In this chapter, another pair of indices was developed on the basis of a PC analysis of SOI, CT, and PDO indices. The resulting PC series were closely related to CT and GR, but they were not formed by application of explicit temporal filters to separate ENSO and decadal ENSO-like variations (as was GR). By construction, the PC series are completely uncorrelated with each other to prevent double counting of climate variations when their respective contributions to regional climate variations are compared. The CT and tropical PC indices are two expressions of the tropically based, dominantly interannual climate variations associated with ENSO. Compared to the interannual ENSO signal, the GR and North Pacific PC indices are dominantly decadal and are expressed most fully in the subtropics and midlatitudes of both hemispheres in the Pacific basin. A number of different mechanisms have been described in the literature to explain the decadal character of the extratropical Pacific SST variations—some directly related to ENSO and others not—and no single one of these explanations is preferred by the scientific community at this time; see Section 1.3 on SSTs earlier in this chapter for some examples of the alternatives under consideration.

At the 500-mbar level in the atmosphere (about 5 km above sea level), positive phases of the CT and GR, and their counterpart PC series, are associated with anomalously low heights (pressures) over the North Pacific at the climatological position of the Aleutian Low, stronger westerly winds, and southward diversion of westerlies and storminess across the North Pacific basin. In the tropics, CT is associated with the entire suite of ENSO teleconnections and atmospheric changes; GR is not expressed so clearly in the tropics (as gauged by SSTs there). In the Southern Hemisphere, positive CT and GR, and PCs, are associated with a dipole of 500-mbar height anomalies that also shifts westerlies and storminess equatorward. The spatial patterns of 500-mbar height anomalies associated with CT and GR are remarkably similar, reflecting the qualitative similarities in SST patterns associated with the indices, despite their differing time series. The patterns of atmospheric circulation changes associated with the indices also are remarkably symmetric about the equator in the Pacific sector.

Over North and South America, the similarities of

the spatial distributions of climatic forcings associated with the interannual and decadal ENSO-like variations in the Pacific basin result in similar precipitation and streamflow patterns that are basically symmetric about the equator in the western Americas on both interannual and decadal timescales. The interannual and decadal ENSO-like patterns of precipitation and streamflow also are similar to each other overall, although more differences arise when the precipitation associations with the tropical PC and North Pacific PC (which are by design completely uncorrelated) rather than with CT and GR (which share some variability) are considered. Surface air temperature patterns associated with the ENSO-like climate forcings may be somewhat less spatially similar from timescale to timescale. Wetter than normal conditions accompany positive phases of CT (El Niños) and GR in the subtropical belts of both continents; drier than normal conditions accompany them in much of tropical South America (east of the Andes), in the midlatitudes of the northwestern United States and perhaps in southernmost South America. At the hemispheric scale, positive phases of the CT (El Niños) and GR indices historically have resulted in more consistent and stronger precipitation and streamflow responses in the Americas than have negative phases, except in the same midlatitude belts of the two continents. Temperatures in the Americas have tended to be warmer overall during positive phases of both CT and GR, except across the southeastern United States, eastern United States (for GR), and parts of the Amazon basin.

Together, the interannual and decadal ENSO-like climate variations are related to more than half the year-to-year variance in one-third to one-fifth of the temperature and streamflow series analyzed here. Current understanding of the processes leaves open the possibility that the interannual and decadal SST variations derive from very different physical processes with different timescales and different vulnerabilities to long-term natural and human-induced changes. The atmospheric variations and climatic consequences for the Americas that are associated with the interannual and decadal SST variations may derive from quick physical responses to overall SST variation that depend relatively little upon which SST process is dominant. Even if the atmosphere and continents prove to be slavish in their responses to SST variations, regardless of the source of those variations, the climatic consequences still would reflect the different vulnerabilities to change of the various SST processes. Thus, improved understanding of the processes represented by CT and GR should be a high priority for present climate, paleoclimate, and climate change researchers. It is unfortunate, then, that the strong spatial similarities between the hy-droclimatic outcomes of historical ENSO-like climate variations will probably impede most efforts to reconstruct, separately, these two dominant modes of climate variation in the Pacific basin by paleoclimatic proxies from the Americas. The results presented here, however, compare ENSO-like SST and climate variations; other modes of SST and climate variations may be important in many locales and may be easier to distinguish from ENSO and ENSO-like behavior of the global climate system. Finally, because this analysis is restricted to variations during the last century, the amplitude of the SST variations at work is only a few degrees large. Extrapolation to the larger SST anomalies of past climates would be speculative, and, indeed, even the close spatial similarities of ENSO and decadal ENSO-like climate variations found here may break down under larger SST forcings or under changing mean temperature conditions of the distant past.

Acknowledgments

We are very grateful to Vera Markgraf for organizing and shepherding the PEP 1 Workshop in Mérida, Venezuela; it was a remarkably informative and useful week. Discussions with Dave Enfield of the National Oceanographic and Atmospheric Administration (NOAA) and Atlantic Oceanographic and Meteorological Laboratory, Jose Marengo of the Instituto Nacional de Pesquisas Espacias; and Henry Diaz of the NOAA Climate Diagnostics Center contributed to the ideas developed here. Comments by Dan Cayan, Scripps Institution of Oceanography, and Michael Evans, Lamont Doherty Earth Observatory, greatly improved this chapter. Access to streamflow, precipitation, and temperature data provided by Dan Cayan and Larry Riddle at the Scripps Institution of Oceanography; to Brazilian streamflow series made available by Jose Marengo, courtesy of Eletrobras and Eletronorte; and to 1948–58 daily 500-mbar height fields by Wayne Higgins of the NOAA Climate Prediction Center made parts of this analysis possible. In addition to National Science Foundation funding for the workshop as a whole, Dettinger's efforts were supported by the U.S. Geological Survey's Global Change Hydrology Program and Battisti's efforts were supported by grants from the NOAA's Office of Global Programs.

References

Allan, R., J. Lindesay, and D. Parker, 1996: *El Niño, Southern Oscillation and Climate Variability*. Collingwood, Australia: CSIRO Publishing, 405 pp.

Barsugli, J. J., and D. S. Battisti, 1998: The basic effects of atmospheric-ocean thermal coupling on midlatitude variability. *Journal of the Atmospheric Sciences*, **55**: 477–493.

Bitz, C. M., and D. S. Battisti, 1999: Interannual to decadal variability in climate and the glacier mass balance in Washington, western Canada, and Alaska. *Journal of Climate*, **12**: 3187–3196.

Cayan, D. R., M. D. Dettinger, H. F. Diaz, and N. Graham, 1998: Decadal variability of precipitation over western North America. *Journal of Climate*, **11**: 3148–3166.

Deser, C., and M. Blackmon, 1993: Surface climate variations over the North Atlantic Ocean during winter: 1900–1989. *Journal of Climate*, **6**: 1743–1753.

Deser, C., and J. M. Wallace, 1990: Large-scale atmospheric circulation features of warm and cold episodes in the tropical Pacific. *Journal of Climate,* **3**: 1254–1281.

Dettinger, M. D., and H. F. Diaz, 2000: Global characteristics of streamflow seasonality and variability. *Journal of Hydrometeorology,* **1**: 289–310.

Dettinger, M. D., and M. Ghil, 1998: Seasonal and interannual variations of atmospheric CO_2 and climate. *Tellus,* **50B**: 1–24.

Dettinger, M. D., D. R. Cayan, H. F. Diaz, and D. Meko, 1998: North-south precipitation patterns in western North America on interannual-to-decadal timescales. *Journal of Climate,* **11**: 3095–3111.

Dettinger, M. D., D. R. Cayan, G. J. McCabe, and J. A. Marengo, 2000: Multiscale hydrologic variability associated with El Niño/Southern Oscillation. *In* Diaz, H. F., and V. Markgraf (eds.), *El Niño and the Southern Oscillation, Multiscale Variability, Global and Regional Impacts.* Cambridge, U.K.: Cambridge Univ. Press, 113–146.

Diaz, H. F., and G. N. Kiladis, 1992: Atmospheric teleconnections associated with the extreme phases of the Southern Oscillation. *In* Diaz, H. F., and V. Markgraf (eds.), *El Niño: Historical and Paleoclimatic Aspects of the Southern Oscillation.* Cambridge, U.K.: Cambridge Univ. Press, pp. 7–28.

Diaz, H. F., and V. Markgraf (eds.), 1992: *El Niño: Historical and Paleoclimatic Aspects of the Southern Oscillation.* Cambridge, U.K.: Cambridge Univ. Press, 476 pp.

Ebbesmeyer, C. C., D. R. Cayan, D. R. McClain, F. H. Nichols, D. H. Peterson, and K. T. Redmond, 1991: 1976 step in the Pacific climate: Forty environmental changes between 1968–75 and 1977–84. *In* Betancourt, J. L., and V. L. Tharp (eds.), *Proceedings of the 7th Annual Pacific Climate (PACLIM) Workshop,* California Department of Water Resources, Interagency Ecological Studies Program Technical Report 26, pp. 115–126.

Eischeid, J. K., C. B. Baker, T. R. Karl, and H. F. Diaz, 1995: The quality control of long-term climatological data using objective data analysis. *Journal of Applied Meteorology,* **34**: 2787–2795.

Eischeid, J. K., H. F. Diaz, R. S. Bradley, and J. D. Jones, 1991: A comprehensive precipitation data set for global land areas. DOE/ER-69017T-H1, TR051, 81 pp.

Enfield, D. B., and A. M Mestas-Nuñez, 2000: Interannual to multidecadal climate variability and its relationship to global sea-surface temperatures. *In* Markgraf, V. (ed.), *Interhemispheric Climate Linkages.* San Diego: Academic Press, Chapter 2.

Folland, C. K., and D. E. Parker, 1990: Observed variations of sea surface temperature. *In* Schlesinger, M. E. (ed.), *Climate-Ocean Interaction.* New York: Kluwer, pp. 21–52.

Folland, C. K., and D. E. Parker, 1995: Correction of instrumental biases in historical sea surface temperature data. *Quarterly Journal of the Royal Meteorological Society,* **121**: 319–367.

Garreaud, R. D., and D. S. Battisti, 1999: ENSO-like interannual to interdecadal variability in the Southern Hemisphere. *Journal of Climate,* **12**: 2113–2123.

Gershunov, A., and T. P. Barnett, 1998: Interdecadal modulation of ENSO teleconnections. *Bulletin of the American Meteorological Society,* **79**: 2715–2725.

Gershunov, A., T. P. Barnett, and D. R. Cayan, 1999: North Pacific interdecadal oscillation seen as factor in ENSO-related North American climate anomalies. *Eos (Transactions of the American Geophysical Union Newsletter),* **80**: 25, 29, 30.

Graham, N. E., 1994: Decadal scale variability in the 1970's and 1980's: Observations and model results. *Climate Dynamics,* **10**: 135–162.

Gu, D. F., and S. G. H. Philander, 1997: Interdecadal climate fluctuations that depend on exchanges between the tropics and extratropics. *Science,* **275**: 805–807.

Hurrell, J. W., 1995: Decadal trends in the North Atlantic Oscillation: Regional temperatures and precipitation. *Science,* **269**: 676–679.

Jacobs, G. A., H. E. Hurlburt, J. C. Kindle, E. J. Metzger, J. L. Mitchell, W. J. Teague, and A. J. Wallcraft, 1994: Decadal-scale trans-Pacific propagation and warming effects of an El Niño anomaly. *Nature,* **370**: 360–363.

Jiang, S., F. F. Jin, and M. Ghil, 1995: Multiple equilibria, periodic and aperiodic solutions in a wind-driven, double-gyre, shallow-water model. *Journal of Physical Oceanography,* **25**: 764–786.

Jones, P. D., S. C. B. Raper, and T. M. L. Wigley, 1986a: Southern Hemisphere surface air temperature variations: 1851–1984. *Journal of Climate and Applied Meteorology,* **25**: 1213–1230.

Jones, P. D., S. C. B. Raper, R. S. Bradley, H. F. Diaz, P. M. Kelly, and T. M. L. Wigley, 1986b: Northern Hemisphere surface air temperature variations: 1851–1984. *Journal of Climate and Applied Meteorology,* **25**: 161–179.

Kalnay, E., M. Kanamitzu, R. Kistler, W. Collins, D. Deaven, L. Gandin, M. Iredelli, S. Saha, G. White, J. Woolen, Y. Zhu, M. Chelliah, W. Ebisuzaki, W. Higgins, J. Janowiak, K. Mo, C. Ropelewski, J. Wang, A. Leetmaa, R. Reynolds, R. Jenne, and D. Joseph, 1996: The NCEP/NCAR Reanalysis Project. *Bulletin of the American Meteorological Society,* **77**: 437–471.

Keeling, C. D., T. P. Whorf, M. Wahlen, and J. van der Plicht, 1995: Interannual extremes in the rate of rise of atmospheric carbon dioxide since 1980. *Nature,* **375**: 666–670.

Kidson, J. W., 1988: Interannual variations in the Southern Hemisphere circulation. *Journal of Climate,* **1**: 1177–1198.

Latif, M., and T. P. Barnett, 1994: Causes of decadal climate variability over the North Pacific and North America. *Science,* **266**: 634–637.

Lees, J., and J. Park, 1995: Multiple-taper spectral analysis: A stand-alone C subroutine. *Computers and Geosciences,* **21**: 199–236.

Mann, M. E., and J. M. Lees, 1996: Robust estimation of background noise and signal detection in climatic time series. *Climatic Change,* **33**: 409–445.

Mantua, N. J., S. R. Hare, Y. Zhang, J. M. Wallace, and R. C. Francis, 1997: A Pacific interdecadal climate oscillation with impacts on salmon production. *Bulletin of the American Meteorological Society,* **78**: 1069–1079.

Miller, A. J., and N. Schneider, 1998: Interpreting the observed patterns of Pacific Ocean decadal variations. *In* Holloway, G., P. Muller, and D. Henderson (eds.), Biotic impacts of extratropical climate variability in the Pacific. Proceedings of the Aha Huliko'a Hawaiian Winter Workshop, Honolulu, Hawaii, School of Ocean and Earth Science and Technology, University of Hawaii at Manoa, pp. 19–27.

Mo, K. C., and M. Ghil, 1987: Statistics and dynamics of persistent anomalies. *Journal of the Atmospheric Sciences,* **44**: 877–901.

Mo, K. C., and R. W. Higgins, 1998: The Pacific South American modes and the tropical intraseasonal oscillation. *Monthly Weather Review,* **126**: 1581–1596.

Percival, D. B., and A. T. Walden, 1993: *Spectral Analysis for Physical Applications—Multitaper and Conventional Univariate Techniques.* Cambridge, U.K.: Cambridge University Press, 580 pp.

Richman, M. B., 1986: Rotation of principal components. *International Journal of Climatology,* **6**: 293–335.

Siegenthaler, U., 1990: El Niño and atmospheric CO_2. *Nature,* **345**: 295–296.

Slack, J. R., and J. M. Landwehr, 1992: Hydro-climatic data network (HCDN): A U.S. Geological Survey streamflow data set for the United States for the study of climate variations, 1874–1988. U.S. Geological Survey Open File Report 92-129, 193 pp. [Available from U.S. Geological Survey, Books and Open File Reports Section, Federal Ctr., Box 25286, Denver, CO 80225.]

Trenberth, K. E., 1990: Recent observed interdecadal climate changes in the Northern Hemisphere. *Bulletin of the American Meteorological Society,* **71**: 988–993.

Trenberth, K. E., and D. A. Paolino, 1980: The Northern Hemisphere sea-level pressure data set trends, errors, and discontinuities. *Monthly Weather Review,* **108**: 855–872.

Troup, A. J., 1965: The "Southern Oscillation." *Quarterly Journal of the Royal Meteorological Society,* **91**: 490–506.

Wallace, J. M., E. M. Rasmusson, T. P. Mitchell, V. E. Kousky, E. S. Sarachik, and H. Storch, 1997: On the structure and evolution of ENSO-related climate variability in the tropical Pacific: Lessons from TOGA. *Journal of Geophysical Research,* **103**: 14241–14259.

Zhang, Y., J. M. Wallace, and D. S. Battisti, 1997: ENSO-like inter-decadal variability: 1900–93. *Journal of Climate,* **10**: 1004–1020.

2

Interannual to Multidecadal Climate Variability and Its Relationship to Global Sea Surface Temperatures

DAVID B. ENFIELD and ALBERTO M. MESTAS-NUÑEZ

Abstract

As a benchmark to help profile paleoclimates across the Americas, we have developed an overview of what is known of modern climate variability on a planetary scale, with emphasis on climate manifestations in the Western Hemisphere. From instrumental observations taken as early as the mid-nineteenth century, we look at both atmospheric and oceanic variables and consider their relationships on timescales ranging from interannual to multidecadal. We focus on three of the most important climate modes: the interannual El Niño / Southern Oscillation (ENSO), the interdecadal Pacific Decadal Oscillation (PDO), and the multidecadal North Atlantic Oscillation (NAO). The variable of greatest interest is the sea surface temperature (SST) because it is arguably the least understood of the atmospheric boundary conditions for prehistoric climates and yet one of the most critical for developing atmospheric model simulations of those climates. The analysis begins by computing a global distribution of the trend in SST, which turns out to be highly nonuniform, with characteristics that may reflect low-frequency changes in shallow water mass formation. We then compute a global, canonical mode for ENSO that preserves the amplitude and phase structures of interannual ENSO variability worldwide. The ranking of the modal ampli-

tudes of ENSO events differs from the absolute amplitudes obtained by indexing SST data directly. This approach reflects the importance of the (non-ENSO) decadal-multidecadal climate modes in modifying the intensity of ENSO-related ocean warmings. Comparing the global mode between the ends of the nineteenth and twentieth centuries, we see essentially no difference in the amplitudes and frequency of ENSO on the global warming timescale, although such changes have occurred on shorter, multidecadal timescales. Upon removal of the global ENSO mode from the data, the residual variability is subjected to two different analyses that extract very similar spatiotemporal patterns of SST for the PDO- and NAO-like climate modes. The climate variations with longer timescales (PDO, NAO) together account for about the same amount of variance as ENSO, globally, and in some regions—e.g., the northeastern North Pacific—may rival ENSO in their climate and marine impacts. The NAO, in particular, involves an Atlantic–Pacific connection that may arise through fluctuations in the polar vortex, an aspect that may also characterize previous climates. In our discussion, we speculate on what might be learned from the instrumental record regarding possible characteristics of ancient climates, especially regarding the possibility that ENSO may have been considerably different or even absent in the mid-Holocene. Copyright © 2001 by Academic Press.

Resumen

Como referencia, para ayudar a perfilar los paleo-climas a través de las Américas, resumimos el cono-cimiento de la variabilidad climática moderna en escala planetaria con énfasis en sus manifestaciones en el hemisferio occidental. Basados en observaciones instrumentales tomadas a partir de mediados del siglo XIX, consideramos tanto las observaciones oceánicas como las atmosféricas, así como sus relaciones mutuas en escalas temporales de interanual a multidecadal. Enfocamos en tres de los modos climáticos más importantes: el modo interanual El Niño/Oscilación del Sur (ENOS), el modo interdecadal Oscilación Decadal Pacífica (ODP), y el modo multidecadal Oscilación del Atlántico Norte (OAN). La variable de mayor interés es la temperatura superficial del mar (TSM), la cual es quizás la menos comprendida de las condiciones de borde atmosféricas para climas prehistóricos y sin embargo es una de las más cruciales para realizar simulaciones de esos climas con modelos atmosféricos. El análisis comienza con el cálculo de la distribución global de la tendencia en la TSM, la que resulta ser altamente no uniforme, con características que podrían reflejar cambios de baja frequencia en la formación de masas de agua someras. Luego calculamos un modo canónico global para el ENOS que preserva la estructura de la amplitud y de la fase de la variabilidad interanual ENOS en forma global. El orden de las amplitudes modales de los eventos ENOS difiere del orden de las amplitudes absolutas obtenidas a partir de los índices calculados utilizando directamente TSM. Esto refleja la importancia de los modos (no ENOS) decadales a multidecadales en modificar la intensidad de los calentamientos oceánicos relacionados con ENOS. Comparando el modo global ENOS al final de los siglos XIX y XX, no vemos diferencia alguna en la amplitud y la frecuencia de ENOS en la escala de calentamiento global, aunque tales cambios han ocurrido en escalas temporales multidecadales más cortas. Luego de extraer de los datos el modo global ENOS, la variabilidad residual es sujeta a dos análisis diferentes que extraen patrones espacio-temporales de TSM muy similares para los modos climáticos ODP y OAN. En conjunto, las variaciones climáticas con escalas temporales más largas (ODP, OAN) representan aproximadamente la misma cantidad de varianza que ENOS globalmente, y en algunas regiones, por ejemplo en el noreste del Pacífico Norte, pueden competir con ENOS en sus impactos climáticos y marítimos. La OAN, particularmente, comprende una conección entre el Atlántico y el Pacífico que puede provenir de fluctuaciones del vórtice polar, un aspecto que también puede ser característico de climas previos. En nuestra discusión, nosotros especulamos en lo que podría aprenderse del registro instrumental con respecto a las posibles características de los climas del pasado, especialmente considerando la posibilidad de que ENOS podría haber sido muy diferente o quizás no haber existido en el holoceno medio.

2.1. INTRODUCTION

Progress in understanding pre-instrumental climate fluctuations requires a complex mix of climate-impact proxy analysis, modeling, reconstruction of currently measured variables for the past, and inferences from modern climate variability. The task of this chapter (ocean emphasis) and Chapter 1 (Dettinger et al., 2000, atmospheric and continental emphasis) is to consider recent progress in understanding the instrumental climate record of the nineteenth and twentieth centuries, how it may relate to previous climate variability, and how the paleoclimatic perspective may relate to future climate. The primary concern of this chapter is to consider the role of the ocean in determining climate through the variability of sea surface temperature (SST). One purpose of this chapter is simply to bring paleoclimate investigators up to date on recent findings on the oceanographic and atmospheric aspects of modern climate. Another aim is to awaken the consciousness of the paleoclimate community to certain issues that must be considered for past climates in view of what we are learning about the present climate.

One of our hopes for projecting the present into the past is that we observe considerable self-similarity in the modern climate across timescales. Therefore, this aspect will be commented upon as we describe the three principal timescales observed in more than a century of SST measurements: interannual (e.g., El Niño), decadal to multidecadal, and secular (global warming). However, as one delves into past climates more remote in time, and farther removed from the recent range of fluctuation, our confidence in projections based on self-similarity inevitably diminishes. Hence, we will put more effort into speculating on the Holocene (mainly 10,000 B.P. to present) than on the pre-Holocene (more than 15,000 B.P.). This approach is appropriate, of course, because the range of variations that occurred in the Holocene is more likely to be representative of what might be observed in the next century.

2.2. DATA AND METHODS

Except in the next section, we shall depend almost exclusively on the SST data taken since the mid-nineteenth century. However, a problem with using summary compilations of SST, such as the Comprehensive Ocean-Atmosphere Data Set (COADS; Woodruff et al., 1987), is that the sampling frequency and spatial coverage of ship observations has increased severalfold over the last 150 years. This increase has resulted in a record that is very inhomogeneous in both space and time. Therefore, we use a reanalyzed version of data on SST anomalies (SSTAs) by Kaplan et al. (1998; henceforth K98) wherein only the variability of large-scale patterns has been retained through a combination of multivariate filtering and inverse analysis. The K98 data are monthly on a 5° × 5° global grid from 1856–1991. The K98 SSTA reanalysis is very similar to the one by which the contemporaneous relationship between tree-ring proxy data and modern, well-measured SSTs are used to reconstruct pre-instrumental SST variability (Chapter 4 by Evans et al.). With the K98 approach, confidence in the resolution of basin-scale features is increased at the expense of resolving small-scale features.

Although the methods used in the K98 reanalysis minimize the effects of poor sampling, we naturally have less confidence in the accuracy of the earlier maps relative to those plotted after about 1950. If one examines decadal plots of ship tracks (Woodruff et al., 1987), it is clear that the historical ship sampling of SSTA is least complete in the southeastern South Pacific and the southwestern South Atlantic. Although the tropical Pacific is more thinly sampled in the early part of the record, the important features are of large zonal scale (retained by the leading multivariate modes used in the K98 reanalysis), and their temporal fluctuations are probably resolved adequately by the mainly meridional ship tracks, east of the date line, after about 1870 (Smith et al., 1998). The mid- and high-latitude regions of both the North Atlantic and North Pacific were comparatively well sampled through most of the reanalyzed period. All of the regions discussed in this chapter were fairly well sampled from 1890 onward, and the Atlantic regions were fairly well sampled from the 1860s onward. We refer the reader to K98 for a more complete discussion of the reanalysis quality. In the SSTA analyses performed here, we use the more reliable 1871–1991 period. We also refer frequently to similar calculations done by Enfield and Mestas-Nuñez (1999) and Mestas-Nuñez and Enfield (1999) for the entire (1856–1991) period of the K98 reanalysis.

Our first task (Section 2.3) will be to describe the secular trend in SSTA, globally, and the canonical global mode of SSTA associated with El Niño/Southern Oscillation (ENSO). Subsequent removal of these two components allows us to concentrate more confidently on the intermediate timescale (decadal-multidecadal) and on variability unrelated to interannual ENSO variations.

To establish a link between decadal to multidecadal SSTA variability and the associated atmospheric climate fluctuations, we will consider how SSTA is related to commonly used atmospheric climate indices based on modern measurements of sea level pressure (SLP). These and similar indices have been used by many authors to characterize the temporal variability of land climate patterns. The analysis method used in this comparison is a simple composite of SSTA based on the positive and negative phases of the anomalous SLP (SLPA) indices. The SLP data used here are the 1899–1997 Northern Hemisphere monthly analyses available at the U.S. National Center for Atmospheric Research (NCAR), denoted as DS010.0.

The third analysis is of the decadal to multidecadal SSTA variability using ordinary and rotated empirical orthogonal function (REOF) analysis after removing the linear trends and global ENSO mode. In this analysis we extract several global modes and compare them to earlier SLPA-SSTA composites. In the final section, we discuss some of the implications of these analyses for previous climates.

2.3. SECULAR TRENDS AND THE GLOBAL ENSO MODE

One reason for considering secular trends and the global ENSO mode first is that they will be removed from the data prior to analysis of decadal-multidecadal variability. It is also appropriate to discuss them together because of current speculation that global warming may affect ENSO variability (Trenberth and Hoar, 1996). More generally, the characteristics of ENSO, or even its existence (or lack of it) during prehistoric periods, may be dependent on the background climate (Sun, 2000). We shall redirect our attention to these timescales when we discuss the paleoclimate implications of modern measurements later in Section 2.6.

2.3.1. The Global Distribution of Secular Trends

Figure 1A is a map of the globally distributed linear trend in SSTA over a 136-year time period (1856–1991), based on a simple, unweighted linear regression of SSTA at each 5° × 5° grid point on the monthly time vector. The associated temporal variation of the global mean SSTA (Fig. 1B) is relatively trendless prior to 1900 and has a more accentuated trend after that time. Therefore, while the trend map is representative of the true secular pattern after 1900, the magnitudes shown tend to underestimate the post-1900 trend by about 40%. We have tried alternate approaches for treating

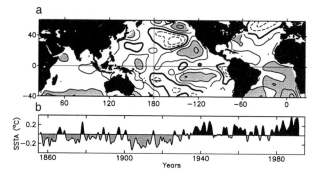

FIGURE 1 (A) Distribution of the 1856–1991 linear (least-squares) trend (°C/century) in the smoothed Kaplan et al. (1998) sea surface temperature anomaly (SSTA) data set. The contour interval is 0.2°C/century, the heavy contour is zero, and the light solid (dashed) contours are positive (negative). Positive trends exceeding 0.2°C/century are shaded. (B) Time variation of the global average SSTA.

the secular change in the data—e.g., division into two periods with different means—and the description of the secular variation is not sensitive to the representation. Therefore, we adopted the simplest scheme (linear trend) rather than making necessarily arbitrary judgments.

At the largest global scale, the overriding impression from the trend map is that warming trends dominate over cooling and that the rate of secular change shows high spatial inhomogeneity. Off the equator, the shaded regions of largest positive trend (>0.4°C/century) are in the South Atlantic, the subtropics of the central and eastern Pacific, the Gulf of Alaska, and the region north of Cape Hatteras where the Gulf Stream separates from the eastern coastline of the continental United States. Regions of negative trend (<−0.2°C/century) are found in the midlatitudes of the east-central North Pacific and in the high-latitude region southeast of Greenland. The equatorial Pacific, of particular interest because of its importance for ENSO variations, shows only a null or weakly positive trend between the Galápagos Islands and the date line, strong cooling off Ecuador and Peru, and warming west of the date line. The equatorial Atlantic shows moderate warming that appears as a low-latitude extension of the stronger warming in the South Atlantic. Other than in the central South Pacific, all other regions are characterized by a mild warming trend of ca. 0.1°–0.3°C/century.

The linear trends in the instrumental record shown here are the closest thing, in timescale, to the slow climate fluctuations that might be detected in previous epochs by less direct, paleoclimatic methods. Examples that come to mind are the Little Ice Age (LIA) and the Medieval Warm Period (MWP). To the extent that the LIA and MWP imply globally warmer or cooler climates in the past, we must keep in mind that those pat-

terns also may have been quite heterogeneous and that certain regions may also have been cooling (warming) when the Earth as a whole was warming (cooling). This lack of homogeneity is confirmed by the 500-year atmospheric model simulation of Hunt (1998), which reproduces ~100-year-long periods of globally warmer or cooler air temperatures similar to the LIA and MWP. Extreme decadal averages in these periods, which occur as natural (externally unforced) climate variability in the model, exhibit large contiguous areas of ocean and land with anomalies opposite to the contemporaneous global mean. This pattern reinforces the likelihood that a period like the mid-Holocene would have had very nonuniform differences from our present climate. These differences could be especially significant if the ancient patterns of SST change were anything like those shown in Fig. 1, because the regions with trends counter to the global averages are in the North Pacific and the North Atlantic, where present-day water masses are formed and subducted (Nakamura et al., 1997; Curry et al., 1998). These processes are currently suspected of playing an essential role in the shifts between climate regimes. The existence of such inhomogeneities at the secular timescale implies that similar feedback mechanisms may play a future role in climate variations with centennial or longer timescales and may also have played a role in the past.

The linear trends were removed from the K98 data prior to the removal of the ENSO mode described in Section 2.3.2. The order in which these removals were done has no effect on the residual (trendless, non-ENSO) variability. All discussions of the SSTA variance explained by the various components of variability have been referred to the detrended data set.

2.3.2. The Global ENSO Mode

The interannual ENSO signal is distributed from the Pacific to other tropical oceans through its associated global tropospheric anomalies (Hastenrath et al., 1987; Latif and Barnett, 1995; Lanzante, 1996; Enfield and Mayer, 1997; Enfield and Mestas-Nuñez, 1999). The characteristic lag of the extended ENSO response in the Atlantic and Indian Oceans is approximately one to three seasons, which is comparable to the lag of the high-latitude North Pacific with respect to the tropical Pacific. This is not a mere curiosity. According to Lau and Nath (1994), if the extended ENSO signal in other ocean basins is not included in a global coupled model, the model cannot accurately replicate the tropospheric response, such as the Pacific–North American (PNA) pressure pattern (Barnston and Livezey, 1987). On the other hand, if one wishes to understand the relationship between land climate and another climate mode such as the North Atlantic Oscillation (NAO), the existence of the ENSO signal in the Atlantic makes unequivocal attribution difficult.

To better characterize the global ENSO mode of variability, we have performed a complex empirical orthogonal function (CEOF) analysis of the monthly K98 data that correctly preserves the phase lags within the Pacific and other oceans. The detrended data are first band-passed to block periodicities larger than 8 years or shorter than 1.5 years. The leading mode of the decomposition is the one that represents the global ENSO. It explains 13.4% of the variability in the detrended, unfiltered data set for the December–January–February (DJF) season. For additional details on the methods used, the reader is referred to Enfield and Mestas-Nuñez (1999). However, we have repeated the Enfield and Mestas-Nuñez calculation here for the shorter but more reliable 1871–1991 period. The differences be-

tween the two analyses are small, and the conclusions are essentially unchanged.

Because the ENSO mode is based on a complex decomposition of the data, it has complex eigenvectors and expansion coefficients which can be transformed into amplitudes and phases in both the spatial and temporal domains. We show the spatial amplitude and phase in Figs. 2A and 2B. The obvious reference rectangle to characterize the ENSO mode is the well-known Niño-3 region in the equatorial Pacific, bounded by 5°N–5°S, 90°–150°W (Fig. 2B). The spatial amplitude (Fig. 2A) is shown as a percent gain with respect to the reconstructed temporal amplitudes for Niño-3 (Fig. 2C). Thus, the average gain in the Niño-3 rectangle is 100, and the contour labeled 30 passes through regions where the amplitude is 30% of the Niño-3 amplitude. Significance tests on the correlation coefficients between the data and the temporal reconstruction show

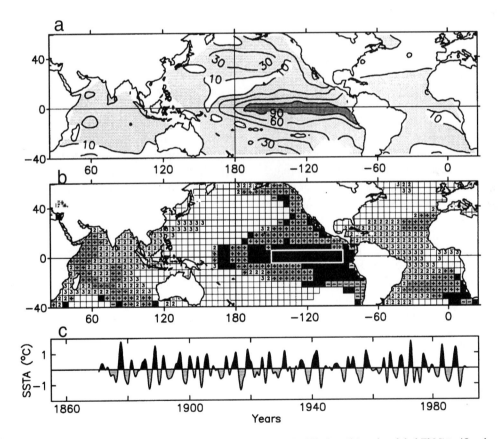

FIGURE 2 First complex empirical orthogonal function (CEOF) describing the global El Niño/Southern Oscillation (ENSO) variability. (A) Spatial distribution of the amplitude with respect to the modal reconstruction over the Niño-3 index region (white rectangle in B). Contours shown are 10, 30, 60, and 90, and values in excess of 10 are shaded. A score of 100 is the average amplitude over the Niño-3 region. (B) Spatial distribution of phase lag (seasons); ± indicates that the data lags/leads Niño-3 by one season, and positive numbers indicate data lags of more than one season. (C) Temporal reconstruction of the mode-related variability averaged over the Niño-3 index region.

that amplitudes in the white regions (<10) are insignificant. For ease of display, the spatial phase (Fig. 2B) has been transformed into a temporal lag at each grid point (units of 3-month seasons) and referenced to the Niño-3 phase at a nominal midband periodicity of 43.7 months (determined from zero upcrossings in the temporal reconstruction).

To represent the temporal behavior of the mode (Fig. 2C), we show the temporal reconstruction (temperature units) for phases and amplitudes combined over an index region rather than the more confusing temporal amplitude and phase functions from which they are derived (i.e., the complex expansion coefficients for the mode). The modal time variability elsewhere is merely the same series lagged by the appropriate spatial phase and with amplitude attenuated in proportion to the spatial amplitude.

The gain and phase maps show the well-known features of mature El Niño events in the Pacific: maximum amplitude within ±10° of the equator (Fig. 2A), opposite phase in the midlatitudes of the central North and South Pacific (Fig. 2B), and near-coastal lags increasing to one to two seasons poleward of the reference region (Fig. 2B). Amplitudes in the tropical Atlantic and Indian Oceans are 10–15% of the Niño-3 amplitude, and the extrema occur one to three seasons after the maximum Niño-3 amplitude. The delayed responses in the Atlantic and Indian Oceans have been noted in previous studies (e.g., Enfield and Mayer, 1997; Latif and Barnett, 1995).

The larger positive peaks in the temporal reconstruction correspond to historical El Niño events (e.g., Quinn et al., 1987). ENSO activity has varied over more than a century of measurements, as evidenced by the relatively quiescent decades following the 1941–42 event. However, there is no visually detectable difference between the first 30 years of the record (1875–1905) and the last 30 years (1960–90); both periods had about seven to nine events of similar intensities. Hence, there is no obvious indication, in this representation, that global warming has had a significant effect on the ENSO activity.

The global ENSO mode is a canonical representation of the spatial and temporal commonalities found in the ENSO extrema of the historical SSTA data. Filtering has removed the modifying effect that long-period anomalies have on ENSO amplitudes, as well as the frequently occurring double peaks and other temporal details caused by fluctuations of annual and shorter timescales. Nor are the unusual aspects of individual ENSO extrema captured; rather, they are relegated to higher modes. As a result, the amplitudes are generally smaller than an unfiltered data average over the Niño-3 region.

Significantly, the reconstructed event amplitudes sort differently than the data-averaged Niño-3 index (not shown). Thus, 1983 has the fifth largest positive amplitude in the reconstructed series, rather than one of the two largest in the data, and 1972 is one of the two largest reconstructed events (along with 1878). The main reason for this change in rank is that 1972 occurred during a two- to three-decade period of generally lower background temperatures, while 1983 occurred after Pacific interdecadal variability at low latitudes had flipped from negative to positive background values in the late 1970s. Much of the 1983 amplitude is captured instead by the leading EOF modes of the non-ENSO residual data (Section 2.5).

2.4. DECADAL-MULTIDECADAL CLIMATE FLUCTUATIONS

In recent years it has been increasingly recognized that smaller (than ENSO) but significant climate shifts occur on decadal to multidecadal timescales. Such fluctuations may be related to ocean–atmosphere interactions, amenable to simulation and prediction by numerical models. An example in the tropics is the interdecadal succession of comparative rainy periods with drought periods in the Sahel region of sub-Saharan Northwest Africa (Nicholson, 1989), along with its strong association with the frequency of Atlantic tropical cyclones (Gray, 1990). Climatic shifts in rainfall and air temperature have also been noted in the Northern Hemisphere extratropics, especially in central and northern Europe and in western North America. These slow climate oscillations correspond to upstream (to the west) fluctuations in the midlatitude westerly wind belts and associated persistent features in atmospheric pressure patterns, such as the Aleutian Low, the Icelandic Low, and the Azores High. Two dominant and quasi-independent oscillations have been noted: the Pacific Decadal Oscillation (PDO) and the NAO. The index time series frequently used to represent the tropospheric variability of the PDO and NAO are based on long records of SLP measured near, or interpolated to, the critical nodal points of SLPA patterns.

We are interested in the non-ENSO patterns of variability in SSTA and how they relate to these climate oscillations. Hence, in this section we will show SLPA-based index series for the PDO and NAO with their composite mean SLPA maps over the Northern Hemisphere, and then we will show the corresponding composite maps for SSTA. The composite SSTA maps can then be used as a reference for the non-ENSO global modes discussed in Section 2.5.

For the PDO, the index we have chosen is the DJF average of SLPA over the Aleutian Low region, 167.5°–177.5°W, 42.5°–52.5°N. Trenberth and Hurrell (1994) and Latif and Barnett (1996) have used similar Pacific indices. In Fig. 3A, we show the PDO index time series and a smoothed version of it using a loess filter, which is a locally weighted quadratic smoother, with a half-span of 8 years. Also shown (Fig. 3B) is a polar-projected NH map of the difference between composite means of the winter SLPA data, based on positive and negative values of the index series. Figure 3C shows the K98 SSTA, similarly composited on the PDO index. The corresponding DJF plots for the NAO are shown in Fig. 4. For the NAO index we have chosen to use the difference of normalized SLPA between Lisbon, Portugal, and Stykkisholmur, Iceland, previously used by Hurrell (1995). Both SLPA indices are signed to be positive

when pressures at higher latitudes are negative and the midlatitude westerlies south of the negative pressure node are anomalously strong.

The positive phase of the PDO is characterized by an enhanced Aleutian Low and strengthened westerly winds over the midlatitude central North Pacific (Fig. 3B). There is also some weakening of the polar vortex, but the pattern is neutral over the rest of the Northern Hemisphere. The smoothed index was positive from 1925–45, negative during 1945–75, and positive again after the mid-1970s (Fig. 3A). The dominant timescale of the smoothed PDO is shorter (one to three decades) than that of the NAO (>25 years). The high- (low-) index state is associated with greater (less) winter storminess over the central North Pacific and a warmer (cooler) winter climate along the west coast of North America (Latif and Barnett, 1994; Wiles et al., 1998).

The SSTA imprint of the PDO's positive (negative) phase is dominated by an extensive zone of cooler (warmer) SSTs over the Kuroshio Extension region (40°N) east of Japan as far as 145°W (Fig. 3C). This is also consistent with enhanced (diminished) evaporation and mixing under the stronger (weaker) westerly winds, with enhanced (diminished) cooling over that region. An SSTA of opposite sign and equally as large in amplitude is found in the northeastern North Pacific, consistent with observed warmer land temperatures over western North America (Wiles et al., 1998). Accompanying an enhanced Aleutian Low (high PDO index), west coast warmer temperatures are consistent (in an advective sense) with an enhanced poleward ocean circulation in the subarctic gyre of the northeast Pacific and a diminished equatorward California Current System to the south. Corresponding to the North Pacific changes, there is also warming south of the equator in the eastern Pacific, extending poleward to the Chilean coast. Most attempts to explain the SSTA signature of the PDO invoke ocean–atmosphere interactions in the North Pacific, but no clear explanation has yet been proposed for the features off South America.

We note that the overall spatial pattern of SSTAs for the PDO looks very much like the ENSO pattern, but with weaker and broader warming in the eastern tropical Pacific and more intense cooling over the central North Pacific. As with the warm phase of ENSO, the warm eastern Pacific observed since the mid-1970s is associated with a migration of salmon stocks northward from Oregon–Washington to the Gulf of Alaska (Mantua et al., 1997) and a replacement of anchovies by sardines and horse mackerel populations off Peru and Chile (Muck, 1989).

In the high-index state of the NAO (Fig. 4B), the polar vortex is intensified with enhanced low pressures

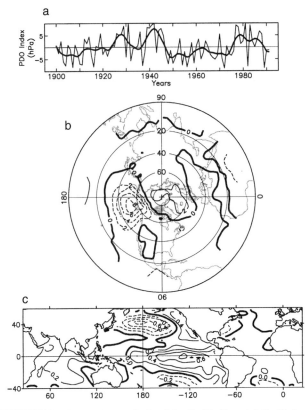

FIGURE 3 Composite analysis of the Pacific Decadal Oscillation (PDO) for the December–January–February (DJF) season. (A) The smoothed (dark) and unsmoothed (light) versions of a PDO temporal index defined by averaging sea level pressure anomaly (SLPA) over an area near the Aleutian Islands and 170°W, signed so as to represent the strength of the midlatitude westerly winds in the North Pacific (see text for details). (B) Distribution of the difference between the composite averages of SLPA over the Northern Hemisphere for positive and negative phases of the unsmoothed PDO index. Units are in hectoPascals (hPa). (D) Distribution of the difference of composites of SSTA, similarly calculated for the phases of the PDO index. Units are in degrees Celsius (°C).

concentrated in the Greenland region, while pressures are high over the Azores. Westerly winds blow more strongly across the North Atlantic between the two nodes. The NAO has been in its high-index state since about 1970, preceded by low values in the 1950s and 1960s (Fig. 4A). A previous high state during 1905–30 was followed by 25 years of near-normal values. Timescales range from less than a decade to multidecadal. Under extremes of the NAO, comparatively severe winters (high rainfall) alternately affect northern (high-index) and southern (low-index) Europe (Hurrell, 1995). The NAO and its climate impacts are consistent with inferences drawn from tree-ring records around the North Atlantic basin (D'Arrigo et al., 1996). In the tropical sector, a high (low) index is associated with less (more) rainfall in the African Sahel (Nicholson, 1989) and less (more) tropical cyclone development west of northwestern Africa (Gray, 1990). That the NAO may now be returning to the low-index phase is suggested by the recent rapid fall in the unfiltered data (Fig. 4A) and is confirmed by a multivariate mode in the updated global SSTA, similar to the Atlantic mode discussed later in Section 2.5 (Landsea et al., 1999).

The winter (DJF) SSTA distribution corresponding to the high phase of the NAO (post-1970) shows moderately warmer conditions in the western midlatitudes of the North Atlantic (35°–40°N), roughly west of the intensified Azores High and sandwiched between enhanced westerlies to the north and easterlies to the south (Fig. 4C). Cooler temperatures are found under the latter zones of enhanced wind speeds (40°–50°N, 15°–20°N). Although this cooling is consistent with greater evaporation rates at the short timescales, the modeling study of Timmerman et al. (1998) argues that the >35-year timescale is primarily controlled by changes in the thermohaline circulation (THC). At intermediate timescales advection may play a greater role, and for the shorter timescales diabatic heating due to surface fluxes and thermocline entrainment becomes increasingly important (e.g., Bjerknes, 1964; Halliwell, personal communication). The tropical South Atlantic is neutral with respect to the NAO, and the high-latitude South Atlantic is moderately warm when the index is positive. One must be cautious in regard to the last feature, however, because sampling is relatively poor south of 30°S and west of the Greenwich meridian. When considered over all months (not shown), the warming region in the midlatitude North Atlantic is found to be stronger and more zonally extensive, while the warming in the extratropical South Atlantic disappears.

The SSTA signature of the NAO includes an interocean feature in the Pacific: when the NAO index is high (low), SSTAs along the U.S. West Coast, in the Gulf of

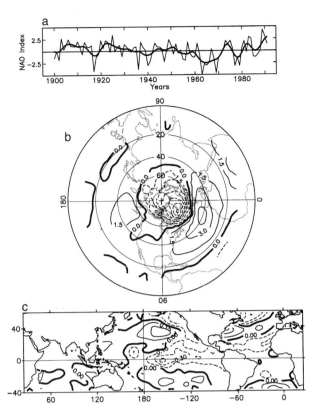

FIGURE 4   Same as in Fig. 3, but for a December–January–February (DJF) version of the North Atlantic Oscillation (NAO) index defined by Hurrell (1995).

Alaska, and in the central equatorial Pacific are significantly cooler (warmer). Again, the pattern in the Pacific region resembles the ENSO pattern, but as we point out in Section 2.5, these climate modes do not necessarily separate cleanly, and alternate representations with differing Pacific signatures may be equally, or more, reasonable.

2.5. GLOBAL MODES OF SSTA VARIABILITY

After removal of the interannual global ENSO mode (Fig. 2) from the K98 SSTA data, we computed the ordinary (not complex) EOFs of the non-ENSO residuals for the DJF season during the period 1871–1991. The first three leading modes explain 6.1, 5.4, and 3.6% of the residual unfiltered variance after removal of trends. The total—15.1%—is comparable to the 13.4% explained by the global ENSO mode for the same season. As frequently occurs when the geographic scale of a data set is much larger than the scales of the physical processes, the leading global EOFs include contributions from different ocean basins. Such regions may be physically linked through teleconnective processes, or they may be associated statistically by accidentally conforming to a common principal axis of variation. Therefore, we applied a varimax rotation (REOF) to the first 10 EOF modes to separate regional processes that are more appropriately considered independently of each other. This was done to preserve spatial orthogonality between REOF modes while relaxing temporal orthogonality, meaning that zero-lag temporal correlations are allowed as the leading (unrotated) modes are reconfigured. The unrotated EOFs are very similar to those of Enfield and Mestas-Nuñez (1999), and the REOF modes are similar to those of Mestas-Nuñez and Enfield (1999). In those analyses, however, all months of the calendar year were used and the longer period (1856–1991) was included. The rotation procedure did not improve on the ability of the first unrotated mode to represent Pacific variability associated with the PDO. However, the third unrotated mode, which best corresponds to the NAO in the North Atlantic, was found to include spatial features in other basins (eastern equatorial Pacific, South Atlantic) that split off into separate modes under rotation. Those REOFs correlate poorly (in time) with the only rotated mode that preserves the essence of the NAO variability in the North Atlantic: REOF 3. REOF 3 also explains almost as much variability as its unrotated parent (EOF 3): 3.4%. Hence, we will discuss only the first unrotated (EOF 1) mode and the third rotated mode (REOF 3), because they best help us to understand the canonical forms of the PDO and NAO, respectively.

Figure 5 shows the spatial and temporal variability for the first, PDO-related EOF of the global SSTA residuals. As with the global ENSO mode (Fig. 2), the spatial pattern (Fig. 5A) is a map of gain (%) for the data with respect to the temporal reconstruction for the plotted rectangular region (Fig. 5B). Most of the dominant features of the PDO/SSTA composite (Fig. 3) are well reproduced in this mode for the Pacific basin. One is an area of strong anticorrelated variability that stretches across the central North Pacific along 30°–50°N. Also seen is the strong in-phase variability along the Pacific eastern boundary north and south of the equator. Another feature of the PDO/SSTA composite is the in-phase variability on and about the equator in the central and eastern tropical Pacific. Only the off-equatorial aspects of the PDO/SSTA composite (Fig. 3) are reproduced by EOF 1 (Fig. 5). The equatorial component was captured by EOF 2 (not shown; see Enfield and Mestas-Nuñez, 1999). The time variabilities in Figs. 3 and 5 correspond well, and both contain the well-documented, recent phases of the PDO. Although the year-to-year (unsmoothed) variations in the PDO are less well correlated with EOF 1 (0.36) than the decadal (smoothed)

fluctuations (0.60), both correlations are about equally significant (95%) after serial correlation in the data is accounted for.

Figure 6 shows the rotated modal variability (REOF 3) corresponding to the NAO composite (Fig. 4). Because the region south of Greenland was chosen as the reference for the temporal reconstruction and the gain map, the signs of features in REOF 3 are opposite to those of the NAO in both the spatial and the temporal sense. Features in REOF 3 that are common with the NAO (but with opposite sign) are in-phase variability south of Greenland, in the tropical North Atlantic, and in a region off the west coast of North America. The reduced gain in the 20°–40°N zonal band of the North Atlantic corresponds qualitatively to the reduced or opposite-sign composite values in the NAO distribution

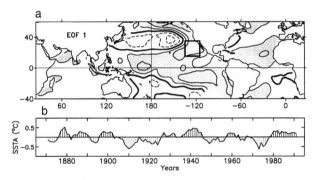

FIGURE 5   First empirical orthogonal function (EOF 1) of the non-ENSO residual data set describing the Pacific interdecadal variability for the December–January–February (DJF) season. (A) Spatial distribution of the response with respect to the modal reconstruction over the index region (square); the contour interval is 40 and dashed contours are negative. A score of 100 is the average response over the index region. (B) Temporal reconstruction of the mode-related variability averaged over the index region. Positive gain in excess of 40 is shaded.

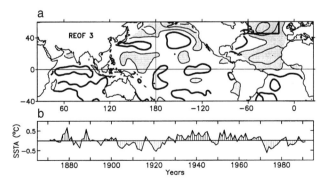

FIGURE 6   Same as in Fig. 3, but for the third rotated empirical orthogonal function (REOF 3) of the non-ENSO residual data set describing the Atlantic multidecadal variability for the December–January–February (DJF) season.

(Fig. 4). A feature from the NAO/SSTA composite that is not well reproduced by REOF 3 is the negative composite mean region in the central equatorial Pacific. The temporal correlation between REOF 3 and the NAO is -0.35 (-0.41) for the unsmoothed (smoothed) series. After serial correlation is accounted for, however, the association for the unsmoothed data is more significant (98%) than that for the smoothed data (90%).

Of special significance in REOF 3 (Fig. 6) is the fact that the Varimax rotation did not separate the positive gains at the high latitudes of the North Pacific (especially the Gulf of Alaska) from the NAO-related features in the North Atlantic. This fact, combined with similar indications from the NAO composite mean distribution, suggests that SSTAs at the high latitudes of both the Pacific and the Atlantic are tropospherically linked through fluctuations in the polar vortex. We can also see a spatial similarity in the North Pacific gain distributions of EOF 1 and REOF 3, especially in the region off the west coast of North America. The temporal reconstructions of EOF 1 and REOF 3 have a correlation of 0.59, suggesting that the two climate oscillations are linked through shared variability in that same west coast region and especially in the Gulf of Alaska.

2.6. DISCUSSION: IMPLICATIONS FOR PALEOCLIMATES

2.6.1. Past Climate Variability at Long Timescales

There are many reasons to doubt that modern records can be used to make inferences about past background climates that lasted hundreds of years or more. The longest timescale in the instrumental record analyzed here is the secular trend, which might be representative of past variability for periods of 100–200 years, but probably not of variability at millennial scales or longer. Moreover, given the similarity of the twentieth-century variations of the global mean SSTA (Fig. 1) to those of globally averaged air temperatures (Houghton et al., 1996), it seems likely that a significant part of the trends and their spatial distribution may be anthropogenic and not necessarily representative of past natural variability, even for short intervals such as the LIA or MWP. It seems even less likely that any of the other climate modes presented here (ENSO, PDO, NAO) can serve as templates for how past climates of centennial or longer timescales differ from the modern climate. The physical processes that would produce significantly longer shifts would probably be quite different. This would be especially true if interactive feedbacks are involved, which we suspect is the case for the shorter timescales, since the timescales of

oscillations such as ENSO are set by the nature of the processes involved. Specifically, the patterns of variability discussed here may not be representative of background climates in the early to mid-Holocene or in Pleistocene glacial epochs.

The spatial similarities among the various modes analyzed here, especially in the Pacific, give us more optimism. The spatial self-similarity seen in the modes cuts across several timescales, despite the possibility that different processes are involved. Thus, even if we postulate that different timescales are governed by different subsurface ocean feedbacks (e.g., subduction at different locations or to different depths), many of the timescales and mechanisms may involve common elements that affect SSTA, such as the strength of the westerly wind belts. Moreover, the westerly wind strength will, in turn, affect the gyre circulations and their associated SST advection in similar ways. Finally, there remains the possibility, still unclarified by research, that interactive feedbacks are not crucial to long climate modes and that they merely arise from the chaotic nature of atmospheric variability with the oceans responding passively (Sarachik et al., 1996).

Present climate modes and their associated SSTA patterns are, in fact, related to fluctuations in the strength of the westerly wind belts overlying major extratropical oceans, to the strength of the polar vortex, and to changes in atmospheric pressure nodes surrounding the vortex. It is not difficult to surmise that the greater land and ice coverage that accompanied lower sea level stands would affect the same atmospheric features and on very long timescales. However, this is only speculation. Our best hope for understanding what those patterns may have been like must depend more on a combination of paleoclimatic proxy indicators and numerical modeling than on the analogies provided by modern climate scenarios. It is especially important that atmospheric models be forced with realistic surface boundary characteristics (sea level stand, ice cover, and SST) as indicated by proxy data.

Just as in the twentieth century, earlier background climates must have been accompanied by a full spectrum of variability on shorter timescales ranging down to the interdecadal and interannual periodicities described in this chapter. How similar that variability would be to the present variability is also best addressed by modeling. Whether or not a forced atmospheric model can accurately portray that variability depends on the role of the ocean in producing and enhancing phase changes (Sarachik et al. 1996). If the ocean is an active and critical participant, coupled ocean–atmosphere models will be required. Present coupled models are precluded from doing this because of limitations having to do with resolution, integration time, and stability. Until the models improve and debates about ocean interactivity are settled, paleoclimatologists should probably focus on improving the accuracy of the surface boundary conditions for past climates, especially with regard to their SST distributions.

2.6.2. Holocene ENSO Variability

A critical question for the modeling of current and future ENSO variability, as well as past variability, is whether or not the characteristics of ENSO vary with the slowly evolving background climate. Our analysis of the global ENSO mode reveals more than a century of variability during which global warming has occurred (Fig. 1). However, we see no apparent differences in the return intervals and intensities of ENSO between one century and the next, at least not in the global ENSO mode derived from band-passed data (Fig. 2C). Enfield and Cid (1991) were similarly unable to detect differences in El Niño return intervals between the LIA and the twentieth century. It is possible, however, that slowly varying, noncanonical aspects of recent ENSO events have been absorbed by the non-ENSO modes of variability. Whatever the case, we are left wondering if differences in background climate over much longer (multimillennial) timescales are perhaps strong enough to trigger changes in ENSO behavior. This question underscores the importance of determining how ENSO interacts with the interdecadal timescale fluctuations and how the longer timescale climate variability has varied in the past.

There is evidence, from both paleoclimate studies and modeling, that ENSO variability has not been ubiquitous throughout the Holocene. Sandweiss et al. (1996) and Rollins et al. (1986) have argued, based on the speciation record of molluscan assemblages along the north-central Peru coast, that El Niño events started to occur at about 5000 B.P., but were absent from the early to mid-Holocene (8000–5000 B.P.). Although the significance of the moluskan data has been debated and the conclusions have been contested (DeVries and Wells, 1990; DeVries et al., 1997; Wells and Noller, 1997; Sandweiss et al., 1997), independent evidence from lacustrine laminations in southwestern Ecuador supports the Sandweiss interpretation (Rodbell et al., 1999).

While the paleogeographic evidence is being sifted and debated, it is nevertheless of interest for climatologists to consider such a climate scenario. It has been recently argued on physical grounds that ENSO variability is required for the aperiodic removal (to polar regions) of excess heat that accumulates in the equatorial Pacific Ocean (Sun and Trenberth, 1998). If this the-

ory is correct, then it may be that the background climates of prior epochs (e.g., the early to mid-Holocene) had smaller pole-to-equator temperature contrasts and no need for the augmented poleward heat transport that is presently accomplished through El Niño events. One can speculate that such a climate would have lacked the equatorial Pacific characteristics that presumably lead to ENSO cycles now, such as intensified trade winds, western Pacific accumulations of warm upper-layer water, large zonal SST gradients, and an exaggerated eastward uptilt of the thermocline. Using a relatively simple model, Sun (2000) illustrates how ENSO variability may be sensitive to the temperature contrast between the eastern and western Pacific and the related availability (through thermocline uptilt) of cold-temperature thermocline water that can be upwelled and mixed into the surface layer in the eastern Pacific. Without the SST gradients that we see in non-El Niño years today, ENSO instabilities would not have arisen, ultimately because they would not have been required for the poleward removal of heat. The equatorial cold tongue would have been less well developed, and the coastal temperatures off Peru would have been warmer (though still relatively cool and characterized by coastal upwelling) and without the alternation between warm and cold years (ENSO).

The previous scenario, though speculative, is consistent with the findings of Sandweiss et al. (1996) and Rodbell et al. (1999) and also with other, independent findings. For example, there is evidence that the contrast in seasonal insolation between the equator and high latitudes has increased between the early and late Holocene (Berger, 1978), which would be consistent with smaller meridional temperature differences before 5000 B.P. and a corresponding lack of ENSO activity. Consistent with a (previously) warmer Peru coastal climate, ice core evidence indicates that the climate of the adjacent north-central Peru highlands was warmer during the 8000–5000 B.P. period in question (Thompson et al., 1995).

What aspect of the background climate might account for a warmer thermocline temperature? A possible clue lies in the self-similarity in SSTA patterns across timescales. In three separate SSTA patterns—secular (Fig. 1), ENSO (Fig. 2), and interdecadal (Fig. 5)—we see common elements that differ mainly in their intensities. One is a general warmth (coolness) in the low latitudes of the eastern Pacific extending poleward off the coasts of North and South America and as far north as the Gulf of Alaska. Also seen in all three patterns is an area of cool (warm) temperatures across the midlatitudes of the central North Pacific. The latter feature, in particular, is important because it extends across the subtropical convergence zone north of

Hawaii, a region of ocean instability where the mixing of upper ocean water by winter storms can entrain surface water to thermocline depths. This water is of an appropriate density to be transported by a circuitous route along the thermocline to low latitudes, where it can upwell on or near the equator (Gu and Philander, 1997). An oscillation presumably occurs because as thermocline water becomes warmer and/or deeper at low latitudes, cooler water is entrained at the high latitudes to replace it, reversing the cycle. Because the transit from the central North Pacific to the equator requires the better part of a decade, the interdecadal timescale is associated with maximum amplitude across the subtropical convergence.

However, there is a problem in applying this self-similar pattern to the millennial timescale. To produce a quasi-stationary (very long timescale) change in the background temperature of the tropical thermocline, the midlatitude thermal anomaly across the subtropical convergence would have to be of the same sign. That, or some entirely different reversal mechanism with a much longer oscillation timescale, would have to be involved: water mass trajectories at tropical thermocline depths do not have millennial (or even centennial) transit times.

This discussion has identified several areas in which further paleoclimate research can be especially fruitful. Although we can see some enticing areas of convergence between paleoclimate scenarios for Holocene changes in ENSO variability and inferences from the instrumental climate record (Figs. 1 and 4), inconsistencies remain. Further paleoclimatic evidence is needed to determine whether a background ocean climate consistent with current theories existed (e.g., Sun 2000) and whether the Peru coastal climate was consistent with a warmer but steady East Pacific (e.g., with increased coastal rainfall). In regard to the former, studies such as Climate Mapping [Climate: Long-Range Investigation, Mapping, and Prediction (CLIMAP)] should consider variations, vis-à-vis the present, in the pole-to-equator SST differences and in the subtropical subduction regions of the North and South Pacific. In regard to the latter, paleogeographers have yet to uncover evidence of the increased vegetation that would be expected of a warmer (albeit semiarid) coastal climate. It must also be hoped that paleoecologists can clarify what kind of an ocean climate (steady, but how much warmer?) would support the observed warm-water molluskan assemblages during the 8000–5000 B.P. period and discourage the cold-water species observed after 5000 B.P. Finally, inconsistencies in the paleomarkers of the early to mid-Holocene climate of Peru, such as those noted by DeVries and Wells (1990), must be addressed.

Acknowledgments

J. Harris prepared all of the basic data sets used in this project. We are grateful to M. Dettinger for his extraordinary thoroughness and his many helpful comments while reviewing the manuscript. This work has been supported by base funding from the Environmental Research Laboratories, by a grant from the Pan-American Climate Studies program, and by a grant from the Inter-American Institute for Global Change Research (IAI).

References

Barnston, A. G., and R. E. Livezey, 1987: Classification, seasonality and persistence of low-frequency atmospheric circulation patterns. *Monthly Weather Review*, **115**: 1083–1126.

Berger, A., 1978: Long-term variations of caloric insolation resulting from the Earth's orbital elements. *Quaternary Research*, **9**: 139–167.

Bjerknes, J., 1964: Atlantic air-sea interaction. *Advances in Geophysics*, **10**: 1–82.

Curry, R. G., M. S. McCartney, and T. M. Joyce, 1998: Oceanic transport of subpolar climate signals to mid-depth subtropical waters. *Nature*, **391**: 575–577.

D'Arrigo, R. D., E. R. Cook, and G. C. Jacoby, 1996: Annual to decadal-scale variations in northwest Atlantic sector temperatures inferred from tree rings. *Canadian Journal of Forest Research*, **26**: 143–148.

Dettinger, M. D., D. S. Battisti, G. M. McCabe, Jr., R. D. Garreaud, and C. M. Bitz, 2000: Interhemispheric effects of interannual and decadal ENSO-like climate variations on the Americas. In Markgraf, V. (ed.), *Interhemispheric Climate Linkages*. San Diego: Academic Press, Chapter 1.

DeVries, T. J., and L. E. Wells, 1990: Thermally-anomalous Holocene molluscan assemblages from coastal Peru: Evidence for palaeogeographic, not climatic change. *Palaeogeography, Palaeoclimatology, Palaeoecology*, **81**: 11–32.

DeVries, T. J., L. Ortlieb, A. Diaz, L. Wells, and Cl. Hillaire-Marcel, 1997: Determining the early history of El Niño. *Science*, **276**: 965–966.

Enfield, D. B., and L. S. Cid, 1991: Low-frequency changes in El Niño—Southern Oscillation. *Journal of Climate*, **4**: 1137–1146.

Enfield, D. B., and D. A. Mayer, 1997: Tropical Atlantic SST variability and its relation to El Niño—Southern Oscillation. *Journal of Geophysical Research*, **102**: 929–945.

Enfield, D. B., and A. M. Mestas-Nuñez, 1999: Multiscale variabilities in global sea surface temperatures and their relationships with tropospheric climate patterns. *Journal of Climate*, **12**: 2719–2733.

Evans, M. N., A. Kaplan, M. A. Cane, and R. Villalba, 2000: Globality and optimality in climate field reconstructions from proxy data. In Markgraf, V. (ed.), *Interhemispheric Climate Linkages*. San Diego: Academic Press, Chapter 4.

Gray, W. M., 1990: Strong association between West African rainfall and U.S. landfall of intense hurricanes. *Science*, **249**: 1251–1256.

Gu, D., and S. G. H. Philander, 1997: Interdecadal climate fluctuations that depend on exchanges between the tropics and extratropics. *Science*, **275**: 721–892.

Hastenrath, S., L. C. de Castro, and P. Aceituno, 1987: The Southern Oscillation in the Atlantic sector. *Contributions in Atmospheric Physics*, **60**: 447–463.

Houghton, J. T., L. G. Mehra Filho, B. A. Callander, N. Harris, A. Kattenberg, and K. Maskell (eds.), 1996: *Climate Change 1995, the Science of Climate Change: Contribution of WGI to the Second Assessment Report of the Intergovernmental Panel on Climate Change*. Cambridge, U.K.: Cambridge Univ. Press.

Hunt, B. G., 1998: Natural climatic variability as an explanation for historical climate fluctuations. *Climatic Change*, **38**: 133–157.

Hurrell, J. W., 1995: Decadal trends in the North Atlantic Oscillation, regional temperatures and precipitation. *Science,* **269**: 676–679.

Kaplan, A., M. A. Cane, Y. Kushnir, A. C. Clement, M. B. Blumenthal, and B. Rajagopalan, 1998: Analyses of global sea surface temperature 1856–1991. *Journal of Geophysical Research,* **103**: 18567–18589.

Landsea, C. W., R. A. Pielke, A. M. Mestas-Nuñez, and J. A. Knaff, 1999: Atlantic basin hurricanes: Indices of climatic changes. *Climatic Change,* **42**: 89–129.

Lanzante, J. L., 1996: Lag relationships involving tropical sea surface temperatures. *Journal of Physical Oceanography,* **9**: 2568–2578.

Latif, M., and T. P. Barnett, 1994: Causes of decadal climate variability over the North Pacific and North America. *Science,* **266**: 634–637.

Latif, M., and T. P. Barnett, 1995: Interactions of the tropical oceans. *Journal of Climate,* **8**: 952–964.

Latif, M., and T. P. Barnett, 1996: Decadal climate variability over the North Pacific and North America: Dynamics and predictability. *Journal of Climate,* **9**: 2407–2423.

Lau, N.-C., and M. J. Nath, 1994: A modeling study of the relative roles of tropical and extratropical anomalies in the variability of the global atmosphere-ocean system. *Journal of Climate,* **7**: 1184–1207.

Mantua, N., S. R. Hare, Y. Zhang, J. M. Wallace, and R. C. Francis, 1997: A Pacific interdecadal climate oscillation with impacts on salmon production. *Bulletin of the American Meteorological Society,* **78**: 1069–1079.

Mestas-Nuñez, A. M., and D. B. Enfield, 1999: Rotated global modes of non-ENSO sea surface temperature variability. *Journal of Climate,* **12**: 2734–2746.

Muck, P., 1989: Major trends in the pelagic ecosystem off Peru and their implications for management. *In* Pauly, D., P. Muck, J. Mendo, and I. Tsukuyama (eds.), *The Peruvian Upwelling System: Dynamics and Interactions.* ICLARM Conference Proceedings, Instituto del Mar del Peru (IMARPE), Callao, Peru; Deutsche Gesellschaft für Technische Zusammenarbeit (GTZ) GmbH, Eschbom, Federal Republic of Germany; and International Center for Living Aquatic Resources (ICLARM), Manila, Philippines, pp. 386–403.

Nakamura, H., G. Lin, and T. Yamagata, 1997: Decadal climate variability in the North Pacific during the recent decades. *Bulletin of the American Meteorological Society,* **78**: 2215–2225.

Nicholson, S. E., 1989: Long-term changes in African rainfall. *Weather,* **44**: 46–56.

Quinn, W. H., V. T. Neal, and S. Antunez de Mayolo, 1987: El Niño occurrences over the past four and a half centuries. *Journal of Geophysical Research,* **92**: 14449–14461.

Rodbell, D., G. O. Seltzer, D. M. Anderson, D. B. Enfield, M. B. Abbott, and J. H. Newman, 1999: A high-resolution ~15,000 year record of El Niño driven alluviation in southwestern Ecuador. *Science,* **283**: 516–520.

Rollins, H. B., J. B. Richardson, and D. H. Sandweiss, 1986: The birth of El Niño: Geoarchaeological evidence and implications. *Geoarchaeology,* **1**: 3–15.

Sandweiss, D. H., J. B. Richardson, III, E. J. Reitz, H. B. Rollins, and K. A. Maasch, 1996: Geoarchaeological evidence from Peru for a 5000 years B.P. onset of El Niño. *Science,* **273**: 1531–1533.

Sandweiss, D. H., J. B. Richardson, III, E. J. Reitz, H. B. Rollins, and K. A. Maasch, 1997: Determining the early history of El Niño. *Science,* **276**: 966–967.

Sarachik, E. S., M. Winton, and F. L. Yin, 1996: Mechanisms for decadal-to-centennial climate variability. *In* Anderson, D. L. T., and J. Willebrand (eds.), *Decadal Climate Variability,* Vol. 44 of NATO ASI Series I: Global and Environmental Change. New York: Springer-Verlag, pp. 157–210.

Smith, T. M., R. E. Livezey, and S. S. Shen, 1998: An improved method for analyzing sparse and irregularly distributed SST data on a regular grid: The tropical Pacific Ocean. *Journal of Climate,* **11**: 1717–1729.

Sun, D.-Z., 2000: Global climate and ENSO: A theoretical framework. *In* Diaz, H. F., and Markgraf, V. (eds.), *El Niño and the Southern Oscillation: Multiscale Variability, Global and Regional Impacts.* Cambridge, U.K.: Cambridge University Press, pp. 443–463.

Sun, D.-Z., and K. E. Trenberth, 1998: Coordinated heat removal from the equatorial Pacific during the 1986–87 El Niño. *Geophysical Research Letters,* **25**: 2659–2662.

Thompson, L. G., E. Mosley-Thompson, M. E. Davis, P.-N. Lin, K. A. Henderson, J. Cole, J. F. Bolzan, and K.-B. Liu, 1995: Late glacial stage and Holocene ice core records from Huascarán, Peru. *Science,* **269**: 46–50.

Timmerman, A., M. Latif, R. Voss, and A. Grotzner, 1998: Northern Hemispheric interdecadal variability: A coupled air-sea mode. *Journal of Climate,* **11**: 1906–1931.

Trenberth, K. E., and T. J. Hoar, 1996: The 1990–1995 El Niño-Southern Oscillation event: Longest on record. *Geophysical Research Letters,* **23**: 57–60.

Trenberth, K. E., and J. W. Hurrell, 1994: Decadal atmosphere-ocean variations in the Pacific. *Climate Dynamics,* **9**: 303–319.

Wells, L., and J. S. Noller, 1997: Determining the early history of El Niño. *Science,* **276**: 966.

Wiles, G. C., R. D. D'Arrigo, and G. C. Jacoby, 1998: Gulf of Alaska atmosphere-ocean variability over recent centuries inferred from coastal tree-ring records. *Climatic Change,* **38**: 289–306.

Woodruff, S. D., R. J. Slutz, R. L. Jenne, and P. M. Steurer, 1987: A comprehensive ocean-atmosphere data set. *Bulletin of the American Meteorological Society,* **68**: 1239–1278.

3

Polar Air Outbreaks in the Americas: Assessments and Impacts During Modern and Past Climates

JOSE A. MARENGO and JEFFREY C. ROGERS

Abstract

Polar outbreaks have long attracted the attention of meteorologists and climatologists. They are described here in terms of their air temperatures, air mass central pressures, and wind characteristics, as well as in relation to the atmospheric circulation and their trajectories over the continents. Outbreaks of polar air into low latitudes tend to organize tropical convection and rain in summer, while the cold air tends to produce cooling in lower latitudes and freezes in the subtropics during winter. Cold waves accompanying polar outbreaks have other significant impacts in the Americas, including adverse effects on coffee production in South America and on citrus production in North America. Synoptic events leading to cold waves are somewhat similar in the Americas, particularly in terms of the upper air patterns and forcing, but the intensity of the outbreaks in terms of temperature and pressure is larger in North America. Typically, they are associated with an amplified ridge lying across the western edge of the continent. Records of freezes in coffee- and citrus-growing areas are used to reconstruct subtropical climate variability and to identify linkages to low-frequency variability mechanisms in the atmospheric circulation. Results suggest that there is little link between the occurrence of El Niño and cold waves and freezes in either hemisphere. Citrus freeze occurrences in Florida are linked to the long-term variability in the Pacific–North American (PNA) teleconnection pattern, and they appear to occur in clusters concentrated near the end of each century. Comparisons are made of possible paleoclimate scenarios for the occurrence of cold waves based on evidence reconstructed from pollen and lake sediment samples collected in southeastern and central Brazil. Copyright © 2001 by Academic Press.

Resumen

Incursiones de aire frío de origen polar han atraído la atención de meteorólogos y climatólogos desde ya hace mucho tiempo. Estas incursiones de aire polar han sido estudiadas y caracterizadas en base a sus características de temperatura, intensidad del centro de alta presión atmosférica, y del comportamiento de la circulación y trayectorias del aire frío una vez que entran al continente. Las irrupciones de aire polar hacia latitudes más bajas tienden a organizar la convección y producir lluvia en verano, mientras que en invierno el aire frío tiende a producir enfriamiento y a veces, hasta heladas en latitudes subtropicales. Ondas de frío que acompañan las incursiones de aire frío tienen un impacto significativo en las Américas, incluyendo efectos adversos en la producción de café en América del Sur, y de naranjas y otros cítricos en América del Norte. Los eventos sinópticos que preceden a las ondas de frío son similares en ambas América del Norte y del Sur, particularmente en relación a los forzantes y a la circulación en altura, pero la intensidad de estas irrupciones y el enfriamiento que producen es mayor en América del Norte. Las ondas de frío muestran como característica general la amplificación de una cresta localizada a lo largo del extremo oeste del continente. Registros de ocurrencias de heladas que afectaron al café y cítricos son usados como indicadores climáticos, para reconstruir la variabilidad climática en la región subtropical, y para identificar los mecanismos de variabilidad climática de baja frecuencia en la circulación atmosférica, asociada a las ondas de frío. Resultados de los estudios realizados sugieren de que hay poca relación entre la ocurrencia del fenomeno El Niño y la presencia de ondas de frío en invierno de ambos hemisferios. Ondas de frío y heladas que afectan a los cítricos en el estado de Florida, U.S.A. estan asociados a la variabilidad de largo plazo del patrón de teleconección identificado como Pacífico–América del Norte (Pacific–North American, PNA), y parece de que se concentran en grupos y eventos al final de cada siglo. Se hacen comparaciones con posibles escenarios paleo-climáticos caracteristicos de episodios fuertes de ondas de frío, basados en reconstrucciones a partir de muestras de granos de pólen y de sedimentos de lagos, colectados en las regiones del sureste y central de Brasil.

3.1. INTRODUCTION

The North and South American atmospheric circulation is characterized by interactions occurring between warm, moist air masses of tropical latitudes and cold, dry air of higher latitudes. An *air mass* is a large dome of air, typically thousands of kilometers across, that has similar horizontal temperature and moisture characteristics acquired from the underlying surface of the Earth. Five basic air masses affect the Americas' weather: continental Arctic (or Antarctic), continental polar, maritime polar, maritime tropical, and continental tropical. North American continental polar air masses originate to the south of the Arctic Circle over northwestern North America and dominate the winter weather across the United States, while continental Arctic masses originate over the Arctic Ocean or even over Siberia. In South America, cold air masses of Antarctic or polar oceanic origin can reach deep into the core of the tropics, affecting the regional agriculture in southern Brazil and sometimes producing cooling that can be detected in the western Amazon region.

Polar outbreaks occur when polar air masses clash

with continental or maritime tropical air. The transition regions between these air masses are *frontal boundaries*, narrow zones where temperature changes substantially within short horizontal distances of between 10 and 150 km. Typically, cold air pushes and lifts the warm air due to its greater density, and the leading *cold front* boundary separating warm and cool air is identified on weather charts in places where the dense air is advancing to replace the warm air at the Earth's surface. In general, *polar outbreaks* represent a thrust of cold air from higher to lower latitudes, affecting the weather and climate in certain regions of the planet. They are noticeable features of the distribution of low-level temperature, circulation, and rainfall over both North and South America. Evidence suggests that these air mass and frontal exchanges existed in climates of the past, but with different intensities than at the present.

We interchangeably use other common phrases for polar outbreaks in this chapter, including *cold waves*, *cold surges*, and *cold spells*. Polar outbreaks have long attracted the attention of forecasters, especially the rapid temperature change or cold waves. Cox (1916) described the basic atmospheric factors defining the severity of polar outbreaks, factors that go beyond temperature changes:

> It is obvious that the degree of cold recorded in a cold wave depends upon the intensity of the cold in the high and the temperature in the low in front of the high, the magnitude of the high and the low and the consequent wind force, the direction of the movement of the high and also the direction and movement of the high and low or lows which it is following, the general character of the sky, the humidity, whether there is snow on the ground, and the season of the year; in a word, all the atmospheric conditions, sometimes even within an area of 3000 or 4000 square miles or more.

Transient incursions of polar and midlatitude air into subtropical and tropical latitudes are a distinctive feature of the synoptic climatology of the Americas. These incursions are a year-round phenomenon, with relatively modest seasonal changes in their structure, and have their largest impact on summer precipitation and on the wintertime temperature field. Meridional exchanges of air masses between lower and higher latitudes in South America are the most intense in the entire Southern Hemisphere, mainly due to the presence of the Andes Mountains. The effects of these polar outbreaks are mainly regional, and studies of cold surges have focused on southeastern Asia bounded by the Himalayan Plateau (Wu and Chan 1995, 1997); the east side of the Rocky Mountains, including impacts on the western side of the Great Plains in North America (Colle and Mass, 1995); and Central America and the Caribbean (Schultz et al., 1997, 1998). In South America, episodic cold surges into the east of the subtropical

Andes sometimes result in freezing conditions in southeastern Brazil during winter (Hamilton and Tarifa, 1978; Fortune and Kousky, 1983; Marengo et al., 1997a,b; Bosart et al., 1998; Seluchi and Marengo, in press) or in enhanced convection and rainfall during summer (Kousky, 1979; Garreaud and Wallace, in press; Liebmann et al., 1998). Cold frontal passages and cold surges have also been documented in Australia (Baines, 1983; Smith et al., 1995).

This chapter examines polar outbreaks in the Americas and their societal impacts, focusing on their meteorological and climatological characteristics, on the proxy climate evidence available in citrus (North America) and coffee (South America) production, and on the freeze statistics dating to the nineteenth century. Citrus and coffee crops are important to the economy of tropical regions, and freezes associated with polar air outbreaks compromise their production and subsequent prices in world markets. Records of damages to these crops also serve as an indicator of the cold wave severity occurring at higher latitudes along the paths of polar air masses. We document (1) weather patterns of wintertime polar outbreaks in the Americas and their interannual and interdecadal variability, (2) circulation features and mechanisms typical of polar outbreaks and their variability in time, and (3) the impact of these polar outbreaks on coffee and citrus production in South and North America. On paleoclimatic scales, information deduced from fossilized pollen and lake sediments (indicators of vegetation distribution) has been used to detect the potential variability of polar outbreaks in the early Holocene. It will be shown that the presence of relicts of *Araucaria* forests in southern Brazil may be an indicator of variability in the frequency of cold fronts and related polar outbreaks during the late Quaternary period. Our topic fits the scope of this book, covering present and past interhemispheric climate linkages between the Americas and their societal impacts. For a more complete review of the meteorological and dynamics aspects of the polar outbreaks, as well as several case studies in North, Central, and South America, the reader should refer to the papers by Schultz et al. (1997, 1998), Garreaud, R. D. (2000), Seluchi and Marengo (in press), Garreaud and Wallace (in press), Marengo et al. (1997a), Konrad (1996), and Colle and Mass (1995), among others.

3.2. REVIEW OF LOWER- AND UPPER-LEVEL CIRCULATION IN THE AMERICAS

In this section, we show the mean seasonal summer and wintertime ensembles of near-surface (925 hPa [hectoPascal]) and upper level (200 hPa) atmospheric circulation for December–February and June–August, drawing upon the studies of Hastenrath (1996) and Kousky and Ropelewski (1997).

The seasonally varying mean circulation features are intimately linked to the horizontal gradients in temperature. The tropics are heated strongly throughout the year, while middle and high latitudes experience considerable variation in heating from summer to winter due to variations in solar insolation. As a result, mean meridional temperature gradients, and consequently the strength of the low-level zonal winds, vary considerably from summer to winter (Figs. 1A and 1B and 2A and 2B). The strongest meridional temperature gradients and strongest zonal winds are generally observed in middle latitudes of the winter hemisphere, especially the Southern Hemisphere. The Southern Hemisphere winter jet stream (maximum in zonal winds, Fig. 2A) is closer to the equator than the corresponding Northern Hemisphere winter jet stream (Fig. 1A), reflecting the asymmetry in heating occurring between the hemispheres. Zonal asymmetries in atmospheric circulation in the Western Hemisphere arise primarily due to the difference in thermal capacity between land and water. Continental areas are often warmer during the summer and cooler during the winter compared to neighboring oceanic regions.

The subtropical highs over the South Pacific and the South Atlantic display very little seasonality in either central pressure or position between winter and summer (Figs. 1B and 2B). However, there is a seasonal cycle in the sea level pressure (SLP) over central South America, with the lowest pressures occurring during December–February and the highest pressures occurring during June–August (Figs. 1B and 2B). This pattern is consistent with the seasonal variation in the low-level meridional flow, which features strong flow from the Amazon basin southward to northern Argentina during austral summer (December–January–February, DJF), and much weaker flow from the north but stronger flow from southern latitudes into lower latitudes during winter (June–July–August, JJA). An important feature of the upper level summertime circulation over South America is the presence of an anticyclonic circulation over the Bolivian Plateau that is absent during winter (Fig. 2A).

Seasonal variations in SLP are evident over the middle and high latitudes of the North Pacific and North Atlantic Oceans. There is a marked northwestward shift from winter to summer (Figs. 1B and 2B) in the central positions of the oceanic anticyclones, and their central pressure increases somewhat during summer in response to the relative coldness of the underlying surface. High pressure dominates continental North America during winter, and lower pressure dominates

a

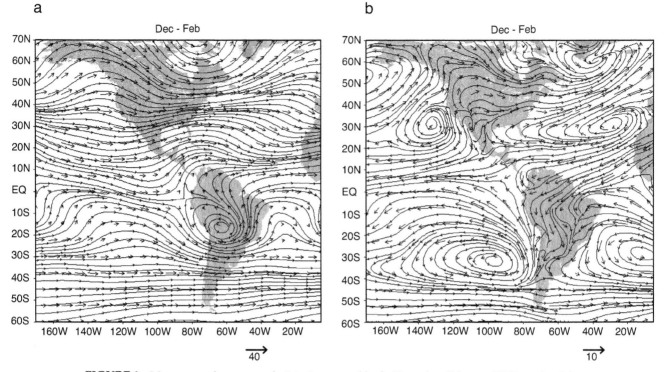

FIGURE 1   Mean seasonal summer and wintertime ensembles for December–February, (A) Upper level circulation (200 hPa) over the Americas, and (B) near-surface circulation (925 hPa) over the Americas.

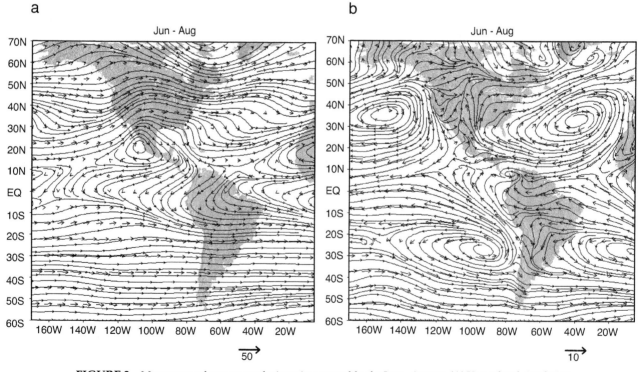

FIGURE 2   Mean seasonal summer and wintertime ensembles for June–August, (A) Upper level circulation (200 hPa) over the Americas, and (B) near-surface circulation (925 hPa) over the Americas.

it during summer; however, the seasonal variations in surface winds are not as apparent as over tropical South America.

Throughout the year a mean westerly (zonal) flow dominates much of the middle latitudes of both hemispheres (Figs. 1A and 2A). The zonal flow is strongest in each hemisphere's winter and weaker in the summer. However, the zonal flow is strongest in both hemispheres in the southern winter (Fig. 2A), and the weakening occurring in the southern summer reaches an intensity that still approximately matches the mean strength of the flow during the northern winter (Fig. 1A). The mean flow during the northern summer is seasonally the weakest on the planet. Comparison between hemispheres in Figs. 1A and 2A also reveals a greater wavelike structure, or meridionality, to the flow in the Northern Hemisphere. Northern winters are occasionally beset by periods in which the westerly flow decreases or becomes blocked, and meridional (north-south) flow begins to dominate the circulation. This flow is accompanied by an amplification of the wave ridges, notably the ridge visible at higher latitudes over western North America and the eastern Pacific (Fig. 1A). During periods of blocked westerlies, unusually cold weather persists over North America for 2–4 weeks. These long cold spells can have cold waves (2–6 days duration) embedded within them, although their severity may not always be nearly as great as those of some of the events described later in this chapter. The persistent cold meridional flow periods have created some extraordinarily cold winters, especially east of the Rocky Mountains, that are long remembered by many people. Some of these in recent memory include the winters of 1962–63, 1968–69, 1976–77, 1978–79, and 1981–82. The 1968–69 and 1978–79 winters did not, however, witness the occurrence of strong, well-known severe polar outbreaks. Blocking ridges can also occur in the Southern Hemisphere around and over the Andes, producing cold waves in South America, but the hemispheric climatological flow (Figs. 1A and 2A) gives little hint of their existence.

3.3. POLAR OUTBREAKS AND THEIR SIGNIFICANCE IN THE CLIMATE OF THE AMERICAS

3.3.1. South American Polar Outbreaks

Synoptic-scale incursions of midlatitude air moving into tropical and subtropical latitudes east of the Andes (leeward side) are observed all year long in two preferred regions: the first is located close to the mountains between 20° and 30°S, and the second region is located in southern Brazil, extending into the adjacent Atlantic.

In summer (December–February), cold air from higher southern latitudes moves into tropical latitudes behind a cold front and organizes tropical convection. This movement has a large influence on climatological rainfall patterns over central Brazil and eastern Amazonia. At this time of the year, there is a strong southward advection of warm and humid air coming from the northern Amazon that creates favorable conditions for the development of mesoscale convective systems linked to abundant rainfall over Paraguay and northern Argentina. On the leeward side of the Andes, incursions of polar air toward lower latitudes accompanying a cold front are favored by the channeling effect of the mountains. During these incursions, cold fronts move northward and merge with the South Atlantic Convergence Zone (SACZ), a region of convergence of moist air from the Amazon and cold air from the south, generating convection that is enhanced by the assimilation of the cold fronts. Because of this convergence, the SACZ is more intense in summer (Garreaud, in press a,b,; Liebmann et al., 1998; Seluchi and Marengo, in press).

In winter (June–August), polar air masses move from higher to lower latitudes and the cold front sometimes becomes stationary, allowing an intense nocturnal cooling of the surface, associated with low humidity and a lack of clouds. This cold air has undergone modifications along its path, but still may be cold enough to produce freezing conditions and snow in the subtropical parts of Brazil. Freezing weather in southeastern Brazil is caused by these polar air outbreaks in May–August. Events like these are known by the Portuguese name of *friagem* (plural: friagems) or *Surazo* in the Amazon region, where they have a marked effect on tropical and extratropical weather. These cold surges occur several times per year (from zero to eight times), producing low temperatures in the midlatitudes, and are sometimes so strong that extensive freezes affect southeastern Brazil (see reviews in Marengo et al., 1997a). In some episodes, the cold surges produce considerable cooling in central and northern Amazonia (Morize, 1992; Serra and Ratisbona, 1942; Parmenter, 1976; Fortune and Kousky, 1983; Satyamurty et al., 1990; Seluchi and Nery, 1992; Marengo et al., 1997a,b; Seluchi and Marengo, in press).

3.3.1.1. Conceptual Model of South American Polar Outbreaks

Most of the literature on South American polar outbreaks has focused on descriptions of the circulation and dynamics of individual episodes (see reviews in Marengo et al. 1997a; Garreaud, 1999, 2000; Marengo, in press), while more recent work has focused on the

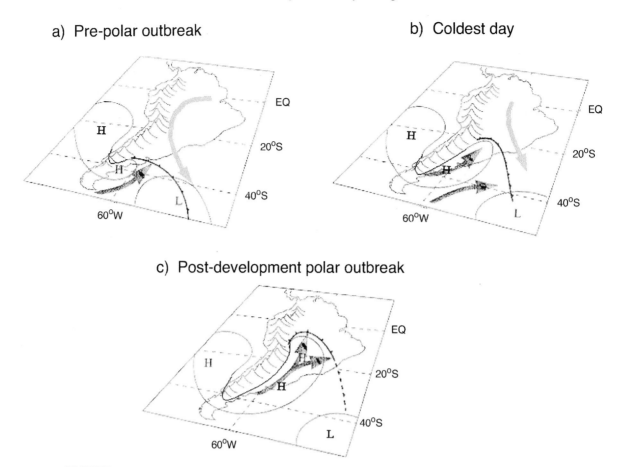

a) Pre-polar outbreak

b) Coldest day

c) Post-development polar outbreak

FIGURE 3 Conceptual model of wintertime polar outbreaks in South America. (Adapted from Marengo et al., 1997a; Garreaud, in press b.)

mean structure, evolution, and dynamics of these phenomena (Garreaud, in press a,b; Seluchi and Marengo, in press; Kousky and Cavalcanti, 1997). Based on literature published from early work by Morize (1922) to the most recent work by Garreaud (1999, 2000) and Marengo et al. (1997a) describing general features and discussing individual case studies, a simple conceptual model for wintertime polar outbreaks is shown in Fig. 3.

Several case studies have shown that cold outbreaks and some freeze events in southern Brazil are preceded by a slow eastward-moving long wave in the South Pacific Ocean, which is amplified greatly some days before (see review in Marengo, in press) and exhibits large meridional displacements of the flow (see Table 1 for some cold events in the state of São Paulo, Brazil). When at maximum amplitude, the wave ridge is usually located near the southern Andes, while the next downstream trough lies over the South Atlantic in Brazilian longitudes. This typical configuration allows air originally from higher latitudes to be channeled equatorward. Frontogenesis is associated with the

wave amplification, and a freeze occurs east of the polar anticyclone. In South American cold waves, an anticyclone–cyclone couplet causes the associated southerlies to transport colder air from high latitudes, contributing substantially to the cooling. The conceptual model (Garreaud, 1999; Marengo et al., 1997a) shows the evolution of near-surface circulation patterns. Key elements include the cold core anticyclone that moves from the southeastern Pacific into southern Argentina near the extreme southernmost region of South America, as well as the cyclone over the southwestern Atlantic. South of 30°S these two pressure systems grow due to upper level vorticity advection. During the pre-polar outbreak period, before the coldest day in southeastern Brazil (Fig. 3A), the geostrophic southerly winds between the high- and low-pressure systems produce low-level cooling along the east coast of South America and farther inland as far north as 25°S.

Closer to the tropical Andes, the low-level flow is blocked during the developmental period, leading to ageostrophic, mountain parallel flow (Fig. 3B) and cold

TABLE 1   List of Cold Events in the State of São Paulo, Brazil and Their Intensity

Date	Intensity (min. temp., °C)	Damage
14 July 1892	0.2	Severe
14 July 1894	1.0	Severe
25 July 1895	1.0	Severe
18 June 1899	1.6	Moderate
19 August 1902	0.2	Very intense
12 August 1904	1.5	Severe
3 September 1912	1.8	Severe
25 June 1918	−1.5	Very intense
12 July 1923	2.0	Weak/absent
4 July 1925	2.0	Weak/absent
29 June 1931	2.0	Moderate
14 July 1933	1.4	Moderate
12 July 1942	−0.2	Severe
15 September 1943	2.0	Moderate
5 July 1953	1.2	Severe
2 August 1955	2.0	Severe
21 July 1957	1.2	Moderate
7 July 1962	2.4	Severe
21 August 1965	0.6	Moderate
11 July 1969	2.4	Severe
9 July 1972	1.6	Moderate
18 July 1975	0.6	Very intense
31 May 1979	0.2	Severe
21 July 1981	0.2	Severe/very intense
8 June 1985	1.4	Moderate
5 June 1988	1.8	Moderate
27 June 1994	0.3	Severe/very severe
10 July 1994	0.1	Severe/very severe

Adapted from Marshall, 1983 and updated by J. Zullo (Centro de Pesquisas Agricolas [CEPAGRI], Universidade de Campinas, São Paulo, Brazil). No cold events that affected coffee were registered after the ones in winter 1994.

air damming. As the cold air moves into lower latitudes (Fig. 3C), the blocking effect of the Andes diminishes (due to a more zonal orientation of the Andes to the north of 18°S). Cold advection produced by the southerly wind produces most of the local cooling at the leading edge of the surge, and the cooling is felt in southeastern Brazil and on the west side of Brazil adjacent to the Andes. Thus, the advance of the polar air outbreak along the subtropical Andes is set up by the topographic blocking of the flow.

Accompanying the cold surge is an area of surface high pressure. The highest pressure detected in central South America with a cold spell was 1044 hPa during the event of 15 July 1975 (Tarifa et al., 1977; Marengo, in press), which was considered the strongest in the twentieth century. Cold wave central pressure intensities in southern Brazil rarely exceed 1038 hPa: e.g., 1035 hPa (June 1994, Marengo et al., 1997a) and 1033 hPa (July 1981, Haddock et al., 1981).

3.3.2. North and Central American Polar Outbreaks

Cold waves have long attracted the attention of North American weather forecasters and climatologists (Garriott, 1906; Cox, 1916). Cold waves have been defined by meteorologists in terms of (1) severity or rapidity of the drop in air temperature (Wendland, 1987), (2) the numerical values of the low temperatures or temperature departures that ultimately occur, and (3) the duration of the cold wave (varying time spans can be used) per the Cox (1916) quotation in the Introduction. To a lesser extent the cold wave can be characterized by the intensity of the polar anticyclone accompanying it (Rogers and Rohli, 1991). In North America, polar outbreaks are frequently observed east of the Rocky Mountains–Mexican Sierras. The most visible feature of the winter polar outbreaks is the surface anticyclone moving from its Alaskan/Yukon source region and associated with the southward transport of cold air across its core and eastern flank (Rogers and Rohli, 1991). SLPs associated with the core of the air mass can exceed 1070 hPa in northwestern Canada, but the central pressure decreases steadily with the southward movement of the polar outbreak, seldom exceeding 1055 hPa once the central high-pressure area enters the United States and 1040 hPa by the time it reaches the United States East Coast or the Gulf of México. The December 1983 and 1989 cold waves and citrus freezes were associated with polar highs with central pressures in excess of 1055 hPa when the anticyclones entered the northern Great Plains of the United States (Rogers and Rohli, 1991). Some examples of anticyclone central pressures in the northern Plains for severe cold events are given in Table 2.

The polar outbreaks are typically associated with record low temperatures (Konrad and Colucci, 1989) and produce numerous adverse socioeconomic impacts across many parts of the United States, including loss of vegetable crops and citrus, damage to homes and plumbing, human mortality, and disruption of transportation systems, among other problems. Air temperatures in the northern Great Plains reached −35°C during the 1989 polar outbreak, and it broke over 250 minimum temperature records across the United States, ultimately producing a citrus freeze in Florida (Rogers and Rohli, 1991). Colle and Mass (1995) have documented northerly surges of cold air that often move equatorward along the eastern side of the Rockies into México and have described the strongest surges that developed in the midwinter of 1994, with temperature decreases of 20°–30°C and pressure rises of 15–30 hPa within 24 hr. These cold surges have several regional names in North and Central America: *Blue*

TABLE 2 Strongest Anticyclones Detected over the Northern Great Plains on 1200 UTC (Universal Time Zone) Synoptic Charts for the Period 1899–1989 and Associated Agricultural Damage in Florida

Date	Intensity (pressure, hPa)	Damage
3 January 1766	Unknown	Severe
7–8 February 1835	Unknown	Very severe
30 December 1880	1048.7	Severe
10–12 January 1886	1047.4	Severe
27 December 1894	1054.5	Severe
7 February 1895	1055.9	Severe
11 February 1899	1059.3	Severe
25 January 1905	1053.8	Moderate
30 December 1909	1032.4	Moderate
1–4 February 1917	1056.9	Severe
28 December 1917	1060.3	Vegetables
20 February 1918	1053.8	None
24 January 1920	1056.5	None
14 February 1923	1056.2	Some
20 December 1924	1057.2	None
27 December 1924	1054.5	None
27 December 1925	1055.2	Vegetables
14 January 1927	1053.2	Citrus
3–4 January 1928	1052.5	Moderate
12–13 December 1934	1030.7	Severe
28–29 January 1940	1051.9	Moderate
3 January 1947	1052.1	None
10–11 February 1947	1052.1	Severe
12–13 December 1957	1046.3	Severe
24 January 1961	1052.3	Severe
10 January 1962	1059.3	Moderate
13–14 December 1962	1045.3	Very severe
9–12 January 1977	1039.9	Moderate
13–14 January 1981	1040.1	Moderate
11–12 January 1982	1054.1	Severe
25 December 1983	1058.7	Severe
21–22 January 1985	1048.0	Severe
5 February 1988	1052.3	None
3 February 1989	1058.5	None
17 February 1989	1057.8	None
23–25 December 1989	1054.9	Severe

Modified from Rogers and Rohli, 1991.

Northers in Texas; *Nortes, Chocolatero,* and *Tehuano* in México; *Papagayo* in Nicaragua and Guatemala; *Atemporalado* in Honduras; and *Invierno de las Chicharras* in Venezuela.

Dalavalle and Bosart (1975) indicate that comparatively weak anticyclones (<1045 hPa) can also produce severe freezes over North America. The cold waves of the 1990s have been among the 10 worst events this century in the midwestern United States in terms of low temperature values (Rogers, 1997), but were associated with anticyclones of only moderate strength. The mid-January 1994 cold wave was associated with two fast-moving highs having intensities under 1040 hPa, the second of which was advected over the Plains behind an intense cyclone that moved across the Great Lakes. The early February 1996 cold wave was ushered in by the cold air advection behind a strong Alberta storm moving across the Midwest, producing an anticyclone that just exceeded 1040 hPa in intensity. Even the Florida freeze of 19–20 January 1997 was associated with an anticyclone of about 1042 hPa in intensity. The trajectory of this anticyclone was characteristic of that for Florida advective freezes (which are the most severe, as described in Section 3.4.2), but its intensity was vastly below normal and the bulk of the freeze damage was probably due to radiative cooling of the air.

As the cold front of a polar outbreak reaches the Gulf of México or the United States East Coast, a strong baroclinic zone is typically established in conjunction with the warm adjacent ocean that is the site of subsequent cyclone development. Thus, as the anticyclone at the core of the polar air mass enters the United States, traveling southward, cyclogenesis begins occurring along the coastal areas. The developing cyclone will frequently evolve into a ferocious *nor'easter* along the eastern coast of the United States (Kocin and Uccellini, 1990), producing tremendous snowfalls along the Atlantic coast and thereby creating many societal impacts of its own. One example is the East Coast storm accompanying what is still regarded as the worst cold wave on record in many parts of the country—that of 8–11 February 1899 (Kocin et al., 1988). The East Coast cyclone developed while the cold wave was advancing across the midsection of the continent, producing a strong baroclinic zone across the southeastern United States. Intensification of the pressure gradient occurred across the anticyclone/cyclone couplet, producing strong cold air advection and a further decline in air temperature and wind chill from the Plains to the East Coast. Schultz et al. (1997) studied the United States superstorm of 12–14 March 1993, which developed after polar air, originating over Alaska and western Canada, brought northerlies exceeding 20 m sec^{-1} and temperature decreases up to 15°C over 24 hr into México and Central America. During this cold surge, topographically channeled northerlies along the Rocky and Sierra Madre Mountains advected cold air equatorward, reaching as far south as 7°N, and the dynamical forcing associated with the low-latitude upper level trough and confluent jet-entrance region over México and Central America, in addition to the topographic channeling, favored the extraordinary equatorward incursion of cold air.

Schultz et al. (1998), Steenburgh et al. (1998), and Klaus (1973) documented midlatitude cold surges affecting México and Central America, based on observations and regional modeling. *A Central American cold surge* is defined as the leading edge of a cold anticyclone originating poleward of México that has penetrated equatorward to at least 20°N. Although the topographies of the Rocky Mountains and the Sierra Madre undoubtedly play an important role in channeling the cold surge equatorward, different planetary- and synoptic-scale flow patterns can also be conducive to longevity of Central American cold surges. From 177 cases studied, they concluded that 75% of the cold surges had durations of 2–6 days, the same timescale as mobile disturbances in the westerlies. There does not appear to be any relationship between the temperature drop and the duration of the event, although cold surges penetrating to low latitudes (7°–10°N) have a weak tendency to persist longer than those that do not penetrate to low latitudes (15°–20°N). In addition, cold surges tend to reach their most equatorward extent where topographic features impede the progress of cold air; the temperature decreases in the postsurge air do not appear to be related to the latitude of the cold outbreak.

3.3.3. Conceptual Model for North-Central American Cold Surges

As with the South American illustration in Fig. 3, the North American conceptual model (Figs. 4A–4C) shows the evolution of the surface circulation components of a polar outbreak, based on the work of Dallavalle and Bosart (1975), Konrad and Colucci (1989), Kocin and Uccellini (1990), Konrad (1996), and Schultz et al. (1998), for severe North American cold waves. Many different synoptic circumstances can lead to polar outbreaks around North America; we illustrate just one here, which is associated with severe cold waves over the eastern United States.

The polar outbreak begins (Fig. 4A) with a strong polar anticyclone moving away from its source region across south-central Canada into the United States. The polar high likely originates over snow-covered plains and forests of northwestern Canada, acquiring its characteristics from radiative cooling during the long polar nights. The dense, cold air is linked to a large positive

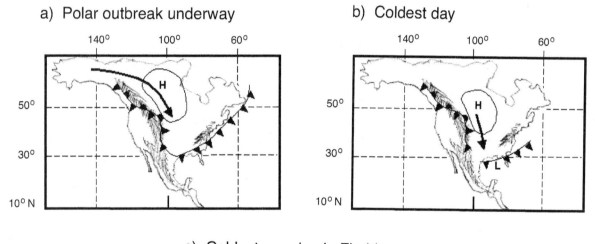

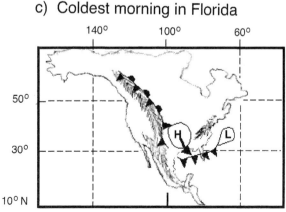

FIGURE 4 Conceptual model of wintertime polar outbreaks in North-Central America. (Adapted from Dallavalle and Bosart, 1975; Kocin and Uccellini, 1990; Schultz et al., 1998; Konrad, 1996; Konrad and Colucci, 1989.)

pressure anomaly at the core of the air mass. The baroclinic zone that is continually present south of the high encourages its south-southeastward motion by the process of cold air advection. The leading edge of the polar air approaches the Gulf of México in Fig. 4A, where strong baroclinicity is also developing. Within 12–24 hr (Fig. 4B), the cold front has entered México and a substantial cold outbreak begins in that country. Low-temperature records are also being set in the northern United States. A frontal wave develops in the Gulf baroclinic zone. The western boundary of the cold air mass is illustrated here as lying along the Rocky Mountains, a common barrier to the shallow cold air mass.

Within another 24–48 hr, the polar high reaches Texas (Fig. 4C) and is no longer supported by cold air advection. It is at its greatest peril of dissipating, especially over the relatively warm environment around the Gulf of México. It has weakened substantially, but is being maintained to some extent by the relatively cold continental surface. The cyclone developing in the Gulf in Fig. 4B has slowly migrated northeastward, undergoing a process of *secondary redevelopment* (Kocin and Uccellini, 1990) that leads to cyclogenesis off the East Coast. The East Coast cyclone will rapidly evolve as an upper air trough of cold polar air (not illustrated) begins to lend support via positive vorticity advection. Severe cold weather and windy conditions now prevail over the southeastern United States. In the final stages (Fig. 4C), both features of the anticyclone–cyclone couplet have received upper level support from the waves in the upper troposphere. The Gulf Coastal anticyclone begins redeveloping eastward, assisted by negative (anticyclonic) vorticity advection occurring ahead of another upper level ridge approaching from the west. It does not greatly reintensify due to its destabilization and proximity to the warm ocean. The cyclone has evolved into a major East Coast winter snowstorm, aided by upper air positive (cyclonic) vorticity advection east of the trough that was, in a large part, created and amplified by the preceding polar outbreak in the Great Plains and the Midwest. The warm front that marks the western boundary of the polar outbreak now tends to move eastward as new western synoptic systems push eastward.

3.3.4. Tracks of Cold-Core Anticyclones and Trajectories of Polar Outbreaks in the Americas

As indicated in the previous sections, there is no direct relationship between the intensity of the cold-core anticyclone and the degree of cooling in regions affected by the polar outbreak associated with that high-

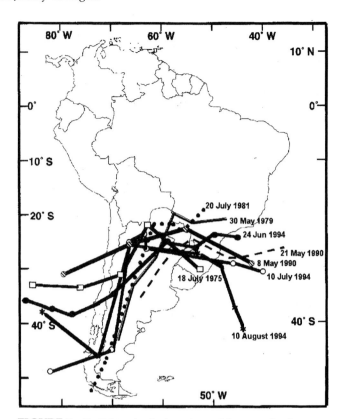

FIGURE 5 Partial track of wintertime cold-core anticyclones related to intense freeze events in southern Brazil. The track indicates the daily paths of the anticyclones from southern Chile, along the Andes, toward southern Brazil (Marengo, submitted).

pressure center. It is also important to consider the track, or path, of the anticyclone and the trajectory of the cold air moving from higher to lower latitudes.

The tracks of some South American anticyclones associated with cold surges (1975, 1979, 1981, 1990, 1993, and 1994), and occasionally freezing conditions in southern and southeastern Brazil, are presented in Fig. 5. Since it was not possible to plot the track of every anticyclone listed in Table 1 due to the lack of weather maps and surface charts for those cases, this figure should not be taken as comprehensive of all cases of major cold surge events. The most prominent characteristics of each path are the eastward trajectory coming from the Pacific, then a turn around the Andes to follow a northward trajectory until it reaches approximately 25°S; there, it turns toward the southeast and southern Brazil–northern Argentina, ultimately reaching the subtropical Atlantic where it is finally assimilated by the subtropical high. In general, the anticyclones take 24–48 hr to travel from the coast of Chile to southern Brazil. There is a tendency for the anticyclones to become somewhat stationary over southern Brazil during severe freeze events, as in 1975, 1981, and 1994.

a) 24 June 1994

b) 25 June 1994

c) 26 June 1994

d) 27 June 1994

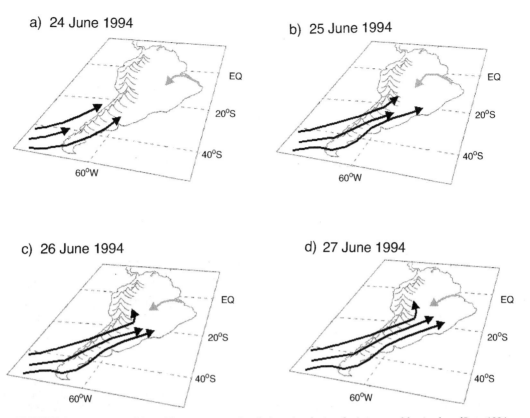

FIGURE 6 Trajectory of the cold air mass over South America during the intense cold episodes of June 1994. (A) 24 June, (B) 25 June, (C) 26 June, and (D) 27 June. Arrows indicate different pressure levels in the atmosphere. (Adapted from Sanchez and Silva Dias, 1996.)

The strong polar outbreaks of June and July 1994 can be analyzed in conjunction with Fig. 5. The case for 24–27 June 1994 is shown in Figs. 6A–6D. On 24 June (Fig. 6A), air coming from the South Pacific reached northern Argentina. On 25 June (Fig. 6B), the trajectory of cold air coming from upper levels above the Pacific turned toward the west and reached western Amazonia, affecting southeastern Peru on 26 June (Fig. 6C). It is interesting to notice that the parcel coming from the middle troposphere above the South Pacific reached central Brazil, while the air parcels from lower levels affected southeast Brazil and eventually reached the coastal region. The 7–10 July 1994 case (Figs. 7A–7D) is significantly different from the June 1994 case, considering the origin of the air parcels. The air parcels that reached western Amazonia and central and southeastern Brazil came from air masses over the South Atlantic. There are indicators that the polar air parcels did not reach such low latitudes as in the June 1994 event. Air came from upper levels of the troposphere and descended until it reached western Amazonia on July 10 (Fig. 7D). This indicates that the air mass reaching southeastern Brazil in July 1994 showed more oceanic characteristics than that of June 1994.

The tracks of surface anticyclones associated with polar outbreaks in North America (listed in Table 2) are presented in Figs. 8A to 8D. Figure 8A shows the path of the strong anticyclones associated with the severe freezes of 1894–95, 1899, 1962, 1983, 1985, and 1989, which produced extensive citrus damage in Florida. It comprises the synoptic setting associated with Fig. 4. The most prominent characteristics of each path are a southward movement across the Plains toward eastern Texas and a subsequent abrupt eastward or northeastward turn toward the Atlantic coast. In general, the anticyclones take 24–48 hr to traverse the distance from the Canadian border to Texas. Rogers and Rohli (1991) indicate that Florida freeze damage generally occurs in the mornings when the high is in Texas. Dalavalle and Bosart (1975) point out that movement of the composite polar anticyclone is linked to changes in forcing mechanisms; cold air advection drives the southward motion, and anticyclonic vorticity advection aloft supports the eastward curvature and reintensification. Figure 8B shows the tracks of slow-moving anticyclones, which generally track east of those in Fig. 8A, crossing over Missouri or Arkansas. This second kind of anticyclone is associat-

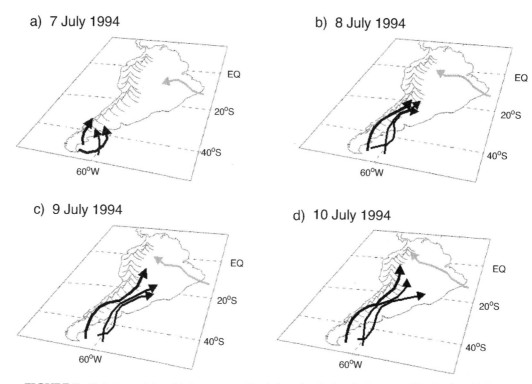

FIGURE 7 Trajectory of the cold air mass over South America during the intense cold episodes of July 1994. (A) 7 July, (B) 8 July, (C) 9 July, and (D) 10 July. Arrows indicate different pressure levels in the atmosphere. (Adapted from Sanchez and Silva Dias, 1996.)

ed with freezes caused by overnight radiative cooling in Florida. They move into the Gulf of México or Atlantic Ocean, becoming part of the warm Atlantic subtropical high. The third set of anticyclone paths (Fig. 8C) is associated with high-pressure areas that move toward Texas, but do not create severe freezes in Florida. Figure 8D shows the paths of a small subset of the many strong anticyclones that are linked to meridional winter flow and unusually cold weather across North America, but do not produce notable impacts across the southeastern United States.

3.3.5. Atmospheric Teleconnection Patterns Associated with Polar Outbreaks in the Americas and Their Interdecadal Variability

North and South American polar outbreaks that produce damaging freezes do not appear to be related to El Niño or La Niña. For southern Brazil, there is no clear signal of El Niño or La Niña impacts at interannual timescales on either the frequency or intensity of polar outbreaks in South America. The coldest events registered in July 1975 and June and July 1994 occurred during transition years between El Niño and La Niña.

What is observed is a tendency for warmer than normal winters during El Niño years with a lower probability of strong polar outbreaks in those winters. However, some of the events listed in Table 1 occurred during some El Niño or La Niña episodes.

Large-scale circulation factors controlling austral winter freeze events over subtropical and tropical latitudes are generally accompanied by major synoptic- to planetary-scale changes in the pressure systems over an extensive part of the Southern Hemisphere. These changes are reflected in a large-amplitude trough of the middle-latitude westerlies. The tropical extension (or penetration) of such large-amplitude troughs appears to be associated with synoptic- to planetary-scale events. The mechanism of downstream amplification across the Pacific into South America generally precedes heavy frost events. This is a well-known wave train of the midlatitudes that is seen to traverse across the Pacific Ocean toward South America, exhibiting the successive intensification of downstream troughs and ridges (Krishnamurti et al., in press).

The occurrence of severe North American freezes in the far southeastern United States has been linked to the positive phase of the Pacific–North American Oscillation (PNA) atmospheric teleconnection pattern

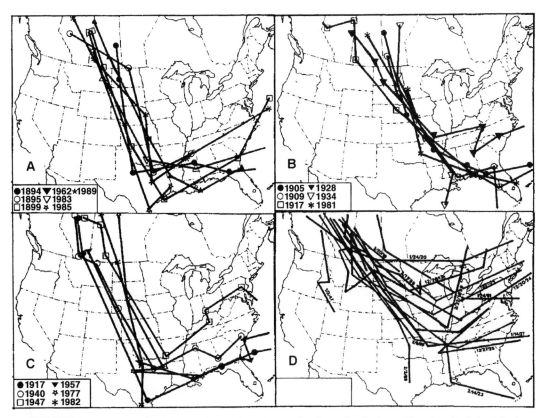

FIGURE 8 Partial track of the strongest wintertime anticyclones, beginning with points in their paths in southern Canada and then across the United States. Positions are based on 12 or 13 October Universal Time Zone (UTC) synoptic charts. The years labeled in (A)–(C) refer to years of Florida citrus freezes, while the tracks in (D) are from the remaining nonfreeze events. (From Rogers and Rolhi, 1991. With permission.)

(Rogers and Rohli, 1991; Rohli and Rogers, 1993; Downton and Miller, 1993). This prominent mode of low-frequency variability in the Northern Hemisphere extratropics is observed in all months of the year except June and July. Its positive phase is characterized by amplification of the upper tropospheric ridge off the west coast of North America (Fig. 1A), which establishes a strong meridional flow of polar air southward toward the United States and Central America. The PNA positive phase can be forced by the tropical heating associated with El Niño in the tropical Pacific. It has been shown, however, that El Niño is not the only forcing mechanism of the PNA, and while the record of freezes in Florida is dominated by the occurrence of the PNA positive mode, the occurrence of El Niño during polar outbreaks and freezes is only occasional. Cold waves are, of course, synoptic events occurring over 2–6 days, and while the presence of a ridge over the western peripheries of the continents is almost mandatory, the long-duration El Niño is not or is merely coincidental. Furthermore, the nature of winters in the eastern United States associated with El Niño has changed since

1977, with relatively mild winters consistently prevailing in events since that year. El Niño is statistically linked to colder, but wetter than normal, winters along the Gulf of México coast and in Florida. The wetter and colder weather is typically linked to the presence of skies that are cloudier than normal, conditions not generally linked to polar outbreaks.

The PNA is linked to variability in the mean surface pressure of the Aleutian Low. Around 1977, the Pacific Ocean sea surface temperatures (SSTs) underwent a remarkable regime shift that has attracted considerable attention (Latif and Barnett, 1994; Mantua et al., 1997). The climate shift produced a deepening of the wintertime Aleutian Low and moved the PNA index into an almost continuously positive phase during subsequent winters. Freezes affecting North and Central America and the Florida citrus-growing regions began occurring with great rapidity, such as in 1977, 1981, 1982, 1983, 1985, 1989, and 1997, whereas they had previously occurred only about once per decade (Rogers and Rohli, 1991). In contrast, 1948–57 was a freeze-free period characterized by anomalously high pressure in the

Aleutian Low center and anomalously low index values of the PNA. A strong, persistent PNA pattern, *within* and *among* winters, substantially increases the probability that Florida will experience a widespread, severe tree-damaging freeze. Further comments on the long-term interdecadal variability of citrus freezes appeared in the earlier discussion.

Strong cold surges in México and Central America usually possess a positive PNA and a confluent subtropical jet over the Gulf of México and southeast United States that is also found during El Niño conditions (Schultz et al., 1998). In fact, Klaus (1973) and Schultz et al. (1998) presented evidence that cold surges in this region tended to be more numerous during the cold season after the El Niño year.

In South America, research on interdecadal variability in cold surges focuses on variability in cold frontal passages. Calbete (personal communication) identified cold fronts at three different latitudinal bands over 1975–98, and noticed that the number of cold fronts reaching these latitudinal bands was larger during the period 1975–84 as compared to the period 1987–98, regardless of the intensity of the cold air masses related to the fronts. Calbete (1996) also showed that cold air masses reach southern Brazil from April to October, and in some regions at high altitudes freezes can occur all year long. Table 3 summarizes the number of cold air masses and snow events during the whole period 1988–96. However, the occurrence of very strong freezes in southern Brazil may be more of a random short-time variability event, occurring anytime even though the number of cold fronts may have been lower in recent decades.

Lemos and Calbete (1996) and Marengo (submitted) identified the number of cold fronts affecting the coastal section of Brazil from 35°–20°S, encompassing the coffee-growing areas of southeastern Brazil. The number reaching this latitudinal band was larger during 1975–84 compared to the period 1987–95. For this latter period, the number of cold fronts that affected this region in wintertime vary considerably: 40 cold fronts during 1987–90, 31 from 1991–94 (period of the extended El Niño of 1991–94), and 34 during 1995–98.

Trends toward warmer climates have been detected in São Paulo and other states in southern and southeastern Brazil (Victoria et al., 1998). In fact, Marengo (submitted) has detected an upward trend since the beginning of the century, in São Paulo, Rio de Janeiro, Londrina, and Curitiba, near the coffee-growing areas in southern and southeastern Brazil. Work by Venegas et al. (1996, 1998) has shown that a positive trend in SSTs and SLP in the South Atlantic may be indicative of a systematic warming in this region.

3.4. HISTORIC POLAR OUTBREAK EVENTS IN THE AMERICAS AND THEIR IMPACT ON REGIONAL AGRICULTURE

This section documents how extreme meteorological events such as polar outbreaks can affect human activities. Statistics on coffee and orange production are used as qualitative indicators of long-term climate variability in the region, as related to interhemispheric indicators of polar outbreaks in present-day climates.

3.4.1. Chronology and Impacts of Polar Outbreaks in Coffee-Growing Regions of South America Since the Late Eighteenth Century

Coffee was introduced to Brazil from French Guiana at the beginning of the eighteenth century. The first seeds, planted in Cayenne in 1718, can be traced back to seedlings offered by the Dutch to the French from the Botanic Gardens of Amsterdam. Brazil is the world's largest coffee producer, supplying approximately 27% of the coffee consumed worldwide. It competes with Colombia as the world's largest exporter, ranking either first or second each year depending upon the volume of its output. It is also the second world consumer of coffee after the United States.

Historically, weather has played a major role in determining the world supply of coffee. For example, production increases after recovery from the 1953 Brazilian frost created big price declines. In another situation, drought is reported to have aggravated the effect of the July 1981 frost because it came after the crop damage had occurred. In many other frost incidents, subsequent abundant rain has helped damaged trees to recover very swiftly.

Freezing temperatures and frost affect a large part of the harvest of wheat, coffee, soybeans, and oranges in

TABLE 3 Number of Cold Air Masses and Number of Snow Events in Southern Brazil during 1988–96 during the Cold Season

Month	Number of cold air masses	Number of snow events
April	10	0
May	30	3
June	33	4
July	28	12
August	28	6
September	25	2
October	13	0

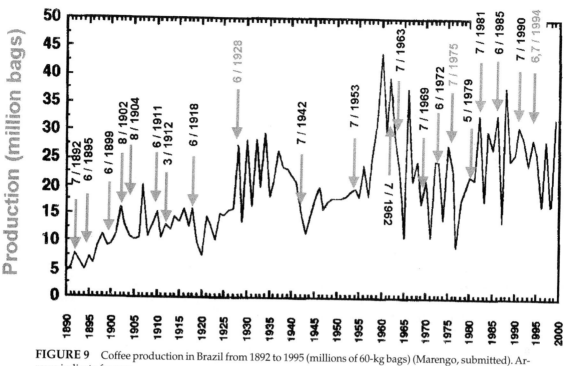

FIGURE 9 Coffee production in Brazil from 1892 to 1995 (millions of 60-kg bags) (Marengo, submitted). Arrows indicate freezes.

the agricultural lands of southeastern Brazil. For southeastern Brazil, reports issued by the United States Department of Agriculture (USDA) indicated that the freeze of July 1975 (perhaps the most intense in this century) reduced the 1976–77 harvest to 9.3 million bags (60 kg/bag), compared to the 1961–80 average of 19 million bags. The damage was severe enough to motivate the moving of some coffee plantations from the former growing region of Paraná (southern Brazil) to the northern states of São Paulo and Minas Gerais. This severe event is comparable to the intense freezes of August 1908 and June 1918. In 1975, some 75% of the trees in Brazil were affected in some degree, and the 1976 crop was a paltry 6 million bags.

Frost in mid-July claimed 60–70% of São Paulo's 1975–76 coffee crop and 5–10% of its sugarcane crop. The intense cold also hurt pastures, tomatoes, wheat, and bananas in that state. Because of the frost damage, virtually no coffee was harvested in the year following the freezes in Paraná, and only 25–33% of its normal harvest was projected for 1977–78.

Frigid air surged through Brazil's coffee regions during the third week of July 1981, killing buds and potentially causing a significant reduction in 1982–83 production (Haddock et al., 1981). The damage was not as severe as in 1975 when coffee trees were killed. Likewise, the Brazilian frosts in June and July 1994 caused a sharp drop in coffee production and dramatic increases in coffee prices (Marengo et al., 1997a).

Statistics on coffee production in the southern states of Brazil, available since the late 1800s, can be used to indirectly assess the presence and intensity of regional cold waves. Figure 9 shows coffee production in southern and southeastern Brazil, provided by the Brazilian Institute of Coffee and *The New York Times*, from 1892–1995. A steady increase is observed in production due to increases in the cultivated area. Large drops in production were observed after the freezes of June 1928, July 1975, June 1981, and June and July 1994, related largely to the cold weather. Estimates of losses in 1995 due to the frosts of June and July 1994 are 50–80% in the states of São Paulo and Paraná (Marengo et al., 1997a). For the 1995–96 production, estimates put the crop at 13.15 million bags. Marengo et al. (1997a) indicate that heavy commodity market speculation after the 24 June 1994 cold event resulted in coffee price increases of almost 70% over a few days, with the New York, London, and Chicago Stock Exchanges reacting more quickly than the Brazilian market. For 1997–98, in Brazil, mild winter weather linked to El Niño kept the top production areas free of frost, but heavier than normal rainfall in June affected the quality of the harvested crop. El Niño–induced drought in Espírito Santo failed to damage the Brazilian coffee harvest, which increased in 1997–98 to 18.86 million bags. Recently, during 16–17 April 1999, the first cold wave entered from higher latitudes and extended over central Brazil and western Amazonia. On the coffee-growing areas of southern and southeastern Brazil (western Paraná and São Paulo), air temperatures dropped to between 0° and 2°C, creating moderate and weak freezes. However, given the occurrence of these freezes during early stages, damage to the coffee plantations was not as severe as one would expect.

Table 1 shows a chronology of events and their intensities since 1892 for the coffee-growing areas of the state of São Paulo (southeastern Brazil); the most devastating events are marked as very intense. The listed frosts in coffee regions of Brazil indicate that the change over the years in temperature range is probably very small, but it takes only marginal variation and a few hours of frost one night to do real damage; as in 1975, the trees can take 2 or 3 years to recuperate. In 106 years, from 1890–1996, 18 intense freeze events brought damage to coffee production. Of these, five were considered catastrophic. On average, there has been one severe event every 6 years and one very severe event every 26 years in the coffee-growing region. Two very severe events were registered in 1892–1925 (in 1902 and 1918) and the other three in 1962–96 (in 1975, 1981, and 1994).

3.4.2. North and Central American Polar Outbreaks and Their Impacts on the Citrus Industry in the Southeastern United States

Climate has played an important role in the development of Florida's citrus industry. Florida's subtropical climate characteristics set the stage for that state to become a major producer of oranges since the eighteenth century, ultimately becoming the world's first major producer of frozen concentrated orange juice. Freeze damage to Florida's citrus industry due to intense polar outbreaks occurring since 1977 has been responsible for launching the Brazilian citrus industry into its current position as the world's leading producer and exporter of orange juice concentrate (Miller and Glantz, 1988). Studies by Rogers and Rohli (1991), Miller and Downton (1993), Rohli and Rogers (1993), and Downton and Miller (1993) describe an increase in freeze frequencies in central Florida since 1977, which have brought about a large dislocation in the Florida citrus industry, similar to the dislocation of the coffee plantations to the north of their normal places after the freeze of July 1975 in southeastern Brazil. Citrus is also grown in the Rio Grande valley region of extreme southern Texas. As in Florida, however, production in Texas has been hampered by freezes, especially those of 1949, 1951, and 1962 (Rohli and Rogers, 1993).

Descriptions of damage to the citrus crop typically focus on freezing of the citrus, defoliation of the trees, reductions in citrus production, and damage to tree limbs or bark leading to injury or death. The most damaging are advective freezes, wherein a powerful polar anticyclone migrates to Texas (Figs. 4C, 8A), and the winds around the eastern edge of the high often reach 50 km/hr, causing excessive defoliation and transpiration and breakage of twigs and branches, in addition to bringing subfreezing air. Damage becomes very severe if temperatures remain below −5.5°C for more than 1 hr. The worst advective freezes were those of 1962, 1983, 1985, and 1989, in addition to those during 1835, 1894, 1895, and 1899 (discussed in more detail later). The advective freeze is that identified in the conceptual model (Section 3.3.2 and Figs. 4A–4C) and in the anticyclone tracks of Fig. 8A. Radiative freezes occur when the anticyclone at the core of the polar air mass begins to settle over the southeastern United States and Florida (Fig. 8B). Radiative cooling in the clear winter night brings air temperatures to below freezing, with the air temperature dropping to the most dangerous levels only near sunrise, after which warming rapidly occurs. While most Florida freezes may combine elements of both the advective and radiative varieties, the less damaging events have lower wind speeds and are primarily radiative.

As was pointed out in Section 3.3.2, a devastating cluster of freezes affected a large portion of Florida citrus-growing areas in January 1977, 1981, and 1982; December 1983; and January 1985 (Table 2). Florida's citrus industry survived the first three freezes without major problems. However, the advective freezes of 1983 and 1985 occurred back-to-back in adjacent winters and sent the industry into a tailspin with serious crop and tree damage. Nearly one-third of the state's commercial citrus trees were lost. The next severe freeze, in December 1989, killed a large number of newly planted tress in areas that already had been affected by the earlier events. By 1989, citrus production in northern Florida had largely been abandoned. Citrus production in northern Florida had originally been encouraged by the relatively mild winters of that region prior to 1977. The industry had suffered only relatively minor setbacks occurring about once a decade, interspersed with some rare but substantial disasters, such as the freezes of 1917 and 1962, for the first three-quarters of the twentieth century.

Figure 10 shows Florida's orange production in thousands of 90-lb boxes since 1965, as shown by Miller and Glantz (1988), and updated to 1997–98. A general rising trend can be seen over the 1970s, but in general it is clear that freezes have had very uneven effects on Florida's orange growers. Major freeze events are noted in Figs. 9 and 10, denoted by arrows. As with coffee production in Fig. 9, the effect of the cold weather is often more pronounced in the year following the freeze. Tree damage is longer lasting, while the fruit frozen on the

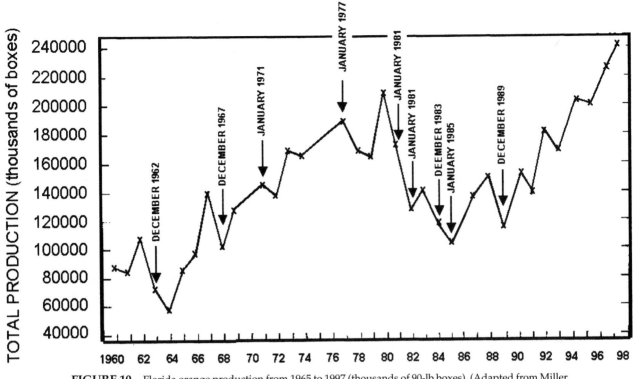

FIGURE 10 Florida orange production from 1965 to 1997 (thousands of 90-lb boxes). (Adapted from Miller and Glantz, 1988; updated with data from the Florida Department of Citrus). Arrows indicate freezes.

tree can often be salvaged, and those freezes occurring in January do not affect the harvest of early-bearing varieties. The dramatic impacts of the December 1983, January 1985, and December 1989 freezes can readily be seen. Miller and Glantz indicate that the December 1962 freeze had a striking effect on orange prices, as did that of 1977. The January 1981 and January 1982 freezes, however, had virtually no effect on prices. By the 1984–85 season, continued reductions in Florida production had succeeded in raising the price of Florida oranges to its highest levels in 20 years.

Northern Florida was abandoned as a citrus-growing region once before, during an earlier cluster of freezes that culminated in the nineteenth century, from 1880–1899. During this spell, the most lethal events were two severe freezes that occurred within ca. 8 weeks of each other during the 1894–95 winter. These freezes were preceded by a freeze in 1886, widely considered the worst since 1835 (see Rogers and Rohli, 1991). The 1894–95 pair of freezes was followed by the extraordinarily severe freeze of 1899 (associated with the famous East Coast blizzard described earlier), which killed both old and new citrus trees in the northern half of Florida. The late nineteenth-century freezes kept Florida citrus production below 1894 levels until 1909–10 and, as with the recent cluster, contributed to

a net southward migration in the citrus production belt (Chen and Gerber, 1985).

The milder decades between the severe citrus freeze clusters appear to have been associated with a gradual northward migration in citrus production (Miller and Glantz, 1988). During these mild periods the severe freezes were sporadic (e.g., those of 1835, 1917, and 1962), and other freezes that occurred were relatively benign radiative events. The December 1962 freeze was described as the "worst since 1899" (Rogers and Rohli, 1991). Very strong cold waves occurred in 1835 and 1857 (Pardue, 1946). Since the beginning of systematic weather observations, severe cold waves have been experienced in January 1866, December 1894, February 1895, 12–13 February 1899, 2–6 February 1917, 3–4 January 1928, 12–13 December 1934, 25–29 January and 16–17 November 1940, 2 March 1941, and 15–16 February 1943. Similarly, in the nineteenth century, the 1880 freeze was the coldest since 1857, but the far more severe freeze of 1886 at the start of the nineteenth century cluster is compared to the event of 1835, which is among the worst known. The data suggest that at least half a century of comparatively milder winters precedes each severe cold-wave cluster.

Tropical fruits had also been uninjured in more than half a century at St. Augustine, FL, at the time of the freeze of 1835 (Blodget, 1857). The *Autobiography of Thomas Douglas* (published in 1856; Chen and Gerber, 1985) indicates that many of the trees destroyed in 1835 around St. Augustine were nearly 100 years old. A cluster of severe winters over the eastern and southeastern United States is also known to have occurred during the last quarter of the eighteenth century, including the winters of 1776–77, 1779–80, 1783–84, during the United States Revolutionary War, and 1786–87, 1796–97, and 1798–99 (Blodget, 1857; Ludlam, 1966). If the St. Augustine, FL, reports are characteristic, however, either any cold waves during these winters may not have greatly affected Florida citrus production or the citrus trees may have been able to withstand them due to their age, the timing of the freeze, or other factors such as abundant soil moisture. It is interesting to note that many of the worst winter freezes in the southeastern United States have been clustered in the final quarter of each of the recent three centuries.

3.5. PALEOCLIMATIC EVIDENCE OF LONG-TERM VARIABILITY IN SOUTH AMERICAN POLAR OUTBREAKS

Our understanding of the climatic history of the South American tropical lowlands is more limited than for other regions of South America. It is not possible to use the available surface meteorological observations (especially those from southeastern tropical South America) to describe variations at century or longer timescales. The statistics for coffee and oranges help to identify years with possible intense freezes since the late nineteenth century, but only qualitatively. Climate proxy data covering longer timescales are available from a few sites in the South American tropics based on paleoenvironmental records, primarily pollen records

(Ledru, 1993; Ledru et al., 1994; Servant et al., 1993; Suguio et al., 1997; Behling and Lichte, 1997; Bush et al., 2000). These records identify possible mechanisms of past atmospheric circulation, including cold air outbreaks.

Today, for the Southern Hemisphere, the Earth is farthest from the Sun in June (winter) and closer to the Sun in December (summer) (Martin et al., 1997). As a consequence, seasonal differences in insolation are strong with warm summers and cold winters, and the seasonal shifts of the Intertropical Convergence Zone (ITCZ) are strong. Equatorward extents of cold advections under present-day conditions are shown in Fig. 11A. In contrast, at ca. 11,000 B.P., for the Southern Hemisphere, the Earth was closer to the Sun in June and farther from it in December, resulting in relatively colder summers, but relatively warmer winters than now and reduced seasonality. As a consequence, the continent was not warming as much then as during today's southern summer, and the ITCZ was probably located farther north than it is today. A weaker ITCZ, on the other hand, would have helped cold advections to penetrate further equatorwards in spring and autumn and possibly even in summer (Fig. 11B).

Numerous paleoecological studies found evidence in the South American tropics of well-developed dense forests during late glacial and early Holocene times (Behling, 1996; Servant et al., 1993; Ledru, 1993; Ledru et al., 1994; Behling and Lichte, 1997; Ledru and Mourguiart, 2000). The main features that characterized the tropical forests during those times were (1) montane Andean forest elements, such as *Podocarpus,* and southern forest elements, such as *Araucaria,* expanded into the tropical lowlands and equatorwards, respectively, and (2) the reduction of climate seasonality prior to 7500 B.P. according to the vegetation composition at that time. All these aspects, inferred from paleoecolog-

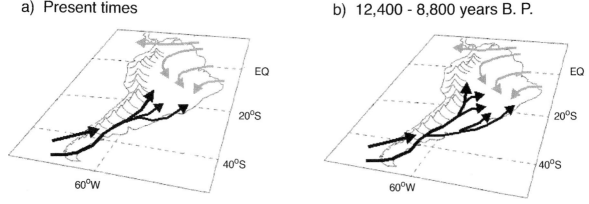

FIGURE 11   Position of the polar outbreaks for (A) present times and (B) 12,400–8800 B.P. (Adapted from Martin et al., 1997 and Servant et al., 1993.)

ical evidence, suggest climates colder than today during the glacial/interglacial transition. The most likely explanation is that polar air masses reached tropical regions more frequently during those times than they do today (Figs. 11A and 11B). Currently, climate in the region is influenced by polar outbreaks from the south only occasionally, but overall the climate is not cold enough to support the presence of either Andean forest elements or *Araucaria* forests in these lowland regions.

Climatic conditions during full glacial times were apparently too dry and too cold for large *Araucaria* populations (Ledru, 1993; Ledru et al., 1994; Behling and Lichte, 1997; however, see discussion in Bush et al., 2000). Dry climatic conditions with reduced cloud cover might have favored frosts during winter nights. Cold air associated with cold fronts and polar outbreaks might have reached farther north and had a significantly stronger influence in southeastern Brazil during the last glacial maximum than it does today (see also Bradbury et al., 2000).

Increases in temperature and moisture content of the air masses after 8000–6000 B.P., interpreted from paleoecological records, are probably due to a reduction in the influence of cold fronts (Suguio et al., 1997). Thus, throughout the late Quaternary, variations in the intensity and trajectory of the polar outbreaks affected the climate of southern and southeastern Brazil.

3.6. CONCLUSIONS

At several timescales, from the interannual to the long term, interpretations of climatic and proxy climatic data have demonstrated that changes in weather and climate patterns can have strong impacts on human societies, exemplified here by the production of coffee and citrus in South and North America, respectively. From timescales of days to decades and beyond, there are associations between the frequencies and intensities of polar outbreaks and freezes and atmospheric circulation anomalies in both hemispheres. Observed negative trends in wintertime temperatures in central Florida and positive trends in southern South America (and less frequent cold fronts) seem to be linked to decadal changes in the PNA pattern and the SST in the South Atlantic. Features such as clusters of freezes in the Florida citrus areas or hard-hitting freezes in southeastern Brazil in the middle 1990s, even though the winter of 1994 was relatively warm, are indicators of the large climate variability in the Americas.

The marked similarity in meteorological aspects of the polar outbreaks in the southeastern United States and Brazil has long been noted. The near-surface and upper level circulations, as well as the role of the mountain ranges, favor the thrust of polar air toward subtropical and tropical latitudes in both the Americas. There are important differences, though, related presumably to geographical differences, outbreak intensities, and impacts. Both regions have different vegetation and agricultural crops that can be threatened by surges of polar air. In that sense, this chapter has used historical data and meteorological evidence to document polar outbreaks, their synoptic setting, and longer term variability.

The extreme weather events cause loss of crop production with inevitable social and economic difficulties. For the past, few observational data can be used directly to indicate climate variability on century timescales in South America, in the way that tree rings do for identifying climate variations in North America. Proxy data such as statistics on coffee and orange production and relics of vegetation distribution are the only possible observational evidence.

In future climate scenarios, consequences of global warming and extreme climate events on many types of agricultural crops must be of major concern in ensuring continuity of the food supply. But besides climate-related effects, agricultural technology, economics, and politics will continue to have profound effects on farming in most societies, and responses to the changing environment will be included in the overall adaptation of farming.

Acknowledgments

Jose A. Marengo was supported by the Brazilian Conselho Nacional de Desenvolvimento Cientifico e Tecnologico (CNPq). The authors would like to thank Drs. Vera Markgraf (Institute of Arctic and Alpine Research [INSTAAR], University of Colorado, Boulder, Colorado, U.S.) and David Schultz (National Oceanic and Atmospheric Administration/National Severe Storm Laboratory [NOAA/NSSL], U.S.) for their input during the preparation of this chapter. Thanks are also due to the Inter American Institute for Global Change (IAI-ISP2) for funding this research, to Dr. Jurandir Zullo (Universidade de Campinas, Campinas, Sao Paulo, Brazil) for providing information on coffee production in the state of São Paulo, and to Mr. David Mendes and Ms. Leticia de Faria for data processing and figures.

References

Baines, P. G., 1983: A survey of blocking mechanisms with application to the Australian region. *Australian Meteorological Magazine*, **31**: 27–36.

Behling, H., 1996: First report on new evidence for the occurrence of *Podocarpus* and possible human presence of the Amazon during the late glacial. *Vegetation History and Archaeobotany*, **5**: 241–246.

Behling, H., and M. Lichte, 1997: Evidence of dry and cold climatic conditions at glacial times in tropical southeastern Brazil. *Quaternary Research*, **48**: 348–358.

Blodget, L., 1857: *Climatology of the United States*. Philadelphia, PA: Lippincott, pp. 145–154.

Bosart, L., R. Lupo, D. Knight, E. Hoffman, and J. Nocera, 1998: Cyclonic and anticyclonic South American cold surges. *8th Conference on Mountain Meteorology*, Flagstaff, AZ, *American Meteorological Society*, pp. 102–103.

Bradbury, J. P., M. Grosjean, S. Stine, and F. Sylvestre, 2000: Full and late glacial lake records along the PEP 1 transect: Their role in developing interhemispheric paleoclimate interactions. *In* Markgraf, V. (ed.), *Interhemispheric Climate Linkages*. San Diego: Academic Press, Chapter 16.

Bush, M. B., M. Stute, M.-P. Ledru, H. Behling, P. A. Colinvaux, P. E. De Oliveira, E. C. Grimm, H. Hooghiemstra, S. Haberle, B. W. Leyden, M.-L. Salgado-Labouriau, and R. Webb, 2000: Paleotemperature estimates for the lowland Americas between 30°S and 30°N at the last glacial maximum. *In* Markgraf, V. (ed.), *Interhemispheric Climate Linkages*. San Diego: Academic Press, Chapter 17.

Calbete, N. O, 1996: Casos de ocorrencia de geada e neve sobre a regiao Sul do Brasil-Periodo de 1988 a 1996. *Climanalise*, special issue. CPTEC/INPE, São Paulo, Brazil, 129 pp.

Chen, E., and J. F. Gerber, 1985: Minimum temperature cycles in Florida. *Proceedings Florida State Horticultural Society*, **98**: 42–46.

Colle, B. A., and C. Mass, 1995: The structure and evolution of cold surges east of the Rocky Mountains. *Monthly Weather Review*, **123**: 2577–2610.

Cox, H. J., 1916: *Cold Waves. Weather and Forecasting in the United States*. Henry, A. J., E. H. Bowie, H. J. Cox, and H. C. Frankenfield (eds.), Washington, DC: U.S. Govt. Printing Office, pp. 143–167.

Dalavalle, J., and L. Bosart, 1975: A synoptic investigation of anticyclogenesis accompanying North American polar air outbreaks. *Monthly Weather Review*, **103**: 941–957.

Downton, M., and K. Miller, 1993: The freeze risk to Florida citrus. Part 2: Temperature variability and circulation patterns. *Journal of Climate*, **6**: 364–372.

Fortune, M., and V. E. Kousky, 1983: Two severe freezes in Brazil: Precursors and synoptic evolution. *Monthly Weather Review*, **111**: 181–196.

Garreaud, R. D., 1999: Cold air incursions over subtropical and tropical South America: A numerical case study. *Monthly Weather Review*, **127**: 2823–2853.

Garreaud, R. D., 2000: Cold air incursions over subtropical and tropical South America: Mean structure and dynamics. *Monthly Weather Review*, **128**: 2544–2559.

Garreaud, R., and J. M. Wallace, in press: Summertime incursions of midlatitude air into subtropical and tropical South America. *Monthly Weather Review*, in press.

Garriott, E. B., 1906: Cold waves and frosts in the United States. *Bulletin P*. Washington, DC: United States Department of Agriculture, Weather Bureau, 22 pp. plus 328 charts.

Haddock, D., R. Motta, R. McInturff, M. Halpert, and R. Smelser, 1981: Freeze hits Brazilian coffee area on July 20 and 21, 1981. *Weekly Weather and Crop Bulletin*, **68**(30): 25–27.

Hamilton, M. G., and J. Tarifa, 1978: Synoptic aspects of a polar outbreak leading to frost in tropical Brazil, July 1972. *Monthly Weather Review*, **106**: 1545–1556.

Hastenrath, S., 1996: *Climate Dynamics of the Tropics*. Dordrecht: Kluwer Academic, 489 pp.

Klaus, D., 1973: Las invasiones de aire frio en los tropicos a sotavento de las Montañas Rocallosas. *Geofisica Internacional*, **13**: 99–143.

Kocin, P. J., and L. W. Uccellini, 1990: Snowstorms along the northeastern coast of the United States. *Meteorological Monographs*, **22**: 44.

Kocin, P. J., A. D. Weiss, and J. J. Wagner, 1988: The great arctic outbreak and East Coast blizzard of February 1899. *Weather and Forecasting*, **3**: 305–318.

Konrad, C. E., 1996: Relationships between the intensity of cold-air

oubreaks and the evolution of synoptic and planetary features. *Monthly Weather Review*, **124**: 1067–1083.

Konrad, C. E., and S. J. Colucci, 1989: An examination of extreme cold air outbreaks over eastern North America. *Monthly Weather Review*, **117**: 2687–2700.

Kousky, V., 1979: Frontal influences on Northeast Brazil. *Monthly Weather Review*, **107**: 1140–1153.

Kousky, V., and I. Cavalcanti, 1997: The principal modes of high-frequency variability over the South American region. 5th International Conference of Southern Hemisphere Meteorology and Oceanography, Pretoria, South Africa, American Meteorological Society, pp. 7.2–7B.3

Kousky, V., and C. Ropelewski, 1997: The tropospheric seasonal varying mean climate over the Western Hemisphere (1979–1995). Atlas No. 3, National Centers for Environmental Prediction (NCEP)/Climate Prediction Center, U.S. Department of Commerce.

Krishnamurti, T. N., M. Tewari, D. Chakrabarty, J. Marengo, P. Silva Dias, and P. Satyamurty, in press: Downstream amplification: A possible precursor to major freeze events over southeastern Brazil. *Weather and Forecasting*, in press.

Latif, M., and T. P. Barnett, 1994: Causes of decadal climate variability over the North Pacific and North America. *Science*, **266**: 634–637.

Ledru, M. P., 1993: Late Quaternary environmental and climatic changes in central Brazil. *Quaternary Research*, **39**: 90–98.

Ledru, M.-P., and P. Mourguiart, 2000: Late glacial vegetation records in the Americas and climatic implications. *In* Markgraf, V. (ed.), *Interhemispheric Climate Linkages*. San Diego: Academic Press, Chapter 20.

Ledru, M.-P., H. Behling, M. Fournier, L. Martin, and M. Servant, 1994: Localisation de la forêt d'Araucaria du Brésil au cours de l'Holocene. Implications paléoclimatiques. *Comptes Rendus Academie Sciences, Paris, Sciences de la Vie*, **317**: 527–521.

Lemos, F., and N. O. Calbete, 1996: Sistemas frontais que atuam no litoral do Brasil (periodo 1987–1995). *Climanalise*, special issue. Centro de Previaso de Tempo e Estudos de Clima/Instituto Nacional de Pesquisas Especiais (CPTEC/INPE), São Paulo, Brazil, pp. 131–135.

Liebmann, B., G. Kiladis, J. Marengo, T. Ambrizzi, and J. Glick, 1998: Submonthly convective variability over South America and the South Atlantic Convergence Zone. *Journal of Climate*, **7**: 1877–1891.

Ludlam, D. M., 1966: *Early American Winters 1604–1820*. Boston, MA: Am. Meteorol. Soc., 285 pp.

Mantua, N. J., S. R. Hare, Y. Zhang, J. M. Wallace, and R. C. Francis, 1997: A Pacific interdecadal climate oscillation with impacts on salmon production. *Bulletin American Meteorological Society*, **78**(6): 1069–1079.

Marengo, J., in press: On the impacts of weather and climate variability and change in agricultural production and prices: Polar air outbreaks and their impact on the coffee growing industry in southern and southeastern Brazil. *Climate Research*, in press.

Marengo, J., A. Cornejo, P. Satyamurty, C. A. Nobre, and W. Sea, 1997a: Cold waves in the South American continent. The strong event of June 1994. *Monthly Weather Review*, **125**: 2759–2786.

Marengo, J., C. Nobre, and A. Culf, 1997b: Climatic impacts of friagems in forested and deforested areas of the Amazon basin. *Journal of Applied Meteorology*, **36**: 1553–1566.

Marshall, C. F., 1983: *The World Coffee Trade*. Cambridge, U.K.: Woodhead-Faulkner, 254 pp.

Martin, L., J. Bertaux, T. Corrége, M. P. Ledru, P. Mourguiart, A. Sifeddine, F. Soubiés, D. Wirrmann, K. Seguio, and B. Turcq, 1997: Astronomical forcing of contrasting rainfall changes in tropical

South America between 12,400 and 8000 yr B.P. *Quaternary Research*, **47**: 117–122.

Miller, K., and M. Downton, 1993: The freeze risk to Florida citrus, Part I: Investment decisions. *Journal of Climate*, **6**: 354–363.

Miller, K., and M. Glantz, 1988: Climate and economic competitiveness: Florida citrus and the global citrus processing industry. *Climatic Change*, **12**: 135–164.

Morize, H., 1922: Contribução ão Estudo do Clima do Brasil. Ministerio da Agricultura, Industria e Commercio, Observatorio Nacional de Rio de Janeiro. (In Portuguese.)

Pardue, L., 1946: Pattern of winter temperature in Peninsular Florida November 1937 through March 1946. Federal-State Horticultural Protection Service, U.S. Dept. of Commerce, Weather Bureau, University of Florida Experiment Stations, Lakeland, FL, 26 pp.

Parmenter, F. C., 1976: A Southern Hemisphere cold front passage at the equator. *Bulletin of the American Meteorological Society*, **57**: 1435–1440.

Rogers, J., 1997: Changes in cold wave frequencies and intensities in the United States. *In* "Workshop on Indices and Indicators for Climate Extremes." Asheville, NC: National Climatic Data Center, pp. 337–352.

Rogers, J., and R. Rohli, 1991: Florida citrus freezes and polar anticyclones in the Great Plains. *Journal of Climate*, **4**: 1103–1113.

Rohli, R., and J. Rogers, 1993: Atmospheric teleconnections and citrus freezes in the southern United States. *Physical Geography*, **14**: 1–15.

Sanchez, O., and P. Silva Dias, 1996: Penetração de ar frio na Amazonia Peruana. Extended Abstracts of the IX Congresso Brasileiro de Meteorologia. Campos de Jordão, Sao Paulo, Brasil, November 1996, pp. 512–516.

Satyamurty, P., P. Etchichury, C. Studzinsky, N. Calbete, R. Lopes, I. A. Glammelsbacher, and E. A. Glammelsbacher, 1990: A primera friagem do 1990: Uma descripção sinótica. *Climanalise*, **5**(5): 43–51. (In Portuguese.)

Schultz, D., W. Bracken, and L. Bosart, 1998: Planetary and synoptic scale signatures associated with Central American cold surges. *Monthly Weather Review*, **126**: 5–27.

Schultz, D., W. Bracken, L. Bosart, G. Hakim, M. Bedrick, M. Dickinson, and K. Tyle, 1997: The 1993 superstorm cold surge: Frontal structure, gap flow, and extratropical-tropical interaction. *Monthly Weather Review*, **125**: 5–39.

Seluchi, M., and J. Marengo, in press: Tropical-midlatitude exchange of air masses during summer and winter in South America: Climatic aspects and examples of intense events. *International Journal of Climatology*, in press.

Seluchi, M., and J. Nery, 1992: Condiciones meteorologicas asociadas a la ocurrencia de heladas en la region de Maringá. *Revista Brasileira de Meteorología*, **7**: 523–534.

Serra, A., and L. Ratisbona, 1942. As massas de ar da America do Sul. Ministerio da Agricultura, Serviço de Meteorologia, Rio de Janeiro, Brazil, 23 pp.

Servant, M., J. Maley, B. Turcq, M. Absy, P. Brenac, and M. P. Ledru, 1993: Tropical rain forest changes during the late Quaternary in African and South American lowlands. *Global and Planetary Changes*, **7**: 25–40.

Smith, R., N. Tapper, and D. Christie, 1995: Central Australian cold fronts. *Monthly Weather Review*, **123**: 16–38.

Steenburgh, W., D. Schultz, and B. Colle, 1998: The structure and evolution of gap flow over the Gulf of Tehuantepec, México. *Monthly Weather Review*, **126**: 2673–2691.

Suguio, K., B. Turcq, L. Martin, J. Abrao, and J. Flexor, 1997: Estudos sobre as flutuaçoes dos climas do Quaternario Tardio como subsidio para o desenvolvimento sustentavel. Abstracts of the Seminario Ciencia e Desenvolvimento Sustentavel. Instituto de Estudos Avançados / Universidade de São Paulo, São Paulo, Brasil, pp. 26–30.

Tarifa, J., H. Pinto, R. Affonsi, and M. Pedro, 1977: A gênese dos episódios meteorológicos de julho de 1975 e a variação espacial de danos causados pelas geadas na cafeicultura no Estado de São Paulo. Sociedade Brasileira para o progresso da Ciência, São Paulo, Brasil. *Ciência y Cultura*, **29**: 1362–1374. (In Portuguese.)

Venegas, S., L. Mysak, and N. Straub, 1996: Evidence for interannual and interdecadal climate variability in the South Atlantic. *Geophysical Research Letters*, **23**: 2673–2676.

Venegas, S., L. Mysak, and N. Straub, 1998: Atmosphere-ocean coupled variability in the South Atlantic. *Journal of Climate*, **10**: 2904–2920.

Victoria, R., L. Martinelli, J. Moraes, M. Ballester, A. Krusche, G. Pellegrino, R. Almeida, and J. Richey, 1998: Surface air temperature variations in the Amazon region and its border during this century. *Journal of Climate*, **1**: 1105–1110.

Wendland, W. M., 1987: Prominent November cold waves in the north-central United States since 1901. *Bulletin of the American Meteorological Society*, **68**: 616–619.

Wu, M., and J. Chan, 1995: Surface features of winter monsoon surges over South China. *Monthly Weather Review*, **123**: 662–680.

Wu, M., and J. Chan, 1997: Upper-level features associated with winter monsoon surges over South China. *Monthly Weather Review*, **125**: 317–340.

Globality and Optimality
in Climate Field Reconstructions
from Proxy Data

M. N. EVANS, A. KAPLAN, M. A. CANE, AND R. VILLALBA

Abstract

A primary objective of paleoclimate research is the characterization of natural climate variability on timescales of years to millennia. In this chapter, we have developed a systematic methodology for the objective and verifiable reconstruction of climate fields from sparse observational networks of proxy data, using reduced space Objective Analysis (OA). In this approach we seek to reconstruct only the leading modes of large-scale variability that are observed in the modern climate and resolved in the proxy data. Given explicit assumptions, the analysis produces climate fields and indices and their associated estimated errors. These parameters may be subsequently checked for consistency with parameter choices and procedural assumptions by comparison with withheld data and results from benchmark experiments.

The methodology is applied to the candidate tree-ring indicator data set described in Chapter 10 by Villalba et al., (2000) for the reconstruction of gridded Pacific basin sea surface temperature (SST) over the interval 1001–1990. We find that one mode of variability may be verifiably distinguished by the tree-ring indicators. This mode may be interpreted as a decadal, El Niño/Southern Oscillation (ENSO)-like pattern that influences climatic conditions in both North and South America. We speculate that this pattern is recovered because large-scale SST variability influences downstream local meteorological conditions sensed by the tree-ring data. Verification statistics for the proxy-reconstructed fields may be compared with those derived from calibrated red noise time series and from historical temperature and precipitation data collected near the proxy sampling sites. The results suggest that while credible reconstructions are possible, resolved variance is small and reconstruction errors are large, limiting interpretation to regions where there is the most skill. Additional proxy data from the Pacific coast of the Americas should improve the resolution and number of reconstructed large-scale SST modes, given that the observations are unbiased, the map describing the connection between SST and the proxy data is well defined, and appropriate observational errors have been prescribed. Copyright © 2001 by Academic Press.

Resumen

Uno de los objetivos principales de la investigación paleoclimática es la caracterización de la variabilidad climática natural en diferente escalas temporales que van desde la anual a la secular. En este estudio se ejemplifica el uso del Análisis Objetivo (OA) para reconstruir en forma confiable los campos climáticos globales a partir de un grupo reducido de registros paleoclimáticos. Sólo se intenta reconstruir aquellos modos dominantes de variabilidad climática de gran escala que están presentes en los datos instrumentales y que pueden ser resueltos a partir de los registros paleoclimáticos. Bajo ciertos supuestos explícitos, este análisis reconstruye campos climáticos, índices de circulación, así como los errores de estimación asociados. La consistencia de los parámetros seleccionados para las reconstrucciones bajo diferentes supuestos puede ser probado posteriormente a través de la comparación con información reservada para la verificación o con los resultados de experimentos de referencia.

En este trabajo, se aplica la metodología de OA a un grupo reducido de cronologías de anillos de árboles, descriptos en Villalba et al. (2000), con el objeto de reconstruir una grilla de temperaturas de la superficie del mar (SST) sobre el Océano Pacífico durante el intervalo 1001–1990. Se observa que uno de los modos de variabilidad en SST puede ser resuelto por el conjunto de cronologías de registros de anillos de árboles. Este modo podría ser interpretado como un modo de variabilidad decenal con un patrón espacial similar al de los eventos El Niño/Oscilación del Sur y que afecta las condiciones climáticas tanto en América del Norte como en América del Sur. El hecho de que este patrón puede ser resuelto por los registros dendrocronológicos se debe a que la variabilidad de gran escala en SST influencia las condiciones climáticas locales donde crecen los árboles muestreados. La verificación de estos campos climáticos reconstruidos a partir de los registros de anillos de árboles pueden llevarse a cabo a través de la comparación con aquellos campos derivados a partir de series temporales de ruido rojo y/o aquellos obtenidos a partir del uso de datos instrumentales de temperatura y precipitación próximos a los sitios de muestreos. Los resultados obtenidos a partir de este reducido número de cronologías sugiere que aún cuando se pueden obtener reconstrucciones confiables, el porcentaje de varianza explicado es pequeño y los errores asociados grandes, limitando la efectividad de las reconstrucciones a aquellas regiones donde los resultados son más significativos. La incorporación de un número mayor de registros paleoambientales a lo largo de la costa Pacífica de las Américas podría incrementar la resolución y el número de modos de variabilidad en SST a ser reconstruidos dado que las observaciones no son sesgadas, los mapas que describen las conecciones entre SST y los registros dendrocronológicos están bien definidos y los errores observaciones correspondientes han sido establecidos.

4.1. INTRODUCTION

4.1.1. Motivation

Much of the work in the rapidly growing field of paleoclimatology has emphasized the need to put recently observed secular climatic trends and shifts within the context of natural low-frequency variability (Martinson et al., 1995). For example, evidence for (1) global warming since the middle of the last century (Hansen and Lebedeff, 1988); (2) variations in the amplitude, duration, and frequency of El Niño/Southern Oscillation (ENSO) events (Trenberth and Shea, 1987; Trenberth and Hoar, 1996); (3) apparent state changes in the Pacific basin climate (Ebbesmeyer et al., 1991) and the timing of the Indian Monsoon (Parthasarathy et al., 1991; Krishna Kumar et al., 1999); and (4) severe mountain glacier retreat (Diaz and Graham, 1996) suggest changes in the base state of the global climate since the beginning of the Industrial Revolution. However, the processes underlying these observations are not very well understood, and some may indeed be part of quasi-periodic or low-frequency phenomena.

It has been hypothesized that many of these observations are tied to a small number of preferred, natural modes of spatial and temporal variability within the ocean–atmosphere system. Together these modes may explain much of the modern large-scale, low-frequency variance in the climate system (Wallace, 1996a,b). Examples of large-scale climatic processes are ENSO, the Pacific–North American Oscillation (PNA), and the North Atlantic Oscillation (NAO) patterns of standing atmospheric pressure waves, decadal variability in surface and intermediate circulation of the subtropical and tropical oceans (Chang et al., 1997), and globally homogeneous trends (Wallace, 1996b; Cane et al., 1997). Since these phenomena have long temporal and spatial scales, variability associated with them may be successfully retrieved from sparse observations (Miller, 1990; Cane et al., 1996). For instance, limited observations in the eastern equatorial Pacific are sufficient to crudely indicate the occurrence of an ENSO event, albeit with large uncertainties (Kaplan et al., 1998). In this chapter, we test implicitly the hypothesis that a sparse observational network composed of proxy paleoclimatic data (Epstein and Yapp, 1976; Fritts, 1976; Evans et al., 1998) may be used to reconstruct the major features of large-scale climate variability. The focus of this chapter is the development of a methodology by which we may seek skillful and verifiable reconstructions of these dominant modes using proxy data sources.

4.1.2. Proxy Climate Data and Their Interpretation

In the absence of direct physical observations, the science of paleoclimatology provides climate information through the measurement of proxies. In a broad sense these are parameters that vary in some understood way with the climatic variable of interest. Proxy measurements made on tree rings, ice cores, reef corals and sponges, sediment cores, and other geological and biological archives provide qualitative to semiquantitative estimates of climate for decades to millennia prior to the rise of direct observations (Bradley and Jones, 1992; Jones et al., 1996). Ideally, proxy observations may be used to produce what we term here a *climate field reconstruction* (CFR): a near globally complete, verifiable, and temporally continuous estimate of climate variability over some period of interest. Once we have the CFR, it is straightforward to obtain temporal indices averaged over some subarea of the global spatial domain.

The broader implications of CFR from paleoclimatic data are difficult to overestimate. Traditionally in paleoclimate research, "reconstruction" indicates recovery of a single time-varying climate index, usually of a local nature (such as seasonal air temperature from tree-ring-width chronologies) or having an immediate physical connection (such as NINO3 reconstruction from any coral site affected by ENSO). The interpretation of proxy reconstructions, usually in terms of a linear or linearized function of desired instrumental quantities, is generally verified by means of comparison with local instrumental data over a short, recent common period, although verification is increasingly made via comparison with regional or synoptic-scale observations over longer time periods. When the proxy data are contemporaneous with direct observations, opportunities are provided for intercomparison of proxy and instrumentally observed data. For example, intercomparison of independently derived proxy time series may be used to determine common features of cross-site variability. Potential biases in proxy-based records of sea surface temperature (SST) variability (age uncertainties, sampling artifacts, or other nonclimatic information) are independent of those in other proxy data or those in the bank of historical observations (instrument calibration, precision, accuracy, measurement method, frequency and spacing of sampling, and analysis procedures). In both cases, we hypothesize that common features observed across proxy realizations and in the direct observations are indicative of large-scale climatic phenomena and not small-scale or nonclimatic effects; if the instrumental and proxy data

describe climatic phenomena, then they should agree (Jones et al., 1998; Evans et al., in revision; Villalba et al., 2000; Briffa and Osborn, 1999). Such intercomparison efforts are essential for development, calibration, validation, and interpretation of individual data sets, but they may also be used to determine the potential for the development and interpretation of CFRs from proxy data (Evans et al., in revision). They provide an important link between research in climate dynamics and paleoclimatology.

Hence, we believe the most effective use of paleoclimatic data for the study of climate dynamics is via CFR based on consideration of all available proxy information, and for the reconstruction of patterns of variability, rather than point measurements. This is the idea of globality (Kaplan et al., 1997, 1998; Mann et al., 1998, 2000; Evans et al., 1998, 2000). Our view of the role of CFR as a natural link between data (proxy or instrumental) analysis and the study of climate dynamics follows the pioneering work of Fritts et al. (1971) and Fritts (1976, 1991) and is illustrated schematically in Fig. 1. Fritts provided a lucid and comprehensive introduction to the CFR concept within the context of dendroclimatological reconstruction of air temperature, sea level pressure (SLP), and precipitation anomalies over North America. More recently, the potential use of CFR on a global scale has been discussed by Jones and Briffa (1996). Bradley (1996) and Evans et al. (1998) have examined the observational array design problem, considering CFR from proxy data of varying quality and quantity. The successful extraction of variability in large-scale, low-frequency climatic phenomena from single point proxy data sources or small observational networks (Cook 1995; Wiles et al., 1998; Stahle et al., 1998; Cook et al., 2000; Evans et al., 2000; and others) has shown the potential for CFR from subsets of the available proxy paleoclimatic database. In a recent application, Mann et al. (1998, 2000) obtained very encouraging CFR results by applying space reduction and statistical techniques related to Smith et al. (1996) and Kaplan et al. (1998) for the reconstruction of surface temperature fields from several proxy data sources (Bradley and Jones 1992).

4.1.3. Globality and Optimality

The field of inverse modeling provides tools particularly suited for extraction of maximum information from a sparsely sampled, smoothly varying field. The term *Objective Analysis* (OA) encompasses a set of inverse methods employed in the estimation of the best-fit field (here in a least-squares sense) to both a sparse observational network of data and a description—a

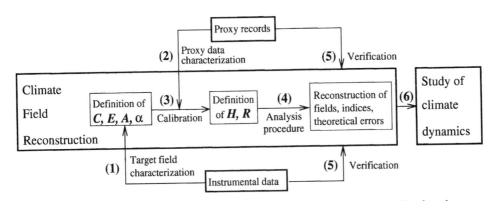

FIGURE 1 Methodology of climate field reconstruction (CFR) from proxy data. Numbered arrows on the diagram correspond to the steps in our CFR procedure (see Section 4.2).

model—of how the field varies. Errors are admitted in both the observations and the model. A cost function, consisting of the appropriately weighted squared errors in the observations and the model, is then formulated. Minimization of the cost function with respect to the field variable produces the analyzed field that is consistent with both the data and the model, given prior estimates of how precisely the model and observations are known. The error in the OA-analyzed field produced in this manner is a function of the observational error, the model error, and the extent to which the field is resolved by the observations and can be compared to either a withheld set of data for validation or to the data used in the assimilation to check the assumed observational error magnitudes.

The OA techniques have become a powerful tool used in the estimation of atmospheric and oceanographic fields of interest from spatially and temporally incomplete observations and imperfect models (Bennett, 1992; Wunsch, 1996). Special *reduced space* analogs of these techniques have been developed to reconstruct fields from sparse sampling in space and time when the fields are sufficiently smooth and have coherent, large spatial, and long temporal-scale patterns (Cane et al., 1996; Kaplan et al., 1997). Here, we apply these same tools to the reconstruction of climate fields from the even sparser but rapidly growing network of annual-resolution paleoclimate records, in order to produce CFRs from proxy data. By the reduced space rationale, we seek reconstruction of only the features of climate variability that are expected to be resolved, given the quality of the observations. Bringing the CFR problem into the context of reduced space OA provides four benefits:

1. The solution is optimal, provided the a priori assumptions hold; that is, if the data are unbiased and the covariance of a priori errors in the observations and

model are correctly estimated, then the solution obtained minimizes the squared error in the reconstructed field.

2. Space reduction emphasizes common features expressed across the proxy data set, which we expect to be climatic in origin, and discounts both climatic and proxy variability that explains little variance and is most likely to be dominated by observational errors.

3. Theoretical error estimates are provided for the solution.

4. The a priori assumptions, space reduction choices, solution, and error estimates can be tested for mutual consistency.

We illustrate this approach using the candidate data set described by Villalba et al. (2000) for the reconstruction of gridded SST for the Pacific basin over the time interval 1001–1990. The procedure is given in detail later. We focus here on the methodology of objective paleoclimatic reconstructions; the actual reconstructions shown here, which are based on a very limited number of tree-ring indicators, are not intended for interpretation. Future work will apply the methodology described here to more extensive sets of proxy indicators. Instead this methodological work is intended to complement the work of others, such as Villalba et al. (2000), who seek to characterize and interpret the climate information content of North and South American Pacific coast tree-ring chronologies. In Section 4.2, we describe the reduced space, OA CFR methodology. Section 4.3 describes the application of the technique to the reconstruction of near-global SST anomaly fields from selected Pacific-American tree-ring-width chronologies. We discuss the character of the climate information provided by tree-ring-width chronologies and conclude (Section 4.4) with some general comments on future applications of the procedure.

4.2. METHODOLOGY OF CFR FROM PROXY DATA

Our approach to CFR from proxy data will be quite similar to that applied previously to historical SST and SLP data (Kaplan et al., 1997, 1998, 2000). At all steps of the procedure we seek globality, optimality, and space reduction. The process involves six steps, which are schematically outlined in Fig. 1 and are described in detail later.

4.2.1. Characterization of the Target Climate Field

We use the modern observed data to provide estimates of the climatological mean and spatial covariance $\mathbf{C}$ of the target climate anomaly field. For the purpose of the study of large-scale phenomena, statistical description of the field can be further improved by first performing an optimal analysis of the historical climate data to provide reconstructions of the target climate field itself. We then determine the leading spatial modes of signal variability $\mathbf{E}$ (expressed as empirical orthogonal functions, or EOFs) and their energy distribution Λ (eigenvalues) through canonical decomposition of the covariance matrix $\mathbf{C}$ and truncation of statistically insignificant modes (Cane et al., 1996; Kaplan et al., 1997, 1998):

$$\mathbf{C} = \, <\mathbf{TT^T}> \, \approx \mathbf{E\Lambda E^T} \tag{1}$$

where boldface variables indicate vectors or matrices, $<\ldots>$ indicates time averaging, and superscript $\mathbf{T}$ denotes the matrix or vector transpose. Here, $\mathbf{E}$ is a matrix whose columns are a relatively small number of the eigenvectors of $\mathbf{C}$, and Λ is a square matrix whose diagonal elements are the eigenvalues corresponding to $\mathbf{E}$. We then assume a reduced space form for the solution

$$\mathbf{T}(t) = \mathbf{E\alpha}(t) + \text{residual} \tag{2}$$

where $\mathbf{\alpha}(t)$ is the low-dimensional vector of amplitudes with which the modes $\mathbf{E}$ contribute to the climate field $\mathbf{T}$ at time t, and $\mathbf{T}$ is the long-term mean zero. For the modern period, the EOFs $\mathbf{E}$ and the vectors $\mathbf{\alpha}(t)$ are known from the analysis of the instrumental data. For the purposes of CFR, a portion of this analyzed data set $\mathbf{T}$ may be used for calibration of the proxy data (see Section 4.2.3). The part of the target climate field not used for calibration purposes can then be used for independent verification of the corresponding period of the proxy-based CFR (see Section 4.2.5).

We presume the analysis of the target modern climate data captures the potential modes of variability reflected in proxy observations. Specifically, we assume that the leading patterns of pre-instrumental SST variability may be adequately represented by some linear combination of the EOFs used in the analysis of the SST data. Although this assumption may be true for the largest scale patterns that are likely to be resolved by proxy data over the past several hundred years, it may be that the climate system has operated in modes that are not spanned by this characterization of the instrumental data. We also assume that the errors in this analysis are small relative to the errors in the proxy-based CFR we seek and that the climatic patterns resolved by the proxy data are a small subset of those resolved in modern historical observations.

4.2.2. Characterization of the Proxy Data

In cases where numerous proxy data realizations are expected to resolve a much smaller dimension of climatic information, we may perform a similar statistical decomposition and analysis of the proxy data, $\mathbf{D}(t)$, to determine the leading modes of variability evident in the proxy data. In doing so, we assume that the leading modes of variability observed in the proxy data set itself are climatic in origin, rather than the expression of small-scale climatic effects or the imprint of nonclimatic influences. If this is the case, the leading eigenvalues of the decomposition may be well separated from the rest, and the corresponding EOFs will explain a large fraction of the variance in the proxy data set. This hypothesis may be checked by examining the corresponding spatial structure in the target climate field or by examining the frequency domain characteristics of the proxy amplitudes (Wallace, 1996a,b). We may also use these results to analyze the proxy data itself, following Eqs. (3)–(12) in Section 4.2.4, to emphasize common cross-site features observed in the proxy data. Such prefiltering of the proxy data reduces the potential for unrelated patterns in the proxy data to falsely project upon the leading modes of climate field variability. Statistical analyses may be checked for consistency with physical or empirical knowledge derived from development and intercomparison of the proxy data with local climatic data.

4.2.3. Calibration

We seek calibration of the proxy data $\mathbf{D}(t)$ in terms of the time amplitudes of the reduced space representation of the climate field, $\mathbf{\alpha}(t)$. More precisely, the low-dimensional representation of $\mathbf{T}$ in terms of $\mathbf{E}$ and $\mathbf{\alpha}(t)$ permits the estimation scheme

$$\mathbf{D}(t) = \mathbf{H\alpha}(t) + \mathbf{\varepsilon}. \tag{3}$$

Equation (3) sets up the Gauss-Markov observational scheme for estimation of $\mathbf{\alpha}(t)$ from the observed proxy vector **D** (Gandin, 1965; Mardia et al., 1979; Rao, 1973). The measurement operator **H** is a linear function that maps from the climate field to the proxy observations. In other words, it represents a linear regression of the proxy data on the temporal amplitudes of the leading patterns of climate variability. For example, if **D** were SST estimates at points, then **H** would be a spatial interpolation of **E** to the observational locations. In this work **H** describes the linear function relating dimensionless tree-ring-width indices from terrestrial locales to leading modes of global SST variability. Locally, it might express the linearized relationship expected from local calibration studies or from principles of dendroclimatology.

We introduce **R** = $<\varepsilon\varepsilon^T>$ as the matrix of observational error covariances. As defined, it is the error in observation of the climate field from the proxy data. Hence, it includes the error in observations of each proxy, the error due to space reduction [Eq. (2)], and the error in estimation of the measurement operator **H**. This last error component is estimated via calibration over a limited time interval with good data coverage. Note that the construction of **H** in this manner permits nonlocal information to be recovered by the proxy data.

We use the proxy data and the instrumental data to estimate **H** and ε over a calibration interval, using the singular value decomposition (SVD). The SVD procedure decomposes the covariance between the proxy data and principal components (PCs) of the climate field into linear combinations of singular values and left and right singular vectors:

$$\mathbf{C} = <\mathbf{D\alpha}^T> = \mathbf{U\Sigma V}^T \qquad (4)$$

where the columns of **U** and **V** are orthogonal vectors describing modes of covariance between the proxy data and the climate field, respectively, and $<...>$ now indicates time averaging over the calibration period.

As a sparse observational network of proxy data may be expected to robustly resolve only a few leading modes of climate variability, we retain for multivariate regression with vector of coefficients **h** only the leading modes of covariance between proxy data and climate field, as defined by the SVD procedure

$$\mathbf{D} \approx \mathbf{hV}_r^T\mathbf{\alpha}. \qquad (5)$$

Here, $\mathbf{V}_r$ is the matrix consisting of the first few columns of **V**. How many columns of **V** to retain is a somewhat subjective choice, though one that may be subsequently tested statistically. The use of such screening techniques to form more robust estimates of the relationships between two fields has long been applied in meteorology (Bretherton et al., 1992) and dendroclimatology (Fritts et al., 1971; Cook et al., 1994).

The vector of regression coefficients **h** is then esti-

mated based on these presumably resolvable and climatic modes:

$$\mathbf{h} = <\mathbf{D\alpha}_r^T><\mathbf{\alpha}_r\mathbf{\alpha}_r^T>^{-1} \qquad (6)$$

where $\mathbf{\alpha}_r = \mathbf{V}_r^T\mathbf{\alpha}$. The error in the map ε is the difference between the proxy observations and their best-fit estimates:

$$\varepsilon = \mathbf{D} - \mathbf{h\alpha}_r, \mathbf{R} = <\varepsilon\varepsilon^T>. \qquad (7)$$

The map from the reduced space representation of the climate field to the proxy data [cf. Eqs. (3) and (5)] is then

$$\mathbf{H} = \mathbf{hV}_r^T. \qquad (8)$$

Equations (7) and (8) specify all parameters necessary for the estimation scheme given in Eq. (3).

In some cases the instrumental data and an understanding of the target climate field and its dynamics can be used to construct a model for climatic time transitions:

$$\mathbf{\alpha}(t+1) = \mathbf{A\alpha}(t) + \varepsilon_t, \mathbf{Q} = <\varepsilon_t\varepsilon_t^T>$$

which can be used in CFR as an additional source of information for OA in an optimal smoother analysis (Kaplan et al., 1997). If no such model is available, and the signal is assumed to have zero mean, then the trivial time transition model **A** = **0** may be employed:

$$\mathbf{\alpha}(t+1) = \mathbf{0\alpha}(t) + \varepsilon_t, <\varepsilon_t\varepsilon_t^T> = <\mathbf{\alpha\alpha}^T> = \mathbf{\Lambda}.$$

This procedure assumes that the proxy measurements may be represented as a linear function of the time variability of large-scale modes of climatic variability and a residual. This functional dependence may include other factors insofar as they are linearly related to the variability exhibited by the climate field **T** within the calibration period. In fact, we have found in studies of coral $\delta^{18}O$ data that nonlocal climatic variability may constitute a significant part of the map to SST (Evans et al., 2000). We further assume that any bias in the linear relationship between climate field and proxy data is removed prior to analysis. Proxy age model uncertainty is included here only through influence on the estimated observational SST error ϵ in the map **H**. We also assume that the map from climate field to proxy data is a robust estimate, which neither overfits nor underrepresents the climatic information in the proxy data, and that this map and its error do not change over the period of the reconstruction. These assumptions may be subsequently subjected to testing and verification (Section 4.2.5).

4.2.4. Analysis

In many observed and gridded meteorological and oceanographic fields, the number of patterns that may be readily discerned above observational noise, data gaps, and other uncertainties is small. For CFR based on proxy data, the number of observations is much smaller and the observational error is far larger, so the dimension of the resolvable climatic *space* is expected to be even smaller.

We seek the field $\mathbf{T(x,t)}$, which is the best fit to the proxy observations constrained to be near the spatial target field covariance. We construct a cost function

$$S(\boldsymbol{\alpha}) = (\mathbf{D} - \mathbf{H}\boldsymbol{\alpha})^{\mathsf{T}}\mathbf{R}^{-1}(\mathbf{D} - \mathbf{H}\boldsymbol{\alpha}) + \boldsymbol{\alpha}^{\mathsf{T}}\boldsymbol{\Lambda}^{-1}\boldsymbol{\alpha} \qquad (9)$$

S evaluated at each time t is a unitless scalar quantity. Here, $\boldsymbol{\alpha}$ is the vector of temporal amplitudes we seek to reconstruct from the proxy data. The first term of Eq. (9) represents the squared residuals between the proxy observations and their $\boldsymbol{\alpha}$-based predictions at each time t. $\mathbf{R}$ is the estimate of their covariance [Eq. (7)] and weights this set of residuals. Similarly, the second term weights the *residuals* of the trivial prediction model $\boldsymbol{\alpha} = \mathbf{0}$ with its inverted error covariance, which is the covariance of the signal [Eq. (1)]. At each time t, the analysis is punished ($S \rightarrow$ large) for putting too much stock either in observations with large error or in patterns that explain little of the energy in the spatial covariance patterns observed in the modern SST field (Kaplan et al., 1997).

Minimization of a quadratic cost function such as Eq. (9) retrieves the generalized least-squares estimate of $\boldsymbol{\alpha}$ (Sokal and Rohlf, 1995). If the proxy observations are unbiased, and if the a priori error covariances $\mathbf{R}$ and $\boldsymbol{\Lambda}$ are well estimated and uncorrelated in time, then minimization of S produces the best linear unbiased estimate (BLUE) (Mardia et al., 1979). For this specific minimization problem, in which S has been written in terms of reduced space variables, minimization at each time t produces the reduced space optimal estimate (Kaplan et al., 1997):

$$\boldsymbol{\alpha}(\mathbf{t}) = \mathbf{P}\mathbf{H}^{\mathsf{T}}\mathbf{R}^{-1}\mathbf{D(t)} \qquad (10)$$

and the field $\mathbf{T}$ is recovered by Eq. (2)

$$\mathbf{T}(\mathbf{t}) = \mathbf{E}\boldsymbol{\alpha}(\mathbf{t}). \qquad (11)$$

$\mathbf{P}$ is the error covariance in the estimate $\mathbf{T}$:

$$\mathbf{P} = (\mathbf{H}^{\mathsf{T}}\mathbf{R}^{-1}\mathbf{H} + \boldsymbol{\Lambda}^{-1})^{-1} \qquad (12)$$

and it represents a weighted average of the error in the observations mapped to the analysis domain and the spatial field covariance estimate. The weights are determined by the observational error covariance matrix $\mathbf{R}$, and the variance $\boldsymbol{\Lambda}$ is resolved by the EOFs $\mathbf{E}$ over the calibration period. The rank of $\mathbf{H}$ is determined as part of the calibration procedure (Section 4.3.3). The matrix $\mathbf{H}$ in Eqs. (10) and (12) consists of the rows of $\mathbf{H}$ defined by Eq. (8) for which the proxy data are available at analysis time t. When the observations are relatively accurate ($\mathbf{R}$ small) or sufficiently numerous ($\mathbf{H}$ has high row dimension), the analysis error is low, and the analysis puts variance into the solution $\mathbf{T}$. When observations are poor ($\mathbf{R}$ large) or absent ($\mathbf{H}$ has low row dimension), the analysis error is large, and the analysis estimate $\mathbf{T}$ approaches climatology (e.g., the trivial solution $\mathbf{T} = \mathbf{0}$). Similarly, if the proxy data do not skillfully describe climatic variance over the calibration period ($\mathbf{H}$ of low rank and/or $\mathbf{R}$ large), the analysis estimate $\mathbf{T}$ will have low amplitude. In either case, the analysis error covariance is formulated to be consistent with our a priori assumptions about the error in the observations, map, and model; our error covariance estimates; and the availability of proxy observations over time. Subsequent verification exercises may show that we have assumed the proxy data are too accurate or the relationship between proxy data and climate field is too well resolved. Iteration of the procedure, with different choices for the rank of $\mathbf{H}$ and more reasonable estimates of $\mathbf{R}$, is required until a priori assumptions match a posteriori results and produce the optimal analysis field estimate $\mathbf{T}$.

4.2.5. Verification

The initial verification of the reconstructed climate field is a comparison with the observed climate data withheld from the calibration procedure (Sections 4.2.1 and 4.3.3). The more complicated issue is testing the consistency of the analysis, as was mentioned earlier. Technically, the analysis solution is optimal, provided the proxy calibration holds for the entire period of reconstruction; the data and model are unbiased estimates; and the observational error ϵ is uncorrelated in time. We may test the validity of these assumptions via cross-validation experiments with withheld data, comparison of the verification residuals with the theoretical error in the analysis, and comparison of the analysis results with other independent data (instrumental or proxy). In addition, to determine the extent to which climatic information is resolved by the proxy data, the proxy-based CFR may be compared to CFRs based on synthetic proxy data chosen to represent benchmark observations or data sets devoid of information.

4.2.6. Study of Climate Dynamics

Once the reconstruction field $\mathbf{T}$ has been satisfactorily validated, we may begin to study it as a source of climatic information within the context of its strengths and limitations. In many cases, the analysis error will be quite large and will limit interpretation to only the largest scale area-average indices. Such indices may be subjected to time series analysis, epoch analysis, frequency domain analysis, joint spatial-temporal mode analysis, and climate change hypothesis testing. Many such studies become more direct and convenient when the reduced space form of the solution [Eqs. (10) and (12)] is utilized.

4.3. APPLICATION: PACIFIC BASIN SST FIELD RECONSTRUCTION FROM PACIFIC AMERICAN TREE-RING INDICATORS

4.3.1. Target Climate Field

SST is an important diagnostic of the global climate. The surface ocean and atmosphere participate (or are suspected to participate) in positive and negative dynamical feedback mechanisms on a wide range of timescales (Namias, 1969, 1980; Cane, 1986; Deser et al., 1996; Latif and Barnett, 1994; Lau and Nath, 1994; Webster, 1994; Trenberth and Hurrell, 1994; Clement et al., 1996; Cane et al., 1997; Clement, 1998). As a statistical field, SST varies relatively smoothly on large time and space scales, and so we hypothesize that its leading temporal and spatial covariance patterns are likely to be captured by a sparse observational network composed of proxy observations.

Our source of SST data for use as a target climate field is the recently produced analysis (Kaplan et al., 1998) of the MOHSST5 (Parker et al., 1994) product of the U.K. Meteorological Office. MOHSST5 is a $5° \times 5°$ monthly compilation of historical (1856–1991) ship-based observations of SST, and the Kaplan et al., (1998) analysis employs the newly developed technique of the reduced space optimal smoother to objectively extract large-scale variations of SST from incomplete spatial and temporal observational coverage. Variance on small spatial scales, amounting to $(0.3–0.4)^2 °C^2$ of the intermonthly variance, which is not constrained by the observations, is filtered out by the reduction of the EOF space. The analysis can reconstruct anomalies associated with ENSO and other large-scale phenomena, including low-frequency variability (Enfield and Mestas-Nuñez, 2000; for several examples and tests, see Kaplan et al., 1998). Since our purpose is the reconstruction of large-scale SST field variability, we chose this product as our basis or *target* climate field for our CFR experiment.

We use 1856–1990 annual means of Pacific basin SST anomaly (87.5°N to 87.5°S, 110°E to 65°W) from this analysis (hereafter termed KaSSTa) as our target climate field. The annual average runs from April of the current year through March of the following year, the natural year for much of the Pacific Ocean (Ropelew-

ski and Halpert, 1987). This choice also serves to integrate dendroclimatological anomalies observed in the Northern Hemisphere growing season (April–October) and those observed during the growing season of the Southern Hemisphere (November–March), which together provide complementary information on annually averaged SST anomaly (Villalba et al., 2000). Following Kaplan et al., (1998), these data are decomposed into a set of EOF weights and temporal loadings, and the leading 30 modes of spatial variability are used to reconstruct approximately 85% of the total variance in the annually averaged gridded $5° \times 5°$ estimates of SST anomaly on which KaSSTa is based (Bottomley et al., 1990). The domain, variance, and error in the analysis of this target climate field are shown in Fig. 2.

4.3.2. Proxy Data

For over a century, those in the field of dendroclimatology have sought estimates of local and regional climate variability from measurements made on trees producing marked annual growth rings (Fritts, 1976, 1991). One of the most common measurements is the width of sequential annual rings. In an individual tree, rings indicate the passage of time, and ring-width variations reflect the combined effects of biophysical and environmental factors, including climate. While the physiological factors controlling ring-width variations within individual trees are complex and difficult to quantify (Fritts et al., 1971), the mean weather and climatic conditions at a given site should have a similar effect on all trees within a homogeneous stand. In addition, using a priori knowledge of local or regional growing conditions, the sampling site may be chosen for its expected sensitivity to a particular, growth-limiting climatic variable of interest (Fritts, 1976). Dendroclimatologists employ statistical approaches to determine the common climatic signal extricable from a population of trees; such a statistical reduction of the data to a well-dated, continuous, standardized, tree-growth index is known as standardization, and its final product is called a chronology.

Many hundreds of tree-ring-width chronologies have now been developed worldwide (ITRDB, 1998), and their relationships to climatic variables have been tested on local, regional, and global scales. In a recent application, Villalba et al. (2000) found that tree-ring-based estimates of air temperature and drought developed from Pacific American tree-ring chronologies share a common, decadal mode of time variability over the last four centuries. They showed that this variability is associated with an ENSO-like pattern of SST anomalies in the Pacific basin. With the hypothesis that this SST pattern forces variability observed in the proxy

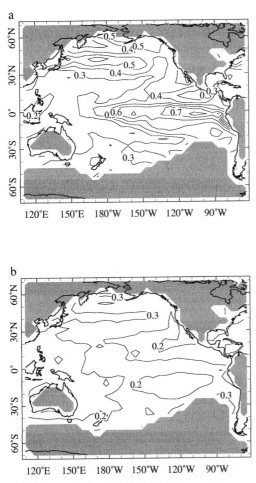

FIGURE 2 (a) The target climate field of the annually averaged Pacific basin sea surface temperature (SST) anomaly. Contours give root-mean-squared (RMS) variance in degrees Celsius (°C). (b) Error in estimation of the field (°C). Gray areas show regions where there are insufficient data for analysis. (Data from Kaplan et al., 1998.)

cal air temperature, precipitation, or both, in the case of drought indicators) and include tree-ring-width chronologies as well as climatic indices derived from tree-ring-width chronologies.

To compare and evaluate the results, we apply the same methodology to reconstruct SST from two additional sets of indicators. The first is a set composed of instrumental climate data from the locations from which the proxy data were collected (Table 1). These are selected to mimic the expected sensitivity (to surface temperature, precipitation, and drought) of the dendroclimatological data employed. The second is a set of red noise processes with lag −1 autoregressive characteristics like those of the dendroclimatological data. These two additional benchmark and noise reconstruction experiments are expected to return estimates of the best possible reconstruction results and of skill expected by chance, respectively. In all cases, the proxy indicators have been normalized, as their original variances relative to one another may or may not have climatic significance.

4.3.3. Calibration

4.3.3.1. Construction of the Relationship between SST and Tree-Ring Indicators

Finding the appropriate mapping **H** from the SST to the proxy data is a somewhat subjective task that we solve in the following manner: we estimate the covariance matrix between the proxy data and the low-dimensional vector of amplitudes known from KaSSTa over the calibration interval 1923–90 [Eqs. (3) to (8)] and we reserve the complementary part of the KaSSTa amplitudes (1856–1922) for verification exercises and as an alternate calibration period (see Section 4.3.5. Following the work of many dendroclimatological calibration studies in general (Fritts, 1976) and employing the findings of Villalba et al. (2000) in particular, we permit the tree-ring data to provide information on SST variability through simultaneous and lag −1 year covariances.

EOF analysis of the tree-ring data suggests that this data set has at least one and perhaps up to three significant EOFs. The first proxy EOF, which explains about 32% of the variance in the time series, is clearly distinguishable above the eigenvalue spectrum retrieved from EOF analysis of red noise time series with the autoregressive characteristics of the tree-ring data (Fig. 3). Similar decomposition of the observed data suggests that, at most, two to three significant EOFs may be retrieved from the sampled regions.

Considering these results and the work of Villalba et al. (2000), we present the analysis with only the leading

data, they inferred increased energy associated with the decadal SST pattern during the period 1600–1850, relative to the present (Villalba et al., 2000).

Following the intriguing and encouraging results of Villalba et al. (2000), we hypothesize that large-scale SST anomalies over the Pacific basin give rise to surface air temperature, precipitation, and drought anomalies across the tropical and extratropical Americas; in turn, dendroclimatological data from affected regions may be used to reconstruct the relevant SST variability. Hence, we apply the approach detailed earlier in Section 4.2 to the problem of reconstructing Pacific basin SSTs from the small subset of dendroclimatological indicators studied by Villalba et al. (2000) (Table 1) for the interval 1001–1990. For our purposes, we term these data *tree-ring indicators* of climate variability, noting that the data have a diversity of sensitivities (lo-

TABLE 1   Tree-Ring Indicators Used in This Analysis[a,b]

Site name	Latitude	Longitude	Interval	Benchmark historical time series
Coast of Alaska—temperature sensitive (Wiles et al., 1998)				
Ellsworth	60.2°N	149.0°W	1580–1991	
Exit Glacier	60.2°N	149.6°W	1580–1988	
Miners Wells	60.0°N	141.7°W	1580–1995	
Tebenkof	60.8°N	148.5°W	1580–1991	SST at (57.5°N, 142.5°W)
Verstovia Ridge	57.0°N	135.3°W	1580–1996	
Whittier	60.8°N	148.6°W	1580–1991	
Wolverine Glacier	60.3°N	148.9°W	1580–1991	
Northern Patagonia—temperature sensitive (Villalba et al., 1997)				
Castaño Overo	41.2°S	71.8°W	1540–1991	
Castaño Overo	41.2°S	71.8°W	1550–1991	SST (42.5°S, 72.5°W)
Co. D. de León	41.3°S	71.7°W	1564–1991	
Southwestern U.S.—drought sensitive (Cook et al., 1999)				
TrPDSI Cell 53	31.0°N	107.5°W	1693–1978	PDSI (31.0°N, 107.5°W)
TrPDSI Cell 63	31.0°N	104.5°W	1690–1978	PDSI (31.0°N, 104.5°W)
TrPDSI Cell 80	39.0°N	98.5°W	1680–1978	PDSI (39.0°N, 98.5°W)
TrPDSI Cell 81	37.0°N	98.5°W	1698–1979	PDSI (37.0°N, 98.5°W)
Northern Patagonia—precipitation sensitive (LaMarche et al., 1979)				
El Asiento	32.7°N	70.8°W	1001–1996	Santiago, Chile, precipitation (33.5°S, 71.5°W)

[a]For more information on the selection of these proxy indicators, see Villalba et al. (2000).

[b]TrPDSI is the gridded estimate of the Palmer Drought Severity Index (PDSI) reconstructed from tree-ring chronologies (Cook et al., 1999).

singular vector of the covariance between SST and proxy data retained. (We leave as a verification exercise [Section 4.3.5] determination of whether additional singular vectors add appreciably to the skill of the reconstruction.) Figure 4 shows the spatial pattern associated with the time series of the leading singular vector for

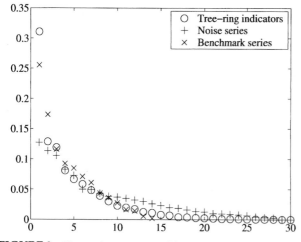

FIGURE 3   Eigenvalue spectrum of the tree-ring data employed in this study (O). For reference, the eigenvalues of red noise time series with similar autocovariance statistics (+) and of historical observations from the proxy sampling sites (×) are also shown.

both the proxy data and benchmark data calibration experiments. Not surprisingly, the proxy calibration captures the same pattern Villalba et al. (2000) observed. Resolution of this pattern is corroborated by a similar (but stronger) pattern from analysis of the benchmark instrumental data set (Table 1). However, the error in the resolution of this pattern [$\mathbf{R}$, Eq. (7)] is large in both cases, due to the difficulty with which these sparse observational networks discern the climatic variability. Results described later show that observational error variances are, on average, about 80% of the signal variance.

4.3.3.2. Contribution of Individual Indicators to the Reconstructed Mode

Table 2 shows the correlations of individual tree-ring indicators with the single mode of reconstructed SST variability as a function of calibration period and time lag. We interpret these correlations as the contribution made by each indicator to the reconstructed pattern. Lag 0 or lag +1 correlations that are significant above the 90% confidence level in both calibration periods are in bold. Seven of the indicators (3, 4, 5, 7, 9, 12, and 15) meet these criteria for simultaneous correlations, and nine (2, 3, 4, 5, 7, 9, 11, 12, and 14) meet these criteria for lag +1 correlations. Six indicators (3, 4, 5, 7, 9, and

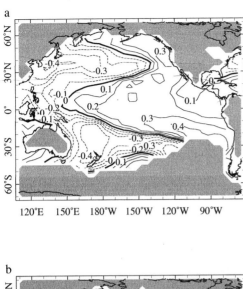

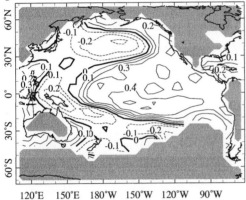

FIGURE 4 (a) Correlation of the first tree-ring principal component (PC) with the sea surface temperature (SST) field. (b) Correlation of the first PC derived from benchmark instrumental observations with the SST field.

TABLE 2   Contribution of Tree-Ring Indicators to Reconstruction

Correlation with time series of reconstructed pattern[a]

Indicator[b]	1923–1990 calibration		1856–1922 calibration	
	Lag 0	Lag +1	Lag 0	Lag +1
1 CAT	0.42 (99)	0.45 (99)	−0.17 (14)	0.06 (64)
2 CAT	0.18 (86)	**0.24** (93)	0.14 (82)	**0.30** (97)
3 CAT	**0.54** (99)	**0.58** (99)	**0.53** (99)	**0.72** (99)
4 CAT	**0.31** (97)	**0.25** (93)	**0.24** (93)	**0.33** (98)
5 CAT	**0.52** (99)	**0.45** (99)	**0.37** (99)	**0.64** (99)
6 CAT	0.31 (96)	0.26 (94)	−0.01 (50)	0.11 (75)
7 CAT	**0.42** (99)	**0.49** (99)	**0.22** (93)	**0.37** (99)
8 NPT	0.37 (99)	0.33 (98)	0.11 (73)	0.17 (86)
9 NPT	**0.51** (99)	**0.31** (98)	**0.21** (91)	**0.21** (92)
10 NPT	0.37 (99)	0.36 (98)	0.13 (83)	0.15 (86)
11 SWP	0.02 (56)	**0.71** (99)	0.28 (98)	**0.71** (99)
12 SWP	**0.20** (94)	**0.76** (99)	**0.47** (99)	**0.71** (99)
13 SWP	−0.24 (92)	0.14 (80)	0.17 (86)	0.54 (99)
14 SWP	−0.18 (88)	**0.29** (97)	**0.22** (92)	**0.64** (99)
15 NPP	**0.25** (94)	0.35 (99)	**0.39** (99)	0.05 (63)

[a]Numbers in parentheses give percent confidence level estimates, which are based on correlations with 1000 synthetic time series having first-order autoregressive characteristics of the tree-ring indicators. Lag 0 or lag +1 correlations significant above the 90% level in both calibration periods are in bold.

[b]Sensitivity of indicator to local conditions as described by Villalba et al. (2000): CAT, coastal Alaskan temperature; NPT, northern Patagonian temperature; SWP, southwest U.S. Palmer Drought Severity Index (PDSI); and NPP, northern Patagonian precipitation.

12) are significantly correlated with the reconstructed mode at both lags and in both calibration periods. These subsets of indicators include precipitation- and temperature-sensitive tree-ring indicators from both hemispheres, in agreement with the hypothesis (Villalba et al., 2000) that the resolved SST pattern represents a common forcing of both Northern and Southern Hemisphere extratropical temperature and precipitation anomalies. Consequently, the reconstructed SST pattern produced independently by using the 1923–90 and 1856–1922 calibration intervals share almost 60% variance. However, the results also show that several indicators (1, 6, 8, and 13) did not contribute consistently within both calibration periods or for either lag relationship to the reconstruction, and two (13 and 14) changed the sign of their simultaneous correlations between calibration periods.

If the assumption is made that the relationships between SST and tree-ring-site climate represented statis-

tically here have not shifted in space, and if the targeted single mode of SST continues to be the desired reconstruction target, then these results may suggest that an additional level of data screening is necessary. For example, Villalba et al. (2000) make this assumption to argue that the tree-ring data indicate different temporal variability before and after 1850. Ten of the fifteen tree-ring indicators support this assumption for the 1856–1990 period by calibrating in the same manner over both halves of the full period (Table 2). However, several indicators (1, 6, 8, 10, and 13) were significantly correlated during one calibration period, but not within the other. This result suggests that either a real change in the relationship between tree-ring indicator and SST or that the observed relationship is weak. We cannot eliminate the influence of real changes in either the proxy SST relationship or the proxy local climate relationship on long timescales. Further steps to test the stationarity assumption may include data-screening experiments with the current proxy data set, incorporation of additional tree-ring indicators from locations sensing the identified mode of climate variability, and quantitative estimation of the sensitivity of tree-ring in-

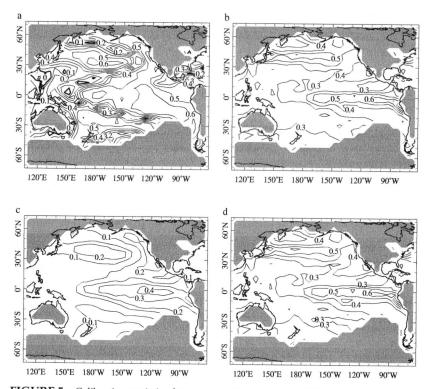

FIGURE 5　Calibration statistics for reconstruction using the 1923–90 calibration interval. All statistics are calculated by using 1923–90 data from Kaplan et al. (1998) (here termed KaSSTa) and the tree-ring-based SST reconstruction (TrSSTa) results. (a) Field correlation (correlation units). (b) Root-mean-squared (RMS) difference between TrSSTa and KaSSTa (°C). (c) RMS variance in TrSSTa (°C). (d) Theoretical error for TrSSTa (°C).

dicators to local climate variables. Analysis of the physical differences between these sites and nearby sites with significant contributions to the reconstruction may indicate reasons for calibration failures and may suggest additional locations from which proxy data may be usefully incorporated or collected. These steps may all serve to reduce or better estimate the observational and mapping error in the reconstruction procedure through better understanding of the physical and biological processes underlying the statistical relationships observed here.

4.3.4. Analysis

We first examine results using 1923–90 as a calibration period, reserving data for 1856–1922 for verification exercises. Four statistics calculated for each point in the analysis grid are shown in Figs. 5 and 6 for calibration and verification periods, respectively. Correlation between reconstructed (TrSSTa) and verification (KaSSTa) fields shows where the reconstruction has skill regardless of signal amplitude (panel a). The root-mean-squared (RMS) difference between TrSSTa and

KaSSTa gives the actual error in the reconstructed fields by comparison of withheld observed data and reconstruction results (panel b). Panel (c) shows the RMS variance in the reconstructed field; this map may be compared to Fig. 2 to assess the extent to which the analysis resolves variance. Panel (d) gives the theoretical error in TrSSTa averaged over the verification period. This map may be compared to the RMS difference described previously to determine consistency of the reconstruction's theoretical error estimate.

4.3.5. Verification

4.3.5.1. Consistency of A Priori Assumptions and A Posteriori Results

Calibration results (Fig. 5) for the 1923–90 period show that correlation between TrSSTa and KaSSTa reaches 0.4–0.6 over much of the eastern tropical Pacific and in the centers of the Pacific subtropical gyres. This result indicates that ca. 20–25% of the variance in KaSSTa was calibrated in these regions, consistent with the map of reconstructed RMS variance in TrSSTa (Fig.

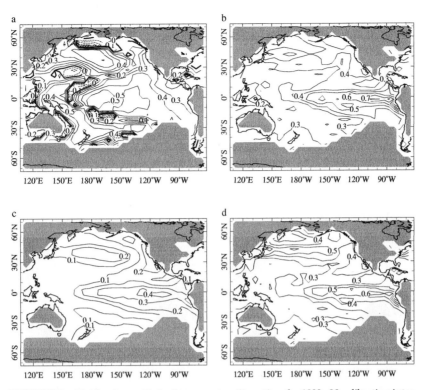

FIGURE 6   Verification statistics for reconstruction using the 1923–90 calibration interval. All statistics are calculated by using 1856–1922 data from Kaplan et al. (1998) (here termed KaSSTa) and the tree-ring-based SST reconstruction (TrSSTa) results. (a) Field correlation (correlation units). (b) Root-mean-squared (RMS) difference (°C). (c) RMS variance in TrSSTa (°C). (d) Theoretical error for TrSSTa (°C).

5c), which has roughly one-fourth to one-fifth of the amplitude of KaSSTa (Fig. 2a) over this period. Areas of minimum correlation correspond to regions in which the map **H** has little amplitude. Retrieval of small-amplitude TrSSTa is consistent with the large error in resolution of **H** shown in Fig. 4; the error plots (Figs. 5b and 5d) indicate that the mapped variance in the eastern equatorial Pacific and in the subtropical gyres is only partially resolved. Comparison of the actual RMS error and average theoretical error estimates (Figs. 5b and 5d) shows that the analysis procedure produces self-consistent errors.

4.3.5.2. Comparison with Withheld Historical SST Data

Results (Fig. 6) computed over the 1856–1922 verification period are similar to those shown in Fig. 5, suggesting that TrSSTa has captured limited but verifiable climatic information. Correlations have shrunk in amplitude by 0.1–0.2 units in the regions in which **H** has nonzero amplitude, but are similar in spatial pattern and strength to the results shown for the calibration period. This result suggests that the resolved pattern is ro-

bustly defined, at least for the interval 1856–1990, and serves as an initial check of analysis assumptions. An overlay of the correlation map shown in Fig. 4 with the verification correlation of TrSSTa and KaSSTa clearly shows that regions with the best reconstruction skill are also regions where the resolved map has a large amplitude; regions of minimal reconstruction skill correspond to regions of minimal map amplitude (Fig. 7 [see color insert]). Actual and theoretical error estimates are approximately equal (Figs. 6b and 6d), suggesting that we have not overcalibrated the reconstruction by retaining modes that cannot be verifiably reconstructed. However, note that the error in the North Pacific region has a slightly different structure than the actual error.

4.3.5.3. Sensitivity to the Calibration Period

We exchange the calibration and verification periods chosen for the previous experiment to estimate the dependence of the map from SST to tree-ring indicators on the calibration interval. The same four statistics are plotted for a 1856–1922 calibration period and a 1923–90 verification period in Figs. 8 and 9, respectively. Calibration results (Fig. 8) give slightly improved skill, but

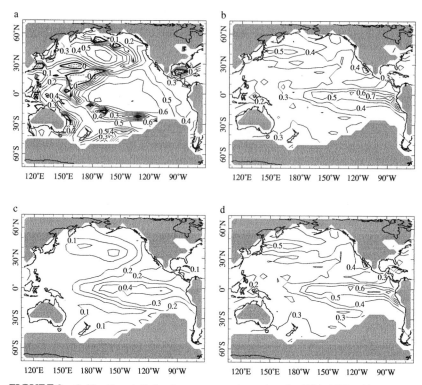

FIGURE 8   Calibration statistics for reconstruction using the 1856–1922 calibration interval (as in Fig. 5).

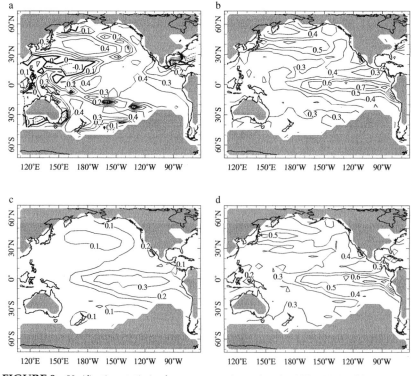

FIGURE 9   Verification statistics for reconstruction using the 1856–1922 calibration interval (as in Fig. 6).

resolve the same pattern of variability. The verification statistics are similar to those shown in Fig. 5. We may also compare the reconstructed time series [Eq. (10) with one mode of variability retained; see also Section 4.3.6] formed by using each of the chosen calibration intervals (1856–1922 and 1923–90) over their respective verification periods (1923–90 and 1856–1922). The reconstructions correlate with r = 0.76 (1923–90) and r = 0.84 (1856–1922); over the pre-observational interval 1001–1855, correlation between the two reconstructions is r = 0.76. For 11-year running averages, the two reconstructions correlate with r ≈ 0.9 over the interval 1856–1990 and the full-time interval 1001–1990. However, there appears to be a tradeoff in skill between the North Pacific and eastern equatorial Pacific regions depending on the calibration period (compare Figs. 5 and 6 to Figs. 8 and 9). This finding suggests some sensitivity of **H** to the calibration interval chosen, which can affect the skill level by about 0.1 correlation unit in regions where TrSSTa has verifiable skill.

4.3.5.4. Comparison with Benchmark and Noise Reconstructions

In addition to checking the sensitivity of the results to map stability, we can also compare TrSSTa to an instrumental data-based reconstruction and a re-construction based on calibrated red noise. Both experiments employ synthetic proxy data. In the first experiment, we reconstruct the SST field using SST, precipitation, and Palmer Drought Severity Index (PDSI) data from locations nearby the tree-ring indicator sampling sites (benchmark). In the second experiment, we perform the reconstruction based on randomly generated time series with normal distribution, unit variance, and lag −1 autocorrelation statistics of the tree-ring indicators (noise). The verification results obtained for these two experiments, using 1923–90 as a calibration period and reconstructing one pattern, are shown in Figs. 10 and 11, respectively. These results may be compared to those already shown in Fig. 6.

The calibrated skill in these experiments (results not shown) has spatial structure similar to that of TrSSTa (Fig. 5). Not surprisingly, the benchmark experiment has higher calibrated skill than TrSSTa, especially along the coasts of the Americas, and there is very little loss of skill (<0.1 correlation unit) between calibration and verification periods (results not shown). More surprisingly, our procedure was content to calibrate the noise proxy data at skill levels very similar to those of TrSSTa (results not shown). The true skill of the experiments, however, becomes apparent in the comparisons with withheld observed SST (Figs. 10a and 11a). While TrSSTa seems to capture the larger scale, open ocean

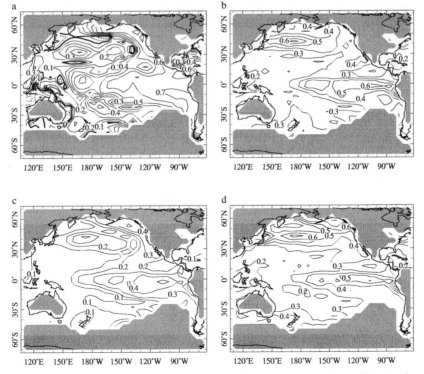

FIGURE 10 Verification statistics for benchmark reconstruction using calibration for 1923–90 (as in Fig. 6).

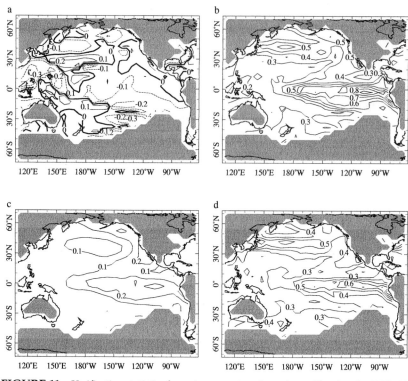

FIGURE 11   Verification statistics for noise reconstruction using calibration for 1923–90 (as in Fig. 6).

signal and has less skill near the proxy sampling sites (Figs. 5–8), the benchmark experiment is more skillful in resolving smaller scale, coastal phenomena, with less skill in the subpolar North Pacific. We speculate that this may be due to temporal signal integration or other smoothing effects introduced by statistical development of the tree-ring chronologies, which may highlight the large-scale oceanographic signal. The noise experiment provides no verifiable skill, as expected (Fig. 11a). Due to the artificial calibrated skill, the reconstruction error is underestimated (Figs. 11b and 11d). However, since the noise reconstruction error (Fig. 11b) is only slightly larger than that of TrSSTa (Figs. 6b and 9b), the magnitudes of the respective reconstructed variances are similar (Figs. 6c, 9c, and 11c) and small relative to those of the benchmark experiment (Fig. 10c). These results suggest that TrSSTa has verifiable skill beyond that expected from red noise. However, the observational and mapping errors are large and must be correctly specified [Eq. (9)] to obtain reconstructions with amplitudes that are consistent with such errors. Verification exercises are required to clearly distinguish artificial skill from climatic information.

4.3.5.5. Rank of the Map H

We have shown results calibrating only the leading pattern of covariance between the SST and the tree-ring indicators. This choice for the number of climatic patterns recoverable from this set of proxy data was motivated by PC and SVD analyses (Section 4.3.3, Fig. 3). Here, we examine the change in the results with increasing rank of **H** to see if additional patterns may be verifiably reconstructed. The calibration and verification skills for retention of two and three patterns are shown in Fig. 12. Over the calibration period 1923–90 (Figs. 12a and 12b), employing two or three patterns improves results marginally in the eastern equatorial Pacific and in the North Pacific. Verification results for the period 1856–1922 (Figs. 12c and 12d) show that skill is not increased by the additional patterns, although some redistribution of skill between the tropics and extratropics occurs. The changes in skill are similar in amplitude to those observed with reversed calibration and verification periods (Figs. 6 and 9). These observations are consistent with the choice of a rank 1 map for the CFR. Similarly, benchmark experiments retaining two and three patterns (results not shown) do not significantly improve on the verification statistics shown in Fig. 10. This result suggests that the number of patterns resolved in the present reconstruction is limited by the scarcity of proxy data employed as well as by the observational error. However, note also that while there is no increase in reconstruction skill with the rank of **H**, there is little or no deterioration either, since additional modes contribute little variance to the reconstruction

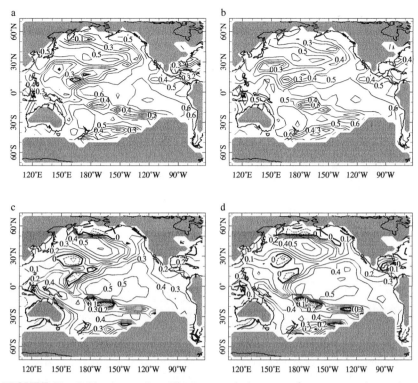

FIGURE 12 Calibration and verification correlation maps for reconstructions using 1923–90 calibration and two and three retained patterns (correlation units). (a) Calibration correlation (1923–90), two patterns retained. (b) Calibration correlation (1923–90), three patterns retained. (c) Verification correlation (1856–1922), two patterns retained. (d) Verification correlation (1856–1922), three patterns retained.

[Eq. (6); second term of Eq. (9)]. Hence, minimization of S makes the choice of the rank of $\mathbf{H}$ not as subjective as it might initially appear.

The preceding results suggest that while TrSSTa has skill over certain parts of the Pacific basin and in regions far removed from the tree-ring indicators, it does not have skill in all parts of the Pacific basin. In addition, as noted in Section 4.2 and shown here, variance resolved is a function of the quality of the map from climate record to proxy, as well as the proxy observational error. The first component is determined in the calibration stage of the procedure; the second component depends on the number of proxy indicators available for analysis as a function of time. In the most recent centuries of the reconstruction, all tree-ring indicators are available, and $\mathbf{H}$ is a full matrix. The analysis reconstructs anomalies with variance similar to that of the calibration period (e.g., Figs. 5c and 6c). However, as we proceed further back in time, fewer chronologies are available for analysis, and $\mathbf{H}$ becomes increasingly sparse. The observational error variance $\mathbf{R}$ grows (Section 4.3.4; Eq. (12)], and the analysis approaches zero or climatology, with the error almost the magnitude of the *true* or calibration period variance. However, since in this example most of the error in the reconstruction is

due to error in $\mathbf{H}$, and since we resolve only a single pattern of climate variability, the reconstruction error remains large even into the relatively well-observed modern period. Additional proxy indicators sensitive to this leading mode of SST variability will be required to reduce the mapping error in the calibration stage, as well as to produce more accurate and precise reconstructions back through time.

4.3.6. Study of Climate Dynamics

4.3.6.1. *Pacific Decadal SST Variability Inferred from Tree-Ring Indicators*

Based on verification results (Figs. 5–8), we form an index from one of the most skillfully resolved regions, NINO3.4 (170°W–120°W, 5°S–5°N). Figure 13a shows the TrNINO3.4 SST anomaly; gray bars indicate 1σ error bars of this index. Since this reconstruction is composed of the time variation of a single mode of spatial variability, all such indices will be simple linear scalings of this time variation. We expect this index to be our best estimate of basin-scale SST variability, since the ratio of signal strength to analysis error is greatest in this region (Figs. 2 and 6). Correlation over the veri-

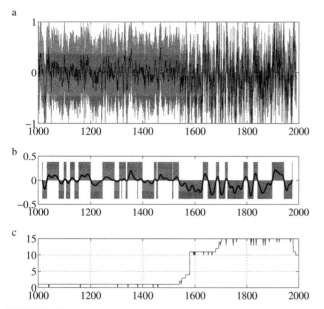

FIGURE 13 (a) NINO3.4 sea surface temperature (SST) anomaly (170°–120°W, 5°N–5°S) from tree-ring-based SST reconstruction (TrSSTa), A.D. 1001–1990. Units are degrees Celsius (°C). Shading indicates 1σ error estimates. (b) TrNINO3.4 index passed with a 31-year Gaussian filter. Overlain gray bars indicate the sign of the filtered index. Units are °C. (c) Number of tree-ring indicators available for analysis over this period.

fication period is 0.53. Figure 13b shows this index passed with a 31-year Gaussian filter. Overlain gray bars on this plot indicate whether the smoothed data show positive or negative anomalies.

Although in reality the NINO3.4 region is dominated by interannual variability, our verifiable reconstruction appears to be of ENSO-like, decadal variability (Figs. 2 and 13; Villalba et al., 2000). Note that the reconstruction prior to the 1590s is based upon only one time series (Fig. 13c), albeit one whose simultaneous contribution to the reconstruction is significant and stable in both calibration periods (Table 2). Given this caveat and the results of verification studies (Section 4.3.5), we interpret the results only qualitatively here. A shift to the positive phase of Pacific decadal variability occurs in the late 1970s. Other shifts occur ca. 1940 and 1895. Prior to 1895, TrNINO3.4 remains predominantly in the negative phase for several centuries, in rough correspondence to the Little Ice Age (LIA) interval noted from historical European climate data. Prior to the seventeenth century, decadal variability in this reconstructed mode was similar to that of the twentieth century; this suggests that twentieth-century Pacific decadal variability is not unprecedented. This result must be considered preliminary until more data can be brought into the analysis and the results are deemed stable to verification exercises.

4.4. CONCLUSIONS

The CFR described here has been derived by using a reduced space optimal estimation technique; the procedure is outlined in this chapter. Multiple benefits arise from placement of high-resolution paleoclimatological data within this context: (1) the solution is optimal (provided the a priori assumptions hold), (2) error estimates are provided for the solution, and (3) the solution and assumptions can be tested for mutual consistency.

As an application of this technique, we have reconstructed a single pattern of Pacific basin SST variability for the period A.D. 1001–1990, employing a set of Pacific American tree-ring indicators. The analysis produces SST field estimates with theoretical errors that are, at best, about twice the size of the historical SST analysis error of Kaplan et al. (1998) for the mid-nineteenth century and, at worst, as large as the RMS variance of modern observed SST. The results are encouraging enough to suggest that the tree-ring data from the Pacific coasts of North and South America may be used to reconstruct and study decadal- to century-scale SST variability. The results also demonstrate the nonlocal nature of the climatic information that may be extracted from this set of tree-ring data. However, benchmark and noise experiments show that verification tests are essential to determine the extent to which the reconstructions may be interpreted. Nevertheless, these reconstructions would benefit greatly from more spatially and temporally extensive proxy data sets, which should produce smaller errors in the reconstruction map and thereby more accurate results.

The results presented here are only a first application of the OA reconstruction method to proxy data (Fig. 1) and should be tuned and elaborated on in a number of ways. The algorithm for construction of the map from SST to tree-ring indicator may be designed to better incorporate frequency domain information likely to be well resolved by these particular proxy data (for example, after Mann and Park, 1994). Careful pretreatment of the tree-ring data, based on expert knowledge of the strengths and weaknesses of the tree-ring indicators, may be used to minimize known sources of bias and nonclimatic information. The resulting maps relating the climate field to the proxy variable will hopefully converge with theoretical or empirical studies of the links between proxy data and local climate and between local climate and large-scale climate variability. In turn, these improvements will result in better observational error covariance estimates, which are essential for proper use of the proxy data to form reconstruction estimates and to determine the true resolvable dimension of the map. Further improvements will exploit resolvable time transition information. Similar intercomparison studies between climate variables and other proxy data sources will provide similar characterizations of the information content of these data with respect to the large-scale climate field of interest. Similar work may also permit seasonal or single-season reconstructions. Subsequently, we may be able to merge the information provided by different proxy data sources to produce a reconstruction that takes no more or less than full advantage of each data source. Finally, a similar approach may be taken toward the reconstruction of other important climate fields such as air temperature and precipitation.

Acknowledgments

This work was funded by National Oceanic and Atmospheric Administration/Earth System History (NOAA/ESH) grant NA86GP 0437. We thank contributors to the International Tree Ring Data Bank, whose data are employed in this study. We are grateful to Vera Markgraf and Rob Dunbar for organizing the PEP 1 Meeting in Mérida, Venezuela, March 1998, and to meeting participants for their many useful comments on this work. Ed Cook and an anonymous reviewer gave very helpful critiques of the manuscript. This is Lamont-Doherty Earth Observatory contribution 6042.

References

Bennett, A. F., 1992: *Inverse Methods in Physical Oceanography.* New York: Cambridge University Press.

Bottomley, M., C. K. Folland, J. Hsiung, R. E. Newell, and D. E. Parker, 1990: *Global Ocean Surface Temperature Atlas.* Norwich, England, U.K.: Her Majesty's Stationery Office.

Bradley, R. S., 1996: Are there optimum sites for global paleotemperature reconstruction? *In* Jones, P. D., R. Bradley, and J. Jouzel (eds.), *Climatic Variations and Forcing Mechanisms of the Last 2000 Years;* Vol. 41 of NATO ASI Series I: Global Environmental Change. New York: Springer-Verlag, pp. 603–624.

Bradley, R. S., and P. D. Jones, 1992: *Climate Since 1500 A.D.* London: Routledge.

Bretherton, C. S., C. Smith, and J. M. Wallace, 1992: An intercomparison of methods for finding coupled patterns in climate data. *Journal of Climate,* 5: 541–560.

Briffa, K. R., and T. J. Osborn, 1999: Seeing the wood from the trees. *Science,* 284: 926–927.

Cane, M. A., 1986: El Niño. *Annual Reviews of Earth and Planetary Sciences,* 14: 43–70.

Cane, M. A., A. C. Clement, A. Kaplan, Y. Kushnir, R. Murtugudde, D. Pozdnyakov, R. Seager, and S.E. Zebiak, 1997: 20th century sea surface temperature trends. *Science,* 275: 957–960.

Cane, M. A., A. Kaplan, R. N. Miller, B. Tang, E. C. Hackert, and A. J. Busalacchi, 1996: Mapping tropical Pacific sea level: Data assimilation via a reduced state space Kalman filter. *Journal of Geophysical Research,* 101: 22599–22617.

Chang, P., L. Ji, and H. Li, 1997: A decadal climate variation in the tropical Atlantic Ocean from thermodynamic air-sea interactions. *Nature,* 385: 516–518.

Clement, A. C., 1998: Mechanisms of tropical climate change: Glacial cycles to greenhouse warming. Ph.D. thesis. Columbia University, New York.

Clement, A. C., R. Seager, M. A. Cane, and S. E. Zebiak, 1996: An ocean dynamical thermostat. *Journal of Climate,* 9: 2190–2196.

Cook, E. R., 1995: Temperature histories from tree rings and corals. *Climate Dynamics,* 11: 211–222.

Cook, E. R., K. R. Briffa, and P. D. Jones, 1994: Spatial regression methods in dendroclimatology—A review and comparison of 2 techniques. *International Journal of Climatology,* 14: 379–402.

Cook, E. R., B. M. Buckley, R. D. D'Arrigo, and M. J. Peterson, 2000: Warm-season temperatures since 1600 B.C. reconstructed from Tasmanian tree rings and their relationship to large-scale sea surface temperature anomalies. *Climate Dynamics,* 16: 79–91.

Cook, E. R., D. M. Meko, D. W. Stahle, and M. K. Cleaveland, 1999: Drought reconstructions for the continental United States. *Journal of Climate,* 12: 1145–1162.

Deser, C., M. A. Alexander, and M. Timlin, 1996: Upper-ocean thermal variations in the North Pacific during 1970–1991. *Journal of Climate,* 9: 1840–1855.

Diaz, H. F., and N. E. Graham, 1996: Recent changes in tropical freezing heights and the role of sea surface temperature. *Nature,* 383: 152–155.

Ebbesmeyer, C. C., D. R. Cayan, D. R. McLain, F. H. Nichols, D. H. Peterson, and K. T. Redmond, 1991: 1976 step in the Pacific climate: Forty environmental changes between 1968–1975 and 1977–1984. *In* Betancourt, J. L., and V. L. Tharp (eds.), *Proceedings of the Seventh Annual Pacific Climate PACLIM Workshop,* California Department of Water Resources, Interagency Ecological Studies Program Technical Report 26, pp. 115–126.

Enfield, D. B., and A. M. Mestas-Nuñez, 2000: Interannual to multidecadal climate variability and its relationship to global sea surface temperatures. *In* Markgraf, V. (ed.), *Interhemispheric Climate Linkages.* San Diego: Academic Press, Chapter 2.

Epstein, S., and C. J. Yapp, 1976: Climatic implications of the D/H ratio of hydrogen in C-H groups in tree cellulose. *Earth and Planetary Science Letters,* 30: 252–261.

Evans, M. N., A. Kaplan, and M. A. Cane, 1998: Optimal sites for coral-based reconstruction of sea surface temperature. *Paleoceanography,* 13: 502–516.

Evans, M. N., A. Kaplan, and M. A. Cane, 2000: Cross-validation of a sparse network of coral data and sea surface temperatures: Potential for coral-based SST field reconstructions. *Paleoceanography,* in press.

Evans, M. N., A. Kaplan, and M. A. Cane, 2000: Intercomparison of coral oxygen isotope data and historical sea surface temperature (SST): Potential for coral-based SST field reconstructions, *Paleoceanography,* 15: 551–563.

Fritts, H. C., 1976: *Tree Rings and Climate.* New York: Academic Press.

Fritts, H. C., 1991: *Reconstructing Large-Scale Climatic Patterns from Tree-Ring Data: A Diagnostic Analysis.* Tucson: University of Arizona Press.

Fritts, H. C., T. R. Blasing, B. P. Hayden, and J. E. Kutzbach, 1971: Mulitvariate techniques for specifying tree-growth and climate relationships and for reconstructing anomalies in paleoclimate. *Journal of Applied Meteorology,* 10: 845–864.

Gandin, L. S., 1965: *Objective Analysis of Meteorological Fields.* Jerusalem: Israeli Program for Scientific Translation, translated from the Russian, *Gidrometeorologicheskoye Izdatel'stvo,* Leningrad, Russia, 1963.

Hansen, J., and S. Lebedeff, 1988: Global surface air temperatures: Update through 1987. *Journal of Geophysical Research,* 92: 13345–13372.

ITRDB, 1998: International Tree-Ring Data Bank. http://www.ngdc.noaa.gov/paleo/ftp-treering.html.

Jones, P. D., and K. R. Briffa, 1996: What can the instrumental record tell us about longer timescale paleoclimatic reconstructions? *In* Jones, P. D., R. Bradley, and J. Jouzel (eds.), *Climatic Variations and*

Forcing Mechanisms of the Last 2000 Years, Vol. 41 of NATO ASI Series I: Global Environmental Change. Berlin: Springer-Verlag, pp. 625–644.

Jones, P. D., R. S. Bradley, and J. Jouzel, 1996: *Climatic Variations and Forcing Mechanisms of the Last 2000 Years,* Vol. 41 of NATO ASI Series I: Global Environmental Change. Berlin: Springer-Verlag.

Jones, P. D., K. R. Briffa, T. P. Barnett, and S. F. B. Tett, 1998: High-resolution palaeoclimatic records for the last millennium: Interpretation, integration and comparison with General Circulation Model control run temperatures. *The Holocene,* 8: 455–471.

Kaplan, A., M. A. Cane, Y. Kushnir, A. C. Clement, M. B. Blumenthal, and B. Rajagopalan, 1998: Analyses of global sea surface temperature 1856–1991. *Journal of Geophysical Research,* **103**: 18567–18589.

Kaplan, A., Y. Kushnir, and M. A. Cane, 2000: Reduced space optimal interpolation of historical marine sea level pressure: 1854–1992. *Journal of Climate,* **13**: 2987–3002

Kaplan, A., Y. Kushnir, M. A. Cane, and M. B. Blumenthal, 1997: Reduced space optimal analysis for historical datasets: 136 years of Atlantic sea surface temperatures. *Journal of Geophysical Research,* **102**: 27835–27860.

Krishna Kumar, K., R. Kleeman, M. Cane, and B. Rajagopalan, 1999: Epochal changes in the Indian monsoon-ENSO precursors. *Geophysical Research Letters,* **26**: 75–78.

LaMarche, V. C., R. L. Holmes, P. W. Dunwiddie, and L. G. Drew, 1979: Chile. *In Tree-Ring Chronologies of the Southern Hemisphere,* Vol. 2 of Chronology Series V. Tucson: University of Arizona Press.

Latif, M., and T. P. Barnett, 1994: Causes of decadal climate variability over the North Pacific and North America. *Science,* 266: 634–637.

Lau, N. -C., and M. J. Nath, 1994: A modeling study of the relative roles of tropical and extratropical SST anomalies in the variability of the global atmosphere-ocean system. *Journal of Climate,* 7: 1184–1207.

Mann, M. E., and J. Park, 1994: Global-scale modes of surface temperature variability in interannual to century timescales. *Journal of Geophysical Research,* 99: 25819–25833.

Mann, M. E., R. S. Bradley, and M. K. Hughes, 1998: Global temperature patterns over the past five centuries: Implications for anthropogenic and natural forcing of climate. *Nature,* 392: 779–787.

Mann, M. E., R. S. Bradley, and M. K. Hughes, 2000: Long-term variability in the El Niño/Southern Oscillation and associated teleconnections. *In* Diaz, H. F., and V. Markgraf (eds.), *El Niño and the Southern Oscillation: Multiscale Variability and Global and Regional Impacts.* Cambridge, U.K.: Cambridge University Press.

Mardia, K. V., J. T. Kent, and J. M. Bibby, 1979: *Multivariate Analysis.* San Diego: Academic Press.

Martinson, D. G., K. Bryan, M. Ghil, M. M. Hall, T. R. Karl, E. S. Sarachik, S. Sorooshian, and L. Talley (eds.), 1995: *Natural Climate Variability on Decade-to-Century Time Scales.* Washington, DC: National Academy of Sciences.

Miller, R. N., 1990: Tropical data assimilation experiments with simulated data: The impact of the tropical ocean and global atmosphere thermal array for the ocean. *Journal of Geophysical Research,* **100**: 13389–13425.

Namias, J., 1969: Seasonal interactions between the North Pacific Ocean and the atmosphere during the 1960s. *Monthly Weather Review,* 97: 173–192.

Namias, J., 1980: Causes of some extreme Northern Hemisphere climatic anomalies from summer 1978 through the subsequent winter. *Monthly Weather Review,* 108: 1334–1346.

Parker, D. E., P. D. Jones, C. K. Folland, and A. Bevan, 1994: Interdecadal changes of surface temperature since the late nineteenth century. *Journal of Geophysical Research,* 99: 14373–14399.

Parthasarathy, B., K. R. Kumar, and A. A. Munot, 1991: Evidence of secular variations in Indian monsoon rainfall-circulation relationships. *Journal of Climate,* 4: 927–938.

Rao, C. R., 1973: *Linear Statistical Inference and Its Applications.* New York: John Wiley & Sons.

Ropelewski, C. F., and M. S. Halpert, 1987: Global and regional scale precipitation patterns associated with the El Niño/Southern Oscillation. *Monthly Weather Review,* **114**: 2352–2362.

Smith, T. M., R. W. Reynolds, R. E. Livezey, and D. C. Stokes, 1996: Reconstruction of historical sea surface temperatures using empirical orthogonal functions. *Journal of Climate,* **9**: 1403–1420.

Sokal, R. R., and F. J. Rohlf, 1995: *Biometry: The Principles and Practice of Statistics in Biological Research,* 3rd ed., New York: W. H. Freeman.

Stahle, D. W., R. D. D'Arrigo, P. J. Krusic, M. K. Cleaveland, E. R. Cook, R. J. Allan, J. E. Cole, R. B. Dunbar, M. D. Therrell, D. A. Gay, M. D. Moore, M. A. Stokes, B. T. Burns, J. Villanueva-Diaz, and L. G. Thompson, 1998: Experimental dendroclimatic reconstruction of the Southern Oscillation. *American Meteorological Society Bulletin,* **79**: 2137–2152.

Trenberth, K. E., and T. Hoar, 1996: The 1990–1995 El Niño Southern Oscillation event: Longest on record. *Geophysical Research Letters,* **23**: 57–60.

Trenberth, K. E., and J. W. Hurrell, 1994: Decadal atmosphere-ocean variations in the Pacific. *Climate Dynamics,* **9**: 303–319.

Trenberth, K. E., and D. J. Shea, 1987: On the evolution of the Southern Oscillation. *Monthly Weather Review,* **115**: 3078–3096.

Villalba, R., J. A. Boninsegna, T. T. Veblen, A. Schmelter, and S. Rubulis, 1997: Recent trends in tree-ring records from high elevation sites in the Andes of northern Patagonia. *Climatic Change,* **36**: 425–454.

Villalba, R., R. D. D'Arrigo, E. R. Cook, G. Wiles, and G. C. Jacoby, 2000: Decadal-scale climatic variability along the extratropical western coast of the Americas: Evidence from tree-ring records. *In* Markgraf, V. (ed.), *Interhemispheric Climate Linkages.* San Diego: Academic Press, Chapter 10.

Wallace, J. M., 1996a: Observed climatic variability: Spatial structure. *In Decadal Climate Variability: Dynamics and Predictability,* Vol. 44 of NATO ASI Series I: Global Environmental Change. Berlin: Springer-Verlag, pp. 31–81.

Wallace, J. M., 1996b: Observed climatic variability: Time dependence. *In* Anderson, D. L. T., and J. Willebrand (eds.), *Decadal Climate Variability: Dynamics and Predictability,* Vol. 44 of *NATO ASI Series I: Global Environmental Change.* Berlin: Springer-Verlag, pp. 1–30.

Webster, P. J., 1994: The role of hydrological processes in ocean-atmosphere interactions. *Reviews of Geophysics,* 32: 427–476.

Wiles, G. C., R. D. D'Arrigo, and G. C. Jacoby, 1998: Gulf of Alaska atmosphere-ocean variability over recent centuries inferred from coastal tree-ring records. *Climatic Change,* 38: 289–306.

Wunsch, C., 1996: *The Ocean Circulation Inverse Problem.* New York: Academic Press.

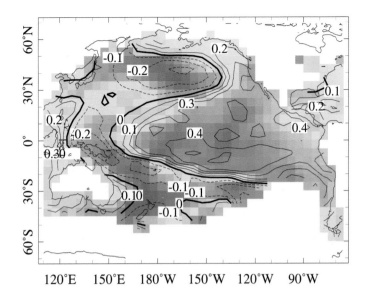

CHAPTER 4, FIGURE 7   Color plot of the verification correlation (Figure 6a) overlain by a contour plot of the correlation with the leading PC of the instrumental data from the tree-ring location (Figure 4b).

CHAPTER 6, FIGURE 4   A Chac (rain god) mask adorns a building at the archaeological site of Labná in the Puuc Region of the Maya lowlands, Yucatán, México. (Photo courtesy of M. Brenner.)

CHAPTER 6, FIGURE 8   Photos of the Pampa Koani, Bolivia, taken in 1986 (a), when Lake Titicaca was at a 20th-century high stand, and in 1996 (b) when the lake level had declined by several meters. Rehabilitated raised agricultural fields are seen in the foreground of (a). The "corduroy" appearance of the landscape is attributed to ancient Tiwanaku raised fields. Small changes in the relation between evaporation and precipitation (E/P) leave the flatlands either inundated or desiccated. (Photos courtesy of M. Brenner.)

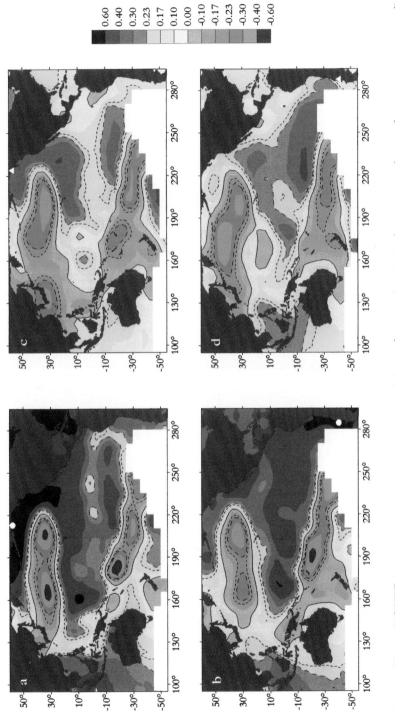

CHAPTER 10, FIGURE 7 Spatial correlation patterns estimated during the interval 1857–1988 between annual sea surface temperature anomalies (SSTAs) over the Pacific Ocean and annual SSTAs on (a) the Alaska coast (57.5°N, 142.5°W) and (b) the Chilean coast (42.5°S, 72.5°W), and tree-ring variations for (c) coastal Alaska and (d) northern Patagonia. Grid cell locations on the Alaska and Chilean coasts are indicated by solid white dots (a and b), whereas tree-ring sites are indicated by solid white triangles (c and d).

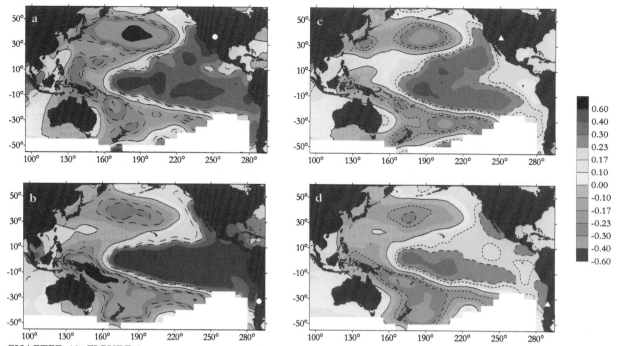

CHAPTER 10, FIGURE 8 Spatial correlation patterns between annual sea surface temperature anomalies (SSTAs) over the Pacific Ocean and (a) Palmer Drought Severity Index (PDSI) variations in the Midwest–southern United States (PC1 amplitudes from cells 53, 63, 80, and 81, interval 1902–78, taken from Cook et al., 1996), (b) total annual precipitation at Santiago de Chile (interval 1861–1988), (c) tree-ring-based PDSI reconstructions for the Midwest–southern United States (PC1 amplitudes from reconstructed cells 53, 63, 80, and 81, interval 1857–1978, taken from Cook et al., 2000), and (d) tree-ring variations from El Asiento in central Chile. PDSI cells and Santiago de Chile locations are indicated by solid white dots (a and b), whereas PDSI-reconstructed cells and the El Asiento site are indicated by solid white triangles (c and d).

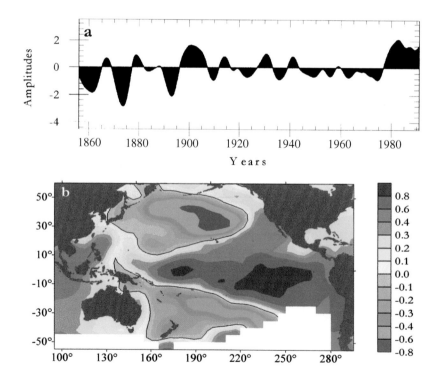

CHAPTER 10, FIGURE 15 The leading rotated principal components (RPCs) of low-pass-filtered sea surface temperatures (SSTs) over the Pacific domain during the interval 1856–1991 (a) and its associated spatial signature (b). Units are dimensionless.

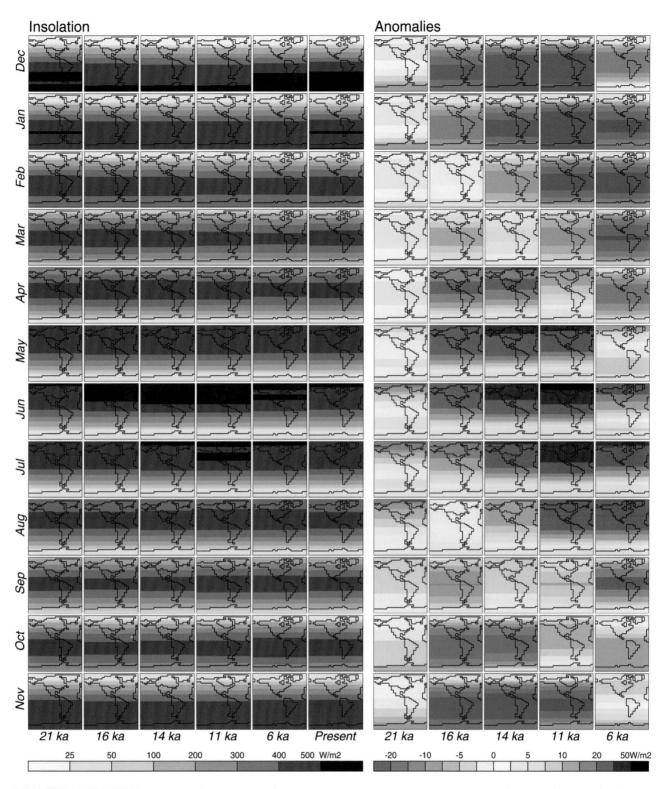

CHAPTER 21, FIGURE 3 Temporal (21 ka to present) and seasonal variations of computed insolation at the top of the atmosphere (Berger, 1978) along with past minus present anomalies.

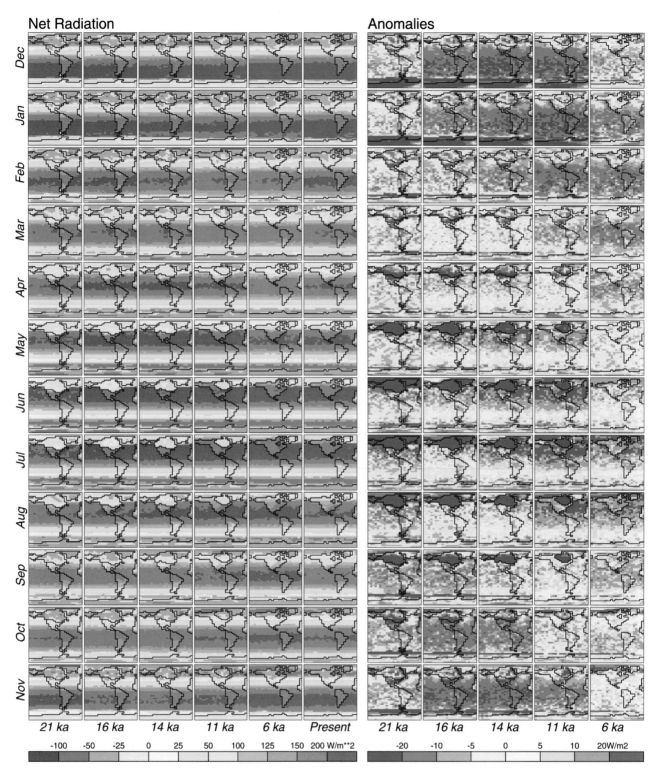

CHAPTER 21, FIGURE 4 Temporal (21 ka to present) and seasonal variations of simulated surface net radiation, along with the simulated past (experiment) minus simulated present (control) anomalies.

Surface Air Temperature

Anomalies

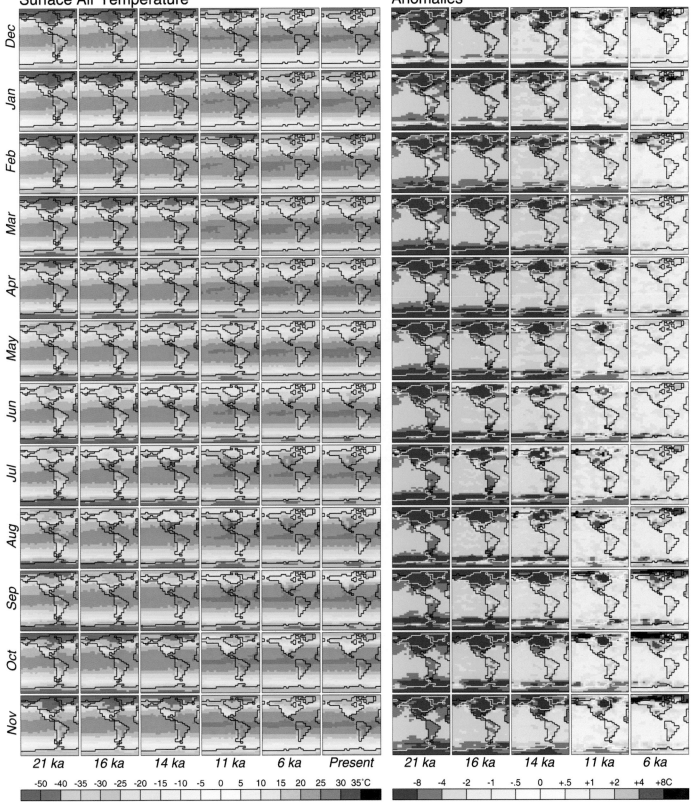

CHAPTER 21, FIGURE 5 As in Figure 4, for near-surface air temperature.

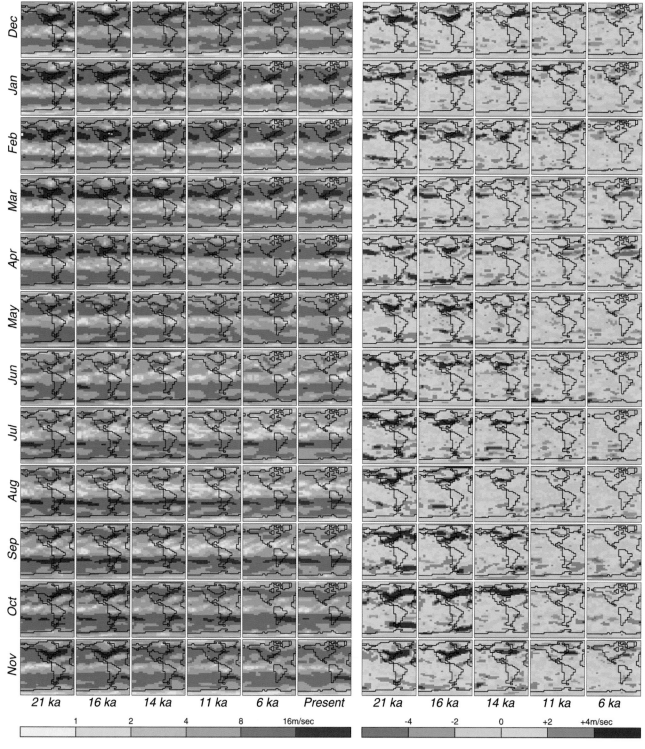

CHAPTER 21, FIGURE 6   As in Figure 4, for 500 mb wind speed.

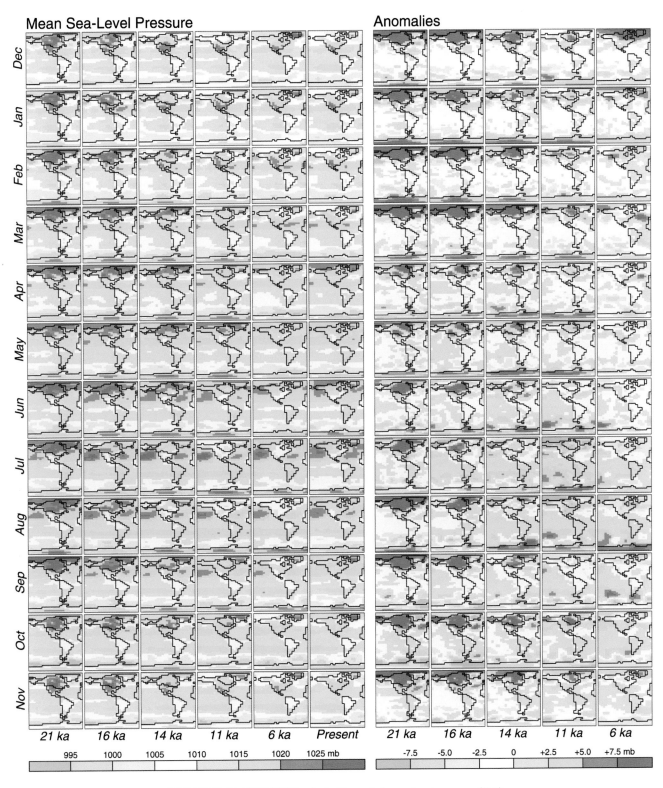

CHAPTER 21, FIGURE 7 As in Figure 4, for sea level pressure (SLP).

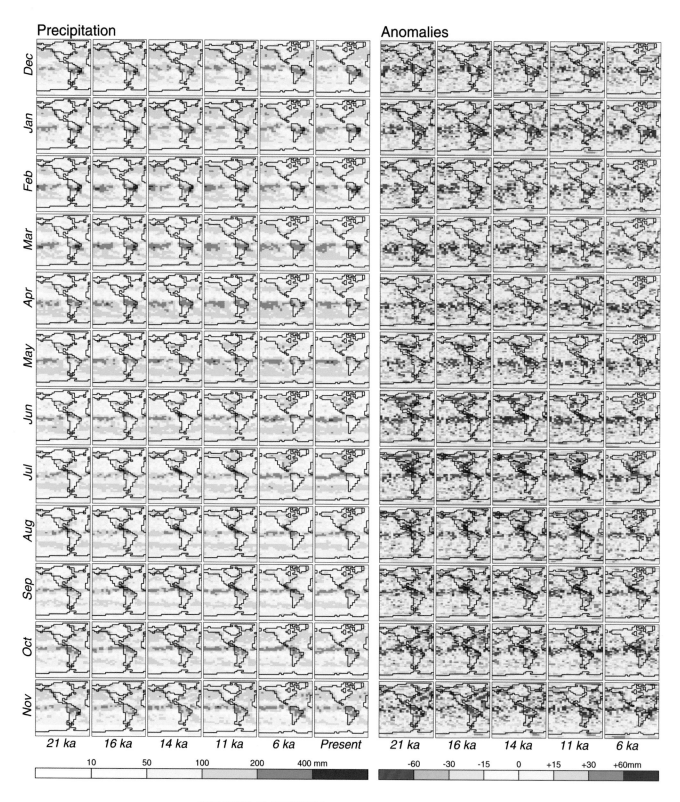

CHAPTER 21, FIGURE 8 As in Figure 4, for precipitation.

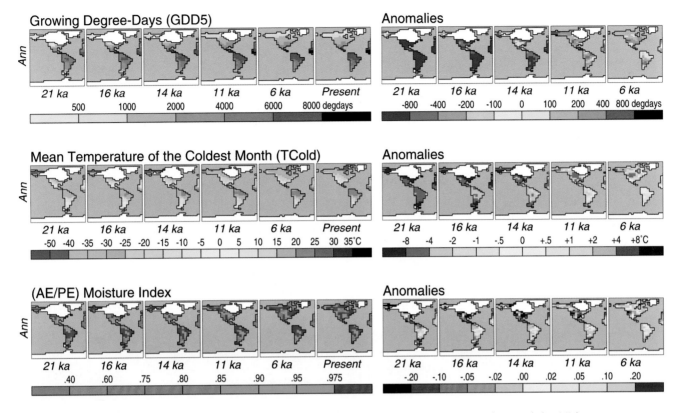

CHAPTER 21, FIGURE 9 Temporal variations in bioclimatic variables: (top) growing degree-days (5°C base); (middle) mean temperature of the coldest month; and (bottom) effective moisture, expressed as the ratio of actual to potential evapotranspiration (AE/PE), and their anomalies.

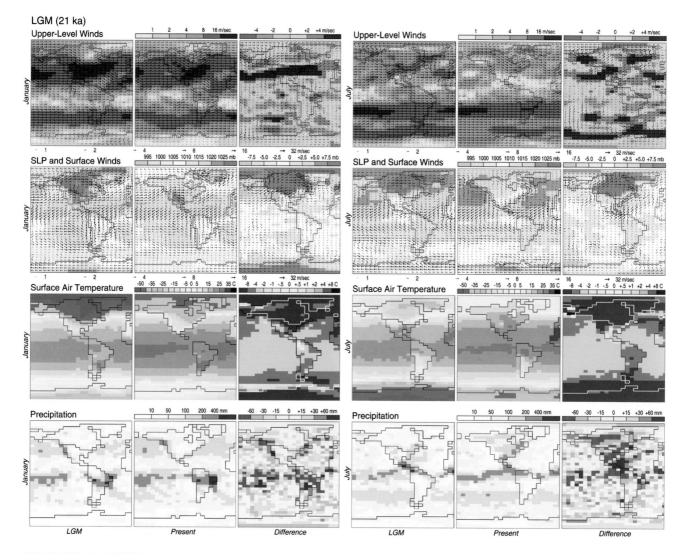

CHAPTER 21, FIGURE 10 Simulated upper-level wind speed and direction, sea level pressure (SLP) and surface winds, surface air temperature, and precipitation for the last glacial maximum (LGM) for January and July.

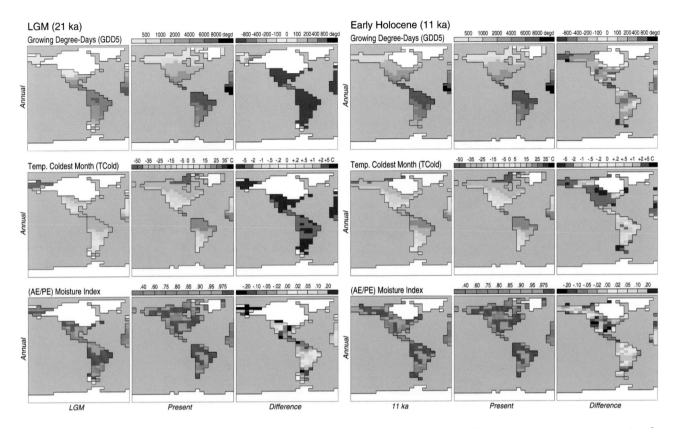

CHAPTER 21, FIGURE 11 Simulated bioclimatic variables for the last glacial maximum (LGM) and early Holocene as compared with present conditions.

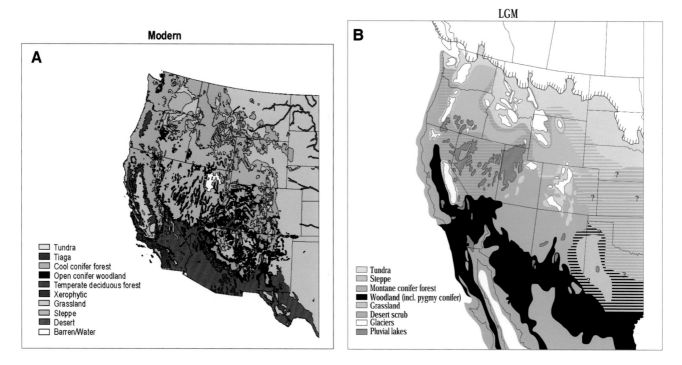

CHAPTER 21, FIGURE 12 Modern vegetation (based on Thompson and Anderson, in press) (A) and full glacial vegetation of the western United States (B). (Based on inferences from pollen data and Betancourt et al., 1990.)

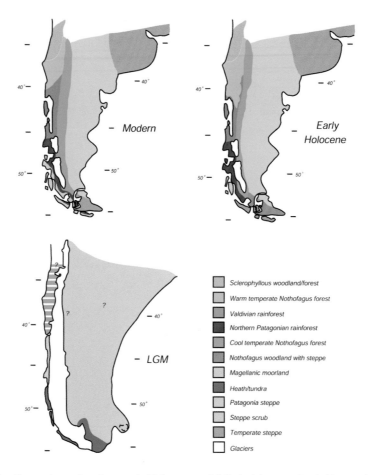

CHAPTER 21, FIGURE 13 Comparison of modern, early Holocene, and full glacial vegetation in Patagonia, based on inferences from fossil pollen data.

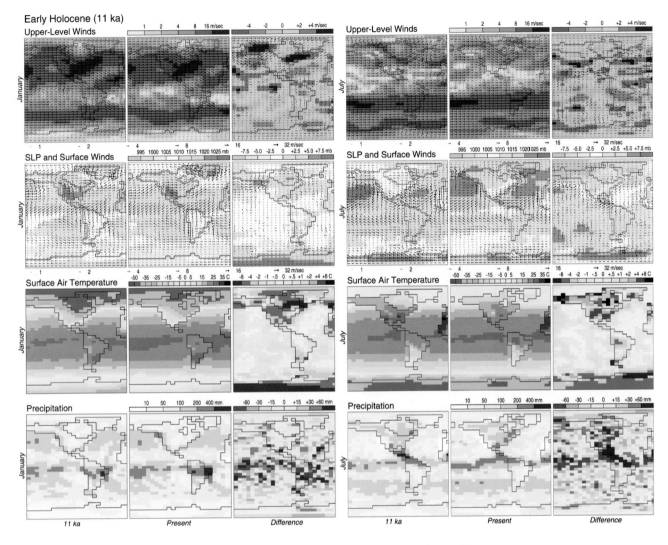

CHAPTER 21, FIGURE 15 As in Figure 10, for 11 cal ka.

Mean July Temperature (°C)

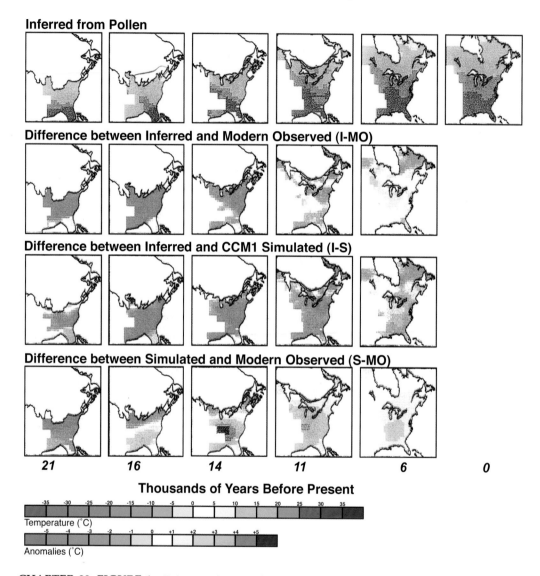

CHAPTER 22, FIGURE 1 Estimates of mean July temperature inferred from pollen data with modern observed values for 0 ka (top panel), differences between the pollen-inferred values and modern observed values (second panel), differences between pollen-inferred values and values simulated by CCM1 (third panel), and differences between CCM1 simulations and modern observed values, which identically are the differences from the model control run (fourth panel, same as panel two in Fig. 1).

HUMAN DIMENSIONS, THE LAST MILLENIUM OF CLIMATE CHANGE

5

The Effects of Explosive Volcanism on Simple to Complex Societies in Ancient Middle America

PAYSON SHEETS

Just as the progress of a disease shows a doctor the secret life of a body, so to the historian the progress of a great calamity yields valuable information about the nature of the society so stricken.

——Marc Bloch

Abstract

The ancient societies of Middle America (México and Central America) were not equally vulnerable to the sudden stresses of explosive volcanic eruptions. Although the sample size of known cases is small, the simple egalitarian societies apparently were more resilient to sudden massive stress, in the long run, than were the more complex chiefdoms or states. The simpler societies often repeatedly reoccupied the recovering landscapes, taking advantage of the short-term benefits of the volcanically active environments, with little or no detectable culture change that can be traced to the volcanic impacts. The more complex societies in this sample had greater difficulties in coping with sudden, massive stresses, apparently because of their greater reliance on the built environment, staple cultigens, long-range trade networks, greater population density, competitive to hostile political landscapes, redistributive economies, and other centralized institutions. An unanticipated result of this study is the real-

ization of the importance of the political context of a society prior to a sudden stress. Copyright © 2001 by Academic Press.

Resumen

Los pueblos Precolombinos de Mesoámerica (México and Central América) mostraron diferentes grados de vulnerabilidad a las rápidas tensiones impuestas por las erupciones volcánicas. Aún cuando los casos estudiados no son numerosos, pareciera que las sociedades menos desarrolladas fueron más resistentes, en el largo plazo, a las tensiones provocadas por las erupciones volcánicas. Por el contrario, los grupos socialmente más desarrollados, tales como las tribus o los estados, fueron más vulnerables a las erupciones volcánicas. Las sociedades más primitivas reocuparon las tierras desvastadas por las erupciones volcánicas, sacando ventajas de los beneficios que en el corto plazo le ofrecían los ambientes volcánicamente activos, con muy pocos o sin cambios culturales que pudieran detectarse a través de los registros arqueológicos. Los estados o tribus tuvieron, por el contrario, mayores dificultades con las erupciones volcánicas debido a su mayor dependencia de grandes construcciones de piedra, el cultivo del maíz como la mayor fuente de comida, la existencia de redes económicas extensas, la alta

73

densidad poblacional y la existencia de instituciones centralizadas. Este estudio pone en evidencia la gran importancia del contexto político y cultural en la respuesta de grupos sociales a las tensiones impuestas por presiones ambientales.

5.1. INTRODUCTION

The objective of this chapter is to compare the effects of explosive volcanic eruptions on the ancient societies of Middle America (México and Central America) and to explore the role of social complexity in determining the vulnerability of societies to sudden, massive stresses. It is clear that some societies were resilient to explosive eruptions, reestablishing their societies, adaptations, and material cultures in the devastated area after natural processes of recovery allowed human resettlement. In other cases, societies were more vulnerable to massive environmental change from explosive eruptions, as they were not able to cope with the changed circumstances. The more complex societies in this small sample were more vulnerable to these stresses, particularly when they were quite competitive or hostile with nearby societies.

Ancient egalitarian societies can be distinguished from complex societies readily with archaeological evidence. Egalitarian societies—whether based on hunting and gathering, marine resources, or agriculture—are characterized by household possessions that are essentially the same throughout the society. The differences that show up in households or burials are due to age or sex differences within the household rather than differential wealth or authority among households. Complex societies exhibit centralized authority in a number of domains, including politics, religion, and economics. An important source of evidence of societal complexity detectable in ancient societies is the degree of centralization of wealth or authority. Differential wealth usually manifests itself in burials, with richly stocked tombs of the elite contrasting with the more plain commoners' burials. Differential wealth is recorded also in architecture and artifacts. Differential resource access within a society can be seen in the archaeological record in storehouses attached to palaces and in the size, construction, and decoration of the palaces themselves as contrasted to the housing of most of the population. The elite control aspects of production and distribution systems and thus accumulate wealth, they live in more permanent and elaborate dwellings, they bury their dead with more luxurious grave goods, they commonly are "closer to the deities" and thus have greater religious power, their social prestige is superior to that of the commoners, and they con-

centrate power in their hands. In the cases considered in this chapter, there is sufficient archaeological evidence to characterize a society as egalitarian or complex, and in many cases finer distinctions can be made. Centralization of authority can be detected in control of labor, in power in military or police functions, or in religious authority and control.

This chapter does not focus on the deaths, injuries, or devastation of the built and utilized environments caused directly by the eruptions. The archaeological and geological records are weak in this regard. Gathering systematic regional data on these topics would require immense research efforts. Rather, the record is more suitable for studying human behavior in the environment in the long term, over the centuries following an eruption, to see recovery or the lack of recovery of the society, culture, and / or adaptation.

Of course, volcanic activity, in general terms, can be beneficial or detrimental to human adaptation to an environment, in different ways at different times. The tapestry of human–volcanic interactions is a rich and varied one, which this chapter can only begin to explore. The cliché of Juan Valdez, of Colombian coffee fame, cultivating his rich volcanic soils, can be true, but only after weathering has formed that soil. A volcanic ashfall that formed a fertile soil after decades of soil genesis can have a chemically similar coeval lava flow beside it that required centuries to form an equally fertile soil. Volcanically active areas often provide key resources such as obsidian that can be fashioned into sharp cutting and scraping tools, basalt and andesite for grinding implements, and hematite and other pigments. Further, volcanically active areas are seismically active as well, and indigenous societies generally develop earthquake-resistant vernacular architecture. Within the range of benign to devastating volcanic phenomena, this chapter focuses on sudden, explosive eruptions and how ancient societies coped with them or failed to cope with them. There are other highly destructive volcanic phenomena such as mudflows that are beyond the scope of this chapter. They are not included here largely because the prehistoric cases rarely have been documented from both natural and social science perspectives.

Because of the thousands of years of human occupation in Middle America and the frequency of volcanic eruptions, the number of times that pre-Columbian societies were affected by explosive volcanism must have been great. The number of cases that have been researched and adequately documented is small, but at least sufficient to begin establishing a comparative framework; those cases are presented later. Ideally, one would wish to have a sample size sufficient to explore a single variable such as societal complexity to see the degree to which the internal organization of societies was a factor in their reaction to sudden, explosive volcanic stress. But, the range of significant variables in volcanological–human societal interactions is considerable, and it is often difficult to determine the relative importance of factors such as the nature and magnitude of the eruption, vulnerability of flora and fauna to stress from tephra (volcanic ash, cinders, etc. that moved through the air), duration and dating of eruptions, chemical and grain size variation, climate, weathering rates, and recovery processes, in addition to political, economic, demographic, and social factors of the affected and the surrounding societies. Thus, anything approaching a statistical analysis is not possible with the present state of knowledge, and this study is best seen as exploratory.

The significant variables to be considered in comparing and contrasting cases where explosive eruptions affected societies range from natural to social and cultural factors. They include the nature of the tephra (depth, chemical and physical properties, and area and nature of emplacement), the climate, and the particular edaphic requirements of cultigens. Sociocultural factors include the regional cultural system, including possible competition for resources and warfare, the adaptation and the economy of the society affected, the political system, residential mobility or sedentism, the density and distribution of population, reliance on fixed facilities and the internal complexity of the society affected, and the religious system.

The most useful cases for this study are those in which investigators have dated the eruption well; have done regional analysis of the extent of the tephra blanket; have conducted regional surveys and excavations to document preeruption population density and distribution; and have documented the timing and nature of the recovery, including soils, flora, fauna, and humans. Societies living beyond the extent of the ashfall should also be known and examined for possible indirect effects, including the housing of refugees or taking advantage of the suddenly changed circumstances in the region by rerouting trading systems or eliminating competitors. Regional studies determine the pattern of diminishing ash thickness with distance from the source and distinguish the zone of devastation from the zone of less deleterious effects from the zone where a slight dusting of ash was beneficial shortly after deposition.

A recent example of the latter was the crop improvement in eastern Washington State by a thin dusting of Mt. St. Helens ash from the 1980 eruption, as the tephra acted as a mulch that improved water retention and the fine grains asphyxiated many insect pests. Soil recovery and plant succession generally are more rapid

on the peripheries of a tephra blanket, allowing for earlier human reoccupation. The chemical and physical characteristics of the tephra have been studied in the ideal case, noting how finer grained and more mafic (basic) tephras weather more rapidly and thus allow for more rapid recovery. The more useful cases here have documented the preeruption subsistence system, including the degree to which it was reliant upon wild or domesticated species, swidden or intensive agriculture, and a staple crop or a wide variety of foods. Beyond subsistence, the regional economy is understood, whether it was regionally integrated and hierarchically organized or characterized by societal, settlement, or household self-sufficiency. Similarly, the political and religious systems in the ideal cases are well understood, so their roles in the impact of the sudden stress as well as the coping and recovery can be studied.

5.2. CASES

Two important variables for comparison of these cases of volcanic impacts (Fig. 1) are the depth of burial of sites and the subsistence sustaining areas of those sites. Burial by explosive volcanism is essentially instantaneous, and the depth of burial is proportional to the deleterious impact. Ideally, we would wish to compare cases of volcanic impacts on regional scales, but many cases are weak in this regard because they focus on a single site. In conducting regional volcanological–ecological–archaeological research in El Salvador, Panama, and Costa Rica, my project members and I have been more successful in searching for, finding, and investigating sites buried by $\frac{1}{2}$ to 2 m of tephra. Sites buried by less than $\frac{1}{2}$ m of tephra are generally poorly preserved because of plowing, bioturbation, erosion, and other postimpact factors. The more deeply buried sites are increasingly difficult to find and excavate. Thus, an aspect of practicality inadvertently assists this study for purposes of comparability, as many of the sites under consideration in this chapter had approximately similar depths of burial with approximately similar tephra impacts on the environment and roughly comparable effects upon those sites and their sustaining areas. Some more deeply buried sites, such as Cuicuilco (México) and Ceren (El Salvador), are also considered here, but in less detail.

5.2.1. Arenal, Costa Rica

People occupied the Arenal area of northwestern Costa Rica for at least the last 10,000 years, most of which was prior to the earliest eruption of the Arenal Volcano in ca. 2000 B.C. (Sheets, 1994). The Arenal Vol-

FIGURE 1 Map of Middle America. Volcanoes and the pre-Columbian societies affected by them that are explored in this chapter are indicated. The dashed line in Honduras separates Mesoamerica to the west from the Intermediate Area (lower Central America).

cano erupted 10 times in four millennia with major explosive eruptions (Melson, 1994), including the 1968 eruption, an average of one per 400 years. These eruptions provide cases of sudden stresses on egalitarian social groups with diversified subsistence economies and relatively low population densities. The eruptions that are relatively well dated and understood are presented in Table 1 within the context of the regional cultural history as discovered by the Arenal Research Project (Sheets, 1994). All of Arenal's pre-Columbian eruptions affected egalitarian societies (Sheets, 1994). Thus, the Arenal area provides a contrast to most of the other cases in Middle America where explosive volcanism affected more complex societies with quite different adaptations and political and economic systems and greater population densities. Therefore, the Arenal eruptions and affected societies are examined more closely than most other cases in this chapter.

The hilly research area spans the Continental Divide, with elevations on both Atlantic and Pacific drainages ranging from 400–1000 m. Mean precipitation varies considerably, with averages as low as 1300 mm at low-elevation western stations to averages over 6000 mm at stations in the east close to the Arenal volcano. That

TABLE 1   Principal Explosive Eruptions of the Arenal Volcano, Costa Rica[a]

Eruption	Date (est.)	Cultural phase	Immediate effects	Long-term effects detected
Unit 10	A.D. 1968	Historic	Initial abandonment	Too early to know
Unit 20	A.D. 1450	Tilaran	Abandonment	None
Units 41 and 40	A.D. 800	Silencio	Abandonment	Population concentration in Rio Piedra
Units 53 and 52	A.D. 1	Arenal	Abandonment	None
Unit 55	800 B.C.	Tronadora	Abandonment	None
Unit 61	1800 B.C.	Tronadora	Abandonment	None

[a]Moderate to small eruptions, or poorly dated eruptions, are not included. "Units" are volcanic ash layers that are numbered sequentially, with the earliest eruption unit having the largest number.

moisture gradient correlates with a vegetational spectrum from a tropical, dry, seasonal forest to a nonseasonal dense rain forest. The erosional potential of some very moist areas is so great that it is against federal law to cut trees.

Paleo-Indian peoples (est. 10,000–7000 B.C.) lived in the area in apparently very low population densities, as is evidenced by the recovery of only one artifact datable to that time. However, population increased significantly during the Archaic period (est. 7000–3000 B.C.), as suggested by datable sites and artifacts. The Arenal area is unusual in Middle America in that regional population density did not continuously rise after plant domestication occurred, but remained relatively constant. The utilization of domesticated foods certainly began by 2000 B.C. and perhaps by 3500 B.C., but the dating is not precise. Maize was cultivated by that 2000–3500 B.C. time horizon, but beans and squash are not confirmed in the record until later. Sedentary villages, with heavy (therefore not portable) metates and manos presumably to grind maize, accompanied the addition of maize. In contrast to most areas of Middle America, domesticated plants supplied only a small fraction of the diet, evidently no more than 12%, as a greater reliance was placed on wild flora and fauna even up to the Spanish conquest. Other aspects of Arenal area culture that showed no change from the Archaic period to the Spanish conquest are the basic core-flake lithic technology and the use of heated stones for cooking in villages and cemeteries. We were not able to detect any changes in the circular houses with activity areas and (presumably) thatch roofs from Early Formative times to the Spanish conquest. Other aspects of culture did change, and the key question considered here is whether any of the changes can be attributed to stresses caused by the biggest explosive eruptions of the Arenal Volcano.

The Arenal Volcano first erupted ca. 1800 B.C. (Melson, 1994), thus causing the first of 10 massive stress events for people and beginning the detailed human–volcanological interaction record. That explosive eruption, which deposited a thick tephra layer we call Unit 61, resulted in the devastation of many square kilometers of rain forest and gardens and certainly eliminated human habitation for a large area surrounding the volcano, particularly on the western (downwind) side. However, survivors on the periphery of the devastated zone, a few generations later, reoccupied the area after the soil and vegetation had recovered. The Tronadora phase, conservatively dated from 2000–500 B.C., was characterized by numerous small villages scattered across the landscape. People lived in circular houses and buried their dead just outside those structures in small rectangular pits with occasional ceramic vessel offerings. The small pit sizes probably indicate the burials were secondary (disarticulated). Maize was cultivated and processed with manos and metates, but was probably a minor part of the diet, as numerous wild species were utilized. The ceramics were surprisingly sophisticated in technology and decoration and show relationships with those in southern Mesoamerica (Hoopes, 1987). The Tronadora phase had been under way for at least two centuries, and perhaps as long as 1700 years, when the Arenal Volcano first erupted.

Toward the end of the Tronadora phase ca. 800 B.C., Arenal's second huge explosive eruption occurred. It deposited Unit 55 throughout the area, resulting in a similar devastation of flora, fauna, and human populations. When we compare pre- and posteruption conditions for both of these eruptions, we are able to detect no cultural changes that could be attributed to the disaster. People certainly had to abandon the area for a few generations, but the reoccupants were culturally indistinguishable from the preeruption occupants. That resilience to volcanic disaster continued to characterize Arenal area society for 3500 years.

It was during the Arenal phase (500 B.C. to A.D. 600) that population density reached its peak in the area. Although never approaching the high densities of Mesoamerica during these same centuries, there were more and larger sites than at any other time in the pre-Columbian period in the Arenal area and much of the rest of Costa Rica. Domestic housing remained unchanged. Human burial was in discrete cemeteries on ridges above and a few hundred meters away from the villages. They shifted from the secondary burials of the Tronadora phase to elaborate primary burials with plots demarcated by outlines of upright stones. Then, a layer of round cobbles was laid between the upright stones, and dozens of whole vessels and decorated metates were smashed on the rocks and left in place. Two volcanic ash layers (Units 53 and 52) fell about the middle of the phase, at about the time of Christ, with very little time separating them. Although individually thinner than the other major tephra blankets considered here, the combined deposits are about as thick as those from the other big eruptions and, therefore, probably had ecological and demographic impacts similar to those of the earlier big eruptions. The eruptions do correlate with a subphase cultural boundary, the shift from the early to the late facet, identified by subtle changes in ceramic decoration. These changes apparently were a local manifestation of the regional culture change in the greater Nicoya cultural area rather than an effect of the volcanic disaster. I believe the important point is that they occurred in spite of the eruption, rather than because of it.

A mild population decline occurred at the beginning

of and during the Silencio phase (A.D. 600–1300), as both large and small sites declined in number, a change that also occurred in most of lower Central America for unknown reasons. That process clearly was not caused by volcanism, but, at most, it could have been accelerated locally by volcanic activity during the ninth century. Chipped stone and ground stone industries show little change, and ceramics indicate considerable continuity, but with the addition of polychrome painting. Cemeteries are located at still greater distances from villages, as exemplified by the Silencio cemetery located on the Continental Divide many kilometers from the villages that utilized it. Burial practices remain elaborate and have adopted the stone box technique: using flat-fracturing andesite slabs to create a stone coffin along the two sides and on top.

In ca. A.D. 800 or 900, two large eruptions (Units 41 and 40) occurred in rapid succession during the Silencio phase. These eruptions must have had ecological and demographic effects as drastic as the earlier big eruptions, as their magnitudes are comparable. As with the earlier recoveries, the culture that reestablished itself is indistinguishable from prior occupations. It is possible that the increase in population in the western end of the study area during the latter part of the phase was a result of the eruptions (Mueller, 1994, p. 63). That area received less volcanic ash from the two eruptions than the rest of the study area and could have served as a refuge area. If this interpretation is correct, it is the sole example of our being able to detect a human change attributable to volcanic pressures in the Arenal area at any time. That demographic shift is detectable for perhaps three centuries.

With the disappearance of the last larger sites, as well as a decline in the number of smaller sites, the population decline of the Silencio phase continued through the Tilaran phase (A.D. 1300–1500). The quality of ceramic manufacture and firing, as well as the elaborateness of painted decoration, also decline dramatically during this period. The primary cultural affiliation of Arenal area peoples during all previous phases had been to the west with greater Nicoya and ultimately with Mesoamerica, but that trend reversed itself as the primary affiliation shifted eastward toward the Atlantic watershed and the Meseta Central. The reasons for that change are not clear, but it does not coincide with any volcanic activity, and, thus, it apparently had nothing to do with volcanism. Toward the end of the phase, in ca. A.D. 1450, Arenal emitted one of its largest eruptions of all time, depositing the thick Unit 20 across the landscape. It appears that the combination of this eruption and the Spanish conquest, with the attendant diseases, was too much stress for local native societies, and they disappeared from the area (Ferrero Acosta, 1981).

In summary, the resilience of Arenal area society to the sudden environmental stresses of explosive volcanism is extraordinary, particularly in contrast to the cases discussed later. Only once can we find a significant change in human occupation, beyond the immediacy of the disaster, that persisted and perhaps was caused by volcanism. Thus, what is striking about Arenal area societies is their resilience to explosive volcanism that occurred on the average of every four centuries. That resilience probably is based on a number of factors, each of which is explored and contrasted to the other more complex societies considered in the rest of this chapter. One is the high degree of Arenal village self-sufficiency, as most food, building materials, clay, and stone for chipped and ground stone implements were available locally. Only the material for stone axes was obtained from a distance, and that was not a very great distance—only a day's walk away. Domesticated foods comprised a small portion of the diet, so the destruction of gardens removed only a fraction of the diet. Relocating survivors could increase their exploitation of wild flora and fauna to compensate for lost cultivated foods in areas beyond the devastated zones. I suspect another major reason Arenal societies were so resilient is that even during the millennia of sedentism following 2000 B.C., they still maintained a reasonably high degree of residential mobility and the technology that supported it.

The archaeological record for the Silencio phase, for instance, clearly indicates that households would establish temporary residences at cemeteries considerable distances from their villages, use boiling stones instead of heavy ceramic cooking vessels, and presumably collect wild flora and fauna as food sources. The presence of maize pollen at cemeteries, with no villages for many kilometers, indicates they even cultivated crops at those distant localities to provide a source of food for people involved in elaborate feasts and ancestor worship. A periodicity of a few hundred years is well within the capability of traditional societies to maintain an oral history of volcanism and their adjustments and that may have aided Arenal area societies in dealing with those massive stresses.

5.2.2. Panama

Research in western Panama (Linares, et al. 1975; Linares, 1977; Linares and Ranere, 1980) uncovered an example of volcanism affecting a society intermediate in complexity between the Arenal and the Mesoamerican cases. The Barriles chiefdom developed in the upper reaches of the Río Chiriquí Viejo from ca. 200 B.C. until the seventh century A.D. Although the Barriles chiefdom is called a chiefdom, it actually comprised a series of chiefdoms along watercourses in western Panama and eastern Costa Rica sharing the same culture and presumably language. In spite of the shared characteristics, they fiercely defended their own individual chiefdoms (polities), and in doing so, they developed an iconography clearly depicting aggression, capture, and human sacrifice by decapitation. The eruption of Volcán Barú during the seventh century A.D. deposited a layer of tephra comparable in thickness to those deposited by Arenal on sites mentioned previously in Section 5.2.1. That explosive eruption evidently caused a permanent abandonment of the Barriles–Cerro Punta area by Barriles peoples, as they or their descendants never reoccupied the area. The other chiefdoms farther downriver were not as heavily impacted and weathered the stress *in situ*. Linares and Ranere (1980, pp. 244–245) believe some of the Barriles residents migrated northward over the Continental Divide and founded sites on the Atlantic drainage such as those in the Bocas del Toro area. In contrast to the facility of recovery of Arenal peoples, the Barú ashfall had a much greater effect on local populations, and centuries later when pre-Columbian people finally did reoccupy the valley, they were not the descendants of the societies that lived there before the eruption, as is evidenced by the dramatic and fundamental differences in artifacts and decorations.

I suspect that the cultural impact of the eruption of Volcán Barú was greater than that of Volcán Arenal because the Barriles chiefdoms relied on maize as a staple and had a more intense adaptation; more fixed facilities; and more complex or rigid political, economic, and social systems. Even more important, the Barriles chiefdoms functioned within a more demographically dense and fully utilized landscape with competitive aggressiveness marking interpolity relationships, indicating a lack of land into which migrants could settle. Related to that ecological scarcity, there probably was a lack of desire and incentive for a relatively intact polity (beyond the zone of negative volcanic impact) to assist the survivors of a suddenly stressed polity that was an adversary. In fact, the degree of competitive aggressiveness between polities is greater than in any other of the cases presented in this chapter and probably is the reason why that eruption had a more dramatic societal impact than would be predicted by ecological factors alone. Survivors in the two uppermost Barriles chiefdoms could have expected hostility if they tried to move farther downriver into other chiefdoms' territories and thus had to migrate over the Continental Divide into the very moist, nonseasonal tropical rain forest to the north, near the Atlantic shoreline. As this eruption was the only one known to have affected western Panamanian societies in the two millennia pri-

or to the Spanish conquest, it may have caught them without any oral history of such phenomena, and, thus, they were less prepared than their Arenal neighbors.

5.2.3. El Salvador

El Salvador provides a few cases of eruptions and their effects on pre-Columbian societies, ranging from the immense eruptions of Coatepeque and Ilopango to the small and highly localized Loma Caldera eruption (Table 2). The huge Coatepeque eruption, dated between 10,000 and 40,000 years ago, may have predated human occupation in Central America. The societies affected by the post-Coatepeque eruptions were more complex than those affected by eruptions in Costa Rica and Panama. Of course, the magnitude of an eruption is correlated with the magnitude of ecological and societal disruption, but it is a major point of this chapter that it is not the only primary variable. It appears that the more complex Salvadoran societies were more vulnerable to explosive volcanic stresses than were the Costa Rican societies. Holocene eruptions in El Salvador, only a few of which are considered here, were sufficiently frequent that we can reasonably expect pre-Columbian Salvadoran societies to have maintained oral histories of them and how to deal with them.

The white volcanic ash layer that buried ancient artifacts in central and western El Salvador has been the subject of intermittent interest by archaeologists and geologists for much of the twentieth century, beginning with Jorge Larde in 1917 (Larde, 1926; Lothrop, 1927; Vaillant, 1934; Porter, 1955; Boggs, 1966; Schmidt-Thome, 1975; Sharer, 1978; Sheets, 1979; Hart and Steen-McIntyre, 1983). Called the "Tierra Blanca Joven" (TBJ), it erupted from a volcanic vent under the west end of Lake Ilopango. It has been radiocarbon dated to A.D. 260 ± 114 (calibrated with a 1 sigma range from

TABLE 2 Four Explosive Eruptions Affecting Societies in Central El Salvador during the Past Two Millennia

Eruption (source)	Date	Area covered (km²)	Regional effects
El Playon	A.D. 1658–1659	30	Slight
El Boqueron (San Salvador Volcano)	est. A.D. 900–1100	300 (deeper than 7 cm)	Moderate
Loma Caldera (Ceren tephra)	A.D. 590 ± 90	Less than 5	Insignificant
Ilopango (Tierra Blanca Joven)	A.D. 260 ± 114 or summer A.D. 175 (by ice cores)	10,000 (deeper than 50 cm)	Massive and lasting

A.D. 146–374), but more precise dating may be provided by comparison to ash layers found in Greenland ice cores. A volcanic ash that began falling in the summer of A.D. 175 in Greenland has a relatively high silica content of 69% and, hence, might be from the Ilopango eruption. The Ilopango ash is 69% silica (Hart and Steen-McIntyre, 1983). Where it is well preserved—at Chalchuapa, 77 km from the source—it is over $\frac{1}{2}$ m thick, indicating a regional disaster of major proportions.

The Ilopango eruption occurred in three phases. First, a great explosion deposited a thin, relatively coarse pumice across the countryside to a radius of about 30 km. This explosion was followed by two colossal pyroclastic flows that headed north to the Río Lempa and west into the Zapotitan Valley. These flows were unusually massive and long, stretching for more than 45 km. The one that headed west had sufficient impetus to flow up and over the Santa Tecla (Nueva San Salvador) pass at 1000-m elevation and then down into the Zapotitan Valley, even though the vent elevation is only 500 m above sea level. Both of these first stages were devastating to flora, fauna, and people in their paths, but the third phase was even more damaging regionally. It was an immense blast of very fine grained volcanic ash that fell over all of central and western El Salvador and must have been significant in adjoining southern Guatemala and southwestern Honduras. The eruption must have blasted tephra well into the troposphere, so finding Ilopango ash in Greenland ice cores should not be surprising.

Some 8000 km² may have been rendered uninhabitable by the Ilopango eruption. Segerstrom (1950) found that 10–25 cm of volcanic ash was too much for traditional agriculturalists to cope with in México after Paricutín erupted, and natural processes of tephra erosion, weathering, soil formation, and plant succession were necessary before the affected area could be reoccupied by cultivators. Depending on the tephra and the environment, these processes can take decades or centuries. Given a conservative estimated regional population density in central El Salvador of 40 people per square kilometer, one can very roughly estimate that some 320,000 people could have been killed or displaced by the eruption (Sheets, 1979). The fact that the TBJ tephra was highly acidic (sialic) rendered the impact more drastic and delayed the recovery, as more mafic materials weather more rapidly. The dozens of sites in the Zapotitan Valley buried by some 1–2 m of Ilopango tephra are comparable in tephra depth to many of the sites in the Arenal area. The contrast is how different the pre- and posteruption societies were in El Salvador, and how similar the pre- and posteruption societies were in Costa Rica. To some degree, that contrast is due

to more rapid cultural change in Mesoamerica compared to the Intermediate Area and to the greater area devastated by Ilopango, and, to some degree, it is due to a greater resilience to perturbation of Arenal societies for a number of inherent reasons.

Many areas of central and western El Salvador were abandoned for a century or two (Sheets, 1979, 1980). Archaeologists have noted the striking population decline along Guatemala's Pacific coastal plain at about this time (e.g., Shook, 1965; Bove, 1989), which probably was a result of the environmental damage caused by the Ilopango ashfall.

Flooding is a common aftermath of a volcanic ash deposit in a river's headwaters, as protective vegetation is killed or suppressed. The flooding reported at so many sites in western Honduras at the Formative–Classic boundary was likely caused by the Ilopango tephra and resultant heavy runoff and redeposition (Sheets, 1987). This includes sites such as Copán, Playa de los Muertos, Pimienta, and Los Naranjos. Even some sites as far away as Barton Ramie in Belize witnessed flooding at about this time, but that is at a considerable distance from Ilopango, and it may have occurred there for quite different reasons. In Middle American areas, where only a few millimeters to centimeters of Ilopango tephra were deposited, the deleterious effects were probably minimal, and the light dusting could have been beneficial in the first year after emplacement.

Central El Salvador was affected at least three more times by explosive volcanism after the Ilopango eruption (Sheets, 1983, 1992). The Loma Caldera volcanic vent opened up underneath the Río Sucio and initiated a series of 14 alternating lateral blasts and airfall volcanic ash and lava bomb units that buried a few square kilometers of the Zapotitan Valley, including the Ceren site. From the evidence of artifact placement and plant maturation, the eruption occurred in the evening, during the month of August, ca. A.D. 590, with sufficient warning that the inhabitants could literally *head south*, but without enough warning that they could remove their possessions. Although it had devastating consequences for the inhabitants of the Ceren village, this eruption was so small that it had no detectable effects on the course of societal change in southern Mesoamerica. The Loma Caldera eruption was followed a few centuries later by a moderately large eruption of a San Salvador volcano (Boqueron eruption) in A.D. 900 or 1000 that did negatively affect some 200–300 km². The Boqueron eruption interfered with cultivation for a few decades and must have had a negative impact on San Andres and surrounding settlements for quite some time, but, unfortunately, the research at that key site remains unpublished.

Playon was the most recent explosive eruption to affect the Zapotitan Valley, beginning late in 1658 and burying 30 km² under lava and a larger area under tephra. Although soils on the Playon lava have barely begun to recover to their preeruption state and support only scrub vegetation, soil recovery on the tephra is now sufficient to support intensive agriculture. All three of these post-Ilopango eruptions caused severe problems for villages and households near the sources, but they cannot be seen as having had long-lasting societal repercussions and certainly were not massive regional natural disasters. Of the post-Ilopango eruptions, only Boqueron probably had significant negative societal effects lasting perhaps for a few decades.

5.3. MEXICAN CASES OF VOLCANISM AND ANCIENT SOCIETIES

5.3.1. Matacapan and the Tuxtla Volcanic Field

Matacapan, in the Tuxtla Mountains on the Gulf of México, has been investigated by Santley (1994) and Reinhardt (1991). The Formative occupation was interrupted for almost a millennium by a large eruption (Table 3), but the population and society recovered and grew to maximum size by the middle of the Classic period, in spite of some smaller eruptions.

Volcanism in the Tuxtla Volcanic Field (TVF), located some 150 km southeast of Veracruz, México, has been studied by Reinhardt (1991). The younger (last 50,000 years) volcanic sequence covers some 400 km²,

with an average frequency of eruptions of at least 4 per 1000 years, an activity rate almost twice that of Arenal during the past 4000 years and comparable to that in central El Salvador during the past 2000 years. The most recent eruptions that are preserved in the stratigraphy at the Matacapan archaeological site date to the Late Archaic to the Middle Classic periods. The earliest of these occurred between ca. 3000 and 2000 B.C., and its effects on surrounding populations are unknown. Societies at that time would have been egalitarian and probably at least semisedentary, with a mixed domesticated and nondomesticated subsistence base. It is likely that village life was emerging in the area at that time, but regional population density would have been quite low compared to Formative and Classic times. The eruption left a relatively thin deposit, some 15 cm, at Matacapan.

The second TVF eruption dates to the Early Formative period, ca. 1250 B.C., and deposited the Cerro Mono Blanco (CMB) tephra. Matacapan was occupied at the time by egalitarian agriculturalists, and tephra buried the maize ridges of an agricultural field. The society affected was relatively simple, as small villages dotted the landscape. The deposit, as measured from Reinhardt's drawn sections, is moderately thick, i.e., over 60 cm. The area was abandoned for some nine centuries, but was reoccupied during the Late Formative period. Thus, the second eruption appears to have been much bigger than the first, at least from the Matacapan perspective, and it had exceptionally long-lasting deleterious effects on settlement.

TABLE 3   Volcanism and Human Settlement in the Matacapan Area, Tuxtla Mountains, Mexico

Period	Tephra	Date (approx.)	Thickness (est. from sections)	Effect on habitation
Middle Classic	Laguna Cocodrilos	A.D. 600	40–45 cm	Moderate impact on complex society; gradual but consistent decline
Early Classic	Laguna Nixtamalapan and Cerro Puntiagudo	A.D. 400	20–30 cm	None detected; occupation continues
Late Formative	Cerro Nixtamalapan	A.D. 150	40 cm	Significant impact, abandonment, rapid recovery
Early Formative	Cerro Mono Blanco	1250 B.C.	60 cm	900-year-long abandonment; major impact on simple society
Late Archaic	n.a.[a]	3000–2000 B.C.	n.a.	Unknown

[a]n.a. = not available.

The third TVF eruption, in ca. A.D. 150, buried Late Formative remains in the Matacapan area under the Cerro Nixtamalapan (CN) tephra. It is about 40 cm thick in Reinhardt's section (1991, p. 86). Given that the decline of the Olmec civilization was largely complete by this time and that the local Classic period societal resurgence had yet to occur, I would presume that the volcanism affected relatively simple societies, probably egalitarian villages. As with the second eruption, the area was abandoned, clearly indicating a significant initial impact of a substantial eruption. However, the abandonment lasted for only a short while.

In ca. A.D. 400, during the Early Classic period, two closely spaced eruptions (no evidence of soil developed between the layers) deposited the Cerro Puntiagudo (CP) and the Laguna Nixtamalapan (LN) tephras. These tephras evidently had less of an impact than the other TVF tephra emplacements, as the site apparently was not abandoned. I estimate their thickness from Reinhardt's sections at only 20–30 cm. The society affected would have been moderately complex, perhaps a chiefdom, but this is uncertain. It appears that the stress caused by this relatively thin deposit was within the resilience of the agroeconomic adaptation, and it is possible that for the cultigens that have roots sufficiently deep to tap into the preeruption soil, the tephra could have been beneficial as a mulch layer. The tephra thickness is barely within what Segerstrom (1950) found to be within the resilience of traditional agriculture in México.

The final eruption occurred in the middle of the Classic period, ca. A.D. 600, and the tephra measures some 40–45 cm thick in the sections. It sealed an agricultural field with maize ridges under it. Compared to the previous Matacapan examples, this was a moderately large eruption, and societal complexity as well as population density were at a peak at this time. Coincident with the ashfall is the reversal in Matacapan's fortune. Matacapan's population began to decline during the seventh century A.D., a part of which I suspect may have been caused by the ecological stress of volcanic ash deposition. It should be noted that Santley and Arnold (n.d.) came to the opposite conclusion; they note that volcanism in the Formative period caused abandonment and major population resettlements, while volcanism in the Classic period did not have the same results. Their suggested explanation is that the state had the ability to harness energy and labor on a larger scale and thus coped better with these sudden massive natural stresses.

The relative ecological–agrarian–societal stresses caused by various volcanic eruptions can be researched only if their magnitudes and distributions are understood and can be compared. Because such information is not yet available for the TVF, and because other indications are lacking, the relative magnitudes can be approximated only very roughly by comparing tephra depths from measured sections. I believe it is significant that the thickness of tephra correlates directly with the societal impacts of the eruptions at Matacapan, as the thinnest of ash layers (CP and LN at ca. 20 cm) did not cause abandonment, but the thicker layers did cause abandonment proportional to their thickness (CMB at about 60 cm and CN at 40–45 cm). The Laguna Cocodrilos (LC) tephra, about 40 cm thick, correlates with demographic and societal declines, and the site never recovered. However, viewing societal impacts from only a single site can be very misleading, and regional research is needed to understand Tuxtla impacts and recoveries. What is needed for the TVF (as well as for other Middle American cases) is isopach (depth and distribution) mapping of each unit, with estimates of the magnitude, volume, and distribution of each tephra and documentation of soil recovery, as well as human reoccupation, within a well-dated natural and cultural framework.

Chase (1981) argued that an eruption of the San Martin volcano in ca. 600 B.C. in the Tuxtlas caused sufficient damage to the utilized landscape in the Tres Zapotes area to have rendered it uninhabitable, and migrations of tens of thousands of people resulted. He relied on stratigraphic information from Drucker's research at Tres Zapotes. Unfortunately, the dating of the eruption is rather speculative, as it is not based on extensive data, and evidence indicating the source of the tephra is weak. This tephra might be from the same eruption as the CMB or CN tephra at Matacapan, or it may be unrelated.

5.3.2. Popocatépetl Volcano and Cholula, Puebla, México

For the past couple of decades, a number of separate discoveries of volcanic ash burying cultural features have been made in Puebla, generally to the northeast, east, or southeast of the Popocatépetl Volcano. Seele (1973) reported well-preserved maize ridges under white airfall tephra at San Buenaventura Nealtican, and others have reported other maize ridges or artifacts buried by primary or secondary volcanic deposits in the general area. Fortunately, Siebe et al. (1996) are working to synthesize these isolated discoveries into a regional interpretive framework. They have found evidence of three major eruptions of Popocatépetl in ca. 3000 B.C., between 800 and 200 B.C., and probably in A.D. 822. The earliest eruption was characterized by a series of large phreatomagmatic (magma in contact with water resulting in steam explosions) eruptions, resulting in multiple surge deposits followed by a thick pumice deposit and a number of ash flows that extended more than 50 km from their sources. The effects on Late Archaic populations are unknown, but must have been devastating. The recovery of soils, vegetation, and human societies resulted in reasonably dense occupation by the first millennium B.C., which were then buried by the *Lower Ceramic Plinian* eruption. This eruption consisted primarily of a thick, widespread andesitic pumice deposit followed by ash flows and lahars (mud flows) radiating from Popocatépetl. These primary and secondary deposits buried agricultural fields (e.g., Seele 1973), artifacts, and other cultural features. Even household groups with talud-tablero (sloping panel with overhanging vertical panel) architecture were buried, dating to a few centuries prior to the appearance of that architectural style in Teotihuacán. The migrations forced by this eruption may have brought talud-tablero architecture to the northern Basin of México. The eruption's effects on habitation on or near the slopes of Popocatépetl were obviously devastating, but its effects on Cholula and the Puebla Valley or the southern Basin of México are not clear.

The third eruption, called the *Upper Ceramic Plinian*, was composed primarily of surge deposits from phreatomagmatic explosions followed by pumice falls and extensive lahars. The age of A.D. 822 is from a Greenland ice core (Siebe et al., 1996), but the detailed geochemical analysis needed to confirm that association has yet to be performed. Massive lahars filled the Puebla basin and Atlixco valley after the eruption, apparently having devastating effects on settlements. Lahar deposits a few meters thick buried architecture at Cholula and are clearly visible to the visitor today. They must have devastated agriculture for many square kilometers around Cholula for many decades. I suspect that the weakened Cholulan society was unable to resist the intrusion of Putun Maya, resulting in settlements such as Cacaxtla.

The cultural, economic, political, and demographic effects of these large eruptions, as well as smaller ones reported by Siebe et al. (1996), remain unknown. Archaeologists have reported cultural and architectural declines and florescences (e.g., Weaver, 1993), and two of the declines Weaver reports do approximately correlate with the latter two large eruptions, especially in the Late Classic period. Correlations are intriguing, but establishing probable causalities will require regional interdisciplinary research. Here is a regional integrated archaeological–volcanological project crying to happen, with the high probability of significant results. Such a project could go far to help us understand the beneficial and detrimental aspects of living in volcanically active areas and to explain demographic and so-

cietal declines and surges in these areas. The Puebla area is one of the most culturally important areas of ancient Mesoamerica, yet it remains one of the most poorly studied on an integrated regional basis.

5.3.3. The Basin of México: Cuicuilco and Xitle Volcano

Niederberger (1979) documented the emergence in the Basin of México of sedentary human communities in the resource-rich lacustrine environments of the Chalco and Xochimilco basins by the sixth millennium B.C. In the southern part of the Basin of México, these communities were devastated by a series of volcanic deposits in ca. 3000 B.C. Because that is the same time as the earliest Popocatépetl eruption documented by Siebe et al. (1996), I suggest that this may have been the northwestern component of the same eruption. Niederberger (1979) refers to the deposits as a series of nuées ardentes (*glowing clouds* or pyroclastic flows of hot volcanic ejecta and gases that flow rapidly downhill), leaving thick, white pumice layers. They are thick enough to have caused ecological and human societal devastation, and because of their sialic composition and the high elevation, I would expect slow weathering rates and thus slow environmental recovery, on the scale of a few centuries. Adams (1991, p. 38) finds Niederberger's interpretation convincing and states that it took some five centuries for human reoccupation of the area.

The case of the Xitle eruption in the southern Basin of México some 1500–2000 years ago remains controversial and probably will remain so until a regional multidisciplinary project is conducted to resolve a number of issues. To date, many individual archaeologists and geologists have conducted small and localized investigations, and the interpretations vary widely. Many scholars have noted that the demise of Cuicuilco, the site that had been the dominant polity of the Late Formative period in the Basin of México; the rise of Teotihuacán; and the eruption of Xitle may have been causally interrelated. Parsons (1974, 1976) suggests that Xitle devastated Cuicuilco in the first century A.D. and thus facilitated the rise of Teotihuacán. Similarly, Diehl (1976) notes the political ascendancy of Teotihuacán and suggests that it was, in part, because of Xitle's elimination of Cuicuilco as a competitor. Adams (1991, p. 109) dates the volcanic destruction of Cuicuilco to ca. 150 B.C., correlating with its decline.

A major problem in exploring human–volcanic interaction in the Cuicuilco area is dating of the eruption of Xitle. Cordova et al. (1994) reviewed the attempts to date the eruption. They note that the early research efforts dated it to ca. 400 B.C., while more recent efforts dated it from the third century A.D. to as late as A.D.

400. They conclude that the most accurate radiocarbon date is 1536 ± 65 B.P., or about A.D. 400. They also suggest that Cuicuilco had been abandoned well before the eruption. If these conclusions are correct, the demise of Cuicuilco certainly was not caused by the Xitle eruption, and its abandonment may have been caused by the expansion of, and competition from, Teotihuacán. This explanation would neatly reverse the causality, pluck it from the volcanological domain, and place it in the cultural domain. Heizer and Bennyhoff (1958) note the deterioration of Cuicuilco architecture prior to burial by the Xitle lava, indicating abandonment prior to the eruption. Even if this later dating of the eruption is substantiated, a study still needs to be done to explore the demographic–ecological–societal implications of burying some 75 km^2 of the most fertile land in the Basin of México under basaltic lava during the Early Classic period. Recent research indicates that the most accurate dating of the Xitle eruption is about the time of Christ (Urrutia-Fucugauchi, 1996; Claus Siebe, personal communication, 1997), thus putting it at about the same time as the large eruption of Popocatépetl that devastated such a large area of Puebla. If these more recent datings are more accurate, then the demise of Cuicuilco and the devastation of the southern basin and Puebla could have facilitated the rise of Teotihuacán.

During a survey conducted shortly after World War II, Gifford (1950, p. 185) found a deep pumice deposit over early (Late Formative period) artifact-bearing layers in the Ixtlan del Rio area of southeast Nayarit. Many sites were under 3–4 m of tephra, and Gifford attributed this to the eruption of the Ceboruco Volcano, but that was not convincingly demonstrated. Certainly the eruption was of sufficient magnitude to be the reason for a decline in cultural complexity in the area, but so little research has been conducted there that the topic remains speculative.

5.4. SUMMARY AND CONCLUSIONS

Although the sample of reasonably well-researched cases where pre-Columbian Middle American societies were affected by explosive volcanism is small, at least some patterns are presented here in a preliminary fashion. The trends or patterns are presented as five factors, each of which is explored for the range of variation seen in the data and is examined for the ways in which it affects the resilience or vulnerability of societies. These factors are demography, experience, adaptation, economy, and politics.

Demography, in terms of the distribution and density of population, played an important role in how

pre-Columbian societies reacted to sudden massive stresses of explosive eruptions. At the low end of the spectrum were the Arenal societies of Costa Rica, with very low regional population densities that manifested themselves as small villages scattered along river courses and lakeshores. The fact that Costa Rican societies maintained remnants of residential mobility by living for significant periods of time in graveyards probably assisted their sudden need for mobility in the face of an eruption from the Arenal volcano. Also, the low regional populations afforded opportunities for refugees without threatening surrounding populations. In contrast, the city of Cholula (México) and the densely populated region around it were highly vulnerable to the large eruptions of the Popocatépetl Volcano. Part of that vulnerability is the fact that surrounding arable lands for many hundreds of kilometers were similarly densely occupied, making migration difficult. Cuicuilco (México) and El Salvador some 2000 years ago were similar to Cholula, but with less population, and those two cultures did not recover.

The factor of experience, and the decision making that results from it, is difficult to research archaeologically in nonliterate societies. However, nonliterate societies commonly retain accurate oral histories for many generations, often for centuries, and, thus, some possibilities should be considered here. Given the high frequencies of eruptions in the Tuxtlas, highland central México, El Salvador, and Costa Rica, it is likely that societies in these areas did maintain oral histories of eruptions. It is much less likely that the Barriles chiefdoms in western Panama had similar oral histories because of the rarity of known eruptions in that area. Similarly, Workman (1979) found that the high frequency of eruptions in southwestern Alaska and the Aleut's oral traditions assisted recovery. That contrasted to the rarity of eruptions in the inland Yukon and the nonrecovery of northern Athabascans from the Mt. St. Elias eruption. According to Boehm (1996), it would be erroneous to assume that simpler societies are less capable of efficient decision making in the face of a crisis than are more complex societies.

The factor of adaptation is very important, ranging from the extensive adaptation of the Arenal societies to the intensive adaptations of Mesoamerican states. The Arenal villages relied on wild (nondomesticated) sources for most of their foods, so when their gardens with the domesticated maize, beans, and squash were devastated, they had only to rely more heavily on already exploited wild sources of food, often in refuge areas outside the zone affected by deep tephra deposits. The Mesoamerican states—as exemplified by Cholula, Cuicuilco, and Late Preclassic El Salvador—relied on maize as their staple crop, supplemented by beans, squash, and chiles. The states relied on intensive agriculture, harvesting and storing of foods during the dry season, and food redistribution as controlled by the elite. Their reliance on intensive production, fixed facilities, and vertical controls on redistribution rendered them vulnerable to massive eruptions.

The cases examined here vary widely with regard to the factor of economics. Anchoring one end of the spectrum is the Arenal area societies, where village self-sufficiency in resources is striking. Each village obtained food, materials for artifacts, and housing construction materials from nearby or contiguous territory and needed only a day's walk to obtain materials for stone axes. The reestablishment of village self-sufficiency by refugees from devastated zones could be achieved without too much trouble. The Arenal societies contrast with the centralized economies of the Mesoamerican states such as Cholula, where the elite and their attached specialists provided ceramics, obsidian tools, and other items in exchange for commoner-produced foods and other items. Many of the elite-controlled items were dependent on trade routes that stretched for hundreds of kilometers. The centralization of economic authority in the elite in a specialized and redistributive economy is vulnerable to sudden, massive stresses, as can be seen in the Mesoamerican cases.

Political factors also played a role in varying degrees of vulnerability. The lack of conflict among the egalitarian Arenal villages assisted in relocation following the large eruptions. The warring Barriles chiefdoms are the most striking example of the other end of the spectrum. The relatively small Barú eruption caused sufficient stress on the uppermost two chiefdoms that they had to migrate over the Continental Divide and change their adaptation to the nonseasonal tropical lowlands. Cholula is another example—following the Late Classic Popocatépetl eruption, the aggressive and competitive Putun Maya established a fortified center overlooking the ruins of Cholula.

In general, the pre-Columbian Middle American societies that shared certain characteristics found recovery from explosive volcanic eruptions more difficult. Those characteristics include sedentary large populations relying on intensive agriculture that focused on a domesticated staple, a high degree of *built environment* of fixed facilities, and a vertical organization of decision making with centralized authority concentrated in the elite. That centralized authority encompassed the realms of economics, politics, and religion. The most resilient societies were those that maintained self-sufficiency at the village level, low regional populations, continuation of some residential mobility, maintenance of a broad-spectrum adaptation, and perhaps an oral tradition of dealing with volcanic eruptions.

Acknowledgments

I wish to thank the field and laboratory crews of my research projects in Panama, Costa Rica, and El Salvador for their large contribution in terms of data and interpretations of what life was like in volcanically active areas of the Middle American tropics. Without their hard work, this chapter would not have been possible. Larry Conyers, Vera Markgraf, and an anonymous reviewer contributed many helpful suggestions. I thank Claus Siebe for numerous conversations about volcanism in highland central México and for helping me understand his important contributions in that area.

References

Adams, R. E. W., 1991: *Prehistoric Mesoamerica* (revised edition). Norman: Univ. of Oklahoma Press, 454 pp.

Boehm, C., 1996: Emergency decisions, cultural-selection mechanics, and group selection. *Current Anthropology*, **37**(5): 763–793.

Boggs, S., 1966: Pottery jars from the Loma del Tacuazin, El Salvador. *Middle American Research Records*, **3**(5):175–185.

Bove, F., 1989: Settlement classification procedures in Formative Escuintla, Guatemala. *In* Bove, F., and F. Heller (eds.), *New Frontiers in the Archaeology of the Pacific Coast of Southern Mesoamerica*. Anthropological Research Papers, No. 39. Tempe: Arizona State University, pp. 65–102.

Chase, J., 1981: The sky is falling: The San Martin Tuxtla volcanic eruption and its effects on the Olmec at Tres Zapotes, Veracruz. *Vinculos*, **7**(2): 53–69.

Cordova, C., A. L. Martin del Pozzo, and J. Lopez Camacho, 1994: Palaeolandforms and volcanic impact on the environment of prehistoric Cuicuilco, southern México City. *Journal of Archaeological Science*, **21**: 585–596.

Diehl, R., 1976: Pre-Hispanic relationships between the Basin of México and North and West México. *In* Wolf, E. (ed.), *The Valley of México: Studies in Pre-Hispanic Ecology and Society*. Albuquerque: Univ. of New Mexico Press, pp. 249–286.

Ferrero Acosta, L., 1981: Ethnohistory and ethnography in the Central Highlands—Atlantic watershed and Diquis. *In* Benson, E. (ed.), *Between Continents/Between Seas: Pre-Columbian Art of Costa Rica*. New York: Abrams, pp. 93–103.

Gifford, E., 1950: *Surface Archaeology of Ixtlan del Rio, Nayarit*. Los Angeles: University of California Publications in American Archaeology and Ethnography, **43**(2):183–302.

Hart, W., and V. Steen-McIntyre, 1983: Tierra Blanca Joven tephra from the A.D. 260 eruption of Ilopango volcano. *In* Sheets, P. (ed.), *Archaeology and Volcanism in Central America: The Zapotitan Valley of El Salvador*. Austin: University of Texas Press, pp. 14–34.

Heizer, R., and J. Bennyhoff, 1958: Archaeological investigations of Cuicuilco, Valley of México, 1956. *Science*, **127**: 232–233.

Hoopes, J., 1987: Early ceramics and the origins of village life in lower Central America. Ph.D. dissertation, Harvard University, Cambridge, MA, 759 pp.

Larde, J., 1926: Arqueologia Cuzcatleca: Vestigos de una Población Pre-Mayica en el Valle de San Salvador, C. A., Sepultados Bajo una Potente Capa de Productos Volcánicos. *Revista de Etnología, Arqueología, y Linguistica*. San Salvador, **1**: 3 and 4.

Linares, O., 1977: *Ecology and the Arts in Ancient Panama*. Dumbarton Oaks Studies in Pre-Columbian Art and Archaeology, No. 17. Washington, DC: Dumbarton Oaks, 77 pp.

Linares, O., and A. Ranere (eds.), 1980: *Adaptive Radiations in Prehistoric Panama*. Peabody Museum, Harvard University, 530 pp.

Linares, O., P. Sheets, and E. Rosenthal, 1975: Prehistoric agriculture in tropical highlands. *Science*, **187**: 137–145.

Lothrop, S., 1927: Pottery types and their sequence in El Salvador. *In*-

dian Notes and Monographs, New York: Museum of the American Indian, Heye Foundation, **1**(4): 165–220.

Melson, W., 1994: The eruption of 1968 and tephra stratigraphy of Arenal volcano. *In* Sheets, P., and B. McKee (eds.), *Archaeology, Volcanism, and Remote Sensing in the Arenal Area, Costa Rica.* Austin: University of Texas Press, pp. 24–47.

Mueller, M., 1994: Archaeological survey in the Arenal basin. *In* Sheets, P., and B. McKee (eds.), *Archaeology, Volcanism, and Remote Sensing in the Arenal Area, Costa Rica.* Austin: University of Texas Press, pp. 48–72.

Niederberger, C., 1979: Early sedentary economy in the Basin of México. *Science,* **203**(4376): 131–142.

Parsons, J., 1974: The development of a prehistoric complex society: A regional perspective from the Valley of México. *Journal of Field Archaeology,* **1**: 81–108.

Parsons, J., 1976: Settlement and population history of the Basin of México. *In* Wolf, E. (ed.), *The Valley of México: Studies in Pre-Hispanic Ecology and Society.* Albuquerque: University of New Mexico Press, pp. 69–100.

Porter, M., 1955: Material Preclasico de San Salvador. *Communicaciones del Instituto Tropical de Investigaciones Científicas, San Salvador,* **3/4**: 105–112.

Reinhardt, B., 1991: Volcanology of the younger volcanic sequence and volcanic hazards study of the Tuxtla Volcanic Field, Veracruz, México. M.A. thesis. Tulane University, New Orleans, LA.

Santley, R., 1994: The economy of ancient Matacapan. *Ancient Mesoamerica,* **5**: 243–266.

Santley, R., and P. Arnold, III, n.d.: Prehispanic settlement patterns in the Tuxtla Mountains, southern Veracruz, México. M.S. thesis, in possession of the authors.

Schmidt-Thome, M., 1975: The geology in the San Salvador area: A basis for city development and planning. *Geologisches Jahrbuch,* **813**(S): 207–228.

Seele, E., 1973: Restos de milpas y poblaciones prehispanicas cerca de San Buenaventura Nealtican, Puebla. Puebla, Fundación Alemana para la Ciencia. *Communicaciones,* **7**: 77–86.

Segerstrom, K., 1950: Erosion studies at Paricutin. U.S. Geological Survey Bulletin 965A, Washington, DC, 164 pp.

Sharer, R. (ed.), 1978: *The Prehistory of Chalchuapa, El Salvador.* Vols. 1–3. Philadelphia: University of Pennsylvania Press, pp. 194, 211, 226.

Sheets, P., 1979: Environmental and cultural effects of the Ilopango eruption in Central America. *In* Sheets, P., and D. Grayson (eds.), *Volcanic Activity and Human Ecology.* New York: Academic Press, pp. 525–564.

Sheets, P., 1980: Archaeological studies of disaster: Their range and value. Natural Hazards Research Working Papers #38, University of Colorado, Boulder, 35 pp.

Sheets, P. (ed.), 1983: *Archeology and Volcanism in Central America: The Zapotitan Valley of El Salvador.* Austin: University of Texas Press, 307 pp.

Sheets, P., 1987: Possible repercussions in western Honduras of the third-century eruption of Ilopango volcano. *In* Pahl, G. (ed.), *The Periphery of the Southeastern Classic Maya Realm.* Los Angeles: UCLA Latin American Center, pp. 41–52.

Sheets, P., 1992: *The Ceren Site: A Prehistoric Village Buried by Volcanic Ash in Central America.* Ft. Worth: Harcourt Brace, 150 pp.

Sheets, P., 1994: Summary and conclusions. *In* Sheets, P., and B. McKee (eds.), *Archaeology, Volcanism, and Remote Sensing in the Arenal Region, Costa Rica.* Austin: Univ. of Texas Press, pp. 312–326.

Shook, E., 1965: Archaeological survey of the Pacific coast of Guatemala. *In Handbook of Middle American Indians,* Vol. 2, Part 1. Austin: University of Texas Press, pp. 180–194.

Siebe, C., M. Abrama, J. Macias, and J. Obenholzner, 1996: Repeated volcanic disasters in Prehispanic time at Popocatépetl, central México: Past key to the future? *Geology,* **24**(5): 399–402.

Urrutia-Fucugauchi, J., 1996: Palaeomagnetic study of the Xitle-Pedregal de San Angel lava flow, southern Basin of México. *Physics of the Earth and Planetary Interiors,* **97**: 177–196.

Vaillant, G., 1934: The archaeological setting of the Playa de los Muertos culture. *Maya Research,* **1**(2): 87–100.

Weaver, M., 1993: *The Aztecs, Maya, and Their Predecessors.* San Diego: Academic Press, 567 pp.

Workman, W., 1979: The significance of volcanism in the prehistory of subarctic Northwest North America. *In* Sheets, P., and D. Grayson (eds.), *Volcanic Activity and Human Ecology.* New York: Academic Press, pp. 339–372.

6

Abrupt Climate Change and Pre-Columbian Cultural Collapse

MARK BRENNER, DAVID A. HODELL, JASON H. CURTIS,
MICHAEL F. ROSENMEIER, MICHAEL W. BINFORD,
AND MARK B. ABBOTT

Abstract

Holocene climate changes in the circum-Caribbean and Andean Altiplano are inferred by using paleolimnological methods. Paleoenvironmental data provide a climatic context in which the Maya (Yucatán Peninsula) and Tiwanaku (Bolivian–Peruvian Altiplano) cultures arose, persisted, and collapsed prior to European contact, ca. A.D. 1500. In the circum-Caribbean, the arid late Pleistocene period (>10,500 [14]C B.P., >10,470 B.C.) was followed by a relatively moist early to middle Holocene period (9000–4000 [14]C B.P., 8030–2490 B.C.), probably related to large differences between summer and winter insolation. The earliest Maya settlement dates to the Middle Preclassic period (1000–300 B.C.) and was associated with reduced seasonality and regional drying. In the northern part of the Yucatán Peninsula, the climate became even drier during the Classic period (A.D. 250–850). The driest episode of the middle to late Holocene occurred in the Maya lowlands at ca. A.D. 800–1000 and coincided with the Maya collapse, ca. A.D. 850.

In contrast to the circum-Caribbean area, the Andean Altiplano was relatively wet in the late Pleistocene period and experienced low seasonality and dry conditions in the early and middle Holocene. The southern basin of Lake Titicaca (Lago Wiñaymarka), currently >40 m deep, displayed a low stage between ca. 7700 and 3600 [14]C B.P. (6470–1930 B.C.). Chiripa culture developed in the Titicaca watershed ca. 1500 B.C. (3210 [14]C B.P.), and was associated with greater seasonality, increased moisture availability, and rising lake level. Tiwanaku culture emerged ca. 400 B.C. (2400 [14]C B.P.) and depended on raised-field agricultural technology from A.D. 600 to 1150. A prolonged dry period began in the Altiplano at ca. A.D. 1100, prompting abandonment of raised fields and cultural decline.

Climate changes in the Northern Hemisphere Maya lowlands and the Southern Hemisphere Andean Altiplano were out of phase on millennial timescales, when climate was apparently forced by shifts in seasonal insolation driven by the precession of the Earth's orbit (Milankovitch forcing). Shorter frequency climate changes in the Maya and Tiwanaku regions during the last ~3000 years may have been in phase and were driven by factors other than Milankovitch forcing. In both areas, population growth and cultural development occurred under favorable conditions for agriculture. Rapid cultural collapses in both regions were associated with protracted droughts. Paleoenvironmental data indicate that cultural development is limited by climatic thresholds and that abrupt, unpredictable climate changes can disrupt agricultural production and have devastating consequences for human populations.

Resumen

Cambios climáticos durante el Holoceno fueron inferidos usando métodos paleolimnológicos en las regiones Caribeñas y el Altiplano Andino. Datos paleoambientales proveen un esquema climático en la cual culturas Maya (Peninsula Yucatán) y Tiwanaku (Altiplano de Bolivia/Peru) surgieron, persistieron, y se desintegraron antes de contacto con los Europeos, ca. D.C. 1500. En la región del Caribe, el período del Pleistoceno tardío (>10,500 ^{14}C años A.P., >10,470 A.C.) fue caracterizado por un clima árido. El clima del Holoceno temprano y medio (9000–4000 ^{14}C años A.P., 8030–2490 A.C.) era mas húmedo, probablemente como consecuencia de amplias diferencias entre la insolación veranal e invernal. La colonización de la región por los Mayas empezó en el Preclásico Medio (1000–300 A.C.), asociado con condiciones mas secas y una reducción en la diferencia estacional. En el norte de la Peninsula Yucatán, el clima fue mas seco durante el Período Clásico (D.C. 250–850). En la región Maya, la época mas seca del Holoceno medio y tardío ocurrío cerca D.C. 800–1000 y coincidío con el colapso de la cultura Maya ~D.C. 850.

Los registros de la región Caribeña demuestran diferencias paleoclimáticas con las condiciones en el Altiplano Andino. El Altiplano fue relativamente húmedo durante el Pleistoceno tardío. Durante el Holoceno medio y tardío, el clima fue seco y diferencias estacionales no fueron muy pronunciadas. La cuenca sur del Lago Titicaca (Lago Wiñaymarka), que en el actual tiene una profundidad >40 m, tenía un nivel muy bajo entre cerca 7700 y 3900 ^{14}C años A.P. (6470–1930 A.C.). La cultura Chiripa se desarrolló en la cuenca de Titicaca ~1500 A.C. (3210 ^{14}C años A.P.), asociado con condiciones mas estacionales, un incremento en precipitación, y una subida en el nivel del lago. La cultura Tiwanaku emergío ~400 A.C. (2400 ^{14}C años A.P.) y dependío de una agricultura intensiva (campos elevados o camellones) entre D.C. 600 y 1150. Una sequía prolongada comenzó en el Altiplano alrededor de D.C. 1100, y incitó el abandonamiento de los camellones y el decaimiento de la cultura.

Cambios climáticos en regiones bajas de los Mayas (Hemisferio Norte) y en el Altiplano Andino (Hemisferio Sur) estaban fuera de fase sobre escalas de tiempo milenial, mientras que el clima era controlado por los cambios en insolación estacional, como consecuencia de precesión en la órbita del globo (ciclos Milankovitch). Durante los ultimos ~3000 años pasados, inferencias sobre cambios climáticos en las regiones Maya y Tiwanaku se aparecen mas o menos en fase y fueron causado por factores diferente a los ciclos Milankovitch. En ambas regiones, el crecimiento de poblaciones y el desarrollo cultural ocurrieron durante un tiempo cuando las condiciones climáticas eran favorables para la agricultura. El decaimiento en ambas culturas ocurrió rapidamente y fue asociado con sequías prolongadas. Datos paleoambientales indican que el desarrollo de una cultura es limitado por el clima. Cambios climáticos bruscos y no predecibles pueden alterar la producción agricola y tener graves consecuencias en las poblaciones humanas.

6.1. INTRODUCTION

6.1.1. Environmental Determinism in Pre-Columbian America

The concept of *environmental determinism* was debated in the anthropological and archaeological literature during the 1950s and 1960s. The fundamental tenet of the hypothesis posits that regional agricultural potential limits cultural development. Meggers (1954) concluded that the Maya lowlands of southern México and Petén, Guatemala (Fig. 1) have limited agricultural potential today. She suggested that the region would not have been conducive to the cultural florescence so evident in the archaeological record. A logical conclusion of the deterministic theory was that lowland Classic Maya culture (A.D. 250–850), with its monumental architecture, art, hieroglyphics, concept of zero, corbeled arches, calendrical system, and stela cult, did not develop *in situ*, but was imported from elsewhere and was destined to decline in its new geographic context. The simple theory apparently accounted for both the origin of the lowland Maya and their mysterious collapse in the ninth century A.D.

Meggers's (1954) claims were contested by those who argued that there was no archaeological basis for claiming that Classic Maya culture was imported from the highlands. Coe (1957) noted that several lowland Maya achievements, such as the corbeled arch, the Long Count calendar, and the stela cult, developed in the lowlands. Excavations in the lowlands yielded Early Preclassic (2000–1000 B.C.), *Formative phase* ceramics, supporting the claim that lowland Classic Maya cultural ontogeny had its roots in the low-elevation, tropical karst environment (Altschuler, 1958; Hammond et al., 1977, 1979). Today, it is generally acknowledged that the origins of lowland Maya civilization began in the Early Preclassic period, were complex, and involved interactions with peripheral coastal and high-

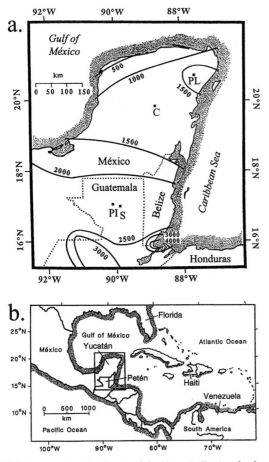

FIGURE 1   (a) Maya lowlands of the Yucatán Peninsula showing isohyets (mm) and locations of lake sediment records discussed in the text (PL = Punta Laguna, C = Chichancanab, PI = Petén-Itzá, S = Salpeten). Rainfall values are based on Wilson (1980). (b) The circum-Caribbean region showing the Yucatán Peninsula (box) and locations of paleoclimate studies referred to in the text, including Petén (Guatemala), Yucatán (México), Florida, Haiti, and Venezuela.

land regions (Sharer, 1994). Coe pointed out that there is no evidence for the *gradual cultural decline* referred to by Meggers. Instead, archaeological data point to abrupt, widespread collapse ca. A.D. 850, after 600 years of population growth and cultural expansion in the Early Classic (A.D. 250–550) and Late Classic (A.D. 550–850) periods.

Ferdon (1959) reassessed Petén's agricultural potential and challenged Meggers's application of *environmental determinism* to the lowland Classic Maya. Using modern temperature, rainfall, soil, and landform criteria, he reclassified the lowlands as a region favorable for agriculture. The analysis did not refute deterministic theory as applied to the Maya, but rather, freed the civilization from expectations dictated by an environment with limited agricultural potential. Ferdon, how-

ever, argued that there is no correlation between natural agricultural potential and cultural development and attributed the cultural decline to invasion of agricultural plots by grasses that interfered with Classic Maya crop cultivation.

Adams (1973) summarized theories that purport to explain the ninth century A.D. Maya collapse. Among the proposed causes that continue to be explored are overpopulation combined with soil erosion and exhaustion (Rice, 1978; Deevey et al., 1979; Paine and Freter, 1996), demographic models that show skewed sex ratios in favor of males (Cowgill and Hutchinson, 1963), health problems and disease (Saul, 1973), social upheaval, warfare, religion (i.e., the collapse was pre-ordained), and insect infestation (Sabloff, 1973). Natural catastrophes such as earthquakes and volcanic eruptions (Ford and Rose, 1995) have also been invoked as factors contributing to the collapse. Others have suggested a correlation between climate and Maya cultural development and decline (Gore, 1992; Gill, 2000; Dahlin, 1983; Gunn and Adams, 1981; Folan et al., 1983; Messenger, 1990; Hodell et al., 1995; Curtis et al., 1996). Several theories for the collapse have been tested by archaeological and paleoenvironmental study, but some of the hypotheses are supported by few, if any, data.

Outside the Maya lowlands, other pre-Columbian cultures also arose, persisted for centuries, and ultimately collapsed in seemingly harsh environments. The Chiripa (1500–200 B.C.) and Tiwanaku (400 B.C. to A.D. 1100) cultures developed in the Lake Titicaca watershed of the Andean Altiplano (Fig. 2), where nitrogen-poor soils on steep slopes are prone to erosion and moisture deficit. Nighttime freezes often kill crops prior to harvest (Kolata and Ortloff, 1989). Flatlands (pampas) near Lake Titicaca are inundated periodically and subject to soil salinization. The prolonged rise and abrupt fall of a great civilization in this region of seemingly limited resources appears to defy environmental determinism.

Human ingenuity, combined with state-level agricultural organization, can overcome natural environmental limitations to food production. Intensive agricultural practices, such as raised-field construction and terracing, were widely used by pre-Columbian civilizations. Remnant raised fields have been found along waterways and in wetlands of the Maya lowlands (Adams, 1980; Adams et al., 1981; Siemens and Puleston, 1972; Matheny, 1976) and in the river drainages and pampas of the Titicaca basin (Kolata, 1991). The ancient agricultural strategy is documented archaeologically throughout the tropics (Denevan, 1970; Denevan and Turner, 1974). In the Maya lowlands, constructed fields in wetlands served to drain land and raise crop

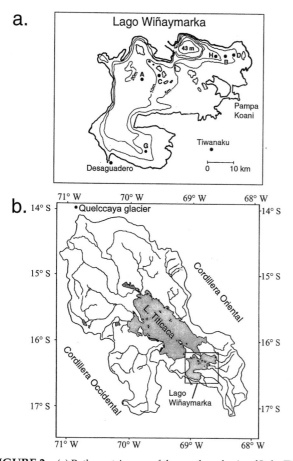

FIGURE 2 (a) Bathymetric map of the southern basin of Lake Titicaca (Lago Wiñaymarka) showing locations of the six sediment cores used to date late Holocene lake level changes. Cores A, B, C, D, G, and H were used to infer lake level shifts over the past ~3500 years. Also indicated are the Río Desaguadero outflow, the archaeological site of Tiwanaku, and the location of the Pampa Koani (see Fig. 8). (b) The Titicaca drainage basin showing the lake and major input rivers (redrawn from Boulangé and Aquize Jaen 1981). High mountain ranges of the Cordillera Oriental and Cordillera Occidental bind the watershed to the east and west, respectively. The Quelccaya ice cap (●) lies at the northwest margin of the drainage.

the day and store heat, protecting crops against nighttime freezes (Kolata and Ortloff, 1989). Cyanobacteria in canals fix atmospheric nitrogen (Biesboer et al., 1999), and the canals retain nutrients and provide fertilizer for N-limited soils (Binford et al., 1996; Carney et al., 1993). Canals between raised fields receive fresh stream water or groundwater from the base of surrounding hills, and low ion concentrations in these waters prevent soil salinization (Sanchez de Lozada, 1996).

Our purpose in revisiting the concept of environmental determinism is not to debate its validity as applied to the Maya and Tiwanaku cultures. Instead, we note that in the Maya and Tiwanaku cases an implicit assumption used to assess natural agricultural potential was indefensible, rendering the discussion moot. Although it is never explicitly stated, evaluations of natural agricultural potential ignored technological innovations and assumed that climate conditions were constant and similar to those of the present. Recent archaeological investigations indicate that many pre-Columbian societies utilized intensive agricultural practices (Denevan, 1970; Pohl, 1990). Paleoclimate studies in the circum-Caribbean (Hodell et al., 1991, 2000; Curtis and Hodell, 1993) and South American Altiplano (Abbott et al., 1997a; Thompson et al., 1985, 1995) provide evidence for late Holocene climate fluctuations. We contend that modern agricultural potential alone cannot establish whether past environments stimulated or constrained cultural development. Instead, we argue that although agricultural potential is, in part, dependent on landscape characteristics such as landforms, soils, and vegetation, crop production varies over time as a function of human social organization, land use practices, and climate change. About 3000 years ago, climate changes in the Maya lowlands and Andean Altiplano generated environmental circumstances that were conducive to human agricultural development. Two millennia later, climatic change leading to droughts exceeded environmental thresholds for Maya and Tiwanaku agricultural sustainability and led to the collapse of both civilizations.

In this chapter, we review paleoclimatological and archaeological results and discuss the correlation between Holocene climate and both Maya and Tiwanaku cultural development and collapse. The two cultural areas were chosen for several reasons. First, they have been subject to intensive archaeological study, so their demographic and agricultural prehistories are well known. Second, both civilizations arose, prospered, and collapsed in low-latitude areas adjacent to lakes that contain paleoclimate archives in their sediments. Third, although the Maya lowlands lie at low elevations (0–300 m above sea level [asl]) and the Tiwanaku

roots above the inundated soil zone (Pohl, 1990). Organic matter that accumulated in canals between raised fields was used to fertilize intensively farmed, nutrient-poor soils. Canals also provided an environment for cultivation of aquatic resources, including edible macrophytes, mollusks, fish, and turtles.

The multiple advantages of raised fields and their contribution to sustainable agricultual production have been demonstrated in the Andean Altiplano by both archaeological excavation (Kolata, 1991) and experimental study of reconstructed raised beds (Erickson, 1988; Kolata et al., 1996). On the pampas, raised fields elevate crop roots above the phreatic zone. Canals between fields absorb solar radiation during

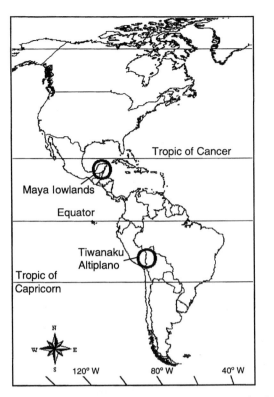

FIGURE 3 Map of the Americas indicating the Maya lowlands and Andean Altiplano where paleoclimate studies were done.

region is located at high altitudes (~3600 to >4000 m asl), both regions are in climatically marginal areas for agriculture, where small changes in available moisture can have profound impacts on crop yields. Finally, the two study areas lie along the Pole-Equator-Pole: Americas (PEP 1) transect (Fig. 3). Combined investigation of the Maya and Tiwanaku cultural areas allows temporal correlation of tropical climate changes north and south of the equator, establishment of interhemispheric climate linkages, and differentiation of forcing factors driving long-term and short-term climate change in the two regions.

6.1.2. The Maya Lowlands

Here, we consider the Maya area of the Yucatán Peninsula encompassed by what is today Belize; the northern department of Petén, Guatemala; and the Mexican states of Campeche, Yucatán, and Quintana Roo (Fig. 1). The region is characterized principally by karst topography that varies in elevation from sea level to ~300 m asl. Annual rainfall across the peninsula is heterogeneous, ranging from <500 mm/year in the extreme northwest to ~2000 mm/year (Fig. 1) in central Petén (Wilson, 1980). The precipitation gradient is reflected in soil development and vegetation (Flores and

Carvajal, 1994). Soils in the flat, northern part of the peninsula are thin, with limestone bedrock exposed at the ground surface. Farther south, in the karst hills of Petén, soils are deeper and more productive. Dry-adapted vegetation in the semiarid northwest is of low stature and diversity. The Petén forest, sometimes referred to as a "quasi-rain forest" (Lundell, 1937), is taller and more diverse.

Rainfall on the Yucatán Peninsula is highly seasonal, with dry conditions from January through April (Deevey et al., 1980). Rainfall is high in May, June, September, and October. A drier interval, referred to as the *canicula,* usually occurs in July or August (Magaña et al., 1999). Interannual variation in the amount and timing of rainfall can be pronounced (Wilson, 1980). Under the influence of the Northeast Trades, summer rains on the peninsula coincide with the northward migration of the Intertropical Convergence Zone (ITCZ) (Hastenrath, 1976, 1984). The Azores–Bermuda high-pressure system also moves northward in summer. Sea surface temperatures (SSTs) are warm in the tropical/subtropical North Atlantic and Caribbean during the Northern Hemisphere summer, providing ample moisture for precipitation. Summertime rainfall is frequently delivered during violent, convective thunderstorms, and total rainfall in any given year can be influenced by tropical storms and hurricanes that contribute substantial precipitation within a short time period.

Dry conditions prevail on the Yucatán Peninsula in the Northern Hemisphere winter, when the ITCZ and Azores–Bermuda high-pressure system move southward (Hastenrath, 1976, 1984). Low winter precipitation is a consequence of relatively low SSTs in the tropical North Atlantic, a steep pressure gradient on the equatorward side of the Azores–Bermuda High, and a strong temperature inversion associated with enhanced trade winds. Differences in precipitation, rather than temperature, define interannual climate variability on the Yucatán Peninsula.

Both pre-Columbian and modern agriculturists on the Yucatán Peninsula have had to contend with strong seasonality and unpredictability of annual precipitation. Although the intraannual pattern of rainfall is known, delayed onset of summer rains or other disruptions of seasonal precipitation can have disastrous consequences for farmers, many of whom still depend on slash-and-burn (swidden) techniques. In rural Yucatán, the contemporary Maya still perform the ancient Cha Cha'ac ceremony to invoke rainfall. Ubiquitous portrayal of the rain god, Chac, on monumental architecture in the northern Maya lowlands (Fig. 4 [see color insert]) demonstrates that the quantity and timing of rainfall were also of great importance to the ancient Maya.

High-resolution Holocene paleoclimate inferences for the Yucatán Peninsula are based on stable isotopic ($\delta^{18}O$) study of multiple ostracod or gastropod shells in samples taken at 1-cm intervals from lake sediment cores. Here, we summarize core results from Lakes Chichancanab (Hodell et al., 1995) and Punta Laguna (Curtis et al., 1996) on the northern part of the peninsula and from Lake Petén-Itzá (Curtis et al., 1998) in Petén, Guatemala (Fig. 1). We are currently studying cores from Lake Salpeten, in the Petén lake district (Fig. 1), to examine the details of climate change in the southern Maya lowlands. The $\delta^{18}O$ of shell carbonate from these lake sediment cores was governed principally by three factors: (1) the $\delta^{18}O$ of the lake water from which the shells were precipitated; (2) taxon-specific fractionation, i.e., *vital effects*; and (3) water temperature. In these stratigraphic studies, we used shells of monospecific, adult snails and ostracods in each core to minimize vital effects. Temperature changes during the Holocene were probably not responsible for the dramatic shifts (>3‰) in shell $\delta^{18}O$, because a 1‰ shift in $\delta^{18}O$ requires a ~4°C change in temperature (Craig, 1965). It is unlikely that long-term temperature fluctuations were solely or largely responsible for the stratigraphic shifts in shell $\delta^{18}O$, as this would require a change in mean water temperature of >12°C. Likewise, intraannual temperature fluctuations probably did not contribute significantly to stratigraphic changes in shell $\delta^{18}O$. Seasonal differences in diurnal water temperatures from these tropical lakes are typically on the order of only 3°–4°C (Covich and Stuiver, 1974; Deevey et al., 1980), and shells collected from 1-cm sample intervals in the cores integrate, on average, material deposited over ~5- to 50-year periods. Although temperature changes may contribute minimally to stratigraphic $\delta^{18}O$ variation, the primary determinant of large shifts in the $\delta^{18}O$ of shell carbonate during the Holocene has been the $\delta^{18}O$ of the lake water from which the carbonate was derived.

In tropical lakes that lack overland outflows, the two major factors that control the $\delta^{18}O$ of lake water over time are the isotopic signature of input waters (rain, runoff, and groundwater) and the relation between evaporation and precipitation (E/P) (Fontes and Gonfiantini, 1967). Rozanski et al. (1993) reported weighted mean isotopic values for several low-elevation, circum-Caribbean sites that were part of the International Atomic Energy Agency (IAEA)/World Meteorological Organization (WMO) precipitation monitoring network. Samples were collected between 1961 and 1987, and in each case, weighted means integrated data obtained over a period of at least 14 years. The isotopic signatures of rainfall from stations at Veracruz, México (−4.13‰, n = 169); Howard Air Force Base, Panama (−5.65‰, n =

165); Barranquilla, Colombia (−5.09‰, n = 85); and Maracay, Venezuela (−4.01‰, n = 64) are similar.

Our mean values for $\delta^{18}O$ in precipitation samples from Petén, Guatemala (−2.86‰) (Rosenmeier et al., 1998) and near Punta Laguna, México (−3.91‰) (Curtis et al., 1996) reflect only a few rainfall events, but are only slightly higher than weighted mean values for the circum-Caribbean reported by Rozanski et al. (1993). Isotopic values for groundwaters in Petén (−3.38‰) and near Punta Laguna (−3.92‰) are similar to their respective rainfall values, suggesting that input waters to Yucatán lakes, whether they are delivered via direct rainfall, runoff, or subterranean infiltration, possess the same isotopic signature.

Lake waters on the Yucatán Peninsula yield isotopic values that are more positive relative to input waters, indicating that ^{18}O is evaporatively concentrated in the water bodies (Covich and Stuiver, 1974; Hodell et al., 1995; Curtis et al., 1996; Rosenmeier et al., 1998). Differences between $\delta^{18}O$ values for precipitation and lake water show enrichment ranging from ~3.6‰ at Lake Petén-Itzá (Curtis et al., 1998) to >7‰ at Lakes Chichancanab (Hodell et al., 1995) and Salpeten (Rosenmeier et al., 1998). Although we cannot demonstrate definitively that the source water $\delta^{18}O$ signature has not changed over the course of the Holocene, we believe that the major process affecting lake water $\delta^{18}O$ on the Yucatán Peninsula for the last ~10,000 years has been shifting E/P. Thus, stratigraphic study of $\delta^{18}O$ in mollusk and crustacean shells in Yucatán sediment cores can be used to reconstruct past changes in moisture availability. Ostracods and snails often occupy different habitats within lakes and display different growth characteristics. Whereas ostracods molt as they pass through instar stages, mollusks simply add shell as they grow. The two taxonomic groups are ecologically and developmentally different, but yield similar stratigraphic $\delta^{18}O$ signals, indicating that the sedimented shell remains faithfully reflect past changes in the oxygen isotope ratio of the lake water.

Shallow lakes (<30 m) and shallow areas of deep lakes on the Yucatán Peninsula were first covered by water and began depositing lacustrine sediments between ca. 8030 and 5840 B.C. (9000–7000 ^{14}C B.P.) (Leyden et al., 1994, 1998). Lakes and sinkholes (cenotes) filled in the early Holocene, after a long, late Pleistocene arid period (Leyden et al., 1994). On the northern part of the peninsula, early Holocene basin filling was a consequence of increasing moisture availability as well as of sea level rise, which raised the local water table. Farther south in the Petén, where the local aquifer lies deep below the land surface (Gill, 2000), increased rainfall at the Pleistocene–Holocene transition filled lakes and supplied moisture for tropical forest synthesis (Leyden, 1984).

Lake Chichancanab, México

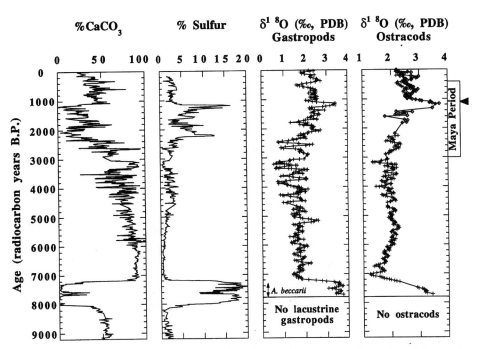

FIGURE 5 High-resolution paleoclimate record from Lake Chichancanab, in the northern Maya lowlands. Percent CaCO$_3$, percent sulfur (gypsum), δ^{18}O of gastropod shells (*Pyrgophorus coronatus*), and δ^{18}O in shells of ostracods *Physocypria* sp. (+) and *Cyprinotus* cf. *salinus* ($\diamond$) at 1-cm intervals, plotted against ^{14}C years B.P. Oxygen isotope results are 3-point running means and are expressed relative to the PDB (PeeDee Belemnite) standard. Arrows between about 7800 and 7300 ^{14}C years B.P. within the gastropod δ^{18}O plot indicate depths at which the foraminifer *Ammonia beccarii* was found. The arrow at the right margin indicates the driest episode of the late Holocene and is dated at 1140 $\pm$ 35 ^{14}C years B.P. (A.D. 922), coinciding with the Classic Maya collapse. (From Hodell, D. A. et al. (1995), with permission.)

The stable isotope (δ^{18}O), geochemical, and microfossil stratigraphies (Fig. 5) of a core from Lake Chichancanab (19°50′ N, 88°45′ W) provide a high-resolution record of changes in E/P over the last ~8200 ^{14}C years (Hodell et al., 1995). Radiocarbon dating of gastropod and ostracod shells from Yucatán lakes is confounded by the effects of hard water lake error (Deevey and Stui-ver, 1964), which makes dates on lacustrine carbonates in Lake Chichancanab ca. 1200 years too old (Hodell et al., 1995). Therefore, the age/depth relation for the core (Fig. 5) was developed by regression using only accelerator mass spectrometer (AMS) ^{14}C dates on terrestrial carbon (Hodell et al., 1995).

Initial filling of Lake Chichancanab ca. 8200 ^{14}C B.P. (7250 B.C.) is marked by gypsum precipitation, relatively high δ^{18}O values for gastropod and ostracod shells, and the presence of the benthic foraminifer *Ammonia beccarii* (Fig. 5). *A. beccarii* can tolerate a wide range of temperatures (10°–35°C) and salinities (7–67 g/L), but is capable of reproducing only at salinities between 13 and 40 g/L (Bradshaw, 1957). Dry conditions in the earliest Holocene are inferred from the biological

and geochemical indicators that point to low lake stage, saline waters, and high E/P. After ~7200 ^{14}C B.P. (6000 B.C.), the lake filled rapidly. Gypsum precipitation was replaced by carbonate deposition, stable isotope values were lower, and *A. beccarii* disappeared from the record, indicating wetter conditions. Relatively moist conditions persisted from ca. 7200 to 3000 ^{14}C B.P. (6000–1250 B.C.) (Fig. 5).

Palynological evidence suggests that swidden activity may have begun in northern Guatemala as early as ~5600 ^{14}C B.P. (4410 B.C.) (Islebe et al., 1996), but this gradual decline of forest in Petén may be attributable to the onset of a regional drying trend. The archaeological record indicates humans did not establish sedentary settlements in the Maya lowlands during the relatively moist early to middle Holocene. Archaeologists date the earliest sedentary populations in the region to the Middle Preclassic period, ca. 1000–300 B.C. (Rice and Rice, 1990; Turner, 1990). Initial settlement in the Maya lowlands corresponds generally to a period of increased regional drying at about 3000 ^{14}C B.P. (1250 B.C.). This pronounced drying trend is reflected in the

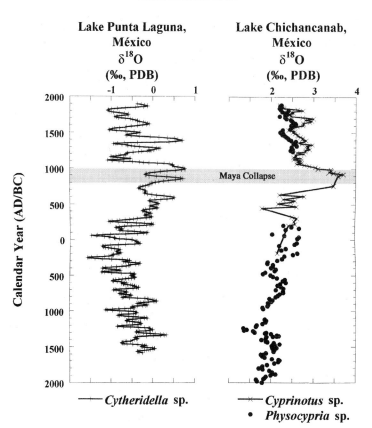

FIGURE 6 Oxygen isotope (δ^{18}O) records based on ostracods from Lakes Punta Laguna and Chichancanab (México), spanning the last 3500 calendar years. Both δ^{18}O records display increasing or relatively high values, i.e., high E/P, within or just following the interval from A.D. 800–1000, corresponding to the period of the Maya collapse.

Chichancanab sediment core by increased sulfur (gypsum) concentrations and higher δ^{18}O values for gastropod and ostracod shells (Fig. 5). Concurrent late Holocene drying has also been documented in Lake Miragoane, Haiti (Hodell et al., 1991), and Lake Valencia in northern Venezuela (Bradbury et al., 1981; Leyden, 1985).

The paleoclimate record from Lake Chichancanab sediments indicates that early Maya agriculturists faced several challenges. Despite the drying trend that began ca. 1250 B.C., general conditions were apparently suitable for shifting agriculture. Nevertheless, in addition to interannual variations in the timing and amount of rainfall, relative E/P on the Yucatán Peninsula between ca. 1250 B.C. and ca. A.D. 1000 varied substantially over decadal to centennial scales. This variability is seen most clearly in the high-resolution, 3500 ^{14}C-year paleoclimate record from Punta Laguna (Fig. 6), which lies about 150 km northeast of Chichancanab at 20°38′ N, 87°37′ W (Fig. 1). Hard water lake dating error at Punta Laguna is also on the order of 1200–1300 years, and the age/depth relation for the core was developed by linear interpolation between five AMS ^{14}C-

dated terrestrial wood samples, assuming linear sedimentation between dated horizons (Curtis et al., 1996). Dates were calibrated by using CALIB 3.0 with a 100-year moving average of the tree-ring data set (Stuiver and Becker, 1993; Stuiver and Reimer, 1993).

In the Chichancanab core, the highest gypsum concentrations and δ^{18}O values of the middle to late Holocene period were found in samples at ~65 cm depth (Fig. 5). An AMS ^{14}C date on a terrestrial seed at 65 cm shows that the protracted dry episode culminated at 1140 ^{14}C B.P. (A.D. 922), coinciding closely with the Classic Maya collapse (Hodell et al., 1995). In the Punta Laguna section, mean δ^{18}O values for the period 1750–940 ^{14}C B.P. (A.D. 300–1100) are higher than mean values for the preceding and following periods (Fig. 6). The δ^{18}O peak at A.D. 862 (1210 ^{14}C B.P.) in the Punta Laguna section may correspond to the Late Classic dry event recorded in the Chichancanab core.

Paleoclimate records from Lakes Chichancanab and Punta Laguna, in combination with the archaeological record, demonstrate a temporal correlation between drought and cultural demise. These water bodies lie in the dry, northern Maya lowlands. This region, however, was apparently least affected by the "collapse" (Lowe, 1985). Demographic consequences of the ninth-century decline were felt most acutely in the wetter, southern lowlands. This pattern may be a consequence of the fact that groundwater is less accessible in the southern lowlands than it is in the north because the depth to the water table increases southward. Therefore, the southern lowlands were more reliant upon surficial water supplies (Gill, 2000). Archaeological survey and test excavation of house mounds in several Petén watersheds yielded estimates of Late Classic Maya population densities in excess of 200 persons per square kilometer (Rice and Rice, 1990). By the Terminal Classic period (A.D. ~900), population densities in nearly all the studied drainage basins had declined to <100 persons per square kilometer, and by the Late Postclassic period (A.D. 1500), many watersheds were abandoned. Drought is implicated in the Classic Maya collapse. Evidence for this climatic change should therefore be found in Petén lake cores.

Recent attention has turned to paleoclimate reconstruction in the Petén lake district (Fig. 1). Lake Petén-Itzá (16°55' N, 89°52' W), the largest Petén water body (area, A = 99.6 km²), yielded a ~9000 ¹⁴C-year paleoenvironmental record based on sediment geochemistry, magnetic susceptibility, pollen, and stable isotopes (Curtis et al., 1998). Chronology for the core was based on AMS ¹⁴C dating of terrestrial wood and charcoal samples (Curtis et al., 1998). Ages between dated horizons were interpolated by assuming constant linear sedimentation. Three $\delta^{18}O$ records from the basin, two based on snail taxa and another on ostracods, display little variability over the past 5000 ¹⁴C years (Fig. 7). Rather than indicating climatic constancy during the past five millennia, however, they may simply reflect the fact that lake water $\delta^{18}O$ in large-volume lakes is insensitive to short-term changes in E/P. In large, deep lakes with long residence times, even protracted droughts may not reduce lake volume sufficiently to alter the lake water isotopic signature.

Lake Salpeten (16°58' N, 89°40' W) lies near Lake Petén-Itzá (Fig. 1), but is smaller (A = 2.6 km²) and more saline (TDS = 4.76 g/L). Lake Salpeten is being studied because its sediments possess abundant ostracod and gastropod shells, it is at saturation with respect to gypsum, and its lakewater $\delta^{18}O$ is demonstrably enriched by >7‰ relative to rainfall $\delta^{18}O$. The relatively greater ¹⁸O enrichment in Lake Salpeten suggests it is more effectively "closed" than Lake Petén-Itzá. Seismic reflection studies in Lake Salpeten were completed in summer 1999. Imaging of the sediment stratigraphy indicates periods of low lake stage during the Holocene, and consequent sedimentation hiatuses at shallow-water locations. Efforts to reconstruct late Holocene cli-

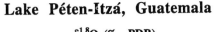

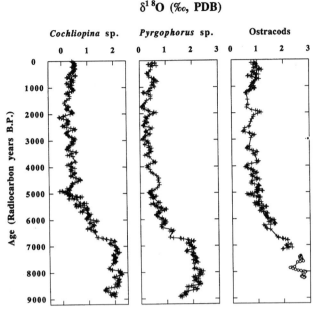

FIGURE 7 Smoothed, 5-point running mean, oxygen isotope ($\delta^{18}O$) records from the Lake Petén-Itzá core, Petén, Guatemala, are plotted against ¹⁴C years B.P., and based on two gastropods (*Cochliopina* sp. and *Pyrgophorus* sp.) and two ostracods (*Cytheridella ilosvayi*, (solid line) and *Candona* sp., (open circles). (From Curtis, J. H., et al. (1998), with permission.)

mate conditions from Lake Salpeten sediments are therefore focusing on cores collected at deep-water sites in the basin.

Lake Salpeten is of particular interest because the history of human occupation in the watershed is well known (Rice and Rice, 1990). A large Maya population occupied the Salpeten drainage basin in the Late Classic Period, and human-mediated vegetation removal in the watershed has been documented palynologically (Leyden, 1987). Massive deforestation can alter watershed hydrology (Bosch and Hewlett, 1982; Stednick, 1996), which may complicate interpretation of isotope records and make it difficult to discern whether drought conditions prevailed in the southern Maya lowlands during the Late Classic.

There is some evidence that dry conditions at the end of the Late Classic period extended beyond the Maya area. Horn and Sanford (1992) reported increased fire frequency in Costa Rica at 1180–1110 ¹⁴C B.P. (A.D. 880–970). Metcalfe et al. (1994) found evidence of a dry episode from 1200 to 1100 ¹⁴C B.P. (A.D. 830–970) in sediments from Lake Pátzcuaro in the central Mexican highlands. A core from La Piscina de Yuriria, Guanajuato, México, also displayed evidence of drying during the period 1500–900 ¹⁴C B.P. (A.D. 570–1070) (Metcalfe et al., 1994), as did the Zacapu basin, ca. 1000 ^{14}C B.P. (A.D. 1020) (Metcalfe, 1995).

6.1.3. The Andean Altiplano

We focus on the region of the Bolivian Altiplano (Fig. 3) that surrounds the southern basin of Lake Titicaca (Lago Wiñaymarka) (16°20′ S, 68°50′ W) (Figs. 2a and 2b) and lies between 3800 and 4200 m asl. Climate conditions in the Lake Titicaca watershed present special challenges for Andean agriculturists. At very high elevations (~5100 m) on peaks at the edge of the Titicaca basin, agriculture is precluded by mean annual temperatures <0°C (Roche et al., 1992). The large area (8560 km²) and volume (~900 × 10⁹ m³) of Lake Titicaca (Wirrmann, 1992) and its relatively warm water temperatures (10°–14°C) have profound thermal effects on the drainage basin, helping maintain mean annual temperatures around the lake between 7° and 10°C (Roche et al., 1992). Nevertheless, the mean minimum monthly temperature near the lake, which occurs in July, is about 1.8°C. Crops are at risk of nighttime freezing during the austral winter.

Long-term, annual rainfall over the entire Titicaca basin averages about 758 mm/year, but is spatially quite variable and influenced by lake effects and orography (Roche et al., 1992). Values range from ~500 mm/year at sites far from the lake to >1000 mm/year over the water body. Precipitation in the Altiplano is derived largely from north-easterly winds that transport moist air from the Amazon basin over the Eastern Cordillera. Rainfall delivery is highly seasonal, with a wet period centered on January and extending from December to March. June marks the middle of the dry season, which runs from May to August. Seasonal rainfall in the Andean Altiplano is out of phase with the timing of precipitation in the Northern Hemisphere Maya lowlands. Intraannual precipitation variability in the Altiplano is best expressed by the percentage of annual rainfall delivered to the Titicaca basin during the 4-month wet season (70%), the 4-month dry season (5%), and the two intermediate 2-month periods (25%) (Roche et al., 1992). Mean monthly evapotranspiration exceeds average rainfall from March to December, leading to soil moisture deficit and salinization (Binford et al., 1997).

Interannual climate change in the Altiplano is expressed as yearly rainfall, which is controlled by the strength of summer monsoon circulation, the position of the ITCZ, and El Niño/Southern Oscillation (ENSO) events. Strong El Niño years are associated with dry conditions in the Altiplano (Roche et al., 1992). Lake Titicaca stage levels have been measured at Puno, Peru, since 1914. Although the relation between lake stage and rainfall is confounded somewhat by inputs of glacial meltwater, the lake level serves as a good proxy for rainfall and illustrates interannual variability in precipitation. The lake outflow sill to the Río Desaguadero is at 3804 m, and the mean lake stage for the period of measurement, 1914–1989, was ~3809 m. The lowest stage over the period of measurement was recorded in 1943 and was 6.37 m below the highest stage, measured in 1986. There was a lake level decline of nearly 5 m between the mid-1930s and mid-1940s (Roche et al., 1992). Such rapid shifts in rainfall and lake level create problems for farmers in the drainage basin. Crops receive insufficient water during very dry periods, but rot or are drowned during wet episodes. Furthermore, extensive, nearshore cultivable flatlands are inundated or exposed by lake level excursions of only a few meters (Fig. 8 [see color insert]). Ion-rich lake waters (conductivity = 1400 µS/cm) retreat during dry periods, leaving soils encrusted with salts.

Rocky, upland soils in the Altiplano are nitrogen deficient (Binford et al., 1996) and prone to erosion on steep slopes. Raised-field canal mucks and intermittently inundated soils have higher total N content (Binford et al., 1996) as a consequence of nitrogen fixation (Biesboer et al., 1999). The construction of raised agricultural fields by the Tiwanaku was a strategy that simultaneously addressed problems of water control, freezes, soil fertility, and salinization. Prehistoric raised fields in the combined Catari basin and Tiwanaku valley covered 130 km² (Kolata and Ortloff, 1996) and bore testimony to the efficacy of this intensive farming practice. Both the widespread use and long persistence (ca. A.D. 600–1100) of this agricultural technique are evidence of its importance.

In the late Pleistocene, 13,000–12,000 ^{14}C B.P. (13,500–12,050 B.C.), Lake Titicaca was deeper and fresher than it is now (P. Baker, personal communication). The extensive paleolake, or Tauca stage, was attributed to a 30–50% increase in precipitation over the Altiplano (Hastenrath and Kutzbach, 1985). At ca. 12,000 ^{14}C B.P., a drying trend began, and water levels in Lake Titicaca consequently fell as much as 85 m (Seltzer et al., 1998). Lago Wiñaymarka essentially desiccated, leaving only scattered, water-filled depressions. The low stage lasted until ca. 3600 ^{14}C B.P. (1930 B.C.) (Mourgiart et al., 1995; Wirrmann et al., 1987, 1992). An increasing E/P trend from the late Pleistocene to the middle Holocene is also inferred from studies of glacial retreat and higher snow lines (Seltzer, 1992), as well as lake sediment records from the Eastern Cordillera (Abbott et al., 1997b). As is predicted by insolation forcing, this long-term history of moisture availability from the Andean Altiplano is out of phase with conditions in the Northern Hemisphere Maya lowlands.

Late Holocene changes in Lake Titicaca's water level were inferred from paleolimnological study of lake sediment cores and serve as proxy estimates of moisture availability in the Altiplano over the last ~3500 ^{14}C years. Cores were taken at six sites throughout Lago Wiñaymarka (Fig. 2). Retrieved lacustrine sediment came from elevations between ~0.5 and 14.5 m below the lake outlet level at 3804 m. Lake stage reconstruction was based on lithostratigraphy of the cores and 60 AMS ^{14}C dates on gastropod shells and sedge (*Schoenoplectus tatora*) achenes found near erosion surfaces (Abbott et al., 1997a). Hard water lake dating error was shown to be about 250 years in Lake Titicaca (Abbott et al., 1997a), and this value was subtracted from ^{14}C ages before dates were calibrated with CALIB 3.0 (Stuiver and Reimer, 1993).

All cores collected from Lago Wiñaymarka displayed laminated lake sediments overlying gray clay soils (Binford et al., 1997). Underlying clay deposits were formed by gleization, implying that the soils developed under anoxic, waterlogged conditions. This conclusion is supported by the presence of littoral *S. tatora* achenes at the clay/organic sediment boundary (Abbott et al., 1997a). At all six coring sites, inception of lacustrine deposition following the protracted low-stage episode dates to between 3560 and 3160 [14]C B.P. (2030–1420 B.C.) (Binford et al., 1997). The timing of this lake level increase is consistent with the earlier findings of Wirrmann et al. (1987). Between the initial late Holocene lake level rise and the current high stand of Lake Titicaca, there have been four dry episodes (Fig. 9) when the lake level declined: 2900–2800, 2400–2200, 2000–1700, and 900–500 cal. B.P. (Abbott et al., 1997a).

In the early and middle Holocene, riparian agriculture in the Altiplano was precluded because of prevailing dry conditions. As the climate ameliorated ca. 1500 B.C. (3210 [14]C B.P.), Chiripa culture emerged in the basin. By ca. 400 B.C., Tiwanaku civilization began to develop, and raised-field cultivation was established by A.D. 600. The agricultural technology was widespread in the region by A.D. 800 and persisted until the Tiwanaku collapse, ca. A.D. 1150. Archaeological excavations in raised-field contexts indicate the technology fell into disuse after A.D. 1150, coincident with drought conditions on the Altiplano (Kolata and Ortloff, 1996) that are documented in sediment records from Lago Wiñaymarka (Abbott et al., 1997a; Binford et al., 1997) and in ice core records from the Quelccaya ice cap in Peru (Thompson et al., 1985).

During the late Tiwanaku IV and Tiwanaku V periods (A.D. 600–1100), dense population centers in the Catari basin and Tiwanaku valley were established near raised-field complexes. Following the state collapse, small populations were dispersed on the landscape and had no apparent relation with raised fields (Kolata, 1993). Populations declined as a consequence of the agricultural collapse (Ortloff and Kolata, 1993), and thereafter, Tiwanaku inhabitants, and later Inca populations, relied on terracing and flatland cultivation.

The protracted dry conditions in the Altiplano lasted more than a century and had negative impacts on raised-field agriculture. Low direct rainfall on crop planting surfaces reduced soil moisture beyond the permanent wilting point. Lack of groundwater recharge reduced or eliminated water flow to input springs and streams that fed canals between raised fields, exacerbating the dry conditions faced by crops. In addition, canal drying removed the benefits of freeze protection and nitrogen fixation. Without continuous flushing, fields were also prone to soil salinization.

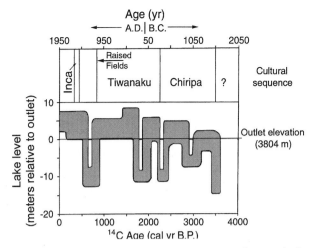

FIGURE 9 Water level record for Lago Wiñaymarka over the last ~3600 calendar years plotted relative to the outlet elevation at the Río Desaguadero (3804 m). The cultural sequence is plotted on a B.C./A.D. scale to show the correlation with the lake-level curve. Tiwanaku raised fields were widely utilized from about A.D. 600–1150. Fields fell into disuse and populations declined coincident with the onset of the last low-water-level event in the record. (Modified from Abbott et al., 1997.)

6.1.4. Climate Forcing in the Circum-Caribbean

Modern studies of interannual rainfall variability in the tropical Atlantic demonstrate that wet years are associated with an enhancement of the annual cycle, driven by greater than normal summertime northward

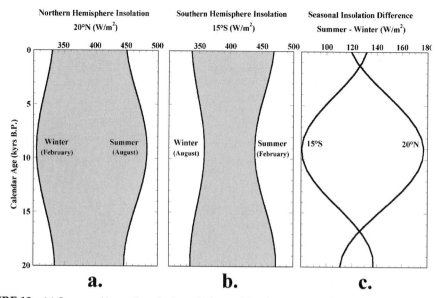

FIGURE 10   (a) Summer (August) and winter (February) insolation curves for the Maya lowlands (20°N) over the last 20,000 calendar years. (b) Summer (February) and winter (August) insolation curves for the Tiwanaku area (15°S) over the last 20,000 years. Values are from Berger and Loutre (1991). (c) Seasonal insolation difference (summer minus winter) at 20°N and 15°S, illustrating that long-term trends in seasonal insolation between the Northern Hemisphere and Southern Hemisphere tropics were out of phase. Since the end of the Pleistocene, the greatest seasonality at 20°N occurred in the early Holocene, whereas the greatest seasonality at 15°S occurred in recent millennia.

migration of the ITCZ and Azores–Bermuda High (Hastenrath, 1984). As a result, wet years occur when sea level pressure (SLP) is low on the equatorward side of the Azores–Bermuda High, the trade winds are displaced poleward (northward), SSTs between 10° and 20°N are warmer, and convergence and cloudiness are enhanced (Hastenrath, 1984). Dry years occur when there is a reduction in the annual cycle, associated with conditions opposite to those described previously (Hastenrath, 1984). Hodell et al. (1991) suggested that long-term E/P changes recorded in the stable isotope record from Lake Miragoane, Haiti, were driven by orbitally forced (Milankovitch) variations in solar insolation, which in turn controlled the intensity of the annual cycle. Long-term changes in seasonal insolation (Fig. 10) may explain the general pattern of a dry late Pleistocene followed by a moister early Holocene in the circum-Caribbean. The dry late Pleistocene–moister early Holocene transition has been documented by paleolimnological studies at sites around the Gulf of México and Caribbean, including Haiti (Hodell et al., 1991); Yucatán, México (Hodell et al., 1995; Whitmore et al., 1996); Guatemala (Leyden et al., 1994); northern Venezuela (Bradbury et al., 1981; Leyden, 1985); and Florida (Watts and Hansen, 1994). Orbital forcing, however, does not fully account for the magnitude of aridity inferred for the late Pleistocene (Hodell et al., 1991), nor the abrupt onset of moister conditions in the early Holocene (e.g., Leyden et al., 1994).

Milankovitch forcing is also thought to be responsible for reduced intensity of the annual cycle and the consequent drying trend in the late Holocene (Fig. 10). Empirical evidence for late Holocene drying around the Caribbean is reported from Lakes Miragoane, Haiti (Hodell et al., 1991); Chichancanab, Yucatán Peninsula, México (Hodell et al., 1995); and Valencia, northern Venezuela (Bradbury et al., 1981). The gradual reduction in the intensity of the annual cycle, however, cannot explain the rapid onset of drier conditions at ca. 3400–3000 [14]C B.P., documented by the δ^{18}O records from Lakes Miragoane (Hodell et al., 1991) and Chichancanab (Hodell et al., 1995). Also, orbital forcing cannot explain the dramatic decadal to centennial E/P fluctuations that are so apparent in the late Holocene portion of the paleoclimate records from Lakes Miragoane, Chichancanab, and Punta Laguna (Curtis et al., 1996). Other, as yet unexplained, forcing factors are responsible for these shorter-term excursions in moisture availability that had such devastating consequences for Maya agriculturists in the ninth century A.D.

6.1.5. Climate Forcing in the Andean Altiplano

Since the late Pleistocene, long-term changes in Lake Titicaca's water level have been driven by insolation forcing that influenced both annual rainfall and glacial advance and retreat. Wet conditions on the northern Altiplano are associated with the Bolivian High, which develops as a consequence of convective precipitation during the Southern Hemisphere summer (Aceituno and Montecinos, 1993; Lenters and Cook, 1997). Perihelion occurred during the Southern Hemisphere winter (July) at 8450 [14]C B.P. (7500 B.C.), and seasonality was reduced relative to the present (Berger, 1988; Kutzbach and Guetter, 1986). Abbott et al. (1997b) argued that reduced summer insolation and increased winter insolation at 8450 [14]C B.P. would have caused cooler summers and warmer winters, decreased moisture transport over the continent, and a net increase in E/P, with the latter caused largely by a decrease in precipitation. Low summertime insolation over the Altiplano lasted from ca. 10,000–7600 [14]C B.P. (9160–6420 B.C.) (Fig. 10). Likewise, the highest June–August insolation at the same latitude occurred from ca. 11,000–6600 [14]C B.P. (11,000–5500 B.C.), probably causing increased ablation of glaciers in the winter dry season. The combined effects of reduced summertime precipitation and increased disappearance of glacial ice in winter contributed to deglaciation and to the mid-Holocene dry episode on the Altiplano (Abbott et al., 1997b).

For the past ~8000 calendar years, austral winter insolation on the Altiplano has decreased and summer insolation has increased (Fig. 10). Higher summertime insolation was accompanied by greater precipitation, which could account for the filling of Lago Wiñaymarka ca. 3600 [14]C B.P. Orbital forcing, however, does not explain the abrupt shifts in water balance that the lake has experienced over the last three millennia. The lake level has been shown to be correlated with rainfall, which tends to be lower during El Niño events (Roche et al., 1992; Binford and Kolata, 1996). Protracted periods of strong El Niño activity, which established its modern periodicity ca. 5000 cal. B.P. (Rodbell et al., 1999), might have been responsible for the documented low stands.

6.1.6. Interhemispheric Correlations

Since the late Pleistocene, long-term patterns of moisture availability in the circum-Caribbean and Andean Altiplano have been out of phase, i.e., negatively correlated. Low-elevation, circum-Caribbean sites were characterized by a dry late glacial period, a moist early and middle Holocene, and a general drying trend over the last ~3000 years. In contrast, the Andean Altiplano was relatively wet during the late Pleistocene period, Tauca stage. Lago Wiñaymarka desiccated when the Altiplano became drier in the early and mid-Holocene. Moist conditions may have returned as early as 4500 [14]C B.P. (Baucom and Rigsby, 1999). Insolation is arguably the driving force behind the general trends, assuming that periods of greater summer insolation and increased seasonality were accompanied by higher rainfall in each of the respective hemispheres. Nonetheless, short-term, secular climatic variations in the two localities north and south of the equator may have been influenced by geographic characteristics or processes such as forest cover, orography, and glacial melting.

Sedentary agriculture developed in the two cultural regions of interest ca. 3000 years ago, when the climate became drier in the circum-Caribbean and wetter in the Altiplano. Although agricultural sedentism may not have been a direct response to climate change, environmental conditions together with agricultural innovations enabled population growth and cultural development in both areas. In the centuries that followed, raised-field technology and terracing in the Altiplano increased crop yields and permitted sustainable farming of planting surfaces. In the Maya lowlands, intensive practices such as raised-field construction, terracing, and perhaps arboriculture (Puleston, 1978) supplemented yields from slash-and-burn activities. In both regions, agriculture was probably under state-level control because intensive techniques required infrastructure and organized labor. Innovative technologies greatly enhanced the natural agricultural potential of the regions. These technologies, in combination with storage and long-distance trade, probably protected human populations against minor disruptions in agricultural output.

6.1.7. Abrupt Climate Change and Cultural Response

Abrupt, but persistent droughts are implicated in the collapse of the Maya (Hodell et al., 1995; Gill, 2000) and Tiwanaku cultures (Binford et al., 1997). General temporal correlations between late Holocene dry episodes in the two cultural areas have been commented upon previously. Curtis et al. (1996) noted that several dry periods between A.D. ~600 and 1400, identified isotopically in the Punta Laguna sediment core (Fig. 6), correlate with periods of high, large (>0.63 μm) microparticle (dust) concentrations in the Quelccaya, Peru, ice core. High concentrations of microparticles in the ice core are attributed to aeolian transport of soils that were exposed during initial raised-field construction (Thompson et al., 1988) as well as long-distance transport of soil and desiccated lake sediment during dry events.

Chepstow-Lusty et al. (1996) noted that palynological and sedimentological shifts in a core from Lake Marcacocha in the central Peruvian Andes (13°3′ S, 72°12′ W) are correlated with Peruvian ice core records from Quelccaya and Huascarán (Thompson et al., 1988, 1995). This suggests that these climatic shifts were regionally significant. Influx of silt and charcoal to Lake Marcacocha between A.D. 600 and 700 may correlate with a major dust event recorded at A.D. 620 in the 1500-year Quelccaya ice core record. The Quelccaya ice core displays a second dust event at A.D. 920, which marks the inception of drier, warmer conditions that prevailed from A.D. 1000–1400, during the Medieval Warm Period (MWP) (Lamb, 1982). This dry episode is correlated with the decline of Tiwanaku agriculture (Ortloff and Kolata, 1993; Binford et al., 1997). At Marcacocha, this climatic episode is recorded by the establishment of *Alnus* after A.D. 1040. A dust event in the 15,000-year Huascarán record occurred 2000 years ago and is correlated with a silty, charcoal-rich layer in the Marcacocha profile that is attributed to the effects of flooding and riparian agriculture. At both Quelccaya and Huascarán, relatively depleted $\delta^{18}O$ values reflect the *Little Ice Age* (LIA) (A.D. 1490–1900). This cool period is marked at Marcacocha by the *Alnus* decline. The LIA is also reflected by relatively depleted $\delta^{18}O$ values between ca. A.D. 1400 and 1800 measured on ostracod (*Limnocythere*) shells in core D (Fig. 2a) from Lago Wiñaymarka (Binford et al., 1996).

Chepstow-Lusty et al. (1996) also point out that changes in the Marcacocha core correlate with events recorded in the profile from Lake Chichancanab, México. During the Maya Late Preclassic period (300 B.C.–A.D. 250), a dry episode is reflected in the Chichancanab core at about A.D. 1 by high gypsum concentrations, relatively enriched $\delta^{18}O$ values, and a change in ostracod taxa (Fig. 5). This dry episode in Yucatán is contemporaneous with the deposition of a silty, charcoal-rich layer in Marcacocha, above which cold-adapted *Plantago* spp. become established. The ninth-century dry episode on the Yucatán Peninsula may be linked to warming in Peru that is reflected by the somewhat delayed expansion of *Alnus*, ca. A.D. 1040. If these climatic events at distant sites on both sides of the equator prove to be synchronous, it would suggest global-scale disruption of atmospheric and oceanic fields.

Temporal correlation between climatic drying and cultural collapses at tropical sites north and south of the equator raises fascinating questions for paleoclimatologists and anthropologists alike. The data indicate that climatic thresholds for agricultural production can be exceeded, with disastrous consequences for human populations. Recent paleoenvironmental studies indicate that climate *surprises* during the Holocene, i.e., significant decadal to centennial climate variability, were more common than was previously believed (Overpeck, 1996). Paleoclimatologists must date late Holocene climate changes accurately, evaluate their causes, assess the geographic extent of areas affected, determine the quantitative reduction in moisture availability that occurred during drought events, and investigate the role of human activity (e.g., deforestation) in compounding the effects of climate change. Social scientists must study factors that make cultures resistant or vulnerable to environmental perturbations such as climate changes. Sheets (2000) notes that simple societies tend to recover from environmental stresses such as explosive volcanism more readily than complex societies. Complex societies, with their state-run trade, agriculture, economies, and other facilities, are unable to cope with sudden, unanticipated environmental perturbations. The Maya and Tiwanaku civilizations instituted intensive agricultural methods that overcame natural limits to food production. These technologies increased yields, promoted sustainability, permitted human populations to reach high densities, and created a vast state-level infrastructure. Thus, intensive agricultural methods may have ultimately contributed to the cultural collapses that occurred as a consequence of sudden, unpredictable climate changes.

Paleoclimatic and archaeological information from the Maya lowlands and Bolivian/Peruvian Altiplano sheds new light on the concept of environmental determinism. Moisture availability in the climatically marginal Maya and Tiwanaku regions fluctuated appreciably over time, illustrating that past regional agricultural potential cannot be assessed by using modern climate variables, but rather should be based on historic climate conditions inferred from paleoclimate proxies. The decadal to centennial E/P variability revealed by the paleoclimate record must have presented serious challenges for pre-Columbian agriculturists. Cultural continuity in light of documented climatic variability provides evidence for both the resilience of the Maya and Tiwanaku food production systems and the ingenuity that went into their development. The multidisciplinary approach to investigating human–environment interactions permitted us to demonstrate a strong temporal correlation between protracted drought and cultural collapse in the high-altitude Tiwanaku region and the Maya lowlands. Drought was a major stressor in both regions, which contributed to agricultural and subsequent cultural declines. The findings suggest that cultural development and survival are profoundly influenced by environmental conditions.

Acknowledgments

This work was supported by National Oceanic and Atmospheric Administration (NOAA) grants NA36GP0304, NA56GP0370, and GC-95-174; National Science Foundation (NSF) grants ATM-9709314, DEB-9207878, DEB-9212641, and BNS-8805490; National Endowment for the Humanities (NEH) grant RO-21806-88; and a grant from the National Geographic Society. We thank G. Seltzer for discussions, and Sarah Metcalfe and Paul Baker for their constructive and helpful reviews of the manuscript. Finally, we are most grateful to Vera Markgraf for organizing the excellent PEP 1 meeting in Mérida, Venezuela, and for inviting us to contribute this chapter.

References

Abbott, M. B., M. W. Binford, M. Brenner, and K. R. Kelts, 1997a: A 3500 ^{14}C yr high-resolution record of water-level changes in Lake Titicaca, Bolivia/Peru. *Quaternary Research,* **47**: 169–180.

Abbott, M. B., G. O. Seltzer, K. R. Kelts, and J. Southon, 1997b: Holocene paleohydrology of the tropical Andes from lake records. *Quaternary Research,* **47**: 70–80.

Aceituno, P., and A. Montecinos, 1993: Circulation anomalies associated with dry and wet periods in the South American Altiplano. Fourth International Conference on the Southern Hemisphere Meteorology and Oceanography, American Meteorological Society, pp. 330–331.

Adams, R. E. W., 1973: The collapse of Maya civilization: A review of previous theories. *In* Culbert, T. P. (ed.), *The Classic Maya Collapse.* Albuquerque: University of New Mexico Press, pp. 21–34.

Adams, R. E. W., 1980: Swamps, canals, and the locations of ancient Maya cities. *Antiquity,* **54**: 206–214.

Adams, R. E. W., W. E. Brown, Jr., and T. P. Culbert, 1981: Radar mapping, archeology, and ancient Maya land use. *Science,* **213**: 1457–1463.

Altschuler, M., 1958: On the environmental limitations of Mayan cultural development. *Southwestern Journal of Anthropology,* **14**: 189–198.

Baucom, P. C., and C. A. Rigsby, 1999: Climate and lake-level history of the northern Altiplano, Bolivia, as recorded in Holocene sediments of the Río Desaguadero. *Journal of Sedimentary Research,* **69**: 597–611.

Berger, A., 1988: Milankovitch theory and climate. *Reviews of Geophysics,* **26**: 624–657.

Berger, A., and M. F. Loutre, 1991: Insolation values for the climate of the last 10 million years. *Quaternary Science Reviews,* **10**: 297–317.

Biesboer, D. D., M. W. Binford, and A. L. Kolata, 1999: Nitrogen fixation in soils and canals of rehabilitated raised-fields of the Bolivian Altiplano. *Biotropica,* **31**: 255–267.

Binford, M. W., and A. L. Kolata, 1996: The natural and human setting. *In* Kolata, A. L. (ed.), *Tiwanaku and Its Hinterland: Archaeological and Paleoecological Investigations in the Lake Titicaca Basin of Bolivia.* Washington DC: Smithsonian Institution Press, pp. 23–56.

Binford, M. W., M. Brenner, and B. W. Leyden, 1996: Paleoecology and Tiwanaku ecosystems. *In* Kolata, A. L. (ed.), *Tiwanaku and Its Hinterland: Archaeological and Paleoecological Investigations in the Lake Titicaca Basin of Bolivia.* Washington DC: Smithsonian Institution Press, pp. 89–108.

Binford, M. W., A. L. Kolata, M. Brenner, J. Janusek, M. T. Seddon, M. B. Abbott, and J. H. Curtis, 1997: Climate variation and the rise and fall of an Andean civilization. *Quaternary Research,* **47**: 235–248.

Bosch, J. M., and J. D. Hewlett, 1982: A review of catchment experiments to determine the effect of vegetation changes on water yield and evapotranspiration. *Journal of Hydrology,* **55**: 3–23.

Boulangé, B., and E. Aquize Jaen, 1981: Morphologie, hydrographie et climatologie du lac Titicaca et de son bassin versant. *Revue D'Hydrobiologie Tropicale,* **14**: 269–287.

Bradbury, J. P., B. W. Leyden, M. Salgado-Labouriau, W. M. Lewis, Jr., C. Schubert, M. W. Binford, D. G. Frey, D. R. Whitehead, and F. H. Weibezahn, 1981: Late Quaternary environmental history of Lake Valencia, Venezuela. *Science,* **214**: 1299–1305.

Bradshaw, J. S., 1957: Laboratory studies of the rate of growth of the foraminifer *Streblus beccarii* (Linné) var. *tepida* (Cushman). *Journal of Paleontology,* **31**: 1138–1147.

Carney, H. J., M. W. Binford, A. L. Kolata, R. R. Marin, and C. R. Goldman, 1993: Nutrient and sediment retention in Andean raised-field agriculture. *Nature,* **364**: 131–133.

Chepstow-Lusty, A. J., K. D. Bennett, V. R. Switsur, and A. Kendall, 1996: 4000 years of human impact and vegetation change in the central Peruvian Andes—with events parallelling the Maya record? *Antiquity,* **70**: 824–833.

Coe, W. R., 1957: Environmental limitation on Maya culture: A re-examination. *American Anthropologist,* **59**: 328–335.

Covich, A. P., and M. Stuiver, 1974: Changes in oxygen 18 as a measure of long-term fluctuations in tropical lake levels and molluscan populations. *Limnology and Oceanography,* **19**: 682–691.

Cowgill, U. M., and G. E. Hutchinson, 1963: Sex-ratio in childhood and the depopulation of the Petén, Guatemala. *Human Biology,* **35**: 90–103.

Craig, H., 1965: The measurement of oxygen isotope paleotemperatures. *In* Tongiorgi, E. (ed.), *Stable Isotopes in Oceanographic Studies and Paleotemperatures.* Consiglio Nazionale delle Richerche, Laboratorio di Geologia Nucleare, Pisa, pp. 161–182.

Curtis, J. H., and D. A. Hodell, 1993: An isotopic and trace element study of ostracods from Lake Miragoane, Haiti: A 10,500 year record of paleosalinity and paleotemperature changes in the Caribbean. *In* Swart, P. K., K. C. Lohmann, J. McKenzie, and S. Savin (eds.), *Climate Change in Continental Isotopic Records.* Washington, DC: American Geophysical Union, Geophysical Monograph 78, pp. 135–152.

Curtis, J. H., M. Brenner, D. A. Hodell, R. A. Balser, G. A. Islebe, and H. Hooghiemstra, 1998: A multi-proxy study of Holocene environmental change in the Maya lowlands of Petén, Guatemala. *Journal of Paleolimnology,* **19**: 139–159.

Curtis, J. H., D. A. Hodell, and M. Brenner, 1996: Climate variability on the Yucatán Peninsula (México) during the past 3500 years, and implications for Maya cultural evolution. *Quaternary Research,* **46**: 37–47.

Dahlin, B. H., 1983: Climate and prehistory on the Yucatán Peninsula. *Climatic Change,* **5**: 245–263.

Deevey, E. S., and M. Stuiver, 1964: Distribution of natural isotopes of carbon in Linsley Pond and other New England lakes. *Limnology and Oceanography,* **9**: 1–11.

Deevey, E. S., M. Brenner, M. S. Flannery, and G. H. Yezdani, 1980: Lakes Yaxha and Sacnab, Petén, Guatemala: Limnology and hydrology. *Archiv für Hydrobiologie,* **Suppl. 57**: 419–460.

Deevey, E. S., D. S. Rice, P. M. Rice, H. H. Vaughan, M. Brenner, and M. S. Flannery, 1979: Mayan urbanism: Impact on a tropical karst environment. *Science,* **206**: 298–306.

Denevan, W. L., 1970: Aboriginal drained field cultivation in the Americas. *Science,* **169**: 647–654.

Denevan, W. L., and B. L. Turner, II, 1974: Forms, functions, and associations of raised fields in the Old World tropics. *Journal of Tropical Geography,* **39**: 24–33.

Erickson, C. L., 1988: Raised field agriculture in the Lake Titicaca basin: Putting ancient agriculture back to work. *Expedition,* **30**: 8–16.

Ferdon, E. N., Jr., 1959: Agricultural potential and the development of cultures. *Southwestern Journal of Anthropology,* **15**: 1–19.

Flores, J. S., and I. E. Carvajal, 1994: Tipos de vegetación de la Peninsula Yucatán. Etnoflora Yucatanense, Fascículo 3. Universidad Autónoma de Yucatán, Mérida, 135 pp.

Folan, W. J., J. Gunn, J. D. Eaton, and R. W. Patch, 1983: Paleoclimatological patterning in southern Mesoamerica. *Journal of Field Archaeology*, 10: 454–468.

Fontes, J. C., and R. Gonfiantini, 1967: Comportement isotopique au cours de l'evapoation de deux bassins sahariens. *Earth and Planetary Science Letters*, 3: 258–266.

Ford, A., and W. I. Rose, 1995: Volcanic ash in ancient Maya ceramics of the limestone lowlands: Implications for prehistoric volcanic activity in the Guatemala highlands. *Journal of Volcanology and Geothermal Research*, 66: 149–162.

Gill, R. B., 2000: The Great Maya Droughts: Water, Life, and Death. Albuquerque: University of New Mexico Press.

Gore, A., 1992: *Earth in the Balance: Ecology and the Human Spirit*. Boston: Houghton Mifflin.

Gunn, J., and R. E. W. Adams, 1981: Climatic change, culture, and civilization in North America. *World Archaeology*, 13: 87–100.

Hammond, N., S. Donaghey, R. Berger, S. de Atley, V. R. Switsur, and A. P. Ward, 1977: Maya Formative phase radiocarbon dates from Belize. *Nature*, 260: 579–581.

Hammond, N., D. Pring, R. Wilk, S. Donaghey, F. P. Saul, E. S. Wing, A. V. Miller, and L. H. Feldman, 1979: The earliest lowland Maya? Definition of the Swasey phase. *American Antiquity*, 44: 92–110.

Hastenrath, S., 1976: Variations in low-latitude circulation and extreme climatic events in the tropical Americas. *Journal of Atmospheric Science*, 33: 202–215.

Hastenrath, S., 1984: Interannual variability and the annual cycle: Mechanisms of circulation and climate in the tropical Atlantic sector. *Monthly Weather Review*, 112: 1097–1107.

Hastenrath, S., and J. Kutzbach, 1985: Late Pleistocene climate and water budget of the South American Altiplano. *Quaternary Research*, 24: 249–256.

Hodell, D. A., M. Brenner, and J. H. Curtis, 2000: Climate change in the northern American tropics and subtropics since the last ice age: Implications for environment and culture. In Lentz, D. (ed.), Imperfect Balance: Landscape Transformations in the Pre-Columbian Americas. New York, Columbia University Press, pp. 13–37.

Hodell, D. A., J. H. Curtis, and M. Brenner, 1995: Possible role of climate in the collapse of Classic Maya civilization. *Nature*, 375: 391–394.

Hodell, D. A., J. H. Curtis, G. A. Jones, A. Higuera-Gundy, M. Brenner, M. W. Binford, and K. T. Dorsey, 1991: Reconstruction of Caribbean climate change over the past 10,500 years. *Nature*, 352: 790–793.

Horn, S. P., and R. L. Sanford, Jr., 1992: Holocene fires in Costa Rica. *Biotropica*, 24: 354–361.

Islebe, G., H. Hooghiemstra, M. Brenner, J. H. Curtis, and D. A. Hodell, 1996: A Holocene vegetation history from lowland Guatemala. *The Holocene*, 6: 265–271.

Kolata, A. L., 1991: The technology and organization of agricultural production in the Tiwanaku state. *Latin American Antiquity*, 2: 99–125.

Kolata, A. L., 1993: *The Tiwanaku: Portrait of an Andean Civilization*. Cambridge, MA: Blackwell Sci.

Kolata, A. L., and C. Ortloff, 1989: Thermal analysis of Tiwanaku raised field systems in the Lake Titicaca basin of Bolivia. *Journal of Archaeological Science*, 16: 233–263.

Kolata, A. L., and C. Ortloff, 1996: Tiwanaku raised-field agriculture in the Lake Titicaca basin of Bolivia. *In* Kolata, A. L. (ed.), *Tiwanaku and Its Hinterland: Archaeological and Paleoecological Investigations in the Lake Titicaca Basin of Bolivia*. Washington DC: Smithsonian Institution Press, pp. 109–151.

Kolata, A. L., O. Rivera, J. C. Ramírez, and E. Gemio, 1996: Rehabilitating raised-field agriculture in the southern Lake Titicaca basin of Bolivia: Theory, practice, and results. *In* Kolata, A. L. (ed.), *Tiwanaku and Its Hinterland: Archaeological and Paleoecological Investigations in the Lake Titicaca Basin of Bolivia*. Washington, DC: Smithsonian Institution Press, pp. 203–230.

Kutzbach, J. E., and P. J. Guetter, 1986: The influence of changing orbital parameters and surface boundary conditions on climate simulations for the past 18,000 years. *Journal of Atmospheric Science*, 43: 1726–1759.

Lamb, H., 1982: *Climate History and the Modern World*. London: Methuen.

Lenters, J., and K. Cook, 1997: On the origin of the Bolivian High and related circulation features of the South American climate. *Journal of Atmospheric Sciences*, 54: 656–677.

Leyden, B. W., 1984: Guatemalan forest synthesis after Pleistocene aridity. *Proceedings of the National Academy of Sciences (U.S.A.)*, 81: 4856–4859.

Leyden, B. W., 1985: Late Quaternary aridity and Holocene moisture fluctuations in the Lake Valencia basin, Venezuela. *Ecology*, 66: 1279–1295.

Leyden, B. W., 1987: Man and climate in the Maya lowlands. *Quaternary Research*, 28: 407–414.

Leyden, B. W., M. Brenner, and B. H. Dahlin, 1998: Cultural and climatic history of Cobá, a lowland Maya city in Quintana Roo, México. *Quaternary Research*, 49: 111–122.

Leyden, B. W., M. Brenner, D. A. Hodell, and J. H. Curtis, 1994: Orbital and internal forcing of climate on the Yucatán Peninsula for the past ca. 36 ka. *Palaeogeography, Palaeoclimatology, Palaeoecology*, 109: 193–210.

Lowe, J. W. G., 1985: *The Dynamics of the Apocalypse*. Albuquerque: University of New Mexico Press.

Lundell, C. L., 1937: *The Vegetation of Petén*. Washington, DC: Carnegie Institution.

Magaña, V., J. A. Amador, and S. Medina, 1999: The midsummer drought over Mexico and Central America. *Journal of Climate*, 12: 1577–1588.

Matheny, R. T., 1976: Maya lowland hydraulic systems. *Science*, 193: 639–646.

Meggers, B. J., 1954: Environmental limitation on the development of culture. *American Anthropologist*, 56: 801–824.

Messenger, L. C., Jr., 1990: Ancient winds of change: Climatic settings and prehistoric social complexity in Mesoamerica. *Ancient Mesoamerica*, 1: 21–40.

Metcalfe, S. E., 1995: Holocene environmental change in the Zacapu basin, México: A diatom-based record. *The Holocene*, 5: 196–208.

Metcalfe, S. E., F. A. Street-Perrott, S. L. O'Hara, P. E. Hales, and R. A. Perrott, 1994: The palaeolimnological record of environmental change: Examples from the arid frontier of Mesoamerica. *In* Millington, A. C., and K. Pye (eds.), *Environmental Change in Drylands: Biogeographical and Geomorphological Perspectives*. Chichester, United Kingdom: Wiley, pp. 131–145.

Mourgiart, P., J. Argollo, and D. Wirrmann, 1995: Evolución paleohidrológica de la cuenca del Lago Titicaca durante el Holoceno. *Bulletin de l'Institut Francais d'Etudes Andines*, 24: 573–583.

Ortloff, C. R., and A. L. Kolata, 1993: Climate and collapse: Agro-ecological perspectives on the decline of the Tiwanaku state. *Journal of Archaeological Science*, 20: 195–221.

Overpeck, J. T., 1996: Warm climate surprises. *Nature*, 271: 1820–1821.

Paine, R. R., and A. C. Freter, 1996: Environmental degradation and the Classic Maya collapse at Copan, Honduras (A.D. 600–1250): Evidence from studies of household survival. *Ancient Mesoamerica*, 7: 37–47.

Pohl, M. D. (ed.), 1990: The Rio Hondo Project in northern Belize. *In* Pohl, M. D. (ed.), *Ancient Maya Wetland Agriculture: Excavations on*

Albion Island, Northern Belize. Boulder, CO: Westview Press, pp. 1–19.

Puleston, D. E., 1978: Terracing, raised field, and tree cropping in the Maya lowlands: A new perspective on the geography of power. *In* Harrison, P. D., and B. L. Turner, II (eds.), *Pre-Hispanic Maya Agriculture.* Albuquerque: University of New Mexico Press, pp. 225–245.

Rice, D. S., 1978: Population growth and subsistence alternatives in a tropical lacustrine environment. *In* Harrison, P. D., and B. L. Turner, II (eds.), *Pre-Hispanic Maya Agriculture.* Albuquerque: University of New Mexico Press, pp. 35–61.

Rice, D. S., and P. M. Rice, 1990: Population size and population change in the central Petén Lake Region, Guatemala. *In* Culbert, T. P., and D. S. Rice (eds.), *Pre-Columbian Population History in the Maya Lowlands.* Albuquerque: University of New Mexico Press, pp. 123–148.

Roche, M. A., J. Bourges, J. Cortes, and R. Mattos, 1992: Climatology and hydrology of the Lake Titicaca basin. *In* Dejoux, C., and A. Iltis (eds.), *Lake Titicaca: A Synthesis of Limnological Knowledge.* Dordrecht: Kluwer Academic, pp. 63–88.

Rodbell, D. T., G. O. Seltzer, D. M. Anderson, M. B. Abbott, D. B. Enfield, and J. H. Newman, 1999: An ~15,000-year record of El Niño-driven alluviation in southwestern Ecuador. *Science,* **283**: 516–520.

Rosenmeier, M. F., D. A. Hodell, J. B. Martin, M. Brenner, and J. H. Curtis, 1998: Late Holocene climatic variability in Central America: Evidence from lacustrine sediments in the Petén, Guatemala, and implications for Maya cultural evolution. Proceedings of the PEP 1 (Pole-Equator-Pole, Paleoclimate of the Americas) Meeting. Mérida, Venezuela, March 16–20, 1998.

Rozanski, K., L. Araguás-Araguás, and R. Gonfiantini, 1993: Isotopic patterns in modern global precipitation. *In* Swart, P. K., K. C. Lohmann, J. McKenzie, and S. Savin (eds.), *Climate Change in Continental Isotopic Records.* Washington, DC: American Geophysical Union, Geophysical Monograph 78, pp. 1–36.

Sabloff, J. A., 1973: Major themes in the past hypotheses of the Maya collapse. *In* Culbert, T. P. (ed.), *The Classic Maya Collapse.* Albuquerque: University of New Mexico Press, pp. 35–40.

Sanchez de Lozada, D., 1996: Heat and moisture dynamics in raised fields of the Lake Titicaca Region, Bolivia. Ph.D. dissertation. Cornell University, Ithaca, NY.

Saul, F. P., 1973: Disease in the Maya area: The Pre-Columbian evidence. *In* Culbert, T. P. (ed.), *The Classic Maya Collapse.* Albuquerque: University of New Mexico Press, pp. 301–324.

Seltzer, G. O., 1992: Late Quaternary glaciation of the Cordillera Real, Bolivia. *Journal of Quaternary Science,* **7**: 87–98.

Seltzer, G. O., P. Baker, S. Cross, R. Dunbar, and S. Fritz, 1998: High-resolution seismic reflection profiles from Lake Titicaca, Peru-Bolivia: Evidence for Holocene aridity in the tropical Andes. *Geology,* **26**: 167–170.

Sharer, R. J., 1994: *The Ancient Maya,* 5th ed. Stanford, CA: Stanford Univ. Press.

Sheets, P., 2000: The effects of explosive volcanism on simple to complex societies in ancient Middle America. *In* Markgraf, V. (ed.), *Interhemispheric Climate Linkages.* San Diego: Academic Press, Chapter 5.

Siemens, A. H., and D. E. Puleston, 1972: Ridged fields and associated features in southern Campeche: New perspectives on the lowland Maya. *American Antiquity,* **37**: 228–239.

Stednick, J. D., 1996: Monitoring the effects of timber harvest on annual water yield. *Journal of Hydrology,* **176**: 79–95.

Stuiver, M., and B. Becker, 1993: High-precision calibration of the radiocarbon time scale A.D. 1950–6000 B.C. *Radiocarbon,* **35**: 35–65.

Stuiver, M., and P. J. Reimer, 1993: Extended ¹⁴C data base and revised CALIB 3.0 ¹⁴C age calibration program. *Radiocarbon,* **35**: 215–230.

Thompson, L. G., M. E. Davis, E. Mosley-Thompson, and K.-B. Liu, 1988: Pre-Incan agricultural activity recorded in dust layers in two tropical ice cores. *Nature,* **307**: 763–765.

Thompson, L. G., E. Mosley-Thompson, J. F. Bolzan, and B. R. Koci, 1985: A 1500-year record of tropical precipitation in ice cores from the Quelccaya ice cap, Peru. *Science,* **229**: 971–973.

Thompson, L. G., E. Mosley-Thompson, M. E. Davis, P.-N. Lin, K. A. Henderson, J. Cole-Dai, J. F. Bolzan, and K.-B. Liu, 1995: Late glacial stage and Holocene tropical ice core records from Huascarán, Peru. *Science,* **269**: 46–50.

Turner, B. L., II, 1990: Population reconstruction for the central Maya lowlands: 1000 B.C. to A.D. 1500. *In* Culbert, T. P., and D. S. Rice (eds.), *Pre-Columbian Population History in the Maya Lowlands.* Albuquerque: University of New Mexico Press, pp. 301–343.

Watts, W. A., and B. C. S. Hansen, 1994: Pre-Holocene and Holocene pollen records of vegetation history from the Florida peninsula and their climatic implications. *Palaeogeography, Palaeoclimatology, Palaeoecology,* **109**: 163–176.

Whitmore, T. J., M. Brenner, J. H. Curtis, B. H. Dahlin, and B. W. Leyden, 1996: Holocene climatic and human influences on lakes of the Yucatán Peninsula, México: An interdisciplinary, palaeolimnological approach. *The Holocene,* **6**: 273–287.

Wilson, E. M., 1980: Physical geography of the Yucatán Peninsula. *In* Moseley, E. H., and E. D. Terry (eds.), *Yucatán: A World Apart.* Tuscaloosa: The University of Alabama Press, pp. 5–40.

Wirrmann, D., 1992: Morphology and bathymetry. *In* Dejoux, C., and A. Iltis (eds.), *Lake Titicaca: A Synthesis of Limnological Knowledge.* Dordrecht: Kluwer Academic, pp. 16–22.

Wirrmann, D., P. de Oliveira, and L. Almeida, 1987: Low Holocene level (7700 to 3650 years ago) of Lake Titicaca (Bolivia). *Palaeogeography, Palaeoclimatology, Palaeoecology,* **59**: 315–323.

Wirrmann, D., J.-P. Ybert, and P. Mourgiart, 1992: A 20,000 year paleohydrological record from Lake Titicaca. *In* Dejoux, C., and A. Iltis (eds.), *Lake Titicaca: A Synthesis of Limnological Knowledge.* Dordrecht: Kluwer Academic, pp. 40–48.

7

Human Dimensions of Late Pleistocene/Holocene Arid Events in Southern South America

LAUTARO NÚÑEZ, MARTIN GROSJEAN, AND ISABEL CARTAJENA

Abstract

Examples from late glacial and early and mid-Holocene archaeological sites in southern South America provide information about the human dimensions of past climate changes. Particularly in semiarid areas with marginal food and water resources (e.g., central and northern Chile), changes from humid to arid conditions coincided with major changes in human occupational and cultural patterns. The end of the first period of human occupation in central Chile (Paleo-Indian period between ca. 11,400 and 9700 ^{14}C B.P.) coincided with rapidly increasing aridity, when trees were replaced by shrubs and herbs, the Pleistocene paleolakes disappeared, and, finally, the Pleistocene megafauna became extinct. The Paleo-Indian hunters occupied *ecological refuges* around lakes, where the resources were concentrated in an environment under general water stress. A similar process was observed 2000 years later in the Atacama Desert and Altiplano of northern Chile, when the first period of human occupation (Early Archaic, between 10,800 and 8000 ^{14}C B.P.) came to an abrupt end, synchronously with the rapid desiccation of early Holocene paleolakes and the disappearance of abundant water, vegetation, and animal resources.

The subsequent extremely arid mid-Holocene environments resulted in a general hiatus of human occupa-
tion in the Atacama basin, whereas widespread resettlement (Tilocalar phase with agriculture and pastoralism, 3200 ^{14}C B.P.) coincided again with a rise of the paleolake levels to modern levels. Copyright © 2001 by Academic Press.

Resumen

Se presentan sitios arqueológicos contemporáneos con el glacial tardío y holoceno temprano y medio del Cono Sur Americano que proveen información acerca de las relaciones entre ocupaciones humanas y cambios paleoclimáticos. Se revisan particularmente las áreas semiáridas con recursos alimentarios y de agua marginales en el centro-norte de Chile, donde se han identificado cambios de condiciones húmedas a áridas en coincidencia con reorientaciones sustanciales en términos de patrones ocupacionales y culturales. Un resumen del rol de los eventos áridos y su efecto en el poblamiento paleoindio tardío, durante la transición pleistoceno-holoceno es presentado a nivel hemisférico y más específicamente en el centro-norte de Chile (11,400–9700 años B.P.). Se evalúa el desecamiento lacustre y la concentración de megafauna en relación a la presión de cazadores especializados en la explotación oportunística de ecorefugios sometidos a estrés paleoambiental. Un proceso similar se ha identificado 2000 años después en la Puna de Atacama (norte de Chile),

105

contemporáneo con la ocupación arcaica temprana (10,800–8000 B.P.). Durante su fase final se afectó por un cambio abrupto de desecación de los lagos altoandinos alterando los recursos de flora y fauna (transición Holoceno Temprano/Medio). Durante el Holoceno Medio el régimen de aridez fue muy intenso, motivando un *hiatus* ocupacional, con ocupaciones excepcionales en ecorefugios de fauna y predadores humanos. Por los 3200 B.P. se ha constatado una recuperación de régimen de humedad, con el alza de los niveles lacustres y el inicio de labores complejas esta vez agrícolas y pastoralistas bajo condiciones modernas (Fase Tilocalar).

7.1. INTRODUCTION

Understanding the relation between late Quaternary climate changes and hunting/gathering societies is fundamental when social processes, spatial expansion and concentration, and resource use of Paleo-Indians are assessed. In this respect, the Americas offer excellent possibilities for studying the human dimensions of climate change because these areas experienced high-amplitude, rapid climatic changes during late glacial/early and mid-Holocene times, when strategies for adaptation of the Paleo-Indian and Archaic hunters to the highly dynamic environments were still limited. This limitation was particularly the case in environments with marginal food and water resources, for instance, as in central Chile, the Atacama Desert, and the highlands of the south-central Andes. Studies of South and Central America also show that in hunting-gathering societies, major cultural changes coincided with climate variations (Irwin-Williams and Haynes, 1970; Byrne, 1988; McMorriston and Hole, 1988). This cultural change was also the case later for complex societies such as the Tiwanaku culture in the Titicaca basin, which broke down coincidentally with the onset of a prolonged dry period that began at ca. A.D. 1100 (Ortloff and Kolata, 1993; Binford et al., 1997; Brenner et al., 2000). Ice core records (Thompson et al., 1988) suggest that climatic variability, reflected in lake level changes, strongly influenced fluctuations in agricultural activities in southern Peru during the last 1500 years.

For the time at the very end of the Pleistocene, climatic fluctuations that seriously affected fauna and vegetation and, ultimately, human resource use are well documented for south-central Chile (e.g., Heusser, 1983, 1990) and Patagonia (Markgraf, 1985). In both areas, rapidly increasing aridity resulted in the extinction of the Pleistocene megafauna (between 11,000 and 10,000 [14]C B.P.), whereas the modern camelids were more competitive (i.e., less specialized in their diet) and

survived. As the megafauna was the primary food resource for the Paleo-Indian hunters in south-central Chile, the Paleo-Indian tradition came to an end synchronously with the extinction of the Pleistocene animals.

In this chapter, we examine environmental aridity stress due to climate changes and its impact on early hunting societies in different parts of the south-central Andes. In particular, two periods of time are interesting: the Pleistocene-Holocene transition and the mid-Holocene arid event, which was a widespread climatic phenomenon in both Americas (Villagrán and Varela, 1990; Martin et al., 1993; Dean et al., 1996; Grosjean et al., 1997b; Villa-Martinez and Villagrán, 1997).

7.2. THE LATE PLEISTOCENE HUNTERS IN CENTRAL CHILE (11,400–9700 [14]C B.P.)

At the current state of knowledge, the earliest well-dated and widely accepted sites of human occupation in southern South America are Monte Verde MV-II (12,500 [14]C B.P.; see Fig. 1) and Piedra Museo (12,890 ± 90 [14]C B.P.) (Meltzer et al., 1997; Miotti and Cattaneo, 1997). In central Chile, the earliest unambiguous evi-

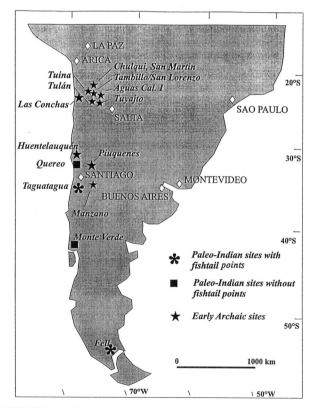

FIGURE 1   Map showing the locations of the archaeological sites in Chile cited in the text.

dence of Paleo-Indian hunters associated with a rich Pleistocene megafauna is documented for the open killing sites of Quereo (32°S, 72°W; Núñez et al., 1994b) and Tagua Tagua (34°30′ S, 71°1′ W; Montané, 1968; Núñez et al., 1994a).

Archaeological strata at the Quereo site are dated between 11,400 ± 155 ¹⁴C B.P. and 10,925 ± 85 ¹⁴C B.P. (Quereo II), possibly as early as 12,000 ± 195 ¹⁴C B.P. (Quereo I; Núñez et al., 1994b; Fig. 2). The butchering site of Quereo I shows bones modified by percussion, a horse skull fractured by a nasofrontal impact, and scarce lithic artifacts. Remains of horse (*Equus* sp.), deer (*Antifer niemeyeri*), mastodon (*Cuvieronius humboldtii*), *Paleolama* sp., *Lama* sp., and *Mylodon* sp. are associated with tree macrofossils (*Dasyphyllum excelsum*), suggesting that the large herbivores were concentrated in favorable sites of limited extent (ecological refuge) with lacustrine facies within a generally semiarid environment. This open killing site provided excellent hunting opportunities. The Paleo-Indian hunters of Quereo II lived at least occasionally on the shoreline of a marsh with peat deposits where Pleistocene herbivores came to drink. At that site, horse (*Equus* sp.), mastodon (*Cuvieronius* sp.), giant sloth (*Mylodon* sp. and *Glossotherium* sp.), deer, and *Lama* sp. were identified. At least the

horse, and most likely the other animals as well, were killed with stones. Laminar flakes, bone fragments with knife marks, and hammered and polished bone artifacts provide evidence for human activities at that site. The remains of animal skeletons were found in sandy matrix, suggesting that the lacustrine environment was already replaced by a fluvial habitat with sandy beaches between 11,100 and 9370 ¹⁴C B.P. Further evidence of increasing aridity with the termination of the Pleistocene is found in pollen records at Quereo (Villagrán and Varela, 1990). Whereas abundant Cyperaceae and *Myriophyllum* pollen suggest the presence of swamps and ponds prior to 11,400 ¹⁴C B.P., the almost complete absence of arboreal and aquatic taxa after 9370 ± 180 ¹⁴C B.P. implies increasing aridity. However, in this changing environment, the concentration of animal, vegetation, and water resources in special places provided the best opportunities for Paleo-Indian hunters. In this sense, increasing aridity at the end of the Pleistocene would have favored animal exploitation near the wetlands, where an *oasis* with abundant resources existed.

Similar processes are found in the late glacial paleo-lake of Tagua Tagua and the associated archaeological open kill sites Tagua Tagua 1 and 2 (Fig. 3; Montané,

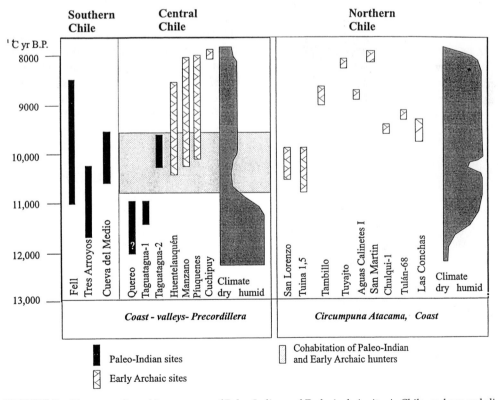

FIGURE 2 Chronostratigraphic sequences of Paleo-Indian and Early Archaic sites in Chile, and general climatic history in central Chile (coast and longitudinal valleys) and northern Chile (circum-Puna area).

1968; Núñez et al., 1994a). Extinct fauna (*Antifer* sp., *Stegomastodon humboldtii*, and *Equus* sp.; Casamiquela, 1970) and numerous lithic artifacts, including *Fell-type* points (fishtail points typical for Pleistocene hunters; Bird, 1951; Massone and Hidalgo, 1981; Flegenheimer and Zárate, 1989; Fig. 3), are dated between 11,380 ± 320 and 9700 ± 90 ^{14}C B.P. (Table 1, Fig. 2). Similar to the record at Quereo I and II, the hunters at Tagua Tagua killed the animals in swamps and lake margins with soft sediment, probably by hitting their skulls with stones. Butchering took place *in situ,* and the presence of charred bones suggests that the hearths were nearby. However, both Tagua Tagua sites show a greater variety of lithic artifacts than Quereo, and strong emphasis was put on killing mastodons. A total of 16 killed individuals was found at both Tagua Tagua sites.

The pollen profile for Tagua Tagua (Heusser, 1983, 1990) shows that Paleo-Indian occupation took place during a phase of rapidly increasing aridity and decreasing lake levels (Fig. 2). Tree taxa (*Podocarpus* and *Nothofagus*) disappeared, and shrubs and herbs increased. The earlier site of Tagua Tagua 1, dated between 11,380 ± 320 ^{14}C B.P. and 11,000 ± 170 ^{14}C B.P. (Montané, 1968), was also located close to the shoreline of the paleolake at its maximum extent, whereas the younger site of Tagua Tagua 2 (dated between 10,190 ± 130 and 9700 ± 90 ^{14}C B.P.) was already located toward the center of the shrinking water body of the paleolake (Fig. 3).

Both sites in central Chile, but also several sites in Patagonia (Markgraf, 1985), have shown that the first humans appeared during times of rapidly changing environments at the end of a long climatic period with relatively stable conditions (at Tagua Tagua since >45,000 ^{14}C years ^{14}C B.P.; Heusser, 1983). Vegetation and megafauna—in particular, animals with specialized diets—were not able to adapt to such rapid changes, and they disappeared or became extinct. In this sense, Tagua Tagua and Quereo are two southern examples for the environmental crisis in many parts of the Americas at the end of the Pleistocene. The climatic variability, rapid warming, and aridification (for instance, in central Chile) particularly affected the last Pleistocene specimen of Proboscidae (mammoths and mastodons) in both Americas (Dreimanis, 1968; Bryan et al., 1978; Correal and van der Hammen, 1977; Bonnichsen et al., 1987; Haynes, 1985a). In northeastern North America, the late glacial Proboscidea were also concentrated in ecological refuges and favorable places and, subsequently, were controlled by the Clovis hunters (between 11,500 and 10,500 ^{14}C B.P.; Saunders, 1980; Haynes, 1985b) at a time of rapidly increasing aridity, water stress, and vegetation changes. Thus, it appears that the *Clovis* projectile points in North America and the *Fell* points in South America are both related to cultures in ecological refuges within rapidly changing environments that resulted in the extinction of Pleistocene fauna.

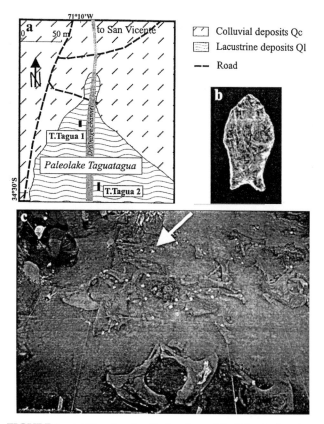

FIGURE 3 (a) Map showing the locations of the Paleo-Indian sites at the Pleistocene paleolake Tagua Tagua and (b) an *in situ* Fell-type (fishtail) point at Tagua Tagua 2 associated with killed mastodons (location arrow in Fig. 3c).

7.3. THE FIRST HUNTERS OF THE PUNA DE ATACAMA (10,800–8000 ^{14}C B.P.)

Although the first colonization of humans in Peru is possibly as early as 12,560 ^{14}C B.P. (Guitarrero Cave; Lynch et al., 1985), the oldest date of Early Archaic hunters in the Puna de Atacama in northern Chile is relatively young at 10,820 ^{14}C B.P. (Table 1, Fig. 2). Small and highly mobile groups occupied the high Puna, the intermediate valleys, the depression of the Salar de Atacama, and the high-elevation sites of the Precordillera. The earliest Archaic sites are found in caves at intermediate elevations (Tuina and San Lorenzo, between 3000 and 3200 m). It is suggested that these were the starting points for a transhumant pattern of resource use, with seasonal access to the paleolakes in the high

Table 1: Selected Radiocarbon Dates Cited in the Text.

Site	Material	Lab #	^{14}C year B.P.	Culture	Reference
Monte Verde (CAE)	Wood	OXA-381	12,450 ± 150	Paleo-Indian	Dillehay, 1989
Piedra Museo (C)	Charcoal	AA-20125	12,890 ± 90	Paleo-Indian	Miotti and Cattaneo, 1997
Quereo I (SCA)	Wood	N-2965	12,000 ± 195	Paleo-Indian	Núñez et al., 1994b
Quereo I (SCA)	Wood	N-2960	11,700 ± 150	Paleo-Indian	Núñez et al. ,1994b
Quereo I (SCA)	Wood	N-2964	11,700 ± 160	Paleo-Indian	Núñez et al., 1994b
Quereo I (SCA)	Wood	N-2963	11,400 ± 155	Paleo-Indian	Núñez et al., 1994b
Quereo I (SCA)	Wood	N-2962	11,400 ± 155	Paleo-Indian	Núñez et al., 1994b
Quereo I (SCA)	Wood	SI-3391	10,925 ± 85	Paleo-Indian	Núñez et al., 1994b
Quereo I (SCA)	Peat	N-2483	9370 ± 200	Post Paleo-Indian	Núñez et al., 1994b
Tagua Tagua 1 (SCA)	Charcoal	GX-1205	11,380 ± 320	Paleo-Indian	Núñez et al., 1994b
Tagua Tagua 1 (SCA)	Charcoal	SR	11,000 ± 170	Paleo-Indian	Montané, 1968
Tagua Tagua 2 (SCA)	Charcoal	Beta-45520	10,190 ± 130	Paleo-Indian	Núñez et al., 1994a
Tagua Tagua 2 (SCA)	Charcoal	Beta-45519	9900 ± 100	Paleo-Indian	Núñez et al., 1994a
Tagua Tagua 2 (SCA)	Charcoal	Beta-45518	9700 ± 90	Paleo-Indian	Núñez et al., 1994a
Cueva Yavi (C)	Charcoal	CSIC-1101	10,450 ± 55	Early Archaic	Kulemeyer and Laguna, 1996
Cueva Yavi (C)	Charcoal	CSIC-908	8320 ± 260	Early Archaic	Kulemeyer and Laguna, 1996
Cueva Yavi	Charcoal	AC-1088	9760 ± 160	Early Archaic	Krapovickas, 1987
Cueva Yavi	Charcoal	AC-1093	9480 ± 220	Early Archaic	Krapovickas, 1987
San Lorenzo-1 (C)	Charcoal	HV-299	10,280 ± 120	Early Archaic	Spahny, 1976
Tuina-1 (C)	Charcoal	SI-3112	10,820 ± 630	Early Archaic	Núñez, 1983
Tambillo-1 (CAE)	Charcoal	Beta-63365	8870 ± 70	Early Archaic	Núñez, this chapter
Tambillo-1(CAE)	Charcoal	Beta-25536	8590 ± 130	Early Archaic	Núñez, this chapter
Tuyajto-1B (CPL)	Charcoal	Beta-105691	8130 ± 110	Early Archaic	Núñez et al., this chapter
Tuyajto-1B (CPL)	Charcoal	Beta-105692	8210 ± 110	Early Archaic	Núñez et al., this chapter
San Martín-4a (CPL)	Charcoal	Beta-116573	8070 ± 50	Early Archaic	Núñez et al., this chapter
Las Conchas (AEC)	Charcoal	P-2702	9680 ± 160	Early Archaic	Llagostera, 1979
Piuquenes (C)	Charcoal	SR	10,160	Early Archaic	Stehberg, 1997
Manzano-1 (C)	Charcoal	SR	10,230	Early Archaic	Saavedra and Cornejo, 1995
Pichasca (C)	Charcoal	IVIC-728	9920 ± 110	Early Archaic	Ampuero and Rivera, 1971
Cuchipuy (Ce)	Charcoal	Beta-1453	8070 ± 100	Early Archaic	Kaltwasser et al., 1985
Tilocalar (Tu-85) (AEC)	Charcoal	Beta-25508	3140 ± 70	Early Formative	Núñez, 1992

Legend: C = cave; Ce = cemetery; AEC = site with architecture; CPL = site at paleolake; CAE = open site with structures; SCA = open hunting site.

Puna (Altiplano between 4000 and 4500 m). The hunters exploited modern fauna (camelids, cervids, birds, and rodents) and used a typical triangular artifact, which is found on paleo-shorelines in numerous lake basins on the Altiplano (Fig. 4). Radiocarbon dates from seven early Archaic sites, of which Tuina, San Lorenzo, and Tambillo are the most important, suggest initial human occupation at 10,800 ^{14}C B.P. and a rapid termination of this occupational phase at ca. 8000 ^{14}C B.P. (Fig. 5). At the time of human arrival, the paleolakes in northern Chile were close to or already at their maximum extent (between 10,800 and <9200 ^{14}C B.P.; Geyh et al., 1998, 1999). Precipitation rates were 2.5–3 times higher than they are today (annual rates >500 mm compared with modern <200 mm; Grosjean, 1994) and provided abundant water resources for vegetation, animals, and ultimately, humans. New radiocarbon dates of open campsites that are related to shorelines of Altiplano paleolakes range between 8700 and 8200 ^{14}C

B.P. (Núñez et al., unpublished data), suggesting that paleolakes were still high at that time. Thus, the link between initial human occupation and favorable environmental conditions as recorded in paleolake sediments, fossil groundwater bodies, and paleosols is evident (Lynch and Stevenson, 1992; Messerli et al., 1993; Grosjean and Núñez, 1994; Grosjean et al., 1995). Further strength to this interpretation is added by the process of how this first occupational phase came to an end. Lake sediment records from the Altiplano suggest that the paleolakes disappeared very rapidly at ca. 8000 ^{14}C B.P. (Grosjean et al., 1995; Geyh et al., 1999), and extremely arid conditions were established from then onwards. The severe decrease in resources in the Atacama Desert at that time is reflected in a synchronous drastic depopulation of the area (*Silencio Arqueológico;* Núñez and Santoro, 1988), suggesting that the favorable humid early Holocene environment was the precondition for early hunting and gathering societies in this area.

FIGURE 4   The Salar Tuyajto shows a typical situation for Early Archaic sites in the Puna de Atacama. The Early Archaic fireplaces (Figs. 4b and 4c) are associated with the early Holocene shorelines of the paleolakes (in this basin 25 m above the modern lake level, Fig. 4a) and usually are far away from the places with water resources under current climatic conditions.

However, the environmental influence over the beginning of human occupation in the Atacama Desert remains debatable. Although megafauna dated to late glacial times existed in the south-central Andes (e.g., Fernandez et al., 1991), a site unequivocally linking hunters and Pleistocene animals in the Atacama Desert has yet to be discovered. At Barro Negro (northwestern Argentina; Fernandez et al., 1991), for instance, evidence of Archaic camelid hunters is found precisely at the time when *Equus* sp. became extinct. Nevertheless, the possibility of Paleo-Indian hunters still exists. For the Chilean part of the Altiplano, we argue that the onset of favorable environmental conditions and initial human occupation was, considering the statistically poor dating control of the first human arrival, broadly synchronous. At that time, the Pleistocene megafauna was possibly already extinct, whereas the modern camelids survived and served as the resource basis for the Early Archaic hunters. The picture drawn for the Puna de At-

acama looks somewhat different from the circum-Titicaca area in Bolivia. In the latter case, paleolakes and most favorable conditions were reported for the time between 15,400 and 11,000 [14]C B.P. (Servant et al., 1995; Sylvestre et al., 1996, 1999), that is long before the first people arrived. Thus, the question remains open of whether environmental conditions or the generally late, but extremely rapid, population of the South American continent (at about 12,500 [14]C B.P.) controlled the timing of the first human arrival in the south-central Andean area.

Indeed, this example from the Atacama Desert shows that the first humans arrived in an environment that was undergoing major changes. However, the changes were directed toward better resources and more favorable environments where hunters and animals coevolved. This is very different from the Paleo-Indian case in central Chile, where the humans intruded into a system that was changing toward unfavorable environ-

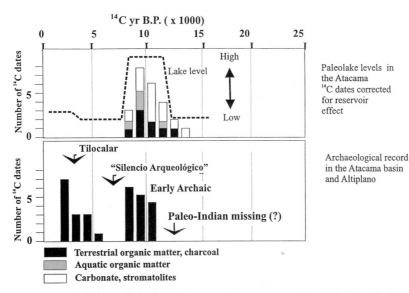

FIGURE 5 Comparison between lake level changes and human occupation in the Puna de Atacama, northern Chile. The Early Archaic period coincided with high paleolake levels, whereas the extremely arid mid-Holocene (low lake levels) resulted in a hiatus of human occupation. Reoccupation (Tilocalar phase) was synchronous with a rise of the lake levels to modern conditions.

mental conditions, and hunters acted possibly as the final blow toward extinction of their own resources.

7.4. HOLOCENE ARIDITY AND HUMAN OCCUPATION

Millennial-scale extremely arid periods during the Holocene provide insight into processes of human adaptation at times of environmental stress and scarcity of resources. The cases of the Paleo-Indian hunters in central Chile and the Early Archaic hunters in northern Chile may again serve as examples.

In central Chile, Holocene arid conditions were established between 11,000 and 10,000 [14]C B.P. (Heusser, 1983; Markgraf, 1985; Villagrán and Varela, 1990), more than 2000 years earlier than in the central Andes of northern Chile (Fig. 2). Much drier conditions than exist today lasted in the coastal areas of central Chile until ca. 3000 [14]C B.P (Villagrán and Varela, 1990; Villa-Martinez and Villagrán, 1997). In central Chile, the Paleo-Indian tradition collapsed together with the extinction of the megafauna and the decreasing potential of the shrinking paleolakes. The youngest date for Paleo-Indian occupation (with Fell points and megafauna) in this area is 9700 ± 100 [14]C B.P. (Núñez et al., 1994a). However, during the time of increasing environmental stress around the shrinking low-elevation lakes, new cultural forms with different resource use and a distinct lithic industry emerged. These Archaic

groups of hunters, gatherers, and fishermen exploited specific ecological zones where resources were still available. These resources included mainly marine resources along the coast of central and northern Chile (dated between 10,490 and 9680 [14]C B.P.; Llagostera, 1979; Jackson, 1993; Weisner and Tagle, 1995) in the Precordillera (El Manzano 10,230 [14]C B.P. and Piuquenes 10,160 [14]C B.P.; Saavedra and Cornejo, 1995; Stehberg, 1997). Increased aridity around the late Pleistocene lakes in the intermediate altitudinal zone—for instance, Laguna de Tagua Tagua—led to a gap of archaeological evidence in this area between the last Paleo-Indians (9700 [14]C B.P.) and the reoccupations after 8070 [14]C B.P. (Cuchipuy) and 6130 ± 115 [14]C B.P. (Tagua Tagua; Montané, 1968; Kaltwasser et al., 1985; Núñez et al., 1994b). It is interesting to note that Paleo-Indian hunters and Archaic hunters coexisted independently in central Chile during a period of at least 500 years, although they used different habitats (Fig. 2).

The swing from humid to arid conditions took place ca. 2000–3000 years later in northern Chile (tropical summer rainfall area) than in central Chile (extratropical winter rainfall area). Whereas Early Archaic hunters in the Atacama Desert still occupied the near-maximum paleoshorelines at ca. 8200 [14]C B.P., the paleolakes in northern Chile and southern Bolivia had disappeared by ca. 8000 [14]C B.P. (Sylvestre, 1997; Sylvestre et al., 1999; Grosjean, 1997; Sifeddine et al., 1998; Geyh et al., 1999), suggesting that the shift from humid to arid conditions was very rapid. Lakes in the Altiplano were extremely shallow or dry, and precipitation rates must have been lower than they are today until ca. 3600–3000 [14]C B.P., when lake levels rose and modern conditions were established (Valero-Garcés et al., 1996; Grosjean et al., 1997b). However, mid-Holocene lake sediments show quite a variable history with repeated wetting and drying cycles, suggesting short-term alternations between humid spells (and/or individual storms) and dry events. Although it appears that the mid-Holocene was a very active period of short-term climatic changes, we must consider that the small lakes with shallow water bodies responded in a much more sensitive way to any moisture changes than the big paleolakes. This phenomenon of climate variability would be masked in the case of the large late glacial/early Holocene paleolakes.

The rapid onset of the arid climate and the harsh conditions during the mid-Holocene affected Early Archaic hunting and gathering societies in the Atacama Desert, where water, plant, and animal resources were likely too scarce for Archaic hunters during this hyperarid period. Such hostile conditions probably resulted in a substantial population decrease or even in a regional hiatus after 8000 [14]C B.P., which is known as the Silencio Arqueológico in the Atacama area (Núñez and Santoro, 1988; Grosjean and Núñez, 1994). All known Early Archaic sites in the Atacama basin (south of the Loa River) were abandoned. In the cave of Tuina-5 (one of the few sites with a continuous sediment stratigraphy), for instance, this hiatus is represented by a 60-cm-deep layer of archaeologically sterile weathered rock material from the roof of the cave. The sterile layer representing the occupational hiatus is bracketed between basal organic-rich layers with bones and lithic artifacts of the Early Archaic period (dated between 10,060 and 9840 [14]C B.P.) and hearths of the post-Archaic period (2240 [14]C B.P.) on top. However, despite the general mid-Holocene Silencio Arqueológico, human occupation continued in a few places with particularly stable resources (ecological *oases* with spring water and vegetation). As has already been observed in the records of the late glacial paleolakes of Tagua Tagua and Quereo, increasing aridity resulted primarily in the formation of pronounced ecological refuges, where resources were concentrated in a limited space, surrounded by hostile environments. One of these places is the Quebrada de Puripica in northern Chile, where 20 individual archaeological campsites (dated between 6200 and 3100 [14]C B.P.) provide detailed insights into human adaptation to the mid-Holocene harsh environment (Grosjean et al., 1997a). The adaptive process of these groups was very different from what was observed on the coast and in the Precordillera in central Chile. The sites in Quebrada de Puripica suggest that the transformation of the early Archaic hunting tradition into Middle and Late Archaic complex societies with camelid domestication, architecture, and semisedentism (between 4800 and 4000 [14]C B.P.) was a stepwise process. The early hunting tradition (prior to 6200 [14]C B.P.) changed by 5900 [14]C B.P. into a cultural system with large campsites, intense exploitation of camelids, and an innovative lithic industry with microliths and perforators. High mobility and the use of exotic lithic materials were still important. At Puripica, this evolution culminated in the Late Archaic period (between 5100 and 4800 [14]C B.P.), which showed hunting and, simultaneously, domestication, the use of local lithic materials, and the development of architecture (Hesse and Hesse, 1979; Núñez, 1982; Grosjean et al., 1997a). However, our currently available paleoclimatic records do not allow us to draw a clear link between climatic change and domestication of camelids at ca. 4800 [14]C B.P., because both the preceding and the succeeding two millennia were similarly arid. This is different from the subsequent major cultural step—the emergence of agriculture and pastoralism (Tilocalar phase in the Atacama basin at ca. 3200 [14]C B.P.; Núñez, 1983). This step coincided with an increase of moisture and lakes to modern levels (Valero-Garcés et al., 1996; Grosjean et al., 1997b) and dense reoccupation of the Altiplano (fireplaces dated between 3390 ± 60 [14]C B.P. at Salar de Capur and 3170 ± 60 [14]C B.P. at Salar de Ollagüe; Fig. 5).

7.5. CONCLUSIONS FOR SOUTHERN SOUTH AMERICA

Paleontological, archaeological, and paleoclimatological records for south-central and northern Chile suggest a strong relation between late Pleistocene and Holocene changing climates, changing environments, and changing human occupational and cultural patterns. In central Chile, the Paleo-Indian hunters specialized in exploiting Pleistocene megafauna, particularly in places around lakes (*ecological refuges*) where natural resources were locally concentrated in an environment under generally increasing aridity and water stress. The termination of the Paleo-Indian hunting tradition in central Chile at ca. 9700 ± 70 [14]C B.P. (Tagua Tagua 2) coincided with the onset of Holocene arid conditions, the disappearance of paleolakes, and the extinction of Pleistocene megafauna (Bird, 1951; Casamiquela, 1970; Massone and Hidalgo, 1981; Núñez et al., 1994a,b). A similar process repeated itself ca. 2000 years later in northern Chile, when rapidly falling lake levels (at ca. 8000 [14]C B.P.) and the onset of extremely arid mid-Holocene conditions were the likely cause for the termination of the Early Archaic period (between

10,800 and ca. 8000 ^{14}C B.P.) and the beginning of the occupational hiatus during the mid-Holocene arid period. In northern Chile, however, the link between Pleistocene megafauna and pre-Archaic Paleo-Indian occupation has yet to be discovered. All the early archaeological sites are associated with modern fauna. Given that the lake levels in northern Chile did not rise before 13,000–12,000 ^{14}C B.P. (Geyh et al., 1999), we suggest that the environmental conditions and vegetational resources on the Atacama Altiplano were not favorable enough to support a dense population of Pleistocene megafauna (mastodons, etc.) during late glacial times, e.g., the time when Paleo-Indians could have been present.

Examples from the Atacama Desert also provide evidence of expanding human occupation at times when environmental conditions became more favorable. The Tilocalar phase (beginning in 3200 ^{14}C B.P.)—with spreading human occupation, pastoralism, and initial agriculture—coincided with the period when lake levels recuperated and modern climatic conditions (i.e., slightly more humid than the mid-Holocene conditions) were established.

It is important to note that the paleoenvironments in the different parts of South America were in different transitional stages when the first people arrived. The Paleo-Indian groups in central Chile intruded into an ecosystem that had experienced a long time of Pleistocene vegetational and faunistic relative stability (>45,000 ^{14}C years B.P.; Heusser, 1983, 1990), became affected by drought stress during late glacial times, declined, and underwent dramatic changes when people appeared. This process resulted in the concentration of animals in ecological refuges around lakes, in intense animal hunting as far as extinction, and, finally, in the termination of the Paleo-Indian tradition. In contrast, the Puna in the central Andes was, until shortly before humans appeared, an *empty,* environmentally hostile place, not suitable for megafauna with large food requirements and specialized diets. With the onset of favorable conditions during the latest late glacial and early Holocene times, modern animals could occupy this *free* space very rapidly, expand their living space, and coevolve with the hunters of the Early Archaic period.

This differential onset of favorable and hostile conditions in south-central and northern Chile resulted in an occupational phase of ca. 1500 years, when Paleo-Indians (near paleolakes) and Archaic hunters (Precordillera, Cordillera, and coast) coexisted, although they exploited different habitats.

Although these examples support an environmental cause and a deterministic interpretation of major changes in early human cultural and expansional patterns, other major cultural and societal changes do not seem to have been triggered by changing environments. This is particularly the case for the process of domestication (at ca. 4800 ^{14}C B.P.). The currently available paleoclimatic archives do not show any major changes around this time. Thus, the concept of the *environmental dictate* must not lead to an oversimplification of the evolutionary processes in human culture, society, and resource use.

7.5.1. Aridity Events and the Early Settlement of the Americas

The human migration into the Americas across the Bering Straits and the subsequent colonization of the whole continent from the Arctic to the subantarctic tip of South America occurred during the final stages of the Pleistocene. There are well-documented sites dating from 14,000 ^{14}C B.P. in Alaska and North America, where megafauna was exploited (Adovasio et al., 1985; Bonnichsen et al., 1987), all the way to the uttermost southern part of the Americas, there dated at 12,500 ^{14}C B.P. Even when different models of migration rates are used (Dillehay, 1989; Meltzer et al., 1997; Miotti and Cattaneo, 1997), all these sites predate both Clovis or Fell cultures of the eleventh and tenth millennia. Although these groups of Pleistocene hunters and gatherers were considered highly mobile, efficient, and adaptable (Kelly and Todd, 1988), they occupied different open and discontinuous environments from the High Plains and Great Basin in North America to the Patagonian steppe wherever faunal resources were abundant. This implies that from the onset of colonization through subsequent times of occupation, populations had to interact with variable environments and variable Pleistocene climates (Markgraf, 1989; Butzer, 1991).

In this context, it is important to identify aridity events during the Pleistocene–Holocene transition in order to understand gaps and convergence as well as divergence of human occupation that depended on subsistence on faunal and plant resources. It is generally accepted that under arid conditions, in the face of an environmental collapse as it has been observed in both hemispheres (Haynes, 1985b; Núñez et al., 1994a), the Pleistocene fauna moved closer to perennial water bodies and springs. This, in turn, must have stimulated the development of specific hunting strategies. Apparently, all biological resources were vulnerable to aridity events. The mammalian megafauna (Proboscids), however, were especially vulnerable to such events. Once the megafauna were restricted by the availability of food, this must have led to predictable hunting strategies in response to the regional change, aiming to intensify the exploitation of resources in those regions with environmental crisis (Minnis, 1983). It has been shown that unspecialized hunting groups avoided exactly those regions where water resources were scarce to avoid the risk of starvation such as that which occurred in southern Africa (Butzer, 1991). In contrast, the specialized hunters in both North and South America, represented during the Pleistocene–Holocene transition by the Clovis and Fell cultures, followed an opportunistic strategy by selectively exploiting regions under environmental stress (Butzer, 1991; Núñez et al., 1994b).

In fact, at that time, various lacustrine environments collapsed due to aridity events in the lowlands of Guatemala as well as in Panama. This affected not only the Paleo-Indian lithic industry (Bird and Cooke, 1978; Wright, 1991), but also the forest cover (Damuth and Fairbridge, 1981). Other isolated cases are known, such as Lake Valencia (Venezuela) where prior to 13,000 ^{14}C B.P. aridity affected the lowland rain forest (Bradbury et al., 1981). For the same time, a mastodon kill site has been found in a megafauna-rich savanna environment near an ecological refugium, which also was affected by aridity (Ochsenius, 1979). It is likely that the Paleo-Indian colonization to South America occurred through such regions where the forests had become less dense because of aridity. Farther south, in the central Andes, high-elevation paleolakes were found, suggesting increased humidity after ca. 15,400 ^{14}C year B.P. (Sylvestre et al., 1999) or after ca. 13,000–12,000 ^{14}C B.P. (Geyh et al., 1999), which actually should have improved the resources. Subsequent geomorphologic processes that may have impacted potential sites, in addition to the lack of detailed excavations, have made it difficult to identify Paleo-Indian occupation in the central Andes (Messerli et al., 1993; Grosjean and Núñez, 1994). On the other hand, lowland lake sites of southern South America, such as those found in the surroundings of Laguna Tagua Tagua, that were colonized by groups representing the Fell lithic culture were quite clearly associated with late Pleistocene aridity events. These events apparently occurred repeatedly throughout the lowland lake region of the central depression of south-central Chile. Because of the low lake levels, the central parts of the former lakes were occupied, as has been shown by open kill sites. These sites were subsequently covered with sediments after precipitation increased and the lakes recovered (Montané, 1968; Núñez et al., 1994a).

Along the eastern slope of the Andes, the paleoclimatic conditions during late Pleistocene times were apparently quite different. Although the Pampas region of Argentina was occupied by Paleo-Indians during times of environmental change, the chronological control is still insufficient to exactly relate both aspects

with each other. In contrast to central Chile, it is known that the occupation by Fell lithic cultures at the Cerro La China site in the Pampas occurred at 10,500 ^{14}C B.P. during a humid climate phase and a period of abundant resources (Zárate, 1997). The effects of drier climates set in only by 8500 ^{14}C B.P., culminating in the extinction of the megafauna at the Holocene transition and limiting human occupation (Tonni and Cione, 1994; Borrero et al., 1998).

7.6. LATE PLEISTOCENE ARIDITY REGIMES IN BOTH HEMISPHERES AND THEIR EFFECTS ON HUMAN OCCUPATION

Faunal remains from late Pleistocene lakes in the lowlands of southern South America show a dominance of mastodons with evidence of human predation dated between 11,000 and 10,000 ^{14}C B.P. (Montané, 1968; Varela, 1976; Núñez et al., 1994a,b), resembling the classical kill-butcher sites in North America (Bryan et al., 1978; Saunders, 1980; Graham et al., 1981). On the other hand, the southeastern and southwestern portions of Ontario (Canada) have the highest frequency of mastodons (*Mammuthus americanum*) and mammoths (*Mammuthus* sp.) in all of the Northern Hemisphere that is associated with lacustrine environments. Also, the transition to a more arid climate regime favored hunting, in this case with fluted Clovis points. As a consequence, Paleo-Indians have been made accountable for the late Pleistocene–Holocene extinction of the megafauna not later than 10,000 ^{14}C B.P. (McAndrews and Jackson, 1988), comparable with the opportunistic hunting technique of the Fell culture in South America, which also has been made responsible for the extinction of the mastodons in the extreme south (Núñez et al., 1994a,b).

In the Northern Hemisphere between 12,000 and 10,000 ^{14}C B.P., ecological refugia for megafauna were also found near lakes together with signs of human predation. Based on the data from central Chile, it appears that the extinction was completed quite late, as has been suggested by Meltzer and Mead (1985), without dramatic abruptness, as has been suggested by Martin and Klein (1984). The presence of mastodons prior to the Clovis occupation in North America (prior to 12,000 ^{14}C B.P.), as well as prior to the Fell occupation in South America, suggests that the faunal extinction was the result of a combination of natural and anthropogenic causes. However, questions about synchroneity of natural changes and types of causes (Martin et al., 1985; Wright, 1991) are not discussed here. The only comment in this context refers to the specific extinction of mastodons in northeastern North America after 11,000

[14]C B.P. Dreimanis (1967) explains their extinction as having been related to increased aridity, interpreted from the replacement of spruce and tundra by pine, a vegetation change that cannot be explained by migrational lags (McAndrews and Jackson, 1988). The increase of aridity is also based on the low abundance of aquatic plants in paleoenvironmental records, suggesting that lake levels were greatly lowered. At the same time, between 11,500 and 10,400 [14]C B.P., there was an abundance of Clovis occupation in the vicinity of these lakes.

On the other hand, the expansion of grasslands between 12,000 and 10,000 [14]C B.P. in the states of Montana and Washington may also have resulted in a reduction of the dense population of mastodons, which earlier were adapted to forested landscapes (King and Saunders, 1984; Wright, 1991). On the Great Plains and in the southwestern United States, the most notable changes in terms of desiccation of the pluvial lakes occurred at ca. 14,000 [14]C B.P. Also related to the increase in aridity, vegetation changed in the vicinity of the desertified basins, including a change in food resources. In other words, with the increase in temperature during hypsithermal times, the amounts of precipitation and evaporation changed so that lake levels in the Great Plains and the southwestern United States became very low or even dried out. Lithic industries related to occupation by Early Man at that time were found on fossil lake beds (Cressman, 1942; Heusser, 1966).

Mastodons are highly vulnerable to prolonged periods of dryness. Hence, in North America as well as in South America, humans could exploit this resource occasionally without even having to kill the prey. The use of Clovis points in North America or of Fell points (*fish-tail points*) in South America was probably not always necessary. In that case, as the ethnographic evidence from Africa suggests, one would expect more dead adult animals, which in fact is shown by archaeological data (Haynes, 1985a). One assumes that during arid periods the hunters concentrated their efforts on those few sites where water was available. Related to the site conditions, this led to a change in the hunting strategy, which included exploitation of those animals that died naturally. It appears that what occurs today in Africa (Dorst and Dandelot, 1973) may have occurred during the late Pleistocene period around the lakes in the southwestern United States and south-central Chile (Haynes, 1985b; Núñez, 1994a,b). In general, there is much information on the role of human predation of the extinct American proboscideans thanks to the kill-butcher sites in both hemispheres (Montané, 1968; Bryan et al., 1978; Saunders, 1980; Graham et al., 1981; Núñez et al., 1994a,b).

Based on this data, North American Clovis hunters developed a specialized and intensive hunting strategy for mastodons and mammoths between 11,600 and 10,800 [14]C B.P., at the same time as paleoenvironmental changes also would have favored extinction of the megafauna. In other words, the human actors preceded the extinction process, but once the megafauna declined in relation to the changing environment and reduction of the ecological refugia, the hunting techniques were also changed. This suggestion is based on the information gained from other earlier cultures, such as the Jobo culture of Venezuela. In other regions and cultures independent from each other, hunting strategies responded to the changing situation in a comparable fashion, including the Clovis hunters in North America and the Fell hunters in southernmost South America (Bird, 1951; Massone, 1985; Bonnichsen et al., 1987; Núñez et al., 1994a,b).

Acknowledgments

This research was financed with grants from the Fondo National de Ciencia y Tecnologia (FONDECYT) 1930022, the National Geographic Society (NGS) 4374-90 and 5836-96, and the Swiss National Fonds 20-36382.92 and 21-50854.97. We thank both reviewers, T. Lynch and P. Sheets, and Vera Markgraf for helpful comments.

References

Adovasio, J. M., C. C. Carlisle, K. A. Cushman, J. Donahue, J. E. Guilday, W. C. Johnson, K. Lord, P. W. Parmalee, R. Stuckenrath, and W. P. Wiegman, 1985: Paleoenvironmental reconstruction at Meadowcroft rockshelter, Washington County, Pennsylvania. *In* Mead, J. I., and D. J. Meltzer (eds.), *Environments and Extinctions: Man in Late Glacial North America.* Orono, ME: Center for the Study of Early Man, University of Maine, pp. 73–120.

Ampuero, G., and M. Rivera, 1971: Secuencia arqueológica del alero rocoso de San Pedro Viejo de Pischasca (Ovalle, Chile). *Boletín del Museo Arqueológico de la Serena,* **14**: 14–45.

Binford, M. W., A. L. Kolata, M. Brenner, J. Janusek, M. B. Abbott, and J. H. Curtis, 1997: Climate variations and the rise and fall of an Andean civilization. *Quaternary Research,* **47**: 235–248.

Bird, J., 1951: South American radiocarbon dates. *Memoirs of the Society for American Archaeology,* **8**: 37–49.

Bird, J., and R. Cooke, 1978: The occurrence in Panama of two types of Paleo-Indian projectile points. *In* Bryan, A. L. (ed.), *Early Man in America from a Circum-Pacific Perspective.* Edmonton, pp. 263–272.

Bonnichsen, R., D. Stanford, and J. L. Fastook, 1987: Environmental change and developmental history of human adaptive patterns: The Paleo-Indian case. *In Geology of North America, Vol. K-3. North America and Adjacent Oceans During the Last Deglaciation.* Boulder, CO: The Geological Society of America, Chapter 18, pp. 403–424.

Borrero, L. A., M. Zárate, L. Miotti, and M. Massone, 1998: The Pleistocene-Holocene transition and human occupations in the southern cone of South America. *Quaternary International,* **49–50**: 191–199.

Bradbury, J. P., B. W. Leyden, M. L. Salgado Labouriau, W. M. Lewis, C. Schubert, M. Binford, D. G. Frey, D. R. Whitehead, and F. H. Weibezahn, 1981: Late Quaternary environmental history of Lake Valencia, Venezuela. *Science,* **214**: 1299–1305.

Brenner, M., D. A. Hodell, J. H. Curtis, M. F. Rosenmeier, M. W. Binford, and M. B. Abbott, 2000: Abrupt climate change and Pre-Columbian cultural collapse. *In* Markgraf, V. (ed.) *Interhemispheric Climate Linkages*, San Diego. Academic Press, pp. 87–103.

Bryan, A. L., R. Casamiquela, J. M. Cruxent, R. Gruhn, and C. Ochsenius, 1978: An El Jobo mastodon kill at Taima Taima, northern Venezuela. *Science*, **20**: 1275–1277.

Butzer, K. W., 1991: An Old World perspective on potential mid-Wisconsinan settlement of the Americas. *In* Dillehay, T. D., and D. Meltzer (eds.), *The First American. Search and Research*. Boca Raton, FL: CRC Press, pp. 137–158.

Byrne, B., 1988: El cambio climático y los orígenes de la agricultura. *In* Manzanilla, L. (ed.), *Coloquio Gordon Childe*. Universidad Nacional Autónoma de México, pp. 27–40.

Casamiquela, R., 1970: Primeros documentos de la Paleontología de vertebrados para un esquema estratigráfico y zoogeográfico del Pleistoceno de Chile. *Boletín de Prehistoria de Chile*, Universidad de Chile, **2**: 2–3.

Correal, G., and T. van der Hammen, 1977: Investigaciónes Arqueológicas en los Abrigos Rocosos de Tebinquiche. Editorial Banco Popular de Colombia, Bogotá, 194 pp.

Cressman, L. S., 1942: *Archaeological Research in the Northern Great Basin*. Washington, DC: Carnegie Institute, Publication 538.

Damuth, J. E., and R. W. Fairbridge, 1981: Equatorial Atlantic deep-sea arkosic sands and ice age aridity in tropical South America. *Geological Society of America Bulletin* **81**: 189–206.

Dean, W. E., T. S. Ahlbrandt, R. Y. Anderson, and J. P. Bradbury, 1996: Regional aridity in North America during the middle Holocene. *The Holocene*, **6**: 145–155.

Dillehay, T. D., 1989: Monte Verde. *Science*, **245**: 1436.

Dorst, J., and P. Dandelot, 1973: *Guía de campo de los mamíferos salvajes de Africa*. Ediciones Omega S.A. Barcelona, Spain.

Dreimanis, A., 1967: Mastodons, geologic age and extinction in Ontario, Canada. *Canadian Journal of Earth Sciences*, **4**: 663–675.

Dreimanis, A., 1968: Extinction of mastodons in eastern North America: Testing a new climatic environmental hypothesis. *Ohio Journal of Science*, **68**: 527–572.

Fernandez, J., V. Markgraf, H. O. Panarello, M. Albero, F. E. Angiolini, S. Valencio, and M. Arriaga, 1991: Late Pleistocene/early Holocene environments and climates, fauna, and human occupation in the Argentine Altiplano. *Geoarchaeology*, **6**: 251–272.

Flegenheimer, N., and M. Zárate, 1989: Paleo-Indian occupation at Cerro El Sombrero locality, Buenos Aires Province, Argentina. *Current Research in the Pleistocene*, **6**: 12–13.

Geyh, M., M. Grosjean, L. A. Núñez, and U. Schotterer, 1999: Radiocarbon reservoir effect and the timing of the late-glacial/early Holocene humid phase in the Atacama Desert, northern Chile. *Quaternary Research*, **52**: 143–153.

Geyh, M. A., U. Schotterer, and M. Grosjean, 1998: Temporal changes of the ^{14}C reservoir effect in lakes. *Radiocarbon*, **40**(2): 921–931.

Graham, R. W., C. V. Haynes, Jr., D. L. Johnson, and M. Kay, 1981: Kimmwick: A Clovis-mastodon association in eastern Missouri. *Science*, **213**: 1115–1117.

Grosjean, M., 1994: Paleohydrology of the Laguna Lejía (North Chilean Altiplano) and climatic implications for late-glacial times. *Palaeogeography, Palaeoclimatology, Palaeoecology*, **109**: 89–100.

Grosjean, M., 1997: Late Quaternary humidity changes in the Atacama Altiplano: Regional, global climate signals and possible forcing mechanisms. *Zentralblatt für Geologie und Paläontologie Teil I*, **3**: 581–592.

Grosjean, M., and L. A. Núñez, 1994: Late glacial, early and middle Holocene environments, human occupation, and resource use in the Atacama (northern Chile). *Geoarchaeology*, **9**: 271–286.

Grosjean, M., M. Geyh, B. Messerli, and U. Schotterer, 1995: Late-glacial and early Holocene lake sediments, groundwater forma-tion and climate in the Atacama Altiplano. *Journal of Paleolimnology*, **14**: 241–252.

Grosjean, M., L. A. Núñez, I. Cartajena, and B. Messerli, 1997a: Mid-Holocene climate and culture change in the Atacama Desert, northern Chile. *Quaternary Research*, **48**: 239–246.

Grosjean, M., B. L. Valero-Garcés, M. Geyh, B. Messerli, U. Schotterer, H. Schreier, and K. Kelts, 1997b: Mid- and late Holocene limnogeology of Laguna del Negro Francisco, northern Chile, and its paleoclimatic implications. *The Holocene*, **7**: 151–159.

Haynes, G., 1985a: Age profiles in elephant and mammoth bone assemblages. *Quaternary Research*, **24**: 333–345.

Haynes, G., 1985b: On watering holes, mineral licks, death, and predation. *In* Mead, J. I., and D. J. Meltzer (eds.), *Environments and Extinctions: Man in Late Glacial North America*. Orono, ME: Center for the Study of Early Man, University of Maine, pp. 53–71.

Hesse, B., and O. Hesse, 1979: *Archaic Animal Exploitation in Inland Northern Chile*. Unpublished report, Smithsonian Institution, Washington, DC.

Heusser, C. J., 1966: Pleistocene climatic variations in the western United States. *In* Blumenstock, D. I. (ed.), *Pleistocene and Post Pleistocene Climatic Variations in the Pacific Area*. Honolulu: Bishop Museum Press, pp. 9–36.

Heusser, C. J., 1983: Quaternary pollen record from Laguna de Tagua Tagua, Chile. *Science*, **219**: 1429–1432.

Heusser, C. J., 1990: Ice age vegetation and climate of subtropical South America. *Palaeogeography, Palaeoclimatology, Palaeoecology*, **80**: 107–127.

Irwin-Williams, C., and V. Haynes, 1970: Climatic changes and early population dynamic in the southwestern United States. *Quaternary Research*, **1**(1): 59–79.

Jackson, D., 1993: Datación radiocarbónica para una adaptación costera del arcaico temprano en el Norte Chico, Comuna de Los Vilos. *Boletín Sociedad Chilena de Arqueología*, **16**: 28–31.

Kaltwasser, J., A. Medina, and J. Munizaga, 1985: Estudio de once fechas de RC-14 relacionadas con el Hombre de Cuchipuy. *Boletín de Prehistoria*, **9**: 9–13.

Kelly, R. L., and L. C. Todd, 1988: Coming into the country: Early Palaeoindian hunting and mobility. *American Antiquity*, **53**: 231–244.

King, J. E., and J. J. Saunders, 1984: Environmental insularity and the extinction of the American mastodon. *In* Martin, P. S., and R. G. Klein (eds.), *Quaternary Extinctions*. Tucson: University of Arizona Press, pp. 315–344.

Krapovickas, P., 1987: Nuevos fechados radiocarbónicos para el sector oriental de la Puna y la Quebrada Humahuaca. *Runa*, **XVII–XVIII**: 207–219.

Kulemeyer, J. A., and L. R. Laguna, 1996: La Cueva de Yavi: Cazadores—Recolectores del borde oriental de la Puna de Jujuy (Argentina) entre los 12,500 y 8000 años ^{14}C B.P. *Ciencia y Tecnología (Jujuy)*, **1**: 37–46.

Llagostera, A. M., 1979: 9,700 years of maritime subsistence on the Pacific: An analysis by means of bioindicators in the north of Chile. *American Antiquity*, **44**: 309–324.

Lynch, T. F., and C. M. Stevenson, 1992: Obsidian hydration dating and temperature controls in the Punta Negra region of northern Chile. *Quaternary Research*, **37**: 117–124.

Lynch, T. F., R. Gillespie, J. A. J. Gowlett, and R. E. M. Hedges, 1985: Chronology of Guitarrero Cave, Peru. *Science*, **229**: 867.

Markgraf, V., 1985: Late Pleistocene faunal extinctions in southern Patagonia. *Science*, **228**: 1110–1112.

Markgraf, V., 1989: Paleoclimates in Central and South America since 18.000 B.P. based on pollen and lake level records. *Quaternary Sciences Reviews*, **8**: 1–24.

Martin, L., M. Fournier, A. Sifeddine, B. Turcq, M. L. Absy, and J.-M. Flexor, 1993: Southern Oscillation signal in South American paleoclimatic data of the last 7000 years. *Quaternary Research*, **39**: 338–346.

Martin, P. S., and R. G. Klein (eds.), 1984: *Quaternary Extinctions*. Tucson: University of Arizona Press.

Martin, L. D., R. A. Rogers, and A. M. Neuner, 1985: The effect of the end of the Pleistocene on man in North America. *In* Mead, J. I., and D. J. Meltzer (eds.), *Environments and Extinctions: Man in Late Glacial North America*. Orono, ME: Center for the Study of Early Man, University of Maine, pp. 15–30.

Massone, M., and E. Hidalgo, 1981: Arqueología de la región volcánica de Palli-Aike (Patagonia Meridionale Chilena). *Anales Instituto de la Patagonia, Punta Arenas*, **12**: 95–124.

McAndrews, J. H., and L. J. Jackson, 1988: Age and environment of late Pleistocene mastodon and mammoth in southern Ontario. *The Buffalo Society of Natural Sciences*, **33**: 161–171.

McMorriston, J., and J. F. Hole, 1988: The ecology of seasonal stress and the origins of agriculture in the Near East. *American Anthropologist*, **93**: 46–69.

Meltzer, D. J., D. K. Grayson, G. Ardila, A. W. Barker, D. F. Dincauze, D. V. Haynes, F. Mena, L. A. Núñez, and D. J. Stanford, 1997: On the Pleistocene antiquity of Monte Verde, southern Chile. *American Antiquity*, **62**: 659–663.

Meltzer, D. J., and J. I. Mead, 1985: Dating late Pleistocene extinctions: Theoretical issues, analytical bias, and substantive results. *In* Mead, J. I., and D. J. Meltzer (eds.), *Environments and Extinctions: Man in Late Glacial North America*. Orono, ME: Center for the Study of Early Man, University of Maine, pp. 145–173.

Messerli, B., M. Grosjean, A. Bürgi, M. Geyh, K. Graf, K. Ramseyer, H. Romero, U. Schotterer, H. Schreier, and M. Vuille, 1993: Climate change and natural resource dynamics of the Atacama Altiplano during the last 18,000 years: A preliminary synthesis. *Mountain Research and Development*, **13**: 117–127.

Minnis, P. E., 1983: *Social Adaptations to Food Stress. A Prehistoric Southwestern Example*. Chicago: University of Chicago Press.

Miotti, L., and R. Cattaneo, 1997: Bifacial/unifacial techonology c. 13000 years ago in southern Patagonia. *Current Research in the Pleistocene*, **14**: 62–65.

Montané, J., 1968: Paleo-Indian remains from Laguna de Tagua Tagua, central Chile. *Science*, **161**: 1137–1138.

Núñez, L. A., 1982: Asentamientos de cazadores recolectores tardíos de la Puna de Atacama: Hacia el sedentismo. *Chungará*, **8**: 110–137.

Núñez, L. A., 1983: Paleo-Indian and Archaic cultural periods in the arid and semiarid regions of northern Chile. *Advances in World Archaeology*, **2**: 161–201.

Núñez, L. A., 1992. Emergencia de complejidad y arquitectura jerarquizada en la Puna de Atacama: las evidencias del sitio Tulán-54. *In* Albeck, M. E. (ed.), *De Costa a Selva. Producción e Intercambio entre los Pueblos Agroalfareros de los Andes Centro Sur*. Buenos Aires: Universidad de Buenos Aires, pp. 85–115.

Núñez, L. A., and C. M. Santoro, 1988: Cazadores de la puna seca y salada del área centro-sur Andina (Norte de Chile). *Estudios Atacameños*, **9**: 11–60.

Núñez, L. A., J. Varela, R. Casamiquela, V. Schiappacasse, H. F. Niemeyer, and C. Villagrán, 1994a: Cuenca de Tagua Tagua en Chile: El ambiente del Pleistoceno y ocupaciones humanas. *Revista Chilena de Historia Natural*, **67**: 503–519.

Núñez, L. A., J. Varela, R. Casamiquela, and C. Villagrán, 1994b: Reconstrucción multidisciplinaria de la ocupación prehistórica de Quereo, Centro de Chile. *Latin American Antiquity*, **5**: 99–118.

Ochsenius, C., 1979: Paleoecology of Taima Taima and its surroundings. *In* Ochsenius, C., and R. Gruhn (eds.), *Taima Taima: A Late Pleistocene Paleo-Indian Kill Site in Northernmost South America*. Final Report of 1976 excavations. South American Documentation Program, pp. 91–103.

Ortloff, C. R., and A. L. Kolata, 1993: Climate and collapse: Agro-ecological perspectives on the decline of the Tiwanaku State. *Journal of Archaeological Science*, **20**: 195–221.

Saavedra, M., and L. Cornejo, 1995: Acerca de la cronología de El Manzano. *Boletín de la Sociedad Chilena de Arqueología*, **21**: 31–34.

Saunders, J. J., 1980: A model for man-mammoth relationship in late Pleistocene North America. *Canadian Journal of Anthropology*, **1**(1): 87–98.

Servant, M., M. Fournier, J. Argollo, S. Servant-Vildary, F. Sylvestre, D. Wirrmann, and J.-P. Ybert, 1995: La dernière transition glaciaire/interglaciaire des Andes tropicales sud (Bolivie) d'après l'étude des variations des niveaux lacustres et fluctuations glaciaires. *Comptes Rendus de l'Académie de Science Paris, Serie IIa*, **320**: 729–736.

Sifeddine, A., J. Bertaux, P. Mourguiart, L. Martin, J.-R. Disnar, F. Laggoun-Défarge, and J. Argollo, 1998: Etude de la sédimentation lacustre d'un site de fôret d'altitude des Andes centrales (Bolivie). Implications paléoclimatiques. *Bulletin Société Géologie France*, **169**: 395–402.

Spahny, J. C., 1976: Gravures et pintures du desert d'Atacama (Chile). *Bulletin de la Societé Suisse des Americanistes*, **40**: 29–36.

Stehberg, R., 1997: El hombre y su medio en el período Holoceno Temprano (5000–10.000 a.P.): Caverna Piuquenes, Cordillera Andina de Chile Central. *Resumenes XIV Congreso Nacional de Arqueología Chilena* (Copiapó), p. 35.

Sylvestre, F., 1997: La dernière transition glaciaire-interglaciaire (18 000–8000 ^{14}C ans B.P.) des Andes tropicales sud (Bolivie) d'après l'étude des diatomées. Unpublished Ph.D. dissertation. Museum Nationale de l' Histoire Naturelle Paris, pp. 1–243.

Sylvestre, F., S. Servant-Vildary, M. Fournier, and M. Servant, 1996: Lake levels in the southern Bolivian Altiplano (19°–21°S) during the late glacial based on diatom studies. *International Journal of Salt Lake Research*, **4**: 281–300.

Sylvestre, F., M. Servant, S. Servant-Vildary, C. Causse, M. Fournier, and J.-P. Ybert, 1999: Lake-level chronology on the southern Bolivian Altiplano (18°–23°S) during late-glacial time and the early Holocene. *Quaternary Research*, **51**: 54–66.

Thompson, L. G., M. E. Davis, E. Mosley-Thompson, and K.-B. Liu, 1988: Pre-Incan agricultural activity recorded in dust layers in two tropical ice cores. *Nature*, **336**: 763–765.

Tonni, E. P., and A. Cione, 1994: Climas en el cuaternario de la región Pampeana y cambio global. II Seminario sobre las geociencias y el cambio global. *Asociación Geológica Argentina*, **2**: 33–35.

Valero-Garcés, B. L., M. Grosjean, A. Schwalb, M. Geyh, B. Messerli, and K. Kelts, 1996: Limnogeology of Laguna Miscanti: Evidence for mid to late Holocene moisture changes in the Atacama Altiplano (northern Chile). *Journal of Paleolimnology*, **16**: 1–21.

Varela, J., 1976: Geología del cuaternario de laguna Tagua Tagua (Provincia de O'Higgins). *Actas del Primer Congreso Geológico Chileno*, Santiago, Chile, pp. 81–112.

Villa-Martinez, R., and C. Villagrán, 1997: Historia de la vegetación de bosques pantanosos de la costa de Chile central durante el Holoceno medio y tardío. *Revista Chilena de Historia Natural*, **70**: 391–401.

Villagrán, C., and J. Varela, 1990: Palynological evidence for increasing aridity on the central Chilean coast during the Holocene. *Quaternary Research*, **34**: 198–207.

Weisner, H. R., and B. D. Tagle, 1995: Paso de las Conchas. Nuevas evidencias acerca del poblamiento costero arcaico de la VI Región. *Actas del XIII Congreso Nacional de Arqueología Chilena* (I—II, Antofagasta), pp. 337–350.

Wright, H. E., 1991: Environmental conditions for Paleo-Indian inmigration. *In* Dillehay, T. D., and D. Meltzer (eds.), *The First American. Search and Research*. Boca Raton, FL: CRC Press, pp. 113–136.

Zárate, M., 1997: Late Pleistocene geoarchaeology of the Southern Pampas, Buenos Aires Province, Argentina. *Anthropologie*, **XXXV/2**: 197–205.

8

Assessing the Synchroneity of Glacier Fluctuations in the Western Cordillera of the Americas During the Last Millennium

B. H. LUCKMAN AND R. VILLALBA

Abstract

This chapter reviews glacier fluctuations of the last millennium from sites along the Pole-Equator-Pole: Paleoclimate of the Americas (PEP 1) transect. The techniques, limitations, and problems of the interpretation and dating of glacier records are discussed. The paleoclimate signal embedded in glacier records is complex; therefore, this chapter is restricted to areas with a reasonable sample base of land-terminating glaciers that provide directly comparable, detailed, and relatively well-dated records. Particular attention is given to sites poleward of 32°S and 47°N, where dating control is generally better and greater numbers of glaciers have been sampled. Regional accounts are given for the subantarctic, Patagonia, the central Andes of Chile and Argentina, Alaska, the Canadian Rockies, and the Coast Ranges of British Columbia. In several cases, the glacier record is compared with available proxy climate records, derived mainly from tree rings, to evaluate possible linkages between glacier and climate fluctuations.

There is a broad synchroneity in the initiation and timing of Little Ice Age (LIA) glacial events along the transect. Initial glacier advances occur in the thirteenth through fourteenth centuries in the Northern Hemisphere and Patagonia, and the "classic" LIA maximum advances are found in most areas during the seventeenth through early twentieth centuries. Only scattered evidence is available for extended glaciers during the fourteenth through sixteenth centuries. However, at the decadal level, there is considerable variability in both the relative extent and timing of events and the dating of the LIA maximum position. Contrasting glacial histories can be observed even for land-based glaciers within relatively short distances, particularly in climatic transition zones. These differences also reflect the varying dominance of precipitation and temperature controls of mass balance, as well as nonclimatic influences on glacier behavior. Copyright © 2001 by Academic Press.

Resumen

Este capítulo reveé las fluctuaciones glaciales durante el último milenio en sitios ubicados a lo largo de la transecta Polo-Ecuador-Polo (PEP 1). Las técnicas, limitaciones y problemas en la interpretación y datación de los registros glaciológicos son examinados. La señal paleoclimática contenida en los registros glaciológicos es compleja y en este estudio nos limitamos a aquellas áreas con un adecuado muestreo de glaciares cuyos frentes terminan sobre tierra-firme y proveen registros directamente comparables, detallados y relativamente bien fechados. Se prestó particular atención a aquellos sitios localizados al sur de los 32°S

119

y al norte de los 47°N donde el control de las dataciones es generalmente mejor y el número de glaciares estudiados es mayor. Cronologías regionales de las fluctuaciones glaciales más significativas se presetan para el Dominio Sub-Antártico, la Patagonia, los Andes Centrales de Chile y Argentina, Alaska, las Rocallosas de Canadá y la Cordillera de la Costa de la Columbia Británica. En varias oportunidades, el registro glaciológico fue comparado con registros paleoambientales de alta resolución, particularmente con anillos de árboles, con el objeto de evaluar las posibles relaciones entre las fluctuaciones glaciales y climáticas. Existe una cierta sincronización en la iniciación y ocurrencia temporal de los eventos glaciales de la Pequeña Edad del Hielo (LIA) a lo largo de la transecta PEP 1. Los primeros avances glaciales durante los últimos 1000 años ocurrieron en los siglos XIII y XIV en el Hemisferio Norte y en la Patagonia. Los máximos avances glaciales asociados a la "clasica" LIA ocurrieron en muchas regiones desde el siglo XVII hasta los comienzos del siglo XX. Sólo existen evidencias fragmentarias de avances glaciales durante los siglos XV y XVI. En una escala más detallada a nivel de décadas, se observa sin embargo, una considerable variabilidad tanto en la extensión relativa, el momento de ocurrencia y el fechado de la máxima posición alcanzada por los glaciares durante la LIA. Glaciares con frentes en tierra-firme pueden presentar historias glaciales contrastantes, aún en sitios separados entre sí por distancias relativamente cortas, particularmente en zonas con transiciones climáticas acentuadas. Estas diferencias también reflejan la alternante dominancia de la precipitación o la temperatura como regulador del balance de masa de los glaciares a lo largo del transecto PEP 1, al igual que la influencia de factores no-climáticos en el comportamiento de los glaciares.

8.1. INTRODUCTION

The study of glacier fluctuations has traditionally provided many of our ideas about the climatic history of the late Quaternary and Holocene periods. The classic terminology of Holocene climate history still largely reflects its glacial origins, e.g., the widespread misuse of the term Little Ice Age (LIA) to describe a climatic rather than a glacial event (see Bradley and Jones, 1992; Luckman, 2000a). However, recent emphasis on and research into the global climate change issue has focused attention on climate variability at much shorter, decade to century timescales. The rapid development of a wide range of paleoclimatology techniques has led to several high-resolution proxy climate records that span the last millennium and emphasize climate variability

rather than change. The increasing spatial coverage of records also demonstrates considerable regional differences in these records, suggesting that well-known Eurocentric scenarios may not be globally applicable. In light of these developments, it is timely to review and re-evaluate the record of glacier fluctuations during the last millennium along the Pole-Equator-Pole: Paleoclimate of the Americas (PEP 1) transect. We first explore briefly the nature and interpretation of the climate signal in the glacier record and the problems of developing a detailed chronology. We then provide a selective review of some of the more detailed regional glacier records in conjunction with other high-resolution paleoenvironmental data and conclude with some recommendations for future studies.

8.1.1. Controls of Glacier Fluctuations

Glacier dynamics are primarily controlled by mass balance relationships that involve various combinations of precipitation and temperature inputs and may be significantly influenced by nonclimatic factors such as surging or calving behavior. Along the PEP 1 transect, glaciers occur in environments that range from the arid, high-elevation tropics to wet, cold maritime temperate and subarctic (subantarctic) regions. As the relative significance of temperature and precipitation controls on mass balance may vary between regions and also at a site over time, the paleoclimate signal in the glacier record is complex. Spectacular calving glaciers occur at both extremes of the transect and have attracted considerable attention because of their extremely large, rapid frontal variations. It is now recognized that, in many cases, this dynamic behavior is controlled by specific morphological and topographical factors unrelated to climate (e.g., Warren, 1992; Warren and Aniya, in press; Wiles et al., 1995). In selecting sites for discussion, we have focused on land-terminating, nonsurging glaciers to avoid these problems. Secure knowledge of the controls of glacier mass balance is necessary to extract an unambiguous climate signal from studies of past glacier fluctuations.

8.1.2. Nature of the Glacial Record

Glacial sediments and landforms can be used to reconstruct the extent and sequence of former glacier fluctuations in a region. However, these records are discontinuous; they contain little information about intervening, nonglacial intervals; and they must be correlated from site to site based on the available dating control. As later glacier advances destroy most evidence of previous, less extensive events, only the most recent history can be reconstructed in great detail.

Therefore, the glacier record of past environmental changes is incomplete, and the available evidence varies in detail, quality, and dating precision, even within the last millennium. The surface record provides information about glacier extent and the dates of the maximum position and recessional history of the most recent events and may be reconstructed by using a variety of techniques. Dating control is provided mainly by ^{14}C dating, dendrochronology and lichenometry, and historic and documentary records, including aerial photographs (e.g., Porter, 1981a; Villalba et al., 1990; Luckman, 1998a). However, as the maximum Holocene glacier advance in many regions was usually attained in the eighteenth or nineteenth century, these surface records are quite short.

Evidence for glacier events prior to the LIA maximum is incompletely preserved and consists of stratigraphic sections within the LIA limits and anomalous sites where older deposits are locally preserved outside the LIA maximum position (e.g., along lateral moraines). Older glacier fluctuations may also be inferred from depositional sequences in former ice-dammed or proglacial lakes (e.g., Clague and Mathewes, 1992). The dating control in these records commonly consists of maximum and/or minimum limiting ages for detrital or *in situ* organic materials associated with deposits an unknown distance upvalley from the terminal position of the glacier advance. Three types of dating uncertainty can be identified in these stratigraphic records. First, there are the inherent limitations associated with the resolution of the specific dating technique used (usually ^{14}C dating). The relatively large errors associated with ^{14}C ages in this time frame (Stuiver and Pearson, 1993) may make it difficult to resolve differences between various dating scenarios where differences of 50–100 years are critical. Second, uncertainties are associated with the relationship between the dated sample and the stratigraphical unit being dated. If, for example, the dated material is a log, was it *in situ* or reworked? What is the difference between the date of the rings sampled and the inception (minimum age) or termination (maximum age) of growth of the tree? How much time elapsed between the death and the burial of a tree or tree establishment and deglaciation of the surface? The third source of uncertainty is the relationship between the dated section and the maximum extent of the glacier advance. In most cases, the downvalley limit of the advance is unknown (as are possible rates of glacier advance or recession), and it is impossible to establish by how much the dated material pre- or postdates the glacial event. With ^{14}C ages of materials, uncertainties of at least ±50–100 years are to be expected, and, even when tree rings provide precise calendar dates for the sample,

the results from either technique remain only limiting dates.

Further uncertainty arises from differences in the available database for different regions. All paleoenvironmental records contain a mixture of local, regional, and global signals. Given a large sample base of glacier histories, local variants from the regional pattern can easily be detected. However, for those regions or periods where data are limited, the regional significance of individual records cannot be evaluated.

Assessing the synchroneity of glacier records depends on both the nature of the dating control and the specific techniques used. Although LIA glacial deposits can be recognized throughout the cordillera along the PEP 1 transect, for many places detailed dating control is lacking, and few detailed historical studies are available. Therefore, in this chapter, we have restricted the data primarily to areas where glaciers have terrestrial, land-based termini (frequently in predominantly forested environments) that may provide directly comparable and relatively well-dated records. We have also selected areas where complementary data may provide some indication of whether temperature or precipitation is the principal control of glacier fluctuations. These constraints considerably truncate what is already a limited data set and preclude a detailed assessment in many areas. We focus on the extremes of the transect, where dating control is generally better and much greater numbers of glaciers have been sampled. In addition, rather than simply summarizing the available history, we have tried to establish the dating of the LIA and/or (where they are different) the Holocene glacial maxima to provide a consistent reference point throughout the transect.

8.2. REGIONAL ACCOUNTS OF GLACIAL HISTORY

The PEP 1 transect spans the cordillera of the western Americas and contains glaciers over most of its length (Fig. 1). However, in the central part of the transect, glaciers are commonly small and are usually widely scattered and restricted to high mountain areas, frequently well above the tree line. The primary research at most of these glaciers has been focused on the deposits and landforms of the last glacial and/or early Holocene. Dating control for many glacial events is poor and/or controversial (see, e.g., Davis, 1988), and, although LIA deposits are readily recognized by morphology, "freshness," and position in the landscape, few absolute age determinations are available. Even in the Sierra Nevada (ca. 38°N), where the LIA was originally defined by Matthes (1939), the most re-

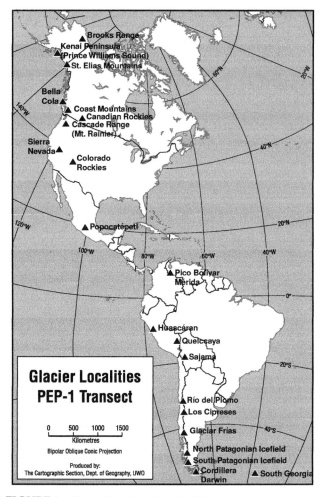

FIGURE 1   Glacier localities along the PEP 1 transect.

ance of tropical glaciers in the high Andes in the late twentieth century, plus the degradation of any potential ice core records they might contain (Thompson, 1995; White, 1981; Granados, 1997); the glaciers on Pico Bolivar near Mérida (see Schubert, 1972, 1992), visited during the PEP 1 meeting, were simply vestigial ice patches. The few available studies of these tropical glaciers indicate that the mass balance–climate relationships are complex (Francou et al., 1995; Ribstein et al., 1995).

Therefore, in presenting regional summaries of the glacier history along the transect, we focus on records from the more heavily glaciated regions poleward of ca. 32°S and 47°N. The relative widths and lengths of the mountain belts in South and North America allow a clearer demonstration of latitudinal variation in the climatic controls of glacier fluctuations in the South American examples. Therefore, we begin with that record.

8.2.1. Southern South America (30°–55°S)

The elevation of the Andes decreases southward from 6000–7000 m in the central Andes of Argentina and Chile to less than 2000 m in southern South America. Along this latitudinal gradient, climate varies from dry subtropical in the north to wet temperate in the south. Precipitation ranges from less than 1 m per year in the central Andes of Chile and Argentina to nearly 10 m on the western coast of southern Patagonia. There are also strong west-east precipitation gradients across the Andes. In the glaciated areas of the central and southern Andes, mountain peaks may be well over 1000 m above the regional snow line (Casassa, 1995).

More than 30 years ago, Mercer (1965) proposed a scheme of three major glacial advances in southern South America during the Holocene. A fourth neoglacial advance during the Holocene has been recently proposed for southern South America (Aniya, 1995). The most recent of these events in southern Patagonia was dated between A.D. 1600 and 1850 (Mercer, 1965, 1968, 1976). Although there has been considerable subsequent interest in Holocene glacier fluctuations in this region (e.g., Clapperton and Sugden, 1988; Grove, 1997), almost all late Holocene glacier chronologies are based on radiocarbon dating of organics (usually wood) in subsurface deposits. The relatively large uncertainties associated with radiocarbon ages in this time frame (see earlier) and the associated problems of limiting dates suggest that most subsurface radiocarbon ages can be resolved only within ±50–100 years. Decadal-annual resolution is rarely attainable, and then only for a few studies that utilize tree-ring dating, lichens, or historical records during the last few cen-

cent account dates only the Matthes (LIA) moraines as younger than 720 ^{14}C years B.P. based on tephrochronology (Clark and Gillespie, 1997). Limited observations of former glacier extent and snow line levels are available from a variety of historical and archival sources in México and Ecuador that extend back 400–500 years (e.g., Hastenrath, 1981, 1997; White, 1981; Vázquez-Selen and Phillips, 1998). Such records indicate, for example, significantly lower historical snow lines near Quito and little change in glacier equilibrium line altitude (ELA) from the 1500s through the early 1800s (Hastenrath, 1997). The most extensive glaciated area in the northern Andes is in Peru, but there are few dated records of LIA glacier fluctuations prior to the twentieth century (see Grove, 1988). Excellent paleoenvironmental records have been recovered from the ice cores at Quelccaya, Huascarán, and Sajama in the high Andes (e.g., Thompson et al., 1984, 1998), but such records are not the focus of this chapter. There is widespread evidence of the rapid recession and disappear-

turies. Synchroneity can be established only in the broadest terms for events predating the local LIA maximum.

An intensive search in the glaciological literature from South America indicates that high-resolution glaciological records for the past 1000 years are extremely rare. Most of these records were obtained during the twentieth century by using aerial photographs (since midcentury) or satellite images (e.g., Warren and Sugden, 1993; Aniya and Enomoto, 1986; Aniya et al., 1997) and are not directly relevant to this chapter. At most localities along the Andes, glaciers reached their LIA maxima prior to 1950. In the following sections, we briefly review the few high-resolution records of glacier fluctuations for the last 1000 years in the central Andes of Argentina and Chile, in northern and southern Patagonia, and in the subantarctic domain. For each of these regions, precisely dated glacial advances are compared with dendrochronological records to cross validate the glacial chronology and to evaluate the possible climate signal represented by these records.

8.2.2. Central Andes of Argentina and Chile

Videla (1997) used old scientific reports, maps, and photographs to document precisely changes in glacier extent on the eastern slope of the central Andes (32°–33°S). The glaciers of the Río del Plomo, Mendoza, reached their maximum LIA extents at the beginning of the twentieth century. According to Videla (1997), no Holocene moraines or other glacial deposits have been observed downstream from the 1910 limit. During the twentieth century, glaciers have generally retreated. Since ca. 1910, the Alto del Río del Plomo Glacier has retreated ca. 400 m in elevation and almost 5000 m in length.

On the western slopes of the Andes, at the same latitude as the Río del Plomo basin (32°S), LaMarche et al. (1979) developed the northernmost tree-ring chronology from *Austrocedrus chilensis* in central Chile. This chronology has been used largely as a proxy record of precipitation variations for central Chile (LaMarche, 1975; Boninsegna, 1988; Villalba, 1990a). Recently, the site was resampled, and the data from the new collection were merged with LaMarche's tree-ring series. The new chronology, which covers the interval A.D. 956–1996, provides a basis to compare glaciological and dendrochronological records from the xeric Andes at 32°S. Replication exceeds 10 samples after A.D. 1200. Except for two short intervals centered in 1860 and 1890, the 1820–1906 interval is the longest period of high sustained growth in this record. Figure 2 shows that *Austrocedrus* ring widths are strongly correlated with precipitation during the period of the instrumental record. This result suggests that the 1820–1910 interval was the longest wet period in the central Andes during the past eight centuries.

Under the extreme dry conditions that prevail in the central Andes, glacier variations are strongly affected by precipitation variations (Leiva et al., 1986). As a consequence, tree-ring variations from the precipitation-sensitive chronology from central Chile support Videla's (1997) observations. At about 32°S, glaciers reached their maximum extent at the beginning of the twentieth century, coincident with the termination of the wettest period in the Andes during the past eight centuries (as inferred from tree-ring variations). In agreement with these results, decadal mean values of rainfall at two stations in central Chile, La Serena (30°S) and Santiago (33°S), show that marked wet years at the end of the nineteenth century were followed by a decreasing trend continuing to the present (Aceituno et al., 1993). At La Serena, decadal mean deviations in precipitation during the period 1880–1900 were greater than 1 SD above the long-term mean.

At 35°S, in a transitional zone to the wetter climatic conditions prevailing southward, glaciers reached their maximum extension during the last millennium sometime earlier than the glaciers located at 32°S. On the western Chilean slopes, radiocarbon ages indicate advances of the Los Cipreses Glacier (34°S) at the beginning of the fourteenth century and around the 1860s (Röthlisberger, 1987). Historical records show that the Ada Glacier extended down to 1800 m in 1858. The elevation of the toe had risen to 1930 m by 1882 and to 2500 m by 1980 (Cobos and Boninsegna, 1983). At about 35°S, in the upper Río Atuel basin in Argentina, the glaciers have undergone almost constant recession since they were observed by Hauthal (1895). At that time, the lower portion of the Humo Glacier, in the Río Atuel basin, consisted of large pieces of dead ice covered by debris (Cobos and Boninsegna, 1983). Cobos and Boninsegna also reconstructed streamflow of the upper Río Atuel from 1534–1968 using three *Austrocedrus* chronologies from central Chile. The reconstruction shows greatly increased runoff between 1820 and 1850, with a second runoff peak in the early 1900s. These high runoff periods could reflect higher precipitation, greater glacier melt and runoff, or a combination of these effects. However, the lack of dating for glacier fluctuations prior to 1895 prevents the exploration of possible linkages between these runoff peaks and glacier fluctuations. It seems probable that glaciers reached their maximum LIA extent in the mid-nineteenth century and underwent significant melting and stagnation by 1895. They have retreated consistently since the 1910s.

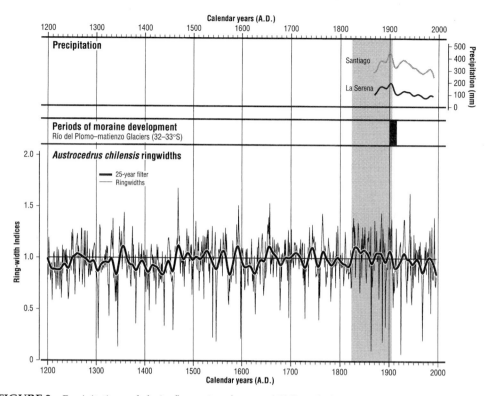

FIGURE 2   Precipitation and glacier fluctuations for central Chile and adjacent Argentina. The upper curves are instrumental precipitation records from Santiago and La Serena, Chile (25-year cubic spline filter). Glacier fluctuations are from Videla, 1997. The lower curve is an indexed tree-ring chronology for *Austrocedrus chilensis* in central Chile, A.D. 956–1996 (LaMarche et al., 1979, revised and updated). This is interpreted as a proxy precipitation record.

8.2.3. Northern Patagonia

For northern Patagonia, most information is available from the glaciers of Monte Tronador, located on the border between Chile and Argentina at 41°S. Radiocarbon ages of tree trunks overridden by the Río Manso Glacier indicate major glacier advances in ca. A.D. 1040, 1330, 1365, 1640, and during the period 1800–50 (Röthlisberger, 1987). For the nearby Frías Glacier, two major glacial advances were identified for the intervals 1270–1380 and 1520–1670 (Villalba et al., 1990). However, only the latter advance has been adequately dated by using tree-ring records. Dating for the massive moraine, which represents the maximum extent of the Frías Glacier during the past 1000 years, is based on ring counts from the oldest tree found on the moraine, plus precise dates from trees damaged by the glacier along its margin during the latest neoglacial advance (Villalba et al., 1990).

Measurements of the terminal position of the Frías Glacier since 1976 indicate a strong relationship between summer climatic conditions (particularly tem-

perature variations) and fluctuations of the position of the Frías Glacier front (Villalba et al., 1990). On the basis of the sparse available data, Warren et al. (1995) also propose that glacier fluctuations on the more continental eastern slopes of the Andes are responding to temperature variations. Figure 3 compares these glacier records with a recently developed summer temperature reconstruction for the eastern side of the northern Patagonian Andes based on 17 multimillennial records for *Fitzroya cupressoides* in northern Patagonia (Villalba et al., 2000). The most striking feature in the 1000-year-long temperature reconstruction is the long, cold interval that extends from ca. 1500 to 1660. Severe, cold summers were particularly frequent from 1630–1650. The occurrence of this century-long cold interval is consistent with the glacial record, which indicates that the maximum extent of the Frías Glacier during the last millennium was between ca. 1660 and 1670 at the end of this cold period. Dating of several of the recessional moraines at the Frías Glacier also coincides with the timing of colder-to-warmer transitions of summer temperatures (see Fig. 3 and discussion in Villalba et al., 1990).

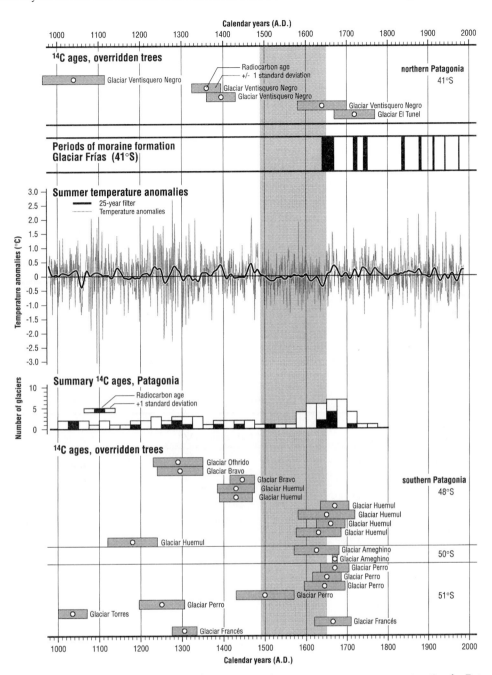

FIGURE 3   Dating of land-based glacier fluctuations and a summer temperature reconstruction for Patagonia. The upper and lower boxes show radiocarbon ages (calendar equivalent ±1 SD) from trees overridden by glaciers in glacier forefields. For sites and sources, see Table 1. Glaciar Frías moraine dates are based on dendrochronology (Villalba et al., 1990). Summer temperature anomalies (25-year cubic spline filter) are reconstructed from 17 *Fitzroya cupressoides* chronologies in the northern Patagonian Andes of Argentina (Villalba et al., 2000). The histogram summarizes the dating based on individual glacier records. Limiting ^{14}C ages ±1 SD are grouped in 20-year intervals for each glacier record. Where multiple ages are available for the same event in a forefield, the mean ^{14}C age is used with the absolute range of all the standard deviation estimates.

8.2.4. Southern Patagonia

Glaciers of the North and South Patagonian ice fields (46°–52°S) have responded to recent climatic changes in a variety of ways. During the twentieth century, some large calving outlet glaciers have advanced to their Holocene maxima, whereas others have fluctuated rapidly (e.g., Warren, 1993; Warren and Sugden, 1993). Glacier Pio XI (Bruggen) in Chile advanced 10 km between 1945 and 1976 and was at its Holocene maximum and advancing in 1993 (Warren and Rivera, 1994; Warren et al., 1997). Such variability is typical of the nonlinear response of calving glaciers to climate variations. These westward flowing glaciers are also thought to be responding primarily to changes in precipitation rather than temperature (Warren et al., 1995). Consequently, climatic inferences from these calving glaciers in southern Patagonia are problematic. In contrast, most land-terminating glaciers of the northern and southern Patagonian ice fields show a consistent, accelerating trend of recession despite individual variations (Aniya and Enomoto, 1986; Aniya, 1992). Although prominent

LIA moraines were developed by many of these glaciers, their precise dating remains uncertain. For the Ameghino Glacier, radiocarbon ages indicate that the culmination of the recent maximum neoglaciation occurred sometime between A.D. 1600 and 1700 (Aniya, 1996). This glacial advance probably began in ca. A.D. 1500, when the Ameghino valley was filled with gravel that killed most of the trees along it (Aniya, 1996).

Few tree-ring records are available for Patagonia between 46° and 52°S. For southern Patagonia, well-replicated tree-ring records start at ca. A.D. 1800, and chronologies are much shorter than those available for northern Patagonia. Due to the lack of meteorological stations in the region, direct comparison of tree-ring variations with climate is difficult.

The sparse data available on glacier advances from sites along the Patagonian Andes show a certain degree of synchronism between northern and southern Patagonia during the past few centuries. Figure 3 and Table 1 summarize the radiocarbon ages associated with overridden trees within selected glacier forefields in Patagonia (land-based termini only). For these data, the

TABLE 1   Radiocarbon Ages from Glacier Forefields in Patagonia

Site	Date (cal. year A.D.)[a]	Source
Glaciar Ventisquero Negro, Argentina	1040 ± 75	Röthlisberger, 1987
	1360 ± 35	Röthlisberger, 1987
	1395 ± 35	Röthlisberger, 1987
	1640 ± 60	Röthlisberger, 1987
Glaciar El Tunel, La Esperanza, Argentina	1720 ± 50	Villalba et al. (unpublished data)
Glaciar Ofthrido, Chile	1290 ± 60	Mercer, 1970
Glaciar Bravo, Chile	1650 ± 70	Röthlisberger, 1987
	1670 ± 35	Röthlisberger, 1987
Glaciar Huemul, Chile	1180 ± 60*[b]	Röthlisberger 1987
	1295 ± 55	Röthlisberger, 1987
	1430 ± 40*	Röthlisberger, 1987
	1430 ± 45*	Röthlisberger, 1987
	1445 ± 45*	Röthlisberger, 1987
	1630 ± 55	Röthlisberger, 1987
	1660 ± 35	Röthlisberger, 1987
Glaciar Ameghino, Chile	1625 ± 55	Aniya, 1996
	1670	Aniya, 1996[c]
Glaciar Perro, Chile	1250 ± 55	Röthlisberger, 1987
	1500 ± 70	Röthlisberger, 1987
	1645 ± 50	Röthlisberger, 1987
	1650 ± 35	Röthlisberger, 1987
Glaciar Torres, Chile	1035 ± 35	Röthlisberger, 1987
Glaciar Francés, Chile	1305 ± 30	Röthlisberger, 1987
	1665 ± 45	Röthlisberger, 1987

[a] All calendar-year equivalent ages are derived from Stuiver and Reimer (1986).
[b] All trees are *in situ* except for those marked with an *.
[c] No correction term is given in the text.

summary histogram shows the number of glacier fore-fields with radiocarbon ages in each 20-year interval, plotting both the actual age and a range of ±1 SD. The major peak in this histogram indicates that most of these land-terminating glaciers in Patagonia reached their maximum extent between 1600 and 1700. There is no strong pattern for the other radiocarbon ages, although four glaciers were advancing between ca. 1260 and 1320. Unfortunately, it is not known whether these specific radiocarbon ages are for the outer rings of trees, but these ages do suggest multiple glacier advances. Patterns for the late 1200–1300s and 1600–1700s are replicated at several sites. The 1600–1700 advance coincides with the longest period of cooler summer temperatures from the tree-ring reconstruction and the dating of the main LIA moraine at Frías Glacier. Villalba (1990b) showed that during the twentieth century, summer temperature variations were relatively uniform on the eastern side of the Andes from about 37° to 50°S, which lends support to the idea of synchroneity for the latest neoglacial maximum for land-terminating glaciers within Patagonia. However, more precise dates for the occurrence of the most recent neoglacial maximum in both northern and southern Patagonia are needed to substantiate this conclusion.

8.2.5. Subantarctic Domain

Holmlund and Fuenzalida (1995) noted contrasting responses of glaciers from the Darwin Cordillera (55°S), Tierra del Fuego, during the twentieth century. Glaciers on the northern and eastern sides of the cordillera have been shrinking since the turn of the century. In contrast, the south-facing glaciers have recently reached positions at or close to their Holocene maxima. Most glaciers from the Darwin Cordillera calve, which complicates the climatic interpretation of their glacial chronologies.

Clapperton et al. (1989) have documented glacier fluctuations on the Sub-Antarctic Islands in the South American sector of the Southern Ocean. The most detailed glacial record is from the island of South Georgia (55°S), which shows a glacial advance between the seventeenth and nineteenth centuries. According to Clapperton et al. (1989), this last neoglacial advance was associated with an ELA lowering of about 50 m and a decrease in temperature of 0.5°C. No more precise dates are available for this event. More recently, Gordon and Timmis (1992) have examined glacier fluctuations in South Georgia during the 1970s and compared the present activity with that over the last 100 years. They observed two different responses of glaciers in relation to their sizes. Small glaciers, which appear to track year-to-year climatic variations, have receded since the

1930s. In contrast, large valley glaciers show a somewhat different response, reaching their most advanced positions since the LIA (seventeenth through nineteenth centuries) during the 1970s (Fig. 4). These large valley glaciers have thinned and receded since the 1970s.

Timmis (1986) compared the areal changes of a group of glaciers of similar size, type, and morphology in three regions around South Georgia. Along the southwest coast, at two of the three glaciers studied, the twentieth-century advance overrode the seventeenth- through nineteenth-century ice limits, whereas at the third glacier the two advances were of comparable extent. This behavior contrasts with that of glaciers on the northeastern coast where the twentieth-century advances terminated some 50–250 m inside the seventeenth- through nineteenth-century limits (Timmis, 1986).

These glaciological observations from both the Darwin Cordillera and South Georgia show contrasting patterns of glacier fluctuations that reflect the topographic locations of the glacier sites. Glaciers in southwest-facing sites are more fully exposed to the prevailing westerly circulation. In South Georgia, glaciers located on the southwest-facing slopes have almost double the accumulation and experience ca. 30% greater cloud cover, and at ca. 0.7°C they experience lower temperatures than do glaciers on the northeast-facing slope (Gordon and Timmis, 1992).

Tree-ring records from the subantarctic domain (Tierra del Fuego and New Zealand) were used to characterize the changes in mean sea level pressure (MSLP) around Antarctica from A.D. 1750 (Villalba et al., 1997). For the South American sector of the Southern Ocean, meteorological records from Grytviken (South Georgia) and Orcades and Signy Islands (both in the South Orkney Islands) were used to develop a regional record of MSLP for the oceanic region. The tree-ring-based MSLP reconstruction for this sector shows that the most persistent interval of low summer pressure during the past 250 years occurred between 1930 and 1970, peaking during the 1960s (Fig. 4). Low pressure in the subantarctic domain is associated with more frequent cyclones and, therefore, higher precipitation, increased cloudiness, and lower temperatures, particularly in summer. Since the 1970s, summer pressure has steadily increased. Long-term changes in pressure patterns at high latitudes in the Southern Ocean (as recorded by tree rings) can be invoked to explain some of the differences in the response of glaciers to climate variations in the subantarctic domain. Concurrent with the long-term negative trend in MSLP from above average during the 1850s to below average in the 1960s, large valley glaciers in South Georgia reached their maximum

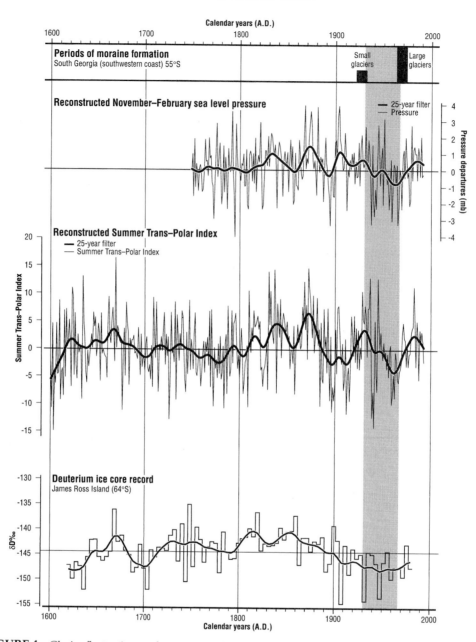

FIGURE 4 Glacier fluctuations and reconstructed atmospheric pressures for South Georgia. Glacier data are for the southwest coast (Gordon and Timmis, 1992). The black bar indicates the timing of the late neoglacial maximum moraines. On the northeast coast, the twentieth-century advance did not overrun the seventeenth-through nineteenth-century moraines. The reconstructed November–February sea level pressure (SLP) is based on five southern beech ring-width chronologies (*Nothofagus betuloides* and *N. pumilio*) from Tierra del Fuego. The reconstructed Summer Trans-Polar Index is based on five New Zealand pine chronologies (three *Holocarpus biformis* and two *Lagarostrobus colensi*) and five southern beech chronologies. Tree-ring data are from Villalba et al. (1997). The lower curve is the oxygen isotope record (4-year mean values) from an ice core on James Ross Island, 64°S (Aristarain et al., 1990).

extent since the seventeenth- through nineteenth-century advances during the 1970s. These observations are also consistent with a 400-year deuterium ice core record from James Ross Island (64°S), which shows a temperature decline of ca. 2°C between 1850 and 1960 (Aristarain et al., 1990).

On the other hand, interannual variations in the

MSLP reconstruction for the South American sector of the Southern Ocean show that the most extreme low annual values of MSLP occurred during the 1930s, when the small year-to-year climatically sensitive glaciers in South Georgia reached their maximum twentieth-century extents. Since the 1970s, these small and larger glaciers have been receding in response to warmer and drier conditions, as reflected by the steady increase in MSLP from 1970 to the present (Fig. 4).

8.2.6. Summary of the South American Record

We have reviewed the scant high-resolution records of glacial variations in relation to the LIA maximum along southern South America, from the subtropical xeric central Andes at 30°S to the wet maritime mountains at 55°S in Tierra del Fuego and the Sub-Antarctic Islands. Far from showing a synchronous pattern along the Andes, glaciers have reached their maximum extents during the last 1000 years at different times from one region to another. Contrasting patterns can be observed between glaciers that are separated by only a few kilometers. This contrast is particularly true for glaciers located in climatic transition zones, for example, between the central Andes and northern Patagonia. Glaciers at 32°S reached their LIA maxima during the first decade of this century; however, at 35°S this event occurred earlier. Contrasting patterns have also been documented for the glaciers in the Darwin Cordillera and the Sub-Antarctic Islands. These glaciers are also located in a climatic transition zone, where steep environmental gradients play an important role in defining the glacial patterns. This marked environmental gradient at about 50°S is also reflected in the tree-ring pattern. Tree growth at Tierra del Fuego (53°–54°S) is not correlated with climate variations in Patagonia north of 51°S, but it is significantly related to climatic variations at 61°S (Orcades and Signy Island).

In contrast, a certain degree of synchronism in glacial variations is observed in climatically uniform regions. The sparse glacial record indicates that land-based glaciers in both northern and southern Patagonia reached their latest maximum extents between A.D. 1600 and 1700. In agreement with these observations, Villalba (1990b) showed that tree-ring records for northern Patagonia at about 41°S explain more than 30% of the total variance in summer temperature variations at Río Gallegos (51.5°S) and 20% at Punta Arenas (53°S), but only 2% at Ushuaia (54.5°S). These results indicate that the relatively uniform pattern of summer temperatures over Patagonia east of the Andes breaks down at 51°–52°S in response to the steep climatic gradient imposed by the interactions between middle and high latitudes in the Southern Hemisphere.

8.3. NORTH AMERICA

The North American glacial record is known in much greater detail than that of South America, but calendar-dated sequences through the last millennium are still rare. In addition, because of the differing latitudinal distribution of landmasses in the two hemispheres, many of the North American sites are poleward of their Southern Hemisphere counterparts. In this chapter, we will concentrate primarily on material that postdates the accounts given in Grove (1988), Calkin (1988), Davis (1988), and Osborn and Luckman (1988).

8.3.1. Alaska

The literature on LIA glacier fluctuations in Alaska is voluminous and comprehensively reviewed in Calkin (1988), Calkin and Wiles (1992), and Calkin et al. (in press). The most interesting recent work has come from coastal Alaska, for which dendrochronological studies have provided calendar dates for forest beds overridden by glacier advances during the last millennium (Wiles et al., 1995; Wiles and Calkin, 1994; Wiles et al., 1999a; Wiles et al., 1999b). In a study of 16 land-based glaciers on the Kenai Peninsula (59°–60°N), Wiles and Calkin (1994) demonstrated slight differences in the timing of periods of glacier advance and recession on the eastern and western sides of the peninsula. Glaciers on the western (cooler and drier) side of the peninsula were advancing and/or building moraines in ca. A.D. 1420–60, 1640–70, 1750, and between 1880 and 1910. Glaciers on the maritime eastern flank were advancing in ca. A.D. 1440–60, 1650–1710, and 1830–60. Recession from LIA maximum positions occurred between the early 1700s and late 1800s on the west and between the early 1800s and the early 1900s on the east. Examination of meteorological and proxy climate records suggests that the glaciers on the western flank expanded during times of lower arctic temperatures, whereas the advance of the eastern maritime glaciers corresponded with warmer temperatures and increased winter precipitation (Wiles and Calkin, 1994).

The record from coastal glaciers in Prince William Sound (ca. 60°–61°N; Wiles et al., 1999a) shows a consistent pattern of decade–century glacier fluctuations. Widespread glacier advances occurred in the twelfth through thirteenth and late seventeenth to early eighteenth centuries. Two major moraine-building episodes (28 moraines, 15 glaciers) have been identified in the early eighteenth century and the late nineteenth century, with the early eighteenth-century event as the most extensive advance for five glaciers. At three glaciers, the greatest advance was in 1874–95, but moraines of this advance are just within the seventeenth-century moraines at several other glaciers. Glaciers were close to their maxima when the earliest scientific visits to the area took place in the late 1800s.

For the Juneau ice field area (58°–59°N), Calkin and Wiles (1992) reported earlier work showing periods of glacier advance based on sheared stumps at ca. A.D.

1265, 1401, and 1752 at Davidson Glacier and ca. 1600 at Bucher Glacier. Nine other glaciers have mid- to late-eighteenth-century maximum positions and re-advanced to form end moraines in the early to late nineteenth century.

Tidewater-calving glaciers in Alaska have shown spectacular fluctuations with a range of maximum ages that may or may not be synchronous with adjacent land-based termini. Major glacier systems dated include: Glacier Bay, ca. A.D. 1700–50 (in Calkin, 1988); Brady Ford, 1886 (Calkin and Wiles, 1992); Lituya Bay, ca. 1600 (Calkin and Wiles, 1992); La Perouse Glacier, 1890s (Calkin and Wiles, 1992); Icy Bay, 1880 (Calkin et al., in press); Taku Glacier, 1750 (maximum, but has advanced ca. 7.3 km between 1890 and 1990; Motyka and Begêt, 1995); and the Yakutat Bay complex, 1250–1500 (Calkin and Wiles, 1992).

Figure 5 (Calkin et al., in press) summarizes glacier history from the forefields of land-based, nonsurging

glaciers in coastal southern Alaska (Prince William Sound and adjacent areas). The subsurface record indicates that glaciers were advancing over forested areas during the late 1200s to the early 1300s and from the early seventeenth century to the early eighteenth century. Limited evidence for advances in the mid-1400s at Beare, Grewingk, and Tustemena Glaciers is omitted from Fig. 5. The surface moraine record indicates that well over half the glaciers reached their LIA maxima between 1800 and 1875, although about one quarter of the glaciers reported had their maxima during the eighteenth century. Many glacier forefields contain re-advance moraines dating from the mid- to late nineteenth century.

Using temperature-sensitive tree-ring chronologies, Wiles (1997) identified 1650–1750 as the main cooler interval in the last 300 years in the Gulf of Alaska and inferred that March–May temperatures are the major control of glacier fluctuations in this area. Figure 6

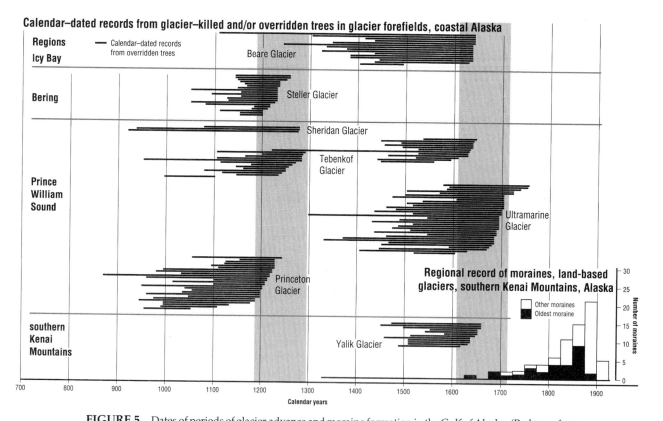

FIGURE 5 Dates of periods of glacier advance and moraine formation in the Gulf of Alaska. (Redrawn from Calkin et al., in press.) The black horizontal bars are calendar-dated records from overridden (subsurface) trees in glacier forefields. At each site, the youngest outermost date provides a minimum-age estimate for glacier overriding at the site. The range of outermost dates provides an indication of the minimum duration of the advance. The histogram shows dates of moraine formation for land-based glaciers in Prince William Sound and the southern Kenai Mountains. Shaded bars are the oldest moraine at each glacier site; open bars represent subsequent (re-advance) moraines.

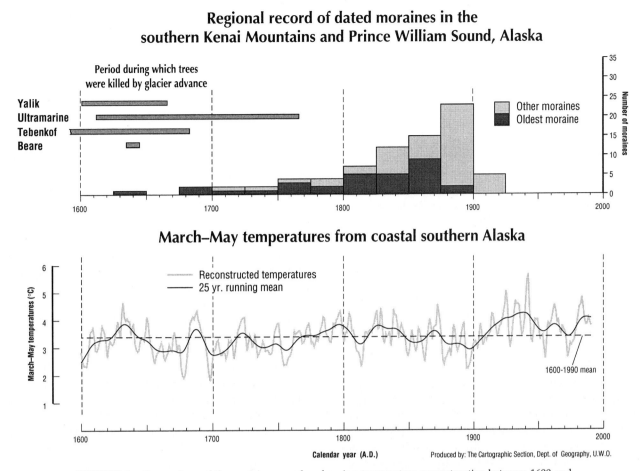

FIGURE 6   Comparison of the moraine record and spring temperature reconstruction between 1600 and 1990 from the southern Alaska coast. The glacial record is from Calkin et al. (in press). The temperature record is based on ring-width chronologies from three temperature-sensitive tree-line sites (*Tsuga mertensiana*; Wiles, 1997; Wiles et al., 1996).

compares the summary record of glacier fluctuations in the southern Kenai Mountains and Prince William Sound regions (Calkin et al., in press) with March–May temperatures reconstructed from coastal tree-ring chronologies (Wiles, 1997; Wiles et al., 1996). The reconstructed periods of lower temperatures correspond with the kill dates for trees of the forests overridden in the 1600s and early 1700s and with the dating of the LIA maximum and re-advance moraines in the nineteenth century.

For interior Alaska and the adjacent Yukon Territory, tree-ring dates are rarely available; dating control is usually provided by lichenometry and radiocarbon dating where suitable materials are present. For the Brooks Range (67°–69°N), Haworth (1988; see Calkin and Wiles, 1992) provides lichenometric data for almost 100 moraines and suggests ages clustering around ca.

AD. 1200, 1570, and 1860 (Fig. 7). The "1570" moraines are the most widely distributed and are found at 85 glaciers, though about 45 glaciers reached their maxima in the 1200s. The 1200 and 1570 events have maximum ELA depressions of 100–200 m (Calkin and Wiles, 1992). These dates indicate that LIA maximum positions were reached much earlier in this interior region than elsewhere along the PEP 1 transect. In the St. Elias-Wrangell Mountains (61°–62°N), 11 glaciers expanded during ca. 1169–959 B.P., and four moraine phases are identified based on lichenometry with dates of ca. A.D. 1500 and the late eighteenth to mid-nineteenth, late nineteenth, and early twentieth centuries (Denton and Stuiver, 1967; Denton and Karlén, 1977; Calkin, 1988). However, the physical expansion of glaciers is limited by earlier Holocene moraines in this region. A similar situation occurs in the Yukon, where mid- to late-nine-

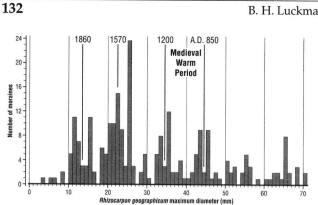

FIGURE 7 Frequency diagram of *Rhizocarpon geographicum* maximum diameters from late Holocene moraines of the Brooks Range, Alaska, with approximate ages. (After Haworth, 1988; Calkin and Wiles, 1992.)

teenth-century LIA maxima were restricted by earlier Holocene deposits. In the St. Elias Range (60°–62°N), White River ash (1230 B.P.) is found in moraines of the Klutlan, Natazhat, Guerin, Giffen, and Russell Glaciers and indicates glacier advance between ca. 1250 and 1000 ^{14}C B.P. (Rampton, 1970, 1978). The LIA maximum of the Kaskawulsh, Donjek, Steele, Silver, and Carne Glaciers occurred in the mid-nineteenth century (Denton and Stuiver, 1967; Sharp, 1951). Dendrochronological dating of driftwood from the beaches of neoglacial Lake Alsek, formerly dammed by the Lowell Glacier, indicate that the 595 and 623 m shorelines postdate A.D. 1848 and 1736, respectively (Clague et al., 1982; Clague and Rampton, 1982). Higher shorelines at 637 and 640 m are probably 400–500 years old (Clague and Rampton, 1982). As higher lake levels probably reflect a larger glacier dam downstream, these dates would suggest that the LIA maximum of Lowell Glacier predates A.D. 1400–1500.

In summarizing the LIA of Alaska, Calkin and Wiles (1992) comment that the LIA was the most extensive Holocene event, with advances of both land- and fjord-based glaciers in the thirteenth century (perhaps later in the interior, ca. fifteenth century). Glaciers reached their Holocene maxima between the sixteenth century and the early to middle eighteenth century. The early and middle 1700s were times of culmination of advances in southern maritime areas. Considerable glacier expansion took place in the nineteenth century, when some glaciers attained their Holocene maxima. During the twentieth century, most glaciers receded, except for those in areas of high precipitation or for some calving or surging glaciers. Present ELAs are 100–200 m higher in the interior and 300–400 m higher in maritime areas than their LIA maximum equivalents.

8.3.2. Rocky Mountains of Canada and the United States

For the continental interior sites of the Canadian Rockies (50°–52°N), there is an extensive database of dated LIA moraine sequences (e.g., Heusser, 1956; Luckman and Osborn, 1979; Smith et al., 1995) most recently reviewed by Luckman (2000a). As with the Alaskan record, the initial dating for earlier events was based on *in situ* or detrital overridden trees that yielded ^{14}C ages between ca. 1100 and 450 B.P. from the forefields of the Kiwa, Robson, Peyto, and Stutfield Glaciers (Heusser, 1956; Luckman, 1986; Luckman et al., 1993; Osborn, 1996). Subsequent tree-ring dating of some of this material has yielded calendar dates that indicate the earliest LIA advances were between ca. A.D. 1150 and 1370 at the Robson (Luckman, 1995), Peyto (Luckman, 1996b, 2000b), and Stutfield (Robinson, 1998; see Fig. 8) Glaciers. At several sites, there has also been dating of much younger overridden and detrital material using dendroglaciological techniques (Luckman, 1996b, 1998b; Robinson, 1998; see Fig. 8). In most cases, this material is associated with a single late-eighteenth-century and early-nineteenth-century glacier advance that culminated in the mid-nineteenth century. However, there is evidence at two sites (Manitoba and Saskatchewan Glaciers; Fig. 8; Robinson, 1998) that trees were also overridden by an earlier advance in the late seventeenth century.

The surface glacial record is based on lichenometric and tree-ring dating of moraines at over 65 glaciers (Luckman, 1996b, 2000a; Fig. 8). Scattered evidence is available for glacier fluctuations between ca. A.D. 1400 and 1600; some moraines are dated to this period by lichenometry or dendrochronology, but the morphological evidence is fragmentary and most are considered limiting dates. Two major subsequent moraine-building periods are identified in the early eighteenth century and the mid to late nineteenth century. The maximum regional extent of ice probably occurred in the first half of the nineteenth century, although at many glaciers extents were slightly greater in the eighteenth century; there are regional differences in this pattern within the Canadian Rockies (Luckman, 2000a). At many sites, glaciers re-advanced several times in the late nineteenth century to positions close to the LIA maximum moraines. During the twentieth century, the dominant pattern has been one of glacier recession with a minor re-advance during the 1960–80 period at several sites (Luckman et al., 1987).

Figure 9 shows a recent spring–summer (April–August) temperature reconstruction based on tree-ring densitometry from *Picea engelmannii* and *Abies lasiocarpa* at the Columbia Icefield (Luckman et al., 1997).

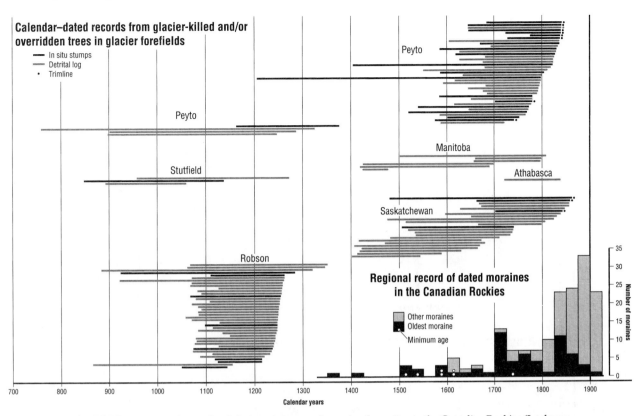

FIGURE 8   Dates of periods of glacier advance and moraine formation in the Canadian Rockies (Luckman, 2000a). Horizontal lines represent calendar-dated records of the minimum life span of glacier overridden or killed trees in six glacier forefields. Outer ring dates from logs at Manitoba and Saskatchewan Glaciers suggest that these assemblages may contain trees killed during more than one glacier advance. The moraine data are based on lichenometric and tree-ring dating in 66 glacier forefields (see Luckman, 1996a, 2000a).

This record shows a strong agreement between periods of glacier advance/moraine building and periods of lower summer temperatures in the 1200s through 1300s, late 1600s through early 1700s, and throughout the nineteenth century. This agreement suggests that summer temperatures are a primary control of glacier fluctuations in this region. Recently, Watson (1998; Luckman and Watson, 1999) has provided the first precipitation reconstruction for this region of the Canadian Rockies. Examination of this record (Fig. 9) shows little obvious correlation with the glacier record, except possibly for the late 1600s and the mid-twentieth-century advance. However, the changes in mass balance at Peyto Glacier since 1976 (Luckman, 1998b; Demuth and Keller, 2000) indicate that increasingly negative mass balances reflect decreasing winter accumulation rather than summer temperature influences.

LIA moraines (late neoglacial or Gannett Peak equivalents) have been extensively reported for sites in the American cordillera (see Davis, 1988), but have rarely been precisely dated. The most commonly utilized dating technique has been lichenometry. In the

Colorado Rockies (ca. 40°N), lichenometric dating suggests the oldest LIA moraines date from the late seventeenth century through the early eighteenth century (Benedict, 1973), similar to those in the Canadian Rockies. For intervening sites in Glacier National Park, Montana (49°N), Carrara and McGimsey (1981) report tree-ring-determined LIA maximum dates of ca. 1860 for two of the larger glaciers. Most other glaciers have single late Holocene (LIA) moraines that are assumed to date from the mid-nineteenth century (Carrara, 1987). Over half of the more than 150 small glaciers that were present in Glacier National Park in the 1850s have subsequently completely disappeared (Carrara, 1987).

8.3.3. Coast Ranges of British Columbia and the Pacific Northwest of the United States

Glacier fluctuations in the Coast Ranges of western British Columbia show a broadly similar pattern to that in the Canadian Rockies. There is a growing body of data from subtill radiocarbon ages and lacustrine

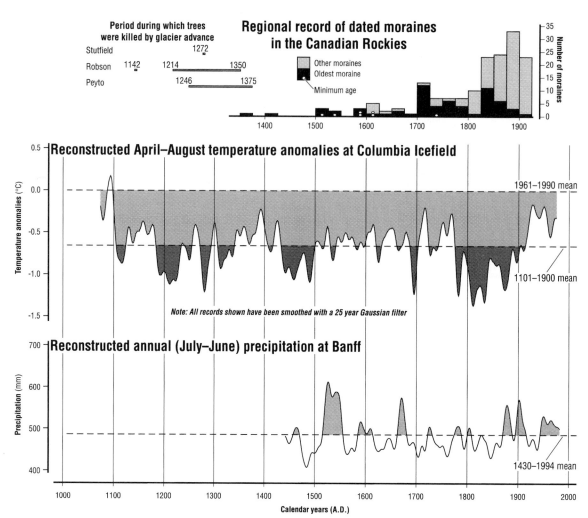

FIGURE 9 Regional glacier, proxy temperature, and precipitation records for the Canadian Rockies. The regional moraine record is from Luckman (2000a). Temperature reconstruction is from Luckman et al. (1997) using ring-width and densitometric data for *Picea engelmannii* and *Abies lasiocarpa*. Precipitation reconstruction is from Watson (1998) using ring-width data for *Pseudotsuga menziesii* at the Powerhouse site near Banff. Both tree-ring records are smoothed with a 25-year filter.

episodes that indicates initial LIA advances between 900 and 400 [14]C B.P. Ryder and Thomson (1986) report subtill maximum ages of ca. 900 ± 40 ^{14}C years B.P. at Klinaklini Glacier, 835 ± 40 at Franklin Glacier, 680 ± 50 and 540 ± 45 at Bridge Glacier, and 460 ± 40 at Sphinx Glacier in the southern Coast Mountains (49°–51°N). Ryder (1987) provides subtill ^{14}C ages of 455 ± 65 and 625 ± 140 at Scud Glacier, 595 ± 60 at Glacier B in the Stikine area (56°–57°N), and 400 ± 60 for the Jacobsen Glacier in the Bella Coola area (52°N; Desloges and Ryder, 1990). The range of these ^{14}C ages indicates the possibility that there may have been several periods of glacier advance between the twelfth and sixteenth centuries, but this can be demonstrated only in the Bella Coola area. Desloges and Ryder (1990) report on a

section from the lateral moraine of Purgatory Glacier indicating that the glacier advanced over an older till at ca. 785 [14]C B.P. (ca. A.D. 1220), re-advanced after 630 [14]C B.P. (A.D. 1410), and remained extensive until after A.D. 1840. At several localities, LIA glacier advances impounded glacial lakes. Recently obtained sections in the deposits of Ape Lake (52°N; Desloges and Ryder, 1990) indicate that the glacier impounded the lake 100–200 years before 770 [14]C B.P. Farther north, the deposits of Tide Lake (ca. 56°N), impounded between the Frank Mackie and Berendon Glaciers, indicate several neoglacial lake phases, the last major impoundment beginning at ca. 1000 [14]C B.P. and finally draining at ca. A.D. 1920–30 (Clague and Mathewes, 1992).

Dating of LIA moraines is based largely on minimum limiting ages for moraines determined by dendrochronology and limited historical observations (e.g., Ryder, 1987). At Tiedemann Glacier (51°N; Ryder and Thomson, 1986) and Berendon Glacier (56°N; Clague and Mathewes, 1996) older neoglacial moraines are locally preserved, dating from ca. 2800–2200 ^{14}C B.P. Figure 10 shows moraine histograms for the southern Coast Ranges (49°–51°N; Ryder and Thomson, 1986) and the Bella Coola region (52°N; Desloges and Ryder, 1990) with comparative data for coastal Alaska (59°–60°N) and the Canadian Rockies (50°–52°N). In the southern Coast Ranges, a small number of glaciers had maxima in the early 1700s, but most had maxima in the nineteenth century with the majority between 1850 and 1900 (Ryder and Thomson, 1986). In the Bella Coola area farther north (Desloges and Ryder, 1990), all the tree-ring-dated moraines were built after 1825, and 12 of the 16 moraines were formed between 1850 and 1900. In discussing results for the Bella Coola area, Desloges and Ryder (1990) suggest that the LIA began slightly later and may have terminated slightly later in the British Columbia Coast Ranges compared with the Rockies. They suggest that the mass balance of these coastal glaciers is strongly controlled by precipitation and is linked to changes in the midlatitude circulation pattern in the adjacent Pacific. However, reconnaissance observations by Smith (personal communication, 1998) do indicate the presence of seventeenth-century

moraines in the Bella Coola area, and Clague and Mathewes (1996) report that the LIA maximum of Berendon Glacier, farther north (56°N), predates 1660. For the Stikine area (57°N), Ryder (1987) estimates the LIA maximum moraines of the Great, Mud, and Flood Glaciers to date from the late seventeenth century to the early eighteenth century, but most other glaciers examined in this area reached their maxima in the mid- to late nineteenth century.

Although many small glaciers occur on Vancouver Island, the only dating presently available comes from the Moving Glacier (49°30′ N; Smith and Larocque, 1996). Cross dating of a subtill detrital and *in situ* assemblage of logs from a gully section, 200–300 m up-ice of the LIA terminal moraine, revealed outermost ring dates of 1713 from the gully and 1718 for an adjacent site, 30 m downvalley. A log protruding from the terminal moraine had an outermost sound ring of A.D. 1818 (the outermost rings were rotted), providing a minimum age for the LIA maximum of this glacier. These data indicate that the glacier was advancing close to the LIA maximum position in 1718 and receded from that limit sometime after 1818 (i.e., as with the adjacent Coast Mountains, the glacial maximum was in the nineteenth century). Aerial photographic evidence indicates that the glacier occupied 77% of its LIA maximum area in 1931 and 20% in 1981.

Results for British Columbia suggest that, with mi-

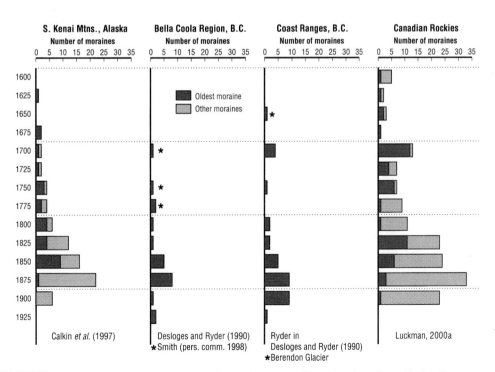

FIGURE 10   Comparison of moraine chronologies for western Canada and southern Alaska. Data sources are indicated on the diagram. Asterisks indicate additions to the original published sources.

nor exceptions, the LIA maxima in the Coast Ranges and on Vancouver Island occurred during the mid- to late nineteenth century. However, available dating from glaciers in the Cascades and Olympic Mountains in the adjacent areas of the United States show sixteenth- to eighteenth-century moraines outside the conspicuous nineteenth-century moraines. At Mount Rainier (47°N), the Nisqually Glacier has been monitored since 1857 (Porter, 1981b), and Burbank (1981) reports minimum tree-ring and lichenometric dates of 1519–28, 1552–76, 1613–24, 1640–66, 1691–95, 1720, 1750, and 1768–77 for pre-nineteenth-century moraines on Mount Rainier. For the Dome Peak area of the Cascades (48°N), Miller (1969) describes a possible thirteenth-century moraine (supporting living trees >680 years old) at Chickamin Glacier and late sixteenth- to early seventeenth-century ages were determined for LIA maximum moraines at the Cascade, La Conte, and Dana Glaciers. At Mount Baker (ca. 49°N), Fuller (1980; Fuller et al., 1983) used dendrochronology to identify sixteenth- (two valleys) and seventeenth-century (three valleys) moraines down-valley of nineteenth-century moraines. Heikkinen (1984) similarly reports moraines from the early 1500s and 1740 outside the nineteenth-century moraines of the Coleman Glacier. For Blue Glacier (ca. 48°N) in the Olympics, Heusser (1957) reports an older moraine fragment dating at ca. 1650, but nineteenth-century moraines (dated by a tilted tree ca. 1850 at Blue Glacier) usually obscure these earlier features at Blue and Hoh Glaciers.

8.3.4. North American Summary

Evidence from throughout the western cordillera suggests that the earliest LIA advances occurred during the twelfth through fourteenth centuries, but the LIA maxima generally occurred during the seventeenth through nineteenth centuries. There are both inter- and intraregional variations in the dating of the LIA maximum position. Mid- through late-nineteenth-century advances occurred throughout the region and appear to have been the most extensive in the more heavily glaciated regions of the British Columbia Coast Mountains and in parts of southern Alaska. However, large numbers of glacier forefields remain uninvestigated. The general absence of multiple till units and/or intervening organic deposits dating from the last millennium suggests that glaciers may have remained at or close to their maximum positions throughout this interval until the mid- through late nineteenth century, although there may have been several intervals of glacier advance during this period. There is no evidence for periods of glacier recession during the last millennium comparable to those of the twentieth century.

Examination of the coastal Alaskan, Rockies, and British Columbian chronologies suggests a strong similarity between the glacier histories of these areas that implies a common climate control. However, contemporary mass balance data suggest more complex patterns and relationships that change over time. Hodge et al. (1998) recently discussed the relationships between Pacific climate indices and mass balance data for Gulkana Glacier (interior Alaska, 63°N), Wolverine Glacier (Kenai Peninsula in Alaska, ca. 60°N), and South Cascade Glacier (Washington, 48°N). The mass balance data for the Peyto Glacier (Canadian Rockies, 51°N) were briefly discussed by Luckman (1998b, 2000b), and all four glaciers have over 30 years of mass balance data. There appears to be no simple relationship between the mass balance records of the two Alaskan glaciers (Gulkana and Wolverine Glaciers), which are only 300 km apart (Hodge et al., 1998). However, the winter balance records for Peyto, Wolverine, and South Cascade Glaciers all show clear "step changes" coincident with the interdecadal 1976–77 regime shift over the North Pacific (Ebbesmeyer et al., 1991). Although, Peyto and South Cascade Glaciers both show sharp decreases in winter balance, Wolverine Glacier shows a marked increase. This negative correlation between the mass balances of Alaskan glaciers and those in southern Canada and the Pacific Northwest was recognized by Walters and Meier (1989) and was attributed to the steering of storm tracks by the Alaskan Low. This teleconnection pattern breaks down after 1989, when all four glaciers show strongly negative balances (Hodge et al., 1998; Luckman, 1998b, 2000b).

8.4. CONCLUSION

The results summarized in this chapter indicate a broad synchroneity in the initiation and timing of LIA glacial events along the PEP 1 transect, with initial glacier advances in the thirteenth through fourteenth centuries in the Northern Hemisphere and Patagonia; scattered evidence for extended glaciers during the fourteenth through sixteenth centuries; and the "classic" LIA advances during the seventeenth to early twentieth centuries. In detail, however, particularly at the decadal level, there is considerable variability in both the relative extent and timing of events and the dating of the LIA maximum position. These differences reflect varying dominance of precipitation and temperature controls of mass balance, as well as nonclimatic influences on glacier behavior. There remains a significant need for detailed, well-dated records of glacier fluctuations, coupled with a better understanding of the factors that control glacier mass balance. Some interpretative problems may be resolved by the development of complementary proxy records (e.g., from tree rings), which are sensitive to similar climatic parameters. For example, Villalba et al. (2000) demonstrate strong correlations between sea surface temperatures (SSTs) and tree-ring chronologies in both the Gulf of Alaska and Patagonia. The Alaskan and the Patagonian tree-ring records clearly show the documented 1976 transition from cold to warm conditions over the tropical Pacific and are significantly correlated over the 1591–1988 period, reflecting a common climatic forcing mechanism that affected past temperature variations in both regions. The Pacific forcing also appears to be reflected in the temperature-sensitive glacier fluctuations of these two regions. Such comparative analyses might ultimately enable large-scale reconstructions of regional patterns of temperature/precipitation variation along the PEP 1 transect at decadal scales for the last several hundred years.

Acknowledgments

We would like to thank the many colleagues who have provided us with material included in this chapter, particularly Jean Grove, Charles Warren, Masamu Aniya, and Greg Wiles. Stephan Hastenrath, John Clague, Greg Wiles, and an anonymous reviewer have provided comments on the manuscript. We thank the Cartographic Section at the University of Western Ontario for the diagrams and Vera Markgraf and Malcolm Hughes for their invitation to present this chapter. This chapter has been partially supported by the Natural Sciences and Engineering Research Council of Canada, the Argentinean Research Council (CONICET) the Argentinean Agency for the Promotion of Science (PICT0703093), and a Lamont-Doherty Earth Observatory Post-Doctoral Fellowship to R.V.

References

Aceituno, P., H. Fuenzalida, and B. Rosenbluth, 1993: Climate along the extratropical west coast of South America. *In* Mooney, H. A., E. R. Fuentes, and B. I. Kronberg (eds.), *Earth System Responses to Global Change: Contrasts Between North and South America.* San Diego: Academic Press, pp. 61–69.

Aniya, M., 1992: Glacier variations in the northern Patagonian Icefield, between 1985/86 and 1990/91. *Bulletin of Glacier Research,* **10**: 83–90.

Aniya, M., 1995: Holocene glacial chronology in Patagonia: Tyndall and Upsala Glaciers. *Arctic and Alpine Research,* **27**: 311–322.

Aniya, M., 1996: Holocene variations of Ameghino Glacier, southern Patagonia. *The Holocene,* **6**: 247–252.

Aniya, M., and H. Enomoto, 1986: Glacier variations and their causes in the northern Patagonian Icefield, Chile, since 1944. *Arctic and Alpine Research,* **18**: 414–421.

Aniya, M., H. Sato, R. Naruse, P. Skvarca, and G. Casassa, 1997: Recent glacial variations in the southern Patagonia Icefield, South America. *Arctic and Alpine Research,* **29**: 1–12.

Aristarain, A. J., J. Jouzel, and C. Lorius, 1990: A 400 years isotope record of the Antarctic Peninsula climate. *Geophysical Research Letters,* **17**: 2369–2372.

Benedict, J. M., 1973: Chronology of cirque glaciation, Colorado Front Range. *Quaternary Research,* **3**: 584–599.

Boninsegna, J. A., 1988: Santiago de Chile winter rainfall since 1220 as being reconstructed by tree rings. *Quaternary of South America and Antarctic Peninsula,* **4**: 67–87.

Bradley, R. S., and P. D. Jones, 1992: When was the "Little Ice Age"? *In* Mikami, T. (ed.), *Proceedings of the International Symposium on the Little Ice Age Climate.* Department of Geography, Tokyo Metropolitan University, Japan, pp. 1–4.

Burbank, D. W., 1981: A chronology of late Holocene glacier fluctuations on Mount Rainier, Washington. *Arctic and Alpine Research,* **13**: 369–386.

Calkin, P. E., 1988: Holocene glaciation of Alaska (and adjoining Yukon Territory, Canada). *Quaternary Science Reviews,* **7**: 159–184.

Calkin, P. E., and G. C. Wiles, 1992: Little Ice Age glaciation in Alaska: A record of recent global climatic change. *In* Weller, G., and C. Wilson (eds.), *The Role of the Polar Regions in Global Change, Proceedings of an International Conference.* Fairbanks: University of Alaska at Fairbanks, Geophysical Institute and the Center for Global Change, Fairbanks, Alaska, pp. 617–625.

Calkin, P. E., G. C. Wiles, and D. J. Barclay, in press: Holocene coastal glaciation of Alaska. *Quaternary Science Reviews.*

Carrara, P. E., 1987: Holocene and latest Pleistocene glacial chronology, Glacier National Park, Montana. *Canadian Journal of Earth Sciences,* **24**: 387–395.

Carrara, P. E., and R. G. McGimsey, 1981: The late-neoglacial history of the Agassiz and Jackson Glaciers, Glacier National Park, Montana. *Arctic and Alpine Research,* **13**: 183–196.

Casassa, G., 1995: Glacier inventory in Chile: Current status and recent glacier variations. *Annals of Glaciology,* **21**: 317–322.

Clague, J. J., and R. W. Mathewes, 1996: Neoglaciation, glacier-dammed lakes and vegetation change in northwestern British Columbia, Canada. *Arctic and Alpine Research,* **28**: 10–24.

Clague, J. J., and W. H. Mathews, 1992: The sedimentary record and neoglacial history of Tide Lake, northwestern British Columbia. *Canadian Journal of Earth Sciences,* **29**: 2382–2396.

Clague, J. J., and V. N. Rampton, 1982: Neoglacial Lake Alsek. *Canadian Journal of Earth Sciences,* **19**: 94–117.

Clague, J. J., L. A. Jozsa, and M. L. Parker, 1982: Dendrochronological dating of glacier-dammed lakes: An example from the Yukon Territory, Canada. *Arctic and Alpine Research,* **14**: 301–310.

Clapperton, C. M., and D. E. Sugden, 1988: Holocene glacier fluctuations in South America and Antarctica. *Quaternary Science Reviews,* **7**: 185–198.

Clapperton, C. M., D. E. Sugden, J. Birnie, and M. J. Wilson, 1989: Late-glacial and Holocene glacier fluctuations and environmental change on South Georgia, Southern Ocean. *Quaternary Research,* **31**: 210–228.

Clark, D. H., and A. Gillespie, 1997: Timing and significance of the late-glacial and Holocene cirque glaciations in the Sierra Nevada, California. *Quaternary International,* **38/39**: 21–38.

Cobos, D. R., and J. A. Boninsegna, 1983. Fluctuations of some glaciers in the upper Atuel River basin, Mendoza, Argentina. *Quaternary of South America and Antarctic Peninsula,* **1**: 61–82.

Davis, P. T., 1988: Holocene glacier fluctuations in the American cordillera. *Quaternary Science Reviews,* **7**: 129–158.

Demuth, M. N., and R. Keller, 2000: An assessment of the mass balance of Peyto Glacier (1966–1995) and its relation to recent and past-century climatic variability. *In* Demuth, M. N., D. S. Munro, and G. J. Young (eds.), *Peyto Glacier: One Century of Science.* National Hydrology Research Institute, Saskatoon, Saskatchewan, Canada, Science Report 8.

Denton, G. H., and W. Karlén, 1977: Holocene glacial and tree-line variations in the White River Valley and Skolai Pass, Alaska and Yukon Territory. *Quaternary Research,* **7**: 69–111.

Denton, G. H., and M. Stuiver, 1967: Late Pleistocene glacial stratigraphy and chronology, northeastern St. Elias Mountains, Yukon Territory, Canada. *Geological Society of America Bulletin*, **78**: 485–510.

Desloges, J. S., and J. M. Ryder, 1990: Neoglacial history of the Coast Mountains near Bella Coola, British Columbia. *Canadian Journal of Earth Sciences*, **27**: 281–290.

Ebbesmeyer, C. R., D. R. Cayan, D. R. McLain, F. H. Nichols, D. H. Peterson, and K. T. Redmond, 1991: 1976 step change in the Pacific climate: Forty environmental changes between 1968–75 and 1977–84. *In* Betancourt, J. L., and J. L. Tharp (eds.), *Proceedings of the Seventh Annual Pacific Climate (PACLIM) Workshop*, Asilomar, California Department of Water Resources, Interagency Ecological Studies Program Technical Report 26, pp. 115–125.

Francou, B., P. Ribstein, R. Saravia, and E. Tiriau, 1995: Monthly balance and water discharge of an inter-tropical glacier: Zongo Glacier, Cordillera Real, Bolivia, 16°S. *Journal of Glaciology*, **41**: 61–67.

Fuller, S. R., 1980: Neoglaciation of the Avalanche Gorge and the Middle Fork Nooksack River valley, Mount Baker, Washington. M.S. thesis. Western Washington University, Bellingham, 68 pp.

Fuller, S. R., D. J. Easterbrooke, and R. M. Burke, 1983: Holocene glacial activity in five valleys on the flanks of Mount Baker, Washington. *Geological Society of America, Abstracts with Program*, **15**: 430–431.

Gordon, J. E., and R. J. Timmis, 1992: Glacier fluctuations on South Georgia during the 1970s and early 1980s. *Antarctic Science*, **4**: 215–226.

Granados, H. I., 1997: The glaciers of Popocatépetel Volcano (México): Changes and causes. *Quaternary International*, **43/44**: 53–60.

Grove, J. M., 1988: *The Little Ice Age*. London: Methuen, 498 pp.

Grove, J. M., 1997: Little Ice Age glacier fluctuations in southern South America, (unpublished manuscript, revision of 1988).

Hastenrath, S., 1981: *The Glaciation of the Ecuadorian Andes*. Rotterdam: Balkema, 159 pp.

Hastenrath, S., 1997: Glacier recession in the high Andes of Ecuador in the context of the global tropics. International Conference on Climatic and Hydrological Consequences of El Niño Events at Regional and Local Scales, Quito, Ecuador, 309 pp.

Hauthal, R., 1895: Observationes generales sobre algunos ventisqueros de la Cordillera de los Andes. *Revista Museo de la Plata*, **6**: 109–116.

Haworth, L. H., 1988: Holocene glacial chronologies of the Brooks Range, Alaska, and their relationship to climate change. Ph.D. thesis. State University of New York at Buffalo, 260 pp.

Heikkinen, O., 1984: Dendrochronological evidence of variation of Coleman Glacier, Mt. Baker, Washington. *Arctic and Alpine Research*, **16**: 53–64.

Heusser, C. J., 1956: Postglacial environments in the Canadian Rocky Mountains. *Ecological Monographs*, **26**: 253–302.

Heusser, C. J., 1957: Variations of Blue, Hoh and White Glaciers during recent centuries. *Arctic*, **10**: 139–150.

Hodge, S. M., D. M. Trabant, R. M. Krimmel, T. A. Heinrichs, R. S. March, and E. G. Josberger, 1998: Climate variations and changes in mass of three glaciers in western North America. *Journal of Climate*, **11**: 2161–2179.

Holmlund, P., and H. Fuenzalida, 1995: Anomalous glacier responses to 20th century climatic changes in Darwin Cordillera, southern Chile. *Journal of Glaciology*, **41**: 465–473.

LaMarche, V. C., 1975: Potential of tree rings for reconstruction of past climate variations in the Southern Hemisphere. *Proceedings of the WMO/IAMAP Symposium on Long Term Climatic Fluctuations*, Norwich, England, World Meteorological Organization, WMO No. 421, pp. 21–30.

LaMarche, V. C., R. L. Holmes, P. W. Dunwiddie, and L. G. Drew, 1979: Chile. *In Tree-Ring Chronologies of the Southern Hemisphere*, Vol. 2 of Chronology Series V. Tucson: Laboratory of Tree Ring Research, University of Arizona Press.

Leiva, J. C., L. E. Lenzano, G. A. Cabrera, and J. A. Suarez, 1986: Variations of Río Plomo Glaciers, Andes Centrales Argentinos. *In* Oerlemans, J. (ed.), *Glacier Fluctuations and Climate Change*. Dordrecht: Kluwer Academic, pp. 143–151.

Luckman, B. H., 1986: Reconstruction of Little Ice Age events in the Canadian Rockies. *Géographie physique et Quaternaire*, **40**: 17–28.

Luckman, B. H., 1995: Calendar-dated, early Little Ice Age glacier advance at Robson Glacier, British Columbia, Canada. *Holocene*, **5**: 149–159.

Luckman, B. H., 1996a: Reconciling the glacial and dendrochronological records of the last millennium in the Canadian Rockies. *In* Jones, P. D., R. S. Bradley, and J. Jouzel (eds.), *Climatic Variations and Forcing Mechanisms of the Last 2000 Years*. Berlin: Springer-Verlag, pp. 85–108.

Luckman, B. H., 1996b: Dendroglaciology at Peyto Glacier, Alberta. *In* Dean, J. S., D. S. Meko, and T. W. Swetnam (eds.), *Tree Rings, Environment and Humanity, Radiocarbon*, pp. 679–688.

Luckman, B. H., 1998a: Dendroglaciologie dans les Rocheuses du Canada. *Géographie physique et Quaternaire*, **52**: 137–149.

Luckman, B. H., 1998b: Landscape and climate change in the central Canadian Rockies during the twentieth century. *Canadian Geographer*, **42**: 319–336.

Luckman, B. H., 2000a: The Little Ice Age in the Canadian Rockies. *Geomorphology*, **32**: 357–384.

Luckman, B. H., 2000b: The neoglacial history of Peyto Glacier. *In* Demuth, M. N., D. S. Munro, and G. J. Young (eds.), *Peyto Glacier: One Century of Science*. National Hydrology Research Institute, Saskatoon, Saskatchewan, Canada, Science Report 8.

Luckman, B. H., and G. D. Osborn, 1979: Holocene glacier fluctuations in the middle Canadian Rocky Mountains. *Quaternary Research*, **11**: 52–77.

Luckman, B. H., and E. Watson, 1999: Precipitation reconstruction in the southern Canadian cordillera. Preprint Volume, 10th Symposium on Global Change Studies, 79th Annual Meeting, American Meteorological Society, Dallas, Texas, 10–15 January 1999, A.M.S., Boston, pp. 296–299.

Luckman, B. H., K. R. Briffa, P. D. Jones, and F. H. Schweingruber, 1997: Tree-ring based reconstruction of summer temperatures at the Columbia Icefield, Alberta, Canada, A.D. 1073–1983. *The Holocene*, **7**: 375–389.

Luckman, B. H., K. A. Harding, and J. P. Hamilton, 1987: Recent glacier advances in the Premier Range, British Columbia. *Canadian Journal of Earth Sciences*, **24**: 1149–1161.

Luckman, B. H., G. Holdsworth, and G. D. Osborn, 1993: Neoglacial glacier fluctuations in the Canadian Rockies. *Quaternary Research*, **39**: 144–153.

Matthes, F. E., 1939: Report of the Committee on Glaciers. *Transactions of the American Geophysical Union*, **20**: 518–523.

Mercer, J., 1965: Glacier variations in southern Patagonia. *Geographical Review*, **55**: 390–413.

Mercer, J., 1968: Variations of some Patagonian glaciers since the late-glacial. *American Journal of Science*, **266**: 91–109.

Mercer, J. H., 1970: Variations of some Patagonian glaciers since the Late-Glacial. *American Journal of Science*, **269**: 1–25.

Mercer, J., 1976: Glacial history of the southernmost South America. *Quaternary Research*, **6**: 125–166.

Miller, C. D., 1969. Chronology of neoglacial moraines in the Dome Peak area, North Cascade Range, Washington. *Arctic and Alpine Research*, **1**: 49–65.

Motyka, R. J., and J. E. Begêt, 1995: Taku Glacier, Southeast Alaska, U.S.A.: Late Holocene history of a tidewater glacier. *Arctic and Alpine Research*, **28**: 41–51.

Osborn, G. D., 1996: Onset of the Little Ice Age at the Columbia Icefield, Canadian Rockies, as recorded in lateral moraines of Stutfield Glacier. American Quaternary Association, Program with Abstracts of the 14th Meeting, Flagstaff, Arizona, p. 181.

Osborn, G. D., and B. H. Luckman, 1988: Holocene glacier fluctuations in the Canadian cordillera (Alberta and British Columbia). *Quaternary Science Reviews,* **7**: 115–128.

Porter, S. C., 1981a: Glaciological evidence of Holocene climatic change. *In* Wigley, T. M. L., M. J. Ingram, and G. Farmer (eds.), *Climate and History.* Cambridge, United Kingdom: Cambridge Univ. Press, pp. 82–110.

Porter, S. C., 1981b: Lichenometric studies in the Cascade Range of Washington: Establishment of *Rhizocarpon geographicum* growth curves at Mount Rainier. *Arctic and Alpine Research,* **13**: 11–23.

Rampton, V. N., 1970: Neoglacial fluctuations of the Natzhat and Klutlan Glaciers, Yukon Territory, Canada. *Canadian Journal of Earth Sciences,* **7**: 1236–1263.

Rampton, V. N., 1978: Holocene tree-line and glacier variations in the White River Valley and Skolai Pass, Alaska and Yukon Territory: A Discussion. *Quaternary Research,* **10**: 130–134.

Ribstein, P., E. Tiriau, B. Francou, and P. Saravia, 1995: Tropical climate and glacier hydrology: A case study in Bolivia. *Journal of Hydrology,* **165**: 221–234.

Robinson, B. J., 1998: Reconstruction of the glacial history of the Columbia Icefield, Alberta. M.S. thesis. Department of Geography, University of Western Ontario, London, Ontario, Canada, 244 pp.

Röthlisberger, F., 1987: *10000 Jahre Gletschergeschichte der Erde.* Aarau, Switzerland: Sauerlander.

Ryder, J. M., 1987: Neoglacial history of the Stikine-Iskut area, northern Coast Mountains, British Columbia. *Canadian Journal of Earth Sciences,* **24**: 1294–1301.

Ryder, J. M., and B. Thomson, 1986: Neoglaciation in the southern Coast Mountains of British Columbia: Chronology prior to the late-neoglacial maximum. *Canadian Journal of Earth Sciences,* **23**: 273–287.

Schubert, C., 1972: Geomorphology and glacier retreat in the Pico Bolivar area, Sierra Nevada de Mérida, Venezuela. *Zeitschrift für Gletscherkunde und Glazialgeologie,* **8**: 189–202.

Schubert, C., 1992: The glaciers of Sierra Nevada de Mérida (Venezuela): A photographic comparison of recent deglaciation. *Erdkunde,* **46**: 517–548.

Sharp, R. P., 1951: Glacial history of Wolf Creek, St. Elias Range, Canada. *Journal of Geology,* **59**: 97–115.

Smith, D. J., and C. P. Larocque, 1996: Dendroglaciological dating of a Little Ice Age glacial advance at Moving Glacier, Vancouver Island, British Columbia. *Géographie physique et Quaternaire,* **50**: 47–55.

Smith, D. J., D. P. McCarthy, and M. E. Colenutt, 1995: Little Ice Age glacial activity in Peter Lougheed and Elk Lakes Provincial Parks, Canadian Rocky Mountains. *Canadian Journal of Earth Sciences,* **32**: 579–589.

Stuiver, M., and G. W. Pearson, 1993: High-precision bidecadal calibration of the radiocarbon time-scale, A.D. 1950–500 B.C. and 2500–6000 B.C. *Radiocarbon,* **35**: 1–24.

Stuiver, M., and P. J. Reimer, 1986: A computer program for radiocarbon age calibration. *In* Stuiver, M., and R. S. Kra (eds.), Proceedings of the 12th International ^{14}C Conference. *Radiocarbon,* **28**(2B): 1022–1030.

Thompson, L. G., 1995: Melting of our Tropical Climate Archives. *PAGES Newsletter,* **3**(2): July 1995.

Thompson, L. G., M. E. Davis, E. Mosley-Thompson, T. A. Sowers, K. A. Henderson, V. S. Zagorodnov, P.-N. Lin, V. N. Mikhalenko, R. K. Campen, J. F. Bolzan, J. Cole-Dai, and B. Francou, 1998: A 25,000-year tropical climate history from Bolivian ice cores. *Science,* **282**: 1858–1862.

Thompson, L. G., E. Mosley-Thompson, W. Dansgaard, and P. M.

Grootes, 1984: The Little Ice Age as recorded in stratigraphy of the tropical Quelccaya ice cap. *Science,* **234**: 361–364.

Timmis, R. J., 1986: Glacier changes in South Georgia and their relationships to climate trends. Ph.D. thesis. University of East Anglia, Norwich, England.

Vázquez-Selen, L., and F. M. Phillips, 1998: Glacial chronology of Iztaccíhuatl volcano, central México, based on cosmogenic ^{36}Cl exposure ages and tephrochronology. American Quaternary Association, Program and Abstracts of the 15th Biennial Meeting, Puerto Vallarta, México, p. 174.

Videla, M. A., 1997: Neoglacial advances in the central Andes, Argentina, during the last centuries. *Quaternary of South America and Antarctic Peninsula,* **10**: 55–70.

Villalba, R., 1990a: Climatic fluctuations in northern Patagonia during the last 1000 years as inferred from tree-ring records. *Quaternary Research,* **34**: 346–360.

Villalba, R., 1990b: Latitude of the surface high-pressure belt over western South America during the last 500 years as inferred from tree-ring analysis. *Quaternary of South America and Antarctic Peninsula,* **7**: 273–303.

Villalba, R., E. R. Cook, R. D. D'Arrigo, G. C. Jacoby, P. D. Jones, J. M. Salinger, and J. Palmer, 1997: Sea-level pressure variability around Antarctica since A.D. 1750 inferred from subantarctic tree-ring records. *Climate Dynamics,* **13**: 375–390.

Villalba, R., R. D. D'Arrigo, E. R. Cook, G. Wiles, and G. C. Jacoby, 2000: Decadal-scale climatic variability along the extratropical western coast of the Americas over past centuries inferred from tree-ring records. *In* Markgraf, V. (ed.), *Interhemispheric Climate Linkages.* San Diego: Academic Press, Chapter 10.

Villalba, R., J. C. Leiva, S. Rubulis, J. C. Suarez, and L. Lenzano, 1990: Climate, tree-ring and glacial fluctuations in the Río Frías valley, Río Negro, Argentina. *Arctic and Alpine Research,* **22**: 215–232.

Walters, R. A., and M. F. Meier, 1989: Variability of glacier mass balances in western North America. *In Aspects of Climate Variability in the Pacific and Western Americas.* Washington, DC: American Geophysical Union, Geophysical Monograph 55, pp. 365–374.

Warren, C. R., 1992: Iceberg calving and the glacioclimatic record. *Progress in Physical Geography,* **16**: 253–282.

Warren, C. R., 1993: Rapid recent fluctuations of the calving San Rafael Glacier, Chilean Patagonia: Climatic or non-climatic? *Geografiska Annaler,* **75A**: 111–125.

Warren, C. R., and M. Aniya, in press: The calving glaciers of southern South America. *Global and Planetary Change.*

Warren, C. R., and A. Rivera, 1994: Non-linear climatic response of calving glaciers: A case study of Pio XI Glacier, Chilean Patagonia. *Revista Chilena de Historia Natural,* **67**: 385–394.

Warren, C. R., and D. E. Sugden, 1993: The Patagonian Icefields. A glaciological review. *Arctic and Alpine Research,* **25**: 316–331.

Warren, C. R., A. Rivera, and A. Post, 1997: Greatest Holocene advance of Glaciar Pio XI, Chilean Patagonia: Possible causes. *Annals of Glaciology,* **24**: 11–15.

Warren, C. R., D. E. Sugden, and C. M. Clapperton, 1995: Asynchronous response of Patagonian glaciers to historic climate change. *Quaternary of South America and the Antarctic Peninsula,* **9**: 85–103.

Watson, E., 1998: Dendroclimatic reconstruction of precipitation in British Columbia and Alberta. M.S. thesis. University of Western Ontario, London, Ontario, Canada, 220 pp.

White, S. E., 1981: Neoglacial to recent glacial fluctuations on the volcano Popocatépetel. *Journal of Glaciology,* **27**: 359–363.

Wiles, G. C., 1997: North Pacific atmosphere-ocean variability over the past millennium inferred from coastal glaciers and tree rings. Preprint Volume, 8th Symposium on Global Change Studies, 2–7 February 1997, Long Beach, CA. American Meteorological Society, pp. 218–220.

Wiles, G. C., and P. E. Calkin, 1994: Late Holocene, high-resolution glacial chronologies and climate, Kenai Mountains, Alaska. *Geological Society of America Bulletin,* **106**: 281–303.

Wiles, G. C., D. J. Barclay, and P. E. Calkin, 1999a: Tree-ring dated Little Ice Age histories of maritime glaciers from western Prince William Sound, Alaska. *The Holocene,* **9**: 163–174.

Wiles, G. C., P. E. Calkin, and A. Post, 1995: Glacier fluctuations in the Kenai Fjords, Alaska, U.S.A.: An evaluation of the controls on iceberg-calving glaciers. *Arctic and Alpine Research,* **27**: 234–245.

Wiles, G. C., R. D. D'Arrigo, and G. C. Jacoby, 1996: Temperature changes along the Gulf of Alaska and the Pacific Northwest coast modeled from coastal tree rings. *Canadian Journal of Forest Research,* **26**: 474–481.

Wiles, G. C., A. Post, E. H. Muller, and B. F. Molnia, 1999b: Dendrochronology and late Holocene history of Bering Piedmont Glacier, Alaska. *Quaternary Research,* **52:** 185–195.

9

Volcanic Signals in Temperature Reconstructions Based on Tree-Ring Records for North and South America

JOSE A. BONINSEGNA AND MALCOLM K. HUGHES

Abstract

High-resolution, well-dated proxy records can extend the climatic information available for analysis for centuries or even millennia. Several reconstructed temperature series covering the past several centuries have been produced for both North and South America. The injection of dust, sulfur dioxide, and other emissions by explosive volcanic eruptions can affect the temperature at regional and, on occasion, global scales. In this chapter, we compare reconstructed temperature series for both hemispheres in a search for signals of volcanic eruptions.

Our results show a rather weak, but discernible pattern of temperature anomalies, probably linked to volcanic activity in the time and spatial domains in the Northern Hemisphere. In the Southern Hemisphere, however, no observable pattern can be related to the set of explosions. Only very few eruptions, particularly those produced by southern volcanoes, seem to have some effect there. The relationships between the sulfate volcanic loads and the reconstructed temperature series are also discussed. Copyright © 2001 by Academic Press.

Resumen

Los registros "proxi" correctamente datados y de alta resolución pueden extender la información climática disponible para análisis en centurias y aún milenios. Tanto en el Hemisferio Norte como en el Hemisferio Sur se han producido varias series de reconstrucciones de temperatura que cubren varias centurias.

La injección de polvo, dioxido de azufre y otros gases por erupciones volcánicas explosivas pueden afectar la temperatura a escalas regionales y en ocasiones a escala global. En este capítulo, nosotros comparamos las series de temperatura reconstruidas para ambos hemisferios buscando en ellas las señales de erupciones volcánicas.

Nuestros resultados muestran un patrón débil pero discernible de anomalias en la temperatura probablemente relacionado a la actividad volcánica en escalas temporales y espaciales en el Hemisferio Norte. Sin embargo, en el Hemisferio Sur no se ha observado ningún patrón que pueda ser relacionado con el conjunto de explosiones estudiadas. Solamente algunas pocas explosiones, en particular aquellas producidas por volcanes ubicados en el Hemisferio Sur, parecen haber provocado perturbaciones en las series de temperatura de dicho hemisferio.

Las relaciones entre la carga de dióxido de azufre emitido por los volcanes y las series reconstruidas de temperatura es también discutida en este capítulo.

141

9.1. INTRODUCTION

Chronology development and climatic reconstruction are the most common products of dendrochronological studies. Networks of tree-ring chronologies at regional or quasi-continental scale have appeared following the exponential increase in the number of tree-ring chronologies. Chronology networks are especially important because they permit the statistical identification of a common regional signal. If the network covers a broad area, spatial variability in climate patterns over time may be detected. These regional differences may reveal information about forcing functions or changes in circulation patterns at quasi-hemispheric scales.

At present, the North America chronology set contains about 2500 chronologies distributed from 70°–23°N and 135°–60°W. For South America, about 150 chronologies are available from 22°–55°S and 78°–68°W.

Despite the difference in the sizes of the North and South American tree-ring record networks, the use of spatial techniques to extract common signals on both sets appears feasible.

Forcing functions operating at annual to centennial scales are possibly recorded in tree-ring series. The most important source of climatic variation at such scales is recognized to be the El Niño/Southern Oscillation (ENSO). Solar influences may also be present in the tree-ring series and in the temperature reconstructions based on tree rings. The climatic effects of anthropogenically increasing atmospheric carbon dioxide concentrations have also been detected in tree-ring series (Mann et al., 1998).

A more elusive signal affecting the temperature at regional and occasionally global scales is the injection of dust, sulfur dioxide, and other emissions by major volcanic explosions. High-resolution, well-dated proxy records can extend the climatic information available for analysis for centuries to even millennia. Coral isotope records, for instance, have been used to chronicle cooling events related to volcanism (Crowley et al., 1997). Several papers have been published dealing with the detection of volcanic influences in both dendrochronological and reconstructed temperature time series for the Northern Hemisphere (Lough and Fritts, 1987; Jacoby and D'Arrigo, 1992; Scuderi, 1990; Briffa et al., 1998; D'Arrigo and Jacoby, 1999), but very few address the Southern Hemisphere (Villalba and Boninsegna, 1992; Palmer and Ogden, 1992; Norton, 1992).

Several authors have reconstructed temperature data using tree-ring chronologies for both North and South America. In this chapter, we compare some of these reconstructions by analyzing possible patterns of climate change in both hemispheres and the influence of volcanic disturbances as a forcing factor of such change.

9.2. DATA AND METHODS

The temperature reconstruction data set is presented in Table 1. We limited the analysis to the period from the year 1602–1960 and selected the set of available reconstructed series to approximately cover a latitudinal gradient from the Arctic regions to Tierra del Fuego. Using singular spectrum analysis (SSA) techniques and, in particular, the additive and filtering properties of the reconstructed components (see Vautard and Ghil, 1989 and Vautard et al., 1992 for a comprehensive approach), we decomposed each series in three wave modes: long-term trend (>50 years), 10- to 49.9-year, and 2- to 9.9-year wave modes. Selection of these three sets of modes results from a previous spectral analysis to establish the significant dominant periods at which variance occurs in the reconstructions (not shown).

We consider only eruptions with a Volcanic Explosivity Index (VEI) of 4 or higher according to the scale of Simkin (Simkin et al., 1981). A list of such explosions is shown in Table 2. The estimate of atmospheric sulfate load due to volcanic activity is from Zielinski (1995).

9.2.1. Mean Differences

We standardized each one of the 2- to 49.9-year combined wave mode reconstructed temperature series to a mean of zero and standard deviation of one over the period from 1602–1960. On these transformed series, we calculated a 7-year mean moving window and used the t-test at the 90% probability level to determine which values differ from the population mean.

9.2.2. Running Correlation Index

Evolving correlations against the sulfate index were calculated for a 50-year window of each reconstructed temperature series.

9.2.3. Event Analysis

Each year in a list of event dates (volcanic explosions in our case) was taken as a key or zero window year. Time series values for event years and for a window of 11 years before and after each event year are superposed and averaged over the series time span. Random simulation determines the significance of the relationship.

TABLE 1 Reconstructed Temperature Series

Site	Label	Description	Ref.
North America			
North Yukon	NYU	Seven tree-ring chronologies from across boreal North America, standardized by using straight or negative exponential line curve fit, were used to reconstruct annual temperature.	D'Arrigo and Jacoby, 1992
Athabasca Glacier	ATH	Ring-width and density chronology mainly derived from *Picea* samples. Series were standardized by using either a negative exponential function or a simple straight line. April–August mean monthly temperatures averaged were estimated as a function of the maximum latewood density and ring width, in the year of ring formation (t) and in the previous year (t−1).	Luckman et al., 1997
Grid13/16 Grid 17/20	G13 G17	Temperatures reconstructed using density chronologies. The chronologies are series of simple arithmetic averages of annual density data for trees at each site. A cubic spline function with a frequency response equal to two-thirds of the series length was employed to detrend each series. Grid 13/16 is the average of the grids 13 to 16 (40°–45° N and 100°–120°W). Grid 17/20 is the average of the grids 17 to 20 (35°–40°N and 100°–120°W).	Briffa et al., 1992
Columbia Intermountain California Southwest	COL IMB CAL SWE	Ring-width index chronologies standardized by using a negative exponential function or by fitting a straight line were used in the annual temperature reconstruction, using a canonical correlation technique.	Fritts, 1991
Grant Grove	GRG	Age-related trend was removed by fitting a negative exponential or linear growth curve. A regression approach was used in reconstruction.	Graybill, D., personal communication
Sierra Nevada	SNE	High-elevation foxtail pine ring-width series. The age-related trend was removed by fitting a negative exponential or linear growth curve. Significant low-order autocorrelation was removed (prewhitening). An approach based on intersecting growth surfaces was used in the reconstruction.	Graumlich, 1993
South America			
Juglans	JUG	Ring-width chronology using *Juglans australis* at Rio Horqueta (27°10′ S, 65°23′ W). The chronology is positively correlated against the November–January monthly mean temperature average (r = 0.53).	Villalba et al., 1998
North Patagonia	NPA	Regional summer temperature reconstruction using eight *Araucaria araucana* ring-width chronologies. A double detrending procedure was employed to remove the growth trend in each tree-ring series.	Villalba et al., 1989
Alerce	ALE	Summer temperature reconstruction using ring width of *Fitzroya cupressoides* chronology from the Alerce River (41°S). Tree-ring series were double detrended by fitting a negative exponential line and in a second step a cubic spline function of two-thirds length of the series.	Villalba, 1990
Lenca	LEN	Summer temperature reconstruction using *F. cupressoides* at Lenca (41°33′ S, 72°36′ W). The tree-ring series were standardized by fitting a 128-year cubic spline and prewhitening before using them in the chronology.	Lara and Villalba, 1993
Ushuaia	USH	Twenty-one chronologies using *Nothofagus pumilio* and *N. betuloides* were produced. The tree-ring series were standardized by applying a cubic spline of 128 years. Four of those chronologies with the strongest positive correlation with summer (November–February) temperature were selected and the first eigenvector scores of a principal components (PC) analysis of these four chronologies were used in a reconstruction of summer temperature.	Boninsegna et al., 1989

(continues)

TABLE 1 (*continued*)

Site	Label	Description	Ref.
Composite reconstructions			
Northern Hemisphere multiproxy	NHM	Annual-resolution, area-weighted multiproxy network.	Mann et al., 1998
Northern Hemisphere summer temperature	NHB	Network of temperature-sensitive, tree-ring density chronologies.	Briffa et al., 1998
Northern Hemisphere multiproxy	NHJ	Average of 10-temperature reconstructions from Northern Hemisphere proxies.	Jones et al., 1998
Southern Hemisphere multiproxy	SHJ	Average of 7-temperature reconstructions from Southern Hemisphere proxies.	Jones et al., 1998

9.3. RESULTS AND DISCUSSION

9.3.1. Spectral Variance Bands

Table 3 shows the relative percent of variance associated with spectral bands for each reconstruction. Important differences are noticeable in the long-term trend, where the North Yukon reconstruction appears with the largest proportion of variance (65.97%) and the Lenca reconstruction appears with the lowest value (2.84%). These results probably reflect the different ways in which the tree-ring chronologies were built. In fact, the Lenca chronology was developed by fitting a rather flexible cubic spline (128 years) and then prewhitening each one of the measurement series. Long-term trends and cycles of periods longer than 128 years are probably underrepresented in this reconstruction.

In general, the percents of variance associated with wave modes of periods from 10–49.9 years do not vary greatly among the reconstructions. The same is true for the 2- to 9.9-year wave modes, with the exception of the North Yukon reconstruction, in which only 8.8% of the total variance is explained by the relatively high frequencies.

Again, the chronologies used in this temperature reconstruction were built by using a very conservative detrending to keep as much as possible of the low-frequency variation (D'Arrigo et al., 1992). Forcing factors acting on decadal to biannual timescales (i.e., volcanic eruptions) are probably well represented in the reconstructions used in this chapter, while those affecting long-term trends are probably not well represented for several of the studied series.

9.3.2. Volcanic Signals

Volcanic eruptions are thought to be responsible for the global cooling that has been observed for a few years after certain major eruptions. The amount and global extent of the cooling depend on the force of an eruption and, possibly, its latitude and seasonal timing. When large masses of gases from an eruption reach the stratosphere, they can produce a large, widespread cooling effect. The effect of each eruption is specific and can be studied as a pulse or event randomly distributed on the time domain. However, as the explosion emissions have a residence time varying from months to several years, the effects of successive eruptions occurring within such an interval are accumulative and probably cannot be separated. As volcanoes erupt, they blast huge clouds into the atmosphere. These clouds are made up of particles and gases, including sulfur dioxide. Millions of tons of sulfur dioxide gas can reach the stratosphere from a major volcano. There, the sulfur dioxide is converted to tiny, persistent sulfuric acid (sulfate) particles, referred to as aerosols. These sulfate particles reflect energy coming from the Sun, thereby preventing the Sun's rays from heating the Earth. The absorption of solar radiation by the volcanogenic stratosphere aerosol layers may lead to a warming of this layer, whereas the decrease of the atmospheric transmission should lead to a cooling of the atmospheric boundary layer near the surface. Moreover, the climate response to explosive volcanic activity should be large scale in space because the stratospheric sulfuric acid aerosols, due to their long residence times, undergo widespread mixing (Schönwiese, 1988).

TABLE 2   List of Volcanic Eruptions with Their Respective Volcanic Explosivity Indexes (VEIs) > 3 for the Period 1602–1963

Year	Volcano	Latitude	Longitude	VEI	Date (month)	Year	Volcano	Latitude	Longitude	VEI	Date (month)
1630	Furnas	37.77 N	25.32 W	4	9	1853	Chikurachki	50.32 N	155.45 E	4	12
1631	Vesuvio	40.82 N	14.42 E	4	12	1854	Sheveluch	56.78 N	161.58 E	4	2
1640	**Komaga-Take**	**42.07 N**	**140.68 E**	**4**	**7**	1856	Komaga-Take	42.07 N	140.68 E	4	9
1640	**Llaima**	**38.7 S**	**71.7 N**	**4**	**2**	1869	Puracé	2.37 N	76.38 W	4	10
1641	**Awu**	**3.67 N**	**125.50 E**	**5**	**1**	1872	Sinarka	48.87 N	154.17 E	4	??
1660	Quilotoa	0.80 S	78.90 W	4	?	1873	Grimsvotn	62.42 N	17.33 W	4	10
1663	**Usu**	**42.53 N**	**140.83 W**	**5**	**8**	1875	Askja	65.03 N	17.65 W	5	3
1667	**Tarumai**	**42.68 N**	**141.38 E**	**5**	**8**	1877	Suwanose	29.53 N	129.72 E	4	??
1671	**San Salvador**	**13.73 N**	**89.28 W**	**4**	**??**	1877	Cotopaxi	0.65 S	78.43 W	4	6
1673	**Gamkonora**	**1.37 N**	**127.52 E**	**4**	**5**	1881	Nasu	37.12 N	139.97 E	4	7
1680	Tongkoko	1.52 N	125.20 E	4	?	**1883**	**Krakatau**	**6.12 S**	**105.42 E**	**6**	**8**
1690	**Chikurachki**	**50.32 N**	**155.45 E**	**4**	**?**	1883	Augustine	59.37 N	153.43 W	4	10
1693	**Hekla**	**63.98 N**	**19.07 W**	**4**	**2**	**1886**	**Tungurahua**	**1.47 S**	**78.45 W**	**4**	**4**
1695?	Long Island	5.38 S	147.12 E	6	???	1886	Tarawera	**38.2 S**	**157.45 W**	**4**	**11**
1707	Fuji	35.35 N	138.73 E	4	12	1889	Suwanose	29.53 N	129.72 E	4	10
1712	Chirpoi	46.52 N	150.87 E	4	8	1899	Doña Juana	1.52 N	76.93 W	4	11
1721	**Öraefajökull**	**64.00 N**	**16.65 W**	**4**	**5**	**1902**	**Soufrière**	**16.05 N**	**61.67 W**	**4**	**5**
1721	**Katla**	**63.63 N**	**10.03 W**	**4**	**5**	**1902**	**Pelée**	**14.32 N**	**61.17 W**	**4**	**5**
1739	**Tarumai**	**42.68 N**	**141.38 E**	**5**	**8**	**1902**	**Santa María**	**14.76 N**	**91.95 W**	**6**	**10**
1741	Oshima	34.73 N	139.38 E	4	8	1903	Thordarhyrna	64.27 N	17.60 W	4	5
1744	Cotopaxi	0.65 S	78.43 W	4	11	**1907**	**Ksudach**	**51.8 N**	**157.5 E**	**5**	**3**
1755	Katla	63.63 N	10.03 W	5	10	1909	Tarumai	42.68 N	141.38 E	5	4
1759	Jorullo	19.03 N	101.67 W	4	9	1911	Taal	14.00 N	121.00 E	4	1
1762	**Peteroa**	**35.25 S**	**70.57 W**	**4**	**12**	**1912**	**Novarupta**	**58.27 N**	**155.16 W**	**6**	**6**
1764	Jorullo	19.03 N	101.67 W	4	9	1913	Colima	19.42 N	103.72 W	4	1
1765	**Komaga-Take**	**42.07 N**	**140.68 E**	**4**	**??**	**1914**	**Sakura Jima**	**38.54 N**	**130.67 E**	**4**	**1**
1766	Hekla	63.98 N	19.07 W	4	4	**1917**	**Agrigan**	**18.77 N**	**145.67 E**	**4**	**4**
1768	Cotopaxi	0.65 S	78.43 W	4	4	1918	Katla	63.63 N	10.03 W	4	10
1772	Papandayan	7.32 S	107.73 E	4	8	1918	Tungurahua	1.47 S	78.45 W	4	4
1778	**Raikoke**	**48.25 N**	**153.25 E**	**4**	**??**	1919	Manam	4.10 S	145.06 E	4	8
1783	**Asama**	**36.40 N**	**138.53 E**	**4**	**5**	**1921**	**Puyehue**	**40.57 S**	**72.10 N**	**4**	**12**
1783	**Lakagigar**	**12.30 N**	**85.73 W**	**4**	**6**	1929	Komaga-Take	42.07 N	140.68 E	4	6
1793	San Martin	18.39 N	95.17 W	4	3	1931	Kliuchesvkoi	56.18 N	160.78 E	4	3
1793	Alaid	50.60 N	155.50 E	4	??	1932	Cerro Azul	35.67 S	70.77 W	5	4
1795	Pogromni	54.6 N	164.7 W	4	??	1932	Fuego	14.48 N	90.88 W	4	1
1812	**Soufrière**	**16.05 N**	**61.67 W**	**4**	**4**	1937	Rabaul	4.27 S	152.20 E	4	5
1814	**Mayon**	**13.35 N**	**123.68 W**	**4**	**2**	1945	Kliuchesvkoi	56.18 N	160.78 E	4	1
1815	**Tambora**	**8.25 S**	**118.00 E**	**7**	**4**	1946	Sarychev	48.09 N	153.53 E	4	11
1818	Beerenberg	71.08 N	8.17 W	4	??	1947	Hekla	63.98 N	19.07 W	4	3
1818	Colima	19.42 N	103.72 W	4	2	1951	Ambrym	16.25 S	168.08 E	4	1
1822	Usu	42.53 N	140.83 W	4	3	1951	Lamington	8.94 S	148.17 E	4	1
1822	Galunggung	7.25 S	108.05 E	5	10	**1952**	**Bagana**	**6.14 S**	**155.19 E**	**4**	**2**
1825	**Isanotski**	**54.75 N**	**163.73 W**	**4**	**3**	1953	Spurr	61.30 N	152.25 W	4	7
1835	**Cosequina**	**12.98 N**	**87.57 W**	**5**	**1**	1955	Nilahue	40.35 S	72.07 W	4	7
1845	Hekla	63.98 N	19.07 W	4	9	1956	Bezymianny	56.07 N	160.72 E	5	3
1849	Puracé	2.37 N	76.38 W	4	12	**1963**	**Agung**	**8.00 S**	**116.00 E**	**4**	**3**

Note: The names of the volcanoes follow the Simkin et al (1981) nomenclature. The boldface indicates those presumed to have emissions rich in sulfur dioxide according to ice core interpretation (Zielinski et al., 1994; Cole-Dai et al., 1997).

However, while the theoretical consequences of specified atmospheric aerosol loading from volcanic events may be reasonably well known, the direct historical record of volcanic aerosol loading is known only poorly (Bertrand et al., 1999). Taking volcanic forcing into account over the last four centuries, variation in volcanic aerosol optical depth has been derived by Zielinski (1995) based on the lower optical depth values from the Greenland Ice Sheet Project 2 (GISP2) and volcanic sulfate time series (Zielinski et al., 1994).

TABLE 3   Percent of Variance of Each Series Associated with Three Wave Modes: Long-Term Trend (>50 Years), 10–49.9 Years, and 2–9.9 Years

Percent of variance	Long-term trend	10- to 49.9-year mode	2- to 9.9-year mode
North America			
NYU	65.97	25.23	8.80
ATH	16.80	28.15	55.05
COL	11.38	38.92	49.70
G13	5.92	30.30	63.78
IVA	14.43	26.82	58.75
G17	5.00	30.44	64.56
GRG	9.02	28.18	62.80
SNE	6.72	17.89	75.39
CAL	35.71	24.67	39.62
SWE	23.65	15.18	61.17
Northern Hemisphere composite			
MAN	53.31	19.80	26.89
BRI	15.10	33.80	51.10
NHJ	42.39	27.21	30.40
South America			
JUG	10.35	40.82	48.83
NPA	3.65	26.91	69.44
ALE	2.84	28.72	68.44
LEN	2.84	26.22	70.94
USH	18.74	35.68	45.58
Southern Hemisphere composite			
SHJ	6.72	28.11	65.17

Note: The decomposition of the series was carried out using singular spectrum analysis (SSA; M = 31).

Several indexes have been proposed to estimate the strength and persistence of the effects of volcanic eruptions on the atmosphere and climate. We used two indexes: the VEI (Newhall and Self, 1982) and the sulfate loads derived from the ice cores. The parameterization of volcanism in the VEI chronology has been criticized as an index of explosivity and not of stratospheric loading. To produce a more accurate index, the residence times of the aerosol and particles ejected must be taken in account. However, we used the VEI only to assess the relative magnitude of each explosion and to select on this basis the explosions included in further analysis.

Ice cores can provide estimates of atmospheric sulfate contents, but are not always dated precisely and can be biased. It is recognized that high-altitude ice cores may overemphasize mid- and high-latitude volcanic eruptions (Zielinski, 1995). They do, however, provide direct evidence of eruptions, although not the nature of the climatic response. While the historical series for these forcing agents are imperfectly known or measured, they do nonetheless represent our best estimates of the time histories of the corresponding forc-

ing. We assumed, as did Sato et al. (1993), that the aerosols were uniformly distributed over the globe.

9.3.3. Anomalies in the Temperature Reconstructed Series

It is possible to find several statistical techniques used for evaluating volcanic signals in the literature proposing volcanic impact on climate. Most of them are applied to monthly surface and ocean temperature series and focus on the impact of selected eruptions (Mass and Portman, 1989; Stothers, 1996; Kelly et al., 1996; Robock and Mao, 1995). Other authors worked with annually resolved proxy series and noted the occurrence of minimum values in the series that are coincident with the dates of major explosive eruptions (Briffa et al., 1998; D'Arrigo and Jacoby, 1999).

We used a simple t-test to determine temporal and spatial anomaly patterns in the series, and we related to such episodes the significant low-temperature anomalies occurring immediately or 1 year after known volcanic explosions. The results are shown in Fig. 1.

It seems rather evident that some eruptions are associated with a coherent spatial pattern in both North America temperature reconstructed series and Northern Hemisphere composite reconstructions. By contrast, few eruptions have affected the South America series and the Southern Hemisphere composite reconstruction in a consistent way.

The eruptions of Komagatake (42.07°N) in 1640 and Awu (3.67°N) in 1641 are reported as especially rich in sulfur dioxide (Briffa et al., 1998) and are probably responsible for the cool spikes observed in most parts of the North America and Northern Hemisphere reconstructed series. No effect is evident on South America. Only the Lenca reconstruction shows some effect. Coincidentally, the eruption of Volcano Llaima, at 38.7°S, occurred in 1640, very close to the site where the Lenca chronology was derived. This result suggests a local, limited effect.

In the decade from 1663–1673, four volcanoes are known to have erupted: Usu, Tarumai, San Salvador, and Gamkonora. The first two were northern midlatitude volcanoes that erupted in 1663 and 1667, respectively. Both explosions were scaled as VEI 5. The two others are equatorial volcanoes and their eruptions were scaled as VEI 4. Cole-Dai et al. (1997) attributed to these volcanoes the spikes of sulfates that appear in two Antarctic ice cores. Some significant cool departures have been observed only in the northern North America series as of this time.

According to Rampino and Self (1982), high northern latitude eruptions (>40°N) usually disperse aerosols

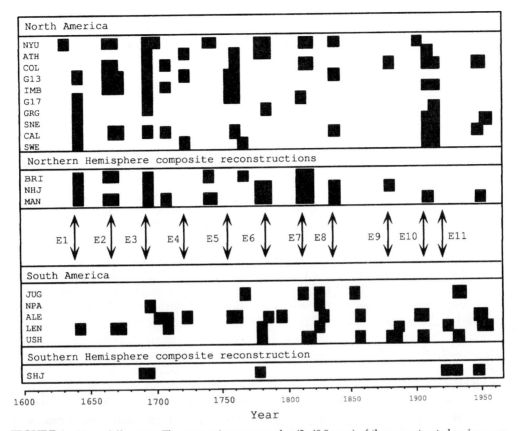

FIGURE 1 Mean differences. The composite wave modes (2–49.9 year) of the reconstructed series were standardized over the entire period to mean 0 and standard deviation 1. A 7-year running mean was calculated and tested for significance against the general mean (0) using a common t-test ($p < 0.1$). Each dark block indicates points where the 7-year mean is significantly lower than 0 and coincides with the occurrence of one of the volcanic explosions listed in Table 2. The arrows indicate the explosions of E1: Komagatake, Llaima, and Awu; E2: Usu, Tarumai, San Salvador, and Gamkonora; E3: Chikurachki, Hekla, and Long Island; E4: Katla-Öraefajökull; E5: Katla and Jorullo; E6: Raikoke, Asama, and Lakagigar; E7: Soufrière, Mayon, and Tambora; E8: Cosequina; E9: Krakatau and Tarawera; E10: Santa María, Pelée, Soufrière, and Ksudach; and E11: Puyehue and Cerro Azul.

over only a portion of the Northern Hemisphere and, thus, have much less impact than low-latitude explosive eruptions, which can affect both hemispheres.

A more extensive cool departure was observed in most of the North America and the composite Northern Hemisphere series in 1690–1700. Three explosive eruptions are known from this decade: Chikurachki (60.3°N), Hekla (63.6°N), and Long Island (5.3°S). The Long Island equatorial volcano produced a powerful explosion between 1695 and 1700 that was scaled as VEI 6. The exact date of the explosion is not yet established. A cold anomaly is also observed on the Southern Hemisphere composite for this period. Coincidentally, the Antarctic ice cores registered an increase in sulfate content (Cole-Dai et al., 1997).

The Jorullo (13.06°N) eruption of 1759 is probably reflected in the cool departures observed in six of the

North America reconstructed series and in Lenca. No signal appears in the composite series. The eruptions of Raikoke (48.2°N), Asama (36.3°N), and Lakagigar (12.3°N) may have affected the north and south high-latitude reconstructed series for the years 1778 and 1783.

The decade between 1810 and 1820 has been noted as especially cool by a number of workers. The eruptions of Soufrière (16°N) in 1812, Mayon (13.3°N) in 1814, Tambora (8.25°S) in 1815, and Beerenberg (71°N) in 1818 are probably responsible for much of the cooling observed. The year 1816 often has been referred to as *the year without a summer*. It was a time of significant weather-related disruptions in New England and in Western Europe, with killing summer frosts in the United States and Canada (1.0°–2.5°C colder than normal). These phenomena were attributed to a major eruption of Tambora (VEI 7) in Indonesia. The volcano threw sulfur dioxide gas into the stratosphere, and the aerosol layer that formed led to striking sunsets seen around the world for several years (Sigurdsson and Carey, 1992).

In our results, this cooling effect is not as widespread as expected. The signal appears in the North Yukon, Columbia, and intermountain basin series and is well defined in the Northern Hemisphere composite reconstructions; it appears in only one of the South American series (Ushuaia). Moreover, the cool anomaly seems to have started sometime earlier, as in 1809. Some evidence (Cole-Dai et al., 1991; Cole-Dai et al., 1997) suggests the occurrence of an unknown volcanic eruption 6 years before Tambora. Rampino et al. (1979) noticed that the average global temperature had already been decreasing since 1810 and then rose again in the 1820s. They also pointed out that the decade 1810–1820 coincided with the most pronounced low in the mean sunspot record of the last 250 years, indicating a lower solar ultraviolet output. This was also a time of low intensity in the solar-terrestrial magnetic wind. These factors plus the volcanic loads may have contributed to the generally cooler climate.

In the South American series, a cool period appears coincidentally with the explosions of Galunggung (7.25°S) in 1822. No signal seems to have been recorded in any of the North American series for that date.

Volcano Cosequina (12.9°N) exploded in 1835 (VEI 5) and probably contributed to the cool pattern observed in three North American reconstructions and two of the Northern Hemisphere composites. No effect was observed in any of the Southern Hemisphere series.

The possible climate effects of the Krakatau (6.12°N) eruption have been the subject of many studies. The consensus seems to be that the depression of Northern Hemisphere temperature was ca. 0.3°–0.4°C for 1 or 2 years following the Krakatau eruption (Jones and Wigley, 1980). The eruptions of Tungurahua (1°S) and Tarawera (38°S) in 1886 may have contributed to the general cooling in the late 1880s (Rampino and Self, 1982). In our data, the signal attributable to these volcanoes is rather weak.

The first two decades of the twentieth century were particularly rich in major volcanic eruptions. Soufrière (16.4°N), Santa María (14.3°N), and Mount Pelée (14.7°N) in 1902; Ksudach (51.2°N) in 1907; Tarumai (42.06°N) in 1909; and Novarupta (58.27°N) in 1912 share a VEI of 5 or higher. The sulfur load has been reported as very high in this period. The reconstructed series show a widespread anomaly, especially in those for North America.

Four explosions have been reported in the Southern Hemisphere during the period from 1918–1932: Tungurahua (1.47°S), Manam (4.1°S), Puyehue (40.57°S), and Cerro Azul (35.6°S). These eruptions could perhaps be linked to the cooling periods observed in some of the South American series and in the Southern Hemisphere composite reconstruction in the same time frame. Hekla (64°N) in 1947, Ambrym (16.2°S) and Lamington (9°S) in 1951, and Bagana (6.14°S) in 1952 are probably connected to the anomalies observed in both North and South American series.

A careful analysis of the results presented here indicates that approximately 38 of the 90 eruptions tested could be linked to significant cool episodes in the reconstructions. Most of the explosions occurred in clusters of years, and due to the window (7 years) used in the analysis, we are not able to separate their effects. The choice of the window has been a compromise between augmenting the noise and reducing the resolution of the estimation. However, several tests using other windows (3, 5, 9, and 11 years) indicate that a 7-year window gives good resolution with more stable means.

The North America and Northern Hemisphere composite reconstructions give more obvious patterns of anomalies linked to the explosions. It is interesting to notice that some of the North America and the Northern Hemisphere composite temperature reconstructions include only tree-ring-density chronologies (Briffa et al., 1998) or a mixture of tree-ring-width and tree-ring-density data (Jones et al., 1998; Mann et al., 1998). No tree-ring-density data have been used in the development of the South America and Southern Hemisphere composite reconstructions. South America and the Southern Hemisphere composite produce a less marked pattern, but apparently more linked to explosions that occurred in the Southern Hemisphere.

According to Mass and Portman (1989), the potential climatic impact of volcanic eruptions is reduced by the inhomogeneous spread of volcanic dust. The largest attenuation of solar radiation does not occur globally, but rather is confined predominantly to the eruption hemisphere.

9.3.4. Sulfate Loads and Climate Relationships

As was expected, the explosions that seem to be more influential in terms of producing cool episodes are those linked to heavy loads of sulfate aerosols in the atmosphere.

In order to investigate this relationship, we ran a superposed event analysis using different combinations of eruption dates (events) and reconstructed temperature series. In a first set of experiments, we tested all the explosion dates against the reconstructed series and

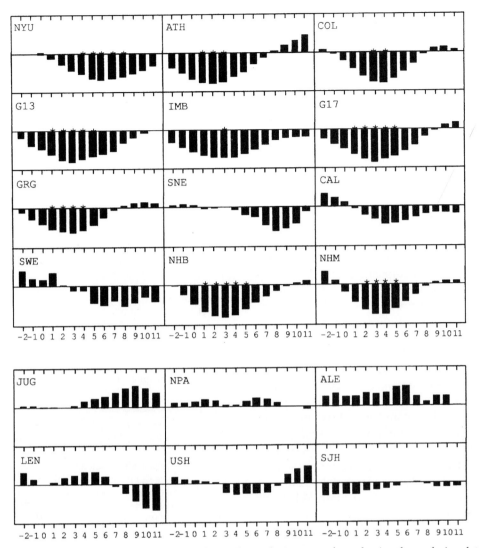

FIGURE 2   Superposed epoch (events) analysis. The analysis was performed using the explosion dates shown in bold in Table 2 as events and the 10- to 49.9-year wave mode series. The window used was 11 years before and after the event date, but the diagrams show only 2 years before the explosion. The ordinates have arbitrary scales, not shown in the diagrams. The significance of lags ($p < 0.05$) is calculated by random simulation and indicated by dots.

against the series decomposed in wave modes of 10–49.9 years and 2–9.9 years (high-frequency band). The results did not show any significant pattern.

Then, we ran a set of similar experiments, but using as events only the dates of the eruptions that reportedly have produced strong sulfate bands in ice cores (Zielinski et al., 1994). Under these conditions, the 10- to 49.9-year wave modes give a coherent pattern (Fig. 2). These results indicate the existence of a significant signal in most of the reconstructions of North America and the Northern Hemisphere composite, but not for the Southern Hemisphere. A delay in the response is apparent, which reaches a maximum of significance at

lag +3. Mass and Portman (1989) suggest two periods of posteruption decline in air temperature: one occurring during the first few months after the eruption key date and another peaking about a year and a half later. Sea surface temperature (SST) holds relatively steady during the first 10 posteruption months and then experiences a 1-year decline. Thus, the primary air temperature perturbation during the second year occurs at a time when appreciable aerosols are still resident in the stratosphere. This consideration and the fact that normally the eruptions occur in temporal clusters could help to explain the observed delay and the presence of a signal in the 10- to 49.9-year wave modes and not in the high-frequency wave modes. As Briffa et al. (1998) pointed out, closely spaced multiple eruptions can reduce hemispheric temperatures on decadal and multidecadal timescales, as occurred in the mid-fifteenth century, 1640s, 1660s, 1670s, and 1810s. The apparent cluster of possible eruptions in the seventeenth century, perhaps in combination with lower solar irradiance, may have been a contributing factor in the extended hemispheric cooling that occurred at that time. Rampino and Self (1982) suggest that a secondary mechanism of climate change, e.g., increased snow and ice cover and perturbations of large-scale atmospheric circulation patterns, might act to enhance and/or prolong the climatic impact of volcanic eruptions.

One possible explanation of the lack of signal observed in all the reconstructions of South America and the Southern Hemisphere composite is that the Southern Hemisphere is the *oceanic hemisphere*. A difference in phasing of the air temperature and SST response is consistent with a variety of modeling studies. Harvey and Schneider (1985), using a global energy balance model, found that air temperature decreased over water with about half the amplitude found over land. McCraken and Luther (1984), using a statistical-dynamical climatic model, found a similar result, with ocean temperatures changing more slowly and with less amplitude than land temperatures. Rampino and Self (1982) pointed out that the 1963 Agung eruption, at 8°S, produced 10 times less optical depth change in the Northern Hemisphere than in the Southern Hemisphere, but was associated with similar temperature deviations in both hemispheres over similar time periods. These differences in responses to the Agung explosion may be a result of a greater sensitivity of the Northern Hemisphere to radiative perturbations. It is also true that there have been only six high southern latitude (>35°S) eruptions with a VEI >3 during the period analyzed.

A simple correlation coefficient with the sulfate series was calculated against each of the reconstructed temperature series. The results are summarized in Table 4. In general, the coefficients indicate a rather weak association, stronger in the case of the Northern Hemisphere composite reconstructions. No association appears in South America or the Southern Hemisphere composite or in the high-frequency (2- to 9.9-year) wave modes in either hemisphere. Caution should be taken in interpreting the correlation coefficients due to the presence of high autocorrelation, especially in the 10- to 49.9-year wave mode series. However, in a previous exploratory analysis, the North America and Northern Hemisphere composite reconstructions and the sulfate series are grouped together over the third component, indicating a link. South America and the Southern Hemisphere composite reconstructions form

TABLE 4   Correlation Coefficients Between the Volcanic Sulfate Series (Zielinski, 1995) and the Reconstructed Temperature Series

	A	B	C	D
North America				
NYU	−0.194	−0.120	−0.283	0.004
ATH	−0.267	−0.172	−0.349	−0.013
COL	−0.085	−0.057	−0.155	−0.003
G13	−0.234	−0.153	−0.406	−0.010
IMB	−0.030	−0.060	−0.110	0.009
G17	−0.307	−0.240	−0.505	−0.020
GRG	−0.160	−0.169	−0.262	−0.007
SNE	−0.013	−0.025	−0.023	−0.001
CAL	−0.087	−0.073	−0.085	−0.004
SWE	−0.095	−0.100	−0.037	0.007
Northern Hemisphere composite				
NHB	−0.416	−0.432	−0.606	−0.032
NHJ	−0.316	−0.310	−0.539	−0.021
NHM	−0.350	−0.264	−0.436	0.004
South America				
JUG	0.105	0.110	0.109	0.001
NPA	0.071	0.043	0.091	0.001
ALE	0.070	0.053	0.093	0.002
LEN	0.021	0.049	0.056	−0.001
USH	0.038	−0.067	−0.085	−0.016
Southern Hemisphere composite				
SHJ	−0.052	−0.083	−0.091	0.001

Note: A, reconstructed temperature series; B, 2- to 49.9-year composite wave mode; C, 10- to 49.9-year wave mode; and D, 2- to 9.9-year wave mode. The 5-year interval measurements of the sulfate series were linearly interpolated to annual values. The boldface coefficients indicate significant values if the series were totally independent. No correction has been made for the degrees of freedom lost due to autocorrelation. Therefore, the significance should be interpreted as indicative only.

a cluster clearly separate from the other variables. The third component explains 11.6% of the total variance.

Running correlation coefficients were calculated by using a window of 50 years between the sulfate series against the reconstructed temperature series. The results are presented in Fig. 3. They show a pattern of negative correlation coefficients spatially distributed toward the northern latitudes in North America. No similar pattern appears in the South America reconstructions. These results also indicate a more consistent and strong negative relation of the sulfate contents with the Northern Hemisphere composite reconstructions. Mann et al. (1998) ran a similar analysis on their series using another index of volcanic activity. They concluded that explosive volcanism exhibits the expected marginally significant negative correlation with temperature during much of 1610–1995 period, most pronounced in the 200-year window centered near 1830, which in-

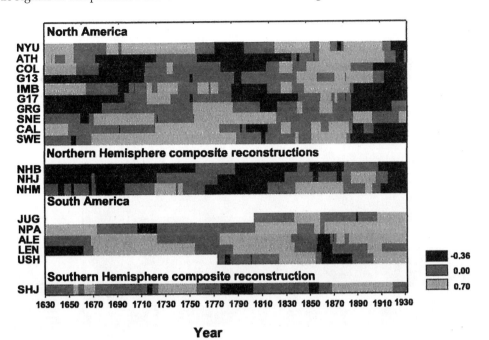

FIGURE 3   Running correlation coefficients (50-year window) between the volcanic sulfate loads and 2- to 50-year wave modes of the reconstructed temperature series. The darkest spots correspond to coefficients lower than −0.36, the limit of significance ($p < 0.05$).

cludes the largest number of known explosive volcanic events.

In Fig. 4, we present a plot of the standardized 5-year running mean of each reconstructed temperature series and a plot of the sulfate loads. Note that high peaks of sulfate content correspond in general, but not always, with cool periods in the series.

Our data suggest a volcanic influence on climate that is not particularly strong, but is discernible in time and space. Several papers have expressed doubts about the existence of a volcanic influence on surface climate (Landsberg and Albert, 1974; Ellsaesser, 1977). After examining the temperature evolution accompanying the largest eruptions since the late eighteenth century, Angell and Korshover (1985) concluded that while volcanic eruptions certainly do not cause warming, the evidence that they cause cooling is not overly impressive. It has become clear in the last decade (Robock and Mao, 1995) that the effect of a volcano on climate is most directly related to the sulfur dioxide content of emissions that reach into the stratosphere and not to the explosivity of the eruption, although the two are highly correlated.

Rampino and Self (1982) signaled that the largest eruptions in terms of total ejecta and eruption column height did not necessarily produce the greatest climatic effects (Self et al., 1981). It seems that significantly smaller, but sulfur dioxide-rich explosive eruptions can produce similar temperature deviations. Past increases in stratospheric optical depth may have been controlled to a large extent by sulfur dioxide-rich explosive eruptions, which, by the nature of their magma bodies, tend to be relative small in volume (VEI 3 or 4). If these eruptions occur in clusters, this might provide an explanation for certain prolonged periods of high stratospheric optical depth (1883–93 and 1902–06). However, a complete historical or volcanological record of these events does not exist for years prior to ca. 1950 (Newhall and Self, 1982). Assigning specific atmospheric opacity variations to known historic eruptions is often not possible. Nevertheless, volcanic aerosols are potentially significant climate forcing mechanisms on timescales of years, decades, and perhaps longer periods (Zielinski, 1995; Overpeck et al., 1997), and the correlation of temperature reductions with volcanic eruptions can be observed even in the instrumental record for the Northern Hemisphere (Bradley, 1988).

Some authors indicated that because the climatic response to ENSO is of the same amplitude and timescale as volcanic responses, it is necessary to separate them to examine the volcano signal (Jones, 1988; Robock and Mao, 1995). Handler (1986) has supposed that tropical explosive volcanoes could cause a warming of the SST, and he assumes a connection as triggering El Niño events. In our series, we have not filtered the ENSO signal, which could mask the volcanic influence. Howev-

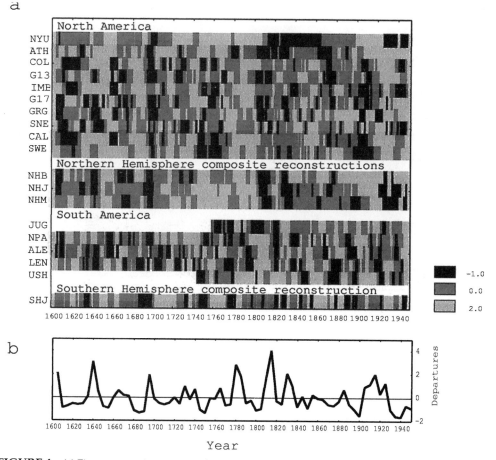

FIGURE 4   (a) Five-year moving average of reconstructed temperature series, standardized (0,1). (b) Standardized volcanic sulfate series.

er, experiments conducted by Robock and Mao (1995) have shown that spatial patterns are clear with or without removing ENSO.

9.4. CONCLUSIONS

The analysis of volcanism–climate relationships on an observational–statistical basis for the last four centuries is possible using proxy reconstructions for both hemispheres. Evidence of volcanic perturbations in the reconstructed temperature series is rather weak and is biased to those explosions that occurred in clusters and with high sulfur dioxide emissions. The North America reconstructed series show a more recognizable pattern associated with eruptions in both the spatial and time domains. The Northern Hemisphere composite reconstructions show a marked signal, perhaps because as they integrate more proxy records, they are capturing more variance.

In contrast, in the South America reconstructed series and in the Southern Hemisphere composite reconstruction, it is hard to define volcanic fingerprints. Some eruptions that occurred in the Southern Hemisphere do seem to have been associated with some observed cold anomalies. However, the relatively large proportion of ocean surface and the relative scarcity of volcanoes are possible explanations for the weakness of the volcanic signal in the Southern Hemisphere.

References

Angell, J. K., and J. Korshover, 1985: Surface temperature changes following six major volcanic episodes between 1780 and 1980. *Journal of Climate and Applied Meteorology*, **24**: 937–951.

Bertrand, C., J. P. van Ypersele, and A. Berger, 1999: Volcanic and solar impacts on climate since 1700. *Climate Dynamics*, **15**: 355–367.

Boninsegna, J., J. Keegan, G. C. Jacoby, R. D'Arrigo, and R. L. Holmes, 1989: Dendrochronological studies in Tierra del Fuego, Argentina. *Quaternary of South America and Antarctica Peninsula*, **7**: 305–326.

Bradley, R., 1988: The explosive volcanic eruption signal in Northern Hemisphere continental temperature records. *Climatic Change,* **12**: 222–243.

Briffa, K., P. D. Jones, and F. H. Schweingruber, 1992: Tree-ring density reconstructions of summer temperature patterns across western North America since 1600. *Journal of Climate,* **5**: 735–754.

Briffa, K. R., P. D. Jones, F. H. Schweingruber, and T. J. Osborn, 1998: Influence of volcanic eruptions on Northern Hemisphere summer temperature over the past 600 years. *Nature,* **393**: 450–455.

Cole-Dai, J., E. Mosley-Thompson, and L. G. Thompson, 1991: Ice core evidence for an explosive tropical volcanic eruption six years preceding Tambora. *Journal of Geophysical Research,* **96**: 17361–17366.

Cole-Dai, J., E. Mosley-Thompson, and L. G. Thompson, 1997: Annually resolved Southern Hemisphere volcanic history from two Antarctica ice cores. *Journal of Geophysical Research,* **102**: 16761–16771.

Crowley, T. J., T. M. Quinn, F. W. Taylor, and P. Joannot, 1997: Evidence for a volcanic cooling signal in a 335-year coral record from New Caledonia. *Paleoceanography,* **12**: 633–639.

D'Arrigo, R., and G. Jacoby, 1999: Northern North American tree-ring evidence for regional temperature changes after major volcanic events. *Climate Change,* **41**: 1–15.

D'Arrigo, R., G. Jacoby, and R. M. Free, 1992: Tree-ring width and maximum latewood density at the North American tree-line: Parameters of climatic change. *Canadian Journal of Forest Research,* **22**: 1290–1296.

Ellsaesser, H. W., 1977: Comments on "Estimate of the global temperature, surface to 100 mb, between 1958 and 1975." *Monthly Weather Review,* **105**: 1200–1201.

Fritts, H., 1991: *Reconstructing Large-Scale Climatic Patterns from Tree-Ring Data: A Diagnostic Analysis.* Tucson: The University of Arizona Press, 286 pp.

Graumlich, L. J., 1993: A 1000-year record of temperature and precipitation in the Sierra Nevada. *Quaternary Research,* **39**(2): 249–255.

Handler, P., 1986: Possible association between the climatic effects of stratospheric aerosols and sea surface temperature in the eastern tropical Pacific Ocean. *Journal of Climatology,* **6**: 31–41.

Harvey, L. D., and S. H. Schneider, 1985: Transient response to external forcing on 10^6–10^4 time scales. 2. Sensitivity experiments with seasonal, hemispherically averaged, coupled atmosphere, land, ocean energy balance model. *Journal of Geophysical Research,* **90**: 2207–2222.

Jacoby, G. C., and R. D'Arrigo, 1992: Spatial patterns of tree-growth anomalies from the North America boreal tree-line in the early 1800's, including the year 1816. *In* Harrington, C. R. (ed.), *The Year Without Summer?: World Climate in 1816.* Ottawa, Canada: National Museum of Science.

Jones, P. D., 1988: The influence of ENSO on global temperature. *Climate Monitor,* **17**: 80–89.

Jones, P. D., and T. M. L. Wigley, 1980: Northern Hemisphere temperatures 1881–1979. *Climate Monitor,* **9**: 43–47.

Jones, P. D., K. Briffa, T. P. Barnett, and S. F. B. Tett, 1998: High resolution palaeoclimatic records for the last millennium: Interpretation, integration and comparison with General Circulation Model control-run temperatures. *The Holocene,* **8**: 455–471.

Kelly, P. M., P. D. Jones, and J. Penguin, 1996: The spatial response of the climatic system to explosive volcanic eruptions. *International Journal of Climatology,* **16**: 537–550.

Landsberg, H. E., and J. M. Albert, 1974: The summer of 1816 and volcanism. *Weatherwise,* **27**: 63–66.

Lara, A., and R. Villalba, 1993: A 3,620-year temperature record from *Fitzroya cupressoides* tree-rings in southern South America. *Science,* **260**: 1104–1106.

Lough, J. M., and H. Fritts, 1987: An assessment of the possible effects of volcanic eruptions on North America climate using tree-ring data: 1602 to 1900 A.D. *Climate Change,* **10**: 219–239.

Luckman, B. H., K. R. Briffa, P. D. Jones, and F. H. Schweingruber, 1997: Tree-ring based reconstruction of summer temperature at the Columbia Icefield, Alberta, Canada, A.D. 1073–1983. *The Holocene,* **7**(4): 375–389.

Mann, M., R. Bradley, and M. Hughes, 1998: Global scale temperature patterns and climatic forcing over the past six centuries. *Nature,* **392**: 779–787.

Mass, C. F., and D. Portman, 1989: Major volcanic eruptions and climate: A critical evaluation. *Journal of Climate,* **2**: 566–593.

McCraken, M. C., and F. M. Luther, 1984: Preliminary estimate of the radiative and climatic effect of the El Chichon eruption. *Geophysical International,* **233**: 385–401.

Newhall, C. G., and S. Self, 1982: The Volcanic Explosivity Index (VEI): An estimate of explosive magnitude for historical volcanism. *Journal of Geophysical Research,* **87**: 1231–1238.

Norton, D., 1992: New Zealand temperatures 1800–30. *In* Harintong, C. R. (ed.), *The Year Without Summer?: World Climate in 1816.* Ottawa, Canada: National Museum of Science.

Overpeck, J., K. Hughen, D. Hardy, R. Bradley, R. Case, M. Douglas, B. Finney, K. Gajewski, G. Jacobi, A. Jennings, S. Lamoureux, G. MacDonald, J. Moore, M. Retelle, S. Smith, A. Wolfe, and G. Zielinski, 1997: Arctic environmental change of the last four centuries. *Science,* **278**: 1251–1256.

Palmer, J., and J. Ogden, 1992: Tree-ring chronologies from endemic Australian and New Zealand conifers 1800–30. *In* Harintong, C. R. (ed.), *The Year Without Summer?: World Climate in 1816.* Ottawa, Canada: National Museum of Science.

Rampino, R., and S. Self, 1982: Historic eruptions of Tambora (1815), Krakatau (1883), and Agung (1963): Their stratospheric aerosols and climatic impact. *Quaternary Research,* **18**: 127–143.

Rampino, R., S. Self, and R. Fairbridge, 1979: Can rapid climatic change cause volcanic eruptions? *Science,* **206**: 826–829.

Robock, A., and J. Mao, 1995: The volcanic signal in surface temperature observations. *Journal of Climatology,* **8**: 1086–1103.

Sato, M., J. E. Hansen, M. P. McCormick, and J. B. Pollack, 1993: Stratospheric aerosol optical depth, 1850–1990. *Journal of Geophysical Research,* **98**: 22987–22994.

Schönwiese, Ch., 1988: Volcanic activity parameters and volcanism-climate relationships within the recent centuries. *Atmosfera,* **1**: 141–156.

Scuderi, L. C., 1990: Tree-ring evidence for climatically effective volcanic eruptions. *Quaternary Research,* **34**: 67–85.

Self, S., M. R. Rampino, and J. Barbera, 1981: The possible effects of large 19th and 20th century volcanic eruptions on zonal and hemispheric surface temperatures. *Journal of Volcanology and Geothermal Research,* **11**: 41–60.

Sigurdsson, H., and S. Carey, 1992: The eruption of Tambora in 1815: Environmental effects and eruption dynamics. *In* Harintong, C. R. (ed.), *The Year Without Summer?: World Climate in 1816.* Ottawa, Canada: National Museum of Science.

Simkin, T., L. Siebert, L. McClelland, D. Bridge, C. Newhall, and J. H. Latter, 1981: *Volcanoes of the World.* Stroudsburg, PA: Hutchinson Ross.

Stothers, R. B., 1996: Major optical depth perturbations to the stratosphere from volcanic eruptions: Pyrheliometric period, 1881–1960. *Journal of Geophysical Research,* **101**(D2): 3901–3920.

Vautard, R., and M. Ghil, 1989: Singular spectrum analysis in nonlinear dynamics with application to paleoclimatic time series. *Physica Dissertation,* **35**: 395–424.

Vautard, R., P. Yiou, and M. Ghil, 1992: Singular-spectrum analysis: A toolkit for short, noise chaotic signals. *Physica Dissertation,* **58**: 95–126.

Villalba, R., 1990: Climatic fluctuations in northern Patagonia during the last 1000 years as inferred from tree-ring records. *Quaternary Research*, **34**: 346–360.

Villalba, R., and J. A. Boninsegna, 1992: Changes in southern South America tree-ring chronologies following major volcanic eruptions between 1750 to 1970. *In* Harintong, C. R (ed.), *The Year Without Summer?: World Climate in 1816*. Ottawa, Canada: National Museum of Science.

Villalba, R., J. A. Boninsegna, and D. R. Cobos, 1989: A tree-ring reconstruction of summer temperature between A.D. 1500 and 1974 in western Argentina. Proceedings of the Third International Conference of the Southern Hemisphere Meteorology and Oceanography, Buenos Aires, Argentina, November 1989, pp. 196–197.

Villalba, R., H. Grau, J. A. Boninsegna, G. C. Jacoby, and A. Ripalta, 1998: Tree-ring evidence for long-term precipitation changes in subtropical South America. *International Journal of Climate*, **5**: 735–754.

Zielinski, G. A., 1995: Stratospheric loading and optical depth estimates of explosive volcanism over the last 2100 years derived from the Greenland Ice Sheet Project 2 ice core. *Journal of Geophysical Research*, **100**: 20937–20955.

Zielinski, G. A., P. A. Mayeski, L. D. Meeker, S. Whitlow, M. S. Twickler, M. Morrison, D. A. Meese, A. J. Gow, and R. B. Alley, 1994: Record of volcanism since 7000 B.C. from the GISP2 Greenland ice core and implications for the volcano-climate system. *Science*, **264**: 948–952.

10

Decadal-Scale Climatic Variability Along the Extratropical Western Coast of the Americas: Evidence from Tree-Ring Records

R. VILLALBA, R. D. D'ARRIGO, E. R. COOK, G. C. JACOBY, AND G. WILES

Abstract

Exactly dated tree-ring chronologies along the western coast of the Americas have been used to track climatic variations that have simultaneously impacted the extratropical regions of North and South America during the past four centuries. Significantly correlated records for the coast of Alaska and northern Patagonia show the existence of common oscillatory modes for temperature variations at 9, 13, and 50 years in both regions. Tree-ring chronologies from precipitation-sensitive regions also reveal the occurrence of decadal-scale oscillations, centered at 8 and 10–18 years, which have simultaneously influenced climatic conditions in the Midwest–southern United States and central Chile. Spatial correlation patterns between tree-ring records and sea surface temperatures (SSTs) over the Pacific show that variations in climate-sensitive records are strongly connected with SST anomalies in the equatorial Pacific and off the western subtropical Americas. Correlations of opposite sign occur for the extratropical central North and South Pacific. These correlation patterns resemble the spatial signature of the decadal mode of SST variability over the Pacific, which has recently been detected in instrumental records. Whereas the spectral decomposition of tree-ring estimates suggests the existence of decadal-scale oscillations common to the climates of western North and South America, the spatial correlation patterns between tree-ring records and SSTs indicate that the Pacific Ocean is the major forcing mechanism for these synchronous oscillations. It is likely that climate variations over the Pacific Ocean result from interactions between interannual, interdecadal, and secular modes of climatic variability. A detailed analysis in time and frequency domains of the temperature- and precipitation-sensitive records for North and South America indicates a major change within the decadal oscillatory mode at ca. 1850. In general, the decadal mode of variability was more energetic from 1600–1850. During this interval, the North and South American records correlated with each other more strongly. The leading modes of the Pacific SST variability indicate that, with the exception of the last two decades, most of the twentieth century has been dominated by the interannual mode of variability. A major reorganization in the Pacific Ocean at ca. 1850 may have altered a persistent decadal mode of oscillation and given rise to a dominant interannual mode. Consequently, what we know of tropical Pacific variability through the analysis of instrumental records has been based largely on a period of predominant interannual variability. Some of the anomalies observed in SSTs over the Pacific during the last two decades may represent a recovery of the decadal mode

---

of SST variability that prevailed in the Pacific before 1850. Copyright © 2001 by Academic Press.

Resumen

La ocurrencia de variaciones climáticas que afectaron simultáneamente las regiones extratropicales de América del Norte y del Sur durante los últimos cuatro siglos fueron investigadas usando cronologías de anillos de árboles existentes a lo largo de las costas oestes de las Américas. Registros dendrocronológicos provenientes de la costa de Alaska y del norte de la Patagonia, correlacionados significativamente entre sí, indican la existencia de oscilaciones comunes en las variaciones de la temperatura en ambas regiones con ciclos alrededor de los 9, 13 y 50 años. Cronologías sensibles a la precipitación, provenientes de las regiones Centrooeste-Sur de los Estados Unidos y Chile Central, también revelan la ocurrencia de oscilaciones comunes en la precipitación con escalas decenales centradas a los 8 y 10–18 años. Los patrones de correlación espacial entre los registros de anillos de árboles y la temperatura de la superficie del mar (SST) sobre el Océano Pacífico, indican que las variaciones en los registros de anillos de árboles están fuertemente relacionadas con las anomalías de las SST en la región Pacífica equatorial y en las vecindades de la costas subtropicales de las Américas. Estos patrones de correlaciones espaciales se asemejan a los patrones que caracterizan el modo de variabilidad decenal en SST sobre el océano Pacífico, el que ha sido recientemente dectectado en los registros instrumentales. Mientras el análisis espectral estaría sugiriendo la existencia de modos comunes de variabilidad climática en escala decenal a lo largo de las costas oestes de las Américas, los patrones espaciales de correlación entre los registros dendrocronológicos y las SST indicarían que el Océano Pacífico es el forzante principal de estos cambios climáticos sincrónicos en regiones geográficamente separadas. Las variaciones climáticas sobre el Océano Pacífico resultan, muy probablemente, de las interacciones entre los modos de variabilidad climática interanual, decenal y secular. Un análisis temporal y de frecuencias muy detallado de los registros dendrocronológicos sensitivos a la temperatura y la precipitación en América del Norte y del Sur indicaría la existencia de un cambio importante en el modo de oscilación decenal alrededor de 1850. En general, las oscilaciones decenales fueron más marcadas entre 1600 y 1850. Durante este intervalo, los registros dendrocronológicos de América del Norte y del Sur estuvieron más fuertemente correlacionados entre sí. Los modos dominantes de variabilidad en SST sobre el Pacífico durante el siglo XX indican que, con excepción de las últimas dos décadas, este siglo ha sido domina-

do por el modo de variabilidad interanual. Una importante re-organización en los modos de variabilidad sobre el Océano Pacífico podría haber tenido lugar alrededor de 1850 modificando el modo de circulación decenal preexistente y dando paso a un modo de variabilidad dominado por oscilaciones de tipo interanual. Consequentemente, nuestro conocimiento sobre la variabilidad climática en el Océano Pacífico tropical a partir de los registros instrumentales, estaría largamente basado en un período dominado fuertemente por la variabilidad interanual. Algunas de las anomalías observadas en SST sobre el Océano Pacífico durante las dos últimas décadas podrían responder a un reforlecimiento del modo de variabilidad decenal de SST, el cúal fue más intenso con anterioridad a 1850.

10.1. INTRODUCTION

An array of 40 atmospheric, terrestrial, and oceanic variables were used by Ebbesmeyer et al. (1991) to demonstrate the remarkable consistency of a major climate change over the Pacific Ocean and the adjacent western coast of the Americas at ca. 1976. This multidimensional data set included dissimilar variables such as air and water temperatures, chlorophyll, salmon, glaciers, atmospheric dust, coral, carbon dioxide, ice cover, ocean transport, and winds from the Bering Strait to the tropical Andes in South America. This compilation by Ebbesmeyer et al. (1991) clearly showed the occurrence of an interhemispheric climatic change that simultaneously affected the western coasts of the Americas.

More recently, on the basis of instrumental records, several studies have reported the existence of decadal-scale oscillatory modes over the Pacific (Trenberth and Hurrell, 1994; Latif et al., 1997; Zhang et al., 1997). Specifically, the climatic change at ca. 1976 reported by Ebbesmeyer et al. (1991) is considered a remarkable manifestation of the decadal-scale climatic variability in the Pacific, which has influenced climatic conditions all along the western Americas. For example, warm climatic conditions have been recorded since ca. 1976 along the Pacific coast of North America, including Alaska (Trenberth and Hurrell, 1994). Regional records also show that temperatures along the Chilean coast have been unusually warm since 1976 compared to the past 80 years (Rosenbluth et al., 1997). One apparent feature of the pattern of changes in sea surface temperatures (SSTs) during the mid-1970s was a wedge-shaped region in the tropical Pacific where SSTs increased by $0.6°–0.8°C$ (Graham, 1994). Interestingly, this region extends eastward from the apex on the equator near the date line to the west coasts of North and South America near 40°N and 40°S, respectively. The results presented by Graham (1994) suggest that the changes in average temperature observed along the extratropical Americas during the 1970s were driven almost entirely by changes in tropical SSTs. Assuming this is to some degree correct, one of the interesting questions concerning this idea is the relative importance of tropical SST in driving temperature and precipitation changes along the Americas during past centuries. Do the climatic changes observed at about the mid-1970s represent a unique manifestation of the tropical Pacific SST as a common forcing of decadal-scale climatic variations along the extratropical western coasts of the Americas, or has this teleconnection pattern been recurrent in the past?

Based on instrumental records, decadal-scale climatic variations have been detected in the North Pacific (Nitta and Yamada, 1989; Trenberth, 1990; Trenberth and Hurrell, 1994; Latif and Barnett, 1994; Graham et al., 1994; Mantua et al., 1997). However, the role of tropical Pacific SSTs as the major forcing mechanism for these decadal oscillations in the North Pacific is still under discussion (Latif and Barnett, 1994, 1996; Trenberth and Hurrell, 1994; Graham, 1994). On the other hand, meteorological records for the Pacific and along the western coasts of the Americas are short, which severely restricts the study of decadal-scale oscillations in climate. Few of these records are longer than 100 years, and most of them are limited to the past 40 years. Temperature- and precipitation-sensitive chronologies from tree rings, which may extend for hundreds of years, can be used to consistently document the existence and stability of interannual- to decadal-scale climatic oscillations along the western coasts of the Americas.

In this study, variations in tree-ring chronologies for western North and South America were compared to investigate the existence of common, decadal-scale climatic forcings affecting mid- to high latitudes across the Americas during the past several centuries. With the increasing number of tree-ring series developed for the western Americas, the use of a spatial approach to study large-scale atmospheric variations connecting mid- to high-latitude climatic changes in these regions now appears to be feasible. Another key incentive for this exercise was the documented change in tree-ring variations, both in Alaska (Wiles at al., 1998) and in northern Patagonia (Villalba et al., 1997), in response to the major decadal-scale climatic change in the tropical Pacific during the mid-1970s. Here, we emphasized the search in the tree-ring records of decadal- to century-scale rather than annual-scale climatic variations for two major reasons. In our study, persistence in tree-ring series was not totally removed, and, consequently, tree rings captured more efficiently decadal- rather than an-

nual-scale climatic variations. Second, we intuitively assumed that simultaneous changes in temperature and precipitation in the extratropical western Americas are primarily modulated by interdecadal variations in the Pacific Ocean, as described by Zhang et al. (1997) and Latif et al. (1997). Temperature- and precipitation-sensitive chronologies for western North and South America were co-spectrally analyzed to establish common modes of climatic variations affecting both regions. Spatial correlation patterns between tree-ring records and SSTs across the Pacific were used to identify the physical links connecting climatic variations at mid- to high latitudes in the Americas with the tropical Pacific. Determining whether climatic changes simultaneously affecting both regions have occurred during the past several centuries will provide some insight on the role of the Pacific Ocean in forcing climatic variations at mid- to high latitudes in the Americas. Finally, past interactions between interannual- and decadal-scale modes of climatic variability over the Pacific Ocean and the adjacent western Americas were evaluated through a detailed study of the tree-ring records in the time and frequency domains.

10.2. TREE-RING RECORDS

Along the western coasts of North and South America, there is a gradual environmental gradient from the dry and warm subtropics to the wet and cold high latitudes. Tree-ring records for subtropical regions, such as southern California and central Chile, are remarkably sensitive to precipitation variations (Haston and Michaelsen, 1994; Boninsegna, 1988). In the transitional zones to higher latitudes, tree-ring responses to climate are largely determined by site conditions. Depending on elevation, aspect, slope, and soil characteristics, tree growth can be influenced by temperature, precipitation, or, more commonly, by a combination of both. At the extreme wet and cold environments in high-latitude upper tree lines, temperature is the limiting factor controlling tree growth (Wiles et al., 1996, 1998; Villalba et al., 1997). These changes in tree response with latitude along the western Americas were instrumental in our setting the strategies for selecting tree-ring records sensitive to temperature and precipitation variations. Upper-elevation chronologies for coastal sites around the Gulf of Alaska and northern Patagonia were selected as proxy records of temperature for North and South America, respectively. Tree-ring records for dry environments in the south-central United States and central Chile were used for the interhemispheric comparison of precipitation-sensitive records (Fig. 1).

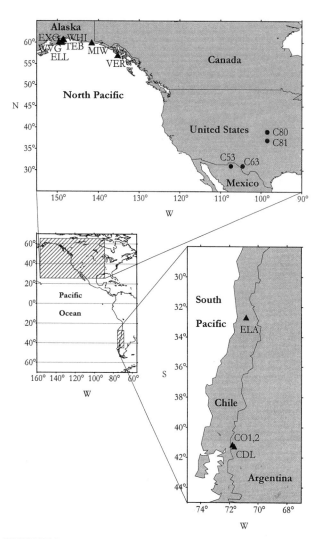

FIGURE 1 Geographical locations of tree-ring chronologies (triangles) and Palmer Drought Severity Index (PDSI) reconstructions (circles) considered in this study. See Table 1 for site names and identification of the PDSI reconstructions.

To facilitate the comparison between the North and South American records, ring-width measurements for all trees at each site were standardized by using similar techniques. Standardization involves fitting an expected growth curve to the observed ring-width series and computing an index of the observed ring widths in relation to the expected values. This procedure reduces the variance among cores and transforms ring widths into dimensionless index values. Thus, standardization permits computation of average tree-ring chronologies without the average being dominated by the faster growing trees with larger ring widths. Residuals from the expected growth curve were calculated. Potential biases introduced by standardizing tree-ring series as ratios between actual and expected ring widths were avoided by taking residuals from the growth curve after appropriate power transformation of tree-ring series (Cook and Peters, 1997).

Tree-ring chronologies were produced by the ARSTAN program (Cook, 1985). ARSTAN generates chronologies by combining standardized tree-ring series with biweighted robust estimation. To preserve the long-term fluctuations in the series, we used a conservative method of detrending. Only negative exponential curves or linear regressions were used.

10.3. TEMPERATURE-SENSITIVE RECORDS

Three long (>400 years), well-replicated tree-ring records were selected from the southern Andes in northern Patagonia (41°S; Table 1, Fig. 1). Chronology sites are located between 1400 and 1500 m elevation near the alpine tree line (Villalba et al., 1997). Chronologies were developed from lenga (*Nothofagus pumilio* Poepp. et Endl.), the dominant subalpine species. Mean intercorrelation among the three chronologies is $r = 0.66$ (range 0.52–0.70), reflecting the large percentage of common variance between tree-ring series.

Tree-line records from the southern Andes have recently been used to characterize past annual temperature variations at upper elevations in northern Patagonia since 1750 (Villalba et al., 1997). Correlations between tree-ring variations and a regional temperature record, consisting of normalized departures from the Bariloche, Mascardi, and Esquel stations, indicate that tree growth is positively correlated with temperatures from November to March (Fig. 2a). The regional temperature record was constructed by computing the monthly standard deviations for each station and then averaging across the stations. As a consequence of using normalized standard deviations, each station, independent of its particular values of temperature, has the same weight in the averaged regional record. The statistical relationships between ring width and each monthly temperature variable were examined over the interval 1949–88 (50 years), which constitutes, in both northern Patagonia and Alaska, the common period for tree rings and instrumental records. The positive response of lenga to increasing temperature in late spring–summer is a typical pattern for trees near the upper tree line and reflects the positive influence of temperature on physiological processes related to tree growth (Tranquillini, 1979).

Seven temperature-sensitive records for North America were selected from a network of tree-ring chronologies for areas along the Gulf of Alaska from 57°–60°N (Table 1, Fig. 1). These chronologies, which

TABLE 1   Site Characteristics of Tree-Ring Chronologies and Palmer Drought Severity Index (PDSI) Reconstructions

Code	Site name	Species	Latitude	Longitude	Elevation (m)
Temperature-sensitive chronologies					
Northern Patagonia					
CO1	Castaño Overo	*Nothofagus pumilio*	41°09′ S	71°48′ W	1480
CO2	Castaño Overo	*N. pumilio*	41°09′ S	71°47′ W	1400
CDL	Cerro Diego de León	*N. pumilio*	41°16′ S	71°40′ W	1500
Coast of Alaska					
ELL	Ellsworth	*Tsuga mertensiana*	60°05′ N	148°58′ W	480
EXG	Exit Glacier	*Picea sitchensis*	60°12′ N	149°35′ W	300
MIW	Miners Wells	*T. mertensiana*	60°00′ N	141°41′ W	650
TEB	Tebenkof	*T. mertensiana*	60°45′ N	148°27′ W	25
VER	Verstovia Ridge	*T. mertensiana*	57°03′ N	135°17′ W	760
WHI	Whittier	*T. mertensiana*	60°50′ N	148°35′ W	300
WVG	Wolverine Glacier	*T. mertensiana*	60°22′ N	148°54′ W	400
Precipitation-sensitive chronologies					
Central Chile					
ELA	El Asiento	*Austrocedrus chilensis*	32°40′ S	70°49′ W	1950
Midwest–southwestern United States					
PDSI Cell 53 (Cook et al., 1996)			31°00′ N	107°30′ W	
PDSI Cell 63 (Cook et al., 1996)			31°00′ N	104°30′ W	
PDSI Cell 80 (Cook et al., 1996)			39°00′ N	98°30′ W	
PDSI Cell 81 (Cook et al., 1996)			37°00′ N	98°30′ W	

span a 700-km zone from the Kenai Peninsula southeastward to Boranof Island, were derived from either mountain hemlock (*Tsuga mertensiana* Bong. Carr.) or Sitka spruce (*Picea sitchensis* Bong. Carr.). The seven chronologies are well replicated over the interval 1589–1990. For the common period of 402 years, the mean correlation between the Alaska ring-width series is $r = 0.43$ (range 0.22–0.72). These correlations, all significant at the 99% confidence level, suggest that trees are responding in a coherent manner to climatic variations across the region.

We compared variations in tree-ring widths with a regional record of temperature for three coastal Alaska stations: Kodiak, Seward, and Sitka. Tree growth is positively correlated with temperature variations over most of the year. The first principal component (PC) scores of the seven tree-ring records are significantly correlated with temperature from January to June and from August to October (Fig. 2b). These results are consistent with those of previous studies showing that coastal Alaska tree rings strongly reflect spring–summer temperature variations in the northeastern Pacific (Briffa et al., 1992; Wiles et al., 1996, 1998).

Both the Patagonia and Alaska tree-ring records clearly show the documented 1976 transition from cold to warm conditions over the tropical Pacific, reflecting a comparable sensitivity to SST changes in the Pacific Ocean (Fig. 3). If decadal-scale climatic variations forced by the tropical Pacific had affected past temper-

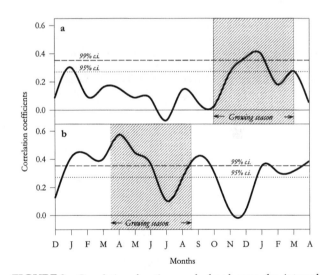

FIGURE 2 Correlation functions, calculated over the interval 1949–88, showing the relationships between regional temperature variations and tree growth in (a) northern Patagonia and (b) coastal Alaska. For northern Patagonia, the leading scores resulting from a principal components (PCs) analysis of three upper elevation chronologies were used for comparison with temperature. For Alaska, the tree-ring record used for comparison is the first PCs resulting from seven chronologies located along the Gulf of Alaska. Positive correlation indicates that above-average tree growth is associated with above-average values of temperature. The meteorological stations included in the temperature regional records are Bariloche, Mascardi, and Esquel in northern Patagonia; and Kodiak, Seward, and Sitka in Alaska. Correlation coefficients greater than 0.27 and 0.35 are significant at the 95 and 99% confidence level, respectively.

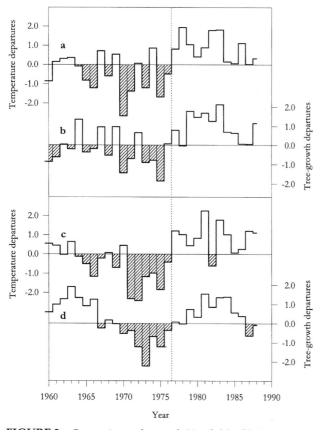

FIGURE 3 Comparison of annual (April–March) temperature fluctuations in (a) northern Patagonia and (c) Alaska with the amplitudes from the leading components resulting from principal components (PCs) analyses of the (b) three northern Patagonian and (d) seven coastal Alaska chronologies listed in Table 1. The simultaneous rise in temperature and tree growth starting in both regions during the mid-1970s is indicated by a vertical dotted line.

ature changes along the extratropical coasts of North and South America, tree-ring records for Alaska and northern Patagonia should present similar oscillatory modes. Indeed, for the common interval 1591–1988, the amplitudes from the first PC of the Alaska chronologies are significantly correlated with those of Patagonia ($r = 0.35$, $p < 0.0001$; Fig. 4), which may be considered a first indication of a common climatic forcing mechanism affecting past temperature variations in both regions.

10.4. PRECIPITATION-SENSITIVE TREE-RING RECORDS

Relationships between SSTs in the equatorial Pacific and precipitation anomalies in central Chile (30°–35°S) have been reported by several authors (Quinn and Neal, 1983; Aceituno, 1988; Rutlland and Fuenzalida, 1991; Aceituno and Montecinos, 1996). Positive rainfall anomalies in central Chile are associated with warmer SSTs in the tropical Pacific. Conversely, cold SSTs correspond quite closely to dry conditions. Tree growth in central Chile is influenced largely by year-to-year variations in precipitation (LaMarche, 1975; Boninsegna, 1988; Villalba, 1990, 1995). A ring-width chronology for ciprés (*Austrocedrus chilensis* D. Don) at El Asiento (32°40′ S), which represents the northernmost extent of this species in central Chile, was used by Boninsegna (1988) to develop a quantitative reconstruction of winter precipitation for Santiago, Chile, during the interval A.D. 1220–1972. Recently, the site was revisited, and the new cores were merged with the original data col-

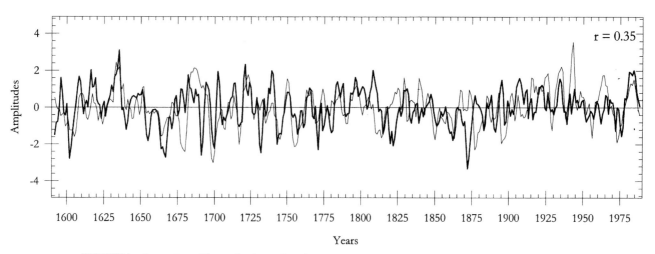

FIGURE 4 Comparison of the amplitudes resulting from principal components (PCs) analyses of temperature-sensitive chronologies from northern Patagonia (thick line) and coastal Alaska (thin line) during the past 400 years. The northern Patagonia record is the leading PC resulting from an analysis of three upper elevation chronologies (year t and t + 1), whereas the coastal Alaska record is the leading PC from seven coastal chronologies (year t and t + 1). The Patagonia and Alaska records are significantly correlated at the 99.99% confidence level.

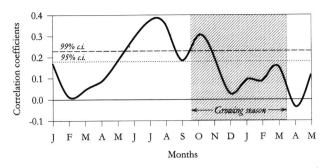

FIGURE 5 The correlation function, calculated over the interval 1861–1988, showing the relationships between Santiago de Chile total monthly precipitation and tree-ring width variations at El Asiento in central Chile. Positive correlation indicates that above-average tree growth is associated with above-average values of precipitation. Correlation coefficients greater than 0.18 and 0.24 are significant at the 95 and 99% confidence level, respectively.

lected in 1974 by LaMarche (1975). For 1861–1988, an interval of time common to the tree-ring and instrumental records, a correlation function between tree-ring variations at El Asiento and Santiago precipitation shows that tree growth is significantly correlated with winter–spring precipitation (May–November; Fig. 5).

A gridded network of drought reconstructions over the continental United States has recently been derived from a large collection of climatically sensitive tree-ring chronologies (Cook et al., 1999). The reconstructions, based on the summer (June–August) Palmer Drought Severity Index (PDSI), were done on a 154-point, 2° latitude × 3° longitude regular grid. For the common time period 1700–1978, we correlated tree-ring variations from El Asiento in central Chile with the 154 PDSI reconstructions for across the United States. Not surprisingly, the strongest correlations between these records came from those PDSI reconstructions located in the Midwest and southern United States, two regions where climate variations are affected by SST changes in the equatorial Pacific.

It is well known that there is a strong teleconnection between SST changes in the tropical Pacific and precipitation anomalies in the southern United States and northern México (Ropelewski and Halpert, 1986; Kiladis and Diaz, 1989; Stahle and Cleaveland, 1993). Warmer SSTs in the tropical Pacific typically result in increased precipitation anomalies in this region. Spatial correlations between different indexes of tropical Pacific circulation and the 154-grid-point PDSI series show that the geographic location of the highest correlation field is the southwestern United States (Cook et al., 2000), a finding that is consistent with the patterns identified using instrumental records.

During the twentieth century, relationships between indexes of the tropical Pacific circulation and precipita-

tion records for the Midwest in the United States have been less obvious. Cook et al. (2000) showed that during the period 1897–1923 the teleconnections between the Southern Oscillation Index (SOI) and precipitation variations across the United States were radically different from those in 1924–1978. The significant SOI teleconnections with PDSI are much more extensive in the early period and penetrate northeastward into the upper Midwest–Great Lakes region. In contrast, the late-period teleconnections are highly contracted in the southwestern United States. Long-term teleconnections across the United States were also investigated based on a winter SOI reconstruction from tree rings (1706–1977; Stahle et al., 1998). For the past three centuries, two contrasting patterns were suggested: one with expanded SOI teleconnections (1708–49 and 1850–99) and one with contracted teleconnections (1750–1999 and 1800–1949; Cook et al., 2000). Consequently, it is highly probable that the significant correlations between the central Chile record and the PDSI reconstructions for the Midwest reflect simultaneous changes in the teleconnections between the tropical Pacific and precipitation anomalies in these regions.

Hence, two PDSI reconstructions for the southwestern United States–northern México (37°N) and two for the Midwest (31°N) were selected for comparison with the precipitation-sensitive records for South America. These four reconstructions explain, on average, more than 55% of the grid-point PDSI variance over the 1928–78 calibration period (Cook et al., 1999).

It is important to note that the strongest teleconnections between precipitation in the southern United States–northern México and SSTs in the Pacific during the twentieth century have been identified for the winter months (Kiladis and Diaz, 1989), whereas the PDSI reconstructions that we used for comparison reflect mainly drought conditions during the summer. Despite this limitation, we found that the scores of the first PC from the four PDSI reconstructions in the United States are significantly correlated with tree-ring variations at El Asiento in central Chile (Fig. 6). For the common interval 1700–1978, the correlation coefficient between them is r = 0.32 (p <0.001), which could be considered as an indication of common modes of variations in these series.

10.5. SPATIAL CORRELATION PATTERNS BETWEEN TREE-RING RECORDS AND PACIFIC SSTs

Precisely dated tree-ring records for North and South America suggest that similar temporal modes of climate variations have affected the extratropical coasts

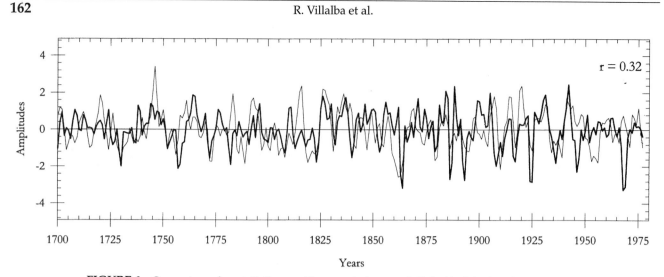

FIGURE 6 Comparison of precipitation-sensitive records for central Chile (thick line) and the Midwest–southern United States (thin line) during the past 300 years. The El Asiento chronology is considered an estimate of past precipitation fluctuations in central Chile. The United States record is the leading principal component (PC) resulting from a PC analysis of the four Palmer Drought Severity Index (PDSI) reconstructions listed in Table 1 (Cook et al. 1996). The central Chile and United States records are significantly correlated at the 99.99% confidence level.

of the western Americas during the past three to four centuries. Assuming this is to some degree correct, the next step in our analysis was to identify the existence of a common forcing mechanism for these climatic oscillations. For comparison across the Americas, tree-ring records were selected for regions where climate variations are, to some degree, associated with SSTs in the tropical Pacific. Then, we proceeded to establish the spatial correlation fields between tree rings and SSTs across the Pacific. If the correlation patterns between the North American tree-ring records and Pacific SSTs are similar to those that result from using the South American chronologies, then they provide an indication of the role of SSTs across the Pacific as a common forcing mechanism of climate variations in the extratropics.

Correlation fields between tree-ring records and SSTs across the Pacific were calculated for the interval 1857–1988. SST records on a 5° latitude × 5° longitude grid were obtained from Kaplan et al. (1997). The SST data set, which consists of 583 points, covers the entire North Pacific from 60°N to the equator. Coverage across the South Pacific is more limited, particularly at higher latitudes. Except for the grids located along the South American coast, there is no information on SST for the southeastern Pacific south of 25°S. At 55°S, the SST coverage is reduced to four points: two south of New Zealand and two off the South American coast. There are no grid points south of 60°S.

Although trees respond primarily to seasonal rather than annual variations in climate, we used annual SSTs

for calculating the correlation fields. The use of seasonal averages of SST may increase the magnitude of the correlation fields between tree rings and SSTs. However, because tree growth responses to climate vary between regions along the Americas, the comparison of spatial patterns based on different seasons could be problematic.

The spatial patterns that result from correlating the Alaska and northern Patagonia tree rings with Pacific SSTs are qualitatively similar (Fig. 7 [see color insert]). In both fields, positive correlations in the tropical Pacific and at mid- to high latitudes along the western coasts of the Americas contrast with negative correlations in the extratropical central and western Pacific. Significant positive correlations penetrate westward into the tropical Pacific region. Interestingly, both sets of temperature-sensitive records are not significantly connected with strong SST anomalies in the eastern equatorial Pacific. Similar spatial patterns were obtained when the tree-ring series were replaced in the calculations by annual SST anomalies from those grid cells located near the chronology sites (Fig. 7), indicating that the tree-ring estimates of temperature are capturing the large-scale spatial pattern of SST.

Correlation fields between SSTs across the Pacific Ocean and the precipitation-sensitive records for the Midwest–southern United States and central Chile were also calculated. In both patterns, significant positive correlations in the central Pacific near the date line extend eastward to the subtropical coasts of North and South America near 30°N and 30°S latitude, respectively. Negative correlations in the extratropical North and South Pacific are also prominent (Fig. 8 [see color insert]).

The spatial patterns obtained by correlating the tree rings with the SST records closely resemble those observed for the decadal mode of Pacific SST variability identified by Latif et al. (1997) and Zhang et al. (1997). In contrast to the interannual mode of El Niño/Southern Oscillation (ENSO) variability, the decadal mode is characterized by less pronounced anomalies in the eastern Pacific (the classic key ENSO region) and is not narrowly confined along the equator. Similar to the tree-ring spatial patterns, the documented decadal oscillatory mode of Pacific SST shows anomalies in the western Pacific that extend to the northeast and southeast into the American subtropics.

10.6. SPECTRAL PROPERTIES OF THE TEMPERATURE- AND PRECIPITATION-SENSITIVE TREE-RING RECORD

Climatically sensitive tree-ring records have been used to identify similar modes of climatic variations in the extratropical regions of North and South America during past centuries. The spatial patterns of correlation between SSTs over the Pacific Ocean and the tree-ring records reveal the importance of the Pacific Ocean in forcing these climatic changes, which simultaneously affect the extratropical Americas. In this section, we use cross-spectral analysis to characterize the oscillatory modes that have contributed, in a large degree, to the occurrence of simultaneous climatic changes in western North and South America.

We used Blackman-Tukey (BT; Jenkins and Watts, 1968) spectral analysis to establish the significant dominant periods at which variance occurs in the tree-ring records. The BT spectrum was estimated from 60 and 80 lags of the autocorrelation function for the precipitation- and temperature-sensitive records, respectively. With this number of lags (~25% of the series lengths), we set a reasonable balance between high resolution and moderate stability. The number of degrees of freedom per spectral estimate is 10. The 95% confidence level of the spectrum was estimated from a *red noise* first-order Markov null continuum based on the lag-1 autocorrelation of the time series (Mitchell et al., 1966).

Figure 9 shows the BT spectra for the temperature-sensitive records from the coast of Alaska and northern Patagonia based on a common period of 397 years (1592–1988). In both sets of records, a large portion of the spectral variance is concentrated at lower (decadal) frequencies. For northern Patagonia, peaks that exceed

the 95% confidence limit are observed at 6.5, 8.4–9.4, and 11.4–13.3 years (Fig. 9). The BT spectra of the Gulf of Alaska record also show a large part of the spectral variance concentrated at similar frequencies: 6.7, 8–10.7, and 15.8–20 years (Fig. 9). Although not statistically significant, obvious peaks are also observed in both sets of temperature-sensitive records at about 50 years. Cross-spectral analysis was used to identify coherent peaks in the Alaska and Patagonia records. Coherent oscillatory modes between the Alaska and northern Patagonia records are observed at 4.8, 5.3, 6.8, 9.0, and 11.1–15.4 years and at temporal scales longer than 50 years (Fig. 9).

The BT spectra for the precipitation-sensitive records for central Chile and the United States (1700–1978) show that both sets of records have peaks in spectral variance at similar frequencies (Fig. 10). Peaks that exceed the 95% confidence limit are observed at 3.2–3.5, 5.1, 7.5–8, and 10.9 years in central Chile and at 4–4.1, 5.7–6.3, 7.5–8, and 10–10.9 years in the Midwest–southern United States records. Cross-spectral analysis of the central Chile and United States records indicates significant coherence at 2.9–3.3, 4.4, 7.4–8, 10, 14–20,

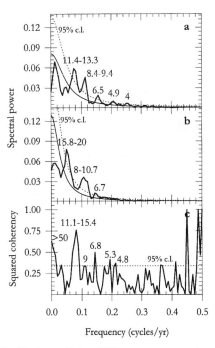

FIGURE 9 Blackman-Tukey (BT) power spectra of temperature-sensitive records for (a) northern Patagonia and (b) the Gulf of Alaska estimated over the interval 1592–1988. The 95% confidence limits are based on a first-order Markov null continuum model. The periods are given in years for each significant peak. The coherency spectrum between these two records is shown in (c). Records are highly coherent at decadal-scale wavelengths at ca. 9, 11.1–15.4, and longer than 50 years.

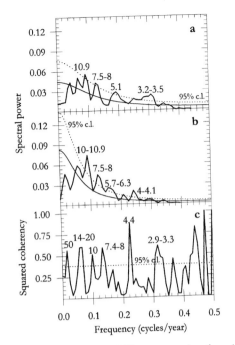

FIGURE 10 Blackman-Tukey (BT) power spectra of precipitation-sensitive records for (a) central Chile and (b) the Midwest–southern United States estimated over the interval 1700–1978. The 95% confidence limits are based on a first-order Markov null continuum model. The periods are given in years for each significant peak. The coherency spectrum between these two records is shown in (c). Records are highly coherent at decadal-scale wavelengths at ca. 10, 14–20, and 50 years.

and 50 years (Fig. 10). These spectral analyses basically reaffirm the existence of common oscillatory modes in temperature- and precipitation-sensitive records for the western coasts of North and South America.

10.7. STABILITY OF THE PAN-AMERICAN TELECONNECTIONS

The cross-spectral analyses of tree-ring records for extratropical regions of North and South America suggest the existence of common decadal modes of climate variations along the western Americas. Over the length of common record, independent tree-ring records for North and South America are significantly correlated in a positive sense (Figs. 4 and 6), implying that similar trends in temperature and precipitation along the western coasts of the Americas have prevailed for the past three to four centuries. However, when correlation coefficients between the tree-ring series from North and South America are calculated for different subintervals, the magnitude of the relationships between these records changes over time, being typically weaker during the twentieth century.

Although local or regional climatic changes may contribute to the lack of correlation during some subperiods, a change in the strength of the tropical Pacific teleconnections is also a valid explanation. We used singular spectral analysis (SSA), a data-adaptive method for extracting signals from noise in a time series (Vautard and Gill, 1989; Vautard, 1995), to determine changes in the oscillatory modes that most contribute to the positive relationships between the independent temperature- and precipitation-sensitive records for North and South America. Similar to other spectral techniques, the choice of lags in SSA is a compromise between the amount of information to be retained (resolution) and statistical significance (stability). We experimented with lags ranging from 5 to 15% of the series length, and we found that a lag equal to 10% of the series length (39 and 29 for the temperature- and precipitation-sensitive records, respectively) adequately resolved most decadal-scale oscillatory modes.

Three major waveforms, representing decadal modes of common variance at 9, 12.8, and >50 years, were isolated from the original temperature-sensitive records (Fig. 11). In general, the temporal evolution of these components shows similar fluctuations in amplitude and intensity from 1600 to ca. 1850. After 1850, relationships between waveforms break up, particularly for the 9- and 12.8-year components. This change is clearly revealed by changes in the correlation coefficients between the northern Patagonia and Alaska waveforms before and after 1850 (Fig. 11). For instance, the correlation coefficient between the components centered on 12.8 years changes from $r = 0.73$ to $r = 0.02$ for the intervals 1592–1849 and 1850–1988, respectively.

As a way of validating the previous results, we split both the northern Patagonia and Alaska records into two independent series (1592–1849 and 1850–1988) and proceeded to spectrally characterize each individual segment using BT spectral analysis. As expected, most of the peaks identified for the interval 1592–1849 agree with those previously reported for the entire series (Fig. 12). In contrast, the spectral estimates for the most recent portion of the series look somewhat different. A significant peak at 10–12 years is observed for the northern Patagonia record, whereas much of the variability for the 1850–1988 part of the Alaska records is confined to a period of 6.6–10 years (Fig. 12). In addition, cross-spectral analyses between the northern Patagonia and Alaska records show large differences in coherence between the early (1592–1849) and late (1850–1988) parts of the series. Tree-ring oscillations for northern Patagonia and Alaska were highly coherent from 1600–1850, particularly at decadal-scale oscillatory modes of 6.7, 8.4–16.5, and >50 years. Yet, except for oscillations longer than 50 years, there is no significant

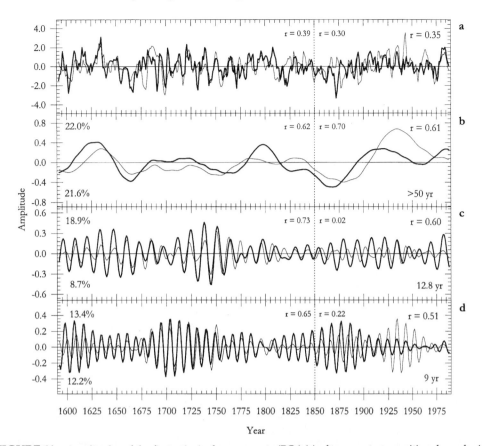

FIGURE 11   Amplitudes of the first principal components (PCs) (a) of temperature-sensitive chronologies for northern Patagonia (thick line) and coastal Alaska (thin line), and their most significant decadal-scale oscillatory modes (b–d) extracted by singular spectrum analysis (SSA). The oscillatory modes have periods of (b) longer than 50 years, (c) ca. 13 years, and (d) 9 years. Percentages of the original variance contributed by each of the Patagonia and Alaska waveforms are indicated in the upper and lower left corners of the figures, respectively. In the upper right corners, r is Pearson's correlation coefficient between the Patagonia and Alaska series. Major changes in the oscillatory modes are observed at about 1850 (vertical dotted line). Correlation coefficients for the intervals 1592–1849 and 1850–1988 between the northern Patagonia and the Alaska oscillatory modes are shown at the left and right of the vertical dotted line, respectively.

coherence at the decadal scale between the records during the most recent interval (Fig. 12).

SSA of the precipitation-sensitive records for central Chile and the Midwest–southern United States in the time and frequency domains also show important shifts in the oscillatory modes at ca. 1825–50 (Fig. 13). Two major temporal patterns, centered at ca. 7.8 and 10–18 years, show contrasting relationships before and after 1825 and 1850, respectively, as indicated by changes in the correlation coefficients between these records during the early and late intervals. The 10- to 18-year waveforms are highly correlated from 1700 to 1849, whereas the 7.8-year modes are strongly correlated from 1825 to 1978.

BT spectral analyses of the precipitation-sensitive records also indicate important changes in the concen-

tration of spectral power at ca. 1850 (Fig. 14). For the early interval 1700–1849, both records show that much of the variability in tree rings was confined principally to frequency bands longer than 5 years. In contrast, the 3.2- to 3.8-year oscillation was the dominant mode in central Chile during the most recent interval, without any significant peak at lower frequencies. Cross-spectral analyses also reveal a shift in the coherence frequency bands from decadal modes at 8.3 and >17 years during the early interval to shorter scale modes at 3.2 and 6.7–9.4 years during the most recent interval (Fig. 14).

Following a procedure that is similar to that used by Zhang et al. (1997), we examined the temporal evolution of the leading decadal-scale oscillatory mode in the Pacific SST field back to 1856. Monthly SST tempera-

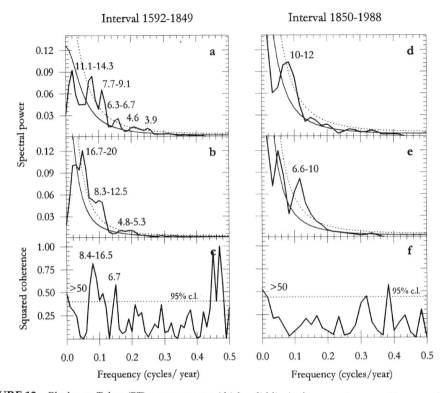

FIGURE 12 Blackman-Tukey (BT) power spectra (thick solid line) of temperature-sensitive records for (a and d) northern Patagonia and (b and e) the Gulf of Alaska estimated over two independent intervals: 1592–1849 (left) and 1850–1988 (right). The 95% confidence limits (dotted line) are based on a first-order Markov null continuum model. The periods are given in years for each significant peak. The coherency spectra between these records are shown for the intervals (c) 1592–1849 and (f) 1850–1988. Note the changes in squared coherency between the two intervals.

ture anomalies for 583 points over the Pacific were taken from Kaplan et al. (1997). Low-pass-filtered time series were generated by applying to the monthly mean SST anomalies a cubic spline (Cook and Peters, 1981) designed to reduce 50% of the variance in a sine wave with a periodicity of 10 years. This procedure preserved 25% of the variance on a timescale up to 7.6 years and only 10% of the variance on timescales up to 5.8 years. Because of our interest in examining decadal-scale modes of variation in SST, the use of a 10-year cubic spline for filtering the SST series was a reasonable choice. Using the Varimax criterion, a rotated PC (RPC) analysis (Richman, 1986) was applied to the low-pass SST series to relieve the orthogonality constraints imposed on unrotated factors. Figure 15a (see color insert) shows the leading Varimax rotated SST factor for the interval 1856–1991. The abrupt change toward warmer SSTs at ca. 1976 is the most obvious feature of the decadal-scale oscillatory mode over the Pacific during the past 130 years. In contrast, low energy appears to be related to the decadal oscillatory frequencies over the

whole Pacific from ca. 1910–1976 (Fig. 15a). The spatial pattern associated with the leading PC is characterized by positive departures in the tropical central and western Pacific and along the subtropical coasts of North and South America. Anomalies of opposite polarity occur in the extratropical central North and South Pacific (Fig. 15b [see color insert]), but they are more pronounced in the North Pacific.

This analysis of both instrumental and tree-ring records shows the occurrence of major changes in the temporal evolution of climate modes over the Pacific and along the western Americas. These records suggest that the patterns of climatic variability over the Pacific have been characterized by interactions between interannual and interdecadal oscillatory modes. An intensification of the interannual ENSO-type forcing from the mid-nineteenth century to ca. 1976, associated with more energy concentrated at high-frequency modes of oscillation, may have played a major role in masking or modifying the decadal-scale mode of oscillation that prevailed in the Pacific during the previous centuries.

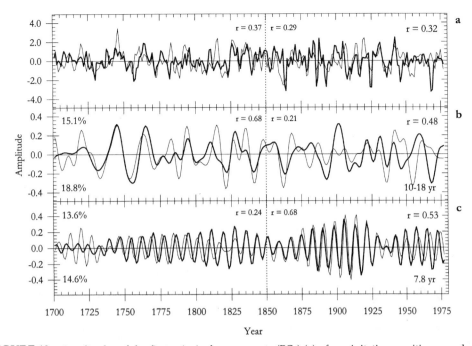

FIGURE 13   Amplitudes of the first principal components (PCs) (a) of precipitation-sensitive records for central Chile (thick line) and the Midwest–southern United States (thin line), and their most significant decadal-scale oscillatory modes (b–c) extracted by singular spectrum analysis (SSA). The oscillatory modes have periods (b) between 10 and 18 years and (c) at ca. 7.8 years. Percentages of the original variance contributed by each of the central Chile and Midwest–southern United States waveforms are indicated in the upper and lower left corners of the figures, respectively. In the upper right corner, r is the Pearson's correlation coefficient between the Chilean and United States series. Major changes in the oscillatory modes are observed at ca. 1850 and 1825 (vertical dotted line) for the 10- to 18- and 7.8-year waveforms, respectively. Correlation coefficients for the intervals before and after the major changes in the central Chile and Midwest–southern United States oscillatory modes are shown at the left and right of the vertical dotted line, respectively.

10.8. DISCUSSION

North-to-south comparative studies along the western Americas, which integrate instrumental and proxy records, provide a unique opportunity for interpreting present and past climatic changes owing to the predominant influence of a single climatic forcing mechanism: the Pacific Ocean. Patterns of modern climate circulation can be used to interpret the dynamics of climatic changes preserved in the paleorecords. On the other hand, high-resolution proxy records allow us to examine changes in phase relationships and magnitude of the dynamic components of the Pacific climate system as they change over time.

The most striking results of this study are the significant relationships between North and South American climate-sensitive records extending back to 1600. Tree-ring records show the occurrence of decadal-scale climatic variations that have simultaneously affected the extratropical western coast of the Americas during the past four centuries. It is very difficult to conceive of any

hypothesis to explain these significant relationships other than the records are responding to large-scale, common climatic variability. This hypothesis also leads to the implication that the temperature- and precipitation-sensitive chronologies have recorded this climatic variability with reasonable fidelity throughout the past four centuries.

It has been shown that decadal-scale climate changes in the western Americas during recent decades have been influenced by the Pacific Ocean. This has been simultaneously reflected in instrumental records and circulation indices such as the Southern Oscillation, Pacific–North America Oscillation (PNA), and the Pacific Decadal Oscillation (PDO) (Trenberth and Hurrell, 1994; Mantua et al., 1997). Many studies suggest that low-frequency climate variability in the North Pacific is caused by tropical forcing (Nitta and Yamada, 1989; Trenberth, 1990; Graham, 1994; Graham et al., 1994; Trenberth and Hurrell, 1994; Lau and Nath, 1994). However, some simulations from global coupled ocean–atmosphere general circulation models (GCMs) sug-

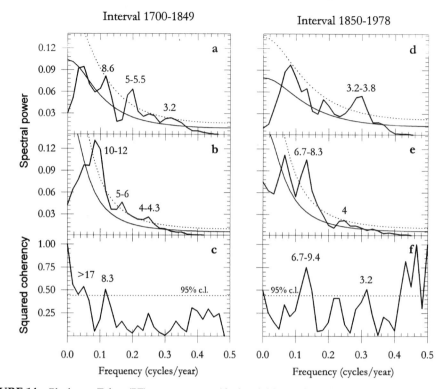

FIGURE 14   Blackman-Tukey (BT) power spectra (thick solid line) of precipitation-sensitive records from (a and d) central Chile and (b and e) the Midwest–southern United States estimated over two independent intervals: 1700–1849 (left) and 1850–1978 (right). The 95% confidence limits (dotted line) are based on a first-order Markov null continuum model. The periods are given in years for each significant peak. The coherency spectra between these records are shown for the intervals (c) 1700–1849 and (f) 1850–1978. Note the changes in squared coherency between the two intervals.

gest that the atmosphere–ocean system over the extratropical North Pacific may generate interdecadal climatic changes by itself (Latif and Barnett, 1994; Robertson, 1996). Recent studies show that the tropical Pacific strongly influences decadal variability in SST within the subtropical gyre, whereas the variability within the subpolar gyre may respond, in a large degree, to a self-maintaining mechanism inherent to ocean–atmosphere interactions in the North Pacific (Nakamura et al., 1997). Significant correlations between tree-ring records for distant sites located in North and South America suggest that the tropical Pacific may have played a major role in forcing common decadal-scale climatic variations in extratropical regions of both North and South America during the past three to four centuries. Additional support for this idea comes from the similarity in the spatial correlation patterns derived from tree-ring records for both North and South America and SSTs over the Pacific. These patterns are characterized by high positive values centered on the tropical western Pacific with up- and downstream regions of positive values along the western coasts of the Amer-

icas. Presently, it is extremely difficult to conceive of a system of atmospheric interactions in which long-term climatic changes in the North Pacific act to force changes in the South Pacific, or vice versa. Although our results support the idea of a common tropical forcing of decadal-scale climate oscillations along the western Americas, Zhang et al. (1998) have recently presented observational evidence that the Pacific warming and the decadal change in ENSO after 1976 may have originated from the midlatitude North Pacific Ocean. In consequence, there is a possibility that thermal anomalies in the North Pacific may indirectly affect the South Pacific by forcing changes in tropical Pacific SSTs, thus leading us to question our hypothesis of a common tropical forcing mechanism for climate variations along the western Americas.

The dominant modes of variability in the tropics during the last few decades have been recently investigated by several authors (Kawamura, 1994; Latif et al., 1997; Zhang et al., 1997). These studies show that the tropical SST variability can be characterized by an interannual mode (*ENSO cycle related*) and a decadal mode (residual independent mode). The presence of a trend or unresolved ultra-low-frequency variability has also been reported by Latif et al. (1997). The spatial structure of the interannual mode is characterized by an equatorial maximum in the eastern Pacific narrowly confined along the equator. In contrast, the spatial signature of the decadal mode is less confined to equatorial regions in the eastern tropical Pacific, with the strongest SST anomalies in the western equatorial Pacific extending to the northeast and southeast into the western American subtropics. Sea surface temperature anomalies (SSTAs) in the extratropical central North and South Pacific are also more prominent in the decadal mode (Latif et al., 1997; Zhang et al., 1997).

The spatial signatures, which result from correlating SST over the Pacific with both the temperature- and precipitation-sensitive records, resemble those of the leading decadal modes of SST variability presented by Latif et al. (1997) and Zhang et al. (1997). Indeed, positive correlations between SST and the tree-ring records are stronger over the western Pacific and along the west coast of the Americas than over the eastern tropical Pacific. In agreement with our spatial correlation patterns, Latif et al. (1997) showed that the SST field associated with the decadal mode of variability over the Pacific originates in the subtropics at the west coast of North and South America (Fig. 10a in Latif et al., 1997), where the variances explained by this mode are the highest (Fig. 10c in Latif et al., 1997). Interestingly, the rotation period proposed by Latif et al. (1997) for this SST variability mode is 13 years, consistent with the 12.8-year oscillation common to the temperature-sensitive records for both Alaska and northern Patagonia during the past 400 years. These observations strongly suggest that interdecadal variations of climate, modulated by Pacific SSTs, appear to be one of the most solid climatic forcing mechanism influencing temperature and precipitation variations along the extratropical western coasts of the Americas.

Additional support indicating the existence of a decadal oscillatory mode associated with the Pacific Ocean comes from a 100-year global spatiotemporal analysis of temperature anomalies by Mann and Park (1994), who detected an interdecadal mode of temperature variability centered on a 15- to 18-year period. The spatial pattern associated with this oscillation resembles that of ENSO, and the time domain signal suggests that the warmth after 1976 was consistent with a large positive excursion of this interdecadal oscillation.

Variations in the temporal evolution of climate-sensitive records for extratropical North and South America show a consistent change in the relationships between modes of decadal-scale climatic variability at about the mid-nineteenth century. Certainly, major

shifts in climate at about the 1850s have been suggested many times and in many places. Starting at ca. 1850, the Arctic warmed to the highest temperatures in four centuries (D'Arrigo and Jacoby, 1993; Overpeck et al., 1997). Since ca. 1850, widespread retreats of mountain glaciers have been reported for the Canadian Rockies (Luckman, 1996), the European Alps (Haeberli et al., 1989), the New Zealand Alps (Gellatly and Norton, 1984), and the tropical Andes (Hastenrath, 1981). Based on the documented relationships between the Pacific SSTs and climate along the western coast of the Americas, we speculated that the recorded changes in the principal modes of extratropical climatic variability at ca. 1850 reflect a major reorganization in the circulation of the Pacific Ocean at that time. This reorganization might have changed the relationships between interannual and interdecadal modes of atmospheric circulation over the Pacific, which in turn may have altered the extratropical teleconnections in time and frequency domains.

Additional support indicating the existence of major changes at ca. 1850 over the tropical Pacific comes from totally independent long records of tropical ocean variability. These records are also related to changes in extratropical teleconnections. Whetton and Rutherfurd (1994) used a set of proxy records from the Eastern Hemisphere (Indonesia, Northeast Africa, North China, and India) to identify interrelationships characteristic of ENSO and how these relationships may have varied over the past 500 years. Based on the information that emerged from the analysis of these proxy records, these researchers noted that ENSO teleconnections, as we know them today, were less active from 1550–1850 than they were during the later period. However, the authors largely attributed the inconsistency between proxy records during the early period to a poorer quality of the data. Documentary records of ENSO manifestations in Peru, central Chile, northeastern Brazil, and central México, in combination with a coral record from the Galápagos, also show poor agreement between ENSO events during the sixteenth, seventeenth, and eighteenth centuries (Ortlieb, 2000). Correspondence between events within the South America sites is better after 1814 and almost perfect in the last quarter of the nineteenth century. Ortlieb (2000) suggests that the changes in event correspondence between documentary records reveal a distinct mode of climatic variability over the Pacific, which in turn might be associated with different ENSO teleconnections before and after the mid-nineteenth century. A composite tree-ring reconstruction of the SOI since 1706 (Stahle et al., 1998) also suggests a significant increase in the interannual ENSO-related oscillatory modes, more frequent cold events, and a stronger sea level pressure

(SLP) gradient across the equatorial Pacific from the mid-nineteenth century. A more energetic interannual mode of ENSO since 1850 may have affected the amplitudes of the decadal-scale oscillatory mode in the Pacific, which according to our analysis has been weaker and less coherent from the mid-nineteenth century to the mid-1970s.

Interactions between different modes of tropical SST variability have also been reported on the basis of instrumental observations (Ji et al., 1996; Latif et al., 1997). In their analysis of the climatic conditions prevailing during the 1990s, Latif et al. (1997) concluded that those anomalies were caused mainly by a prominent decadal mode with some weak interannual fluctuations superimposed on it. In contrast, they noted that the amplitude of the interannual (ENSO) mode was stronger than that of the decadal mode before 1990. Our results suggest that the decadal mode of variability over the Pacific was more prominent from 1600 to ca. 1850; after that, the interannual mode has dominated. Based on instrumental SST records for the past 130 years, the Pacific decadal mode was apparently weaker during most of the twentieth century (Fig. 15a). The widely reported transition from cold to warm winter conditions in the tropical Pacific during 1976 is the most significant decadal-scale oscillation in the instrumental record.

These observations have important implications in relation to our knowledge of ENSO and ENSO-related climate variability. Have the Pacific modes of variability always been dominated by oscillations of between 3 and 7 years as has been indicated by the traditional analysis of the instrumental records during the past 130 years? Or is the decadal mode, of which the shift in the mid-1970s is the most significant manifestation in this century, another important mode of variability inherent to the Pacific? If the idea of a decadal mode of oscillation were incorporated into the natural climatic variability of the Pacific Ocean, it would not be necessary to refer to the 1980s and 1990s as periods of *permanent* or *extended* El Niño conditions, nor would it be necessary to invoke anthropogenic influences (Trenberth and Hoar, 1996) to explain *anomalous* ENSO such as the long 1991–94 event. Although the impact of the decadal mode on the interannual (ENSO) mode is still unclear, Latif et al. (1997) suggest that the decadal mode is a source of irregularities in the interannual (ENSO) mode of variability. Certainly, the short-lived warmings in the tropical Pacific in the early 1990s were poorly simulated by both statistical and dynamic forecast models, which were developed for predicting the strong interannual variability in the tropical Pacific during the previous decades (Ji et al., 1996).

Although tree-ring records provide insight into the temporal evolution of the relationships between the tropical ocean and higher latitudes in the Americas, it is important to note that the variability in the records is related to tropical teleconnections along the western coasts of the Americas and not to direct forcing from the equatorial Pacific. Additional instrumental and proxy records will aid efforts to understand the interactions between different modes of variability over the Pacific Ocean and the adjacent western Americas. As a follow-up step to investigate temporal changes in the oscillatory modes over the whole Pacific, we suggest comparing tree-ring and historical records along the western Americas with high-resolution coral records from the tropical Pacific.

Acknowledgments

This research was supported by a Lamont-Doherty postdoctoral fellowship to Ricardo Villalba by grant ATM-96-16975 of the Climate Dynamics Program of the National Science Foundation (NSF), by National Oceanic and Atmospheric Administration (NOAA) grant NA56 GPO235, by the Argentinean Agency for the Promotion of Science (PICT 0703093), and by the Argentinean Council of Research and Technology (CONICET). Lamont-Doherty Earth Observatory Contribution Number 5916.

References

Aceituno, P., 1988: On the functioning of the Southern Oscillation in the South American sector. Part I: Surface climate. *Monthly Weather Review*, **116**: 505–524.

Aceituno, P., and A. Montecinos, 1996: Assessing upper limits of seasonal predictability of rainfall in central Chile based on SST in the equatorial Pacific. *Experimental Long-Lead Forecast Bulletin*, **5**: 37–40.

Boninsegna, J. A., 1988: Santiago de Chile winter rainfall since 1220 as being reconstructed by tree rings. *Quaternary of South America and Antarctic Peninsula*, **6**: 67–87.

Briffa, K. R., P. D. Jones, and F. H. Schweingruber, 1992: Tree-ring reconstructions of summer patterns across western North America since 1600 A.D. *Journal of Climate*, **5**: 735–754.

Cook, E. R., 1985: A time series analysis approach to tree-ring standardization. Ph.D. dissertation. University of Arizona, Tucson.

Cook, E. R., and K. Peters, 1981: The smoothing spline: A new approach to standardizing forest interior ring-width series for dendroclimatic studies. *Tree-Ring Bulletin*, **41**: 45–53.

Cook, E. R., and K. Peters, 1997: Calculating unbiased tree-ring indices for the study of climatic and environmental change. *The Holocene*, **7**: 359–368.

Cook, E. R., D. M. Meko, D. W. Stahle, and M. K. Cleaveland, 1996: Tree-ring reconstructions of past drought across the conterminous United States: Tests of a regression method and calibration/verification results. *In* Dean, J. S., D. M. Meko, and T. Swetnam (eds.), *Tree Rings, Environment, and Humanity. Radiocarbon*, pp. 155–169.

Cook, E. R., J. E. Cole, R. D. D'Arrigo, D. M. Stahle, and R. Villalba, 2000: Tree-ring records of past ENSO variability and forcing. *In* Diaz, H. F., and V. Markgraf (eds.), *El Niño and the Southern Oscillation, Multiscale Variability, Global and Regional Impacts*. Cambridge, U.K.: Cambridge University Press.

Cook, E. R., D. M. Meko, D. W. Stahle, and M. K. Cleaveland, 1999: Drought reconstructions for the continental United States. *Journal of Climate*, **12**: 1145–1162.

D'Arrigo, R. D., and G. C. Jacoby, 1993: Secular trends in high northern latitude temperature reconstructions based on tree rings. *Climate Change*, **25**: 163–177.

Ebbesmeyer, C. C., D. R. Cayan, D. R. McLain, F. H. Nichols, D. H. Peterson, and K. T. Redmond, 1991: 1976 step in the Pacific climate: Forty environmental changes between 1968–75 and 1977–84. *In* Betancourt, J. L., and V. L. Tharp (eds.), Proceedings of the 7th Annual Pacific Climate Workshop, California Department of Water Resources, Interagency Ecological Studies Program Technical Report 26, pp. 115–126.

Gellatly, A. F., and D. A. Norton, 1984: Possible warming and glacial recession in the South Island, New Zealand. *New Zealand Journal of Science*, **27**: 381–388.

Graham, N. E., 1994: Decadal-scale climate variability in the 1970s and 1980s: Observations and model results. *Climate Dynamics*, **10**: 135–162.

Graham, N. E., T. P. Barnett, R. Wilde, M. Ponater, and S. Schubert, 1994: Low-frequency variability in the winter circulation over the Northern Hemisphere: On the relative role of tropical and midlatitude sea surface temperatures. *Journal of Climate*, **7**: 1416–1442.

Haeberli, W., P. Muller, P. Alean, and H. Bosch, 1989: Glacier changes following the Little Ice Age: A survey of the international data basis and its perspectives. *In* Oerlemans, J. (ed.), *Glacier Fluctuations and Climate Change*. Dordrecht, The Netherlands: Kluwer Academic, pp. 77–101.

Haston, L., and J. Michaelsen, 1994: Long-term central coastal California precipitation variability and relationships to El Niño—Southern Oscillation. *Journal of Climate*, **7**: 1373–1387.

Hastenrath, S., 1981: *The Glaciation of the Ecuadorian Andes*. Rotterdam: Balkema, 159 pp.

Jenkins, G. M., and D. G. Watts, 1968: *Spectral Analysis and Its Applications*. San Francisco: Holden–Day, 525 pp.

Ji, M., A. Leetmaa, and V. E. Kousky, 1996: Coupled model forecasts of ENSO during the 1980s and 1990s at the National Meteorological Center. *Journal of Climate*, **9**: 3105–3120.

Kaplan, A., M. A. Cane, Y. Kushnir, B. Blumenthal, and B. Rajagopalan, 1997: Analyses of global sea surface temperature 1856–1991. *Journal of Geophysical Research*, **101**: 22599–22617.

Kawamura, R., 1994: A rotated EOF analysis of global sea surface temperature variability with interannual and interdecadal scales. *Journal of Physical Oceanography*, **24**: 707–715.

Kiladis, G. N., and H. F. Diaz, 1989: Global climatic anomalies associated with extremes of the Southern Oscillation. *Journal of Climate*, **2**: 1069–1090.

LaMarche, V. C., Jr., 1975: Potential of tree-rings for reconstruction of past climatic variations in the Southern Hemisphere. *Proceedings of the World Meteorological Organization/International Association of Meteorology and Atmospheric Physics (WMO/IAMAP) Symposium on Long-Term Climatic Fluctuations*, Norwich, England, World Meteorological Organization, Geneva, WMO No. 421, pp. 21–30.

Latif, M., and T. P. Barnett, 1994: Causes of decadal climate variability over the North Pacific and North America. *Science*, **266**: 634–637.

Latif, M., and T. P. Barnett, 1996: Decadal climate variability over the North Pacific and North America: Dynamics and predictability. *Journal of Climate*, **9**: 2407–2433.

Latif, M., R. Kleeman, and C. Eckert, 1997: Greenhouse warming, decadal variability, or El Niño? An attempt to understand the anomalous 1990s. *Journal of Climate*, **10**: 2221–2239.

Lau, N.-C., and M. J. Nath, 1994: A modeling study of the relative roles of the tropical and extratropical SST anomalies in the variability of the global atmosphere-ocean system. *Journal of Climate*, **7**: 1184–1207.

Luckman, B. 1996: Reconciling the glacial and dendrochronological records for the last millennium in the Canadian Rockies. *In* Jones, P. D., R. S. Bradley, and J. Jouzel (eds.), *Climatic Variations and Forc-ing Mechanisms of the Last 2000 Years*, Vol. 41 of NATO ASI Series, Series I: Global Environmental Change, pp. 85–108.

Mann, M. E., and J. Park, 1994: Global-scale modes of surface temperature variability on interannual and century timescales. *Journal of Geophysical Research*, **99**: 25819–25833.

Mantua, J. N., S. R. Hare, Y. Zhang, J. M. Wallace, and R. C. Francis, 1997: A Pacific interdecadal climate oscillation with impacts on salmon production. *Bulletin of the American Meteorological Society*, **78**: 1069–1080.

Mitchell, J. M., Jr., B. Dzerdseevskii, H. Flohn, W. L. Hofmeyr, H. H. Lamb, K. N. Rao, and C. C. Wallen, 1966: Climatic Change. World Meteorological Organization, Geneva, Technical Note 79, 79 pp.

Nakamura, H., G. Lin, and T. Yamagata, 1997: Decadal climate variability in the North Pacific during the recent decades. *Bulletin of the American Meteorological Society*, **78**: 2215–2225.

Nitta, T., and S. Yamada, 1989: Recent warming of tropical sea surface temperature and its relationship to the Northern Hemisphere circulation. *Journal of the Meteorological Society of Japan*, **67**: 375–382.

Ortlieb, L., 2000: The documentary historical record of El Niño events in Peru: An update of the Quinn record (sixteenth through nineteenth centuries). *In* Diaz, H. F., and V. Markgraf (eds.), *El Niño and the Southern Oscillation, Multiscale Variability, Global and Regional Impacts*. Cambridge, U.K.: Cambridge University Press.

Overpeck, J., K. Hughen, D. Hardy, R. Bradley, R. Case, M. Douglas, B. Finney, K. Gajewski, G. Jacoby, A. Jennings, S. Lamoureux, A. Lasca, G. MacDonald, J. Moore, M. Retelle, S. Smith, A. Wolfe, and G. Zielinski, 1997: Arctic environmental change of the last four centuries. *Science*, **278**: 1251–1256.

Quinn, W. H., and V. T. Neal, 1983: Long-term variations in the Southern Oscillation, El Niño, and Chilean subtropical rainfall. *Fisheries Bulletin, U.S.*, **81**: 363–374.

Richman, M. B., 1986: Rotation of principal components. *Journal of Climatology*, **6**: 293–335.

Robertson, A. W., 1996: Interdecadal variability over the North Pacific in a multi-century climate simulation. *Climate Dynamics*, **12**: 227–241.

Ropelewski, C. F., and M. S. Halpert, 1986: North American precipitation and temperature patterns associated with El Niño/Southern Oscillation (ENSO). *Monthly Weather Review*, **115**: 1606–1626.

Rosenbluth, B., H. A. Fuenzalida, and P. Aceituno, 1997: Recent temperature variations in southern South America. *International Journal of Climatology*, **17**: 67–85.

Rutlland, J., and H. Fuenzalida, 1991: Synoptic aspects of the central Chile rainfall variability associated with the Southern Oscillation. *International Journal of Climatology*, **11**: 63–76.

Stahle, D. W., and M. K. Cleaveland, 1993: Southern Oscillation extremes reconstructed from tree rings of the Sierra Madre Occidental and southern Great Plains. *Journal of Climate*, **6**: 129–140.

Stahle, D. W., R. D. D'Arrigo, M. K. Cleaveland, P. J. Krusic, R. J. Allan, E. R. Cook, J. E. Cole, R. B. Dunbar, M. D. Therrell, D. A. Gay, M. Moore, M. A. Stokes, B. T. Burns, and L. G. Thompson, 1998: Experimental multiproxy reconstructions of the Southern Oscillation. *Bulletin of the American Meteorological Society*, **79**: 2137–2152.

Tranquillini, W., 1979: *Physiological Ecology of the Alpine Timberline*. New York: Springer-Verlag, 137 pp.

Trenberth, K. E., 1990: Recent observed interdecadal climate changes in the Northern Hemisphere. *Bulletin of the American Meteorological Society*, **71**: 988–993.

Trenberth, K. E., and T. J. Hoar, 1996: The 1990–1995 El Niño–Southern Oscillation event: Longest on record. *Geophysical Research Letters*, **23**: 57–60.

Trenberth, K. E., and J. W. Hurrell, 1994: Decadal atmosphere-ocean variations in the Pacific. *Climate Dynamics*, **9**: 303–319.

Vautard, R., 1995: Patterns in time: SSA and MSSA. *In* von Storch, H., and A. Navarra (eds.), *Analysis of Climate Variability: Applications of Statistical Techniques.* Berlin: Springer-Verlag, pp. 259–279.

Vautard, R., and M. Ghil, 1989: Singular spectrum analysis in nonlinear dynamics, with applications to paleoclimatic time series. *Physica,* **D 35**: 395–424.

Villalba, R., 1990: Latitude of the surface high-pressure belt over western South America during the last 500 years as inferred from tree-ring analysis. *Quaternary of South America and Antarctic Peninsula,* **7**: 273–303.

Villalba, R., 1995: Geographical variation in tree-growth responses to climate in the southern Andes. *In* Argollo, J., and P. Mourguiart (eds.), *Cambios Cuaternarios en América del Sur,* Organisation Scientifique et Technique d'Outre Mer (ORSTOM) (France)-Bolivia, pp. 307–317.

Villalba, R., J. A. Boninsegna, T. T. Veblen, A. Schmelter, and S. Rubulis, 1997: Recent trends in tree-ring records from high-elevation sites in the Andes of northern Patagonia. *Climatic Change,* **36**: 425–454.

Whetton, P., and I. Rutherfurd, 1994: Historical ENSO teleconnections in the Eastern Hemisphere. *Climatic Change,* **28**: 221–253.

Wiles, G. C., R. D. D'Arrigo, and G. C. Jacoby, 1996: Temperature changes along the Gulf of Alaska and the Pacific Northwest coast modeled from coastal tree rings. *Canadian Journal of Forest Research,* **26**: 474–481.

Wiles, G. C., R. D. D'Arrigo, and G. C. Jacoby, 1998: Gulf of Alaska atmosphere-ocean variability over recent centuries inferred from coastal tree-ring records. *Climatic Change,* **38**: 289–306.

Zhang, Y., J. M. Wallace, and D. S. Battisti, 1997: ENSO-like interdecadal variability: 1900–93. *Journal of Climate,* **10**: 1004–1020.

Zhang, R.-H., L. M. Rothstein, and A. J. Busalacchi, 1998: Origin of upper-ocean warming and El Niño change on decadal scales in the tropical Pacific Ocean. *Nature,* **391**: 879–883.

LONG-TERM CLIMATE VARIABILITY

Geomorphic Evidence

11

Glaciation During Marine Isotope Stage 2 in the American Cordillera

CHALMERS CLAPPERTON AND GEOFFREY SELTZER

Abstract

Alpine glaciers respond sensitively to climate change, and records of glacial advance and retreat can be used to assess patterns of climate change in widely disparate areas. Along the Pole-Equator-Pole: Americas (PEP 1) transect, maximum glaciation during Marine Isotope Stage 2 (MIS 2) was reached ca. 24,000–20,000 [14]C B.P. in regions receiving precipitation sufficient to produce an increase in glacial mass balance as global temperatures cooled. Exceptions to this scenario were glaciers that either responded to dynamic changes or were in areas sufficiently arid that cooler temperatures did not result in glacial expansion. However, these glaciers did reach maxima later in MIS 2, when global temperatures were still cool but an increase in moisture resulted in glacier advance. In addition to recording glacial maximum conditions, detailed chronologies of Sierra Nevadan and Chilean Lake District glaciers show high-resolution events that appear to be in synchrony with the Heinrich Events of the North Atlantic. Thus, glaciers along the PEP 1 transect have responded sensitively, and generally in synchrony, to climatic changes on both millennial and longer timescales. Copyright © 2001 by Academic Press.

Resumen

Glaciares alpinos responden sensitivamente a los cambios climáticos y los registros de avance y retroceso glaciar pueden ser usados para determinar parámetros del cambio climático en áreas ampliamente alejadas entre si. A lo largo de la transecta PEP1 la máxima glaciación ocurrió entre ca. 24–20 [14]C B.P., durante el estadio isotópico marino 2, en regiones que recibieron suficiente precipitación y que produjeron un incremento en el balance de masas de los glaciares mientras la temperatura global era mas baja que en el actual. La excepción fueron estos glaciares que por una parte no respondieron a cambios dinámicos o que por otra parte se encontraron en areas suficiente áridos, donde las bajas temperaturas no produjeron la expansion de hielos. Sin embargo, estos glaciares alcanzaron su máxima expansion posteriormente al estado isotópico marino 2, cuando las temperaturas globales eran todavia frías (o bajas) pero que un aumento en humedad provocó el avance de los hielos. Ademas de registrar las condiciones de la máxima glaciación, detalladas cronologías de los glaciares de la Sierra Nevada (EEUU) y de los glaciares del distrito de los lagos en Chile austral, muestran eventos de movimiento de glaciares con alta

173

resolución temporal que parecen ser sincrónicos con los "Heinrich Events" de el Atlantico Norte. De esta manera, los glaciares a lo largo de la transecta PEP 1 han respondido sensitivamente, y generalmente sincrónicos, a los cambios climáticos en escalas milenniales y aún mas largos.

11.1. INTRODUCTION

Moraines in mountain ranges along the Pole-Equator-Pole: Americas (PEP 1) transect document positions reached by glaciers during the last glacial cycle. A common perception has been that the outermost moraines define the CLIMAP (Climate: Long-Range Investigation, Mapping, and Prediction) (1981) last glacial maximum (LGM, 18,000 ^{14}C B.P.), and moraines within this zone were formed during recessional stages. However, it now seems likely that mountain glaciers advanced and retreated many times during the last glacial cycle (e.g., Lowell et al., 1995; Benson et al., 1996; Phillips et al., 1996), perhaps in synchrony with glacial age climate variability indicated by analyses of Greenland ice cores (Dansgaard et al., 1993) and North Atlantic sediments (Bond et al., 1993). Moreover, maximum glaciation rarely occurred at 18,000 ^{14}C B.P., which is important when the relationship between snow line depression and CLIMAP's (CLIMAP, 1981) estimates of sea surface temperature (SST) at the LGM (e.g., Klein et al., 1999; Thompson et al., 1995) is being considered. The extent reached by glaciers during Marine Isotope Stage 2 (MIS 2) depended on the available precipitation as well as on the degree of thermal decline. For this reason, it may be inappropriate to assume that all mountain glaciers reached their maximum extents at the CLIMAP LGM, or that they all reached their maximum limits at the same time.

This chapter presents a brief review of published chronologies for MIS 2 (~32,000–15,000 ^{14}C B.P.) glacier advances along the PEP 1 transect (Fig. 1). The analysis is based largely on overviews of glacial chronologies published in the last 15 years, but incorporates information from recent papers on specific areas where appropriate. This chapter discusses the last glaciation in the cordilleras of North and South America prior to a discussion of the patterns that emerge (Fig. 2).

11.2. NORTH AMERICA

In Alaska, studies of last glaciation advances have been made in the Brooks Range, the Alaska Range, the Alaska Peninsula, and Kodiak Island (Porter et al., 1983; Hamilton and Thorson, 1983; Mann and Peteet,

1994). Radiocarbon dating of organic material associated with glacial and glaciofluvial deposits supports a time-distance diagram presented by Porter et al. (1983, Fig. 4.7) showing two glacial maxima in the Brooks Range during MIS 2, at ca. 24,000–22,000 ^{14}C B.P. and 19,500–17,000 ^{14}C B.P. Glaciers in other parts of Alaska appear to have reached their maxima at about the same time as those in the Brooks Range. Deglaciation in coastal southern Alaska began by ca. 16,000–15,000 ^{14}C B.P. and continued for at least 1500 ^{14}C years, as is shown by the radiocarbon ages of marine fossils in the Bootlegger Cove formation in Cook Inlet (Hamilton and Thorson, 1983).

Farther south, glaciers in the coastal mountains of southwestern British Columbia and the northern Cascade Range advanced to their late Wisconsin maximum limits (Coquitlam advance) at ca. 24,800–18,000 ^{14}C B.P. (Porter et al., 1983). The mountain glaciers then receded for about a millennium during the Olympia interstade before the Vashon advance led to maximum expansion of western outlet glaciers of the Cordilleran ice sheet. Although the Puget ice lobe terminated its maximum late Wisconsin advance south of Seattle shortly after ca. 15,000 ^{14}C B.P. (Mullineaux et al., 1965; Porter and Swanson, 1998), eastern outlets re-advanced less distance (~50%) than they had at ca. 22,000 ^{14}C B.P. (Richmond, 1986). This difference suggests a remarkable influence of either glaciological control (internal flow dynamics, calving) or west-east differences in precipitation supply across the Cordilleran ice sheet during the late Wisconsin.

Moraines in many ranges of the Rocky Mountains mark limits of late Wisconsin (Pinedale stage) glacier advances, but chronologies exist only for those in the Yellowstone Plateau area, the Wind River Range (Wyoming), and the Colorado Front Range. Pierce (1979) mapped three major moraine stages in the Yellowstone River Valley—the Eightmile, Chico, and Deckard Flats moraines (oldest to youngest)—but had little dating control. However, U-series ages from travertine deposits associated with these moraines (Sturchio et al., 1994) now suggest that the Chico stage, with an age in the range of ca. 30,000–22,500 cal. B.P. (ca. 25,000–19,000 ^{14}C B.P.) represents the late Wisconsin maximum, whereas the Deckard Flats moraine (ca. 16,500–13,300 ^{14}C B.P.) may have formed coevally with the Vashon advance of the Pacific Northwest. The most extensive advance (Eightmile) has an age in the range of 47,000–34,000 cal. B.P. (ca. 40,000–28,000 ^{14}C B.P.) and indicates substantial glaciation of the Yellowstone Plateau well before the CLIMAP LGM.

The classic Pinedale sequence of moraines for the Rocky Mountains was first established in the Fremont Valley of the Wind River Range, initially based on

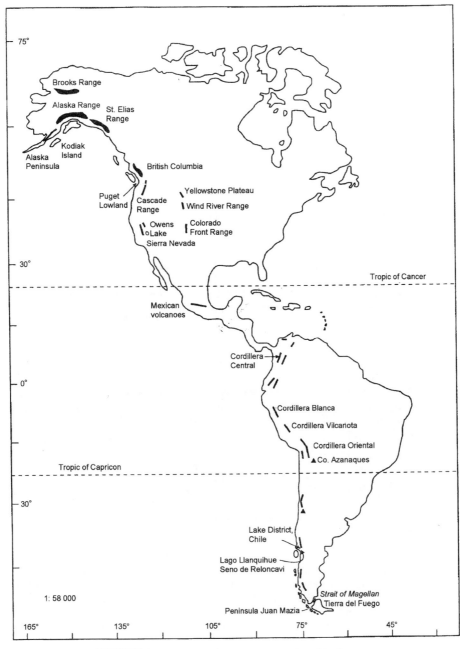

FIGURE 1 Locations of major sites mentioned in the text.

relative weathering criteria (Blackwelder, 1915; Richmond, 1986). Because of the absence of any radiocarbon chronology, surface boulders on the moraines have been dated by cosmogenic exposure methods using [10]Be (Gosse et al., 1995). These results generally suggest that glaciers reached their Pinedale maxima between 22,000 and 16,500 cal. B.P. (ca.18,400 and 13,800 [14]C B.P.). Results from the northern side of the Wind River Range generally confirm the results from Fre-

mont Valley (Phillips et al., 1997). In the Colorado Front Range, three moraines (Winter Park) document the Pinedale glacier maximum, and although it is believed they formed after ca. 30,000 [14]C B.P. (Nelson et al., 1979), their precise ages are not known.

Late Wisconsin moraines in the Sierra Nevada of California are commonly described as those of the Tenaya (older) and Tioga (younger) glaciations (Porter et al., 1983; Fullerton, 1986). Bursik and Gillespie (1993)

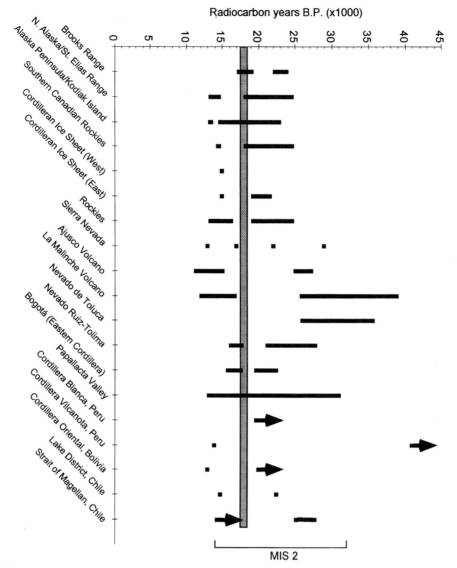

FIGURE 2 Summary of radiocarbon ages bracketing the Marine Isotope Stage 2 (MIS 2)glacial maxima in the American Cordillera. The shaded bar indicates the CLIMAP (Climate: Long-Range Investigation, Mapping, and Prediction) (1981) last glacial maximum (LGM) at 18,000 [14]C B.P. The arrows indicate minimum ages for glaciation. Based on data from Porter et al. (1983), Hamilton and Thorson (1983), and Mann and Peteet (1994) for Alaska; from Porter et al. (1983) and Richmond (1986) for the southern Canadian Rockies, the Puget lowland, and the Cordilleran ice sheet; from Pierce (1979), Sturchio et al. (1994), Zreda and Phillips (1995), Gosse et al. (1995), Phillips et al. (1997), and Nelson et al. (1979) for the Rockies; from Phillips et al. (1990, 1996), Bursik and Gillespie (1993), and Benson et al. (1996) for the Sierra Nevada; from White (1986) for the Ajusco Volcano and the La Malinche and Nevado La Toluca Volcanoes (as interpreted from Heine 1988); from Thouret et al. (1996) and Helmens et al. (1997) for Colombia; from Clapperton (1998) and Clapperton et al. (1995) for Ecuador; from Rodbell (1993), Mercer and Palacios (1977), and Goodman (1999) for Peru; from Seltzer (1994) and Clayton and Clapperton (1997) for Bolivia; from Denton et al. (1999) for the Chilean Lakes Region; and from Clapperton et al. (1995) for the Strait of Magellan.

assigned an age of ca. 26,400 [14]C B.P. to the Tenaya moraine stage, implying that the maximum glacier extent in the Sierra Nevada had occurred well before the CLIMAP LGM. The Tioga glaciation is generally assumed to have peaked at ca. 22,000–21,000 [14]C B.P.

(Fullerton, 1986; Gillespie and Molnar, 1995). Phillips et al. (1996) revised this chronology using [36]Cl analysis of boulder surfaces to determine four ages of MIS 2 glaciation: Tioga 1, 33,000 [36]Cl B.P. (29,000 [14]C B.P.); Tioga 2, 26,000 [36]Cl B.P. (22,000 [14]C B.P.); Tioga 3, 20,000 [36]Cl B.P.

(17,000 ^{14}C B.P.); and Tioga 4, 16,000 ^{36}Cl B.P. (13,000 ^{14}C B.P.).

A good proxy record for glacier advances in the Sierra Nevada comes from analyses of lake sediments taken from Owens Lake, which lies in an intermontane valley immediately north of the range (Benson et al., 1996). Magnetic susceptibility variations in the sediments document intervals when glaciers expanded in the adjacent mountains and delivered rock flour in meltwater to Owens Lake. Twenty-six accelerator mass spectrometry (AMS) radiocarbon dates constrain the sediment accumulation rate, and, from this, it is evident that brief glacier advances occurred at ca. 27,000 and 25,000 ^{14}C B.P. (Tenaya stages?) prior to a prolonged interval of expanded ice cover at ca. 23,000–15,000 ^{14}C B.P. (Tioga glaciation).

11.2.1. Conclusion for North American Ranges

The best chronological data for North American ranges (Fig. 2) suggest that maximum glacier expansion during MIS 2 culminated in most places between ca. 22,000 and 19,000 ^{14}C B.P. After a cool interstade, when glaciers receded an unknown distance, many readvanced by 50–80% of their earlier extents at ca. 15,000 ^{14}C B.P. Exceptions to this pattern include the continuous deglaciation in coastal Alaska from ca. 16,000 ^{14}C B.P., the Vashon advance of the Cordilleran ice sheet in which western outlets reached their greatest Wisconsin limits, and the suggestion that glaciers in the Sierra Nevada may have reached their maximum extents somewhat earlier in MIS 2 at ca. 26,400 ^{14}C B.P.

11.3. MÉXICO

Some chronology for late Quaternary glacier advances has been obtained for the Mexican volcanoes. Although 10 of the volcanic massifs supported glaciers during MIS 2, the best dated sequences are on Nevado de Toluca, Ajusco, Iztaccíhuatl, La Malinche, and Pico de Orizaba. A sequence of five glacial stages identified on Iztaccíhuatl was proposed as the standard for México and was correlated with the pre-Bull Lake, Bull Lake, and Pinedale stages in the Rocky Mountains of North America (White and Valastro, 1984). White (1986) presented a correlation chart for Mexican glaciations in a synthesis that also included the work of Heine (1984). Although few radiocarbon dates constrain the ages of the proposed glacier advances, the most extensive occurred between ca. 27,200 and 25,100 ^{14}C B.P. and was followed by a less extensive glacial advance; a third advance was assigned to the interval of ca.

15,100–11,400 ^{14}C B.P. Heine (1988) disagreed with White's scheme, and, on the basis of mapped moraine limits, marker pumice beds, and radiocarbon dating at five of the volcanoes, he concluded that the only MIS 2 glacier advance in México occurred at ca. 12,000 ^{14}C B.P. He assigned glacial maximum moraines to the interval of ca. 36,000–32,000 ^{14}C B.P. The uncertainty raised by this mismatch of opinions, and the unusual sequence proposed by Heine, will not be resolved until further detailed work is undertaken at these glaciated volcanoes. However, recent unpublished work on Iztaccíhuatl using ^{36}Cl dating suggests that at least two MIS 2 glacial maxima occurred between 20,000 and 14,000 ^{36}Cl B.P. (Phillips, personal communication; Vázquez-Selem and Phillips, 1998), which appears to support the prior work of White.

11.4. SOUTH AMERICA

As the Andes extend for more than 7000 km between latitudes 10°N and 55°S, they cross most of the zonal climate belts between equatorial and cool temperate regions of an entire hemisphere. Because mountain glaciers and ice caps formed throughout this mountain transect during the last glacial cycle (Clapperton, 1983, 1993a,b), chronologies of morainic deposits obtained in key parts of the Andes provide an opportunity to examine the interhemispheric synchrony of glacier advances and, thus, climate change (see, e.g., Lowell et al., 1995). Although spectacular moraines are present at all latitudes, only a few sequences have a series of radiocarbon ages to constrain the timing of glaciation (e.g., Helmens, 1988; Helmens et al., 1996, 1997; Thouret et al., 1996; Lowell et al., 1995; Clapperton et al., 1995; Denton et al., 1999).

Radiocarbon ages for MIS 2 glacier advances in the northern Andes were obtained in the Ruiz-Tolima volcanic massif of the Central Cordillera (Thouret et al., 1996) and in the currently ice-free ranges south of Bogotá in the Eastern Cordillera (Helmens et al., 1997). Although the advances are constrained by rather broad age ranges, it is apparent that the maximum MIS 2 advance occurred between ca. 23,500 and 19,500 ^{14}C B.P. The glaciers reached more extensive limits earlier in the last glacial cycle, prior to MIS 2 (Helmens et al., 1996). Farther south in Ecuador, an outlet glacier from ice fields covering the Eastern Cordillera close to the equator advanced eastward down the Papallacta Pass on two different occasions at some time after ca. 31,000 ^{14}C B.P., but before ca. 13,200 ^{14}C B.P., leaving moraines at altitudes of 3000 and ~3250 m. Last glacial maximum moraines in the Andes of north-central Peru (Cordillera Blanca) were deposited before ca. 19,700 ^{14}C B.P. (Rodbell, 1993), but in southeast Peru the maximum extent of glaciation occurred before ca. 41,000 [14]C B.P. (Mercer and Palacios, 1977; Goodman, 1999). The next most extensive advance was not until shortly after ca. 13,900 [14]C B.P. in southeastern Peru and reached less than half the distance of the earlier event. These data imply that glaciers in this part of the Andes did not advance very far during the CLIMAP LGM. One site in the Eastern Cordillera of Bolivia does suggest that the last glacial maximum was attained prior to 20,000 [14]C B.P. (Seltzer, 1994). However, it appears that glaciers were small in many areas of the Eastern Cordillera of Bolivia during the LGM and may not have advanced to their maximum MIS 2 limits until the interval of ca. 14,000–12,000 [14]C B.P. (Seltzer, 1992; Clapperton et al., 1997). This inference is compatible with evidence from the Cerro Azanaque massif, which rises to 5200 m above the salt basins of Poopó and Uyuni in the southern Altiplano, showing that the maximum MIS 2 advance impacted peat beds at about 13,000 [14]C B.P. Meltwater from this advance discharged sediment to form Gilbert-type deltas in a high stand of paleolake Tauca (Clayton and Clapperton, 1997).

Although small glaciers may have formed in the currently arid region between latitudes 18° and 29°S, where modern glaciers do not exist (Messerli et al., 1993; Jenny and Kammer, 1996), Quaternary glaciers seem to have been absent south of here to about latitude 32°S because of arid conditions. South of 33°S, however, sharp-crested moraines are present in most drainages all the way to Tierra del Fuego, but the only well-dated sequences are in the Chilean Lake District (and adjacent Isla de Chiloé) and along the central Strait of Magellan (Porter, 1981; Porter et al., 1992). The remarkably detailed morphological, stratigraphical, and chronological investigation of last glacial deposits around Lago Llanquihue and the marine embayment of Seno Reloncavi has revealed that outlet glaciers from the Andean ice cap advanced at least four times during MIS 2 (Lowell et al., 1995; Denton et al., 1999). More than 450 radiocarbon dates underpin the current conclusion that these advances culminated at ca. 29,400, 26,800, 22,400, and 14,600 [14]C B.P., all reaching limits within the ca. 5- to 10-km wide Llanquihue moraine belt. Whereas the Llanquihue ice lobe was most extensive at ca. 22,400 [14]C B.P., the Reloncavi lobe farther south, which advanced onto Isla de Chiloé, reached its maximum limit later at ca. 14,600 [14]C B.P., a pattern reminiscent of that in the South Canadian Rockies–Puget lowlands area of North America.

The southernmost part of the conterminous Patagonian ice cap drained to the east and north as large lobate outlet glaciers (Caldenius, 1932). The lobe that flowed along the central Strait of Magellan terminated its maximum Stage 2 advance at Peninsula Juan Mazia, some 180 km from the ice divide (Porter et al., 1992; Clapperton et al., 1995; Anderson and Archer, 1999). The radiocarbon ages of marine shells in deformation till associated with the outermost moraine are in the range of ca. 27,700–44,000 [14]C B.P. This range indicates a long marine interval in the Strait of Magellan during MIS 3 and implies extensive deglaciation early in MIS 2. A minimum age for the maximum advance is given by AMS radiocarbon dates of ca. 25,200 and 23,600 [14]C B.P. on organic matter from the base of an interdrumlin hollow within the moraine limit. Following a brief interval of recession, during which ice-marginal and proglacial lacustrine and outwash sediments accumulated, the Magellan glacier re-advanced over these deposits to limits approximately 2–3 km short of the maximum advance. This event is constrained by a minimum radiocarbon age of ca. 17,700 [14]C B.P. from basal organic matter within the moraine limit. The Magellan glacier then receded an unknown distance from this re-advance limit, but not far enough south to allow penetration of Pacific seawater into the strait. A subsequent advance formed a complex of bouldery moraines along the very edge of both sides of the central Strait of Magellan and terminated about 25 km short of the earlier limits. This event is constrained by minimum radiocarbon dates of ca. 14,000–13,500 [14]C B.P. from the base of kettle holes inside the moraine limit (see McCulloch et al., 2000).

11.4.1. Conclusion for South American Ranges

It is evident that at least three advances occurred between ca. 27,000 and 4500 [14]C B.P., reaching almost to the same limits. Advances in Colombia and along the Strait of Magellan are bracketed by radiocarbon dates that imply they may have occurred coevally with advances dated in the Chilean Lake District to ca. 26,800, 22,400, and 14,800 [14]C B.P. Evidence for additional advances has not yet been observed, other than that dated to ca. 29,400 [14]C B.P. in the Chilean Lake District. Notable exceptions to this pattern of MIS 2 fluctuations are in southern Peru and Bolivia, where the maximum MIS 2 advance peaked between ca. 14,000 and 12,000 [14]C B.P.

11.5. DISCUSSION

The initial cooling of MIS 2 appears to have caused glaciers to advance in Alaska (29,000–27,500 [14]C B.P.), in the Pacific Northwest (35,000–28,000 [14]C B.P.), and in the southern Andes (29,300 [14]C B.P.), suggesting that regions with abundant precipitation responded rapidly to the thermal decline. The Owens Lake data indicate relatively wet conditions at ca. 28,500 [14]C B.P., which may have caused glaciers to advance in the Sierra Nevada at this time also. A second MIS 2 advance possibly occurred in several mountain ranges of the PEP 1 transect during the interval of ca. 27,000–25,000 [14]C B.P. For example, advances are documented along the southern margin of the Cordilleran ice sheet (the Flathead lobe, western Montana at ca. 29,000–25,000 [14]C B.P.), the Sierra Nevada of California (27,000–26,000 [14]C B.P.), the Chilean Lake District (26,700 [14]C B.P.), and the Strait of Magellan (27,700–25,500 [14]C B.P.). The third MIS 2 advance has been radiocarbon dated to the broad interval of ca. 24,000–20,000 [14]C B.P. in many mountain ranges of the PEP 1 transect. For example, in Alaska the event is constrained by ages of ca. 24,300 and 22,700 [14]C B.P. (Mann and Peteet, 1994), and southern lobes of the Cordilleran ice sheet are alleged to have reached their maximum limits at ca. 21,000 [14]C B.P. (Richmond and Fullerton, 1986); the Tioga series of glacier advances in the Sierra Nevada of California began at ca. 24,000 [14]C B.P. (Benson et al., 1996). The maximum MIS 2 advance in Colombia is bracketed by ages of ca. 22,000–19,500 ^{14}C years B.P. (Helmens et al., 1997), and in the Chilean Lake District it culminated at ca. 22,400 [14]C B.P. (Denton et al., 1999). This evidence is suggestive that the MIS 2 maximum glacier extent culminated between ca. 24,000 and 20,000 [14]C B.P. in ranges receiving relatively high amounts of precipitation. It is possible that glaciers in different regions did not reach their maximum limits precisely at the same time, but a broad interhemispheric synchrony is implied by the existing data.

Whether or not strong glacier advances occurred anywhere along the Americas transect between ca. 19,000 and 16,000 [14]C B.P. remains uncertain, but glaciers expanded by 50 to >100% of their earlier maximum extents at ca. 15,000 [14]C B.P. in ranges as far apart as the Canadian Rockies/North Cascades and the southernmost Andes, again implying broad interhemispheric synchrony of climatic forcing. Alpine glaciers apparently advanced worldwide at about 15,000 [14]C B.P. due to atmospheric cooling (Clapperton, 1995), but the reason why western outlets of the Cordilleran ice sheet reached their greatest limits of all at this time may have been due to increased precipitation resulting from a northerly retreat of the Polar Front as the Laurentide ice sheet shrank and relaxed its influence. A similar explanation may apply to some outlets of the Chilean ice cap, which also reached their greatest limits at about 14,800 [14]C B.P. This implies that the belt of high precipitation from the Pacific westerlies may have become located around latitudes 41°–42°S at this time prior to migrating south to its interglacial position.

An outstanding anomaly is the tropical region of southeastern Peru and Bolivia, where glaciers appear to have been smaller at the ca. 24,000–20,000 [14]C B.P. LGM stage than they became after Termination 1, advancing to their maximum late Quaternary limits at ca. 13,000 [14]C B.P. This anomaly is best explained as a consequence of reduced precipitation at the LGM because of cooler temperatures. The dramatic rise in atmospheric moisture after Termination 1 presumably resulted from a greater capacity of the atmosphere to hold more moisture as the Bølling warm interval began, and in the tropical Andes this effect was enhanced by the warming Atlantic Ocean and expansion of the tropical rain forests. As atmospheric temperature was possibly still 2°–3°C cooler than it is now immediately after Termination 1, glaciers in southeastern Peru and Bolivia expanded to their maximum limits.

Acknowledgments

We are grateful to Vera Markgraf and the PEP 1 Organization for inviting our participation in the Mérida conference. C.M. Clapperton acknowledges the National Environment Research Council (NERC, United Kingdom), The Royal Society of London, and The Carnegie Trust for the Universities of Scotland for research grants to work in the Andes. G.O. Seltzer acknowledges the continued support of the National Science Foundation (NSF, United States) for research in the Andes. We also are grateful to Fred Phillips and Tom Lowell for detailed reviews, which helped greatly in improving the manuscript.

Editor's Note

A draft of this chapter was received from C. M. Clapperton prior to his sudden illness, which prevented him from submitting a final version for publication in the PEP 1 volume. G. Seltzer kindly updated the draft and finalized the chapter based on reviewers comments. A more detailed version of this manuscript was recently published in a special tribute volume of the *Journal of Quaternary Science* (Clapperton, C. 2000. Interhemispheric synchroneity of Marine Oxygen Isotope Stage 2 glacier fluctuations along the American cordilleras transect. *Journal of Quaternary Science*, 13: 435–468).

References

Anderson, D. M., and R. B. Archer, 1999: Preliminary evidence of early deglaciation in southern Chile. *Palaeogeography, Palaeoclimatology, Palaeoecology*, 146: 295–301.

Benson, L. V., W. B. James, M. Kashgarian, S. P. Lund, F. M. Phillips, and R. O. Rye, 1996: Climatic and hydrologic oscillations in the Owens Lake basin and adjacent Sierra Nevada of California. *Science*, 274: 746–749.

Blackwelder, E., 1915: Post-Cretaceous history of the mountains in central western Wyoming. *Journal of Geology*, 23: 97–117, 193–217, 307–340.

Bond, G., W. Broecker, S. Johnsen, J. McManus, L. Labeyrie, J. Jouzel, and G. Bonani, 1993: Correlation between climate records from North Atlantic sediments and Greenland ice. *Nature*, 365: 143–147.

Bursik, M., and A. R. Gillespie, 1993: Late Pleistocene glaciation in the Mono basin, California. *Quaternary Research*, 39: 24–35.

Caldenius, C. C., 1932: Las glaciaciones cuaternarías en la Patagonia y Tierra del Fuego. *Geografiska Annaler*, **14**: 1–64.

Clapperton, C. M., 1983: The glaciation of the Andes. *Quaternary Science Reviews, 2/3*: 1–83.

Clapperton, C. M., 1993a: *Quaternary Geology and Geomorphology of South America.* Amsterdam: Elsevier, 779 pp.

Clapperton, C. M., 1993b: Nature of environmental changes in South America at the last glacial maximum. *Palaeogeography, Palaeoclimatology, Palaeoecology*, **101**: 189–208.

Clapperton, C. M., 1995: Fluctuations of local glaciers at the termination of the Pleistocene: 18–8 ka B.P. *Quaternary International*, **28**: 41–50.

Clapperton, C. M., 2000: Interhemispheric synchroneity of Marine Oxygen Isotope Stage 2 glacier fluctuations along the American cordilleras transect. *Journal of Quaternary Science* **15**: 435–468.

Clapperton, C. M., M. Hall, P. Motlies, M. J. Hole, J. W. Still, K. F. Helmens, P. Kuhry, and A. M. D. Gemmell, 1997: A Younger Dryas ice cap in the equatorial Andes. *Quaternary Research*, **47**: 13–28.

Clapperton, C. M., D. E. Sugden, D. Kaufman, and R. D. McCulloch, 1995: The last glaciation in central Magellan Strait, southernmost Chile. *Quaternary Research*, **44**: 133–148.

Clayton, J. D., and C. M. Clapperton, 1997: Broad synchrony of a late-glacial glacier advance and highstand of palaeolake Tauca in the Bolivian Altiplano. *Journal of Quaternary Science*, **12**: 169–182.

CLIMAP Project Members, 1981: Seasonal reconstruction of the Earth's surface at the last glacial maximum. *Geological Society of America Map and Chart Series*, **36**.

Dansgaard, W., S. Johnsen, H. Clausen, D. Dahl-Jensen, N. Gundestrup, C. Hammer, C. Hvidberg, J. Steffensen, A. Sveinbjornsdottir, J. Jouzel, and G. Bond, 1993: Evidence for general instability of past climate from a 250-kyr ice-core record. *Nature*, **364**: 218–220.

Denton, G. H., T. V. Lowell, P. I. Moreno, B. G. Andersen, and C. Schlüchter, 1999: Interhemispheric linkage of paleoclimate during the last glaciation. *Geografisker Annaler*, **81A**: 167–229.

Fullerton, D. S., 1986: Chronology and correlation of glacial deposits in the Sierra Nevada, California. *Quaternary Science Reviews*, **5**: 161–170.

Gillespie, A. R., and P. Molnar, 1995: Asynchronous maximum advances of mountain and continental glaciers. *Reviews of Geophysics*, **33**: 311–363.

Goodman, A. Y., 1999: Subdivision of glacial deposits in southeastern Peru based on pedogenic development and radiometric ages. M.S. thesis. Syracuse University, New York, 143 pp.

Gosse, J. C., J. Klein, E. B. Evenson, B. Lawn, and R. Middleton, 1995: Beryllium-10 dating of the duration and retreat of the last Pinedale glacial sequence. *Science*, **268**: 1329–1333.

Hamilton, T. D., and R. M. Thorson, 1983: The Cordilleran ice sheet in Alaska. *In* Wright, H. E., Jr. (ed.), *Late Quaternary Environments of the United States, Vol. 1, The Late Pleistocene* (Porter, S. C., ed.). Minneapolis: University of Minnesota Press, pp. 38–41.

Heine, K., 1984: The classical late Weichselian climatic fluctuations in México. *In* Mörner, N.-A., and W. Karlén (eds.), *Climatic Changes on a Yearly to Millennial Basis.* Dordrecht: Reidel, pp. 95–115.

Heine, K., 1988: Late Quaternary glacial chronology of Mexican volcanoes. *Die Geowissenschaften*, **6**: 197–205.

Helmens, K. F., 1988: Late Pleistocene glacial sequence in the area of the high plain of Bogotá, Eastern Cordillera, Colombia. *Palaeogeography, Palaeoclimatology, Palaeoecology*, **67**: 263–283.

Helmens, K. F., P. Kuhry, N. W. Rutter, K. Van der Borg, and A. F. M. de Jong, 1996: Warming at 18k yr B.P. in the tropical Andes. *Quaternary Research*, **45**: 289–299.

Helmens, K. F., N. W. Rutter, and P. Kuhry, 1997: Glacier fluctuations in the eastern Andes of Colombia, South America, during the last 45k radiocarbon years. *Quaternary International*, **38/39**: 39–48.

Jenny, B., and K. Kammer, 1996: Climate change in den trocken Anden: Jungquartäre Vergletscherungen. *Geographica Bernensis*, **G46**: 1–80.

Klein, A. G., G. O. Seltzer, and B. L. Isacks, 1999: Modern and last glacial maximum snowlines in the central Andes of Peru, Bolivia, and northern Chile. *Quaternary Science Reviews*, **18**: 63–84.

Lowell, T. V., C. J. Heusser, B. Andersen, P. I. Moreno, A. Hauser, L. E. Heusser, C. Schlüchter, D. R. Marchant, and G. H. Denton, 1995: Interhemispheric correlation of late Pleistocene glacial events. *Science*, **269**: 141–149.

Mann, D. H., and D. M. Peteet, 1994: Extent and timing of the last glacial maximum in southwestern Alaska. *Quaternary Research*, **42**: 136–148.

McCulloch, R. D., M. J. Bentley, R. S. Purves, N. R. J. Hulton, D. E. Sugden, and C. M. Clapperton, 2000: Climatic inferences from glacial and palaeoecological evidence at the last glacial termination, southern South America. *Journal of Quaternary Science*, **15**: 409–417.

Mercer, J. H., and O. Palacios, 1977: Radiocarbon dating of the last glaciation in Peru. *Geology*, **5**: 600–604.

Messerli, B., M. Grosjean, G. Bonani, A. Bürgi, M. A. Geyh, K. Graf, K. S. Ramseyer, H. Romero, U. Schotterer, H. Schreier, and M. Vuille, 1993: Climate change and natural resource dynamics of the Atacama Altiplano during the last 18k years: A preliminary synthesis. *Mountain Research and Development*, **13**: 117–127.

Mullineaux, D. R., H. H. Waldron, and M. Rubin, 1965: Stratigraphy and chronology of late interglacial and early Vashon time in the Seattle area, Washington. *U.S. Geological Survey Bulletin*, **1194-O**: 1–10.

Nelson, T. G., A. C. Millington, J. T. Andrews, and H. Nichols, 1979: Radiocarbon dated upper Pleistocene glacial sequence, Fraser Valley, Colorado Front Range. *Geology*, **7**: 410–414.

Phillips, F. M., M. G. Zreda, L. V. Benson, M. A. Plummer, D. Elmore, and P. Sharma, 1996: Chronology for fluctuations in late Pleistocene Sierra Nevada glaciers and lakes. *Science*, **274**: 749–751.

Phillips, F. M., M. G. Zreda, J. C. Gosse, J. Klein, E. B. Evenson, R. D. Hall, O. A. Chadwick, and P. Sharma, 1997: Cosmogenic ^{36}Cl and ^{10}Be ages of Quaternary glacial and fluvial deposits of the Wind River Range, Wyoming. *Geological Society of America Bulletin*, **109**: 1453–1463.

Phillips, F. M., M. G. Zreda, S. S. Smith, D. Elmore, P. W. Kubik, and P. Sharma, 1990: Cosmogenic chlorine-36 chronology for glacial deposits at Bloody Canyon, eastern Sierra Nevada. *Science*, **248**: 1529–1532.

Pierce, K. L., 1979: History and dynamics of glaciation in the northern Yellowstone National Park area. *U.S. Geological Survey Professional Paper* 729-F, pp. Fl–F90.

Porter, S. C., 1981: Pleistocene glaciation in the southern Lake District of Chile. *Quaternary Research*, **16**: 263–292.

Porter, S. C., C. M. Clapperton, and D. E. Sugden, 1992: Chronology and dynamics of deglaciation along and near the Strait of Magellan, southernmost South America. *Sveriges Geologiska Undersøkning*, Ser. Ca., **81**: 233–239.

Porter, S. C., K. L. Pierce, and T. D. Hamilton, 1983: Late Pleistocene glaciation in the western United States. *In* Wright, H. E., Jr. (ed.), *Late Quaternary Environments of the United States, Vol. 1, The Late Pleistocene* (Porter, S. C., ed.). Minneapolis: Univ. of Minnesota Press, pp. 71–111.

Porter, S. C., and T. W. Swanson, 1998: Radiocarbon age constraints on rates of advance and retreat of the Puget Lobe of the Cordilleran Ice Sheet during the last glaciation. *Quaternary Research*, **50**: 205–213.

Richmond, G. M., 1986: Stratigraphy and chronology of glaciations in Yellowstone National Park. *Quaternary Science Reviews*, **5**: 83–98.

Richmond, G. M., and D. S. Fullerton, 1986: Summation of Quaternary glaciations in the United States of America. *Quaternary Science Reviews,* **5**: 183–196.

Rodbell, D. T., 1993: Subdivision of late Pleistocene moraines in the Cordillera Blanca, Peru, based on rock-weathering features, soils and radiocarbon dates. *Quaternary Research,* **39**: 133–143.

Seltzer, G. O., 1992: Late Quaternary glaciation of the Cordillera Real, Bolivia. *Journal of Quaternary Science,* **7**: 87–98.

Seltzer, G. O., 1994: A lacustrine record of late-Pleistocene climatic change in the subtropical Andes. *Boreas,* **23**: 105–111.

Sturchio, N. C., K. L. Pierce, M. T. Murrell, and M. Sorey, 1994: Uranium-series ages of travertines and timing of the last glaciation in the northern Yellowstone area, Wyoming-Montana. *Quaternary Research,* **41**: 265–277.

Thompson, L. G., E. Mosley-Thompson, M. E. Davis, P.-N. Lin, K. A. Henderson, J. Cole-Dai, J. F. Bolzan, and K.-B. Liu, 1995: Lateglacial stage and Holocene tropical ice core records from Huascarán, Peru. *Science,* **269**: 46–50.

Thouret, J.-C., T. van der Hammen, and B. Salomons, 1996: Paleoenvironmental changes and glacial stades of the last 50k years in the Cordillera Central, Colombia. *Quaternary Research,* **46**: 1–18.

Vázquez-Selem, L., and F. M. Phillips, 1998: Glacial chronology of Iztaccíhuatl Volcano, central México, based on cosmogenic ^{36}Cl exposure ages and tephrochronlogy. American Quaternary Association, 15th Biennial Meeting, Puerto Vallarta, México, p. 174.

White, S. E., 1986: Quaternary glacial stratigraphy and chronology of México. *Quaternary Science Reviews,* **5**: 201–206.

White, S. E., and S. Valastro, Jr., 1984: Pleistocene glaciation of volcano Ajusco, central México, and comparison with standard Mexican glacial sequence. *Quaternary Research,* **21**: 21–35.

Zreda, M. G., and F. M. Phillips, 1995: Insights into alpine moraine development from cosmogenic ^{36}Cl buildup dating. *Geomorphology,* **14**: 149–156.

Late Quaternary Eolian Records
of the Americas and Their
Paleoclimatic Significance

DANIEL R. MUHS AND MARCELO ZÁRATE

Abstract

Eolian deposits, consisting of both dune and sheet sand as well as loess, are extensive in South and North America. Much of the loess present on both continents was derived from glacial outwash sources and deposited during the last glacial maximum (LGM). However, both continents have loess deposits that are nonglaciogenic; e.g., derived from volcaniclastic sources which are unrelated to glaciation. Nevertheless, cold, dry, windy conditions during the last glacial period may have favored loess deposition from a variety of sources. Eolian sand is also extensive on both continents, and some was deposited during the last glacial period as well, particularly in South America, the midcontinent and deserts of North America, and Alaska. Mid-Holocene eolian sand was deposited in parts of the midcontinent of North America under warmer and drier conditions. However, in both the midcontinent of North America and the southwestern deserts, the most recent and extensive eolian sand units were deposited during the late Holocene. Much of the late Holocene eolian sand in midcontinental North America may have been reworked from older sand deposited during the last glacial and mid-Holocene. Late Holocene eolian sand is also abundant in the Pampas region of South America and also may have been reworked from older deposits.

Far-traveled, fine-grained dust is transported to and from the Americas. Deep-sea sediment cores, ice cores, and soils contain long-term records of this dust flux. Deep-sea cores from off the coast of South America and ice cores from the Andes indicate that the last glacial period was characterized by much greater fluxes of dust than the Holocene. Both deep-sea cores and soil chronosequences suggest that the flux of dust from Africa to the Americas via the trade winds has been an important process for much of the Quaternary period. Isotopic data indicate that dust in Antarctic ice cores, also found in greater abundance during glacial periods, has its sources in the Pampas region of South America.

Deposition of loess and eolian sand, from multiple sources during the last glacial period, agrees with pollen records, lake records, and COHMAP; Cooperative Holocene Climate Mapping) climate simulations which suggest that the last glacial period was cold, dry, and windy (relative to now) in the mid- and high-latitude regions of both continents. In the Pampas region of South America, there is at least fair agreement between summer surface winds simulated by COHMAP and paleowinds as recorded by eolian sand and loess. In midcontinental North America, however, the presence of a strong glacial anticyclone centered over the Laurentide ice sheet (as derived from COHMAP models) would have generated northeasterly winds in the region immediately to the south of the ice sheet. Loess paleowind records are not in agreement with this model and suggest westerly or northwesterly winds. The reason for this disagreement is not understood and needs further study. Mid-Holocene eolian activity in parts of the mid-continent of North America agrees with pollen data and COHMAP simulations, both of which suggest that this may have been a warm and dry period. Late Holocene eolian activity on both continents cannot yet be dated precisely enough to be correlated with other records, such as those from diatoms and pollen. However, both eolian records and lacustrine records indicate that there were multiple periods of relatively dry climate during the late Holocene.

Resumen

Los depósitos eólicos que comprenden tanto dunas y mantos de arenas como loess, son extensos en Norte y Sudámerica. Ambos continentes, donde el loess ha sido depositado en gran parte durante el último máximo glacial, presentan depósitos loéssicos que no son glacigénicos, i.e, derivados de fuentes volcaniclásticas, no relacionadas con la glaciación. Sin embargo, las condiciones frías, secas y ventosas del último período glacial pueden haber favorecido la depositación de loess procedente de fuentes diversas. La arena eólica, también de amplia extensión en ambos continentes, en parte se depositó durante el último período glacial, particularmente en Sudamérica, la región continental media y los desiertos de Norteamérica, y Alaska. Arenas eólicas del Holoceno medio se depositaron en sectores de la región continental media de Norteamérica en condiciones más cálidas y más secas. Sin embargo, tanto en esta última región como en los desiertos del sudoeste, las unidades de arena eólica más recientes se depositaron durante el Holoceno tardío. Gran parte de la arena eólica del Holoceno tardío en la región continental media de Norteamérica puede haber sido retrabajada de arenas más antiguas depositadas durante los períodos del último glacial y del Holoceno medio. La arena eólica del Holoceno tardío también es frecuente en las Pampas de Sudamérica, donde puede haber sido retrabajada a partir de depósitos más antiguos como en Norteamérica.

Polvos de grano fino procedentes de fuentes lejanas, son transportados hacia y desde las Américas. Los testigos de sedimento de los fondos marinos, los testigos de hielo, y los suelos contienen registros temporalmente prolongados de este flujo de polvo. Los testigos de fondo marino costa afuera de Sudamérica y de hielo en los Andes indican que el último período glacial estuvo caracterizado por flujos de polvo mucho mayores que durante el Holoceno. Tanto los testigos de fondo marino como las cronosecuencias de suelos sugieren que el flujo de polvo desde Africa a las Américas a través de los vientos alisios ha sido un proceso importante durante la mayor parte del Cuaternario. Los datos isotópicos indican que el polvo en los testigos de hielo de Antártida, también encontrados con mayor abundancia durante los períodos glaciales, tienen sus áreas de procedencia en la región pampeana de Sudamérica.

La depositación de loess y arenas eólicas a partir de múltiples áreas de procedencia durante el último período glacial concuerda con los registros de polen, lagos y las simulaciones climáticas de COHMAP (Cooperative Holocene Mapping Project), las que sugieren que el último período glacial fue frío, seco y ventoso (en relación con la actualidad) en las regiones de latitudes medias y altas de ambos continentes. En las Pampas de Sudamérica, existe al menos un claro acuerdo entre los vientos estivales de superficie simulados por COHMAP y los paleovientos inferidos a partir de las arenas eólicas y el loess. En la región continental media de Norteamérica, sin embargo, la presencia de un fuerte anticiclón glacial centrado sobre el Manto de Hielo Laréntico (tal como se deriva de los modelos COHMAP) habría generado vientos del noreste en la región inmediatamente al sur del manto de hielo. Los paleovientos indicados por los registros de loess no están de acuerdo con este modelo y sugieren vientos del oeste o noroeste. Aún no se entiende la causa de este desacuerdo que requiere más estudios. La actividad eólica en sectores de la región continental media de Norteamérica concuerda con los datos de polen y las simulaciones COHMAP; ambos sugieren que este intervalo pudo haber sido un período cálido y seco. La actividad eólica del Holoceno tardío en ambos continentes no puede ser aún datada con precisión suficiente para ser correlacionada con otros registros tales como aquellos de diatomeas y polen. Sin embargo, tanto los registros eólicos como lacustres indican que hubo múltiples períodos de clima relativamente seco durante el Holoceno tardío.

12.1. INTRODUCTION

The Americas contain rich late Quaternary eolian records in the form of loess (Fig. 1), eolian sand sheets, and dunes. Eolian deposits are important in paleoclimatic studies because they (1) put limits on the moisture balance of a region at the time of formation, (2) can be sensitive indicators of the amount of vegetation cover, and (3) are records of wind direction. The record of

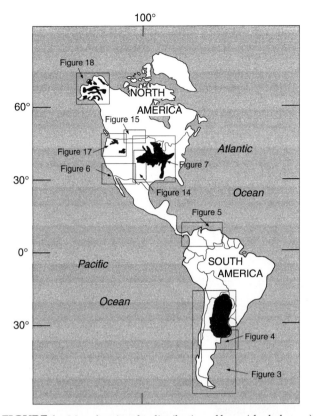

FIGURE 1 Map showing the distribution of loess (shaded areas) in North and South America and locations of detailed figures: loess distribution in North America from Thorp and Smith (1952) and Péwé (1975), and loess distribution in South America from Sayago (1995).

loess and eolian sand is summarized for South and North America with emphasis on three time periods: the last glacial maximum (LGM), the mid-Holocene, and the late Holocene. Throughout this chapter, all ages given, unless otherwise noted, are in uncalibrated radiocarbon years before present (^{14}C years B.P.).

The evolution of concepts about the genesis of loess make for a fascinating history of scientific thought (see Follmer, 1996). A now classical and generally accepted model of loess formation near glaciated terrain (Fig. 2) is that silt-sized particles are formed by glacial grinding and are delivered as outwash to valley trains (Ruhe, 1983). Wind entrains the particles from floodplains of the valley trains and deposits them in downwind-thinning blankets that also show systematic downwind decreases in particle size and carbonate content. The classical concept of loess stratigraphy, therefore, is that thick, unaltered loess deposits record glacial periods and intercalated paleosols record interglacial or interstadial periods (Fig. 2). In this chapter, certain aspects of the classical loess model are supported by loess studies in the Americas, and many aspects are not.

Although eolian sand deposits of the Americas have not received the attention that loess deposits have, there is also a traditional interpretation that such deposits are related to glacial periods. Several investigators (e.g., Sarnthein, 1978; Kutzbach and Wright, 1985) have considered large dune fields in the midlatitudes to be dominantly relict features that were last active during the LGM, when conditions are thought to have been cold, dry, and windy. In addition, eolian sand and loess were thought to be genetically related: i.e., essentially one deposit with a coarse, proximal facies (sand) and fine, distal facies (loess). In some regions of the Americas, this model seems to apply, and in other regions, it does not. New studies from both South and North America are now demonstrating that many dune fields have a rich record of Holocene activity and that dunes and loess may have distinctly different sources.

12.2. EOLIAN RECORDS OF THE SOUTH AMERICAN PAMPAS

The late Quaternary record of the Pampas (Fig. 3) is mainly composed of eolian deposits that include loess, loessial sands, eolian sand mantles, and dunes. Although several new contributions have been recently published (Iriondo, 1997, 1999; Sayago, 1995, 1997), controversial explanations emerge when specific aspects are considered, such as loess distribution, source areas, and origin of the eolian particles. First, the Pampas are a heterogeneous environment with different climatic, geomorphologic, and vegetative characteristics; key stratigraphic localities are found in different set-

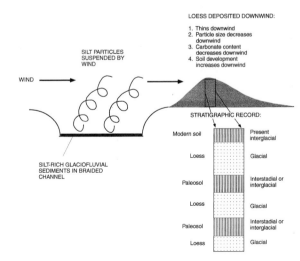

FIGURE 2 Classical model of midlatitude loess formation by eolian deflation of glaciogenic silt, downwind trends, and traditional climatic interpretation of the loess stratigraphic record.

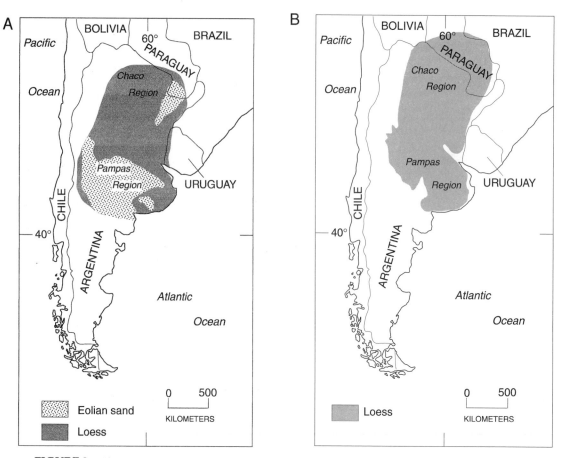

FIGURE 3 Alternative distributions of eolian sand (stippled) and loess (shaded) in Argentina, Paraguay, and Uruguay. (a) Redrawn from Iriondo (1997); (b) redrawn from Sayago (1995).

tings. Pioneering contributions on Pampean loess composition and distribution came from one key locality (Mar del Plata; see Figs. 4a and 4b) on the coast of the southern Pampas, from which results were extrapolated to the rest of Argentina (Teruggi et al., 1957; Teruggi, 1957). Second, inadequate chronological control, based on few ages, hinders stratigraphic correlation of eolian records between different areas of the Pampas, making even the present interpretations tentative and speculative.

12.2.1. Full Glacial and Late Glacial Times

Iriondo (1990a, 1997) proposed a regional model of eolian deposition that he named the Pampean Eolian System. The eolian source areas are thought to be the floodplains and fans of the combined fluvial system of the Bermejo, Salado, Desaguadero, and Colorado Rivers (Fig. 4a), which drained the glaciated Andes Mountains from approximately 28°–38°S latitude. A minor amount of material was supplied locally from the mountain ranges in central Argentina (Córdoba and

San Luis). Volcanic ashfalls were a secondary source of eolian particles transported in high suspension, which came from volcanoes south of 38°S (southern Mendoza and Neuquén; Fig. 4a) and from volcanoes from the Puna region (Bloom, 1990), and were deposited in the extensive Chaco-Pampean plains.

The initiation of late Quaternary eolian deposition has not been extensively dated. Traditionally, the latest episodes of major deposition of the loess and sandy eolian facies have been related to the LGM (Groeber, 1936; Iriondo, 1997; Zárate and Blasi, 1991), and this is supported broadly by magnetostratigraphic data from the classic Mar del Plata section (see Zárate and Fasano, 1989). Recently, Kröhling (1999) reported thermoluminescence (TL) ages of ~36,000 cal. B.P. near the base of loess assigned to the Tezanos Pinto formation and ~9000 cal. B.P. near the top of this unit, in the Carcarañá River basin. These bracketing ages, if correct, broadly support a last glacial age for major loess deposition in this area, although Kröhling (1999) also shows eolian sand deposited prior to the loess deposition, with TL ages going back to at least 52,000 cal. B.P. During the

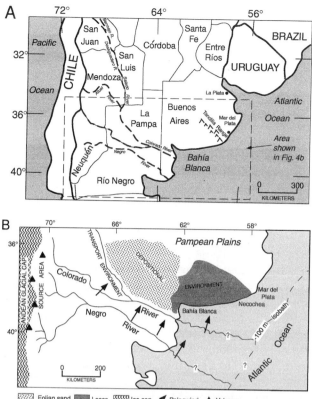

FIGURE 4 (a) Location map of provinces within central Argentina and surrounding countries. (b) Sediment sources and transport directions during the last glacial maximum (LGM) in southern Buenos Aires Province, Argentina. (Redrawn from Zárate and Blasi, 1993.)

LGM, the aggradation of fluvial valleys was mostly related to eolian deposition under dry conditions, with eolian material partially reworked and redeposited by fluvial processes.

The eolian sediments are of volcano-pyroclastic composition, characterized by an abundance of volcanic glass shards and plagioclase, with quartz as a minor component. A relatively coarse grain size is characteristic of these eolian deposits, which are classified as loessial sands, sandy loess, and loess (Zárate and Blasi, 1991, 1993; Bidart, 1996). Mobilization of the eolian grains and particles is attributed to short-term suspension and modified saltation. These mechanisms are related to the occurrence of low-level dust storms, which account for the bulk of the eolian deposits. The source areas of the loessial sands and sandy loess are thought to be the distal segments of the Colorado and Negro Rivers, located ca. 200 km south of the southern Pampas, when sea level was about 120 m below present levels at the LGM (Fig. 4b). Consequently, the eolian deposits of the southern Buenos Aires Province can be regarded as proximal facies and ascribed to a medium-distance transport of dust. Only a minor amount of sed-

iment (estimated to be less than 10%) was transported in high suspension during volcanic ashfalls (Zárate and Blasi, 1991).

In the southern Pampas, no evidence has been found to support the physical weathering suggested by Iriondo (1990a) as the main mechanism for particle formation. The dominant explosive vulcanism of the Andes seems to have played an important role in particle formation, generating extensive pyroclastic deposits along the Andean piedmont and northern Patagonia, which were easily reworked by glacial, fluvial, and eolian processes (Zárate and Blasi, 1993). This interpretation is supported by recent geochemical and isotopic studies, which confirm the importance of volcanic-source sediments for loess in this region (Gallet et al., 1998).

The southern Pampas facies of fine sand sheets grade into dune fields toward La Pampa (Gardenal, 1986; Salazar Lea Plaza, 1980), southern San Luis (Ramonell et al., 1992), northwest of Buenos Aires, and southern Córdoba (Giménez, 1990; Iriondo, 1990a). According to Iriondo (1997, 1999), this central region of Argentina is characterized by a sand sea that covers an area of 180,000 km^2. Hence, the loess distribution is much more restricted than what was mapped by Teruggi (1957) and slightly modified by Sayago (1997) (compare Figs. 3a and 3b).

The eolian facies of the southern Pampas have been deposited by westerly and southwesterly winds. Geomorphological analysis revealed the occurrence of southwest-to-northeast- and west-to-east-oriented eolian landforms, partially superposed in the western Buenos Aires Province (Gardenal, 1986). The inferred south-southwesterly and westerly paleowinds have been related to high-pressure cells situated on the ice fields of the Patagonian Andes (Groeber, 1936; Tricart, 1973; Iriondo, 1997). The subhumid-semiarid climate of the late glacial inferred from pollen analysis in this region could have been due to a more northerly location of the subtropical circulation or a stronger Atlantic anticyclone. An asymmetry with the Pacific anticyclone would imply a meridional circulation predominantly from the southwest (Prieto, 1996).

In the northern Pampas (northern Buenos Aires, Santa Fe, and northern Córdoba; see Fig. 4a), silty eolian facies are dominant. In these areas, Iriondo (1997) located a loess belt that rims the sandy facies of central Argentina (Fig. 3b), with TL ages spanning the late glacial period (Kröhling and Iriondo, 1998; Kröhling, 1999), as was discussed earlier. This distribution also suggests a general trend of decreasing grain size northward and northeastward.

The rate of eolian deposition decreased remarkably between 11,000 and 10,000 B.P. and was followed by an interval of pedogenesis under more humid conditions. Instead, Iriondo (1999) and Tonni et al. (1999) suggested that cold and dry conditions extended until ca. 8500 B.P. This humid period gave rise to most of the present cultivated soils in the southern Pampas (Zárate and Flegenheimer, 1991). The change to more humid conditions could be related to a poleward shift of the Atlantic Convergence Zone, with a predominant circulation from the northeast, or alternatively to a weakening of the southern Atlantic anticyclone (Prieto, 1996).

12.2.2. Mid- and Late Holocene

The lack of chronological control does not permit a detailed reconstruction for the mid- and late Holocene, but a broad approximation of the environmental conditions can be made. The eolian activity, which was very much reduced or negligible during the early and mid-Holocene, increased sometime between 5000 and 4000 ^{14}C B.P. and continued during the late Holocene period. In the southern Pampas, after 5000 B.P., soils were partially truncated at some locations, whereas small ponds and swampy areas of floodplains dessicated and were somewhat modified by soil formation (Zárate et al., in press). Sandy loess accumulation started at ca. 5200 B.P. in Bahía Blanca (González and Weiler, 1987/1988) and at ca. 4600 B.P. in the Tandilia Range (Fig. 4a), where localized eolian activity mostly reworked and redeposited late Pleistocene sandy loess (Zárate and Flegenheimer, 1991); this eolian activity continued for an unknown span of time. In a fluvial setting, loessial sand up to 2.5 m thick with an A/C soil horizonation gave a TL age of ~4300 cal. years B.P. in its lower section (Prieto, 1996). At another locality (also in a fluvial setting), 1-m thick sandy loess with a cumulic A horizon is bracketed between alluvial deposits (which have been modified by pedogenesis) dated at ca. 2500 B.P. and modern alluvium (Zárate et al., in press). A more recent and localized episode of eolian redeposition in the Tandilia Range occurred after the Spanish arrival in the sixteenth century (Zárate and Flegenheimer, 1991).

In the northeastern Pampas of Buenos Aires, several indicators of much drier conditions during the late Holocene also have been reported. Deflation basins were reexcavated and lunettes formed, whereas soils were truncated and then covered by a thin blanket of loess, which in turn was modified by pedogenesis (Tricart, 1973). According to Iriondo and García (1993), further northwest, in Santa Fe (Fig. 4a), a thin eolian layer was deposited regionally during a dry interval between 3500 and 1000 B.P. Recently, a basal TL age of 2050 ± 100 years B.P. was reported by Kröhling and Iriondo (1998). Although soil truncation certainly oc-

curred at some localities, the ubiquity and regional impact of soil erosion processes currently assumed by some investigators pose critical questions on the intensity and magnitude of the hypothesized dry interval, the vegetation changes associated with it, and the processes involved in the transport and deposition of the eroded soil material. These assumptions are untested by the available information.

Several dune fields are assumed to have formed during the late Holocene in southern San Luis, Córdoba, Buenos Aires (Fig. 4a), and along the distal eastern piedmont of the Andes (Ramonell et al., 1992; Iriondo, 1997, 1999). In northwestern Buenos Aires, included in the so-called sandy Pampa (Pampa arenosa) are longitudinal dunes over 100 km long, 2–5 km wide, and up to 6 m high. These dunes are arranged as extensive southwest-to-northeast archlike forms and are presently stabilized by vegetation; Iriondo (1999) considers them to be of late Pleistocene age. However, Iriondo (1999) also maps smaller dunes, both longitudinal and parabolic, that he considers to be of late Holocene age. A field of parabolic dunes, most of them also stabilized and oriented in a southwest-to-northeast direction, extends south of the longitudinal dunes. The parabolic dunes have developed from the reworking of preexisting sandy deposits. Their dimensions are variable, between 0.2 and 6 km long and from 0.2–3 km wide (Cabral et al., 1988; Giménez, 1990). Soils in both the longitudinal and the parabolic dunes show weak development (Entisols) with A/C horizon sequences, including buried A horizons that indicate previous intervals of stability (Hurtado and Giménez, 1988).

The late Holocene eolian reactivation was apparently triggered by a climatic shift to drier conditions at about 5000 B.P. that may have lasted until ca. 1000 B.P. (Tonni 1992) when subhumid-semiarid conditions returned (Prieto, 1996). This climatic change is interpreted as a northward shift of the anticyclonic centers (Prieto 1996). The shift may have been seasonal, equatorward during the winter and poleward during the summer (Markgraf, 1993a). A seasonal anticyclonic anomaly that covered most of the region and generated a warm and semiarid climate is proposed by Iriondo (1990b, 1997). His estimated paleowind trajectories are based on an assumption that eolian deposits in the region are synchronous. In the area of southern Córdoba, geomorphological studies do not support the inferred paleowind trajectories from this anticyclonic anomaly, suggesting northeasterly winds for supposedly late Holocene eolian deposits (Cantú and Degiovanni, 1984). Little evidence is currently available to support the notion of synchronous, regionwide dune formation. It is likely, as the few available ages suggest, that several pulses of eolian reactivation may have occurred during the mid- and late Holocene that did not necessarily affect the heterogeneous Pampean region in the same manner. Factors such as geomorphic setting, vegetation, and soils may have exerted local controls, resulting in responses of different magnitudes. For the most recent dry interval after the fifteenth century, Markgraf (1993a) proposed a transitional or intermediate stage of the circulation patterns that remained between the extreme positions.

The distribution, age, and thickness of the mid- and late Holocene eolian deposits show that the Pampean region may have presented a diverse sensitivity to climate change, with much more intense eolian activity at the southern and western margins. These areas are currently characterized by transitional climatic conditions toward the semiarid/subhumid environments of central Argentina, where sandy eolian deposits were remobilized and dunes formed or were reactivated.

12.3 EOLIAN RECORDS OF THE CHACO REGION

For the Chaco region (Fig. 3), Iriondo (1997) has also proposed a model of eolian deposition. In his conception, glacial and periglacial processes occurred during the LGM in the upper basins of the Parapetí, Pilcomayo, and Bermejo Rivers, which drain the Bolivian Andes. Northern winds were the transport agents of these materials, giving rise to the formation of parabolic dunes in the Bolivia–Paraguay low plain and loess deposits along the sub-Andean ranges and the Argentine Chaco plain. The mineral composition of these sediments is markedly different from that of eolian deposits of the Pampas; quartz is dominant (60–80%) with lesser amounts of hornblende and altered plagioclase. The heavy mineral assemblage is composed of more than 50% magnetite, zircon, tourmaline, and apatite; volcanic glass is completely absent. Only one ^{14}C age of 16,900 ± 270 years B.P. is available from the whole section of the type locality. According to Iriondo (1997), the large dune fields in southeastern Bolivia and northwestern Paraguay, consisting of well-preserved parabolic dunes, were formed by northern winds during the later Holocene. Clapperton (1993) also suggested that the southern Altiplano of Bolivia could have been a potential source area for the eolian sediments of the Chaco region. The subtropical jet stream may have entrained the fine particles, which were then carried eastward where they were eventually reworked by the strong seasonal northerly winds.

A different mode of origin for sediments of the Chaco region was postulated by Sayago (1995). He suggested that the *neotropical loess* of northwestern Ar-

gentina was transported by southerly winds during cool and dry periods due to a northward shift of the Polar Front and a weakening of the midlatitude South Pacific anticyclone. The significant differences in mineral composition between the Pampean loess and eolian deposits of the Chaco region might be attributed to transport processes, local lithological influences, or postdepositional processes driven by the regional climate (Sayago, 1995, p. 758).

12.4. LATE PLEISTOCENE EOLIAN RECORDS OF NORTHERN SOUTH AMERICA

Although fewer studies of the late Quaternary eolian record have been conducted in northern South America, one spectacular example of a large, stabilized dune field is discussed here. In the Llanos del Orinoco of Venezuela and extending into Colombia, Roa (1980) mapped a large dune field of geomorphically well-expressed linear and parabolic dunes (Fig. 5). A minimum limiting radiocarbon age of ~11,000 years B.P. was obtained by Roa (1979) from a paleosol capping one of these dunes, which in turn is overlain by younger eolian sand. Iriondo (1997) also described a finer grained loess that occurs downwind of the dune field. Roa (1979), Iriondo (1997), and Clapperton (1993) interpreted this sand to represent deposition during a time when there was a migration of the Intertropical Convergence Zone (ITCZ) to the south during the last (?) glacial period, with the result that the Llanos del Orinoco region became drier and favored dune formation. If this interpretation is correct, it is consistent with the paleontological, geochemical, and mineralogical record of relatively dry conditions in this part of Venezuela during the late glacial period (~13,000–

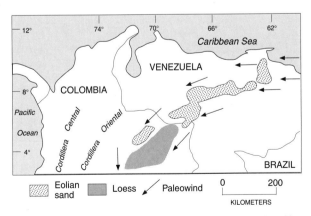

FIGURE 5 Map showing the distribution of eolian sand and loess (silt) and paleowind directions in Venezuela and Colombia. (Redrawn from Iriondo, 1997.)

10,000 B.P.) as is seen in the sediments of Lake Valencia (Bradbury et al. 1981).

12.5. EOLIAN RECORDS IN THE DESERTS OF THE SOUTHWESTERN UNITED STATES AND NORTHERN MÉXICO

Dune fields are widespread over the desert regions of the southwestern United States and northern México, although these features are generally smaller than those found in the Great Plains to the east (Fig. 6). The largest dune fields, mostly stabilized, are found on the Colorado Plateau in northeastern Arizona and northwestern New Mexico (Hack, 1941; Wells, et al. 1990). Elsewhere in this mostly arid region, some dune fields are active, such as the Gran Desierto in México, most of the Algodones and parts of the Kelso dunes in California, and White Sands in New Mexico.

Detailed studies on the Colorado Plateau show that these dune fields have a long and complex history. Wells et al. (1990) conducted stratigraphic and soil geomorphic studies in the Chaco dune field of northwestern New Mexico and identified three main periods of eolian sand activity. Radiocarbon ages indicate that while the oldest eolian sand unit in the area is of late Pleistocene age (~16,000–12,000 ^{14}C years B.P.), the other two units are both of late Holocene age, ca. 5600–2800 ^{14}C years B.P. and less than ~1900 ^{14}C years B.P. These ages agree with late Holocene luminescence ages of dunes elsewhere on the Colorado Plateau in northeastern Arizona (Stokes and Breed, 1993). Results of both studies indicate that, in contrast to earlier concepts, the mid-Holocene *Altithermal* warm period was

not a particularly important period for eolian sand activity. Wells et al. (1990) point out that late Holocene eolian sand is volumetrically far more important than mid-Holocene eolian sand.

Although they are of smaller extent than elsewhere in North America, the dunes of the Mojave and Sonoran Desert regions of southeastern California, Arizona, and northern México have received the most attention for studies of geomorphology, sediment transport, and origins (Lancaster et al., 1987; Tchakerian, 1991; Lancaster, 1994; Zimbelman et al., 1995; Muhs et al., 1995; Lancaster and Tchakerian, 1996), as well as geochronology (Clarke, 1994; Wintle et al., 1994; Rendell et al., 1994; Clarke et al., 1996; Rendell and Sheffer, 1996; Clarke and Rendell, 1998). Despite the evidence that Great Plains dunes seem to record relatively arid periods (see discussion later), a general concept that is emerging from studies in the Mojave and Sonoran Deserts is that the degree of activity of many dunes is only indirectly a function of climate. Many of the dune fields in southeastern California seem to be supply-limited systems, and therefore, the degree of dune activity may not be related to overall moisture balance per se, but rather to the effect climate has on supplying new sediment from fluvial or lacustrine sources (Lancaster, 1994; Lancaster and Tchakerian, 1996; Muhs et al., 1995; Clarke and Rendell, 1998). In an overall summary of luminescence dating of Mojave Desert dunes and sand sheets, Clarke and Rendell (1998) showed that ages range from as old as ~23,000 cal. years B.P. to the late Holocene, and are similar to ages of nearby pluvial lake high stands, whether of late Pleistocene or Holocene age. They suggested that the main effect of climate change on dune activity is to bring about conditions whereby new supplies of sediment become available to dune fields, such as from pluvial lake beaches or alluvial sources. An arid climatic regime with a minimal vegetation cover almost certainly plays some role in degree of dune activity, however. For example, Muhs et al. (1995) showed that while sediments of the Algodones dunes had a pluvial lake source in the past, the region is sufficiently arid now that the dunes are active, despite no new supplies of pluvial lake beach sand.

12.6. EOLIAN RECORDS IN THE NORTH AMERICAN MIDCONTINENT

12.6.1. Full Glacial and Late Glacial Times: Loess

During the LGM, loess was deposited over much of the North American midcontinent, particularly in the central United States (Fig. 7). Little loess was deposited in Canada during the last glacial period because the

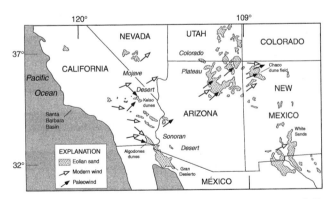

FIGURE 6 Map showing the distribution of eolian sand (stippled) in desert regions of northern México and the southwestern United States, modern sand transport directions, and late Holocene paleowinds from stabilized dunes. Data compiled mainly from Hack (1941), Jennings (1967), Hunt (1977), Lancaster et al. (1987), Wells et al. (1990), Lancaster (1994), Muhs and Holliday (1995), and Muhs et al. (1995); modern wind data are compiled by the authors.

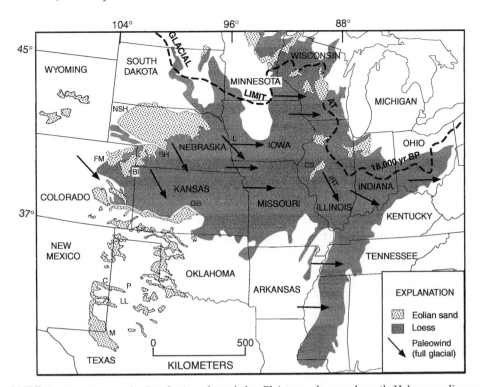

FIGURE 7 Map showing the distribution of mostly late Pleistocene loess and mostly Holocene eolian sand in the midcontinent region of North America and inferred paleowinds during the last glacial maximum (LGM). Loess distribution data from Thorp and Smith (1952), eolian sand distribution data from Muhs and Holliday (1995), and inferred LGM paleowinds data from Muhs and Bettis (2000). NSH, Nebraska Sand Hills localities (Gudmundsen Ranch and Swan Lake, see Fig. 10; Whitman and Snake River, see Fig. 13); FM, Fort Morgan dune field localities (Logan County, see Fig. 10; Hudson and Coors Quarry, see Fig. 11; Friehaufs Hill and Hillrose, see Fig. 13); GB, Great Bend Sand Prairie localities (Cullison Quarry and Reno County, see Fig. 13); C, Clovis, NM (see Fig. 10); M, Midland, TX (see Fig. 10); BI, Beecher Island, CO (see Fig. 8); BH, Bignell Hill, NE (see Fig. 8); L, Loveland, IA (see Fig. 8); CS, Cottonwood School, IL (see Fig. 8); IRT, Illinois River transect (see Fig. 9).

Laurentide ice sheet covered most of the region during the glacial maximum and extensive proglacial lakes covered the region and served as sediment sinks as ice receded. Eolian silt dating to the LGM in the United States is referred to as Peoria Loess. Peoria Loess is as thick as ~50 m in western Nebraska (Maat and Johnson, 1996) and as thick as ~40 m in western Iowa (Muhs and Bettis, 2000). From western Iowa to the east, loess is thickest adjacent to major valleys, such as those of the Missouri, Mississippi, Illinois, and Wabash Rivers.

Numerous studies show that while the bulk of late Quaternary loess in midcontinental North America dates broadly to the LGM, there are important differences in the timing of deposition within the region. Maximum limiting radiocarbon ages of Peoria Loess range from ~30,000 to ~20,000 years, and minimum limiting ages range from ~14,000 to ~10,000 years (Fig. 8). Localities near major loess sources record deposition earlier than those farther away. For example, in west-

ern Tennessee, immediately east of the Mississippi River, maximum limiting ages for Peoria Loess are as old as ~29,000 ^{14}C years B.P., and an age of 21,800 ^{14}C years B.P. has been reported in the lower part of the loess (Markwich et al., 1998). Thermoluminescence and [10]Be ages of Peoria Loess from this locality and elsewhere in Tennessee agree with these ages (Markewich et al., 1998; Rodbell et al., 1997). In much of Illinois, Peoria Loess deposition probably began sometime after ca. 25,000 [14]C B.P. (Curry and Follmer, 1992) and was still in progress at ca. 12,000 [14]C B.P. (Grimley et al., 1998). In central Illinois, however, maximum limiting radiocarbon ages for Peoria Loess decrease from ~25,000 to ~21,000 ^{14}C years B.P. along a 20-km long traverse away from the source (Kleiss and Fehrenbacher, 1973). Similarly, in Iowa, the maximum period of Peoria Loess deposition was from ca. 24,000 to 14,000 [14]C B.P. (Ruhe, 1983). Ruhe (1983) points out, however, that initiation of Peoria Loess deposition in Iowa, like Illinois, was

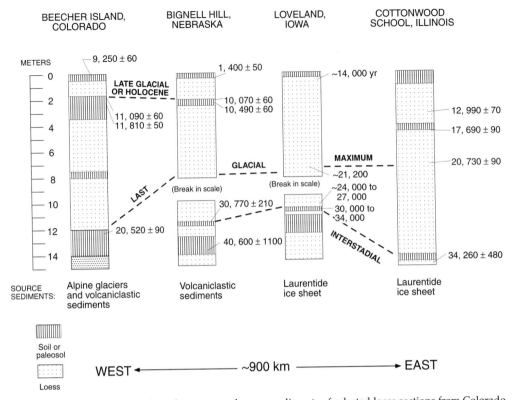

FIGURE 8 Stratigraphy, radiocarbon ages, and source sediments of selected loess sections from Colorado to Illinois in the North American midcontinent (locations of sections given in Fig. 7). Data from Colorado and Nebraska from Muhs et al. (1999), Iowa data from Ruhe (1983) and Forman et al. (1992a), and Illinois data from Grimley et al. (1998).

time transgressive; maximum limiting radiocarbon ages are ~24,500 ^{14}C years B.P. near the Missouri River source, whereas they are ~19,000 ^{14}C years B.P. ~280 km to the east of the river.

Loess deposition in Tennessee, Illinois, Iowa, and other areas east of the Missouri River was a function of source sediment availability from the Laurentide ice sheet via the Ohio, Wabash, Illinois, Mississippi, and Missouri Rivers, and the timing of loess deposition probably followed the history of movement of the ice sheet. However, in the Great Plains region of Nebraska, Kansas, and Colorado, west of the area that drained outwash from the Laurentide ice sheet, extensive loess sheets were derived from Rocky Mountain glaciers and, more importantly, Tertiary volcaniclastic sediments (Aleinkoff et al., 1998, 1999). Interestingly, in some places, the timing of loess deposition seems to differ little from that of Laurentide outwash-derived loess found farther to the east (May and Holen, 1993; Martin, 1993; Pye et al., 1995; Maat and Johnson, 1996; Muhs et al., 1999). For example, in eastern Colorado, loess deposition began after ca. 20,000 [14]C B.P., but ended by ca. 11,800 [14]C B.P. (Fig. 8). Elsewhere, however,

loess deposition differs significantly from that found farther east. In western Nebraska, Peoria Loess was deposited sometime after ca. 30,000 [14]C B.P., but continued until sometime just before ca. 10,500 [14]C B.P. (Maat and Johnson, 1996; Muhs et al., 1999), indicating a much longer period of eolian sedimentation.

Loess deposition was not continuous in midcontinental North America during the LGM. In western Illinois, multiple, poorly developed buried soils are found within Peoria Loess, indicating brief periods of landscape stability with little or no loess deposition (Wang et al., 1998). Radiocarbon ages indicate as many as five cycles of loess sedimentation and soil formation in the period from 20,710 to 17,630 [14]C B.P. Ruhe et al. (1971) also report multiple buried soils in Peoria Loess in western Iowa, and at least one possible buried soil has been identified in Peoria Loess in eastern Colorado (Muhs et al., 1999).

Multiple lines of evidence indicate that loess sources changed during the LGM over midcontinental North America, even within the area draining the Laurentide ice sheet. The most detailed work has been done in Illinois, where Peoria Loess shows distinctive zonations, based on clay mineralogy (Frye et al., 1968), dolomite content (McKay, 1979), and magnetic susceptibility (Grimley et al., 1998). The compositional changes of the loess in this area are clearly linked to different lobes of the Laurentide ice sheet that provided outwash to the Mississippi and Illinois River systems (Grimley et al., 1998). In western Iowa, Muhs and Bettis (2000) found three zones within Peoria Loess: (1) a lower zone characterized by low carbonate, probably related to syndepositonal leaching; (2) a middle zone, probably derived dominantly from the Missouri River; and (3) an upper zone that could be derived from both the Missouri River and sources in Nebraska farther west. In eastern Colorado, Aleinikoff et al. (2000) used Pb isotopes in K-feldspars to show that Peoria Loess was derived from two competing sources: outwash derived from Front Range glaciers and volcaniclastic Tertiary siltstone. At least two cycles of alternating dominance by these competing sources are evident in the period of Peoria Loess deposition.

Loess distribution, as well as thickness, particle size, carbonate, and geochemical trends, indicates that most loess in North America was deposited by northwesterly winds (Figs. 7 and 9). There were apparently some episodes of easterly winds, because areally limited tracts of loess occur to the west of some source valleys (see Muhs and Bettis, 2000, for a review of these localities). Nevertheless, such areas are of minor extent, and the picture that emerges is one of dominantly northwesterly winds.

12.6.2. Last Glacial Period: Eolian Sand

Eolian sand deposits are extensive in North America, but until recently were much less studied than loess deposits. The most widespread eolian sands in North America are found in the central and southern Great Plains region, from Nebraska south to Texas (Fig. 7). There are few active dunes in this region at present. However, stabilized dunes and sand sheets cover tens of thousands of square kilometers. Recent studies show that the most critical factor in the amount of eolian activity in this region—assuming that adequate winds and sand supplies are available—is the degree of vegetation cover, which in turn is a function of the balance between precipitation and evapotranspiration (Muhs and Maat 1993; Muhs and Holliday 1995; Wolfe 1997; Muhs and Wolfe, 1999).

Limited radiocarbon ages, archaeology, and degree of soil development suggest that eolian sand sheet, lunette, and dune deposition took place during the last glacial period in Nebraska, Colorado, New Mexico, and Texas (Fig. 10). Radiocarbon results from the Nebraska Sand Hills indicate that there was sufficient

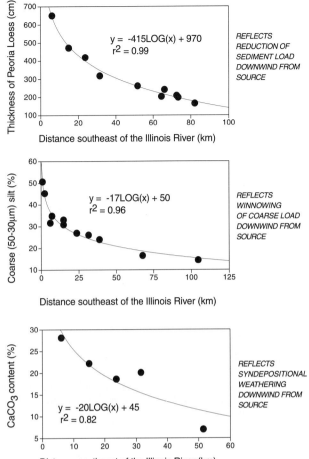

FIGURE 9 Thickness, coarse silt content, and carbonate content of Peoria (last glacial) loess as a function of distance southeast of the Illinois River (location of transect given in Fig. 7). (Data from Smith, 1942.)

dune building to dam sizable rivers and create lakes during late glacial time (Loope et al., 1995; Mason et al., 1997). These data are supported by an earlier report of maximum limiting ages of ~13,000 years B.P. for some of the largest barchanoid ridges (see Swinehart, 1990) in the Nebraska Sand Hills (Swinehart and Diffendal, 1990). In Colorado, stratigraphic, radiocarbon, and soil evidence indicate that eolian sheet sands were deposited over large areas during the last glacial period (Forman and Maat, 1990; Forman et al., 1992a, b, 1995; Madole, 1995; Muhs et al., 1996). Stratigraphic studies show that eolian sheet sands in eastern Colorado have maximum limiting radiocarbon ages of ~27,000 ^{14}C years B.P. and, where exposed at the surface, have a degree of soil development similar to that in river terrace sands dated to ca. 11,000–10,000 [14]C B.P., suggesting that sand sheet deposition took place between ca. 27,000 and 10,000 [14]C B.P. (Muhs et al., 1996). In Texas,

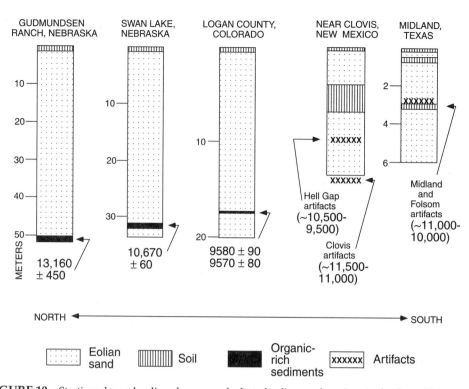

FIGURE 10 Stratigraphy and radiocarbon ages of selected eolian sand sections in the Great Plains region of North America that show evidence of possible late glacial eolian activity. Nebraska data from Loope et al. (1995), Colorado data are from Muhs et al. (1996), and New Mexico and Texas data from Holliday (1997b). Locations are shown in Fig. 7.

there was widespread lunette formation during full glacial time (Holliday, 1997a), and extensive eolian sand sheet deposition is dated to ca. 11,000–10,000 [14]C B.P. (Holliday 1997b, 2000). Extensive eolian sheet sands in Texas and adjacent parts of New Mexico are confidently correlated to the late glacial period because of their association with well-dated artifacts (Fig. 10). Forman et al. (1995) stated that eolian activity in the midcontinent ceased between ca. 12,000 and 9000 [14]C B.P., based on examination of a few sections in a limited area of northeastern Colorado. However, the studies from Nebraska (Loope et al., 1995; Mason et al., 1997), Texas, and New Mexico (Holliday, 2000) indicate that the late glacial period was an important time of eolian sand deposition in the midcontinent.

Because few dunes (as opposed to sand sheets) can be confidently linked to the last glacial period in the Great Plains, it is difficult to reconstruct paleowinds. However, distribution of last glacial eolian sands relative to probable source sediments indicates that winds were probably northwesterly on the central Great Plains and westerly on the southern Great Plains during last glacial time (Loope et al., 1995; Mason et al., 1997; Muhs et al., 1996; Holliday, 1995a, 1997a), similar to the present.

12.6.3. Mid-Holocene: Loess and Sand

Mid-Holocene eolian sediments have been reported in some parts of midcontinental North America. A thin (1–4 m) eolian silt called the Bignell Loess is found in a patchy distribution west of the Missouri River in the Great Plains region of Nebraska, Kansas, and Colorado (Johnson and Willey, 2000; Mason and Kuzila, 2000). Radiocarbon ages indicate that it was deposited between ca. 11,000 and 9000 B.P. at some localities and possibly during the mid-Holocene, between ca. 10,000 and 1500 B.P., at others (Fig. 8). Direct dating of Bignell Loess using TL methods gives ages ranging from ~9000 to ~3000 cal. years B.P. (Pye et al., 1995; Maat and Johnson, 1996). A recent study by Olson et al. (1997) indicates that a post-mid-Holocene loess covers extensive parts of southwestern Kansas and may be younger than the Bignell Loess found farther north. Holocene loess has also been described in Saskatchewan (David, 1970; Vreeken, 1996) and, as with Holocene loess farther south, has a patchy distribution.

Although there is little direct geomorphic evidence for it, some stratigraphic evidence suggests mid-Holocene eolian sand activity in midcontinental North

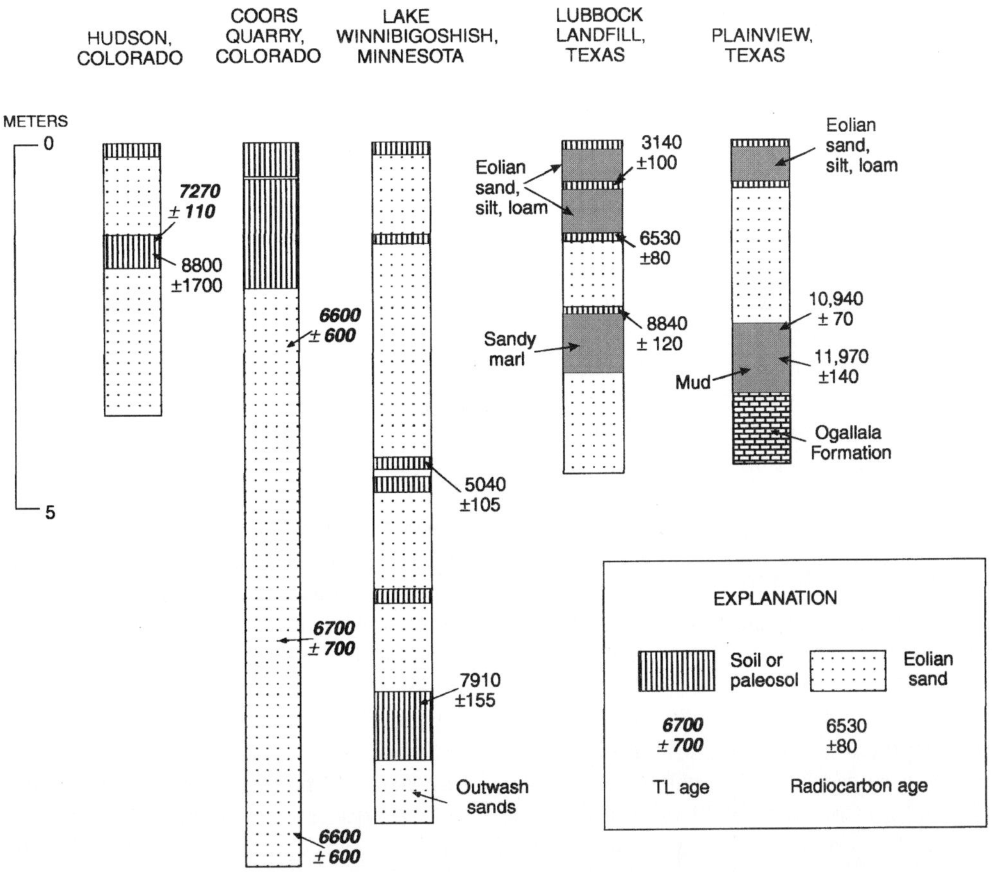

FIGURE 11 Stratigraphy and ages of selected eolian sand sections in Colorado, Minnesota, and Texas that show evidence of mid-Holocene eolian activity. Colorado data are from Forman and Maat (1990) and Forman et al. (1995), Minnesota data are from Grigal et al. (1976), Texas data are from Holliday (1995b). Locations are shown in Fig. 7.

America (Fig. 11). Radiocarbon ages and stratigraphic data from the western Nebraska Sand Hills indicate that mid-Holocene eolian activity probably created a large dune dam that blocked a major drainage in the area, as was the case during the late glacial period (Loope et al., 1995; Mason et al., 1997; Muhs et al., 2000). In eastern Colorado, there are a few closely spaced localities where evidence of mid-Holocene eolian sand movement has been documented (Forman and Maat, 1990; Forman et al., 1995). Mid-Holocene dune activity took place in parts of the intermountain basins of Wyoming to the west of the Great Plains (Gaylord, 1990). Holliday (1995b, 1997a,b) reported radiocarbon and stratigraphic evidence of mid-Holocene lunette formation in the southern Great Plains, and dry valleys or *draws* in this region contain thick eolian sands that are of mid-Holocene age. However, the upland dunes in this region, mostly parabolic forms, do not date from the mid-Holocene and are largely of late Holocene age (Holliday, 1995c). Preservation of mid-Holocene eolian

sand in draws and not on uplands may be a function of the more protected sedimentary setting in the draws, where late Holocene reworking of sediment would be less likely. Elsewhere, the record for mid-Holocene eolian sand deposition is scanty and is based on indirect lines of evidence such as dunes with a degree of soil development that is greater than that on late Holocene eolian sand, but less than that on late Pleistocene sand (Madole, 1995; Muhs et al., 1996). In some parts of the Great Plains, dunes are depleted in carbonate minerals, thought to be the result of physical reduction of sand-sized grains from extended periods of eolian activity in the mid-Holocene period (Muhs et al., 1997b; Arbogast and Muhs, 2000). Although it seems likely, from other proxy paleoclimate evidence (discussed later), that conditions were optimal for eolian sand activity over the semiarid Great Plains during the mid-Holocene extensive late Holocene, eolian activity may have removed much of the record.

Interestingly, subhumid areas around the margins of

the Great Plains may have a better record of mid-Holocene eolian activity, such as that documented for northern Minnesota (Fig. 11) by Grigal et al. (1976). It can be hypothesized that these currently subhumid to humid areas were affected by a mid-Holocene dry period, but, unlike the drier Great Plains to the west, were unaffected by reworking during the late Holocene. Stabilized eolian sand is extensive in parts of Minnesota, Wisconsin, Illinois, and Indiana (Fig. 7) and in smaller areas in Kansas and Michigan. It would be worthwhile to conduct detailed stratigraphic and geochronologic studies of these areas to see if the last period of eolian activity dates to the mid-Holocene. Some of the best records of mid-Holocene eolian activity may be found in lake sediments and in areas marginal to the Great Plains (Dean et al., 1996).

12.6.4. Late Holocene

In the past, a number of investigators assumed that dunes in the Great Plains were last active during full glacial time, at ca. 22,000–16,000 ^{14}C B.P. (Watts and Wright, 1966; Wright, 1970; Warren, 1976; Sarnthein, 1978; Wells, 1983; Kutzbach and Wright, 1985), or at least no later than during what is perceived to be a warm and dry mid-Holocene, at ca. 8000–5000 B.P. Recent studies have confirmed (as discussed earlier) that the last glacial period was indeed a time of eolian sand activity over much of the region. However, a number of recent studies also show that much eolian sand in the Great Plains region is of late Holocene age. Indeed, in the Great Plains of both Canada and the United States, the areal extent of late Holocene eolian sand is much greater than that of last glacial sand and is a more important part of the record than previously thought. For example, dunes of the Nebraska Sand Hills cover ~50,000 km^2, yet all have only simple A/AC/C soil profiles (Entisols), similar to the young dune fields of the Pampas in Argentina discussed earlier. This observation, along with late Holocene radiocarbon ages at numerous localities, suggests that most or all of the Nebraska Sand Hills region has been active at some time in the late Holocene, though not all parts of the dune field necessarily were active simultaneously (Muhs et al., 1997a).

An indirect piece of evidence supporting late Holocene ages for eolian sand comprises new data that show that sand and loess in the Great Plains are geochemically and mineralogically distinct, indicating different source sediments. This is a departure from the long-held concept, articulated by Lugn (1935, 1939, 1962, 1968), that eolian sand and loess in the Great Plains are essentially the same deposit, with eolian sand a coarse-grained, proximal facies and loess a fine-grained, distal facies. In both eastern Colorado and Nebraska, loess is highly calcareous, whereas eolian sand is not. This mineralogical difference is reflected in the higher Ca concentrations in loess compared to eolian sand; Ca/Sr values are also significantly different between the two sediment types (Fig. 12). Eolian sand has concentrations of both Ti and Zr that are lower than in loess that occurs to the south and east of the dune fields. Furthermore, ratios of Ti to Zr in eolian sand are lower than these values in loess in both areas (Fig. 12). These compositional differences, which reflect both carbonate mineral and heavy mineral assemblages, cannot be explained by differences in transport processes alone, and they suggest that eolian sand and loess have different sources. Muhs et al. (1996) presented evidence to show that eolian sand in Colorado is derived mainly from South Platte River sediments. However, loess in Colorado is derived partly from South Platte River sediments, but also from Cretaceous and Tertiary bedrock sources (Aleinikoff et al., 1998, 1999). Thus, because different source sediments supply dune fields and loess, eolian sand and loess entrainment need not be synchronous in time.

Stratigraphic, soil geomorphic, radiocarbon, and luminescence methods demonstrate that eolian sand over a wide range of midcontinental North America has been active in the past 3000 years (Ahlbrandt et al., 1983; Muhs, 1985; Swinehart and Diffendal, 1990; Madole, 1994, 1995; Holliday, 1995a,c, 1997a,b; Forman et al., 1995; Loope et al., 1995; Arbogast, 1996; Muhs and Holliday, 1995; Muhs et al., 1996, 1997a,b; Wolfe et al., 1995, 2000; Stokes and Swinehart, 1997). In addition, most of these studies have stratigraphic data indicating multiple periods of eolian activity in the late Holocene (Fig. 13). The number of radiocarbon ages and their analytical uncertainties do not yet make it possible to test the hypothesis of regional synchroneity of activity. However, these observations indicate that, contrary to earlier beliefs, eolian sand in this region can be active under an essentially modern climatic regime. In fact, observations by explorers in the nineteenth century indicate that many parts of the Great Plains had active dune sand where it is now stable (Muhs and Holliday, 1995; Muhs and Wolfe, 1999). The evidence for multiple periods of activity and stability in the past few thousand years indicates that Great Plains eolian sand is quite sensitive to small changes in the overall moisture balance and degree of vegetation cover. Most of the late Holocene dunes in the midcontinent of North America are parabolic forms, which are excellent paleowind indicators. Orientations of late Holocene dunes indicate northwesterly winds in the central Great Plains and southwesterly winds in the southern Great Plains, similar to modern wind regimes (Fig. 14).

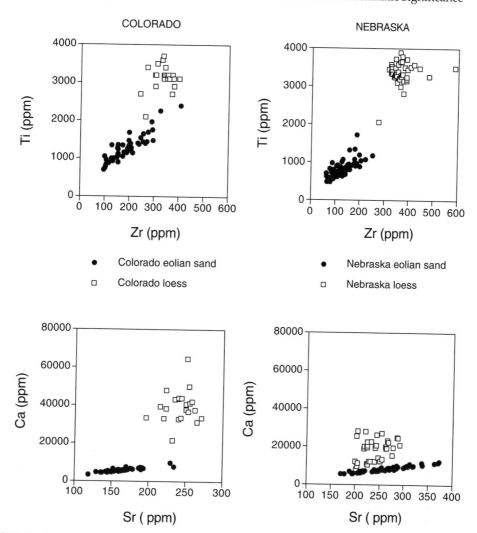

FIGURE 12   Plots of Ca, Sr, Ti, and Zr concentrations in eolian sand and loess in Colorado and Nebraska. Data are from Muhs et al. (1996, 1999) and previously unpublished Nebraska eolian sand data of the authors.

Stable and partially active dune fields, dominated by parabolic dunes, are widespread over the subhumid to semiarid northern Great Plains of southern Canada and the northern United States (Fig. 15). Although winds in the northern Great Plains are among the strongest in the world (compared to other areas where eolian sand is found), most dunes are presently inactive because of relatively high ratios of precipitation to potential evapotranspiration, which has the dual effects of increasing moisture content within dunes and maintaining a vegetation cover (Wolfe, 1997). Dune fields in the region are smaller than those in the central and southern Great Plains, as most are derived from finite supplies of glaciofluvial or glaciolacustrine sediments deposited during the last deglaciation (David, 1971; Wolfe et al., 1995, 2000; Wolfe, 1997; Muhs et al., 1997b; Muhs and

Wolfe, 1999). Although some dune fields in Canada appear to have been generated from glaciofluvial and glaciolacustrine sediments by northeasterly glacial anticyclonic winds as the ice receded (David, 1981, 1988), stratigraphic and geochronologic studies indicate that many northern Great Plains dunes are not relict features from the last deglaciation. The last episodes of eolian activity were during the late Holocene, and several localities show evidence of having been active in the past millennium (Fig. 16), as is the case with the central and southern Great Plains. Although several dune fields in the northern Great Plains of Canada and the northern United States show evidence of multiple episodes of activity in the past few thousand years, as yet, there is too little evidence to ascertain whether there is regional synchroneity of dune sand movement.

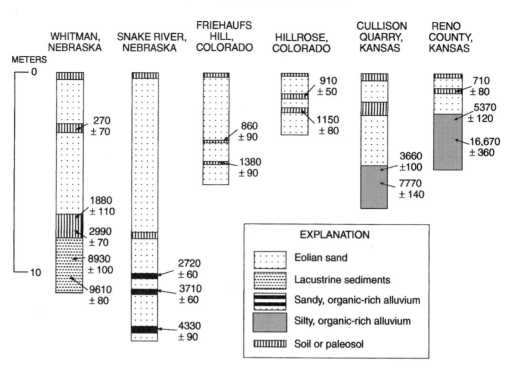

FIGURE 13 Stratigraphy and ages of selected eolian sand sections in Nebraska, Colorado, and Kansas that show evidence of late Holocene eolian activity. Data are from Stokes and Swinehart (1997), Muhs et al. (1997b), Madole (1994), and Arbogast (1996). Locations are shown in Fig. 7.

Historic accounts indicate that at least four dune fields had some degree of activity in the mid-nineteenth century before major European settlement of the region took place (Muhs and Wolfe, 1999). Orientations of stabilized parabolic dunes indicate late Holocene paleowinds from the west or northwest, similar to dune-forming winds of the present (Fig. 15).

12.7. EOLIAN RECORDS IN THE NORTHWESTERN UNITED STATES

Loess is extensive over parts of the northwestern United States, including a small part of western Wyoming, a large part of eastern Idaho, a small part of northern Oregon, and much of eastern Washington (Fig. 17). Loess in this region has not been studied for as long as loess in the midcontinent of North America. However, recent work indicates that a long stratigraphic record is present and that loess origins in parts of the region differ from those found elsewhere.

In Idaho and adjacent parts of westernmost Wyoming, loess covers large areas to the north and south of the Snake River Plain, which is probably one of its major sources (Lewis et al., 1975; Scott, 1982; Pierce et al., 1982; Glenn et al., 1983). Loess in this region has a thick-

ness of up to 12 m in places, but most is 2 m or less (Lewis et al., 1975). Thickness and particle size data (Lewis et al., 1975; Lewis and Fosberg, 1982; Pierce et al., 1982; Glenn et al., 1983) suggest that loess deposition took place under northwesterly or westerly winds (Fig. 17).

Stratigraphic studies of loess in Idaho have been conducted by Scott (1982), Pierce et al. (1982), and Forman et al. (1993). Pierce et al. (1982) studied the loess stratigraphy at numerous localities over an ~400-km long transect from western Wyoming to central Idaho and found two major units, informally designated loess *A* and loess *B*. The two loesses are separated by a well-developed paleosol. Based on the degree of soil development and stratigraphic relations with K/Ar-dated lava flows and well-dated Lake Bonneville flood deposits, they suggested that loess A was deposited during the last glacial period and loess B was deposited during the penultimate glacial period. Forman et al. (1993) studied loess at two closely spaced sites in eastern Idaho and reported TL data that support correlation of loess A with the last glacial period (late Wisconsin time). However, they suggested, on the basis of TL data, that loess B was also deposited during the last glacial period (early Wisconsin time). At present, probably too few data exist to determine which age estimate is correct.

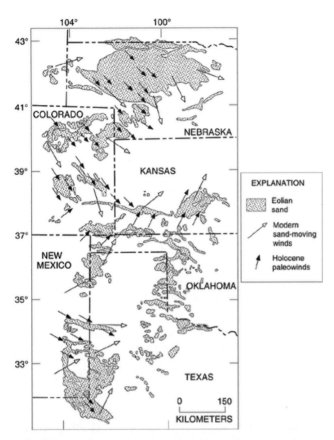

FIGURE 14 Map showing the distribution of eolian sand (stippled) in the central and southern Great Plains of the United States, modern potential drift directions, and inferred late Holocene paleowinds based on dune orientations. Sand distribution data from Muhs and Holliday (1995); drift directions and paleowinds data from Madole (1995), Muhs et al. (1996, 1997a), Arbogast (1996), and unpublished data of D.R. Muhs and V.T. Holliday.

In eastern Washington, detailed studies by A.J. Busacca and colleagues (Busacca, 1991; Busacca et al., 1992; McDonald and Busacca, 1992; Busacca and McDonald, 1994; Berger and Busacca, 1995; Richardson et al., 1997) have added tremendously to our knowledge of loess history in this region. Loess in eastern Washington and adjacent parts of Idaho and Oregon may cover as much as 60,000 km² (McDonald and Busacca, 1992). It is as thick as 75 m (Ringe, 1970), and Busacca (1991) estimates that loess deposition may have begun as early as 2 Ma (million years ago). There are dozens of paleosols within eastern Washington loess, indicating many periods of nondeposition and surface stability between times of loess deposition. The loess itself is thought to be derived primarily from fine-grained slackwater sediments, which in turn are derived from cataclysmic floods of proglacial Lake Missoula (McDonald and Busacca, 1992).

Busacca and McDonald (1994) designated the two youngest loess units in eastern Washington as *L1* (upper) and *L2* (lower). Well-dated Holocene and late glacial tephras from the volcanically active Cascade Range to the west are found within eastern Washington loess and provide valuable time lines for dating and correlation (Busacca et al., 1992; Richardson et al., 1997). In addition, TL ages by Berger and Busacca (1995) and TL and infrared stimulated luminescence (IRSL) ages by Richardson et al. (1997) confirm correlations inferred from tephra data that L1 is of late Wisconsin age and L2 is of early to middle Wisconsin age. Agreement between TL and IRSL ages is good back to ~70,000 years ago, which strengthens the age estimates made by these investigators. However, there are differences between the two studies; TL ages reported by Richardson et al. (1997) for the same sections are generally younger than those reported by Berger and Busacca (1995).

Stratigraphic and geochronologic data show when in a glacial cycle most loess deposition takes place in eastern Washington. Because the loess is thought to be derived primarily from proglacial Lake Missoula flood sediments, most loess deposition probably takes place late in each glacial cycle, and some deposition has continued into the Holocene. Eastern Washington loess becomes thinner and finer grained from southwest to northeast, indicating that late glacial winds were from the southwest (Busacca, 1991; Busacca and McDonald, 1994).

12.8. EOLIAN RECORDS IN ALASKA AND NORTHWESTERN CANADA

Loess is areally the most extensive surficial deposit in Alaska and adjacent parts of the Yukon Territory in Canada and is found over most parts of this region where rugged mountains, such as the Alaska Range and the Brooks Range, are absent (Fig. 18). As with South America and the North American midcontinent, the classical concept has been that most loess deposition in Alaska took place during glacial periods. However, extensive Holocene loess is found in Alaska (Péwé, 1975; Begét, 1990, 1996; Muhs et al., 1997c), and, at localities near Fairbanks (Eva Creek and Fox), Holocene deposits are almost as thick or thicker than the best candidates for deposits of last-glacial age (Fig. 19). Because glaciers still exist in Alaska and northwestern Canada, it is not surprising that abundant glaciogenic silt is available in many of the region's major river systems, and contemporary loess deposition is well documented (Péwé, 1951, 1975; Nickling, 1978).

Eolian sand is also extensive in Alaska (Fig. 18), but

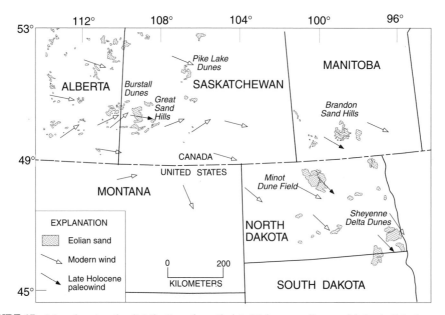

FIGURE 15 Map showing the distribution of mostly late Holocene eolian sand (stippled) in the northern Great Plains of southern Canada and the northern United States, modern sand-moving winds, and late Holocene paleowinds based on dune orientations. Data compiled from Running (1996), Wolfe (1997), Muhs et al. (1997b), and Muhs and Wolfe (1999).

with few exceptions is largely stabilized by boreal forest or tundra vegetation. Geomorphic and stratigraphic evidence suggests that most eolian sand was derived from glaciofluvial sediments and was deposited originally in sand sheets rather than dunes, at least in part because of limited sand supply (Lea and Waythomas, 1990; Lea, 1996). Later dune building is thought to have occurred mainly as a result of reworking of previously

deposited sheet sand. It is not clear at all localities, however, whether most of this dune building took place during latest Pleistocene, early Holocene, or mid-Holocene time (Lea and Waythomas, 1990). Available radiocarbon ages suggest that much eolian sand deposition, whether of sand sheets or dunes, took place in the interval from ca. 18,000 to ca. 10,000 [14]C B.P. However, Holocene eolian sand deposition is also docu-

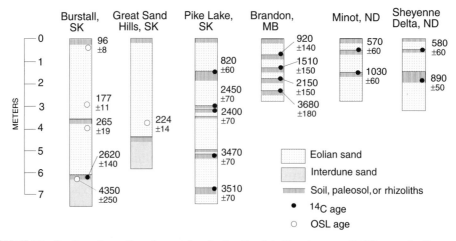

FIGURE 16 Stratigraphy, radiocarbon, and optically stimulated luminesence (OSL) ages of eolian sand in the northern Great Plains of the United States and Canada, showing evidence of multiple episodes of late Holocene eolian activity. Localities are shown in Fig. 15. Data from David (1971), Wolfe et al. (1995), Running (1996), Muhs et al. (1997b), and sources in Muhs and Wolfe (1999).

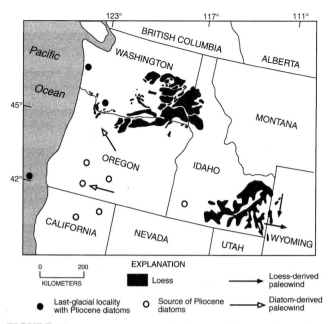

FIGURE 17   Map showing the distribution of loess in the northwestern United States (from Thorp and Smith, 1952), inferred paleowinds from loess and wind flute data (from Lewis et al., 1975; Lewis and Fosberg, 1982; Pierce et al., 1983; Glenn et al., 1983; and Busacca and McDonald, 1994), and inferred paleowinds from offshore and lacustrine diatom data (from Sancetta et al., 1992). Locations of last glacial maximum (LGM) localities where exotic Pliocene diatoms are found are from Sancetta et al. (1992); locations of Pliocene diatom-bearing beds are from Bradbury (1982, 1992).

mented in both the interior of Alaska and on the Arctic Coastal Plain (Fig. 20). Distribution of eolian sand relative to probable source sediments and, in some places, orientations of dunes indicate northeasterly paleowinds in most parts of Alaska (Fig. 18).

12.9. LONG-DISTANCE TRANSPORT OF FINE-GRAINED AIRBORNE DUST

A third category of eolian sediment (in addition to sand and loess) is fine-grained aerosolic dust that is capable of traveling long (intercontinental) distances (Prospero, 1981; Pye, 1987; Rea, 1994). The particle size range of this dust overlaps the finest particle size range of loess, but has mean values less than ca. 5 μm. Fine-grained dust is of considerable interest to climate modelers because studies have demonstrated that it can play a role in climate change, both through changes in the Earth's radiative balance and as a nutrient for primary productivity in the oceans (Harvey, 1988; Joussaume, 1993; Li et al., 1996; Tegan et al., 1996; Overpeck et al., 1996; Andersen et al., 1998; Mahowald et al., 1999). Where they can be found, geologic records of far-

traveled dust are also important paleoclimate indicators because they yield information about synoptic-scale circulation. Records in deep-sea sediment cores, ice cores, and soils indicate that the Americas have been both a source and a sink for far-traveled, fine-grained dust.

Darwin (1846) was one of the first investigators to observe airborne dust over the Atlantic Ocean enroute to the Americas. A century later, numerous researchers have confirmed that African dust travels to the Americas via the northeasterly trade winds (Fig. 21). Delany et al. (1967), Prospero et al. (1970, 1981), and Prospero and Nees (1977, 1986) demonstrated, through analysis of satellite imagery and dust trap collections, that African dust falls on Florida, Barbados, and northern South America. By the time it reaches the western Atlantic Ocean, airborne dust from Africa is composed mostly of particles less than 20 μm in diameter, and about half of this is clay-sized material less than 2 μm in diameter (Prospero et al., 1970). The most important mineral in the silt-sized fraction is quartz, and mica dominates the clay-sized fraction (Delany et al., 1967; Glaccum and Prospero, 1980). The modern flux of dust from Africa is, to a great extent, a function of the degree of aridity in the southern Sahara and Sahel (Prospero

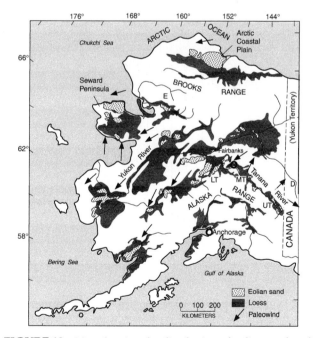

FIGURE 18   Map showing the distribution of eolian sand and loess and paleowinds inferred from eolian sand data in Alaska and northwestern Canada. E, Epiguruk (see Fig. 20); UT, upper Tanana River (see Fig. 20); MT, middle Tanana River (see Fig. 20); LT, lower Tanana River (see Fig. 20); D, Dawson, Yukon Territory, Canada (see Fig. 19). Eolian sand and paleowinds data are from Lea and Waythomas (1990); loess distribution is redrawn from Péwé (1975).

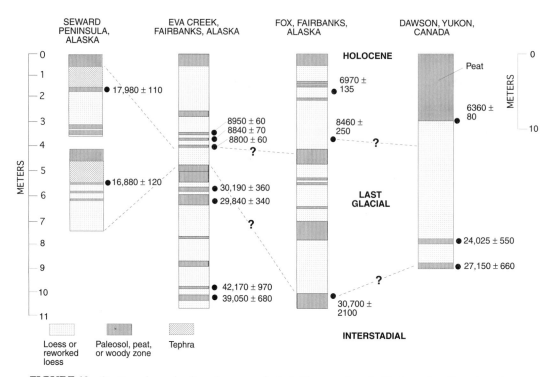

FIGURE 19 Stratigraphy and radiocarbon ages of selected loess sections in Alaska and northwestern Canada. Localities are shown in Fig. 18. Generalized from Hamilton et al. (1988), Höfle and Ping (1996), Muhs et al. (1997c), and Fraser and Burn (1997).

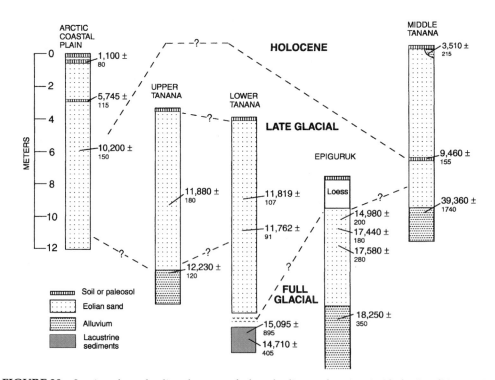

FIGURE 20 Stratigraphy and radiocarbon ages of selected eolian sand sections in Alaska. Localities are given in Fig. 18. Generalized from Weber et al. (1981), Carter and Galloway (1984), Galloway and Carter (1993), Hamilton and Ashley (1993), and Lea (1996).

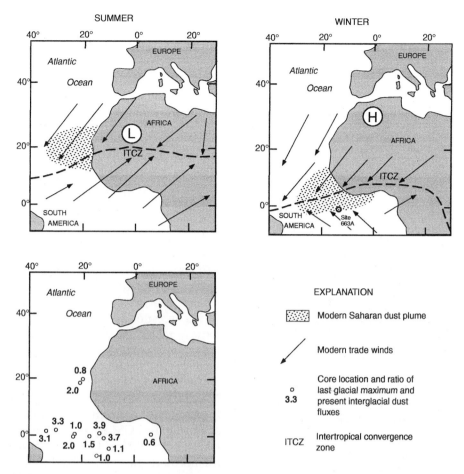

FIGURE 21 Modern summer and winter circulation patterns over a portion of the Atlantic Ocean, the location of modern Saharan dust plumes, and the ratio of last glacial maximum (LGM) to modern dust fluxes as recorded in deep-sea cores between Africa and South America. (Redrawn from Ruddiman 1997.)

and Nees, 1977), but measurable dust has been collected in Florida and Barbados most years since collections began in the 1960s (Prospero, 1981).

It has been known for some time that deep-sea sediments in the Atlantic Ocean that date to the LGM are enriched in quartz relative to Holocene sediments (Bowles, 1975; Kolla et al., 1979). Quartz enrichment in the North Atlantic is due to ice rafting, but in the equatorial Atlantic it is interpreted to be the result of westward eolian transport of quartz-rich particles from Africa via the trade winds. Deep-sea sediment records and soils indicate that African dust transport to the Atlantic Ocean and the Americas have been an important process for hundreds of thousands of years, and show distinctive maxima (Fig. 22). At site 663A in the Atlantic Ocean, maximum fluxes of African dust to the Atlantic Ocean generally occur during glacial or stadial periods, whereas peak interglacial periods have low dust fluxes (deMenocal et al., 1993; Ruddiman, 1997). Studies of

soils on high-purity carbonate reef terraces and eolianites on Barbados, the Bahamas, and the Florida Keys show that they are unlikely to have formed as residual products by carbonate dissolution (Muhs et al., 1987, 1990). Lesser Antilles island arc volcanic ash deposition on these islands has been an important process in the Quaternary period (Carey and Sigurdsson, 1980). It is possible that soils have developed primarily from this source. However, geochemical analyses show that African dust is a much more likely parent material and has been so for the last several glacial–interglacial cyles (Fig. 22).

Some of the highest resolution paleoclimate records of the Americas are ice cores from glaciers in the high Andes of South America. Dust flux is one of the records available from these cores, and there are differences over both time and space. The Huascarán, Peru, ice core shows a dust flux during the LGM that is about 200 times greater than most Holocene values (Thompson et

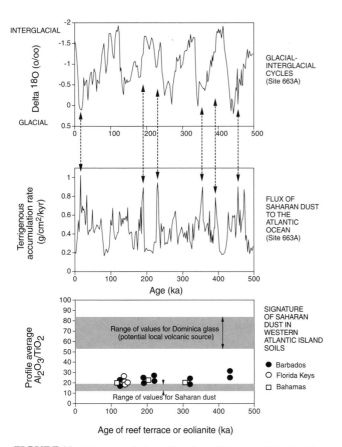

FIGURE 22 Records of dust flux from Saharan Africa to the Americas over glacial–interglacial cycles of the past 500,000 years. Top: Oxygen isotope record of foraminifera in deep-sea sediments at site 663A in the Atlantic Ocean showing glacial–interglacial cycles. Middle: Accumulation rates of terrigenous sediments at site 663A, interpreted to be a measure of Saharan dust flux to the Atlantic Ocean. Arrows show correspondence between generally cold (glacial or stadial) periods with periods of maximum dust flux. Bottom: Soil profile average values of Al_2O_3/TiO_2 on Barbados, the Bahamas, and the Florida Keys compared to average values (plus or minus one sigma) for soil parent material sources. Site 663A data from deMenocal et al. (1993); soil data from Muhs et al. (1990).

al., 1995). The timing of this high dust flux is remarkably similar to that found in Greenland and Antarctic ice cores. Thompson et al. (1995) interpreted the high dust flux during the LGM to reflect an overall decrease in humidity, precipitation, and vegetation cover in South America at this time, as well as enhanced eolian transport processes (i.e., stronger winds). In contrast, an ice core taken from the summit of Sajama mountain in Bolivia shows much higher values in the Holocene than during the LGM, which is inconsistent with both the Peruvian record and that of polar ice cores (Thompson et al., 1998). Thompson et al. (1998) interpreted the enhanced Holocene dust flux in Bolivia to represent increased volcanic activity, elevated snow lines, and decreased net accumulation since the LGM.

The Americas have been studied not only as the endpoints for far-traveled dust, such as that from Africa, but also as sources for dust deposited elsewhere. Petit et al., (1990) showed that the flux of dust into Antarctic ice was at a maximum during glacial periods of the past two glacial–interglacial cycles. Potential sources for this dust include southern Africa, Australia, New Zealand, and South America. Basile et al. (1997) demonstrated, using Sr and Nd isotopes, that the Pampas and Patagonia regions of South America are the most likely source areas for this dust. Other studies identifying South America as a dust source have been questioned. Molina-Cruz (1977) and Molina-Cruz and Price (1977) interpreted the mineral fraction of deep-sea sediments off the west coast of South America as having an eolian origin. Rea (1994) challenged this hypothesis and suggested instead that the increased mineral mass accumulation rates recorded in these cores during glacial periods are due to hemipelagic processes.

The question is open as to whether there is a Quaternary record of dust flux from arid regions of western North America to the eastern Pacific Ocean. Hemipelagic processes dominate much of the mineral accumulation recorded in cores off the west coast of North America. In addition, winds are predominantly from the west or northwest and are likely to have been so during the LGM as well (Johnson, 1977). There is little question that most airborne dust that is delivered to the northern part of the Pacific Ocean comes from the loess regions of Asia (Rea, 1994), and eolian quartz has been found in soils as far east of Asia as the Hawaiian Islands (Jackson et al., 1971). Nevertheless, some data suggest eolian inputs from North America to the eastern Pacific Ocean, at least locally. A high-resolution core from the Santa Barbara basin (for location, see Fig. 6) is dominated by silt- and clay-sized particles, with a few thin sandy layers present, over the past two interglacial–glacial cycles (Stein, 1995). Quartz abundances in these sediments show distinct maxima during three glacial periods: oxygen isotope stages 2, 4, and 6. Stein (1995) speculated that the quartz enrichments at these times may have been due to greater eolian flux from North American desert regions. Other data also suggest that the Mojave Desert of California periodically supplies dust to the eastern Pacific. Although prevailing winds are from the northwest on the California coast, easterly winds known as "Santa Anas" occur several times per year. After the passage of a frontal system, high pressure settles in the Great Basin and winds from this high move to the south and southwest over the Mojave Desert. In areas where there are loose, dry, silt- and clay-sized particles, Santa Ana winds are strong enough to carry the sediments at least as far west as the California Channel Islands, based on examination of satellite imagery, dust trap data, and reports of aircraft pilots (Muhs, 1983). Soils on San Clemente Island, CA, rest on uplifted marine terraces cut primarily into Miocene andesite, which has no mica and only scarce quartz. Nevertheless, soils on alluvial fans and marine terraces contain abundant silt- and clay-sized quartz and mica at all depths in all soils of deposits ranging in age from the Holocene to greater than 1.0 Ma (Muhs, 1982). These observations support the hypothesis of Stein (1995) that the Mojave Desert may deliver eolian dust to the eastern Pacific Ocean and that the process may have been important for much of the Quaternary period.

12.10. COMPARISONS BETWEEN SOUTH AMERICAN AND NORTH AMERICAN EOLIAN RECORDS

There are both similarities and differences in the late Quaternary eolian records of South and North America. Both continents have extensive loess and eolian sand deposits. Loess is widespread over much of Argentina and Paraguay (Zárate and Blasi, 1991, 1993; Sayago, 1995; Iriondo, 1997, 1999). Similar to North American loess deposits, much of this sediment was deposited during the LGM under cold, arid, and windy conditions. In contrast, however, South American loesses have a much higher contribution from volcaniclastic sources delivered by fluvial transport and even directly from explosive eruptions (Zárate and Blasi, 1993). In North America, only Great Plains loesses have a significant volcaniclastic component, and this is mostly reworked from Tertiary rocks (Aleinikoff et al., 1998, 1999). On both continents, however, fluvial transport is an important step in the sediment cascade before eolian entrainment (Zárate and Blasi, 1993). One difference is that there was significant Holocene loess deposition in South America (Zárate and Blasi, 1993), but it was rare in North America outside of Alaska and isolated parts of the Great Plains.

In South America, eolian sand is found on the Llanos del Orinoco in Venezuela and Colombia, in the Pantanal and Beni basins of Paraguay and southwestern Amazonia, and on the Pampas and Patagonian plains of Argentina (Zárate and Blasi, 1993; Roa, 1979, 1980; Clapperton, 1993; Iriondo, 1997, 1999). As is the case with its North American counterparts, almost all of this eolian sand is stabilized by vegetation. As in North America, active dunes in South America are rare, an important exception being the field of active barchan dunes along the coast of Peru (Finkel, 1959). Most South American eolian sand is undated (see Kröhling, 1999 for an exception to this), but much is thought to have

been deposited during the LGM. However, parabolic dunes with minimally developed soils are reported to have formed in Argentina during the late Holocene (Cabral et al., 1988; Hurtado and Giménez, 1988; Giménez, 1990; Ramonell et al., 1992; Iriondo and García, 1993; Iriondo, 1999), similar to the episodes of parabolic dune formation during the late Holocene in the midcontinent of North America (David, 1971; Ahlbrandt et al., 1983; Muhs, 1985; Swinehart, 1990; Madole, 1994, 1995; Wolfe et al., 1995; Arbogast, 1996; Muhs et al., 1996, 1997a,b). Buried soils in these dunes, on both continents, indicate that there were multiple periods of Holocene eolian activity.

12.11. PALEOCLIMATIC IMPLICATIONS OF EOLIAN RECORDS

Eolian deposits are potentially important paleoclimate proxies because of their sensitivity to overall moisture balance and degree of vegetation cover and because they are direct records of atmospheric circulation. Worldwide, many long-term and quasi-continuous records such as deep-sea cores, ice cores, and certain loess sections (such as those in China) suggest that during the LGM, the atmosphere contained more dust than during the Holocene. Mahowald et al. (1999), who reviewed this evidence, summarized three reasons why dust deposition during the LGM may have been greater than it is now: (1) increased glacial age wind intensities could entrain more dust and carry it farther from sources, (2) reduced intensity of the hydrological cycle could allow dust to remain suspended in the atmosphere longer (i.e., fewer *rain-out* periods), and (3) increased source areas for dust existed, due to reduced vegetation cover and decreased soil moisture. In this section, we compare paleoclimatic inferences made from eolian records with paleoclimate models and independent records of paleoclimate such as pollen, faunas, and lake levels. Climate models and synoptic-scale paleoclimatic reconstructions for the past 18,000 years provided by the COHMAP effort (COHMAP Members, 1988; Kutzbach et al., 1993; Markgraf et al., 1993b; Thompson et al., 1993; Webb et al., 1993a,b), as well as later modeling efforts (Kutzbach et al., 1998; Bartlein et al., 1998; Mahowald et al., 1999), are particularly well suited to comparisons with eolian records of South and North America.

12.11.1. Low Latitudes

The greater dust flux at low latitudes during the LGM is well recorded in Atlantic deep-sea sediments from Africa to South America (Ruddiman, 1997) and in

Huascarán ice cores from Peru (Thompson et al., 1995). Ruddiman (1997) concluded that glacial age aridity in the source areas of Africa cannot explain the greater flux of dust to the Atlantic (and the Americas) during the LGM because there was no increase in dust flux following the early Holocene monsoon maximum. Instead, he argued that stronger winter trade winds were the most likely cause for the greater dust flux during the LGM and that there was also a southward shift in these winds by several degrees of latitude. An atmospheric general circulation model simulation by Rind (1987) demonstrated that a cooler North Atlantic Ocean sea surface would have intensified the winter subtropical high over the Atlantic, which in turn would have increased the strength of the winter trade winds. Countering Ruddiman's (1997) viewpoint, however, is the notion that the amount of dust transport is not a simple function of average wind speed. Rea (1994) pointed out that, at the present time, virtually all dust transport from Africa to Barbados or from Asia to the Pacific Ocean takes place in spring, when both the northeasterly trade winds (Africa to Barbados) and the westerlies (Asia to the North Pacific Ocean) are weaker than during winter. In addition, he noted that there is a good correlation between wind speed and size of particles transported, but little correlation between accumulation rates and size of particles. The simulations of Mahowald et al. (1999), based on both source area conditions *and* transport processes, show the Sahara and Sahel of Africa as important dust sources during the LGM; so perhaps it is a combination of source sediment conditions and wind strength that is critical to explaining the Atlantic Ocean dust flux record.

The record of formerly active dunes on the Llanos del Orinoco in Venezuela has been inferred to represent a drier climate in this region during the LGM (Roa, 1979; Iriondo, 1997; Clapperton, 1993). These latter workers suggested that the drier conditions could have resulted from a southward migration of the ITCZ during the LGM, which agrees with the reconstruction made by Ruddiman (1997) from data on dust flux to the east of South America. Although most lake and pollen records in this region go back only to ca. 13,000 ^{14}C B.P., they support the notion that this region was drier during the late glacial period (Bradbury et al., 1981; Markgraf, 1993b).

12.11.2. Midlatitudes: Last Glacial Period

Many of the paleoclimate records derived from pollen and lake-level data (Markgraf, 1993b; Thompson et al., 1993; Webb et al., 1993a,b) suggest that during the LGM, midlatitude regions such as the North American midcontinent south of the Laurentide ice

sheet and the Pampas region of Argentina (east of the Andes glacial cap) were very cold, dry, and windy. Virtually all investigators who have studied last glacial eolian deposits in Argentina agree that paleoclimatic conditions at the time of deposition were cold, dry, and probably windy (Zárate and Blasi, 1991, 1993; Iriondo, 1999; Kröhling and Iriondo, 1999; Kröhling, 1999; Carignano, 1999). These interpretations are in agreement with pollen and faunal data for the last glacial period (Markgraf, 1993b; Prieto, 1996; Prado and Alberdi, 1999; Tonni et al., 1999). They are also in agreement with COHMAP model results, which indicate that midlatitudes in the Southern Hemisphere were colder and drier during the LGM and the late glacial period (Kutzbach et al., 1993; Markgraf et al., 1993b; Webb et al., 1993b). Because South American glaciers were of much smaller extent than those in the Northern Hemisphere (Clapperton, 1993), it can be inferred that an important reason for greater eolian activity was a drier climate and diminished vegetation cover, rather than solely an increased sediment supply from glaciogenic silt. The studies of Zárate and Blasi (1993) and Gallet et al. (1998), which demonstrate the importance of volcaniclastic sediments as sources, suggest that loess in the Pampas region is not primarily dependent on glaciogenic silt. Identification of the Pampas as the most important source area for dust found in Antarctic ice cores (Basile et al., 1997) is also consistent with COHMAP simulations of this region as drier during the LGM (Markgraf et al., 1993b; Webb et al., 1993b). Transport of fine-grained loess from the Pampas to Antarctica also agrees with the simulation of Mahowald et al. (1999) showing that this part of South America was an important dust source region.

In contrast, the causes of abundant loess deposition during the LGM and the late glacial period in North America vary from region to region. East of the Missouri River, where the headwaters of major drainages were glaciated, the Laurentide ice sheet generated abundant silt-sized material that was carried in major outwash valleys. Timing of loess deposition shows that shortly after the ice left the headwaters of major drainages to the south, loess deposition ceased (Ruhe, 1983). This concept is supported by the studies of Frye et al. (1968), McKay (1979), and Grimley et al. (1998), which show the linkages between loess composition and various ice lobes, as discussed earlier. These observations indicate that the reason for much loess deposition during the LGM is that the region was a supply-dominated system. The dust model of Mahowald et al. (1999) does not show the area south of the Laurentide ice sheet as a significant dust source during the LGM, although their model does not account for extraordinary sediment supplies due to glacial meltwaters.

Similarly, the northwestern United States is not simulated as a major loess source during the LGM (Mahowald et al., 1999), despite the abundance of loess deposits found in Idaho and Washington (Busacca, 1991). However, as described earlier, eolian entrainment of silt in this region is a function of a particular sediment supply, namely, slackwater deposits of proglacial Lake Missoula outbreak floods. Therefore, loess deposition is a function of a short-lived, supply-dominated system and not primarily a function of strong winds or lack of vegetation cover (although these factors may enhance silt entrainment). Furthermore, loess deposition in this region occurs during both late glacial and Holocene times, rather than primarily during the LGM.

On the other hand, areas of North America not immediately adjacent to glaciated terrain, in the Great Plains, have some of the thickest LGM loess deposits on the continent. As was described earlier, the isotopic studies of Aleinikoff et al. (1998, 1999) demonstrate that loess sources in this region are mostly nonglaciogenic. Thus, the COHMAP and later climate models (COHMAP Members, 1988; Thompson et al., 1993; Kutzbach et al., 1993, 1998; Bartlein et al., 1998) that simulate cold, dry, windy conditions in this region agree well with the loess record. Vegetation cover on the Great Plains is not abundant at present and likely was even more sparse under the cold, dry conditions of the LGM. Lack of vegetation cover may have been one of the most important factors in allowing sediment entrainment in this region during full glacial times. However, the Mahowald et al. (1999) simulation does not suggest that there were significant dust sources in the Great Plains region during the LGM.

There is a major disagreement between LGM circulation modeled by COHMAP studies and the loess record of North America. COHMAP and later simulations (COHMAP Members, 1988; Kutzbach et al., 1993, 1998; Bartlein et al., 1998) indicate that the Laurentide ice sheet generated a glacial anticyclone that would have produced northeasterly winds in the region to the south of the ice sheet. In contrast, the well-dated LGM loess paleowind record of midcontinental North America (Fig. 7) indicates a westerly or northwesterly wind during full glacial times (Muhs and Bettis, 2000). If a zone of northeasterly winds existed, then the loess distribution pattern described earlier indicates that it was extremely limited in geographic extent and may have affected only a narrow zone near the ice sheet. An alternative explanation is that loess deposition occurred during short, exceptionally strong northwesterly winds that occurred infrequently, but transported much sediment (Muhs and Bettis, 2000).

In the northwestern United States, there is also an apparent disagreement between LGM paleowinds derived from the loess record and the COHMAP simulations. As with the midcontinent, LGM paleowinds in unglaciated portions of the northwestern United States are simulated by COHMAP to have been easterly or northeasterly, in this case due to glacial anticyclonic circulation over both the Laurentide and Cordilleran ice sheets (COHMAP Members, 1988; Kutzbach et al., 1993, 1998; Bartlein et al., 1998). Based on eastern Idaho loess thickness and particle size trends, westerly or northwesterly paleowinds can be inferred (Fig. 17) and are in conflict with model simulations.

In contrast, marine and terrestrial records of exotic diatoms provide an independent and novel method to infer LGM paleowinds in the northwestern United States. Sancetta et al. (1992) studied sediments, pollen, and diatoms in a core taken 120 km off the coast of southern Oregon (Fig. 17), with a record that goes back to ca. 30,000 [14]C B.P. In the interval from ca. 25,000–11,000 B.P. (but not in the Holocene), they reported abundant extinct diatoms that occur only in Pliocene lacustrine sediments found in interior Oregon and California and also in southwestern Idaho (Bradbury, 1982, 1992). Although a fluvial mode of transport cannot be completely ruled out, an eolian origin is far more likely, based on the lack of certain Quaternary taxa that would also be expected if streams were the primary means by which the Pliocene diatoms were delivered to the ocean. This finding implies at least occasional easterly paleowinds in the northwestern United States over the period from ca. 25,000–11,000 B.P. Sancetta et al. (1992) suggested that winter was the most likely time that easterly transport might have occurred and that easterly winds need not have been constant. Nevertheless, they pointed out that their findings support the concept of a glacial anticyclone, as modeled then by Kutzbach and Guetter (1986). Perhaps easterly, diatom-bearing winds reflect winter conditions during the LGM, and westerly, loess-bearing winds reflect summer conditions. This intriguing possibility needs to be investigated more thoroughly with additional loess and exotic diatom studies.

There is less disagreement between modeled circulation and the loess and eolian sand record for the Pampas region of Argentina. COHMAP results (Kutzbach et al., 1993) suggest that during full glacial southern summers (January), surface winds would have been westerly, which agrees reasonably well with southwesterly paleowind data derived from loess and eolian sand studies as described by Zárate and Blasi (1993).

During the late glacial period and transition into the Holocene (ca. 12,000–9000 [14]C B.P.), Forman et al. (1995) suggested that the midcontinent of North America experienced little eolian activity. They presented a model wherein the Gulf of México (cooled by glacial meltwaters) and the continental interior (warmed by higher summer insolation) generated a stronger monsoonal flow of air from the Gulf of México to the interior. Because the Gulf of México is the major moisture source for midcontinental North America, they hypothesized that this should have been a time of relatively high precipitation and little eolian activity and cited supportive evidence from some sections studied in eastern Colorado. This model has appeal, but there are numerous examples from the region that do not support it. Eolian activity in the Nebraska Sand Hills is inferred to have been important at this time, sufficient to build dunes that dammed drainages (Loope et al., 1995; Mason et al., 1997; Muhs et al., 2000). Furthermore, maximum limiting radiocarbon ages suggest that some of the largest barchanoid ridge dunes in the Nebraska Sand Hills (see Swinehart, 1990) may have been built at this time (Swinehart and Diffendal, 1990). For the Southern High Plains, Holliday (2000) has compiled numerous well-characterized and well-dated sections that indicate that the Folsom period, which occurred in this late glacial interval, was a time of significant eolian sand movement.

12.11.3. Midlatitudes: Holocene

For South America, some workers who have studied eolian deposits in the Pampas region note that the early to mid-Holocene period appears to have been a period of soil formation rather than eolian activity (Zárate and Flegenheimer, 1991; Kröhling and Iriondo, 1999; Kröhling, 1999). Therefore, these workers infer a warm, but humid, period at this time. Pollen data for the region also seem to suggest a warm, humid period (Prieto, 1996). Faunal data suggest that the early to mid-Holocene from ca. 9000–7000 B.P. was relatively dry (Prado and Alberdi, 1999; Tonni et al., 1999). However, these latter workers also infer a more humid period between ca. 7500 and 5000 B.P. COHMAP model results suggest that this part of South America may have been warmer, but drier, during the mid-Holocene, with a suppressed monsoon (Markgraf, 1993b). More pollen, faunal, and eolian studies with good age control are probably necessary before a clear picture emerges for the mid-Holocene of the Pampas region.

Several workers (Zárate and Blasi, 1993; Iriondo, 1997, 1999; Kröhling and Iriondo, 1999) infer a drier period in the late Holocene for the Pampas region, although not as dry as during the LGM. Loess deposition and dune formation are both recorded for parts of the Pampas region, and there may have been numerous, short-lived episodes of relatively dry conditions. This interpretation is in agreement with pollen (Prieto, 1996)

and faunal data, which suggest a predominantly humid late Holocene period punctuated by arid phases (Prado and Alberdi, 1999).

In North America, there is limited evidence of mid-Holocene eolian activity. The Bignell Loess of the Great Plains may have been deposited at this time (Pye et al., 1995; Maat and Johnson 1996; Johnson and Willey, 2000; Mason and Kuzila, 2000), and there are localities within the Great Plains where eolian sand was active, such as in Nebraska (Loope et al., 1995), Colorado (Forman and Maat, 1990; Forman et al., 1995), and Texas (Holliday, 1995a,b). These periods of eolian sand movement are consistent with pollen data, suggesting that at least parts of the midcontinent were relatively warm and dry during the mid-Holocene (Webb et al., 1993a; Fredlund, 1995; Baker et al., 1998). Mid-Holocene eolian activity is also consistent with COHMAP reconstructions of a relatively warm and dry midcontinental North America (Kutzbach et al., 1993, 1998; Thompson et al., 1993; Webb et al., 1993b; Bartlein et al., 1998). However, there is some disagreement with COHMAP models of overall moisture balance and the eolian record in the Southern High Plains of Texas and New Mexico. The COHMAP simulations suggest positive moisture balances for both summer and on an annual basis ca. 6000 B.P. (Thompson et al., 1993), perhaps due to an enhanced monsoon that affected the southern Great Plains but not the central and northern Great Plains. Packrat midden data from southern New Mexico, just to the west of the southern Great Plains, also suggest wetter conditions at both 9000 and 6000 B.P. (Thompson et al., 1993). In contrast, the eolian record of the southern Great Plains suggests that there was a negative moisture balance at this time (Holliday, 1995a,b).

There is excellent agreement, however, between the eolian record, the vegetation record (primarily from packrat midden data), and model simulations for the Colorado Plateau region during the mid-Holocene. The geomorphic, stratigraphic, and soil data of Wells et al. (1990) suggest there was little eolian activity during the mid-Holocene, and this conclusion agrees well with vegetation records that imply a more positive moisture balance than at present (Thompson et al., 1993). The latest climate model experiments for the southwestern United States for this period (Bartlein et al., 1998) also suggest a much stronger monsoonal flow and a more positive moisture balance.

There is abundant evidence of late Holocene eolian activity in all parts of the Great Plains, from southern Texas to the Canadian prairie provinces. Holocene eolian activity was so widespread that it is likely that much eolian sand active during the mid-Holocene was reworked, and this may explain why the mid-Holocene record of sand movement is sparse. In addition, many localities show stratigraphic evidence of multiple episodes of late Holocene eolian activity, suggesting alternating periods of relatively dry and wet conditions. Independent paleoclimate data for the late Holocene are mostly limited to the northern Great Plains, where there are abundant lacustrine records of diatoms, ostracodes, and pollen (e.g., Fritz et al., 1991, 1994; Laird et al., 1996). Although limits on dating of eolian sediments preclude precise timing of short dry periods, the lacustrine records indicate that there were many alternations of relatively dry and moist periods in the late Holocene. Eolian sand was active during the late Holocene over much of the Colorado Plateau as well (Wells et al., 1990; Stokes and Breed, 1993). Orientations of late Holocene dunes in both the Great Plains and the Colorado Plateau regions indicate that dune-forming winds were very similar to modern winds (Figs. 6, 14, and 15). This observation suggests that dune activation in these regions can take place under climatic regimes that have circulation patterns similar to those of modern climates.

12.11.4. High Latitudes

The loess record of Alaska and adjacent parts of Canada has been used to interpret paleoclimate in a number of ways. The traditional interpretation is that thick loess deposits represent glacial periods, as in the North American midcontinent (Péwé, 1975). However, near Fairbanks, some Holocene loess deposits with good radiocarbon control are as thick or thicker than what could be LGM loess deposits. There is no question that glaciers were more extensive in Alaska and northwestern Canada during the LGM (Péwé, 1975), and it is very likely that more glaciogenic silt was produced at that time compared to the present. However, the relatively thick Holocene loess compared to last glacial loess may be a consequence of the vegetation cover at the time of loess deposition. Since the early Holocene, there has been a boreal forest cover in interior Alaska, whereas during full glacial time, herb tundra was the dominant vegetation (Ager and Brubaker, 1985). The Boreal forest may have served as a more efficient trap of airborne sediment compared to tundra vegetation, even if overall glaciogenic silt production during full glacial time was greater (Begét, 1988, 1991). If this hypothesis is correct, then extreme care must be taken in interpreting Alaskan loess stratigraphy for paleoclimatic records.

Béget (1990, 1991), Béget and Hawkins (1989), and Béget et al. (1990) used magnetic susceptibility variations in Alaskan loess–paleosol sequences to infer glacial to interglacial changes in late Quaternary

wind strengths. Loess–paleosol sequences in Alaska are analogous to those in China in that loess deposition and pedogenesis occur during both glacial and interglacial periods and are competing processes. In this model, loess deposition rates exceed rates of pedogenesis in glacial periods and the reverse occurs in interglacial periods. Thus, although a paleosol may represent a slowing of loess sedimentation, particles are still added to what becomes a cumulic soil surface. Bégét and co-workers further assumed that with lower deposition rates, wind strengths should be lower; therefore, mean particle size should be finer, and heavy mineral content should be lower. Using magnetic susceptibility as a proxy for heavy mineral content (and therefore wind strength), they demonstrated that values are higher in loesses (assumed to represent glacial or stadial periods) and lower in paleosols (assumed to represent interglacials or interstadials). These observations suggest that winds were stronger during the last glacial period compared to the Holocene.

Much still remains to be learned about the paleoclimatic significance of Alaskan loess. Loess source areas in Alaska are not well established, nor is the amount of loess deposition that took place during the LGM. Paleowinds during the LGM could be reconstructed from loess data if well-dated loess sequences were convincingly linked to source areas by using mineralogy, geochemistry, and isotopic composition. However, even apart from the larger glaciers that were present during the LGM, the dust-source model of Mahowald et al. (1999) suggests that interior Alaska was an important source at this time, consistent with the pollen record of cold, dry conditions and a sparsely vegetated landscape (Ager and Brubaker, 1985).

In interior Alaska, eolian sand sheets, rather than dunes, were deposited during the LGM (Lea and Waythomas, 1990). Although mechanisms for the formation of sand sheets are still not entirely understood, it is known that sheet sand tends to be deposited in favor of dune sand in settings where there are sediment supply limitations. Although there was abundant glacial outwash source material at the LGM, Lea and Waythomas (1990) infer that sediment supplies were effectively limited because of ice cementation, snow cover, high groundwater tables, and vegetation cover. Partial vegetation cover during sand sheet formation is inferred from vertebrate tracks that are common in many eolian deposits in Alaska (Lea, 1996). Thus, sheet sand formation during the LGM may imply a sparsely vegetated, permafrost-dominated landscape, which agrees well with full glacial climate reconstructions based on pollen (Ager and Brubaker, 1985). Dunes seem to have formed mainly during the late glacial or

Holocene period, at least in interior and southern Alaska. Lea and Waythomas (1990) hypothesize that the shift from sheet sand deposition to dune formation resulted, at least in part, from deeper seasonal thaws that occurred at the close of the last glacial period, which would have resulted in greater sediment availability. Countering this view are pollen data, which indicate that vegetation in late glacial time would have been more abundant than during the LGM, including an increase in the amount of dwarf birch (Ager and Brubaker, 1985). However, Lea and Waythomas (1990) infer that the deeper thaws and increased sediment supply of the late glacial and early Holocene periods may have outweighed the effect of greater precipitation and more abundant vegetation, with the result that dune formation could take place.

On the Arctic Coastal Plain of Alaska, a different style of eolian deposition took place compared with that of interior Alaska. Longitudinal dunes formed during the LGM, whereas sand sheets formed on top of these after the last glacial period (Carter, 1981). However, some Holocene sands were also worked into dunes, both longitudinal and parabolic types, during at least four late Holocene periods of activity (Galloway and Carter, 1993, 1994). These workers correlated the periods of dune activity with neoglacial expansions of cirque glaciers in the Brooks Range and inferred that dune activity resulted from cooler and drier conditions.

As is the case with the North American midcontinent and the northwestern United States, there is a conflict with paleowinds inferred from eolian sands and model-derived paleowinds in Alaska (Kutzbach et al., 1993, 1998; Bartlein et al., 1998). The model results suggest that paleowinds during the LGM and the late glacial period were dominantly from the southwest. In contrast, both dune orientations and locations of most sand bodies relative to probable sources (note association of dune fields with major drainages on Fig. 18) suggest paleowinds from the northeast. Although Lea and Waythomas (1990) point out that the dunes probably date to the late glacial or early Holocene periods, as was discussed earlier, the locations of the sand bodies from which the dunes were derived probably have not changed since the LGM and still imply mostly northeasterly paleowinds.

12.12. CONCLUSIONS

The Americas contain a rich record of late Quaternary eolian sand, loess, and aerosolic dust. Full glacial time in the midlatitudes of both continents was likely cold, dry, and windy, based both on the eolian record of sand and loess and on other paleoclimate proxies such as pollen. This inference is supported, in part, by the observation that much loess on both continents is nonglaciogenic, implying dry, windy conditions with minimal vegetation cover, prerequisites for eolian entrainment. During full glacial time, midlatitude regions of South America were probably the source for dust in Antarctica, and low-latitude regions of South America received dust from Africa. Paleoclimatic reconstructions of cold, dry conditions from eolian sediments are in agreement with climate model results for the LGM.

There are both agreements and disagreements between last glacial circulation patterns derived from eolian sediments and patterns derived from climate models. In South America, inferred westerly and southwesterly directions of last glacial eolian sand and loess movement are in agreement with westerly paleowind simulations by climate models. However, in North America, including the midcontinent, the northwestern United States, and Alaska, paleowinds derived from eolian sediments do not agree with climate model simulations. The reason for this difference is not understood and needs additional study by both modelers and field scientists.

The mid-Holocene is a time for which eolian records are not as well understood on both continents. In South America, little eolian sediment movement seems to have taken place and warm, humid conditions are inferred; this agrees with pollen and some faunal data, but is not entirely in agreement with climate model simulations. In midcontinental North America, mid-Holocene eolian records are sparse, but may be partly the result of late Holocene reworking. Pollen data and climate model simulations suggest that, at least in the central and northern Great Plains, this should have been an important period of eolian activity, and limited data support that interpretation. On the Colorado Plateau, pollen and climate model data suggest that the mid-Holocene was more moist than at present, and the eolian record agrees with this interpretation.

The late Holocene records eolian sand movement in the midlatitudes of both continents. Too few data currently exist to assume that eolian sand movement was regionally synchronous within dune fields in both South and North America. However, both continents show evidence of widespread late Holocene eolian sand activity and multiple episodes of eolian sand movement during the late Holocene, based on the presence of paleosols within sections. These observations indicate that eolian sediments are highly sensitive to slight shifts in climate and vegetation cover and that widespread eolian sand movement can take place in the midlatitudes of both continents under essentially modern climatic conditions.

Acknowledgments

Eolian studies by Muhs were supported by the Earth Surface Dynamics Program of the U.S. Geological Survey. We thank Platt Bradbury, Vance Holliday, Art Bloom, Steve Forman, and Vera Markgraf for helpful comments on an earlier draft of the chapter. Aldo Prieto, Adriana Blasi, and Jorge Giménez kindly provided information on climate and eolian deposits of the Pampas.

References

Ager, T. A., and L. Brubaker, 1985: Quaternary palynology and vegetational history of Alaska. *In* Bryant, V. M., Jr., and R. G. Holloway (eds.), *Pollen Records of Late-Quaternary North American Sediments.* Dallas, TX: American Association of Stratigraphic Palynologists Foundation, pp. 353–383.

Ahlbrandt, T. S., J. B. Swinehart, and D. G. Maroney, 1983: The dynamic Holocene dune fields of the Great Plains and Rocky Mountain basins, U.S.A. *In* Brookfield, M. E., and T. S. Ahlbrandt (eds.), *Eolian Sediments and Processes.* New York: Elsevier, pp. 379–406.

Aleinikoff, J. N., D. R. Muhs, and C. M. Fanning, 1998: Isotopic evidence for the sources of late Wisconsin (Peoria) loess, Colorado and Nebraska: Implications for paleoclimate. *In* Busacca, A. J. (ed.), *Dust Aerosols, Loess Soils and Global Change.* Pullman: Washington State University College of Agriculture and Home Economics, Miscellaneous Publication No. MISC0190, pp. 124–127.

Aleinikoff, J. N., D. R. Muhs, R. Sauer, and C. M. Fanning, 1999: Late Quaternary loess in northeastern Colorado: Part II–Pb isotopic evidence for the variability of loess sources. *Geological Society of America Bulletin,* **111**: 1876–1883.

Andersen, K. K., A. Armengaud, and C. Genthon, 1998: Atmospheric dust under glacial and interglacial conditions. *Geophysical Research Letters,* **25**: 2281–2284.

Arbogast, A. F., 1996: Stratigraphic evidence for late-Holocene eolian sand mobilization and soil formation in south-central Kansas, U.S.A. *Journal of Arid Environments,* **34**: 403–414.

Arbogast, A. F., and D. R. Muhs, 2000: Geochemical and mineralogical evidence from eolian sediments for northwesterly mid-Holocene paleowinds, central Kansas, USA. *Quaternary International,* **67**: 107–118.

Baker, R. G., L. A. González, M. Raymo, E. A. Bettis, III, M. K. Reagan, and J. A. Dorale, 1998: Comparison of multiple proxy records of Holocene environments in the midwestern United States. *Geology,* **26**: 1131–1134.

Bartlein, P. J., K. H. Anderson, P. M. Anderson, M. E. Edwards, C. J. Mock, R. S. Thompson, R. S. Webb, T. Webb, III, and C. Whitlock, 1998: Paleoclimate simulations for North America over the past 21,000 years: Features of the simulated climate and comparisons with paleoenvironmental data. *Quaternary Science Reviews,* **17**: 549–585.

Basile, I., F. E. Grousset, M. Revel, J. R. Petit, P. E. Biscaye, and N. I. Barkov, 1997: Patagonian origin of glacial dust deposited in East Antarctica (Vostok and Dome C) during glacial stages 2, 4 and 6. *Earth and Planetary Science Letters,* **146**: 573–589.

Begét, J. E., 1988: Tephras and sedimentology of frozen loess. *In* Senneset, K. (ed.), *Fifth International Permafrost Conference Proceedings,* Tapir, Trondheim, **1**: 672–677.

Begét, J. E., 1990: Middle Wisconsin climate fluctuations recorded in central Alaskan loess. *Géographie Physique et Quaternaire,* **44**: 3–13.

Begét, J. E., 1991: Paleoclimatic significance of high latitude loess deposits. *In* Weller, G. (ed.), *International Conference on the Role of the Polar Regions in Global Change,* Fairbanks: University of Alaska,

Geophysical Institute and the Center for Global Change, **2**: 594–598.

Begét, J. E., 1996: Tephrochronology and paleoclimatology of the last interglacial-glacial cycle recorded in Alaskan loess deposits. *Quaternary International,* **34–36**: 121–126.

Begét, J. E., and D. B. Hawkins, 1989: Influence of orbital parameters on Pleistocene loess deposition in central Alaska. *Nature,* **337**: 151–153.

Begét, J. E., D. B. Stone, and D. B. Hawkins, 1990: Paleoclimatic forcing of magnetic susceptibility variations in Alaskan loess during the Quaternary. *Geology,* **18**: 40–43.

Berger, G. W., and A. J. Busacca, 1995: Thermoluminescence dating of late Pleistocene loess and tephra from eastern Washington and southern Oregon, and implications for the eruptive history of Mount St. Helens. *Journal of Geophysical Research,* **100**: 22361–22374.

Bidart, S., 1996: Sedimentological study of aeolian soil parent materials in the Rio Sauce Grande basin, Buenos Aires Province, Argentina. *Catena,* **27**: 191–207.

Bloom, A., 1990: Some questions about the Pampean loess. *In* Zarate, M. (ed.), *Properties, Chronology and Paleoclimatic Significance of Loess.* International Symposium on Loess, 25 Nov.–1 Dec. 1990, Mar del Plata, Argentina, pp. 29–31.

Bowles, F. A., 1975: Paleoclimatic significance of quartz/illite variations in cores from the eastern equatorial North Atlantic. *Quaternary Research,* **5**: 225–235.

Bradbury, J. P., 1982: Neogene and Quaternary lacustrine diatoms of the western Snake River Basin Idaho-Oregon, U.S.A. *Acta Geologica Academiae Scientiarum Hungaricae,* **25**: 97–122.

Bradbury, J. P., 1992: Late Cenozoic lacustrine and climatic environments at Tule Lake, northern Great Basin. *Climate Dynamics,* **6**: 275–285.

Bradbury, J. P., B. Leyden, M. Salgado-Labouriau, W. M. Lewis, Jr., C. Schubert, M. W. Binford, D. G. Frey, D. R. Whitehead, and F. H. Weibezahn, 1981: Late Quaternary environmental history of Lake Valencia, Venezuela. *Science,* **214**: 1299–1305.

Busacca, A. J., 1991: Loess deposits and soils of the Palouse and vicinity. *In* Baker, V. R. (ed.), Quaternary Geology of the Columbia Plateau, *In* Morrison, R. B. (ed.), *Quaternary Non-Glacial Geology: Conterminous United States, Vol. K-2. The Geology of North America,* Boulder, CO: Geological Society of America, pp. 215–250.

Busacca, A. J., and E. V. McDonald, 1994: Regional sedimentation of late Quaternary loess on the Columbia Plateau: Sediment source areas and loess distribution patterns. *Washington Division of Geology and Earth Resources Bulletin,* **80**: 181–190.

Busacca, A. J., K. T. Nelstead, E. V. McDonald, and M. D. Purser, 1992: Correlation of distal tephra layers in loess in the Channeled Scabland and Palouse of Washington State. *Quaternary Research,* **37**: 281–303.

Cabral, M., G. Larroza, and V. Ferreiro, 1988: Estudio hidrogeomorfológico de los partidos de Bolívar e Hipólito Yrigoyen. Comisión de Investigaciones Científicas de la Provincia de Buenos Aires. Unpublished report, 80 pp.

Cantú, M. P., and S. Degiovanni, 1984: Geomorfología de la región centro-sur de la provincia de Córdoba. *Noveno Congreso Geológico Argentino,* San Carlos de Bariloche, ACTAS IV, pp. 76–92.

Carey, S. N., and H. Sigurdsson, 1980: The Roseau ash: Deep-sea tephra deposits from a major eruption on Dominica, Lesser Antilles arc. *Journal of Volcanology and Geothermal Research,* **7**: 67–86.

Carignano, C. A., 1999: Late Pleistocene to recent climate change in Córdoba Province, Argentina: Geomorphological evidence. *Quaternary International,* **57/58**: 117–134.

Carter, L. D., 1981: A Pleistocene sand sea on the Alaskan Arctic Coastal Plain. *Science,* **211**: 381–383.

Carter, L. D., and J. P. Galloway, 1984: Lacustrine and eolian deposits of Wisconsin age at Riverside Bluff in the upper Tanana River valley, Alaska. *U.S. Geological Survey Circular,* **868**: 66–67.

Clapperton, C., 1993: *Quaternary Geology and Geomorphology of South America.* Amsterdam: Elsevier, 779 pp.

Clarke, M. L., 1994: Infra-red stimulated luminescence ages from aeolian sand and alluvial fan deposits from the eastern Mojave Desert, California. *(Quaternary Geochronology) Quaternary Science Reviews,* **13**: 533–538.

Clarke, M. L., and H. M. Rendell, 1998: Climate change impacts on sand supply and the formation of desert sand dunes in the southwest U.S.A. *Journal of Arid Environments,* **39**: 517–531.

Clarke, M. L., A. G. Wintle, and N. Lancaster, 1996: Infra-red stimulated luminescence dating of sands from the Cronese Basins, Mojave Desert. *Geomorphology,* **17**: 199–205.

COHMAP Members, 1988: Climatic changes of the last 18,000 years: Observations and model simulations. *Science,* **241**: 1043–1052.

Curry, B. B., and L. R. Follmer, 1992: The last interglacial-glacial transition in Illinois: 123–25 ka. *In* Clark, P. U., and P. D. Clark (eds.), *The Last Interglacial-Glacial Transition in North America.* Boulder, CO: Geological Society of America Special Paper 270, pp. 71–88.

Darwin, C., 1846: An account of the fine dust which falls upon vessels in the Atlantic Ocean. *Quarterly Journal of the Geological Society of London,* **2**: 26–30.

David, P., 1970: Discovery of Mazama ash in Saskatchewan, Canada. *Canadian Journal of Earth Sciences,* **7**: 1579–1583.

David, P., 1971: The Brookdale road section and its significance in the chronological studies of dune activities in the Brandon Sand Hills of Manitoba. Geological Association of Canada Special Paper 9, pp. 293–299.

David, P. P., 1981: Stabilized dune ridges in northern Saskatchewan. *Canadian Journal of Earth Sciences,* **18**: 286–310.

David, P. P., 1988: The coeval eolian environment of the Champlain Sea Episode. *In* Gadd, N. R. (ed.), The late Quaternary development of the Champlain Sea Basin. Geological Association of Canada Special Paper 35, pp. 291–305.

Dean, W. E., T. S. Ahlbrandt, R. Y. Anderson, and J. P. Bradbury, 1996: Regional aridity in North America during the middle Holocene. *The Holocene,* **6**: 145–155.

Delany, A. C., D. W. Parkin, J. J. Griffin, E. D. Goldberg, and B. E. F. Reimann, 1967: Airborne dust collected at Barbados. *Geochimica et Cosmochimica Acta,* **31**: 885–909.

deMenocal, P. B., W. F. Ruddiman, and E. M. Pokras, 1993: Influences of high- and low-latitude processes on African terrestrial climate: Pleistocene eolian records from equatorial Atlantic Ocean Drilling Program site 663. *Paleoceanography,* **8**: 209–242.

Finkel, H. J., 1959: The barchans of southern Peru. *Journal of Geology,* **67**: 614–647.

Follmer, L. R., 1996: Loess studies in central United States: Evolution of concepts. *Engineering Geology,* **45**: 287–304.

Forman, S. L., and P. Maat, 1990: Stratigraphic evidence for late Quaternary dune activity near Hudson on the Piedmont of northern Colorado. *Geology,* **18**: 745–748.

Forman, S. L., E. A. Bettis, III, T. J. Kemmis, and B. B. Miller, 1992a: Chronologic evidence of multiple periods of loess deposition during the late Pleistocene in the Missouri and Mississippi River Valley, United States: Implications for the activity of the Laurentide ice sheet. *Palaeogeography, Palaeoclimatology, Palaeoecology,* **93**: 71–83.

Forman, S. L., A. F. H. Goetz, and R. H. Yuhas, 1992b: Large-scale stabilized dunes on the High Plains of Colorado: Understanding the landscape response to Holocene climates with the aid of images from space. *Geology,* **20**: 145–148.

Forman, S. L., R. Oglesby, V. Markgraf, and T. Stafford, 1995: Paleoclimatic significance of late Quaternary eolian deposition on the Piedmont and High Plains, central United States. *Global and Planetary Change,* **11**: 35–55.

Forman, S. L., R. P. Smith, W. R. Hackett, J. A. Tullis, and P. A. McDaniel, 1993: Timing of late Quaternary glaciations in the western United States based on the age of loess on the eastern Snake River Plain, Idaho. *Quaternary Research,* **40**: 30–37.

Fraser, T. A., and C. R. Burn, 1997: On the nature and origin of "muck" deposits in the Klondike area, Yukon Territory. *Canadian Journal of Earth Sciences,* **34**: 1333–1344.

Fredlund, G. G., 1995: Late Quaternary pollen record from Cheyenne Bottoms, Kansas. *Quaternary Research,* **43**: 67–79.

Fritz, S. C., D. R. Engstrom, and B. J. Haskell, 1994: "Little Ice Age" aridity in the North American Great Plains: A high-resolution reconstruction of salinity fluctuations from Devils Lake, North Dakota, U.S.A. *The Holocene,* **4**: 69–73.

Fritz, S. C., S. Juggins, R. W. Battarbee, and D. R. Engstrom, 1991: Reconstruction of past changes in salinity and climate using a diatom-based transfer function. *Nature,* **352**: 706–708.

Frye, J. C., H. D. Glass, and H. B. Willman, 1968: Mineral zonation of Woodfordian loesses of Illinois. *Illinois State Geological Survey Circular,* **427**: 1–44.

Gallet, S., B. Jahn, B. van Vliet-Lanoë, A. Dia, and E. Rossello, 1998: Loess geochemistry and its implications for particle origin and composition of the upper continental crust. *Earth and Planetary Science Letters,* **156**: 157–172.

Galloway, J. P., and L. D. Carter, 1993: Late Holocene longitudinal and parabolic dunes in northern Alaska: Preliminary interpretations of age and paleoclimatic significance. *U.S. Geological Survey Circular,* **2068**: 3–11.

Galloway, J. P., and L. D. Carter, 1994: Paleowind directions for late Holocene dunes on the western Arctic Coastal Plain, northern Alaska. *U.S. Geological Survey Bulletin,* **2107**: 27–30.

Gardenal, M., 1986: Geomorfologia del partido de Salliquelo, provincia de Buenos Aires. Comisión de Investigaciones Científicas de la Provincia de Buenos Aires. Unpublished report, 60 pp.

Gaylord, D. R., 1990: Holocene paleoclimatic fluctuations revealed from dune and interdune strata in Wyoming. *Journal of Arid Environments,* **18**: 123–138.

Giménez, J., 1990: The sandy Pampa region. *In* Zárate, M., and N. Flegenheimer (eds.), *Loess Stratigraphy and Geomorphology of the Pampas,* Field Guide for the International Symposium on "Properties, Chronology and Paleoclimatic Significance of Loess," 25 November–1 December, 1990, Mar del Plata, Argentina, pp. 34–42.

Glaccum, R. A., and J. M. Prospero, 1980: Saharan aerosols over the tropical North Atlantic—Mineralogy. *Marine Geology,* **37**: 295–321.

Glenn, W. R., W. D. Nettleton, C. J. Fowkes, and D. M. Daniels, 1983: Loessial deposits and soils of the Snake and tributary river valleys of Wyoming and eastern Idaho. *Soil Science Society of America Journal,* **47**: 547–552.

González, M. A., and N. Weiler, 1987/1988: Sitio arqueológico Fortín Necochea. Informe Geológico Preliminar. *Paleoentológica,* **4**: 55–63.

Grigal, D. F., R. C. Severson, and G. E. Goltz, 1976: Evidence of eolian activity in north-central Minnesota 8000–5000 yr ago. *Geological Society of America Bulletin,* **87**: 1251–1254.

Grimley, D. A., L. R. Follmer, and E. D. McKay, 1998: Magnetic susceptibility and mineral zonations controlled by provenance in loess along the Illinois and central Mississippi River valleys. *Quaternary Research,* **49**: 24–36.

Groeber, P., 1936: Oscilaciones de clima en la Argentina desde el Plioceno. *Revista Centro Estudiantes Doctorado Ciencias Naturales,* Univ. Nacional de Buenos Aires, **1**(2): 71–84.

Hack, J. T., 1941: Dunes of the western Navajo Country. *Geographical Review,* **31**: 240–263.

Hamilton, T. D., and G. M. Ashley, 1993: Epiguruk—A late Quaternary environmental record from northwestern Alaska. *Geological Society of America Bulletin*, **105**: 583–602.

Hamilton, T. D., J. L. Craig, and P. V. Sellmann, 1988: The Fox permafrost tunnel: A late Quaternary geologic record in central Alaska. *Geological Society of America Bulletin*, **100**: 948–969.

Harvey, L. D. D., 1988: Climatic impact of ice-age aerosols. *Nature*, **334**: 333–335.

Höfle, C., and C.-L. Ping, 1996: Properties and soil development of late-Pleistocene paleosols from Seward Peninsula, northwest Alaska. *Geoderma*, **71**: 219–243.

Holliday, V. T., 1995a: Late Quaternary stratigraphy of the Southern High Plains. *In* Johnson, E. (ed.), *Ancient Peoples and Landscapes*. Lubbock: Museum of Texas Tech University, pp. 289–313.

Holliday, V. T., 1995b: Stratigraphy and Paleoenvironments of Late Quaternary Valley Fills on the Southern High Plains. *Geological Society of America Memoir*, **186**: 1–136.

Holliday, V. T., 1995c: Stratigraphy and geochronology of dune fields on the Southern High Plains, Texas and New Mexico. *Geological Society of America Abstracts with Programs*, **27**(3): 59.

Holliday, V. T., 1997a: Origin and evolution of lunettes on the High Plains of Texas and New Mexico. *Quaternary Research*, **47**: 54–69.

Holliday, V. T., 1997b: *Paleo-Indian Geoarchaeology of the Southern High Plains*. Austin: Univ. of Texas Press, 297 pp.

Holliday, V. T., 2000: Folsom drought and Episodic drying on the Southern High Plains from 10,900–10,200 ^{14}C yr B.P. *Quaternary Research*, **53**: 1–12.

Hunt, C. B., 1977: Surficial geology of southeast New Mexico. New Mexico Bureau of Mines and Mineral Resources Geologic Map 41, scale 1:500,000.

Hurtado, M., and J. Giménez, 1988: Entisoles de la región pampeana. Génesis, clasificación, cartografía y mineralogía. *Relatos 2das Jornadas de Suelos de la Región Pampeana, La Plata*, 97–137.

Iriondo, M. H., 1990a: Map of the South American plains—Its present state. *Quaternary of South America and Antarctic Peninsula*, **6**: 297–308.

Iriondo, M. H., 1990b: A late dry period in the Argentine plains. *Quaternary of South America and Antarctic Peninsula*, **7**: 197–218.

Iriondo, M. H., 1997: Models of deposition of loess and loessoids in the upper Quaternary of South America. *Journal of South American Earth Sciences*, **10**: 71–79.

Iriondo, M., 1999: Climatic changes in the South American plains: Records of a continental-scale oscillation. *Quaternary International*, **57/58**: 93–112.

Iriondo, M. H., and N. O. García, 1993: Climatic variations in the Argentine plains during the last 18,000 years. *Palaeogeography, Palaeoclimatology, Palaeoecology*, **101**: 209–220.

Jackson, M. L., T. W. M. Levelt, J. K. Syers, R. W. Rex, R. N. Clayton, G. D. Sherman, and G. Uehara, 1971: Geomorphological relationships of tropospherically derived quartz in the soils of the Hawaiian Islands. *Soil Science Society of America Proceedings*, **35**: 515–525.

Jennings, C. W., 1967: Geologic map of California Salton Sea sheet. California Division of Mines and Geology, Sacramento.

Johnson, D. L., 1977: The late Quaternary climate of coastal California: Evidence for an Ice-Age refugium. *Quaternary Research*, **8**: 154–179.

Johnson, W. C., and K. L. Willey, 2000: Isotopic and rock magnetic expression of environmental change at the Pleistocene-Holocene transition in the central Great Plains. *Quaternary International*, **67**: 89–106.

Joussaume, S., 1993: Paleoclimatic tracers: An investigation using an atmospheric general circulation model under ice age conditions 1. Desert dust. *Journal of Geophysical Research*, **98**: 2767–2805.

Kleiss, H. J., and J. G. Fehrenbacher, 1973: Loess distribution as revealed by mineral variations. *Soil Science Society of America Proceedings*, **37**: 291–295.

Kolla, V., P. E. Biscaye, and A. F. Hanley, 1979: Distribution of quartz in late Quaternary Atlantic sediments in relation to climate. *Quaternary Research*, **11**: 261–277.

Kröhling, D. M., 1999: Upper Quaternary geology of the lower Carcarañá Basin, North Pampa, Argentina. *Quaternary International*, **57/58**: 135–148.

Kröhling, D., and M. Iriondo, 1998: Loess in Argentina. Temperate and tropical. International Joint Field Meeting, Guía de Campo 1 (Pampa Norte, Cuenca del Carcarañá), INQUA, Universidad Nacional de Entre Ríos, 58 pp.

Kröhling, D. M., and M. Iriondo, 1999: Upper Quaternary palaeoclimates of the Mar Chiquita area, North Pampa, Argentina. *Quaternary International*, **57/58**: 149–163.

Kutzbach, J. E., and P. J. Guetter, 1986: The influence of changing orbital parameters and surface boundary conditions on climate simulations for the past 18,000 years. *Journal of Atmospheric Science*, **43**: 1726–1759.

Kutzbach, J. E., and H. E. Wright, Jr., 1985: Simulation of the climate of 18,000 years B.P.: Results for the North American/North Atlantic/European sector and comparison with the geologic record of North America. *Quaternary Science Reviews*, **4**: 147–187.

Kutzbach, J. E., R. Gallimore, S. Harrison, P. Behling, R. Selin, and F. Laarif, 1998: Climate and biome simulations for the past 21,000 years. *Quaternary Science Reviews*, **17**: 473–506.

Kutzbach, J. E., P. J. Guetter, P. J. Behling, and R. Selin, 1993: Simulated climatic changes: Results of the COHMAP climate-model experiments. *In* Wright, H. E., Jr., J. E. Kutzbach, T. Webb, III, W. F. Ruddiman, F. A. Street-Perrott, and P. J. Bartlein (eds.), *Global Climates Since the Last Glacial Maximum*. Minneapolis: Univ. of Minnesota Press, pp. 24–93.

Laird, K. R., S. C. Fritz, K. A. Maasch, and B. F. Cumming, 1996: Greater drought intensity and frequency before A.D. 1200 in the northern Great Plains, U.S.A. *Nature*, **384**: 552–554.

Lancaster, N., 1994: Controls on aeolian activity: Some new perspectives from the Kelso Dunes, Mojave Desert, California. *Journal of Arid Environments*, **27**: 113–125.

Lancaster, N., and V. P. Tchakerian, 1996: Geomorphology and sediments of sand ramps in the Mojave Desert. *Geomorphology*, **17**: 151–165.

Lancaster, N., R. Greeley, and P. R. Christensen, 1987: Dunes of the Gran Desierto sand-sea, Sonora, México. *Earth Surface Processes and Landforms*, **12**: 277–288.

Lea, P. D., 1996: Vertebrate tracks in Pleistocene eolian sand-sheet deposits of Alaska. *Quaternary Research*, **45**: 226–240.

Lea, P. D., and C. F. Waythomas, 1990: Late-Pleistocene eolian sand sheets in Alaska. *Quaternary Research*, **34**: 269–281.

Lewis, G. C., and M. A. Fosberg, 1982: Distribution and character of loess and loess soils in southeastern Idaho. In Bonnichsen, B., and R. M. Breckinridge (eds.), *Cenozoic Geology of Idaho. Idaho Bureau of Mines and Geology Bulletin*, **26**, pp. 705–716.

Lewis, G. C., M. A. Fosberg, R. E. McDole, and J. C. Chugg, 1975: Distribution and some properties of loess in south-central and southeastern Idaho. *Soil Science Society of America Proceedings*, **39**: 1165–1168.

Li, X., H. Maring, D. Savoie, K. Voss, and J. M. Prospero, 1996: Dominance of mineral dust in aerosol light-scattering in the North Atlantic trade winds. *Nature*, **380**: 416–419.

Loope, D. B., J. B. Swinehart, and J. P. Mason, 1995: Dune-dammed paleovalleys of the Nebraska Sand Hills: Intrinsic versus climatic controls on the accumulation of lake and marsh sediments. *Geological Society of America Bulletin*, **107**: 396–406.

Lugn, A. L., 1935: The Pleistocene geology of Nebraska. *Nebraska Geological Survey Bulletin*, **10**(2): 1–223.

Lugn, A. L., 1939: Nebraska in relation to the problems of Pleistocene stratigraphy. *American Journal of Science*, **237**: 851–884.

Lugn, A. L., 1962: The origin and sources of loess. *University of Nebraska Studies,* **26**: 1–105.

Lugn, A. L., 1968: The origin of loesses and their relation to the Great Plains of North America. *In* Schultz, C. B., and J. C. Frye (eds.), *Loess and Related Eolian Deposits of the World.* Lincoln: Univ. of Nebraska Press, pp. 139–182.

Maat, P. B., and W. C. Johnson, 1996: Thermoluminescence and new ^{14}C age estimates for late Quaternary loesses in southwestern Nebraska. *Geomorphology,* **17**: 115–128.

Madole, R. F., 1994: Stratigraphic evidence of desertification in the west-central Great Plains within the past 1000 years. *Geology,* **22**: 483–486.

Madole, R. F., 1995: Spatial and temporal patterns of late Quaternary eolian deposition, eastern Colorado, U.S.A. *Quaternary Science Reviews,* **14**: 155–177.

Mahowald, N., K. Kohfeld, M. Hansson, Y. Balkanski, S. P. Harrison, I. C. Prentice, M. Schulz, and H. Rodhe, 1999: Dust sources and deposition during the last glacial maximum and current climate: A comparison of model results with paleodata from ice cores and marine sediments. *Journal of Geophysical Research,* **104**: 15895–15916.

Markewich, H. W., D. A. Wysocki, M. J. Pavich, E. M. Rutledge, H. T. Millard, Jr., F. J. Rich, P. B. Maat, M. Rubin, and J. P. McGeehin, 1998: Paleopedology plus TL, ^{10}Be, and ^{14}C dating as tools in stratigraphic and paleoclimatic investigations, Mississippi River Valley, U.S.A. *Quaternary International,* **51/52**: 143–167.

Markgraf, V., 1993a: Paleoenvironments and paleoclimates in Tierra del Fuego and southernmost Patagonia, South America. *Palaeogeography, Palaeoclimatology, Palaeoecology,* **102**: 53–68.

Markgraf, V., 1993b: Climatic history of Central and South America since 18,000 yr B.P.: Comparison of pollen records and model simulations. *In* Wright, H. E., Jr., J. E. Kutzbach, T. Webb, III, W. F. Ruddiman, F. A. Street-Perrott, and P. J. Bartlein (eds.), *Global Climates Since the Last Glacial Maximum.* Minneapolis: Univ. of Minnesota Press, pp. 357–385.

Martin, C. W., 1993: Radiocarbon ages on late Pleistocene loess stratigraphy of Nebraska and Kansas, central Great Plains, U.S.A. *Quaternary Science Reviews,* **12**: 179–188.

Mason, J. A., and M. S. Kuzila, 2000: Episodic Holocene loess deposition in Central Nebraska. *Quaternary International,* **67**: 119–131.

Mason, J. P., J. B. Swinehart, and D. B. Loope, 1997: Holocene history of lacustrine and marsh sediments in a dune-blocked drainage, southwestern Nebraska Sand Hills, U.S.A. *Journal of Paleolimnology,* **17**: 67–83.

May, D. W., and S. R. Holen, 1993: Radiocarbon ages of soils and charcoal in late Wisconsinan loess, south-central Nebraska. *Quaternary Research,* **39**: 55–58.

McDonald, E. V., and A. J. Busacca, 1992: Late Quaternary stratigraphy of loess in the Channeled Scabland and Palouse regions of Washington State. *Quaternary Research,* **38**: 141–156.

McKay, E. D., 1979: Wisconsinan loess stratigraphy of Illinois. In Follmer, L. R., E. D. McKay, J. A. Lineback, and D. L. Gross (eds.), *Wisconsinan, Sangamonian, and Illinoian Stratigraphy in Central Illinois.* Illinois State Geological Survey Guidebook 13, pp. 95–108.

Molina-Cruz, A., 1977: The relation of the southern trade winds to upwelling processes during the last 75,000 years. *Quaternary Research,* **8**: 324–338.

Molina-Cruz, A., and P. Price, 1977: Distribution of opal and quartz on the ocean floor of the subtropical southeastern Pacific. *Geology,* **5**: 81–84.

Muhs, D. R., 1982: A soil chronosequence on Quaternary marine terraces, San Clemente Island, California. *Geoderma,* **28**: 257–283.

Muhs, D. R., 1983: Airborne dust fall on the California Channel Islands, U.S.A. *Journal of Arid Environments,* **6**: 223–238.

Muhs, D. R., 1985: Age and paleoclimatic significance of Holocene sand dunes in northeastern Colorado. *Annals of the Association of American Geographers,* **75**: 566–582.

Muhs, D. R., and E. A. Bettis, III, 2000: Geochemical variations in Peoria Loess of western Iowa indicate paleowinds of midcontinental North America during last glaciation. *Quaternary Research,* **53**: 49–61.

Muhs, D. R., and V. T. Holliday, 1995: Evidence of active dune sand on the Great Plains in the 19th century from accounts of early explorers. *Quaternary Research,* **43**: 198–208.

Muhs, D. R., and P. B. Maat, 1993: The potential response of Great Plains eolian sands to greenhouse warming and precipitation reduction on the Great Plains of the U.S.A. *Journal of Arid Environments,* **25**: 351–361.

Muhs, D. R., and S. A. Wolfe, 1999: Sand dunes of the northern Great Plains of Canada and the United States. *Geological Survey of Canada Bulletin,* **534**: 183–197.

Muhs, D. R., T. Ager, T. W. Stafford, Jr., M. Pavich, J. E. Begét, and J. P. McGeehin, 1997c: The last interglacial-glacial cycle in late Quaternary loess, central interior Alaska. *In* Elias, S. A., and J. Brigham-Grette (eds.), *Program and Abstracts, Beringian Paleoenvironments Workshop,* Florissant, Colorado, pp. 109–112.

Muhs, D. R., J. N. Aleinikoff, T. W. Stafford, Jr., R. Kihl, J. Been, S. Mahan, and S. D. Cowherd, 1999: Late Quaternary loess in northeastern Colorado: Part I–Age and paleoclimatic significance. *Geological Society of America Bulletin,* **111**: 1861–1875.

Muhs, D. R., C. A. Bush, S. D. Cowherd, and S. Mahan, 1995: Geomorphic and geochemical evidence for the source of sand in the Algodones dunes, Colorado Desert, southeastern California. *In* Tchakerian, V. P. (ed.), *Desert Aeolian Processes.* London: Chapman & Hall, pp. 37–74.

Muhs, D. R., C. A. Bush, K. C. Stewart, T. R. Rowland, and R. C. Crittenden, 1990: Geochemical evidence of Saharan dust parent material for soils developed on Quaternary limestones of Caribbean and western Atlantic islands. *Quaternary Research,* **33**: 157–177.

Muhs, D. R., R. C. Crittenden, J. N. Rosholt, C. A. Bush, and K. C. Stewart, 1987: Genesis of marine terrace soils, Barbados, West Indies: Evidence from mineralogy and geochemistry. *Earth Surface Processes and Landforms,* **12**: 605–618.

Muhs, D. R., T. W. Stafford, Jr., J. Been, S. Mahan, J. Burdett, G. Skipp, and Z. M. Rowland, 1997b: Holocene eolian activity in the Minot dune field, North Dakota. *Canadian Journal of Earth Sciences,* **34**: 1442–1459.

Muhs, D. R., T. W. Stafford, Jr., S. D. Cowherd, S. A. Mahan, R. Kihl, P. B. Maat, C. A. Bush, and J. Nehring, 1996: Origin of the late Quaternary dune fields of northeastern Colorado. *Geomorphology,* **17**: 129–149.

Muhs, D. R., T. W. Stafford, Jr., J. B. Swinehart, S. D. Cowherd, S. A. Mahan, and C. A. Bush, 1997a: Late Holocene eolian activity in the mineralogically mature Nebraska Sand Hills. *Quaternary Research,* **48**: 162–176.

Muhs, D. R., J. B. Swinehart, D. B. Loope, J. Been, S. A. Mahan, and C. A. Bush, 2000: Geochemical evidence for an eolian sand dam across the North and South Platte Rivers in Nebraska. *Quaternary Research,* **53**: 214–222.

Nickling, W. G., 1978: Eolian sediment transport during dust storms: Slims River Valley, Yukon Territory. *Canadian Journal of Earth Sciences,* **15**: 1069–1084.

Olson, C. G., W. D. Nettleton, D. A. Porter, and B. R. Brasher, 1997: Middle Holocene aeolian activity on the High Plains of west-central Kansas. *The Holocene,* **7**: 255–261.

Overpeck, J., D. Rind, A. Lacis, and R. Healy, 1996: Possible role of dust-induced regional warming in abrupt climate change during the last glacial period. *Nature,* **384**: 447–449.

Petit, J. R., L. Mounier, J. Jouzel, Y. S. Korotkevich, V. I. Kotlyakov, and

C. Lorius, 1990: Palaeoclimatological and chronological implications of the Vostok core dust record. *Nature,* **343:** 56–58.

Péwé, T. L., 1951: An observation on wind-blown silt. *Journal of Geology,* **59:** 399–401.

Péwé, T. L., 1975: Quaternary geology of Alaska. U.S. Geological Survey Professional Paper 835, 145 pp.

Pierce, K. L., H. R. Covington, P. L. Williams, and D. H. McIntyre, 1983: Geologic map of the Cotterel Mountains and the northern Raft River valley, Cassia County, Idaho. U.S. Geological Survey Miscellaneous Investigations Series Map I-1450, scale 1:48,000.

Pierce, K. L., M. A. Fosberg, W. E. Scott, G. C. Lewis, and S. M. Colman, 1982: Loess deposits of southeastern Idaho: Age and correlation of the upper two loess units. In Bonnichsen, B., and R. M. Breckinridge (eds.), Cenozoic Geology of Idaho. *Idaho Bureau of Mines and Geology Bulletin,* 26, pp. 717–725.

Prado, J. L., and M. T. Alberdi, 1999: The mammalian record and climatic change over the last 30,000 years in the Pampean region, Argentina. *Quaternary International,* **57/58:** 165–174.

Prieto, A., 1996: Late Quaternary vegetational and climatic changes in the Pampa grassland of Argentina. *Quaternary Research,* **45:** 73–88.

Prospero, J. M., 1981: Arid regions as sources of mineral aerosols in the marine atmosphere. In Péwé, T. L. (ed.), Desert Dust: Origin, Characteristics, and Effect on Man. Geological Society of America Special Paper 186, pp. 71–86.

Prospero, J. M., and R. T. Nees, 1977: Dust concentration in the atmosphere of the equatorial North Atlantic: Possible relationship to the Sahelian drought. *Science,* **196:** 1196–1198.

Prospero, J. M., and R. T. Nees, 1986: Impact of the North African drought and El Niño on mineral dust in the Barbados trade winds. *Nature,* **320:** 735–738.

Prospero, J. M., E. Bonatti, C. Schubert, and T. N. Carlson, 1970: Dust in the Caribbean atmosphere traced to an African dust storm. *Earth and Planetary Science Letters,* **9:** 287–293.

Prospero, J. M., R. A. Glaccum, and R. T. Nees, 1981: Atmospheric transport of soil dust from Africa to South America. *Nature,* **289:** 570–572.

Pye, K., 1987: *Aeolian Dust and Dust Deposits.* San Diego: Academic Press, 334 pp.

Pye, K., N. R. Winspear, and L. P. Zhou, 1995: Thermoluminescence ages of loess and associated sediments in central Nebraska, U.S.A. *Palaeogeography, Palaeoclimatology, Palaeoecology,* **118:** 73–87.

Ramonell, C., M. Iriondo, and R. Krîmer, 1992: Guía de campo 1 centro-este de San Luis. *5ta reunión de Campo Comité Argentino para la Investigación del Cuaternario,* 18–19 Octubre, Depto de Geología y Minería. Universidad Nacional de San Luis.

Rea, D. K., 1994: The paleoclimatic record provided by eolian deposition in the deep sea: The geologic history of wind. *Reviews of Geophysics,* **32:** 159–195.

Rendell, H. M., and N. L. Sheffer, 1996: Luminescence dating of sand ramps in the eastern Mojave Desert. *Geomorphology,* **17:** 187–197.

Rendell, H. M., N. Lancaster, and V. P. Tchakerian, 1994: Luminescence dating of late Quaternary aeolian deposits at Dale Lake and Cronese Mountains, Mojave Desert, California. *(Quaternary Geochronology) Quaternary Science Reviews,* **13:** 417–422.

Richardson, C. A., E. V. McDonald, and A. J. Busacca, 1997: Luminescence dating of loess from the northwest United States. *(Quaternary Geochronology) Quaternary Science Reviews,* **16:** 403–415.

Rind, D., 1987: Components of the Ice Age circulation. *Journal of Geophysical Research,* **92:** 4241–4281.

Ringe, D., 1970: Sub-loess basalt topography in the Palouse Hills, southeastern Washington. *Geological Society of America Bulletin,* **81:** 3049–3060.

Roa, P., 1979: Estudio de los médanos de los Llanos Centrales de Venezuela: Evidencias de un clima desértico. *Acta Biológica Venezolana,* **10:** 19–49.

Roa, P., 1980: Algunos aspectos de la evolución sedimentológia y geomorfológica de la llanura aluvial de desborde en la Bajo Llano. *Boletín de la Sociedad Venezolana de Ciencias Exactas Naturales,* **35:** 31–47.

Rodbell, D. T., S. L. Forman, J. Pierson, and W. C. Lynn, 1997: Stratigraphy and chronology of Mississippi Valley loess in western Tennessee. *Geological Society of America Bulletin,* **109:** 1134–1148.

Ruddiman, W. F., 1997: Tropical Atlantic terrigenous fluxes since 25,000 yrs B.P. *Marine Geology,* **136:** 189–207.

Ruhe, R. V., 1983: Depositional environment of late Wisconsin loess in the midcontinental United States. *In* Wright, H. E., Jr., and S. C. Porter (eds.), *Late-Quaternary Environments of the United States, Vol. 1: The Late Pleistocene.* Minneapolis: University of Minnesota Press, pp. 130–137.

Ruhe, R. V., G. A. Miller, and W. J. Vreeken, 1971: Paleosols, loess sedimentation and soil stratigraphy. *In* Yaalon, D. H. (ed.), *Paleopedology—Origin, Nature and Dating of Paleosols.* Jerusalem: Israel Universities Press, pp. 41–59.

Running, G. L., IV, 1996: The Sheyenne Delta from the Cass phase to the present: Landscape evolution and paleoenvironment. *In* Harris, K. L., M. R. Luther, and J. R. Reid (eds.), Quaternary Geology of the Southern Lake Agassiz Basin, Field Trip Guidebook, Midwest Friends of the Pleistocene, 43rd Annual Meeting. North Dakota Geological Survey Miscellaneous Series 82, pp. 136–152.

Salazar Lea Plaza, J. C., 1980: Regiones fisiográficas. *In* Inventario Integrado de los Recursos Naturales de la Provincia de la Pampa. Clima, Geomorfología, Suelo y Vegetación. *Instituto Nacional de Tecnología Agropecuaria,* Provincia de la Pampa-Universidad Nacional de la Pampa, 89, 444 pp.

Sancetta, C., M. Lyle, L. Heusser, R. Zahn, and J. P. Bradbury, 1992: Late-glacial to Holocene changes in winds, upwelling, and seasonal production of the northern California Current system. *Quaternary Research,* **38:** 359–370.

Sarnthein, M., 1978: Sand deserts during glacial maximum and climatic optimum. *Nature,* **272:** 43–46.

Sayago, J. M., 1995: The Argentine neotropical loess: An overview. *Quaternary Science Reviews,* **14:** 755–766.

Sayago, J. M., 1997: Distribución y génesis del loess argentino. *VI Congresso da Associaáão Brasileira de Estudos do Quaternçrio e Reunion sobre o Quaternario da America do Sul.* Resumos Expandidos, Curitiba, Paraná, Brasil, 27 Jul-3 ago 1997, pp. 254–260.

Scott, W. E., 1982: Surficial geologic map of the eastern Snake River Plain and adjacent areas, 111° to 115°W, Idaho and Wyoming. U.S. Geological Survey Miscellaneous Investigations Series Map I-1372.

Smith, G. D., 1942: Illinois loess: Variations in its properties and distribution, a pedologic interpretation. *University of Illinois Agricultural Experiment Station Bulletin,* **490:** 139–184.

Stein, R., 1995: Clay and bulk mineralogy of late Quaternary sediments at site 893A, Santa Barbara Basin. *In* Kennett, J. P., J. G. Baldauf, and M. Lyle (eds.), *Proceedings of the Ocean Drilling Program, Scientific Results,* **146**(2): 89–102.

Stokes, S., and C. Breed, 1993: A chronostratigraphic re-evaluation of the Tusayan Dunes, Moenkopi Plateau and southern Ward Terrace, northeastern Arizona. In Pye, K. (ed.), The Dynamics and Environmental Context of Aeolian Sedimentary Systems. Geological Society Special Publication 72, pp. 75–90.

Stokes, S., and J. B. Swinehart, 1997: Middle and late Holocene dune reactivation in the Nebraska Sand Hills. *The Holocene,* **7:** 263–272.

Swinehart, J. B., 1990: Wind-blown deposits. *In* Bleed, A., and C. Flowerday (eds.), *An Atlas of the Sand Hills.* Lincoln: University of Nebraska Press, Resource Atlas No. 5a, pp. 43–56.

Swinehart, J. B., and R. F. Diffendal, Jr., 1990: Geology of the pre-dune strata. *In* Bleed, A., and C. Flowerday (eds.), *An Atlas of the Sand Hills.* Lincoln: University of Nebraska Press, Resource Atlas No. 5a, pp. 29–42.

Tchakerian, V. P., 1991: Late Quaternary aeolian geomorphology of the Dale Lake sand sheet, southern Mojave Desert, California. *Physical Geography,* **12**: 347–369.

Tegan, I., A. A. Lacis, and I. Fung, 1996: The influence on climate forcing of mineral aerosols from disturbed soils. *Nature,* **380**: 419–422.

Teruggi, M. E., 1957: The nature and origin of Argentine loess. *Journal of Sedimentary Petrology,* **27**: 322–332.

Teruggi, M. E., M. C. Etchichury, and J. R. Remiro, 1957: Estudio sedimentologico de los terrenos de las barrancas de la zona Mar del Plata-Miramar. *Revista del Museo Argentino de Ciencias Naturales Bernardino Rivadavia e Instituto Nacional de Investigacioenes de las Ciencias Naturales, Ciencias Geologicas,* Buenos Aires, **IV**(2): 167–250.

Thompson, L. G., M. E. Davis, E. Mosley-Thompson, T. A. Sowers, K. A. Henderson, V. S. Zagorodnov, P.-N. Lin, V. N. Mikhalenko, R. K. Campen, J. F. Bolzan, J. F. Cole-Dai, and B. Francou, 1998: A 25,000-year tropical climate history from Bolivian ice cores. *Science,* **282**: 1858–1864.

Thompson, L. G., E. Mosley-Thompson, M. E. Davis, P.-N. Lin, K. A. Henderson, J. Cole-Dai, J. F. Bolzan, and K.-B. Liu, 1995: Late glacial stage and Holocene tropical ice core records from Huascartn, Peru. *Science,* **269**: 46–50.

Thompson, R. S., C. Whitlock, P. J. Bartlein, S. P. Harrison, and W. G. Spaulding, 1993: Climatic changes in the western United States since 18,000 yr B.P. *In* Wright, H. E., Jr., J. E. Kutzbach, T. Webb, III, W. F. Ruddiman, F. A. Street-Perrott, and P. J. Bartlein (eds.), *Global Climates Since the Last Glacial Maximum.* Minneapolis: Univ. of Minnesota Press, pp. 468–513.

Thorp, J., and H. T. U. Smith, 1952: Pleistocene eolian deposits of the United States, Alaska, and parts of Canada, National Research Council Committee for the Study of Eolian Deposits, Geological Society of America, scale 1:2,500,000.

Tonni, E. P., 1992: Mamíferos y clima del Holoceno en la provincia de Buenos Aires. *In* Iriondo, M. (ed.), *Holoceno* (Comité Argentino para la Investigación del Cuaternario), **1**: 64–78.

Tonni, E. P., A. L Cione, and A. J. Figini, 1999: Predominance of arid climates indicated by mammals in the Pampas of Argentina during the late Pleistocene and Holocene. *Palaeogeography, Palaeoclimatology, Palaeoecology,* **147**: 257–281.

Tricart, J. L., 1973: Geomorfología de la Pampa Deprimida. Base para los estudios Edafológicos y Agronómicos. *Plan Mapa de Suelos de la región Pampeana.* Instituto Nacional de Tecnología Agropecuaria., Colección Científica XII, 202 pp.

Vreeken, W. J., 1996: A chronogram for postglacial soil-landscape change from the Palliser Triangle, Canada. *The Holocene,* **6**: 433–438.

Wang, H., L. R. Follmer, and J. C. Liu, 1998: Abrupt climate oscillations in the loess record of the Mississippi Valley during the last glacial maximum. *In* Busacca, A. J. (ed.), *Dust Aerosols, Loess Soils and Global Change.* Pullman: Washington State University College of Agriculture and Home Economics, Miscellaneous Publication No. MISC0190, pp. 179–182.

Warren, A., 1976: Morphology and sediments of the Nebraska Sand Hills in relation to Pleistocene winds and the development of eolian bedforms. *Journal of Geology,* **84**: 685–700.

Watts, W. A., and H. E. Wright, Jr., 1966: Late-Wisconsin pollen and seed analysis from the Nebraska Sand Hills. *Ecology,* **47**: 202–210.

Webb, T., III, P. J. Bartlein, S. P. Harrison, and K. H. Anderson, 1993a: Vegetation, lake levels, and climate in eastern North America for the past 18,000 years. *In* Wright, H. E., Jr., J. E. Kutzbach, T. Webb, III, W. F. Ruddiman, F. A. Street-Perrott, and P. J. Bartlein (eds.), *Global Climates Since the Last Glacial Maximum.* Minneapolis: University of Minnesota Press, pp. 415–467.

Webb, T., III, W. F. Ruddiman, F. A. Street-Perrott, V. Markgraf, J. E. Kutzbach, P. J. Bartlein, H. E. Wright, Jr., and W. L. Prell, 1993b: Climatic changes during the past 18,000 years: Regional syntheses, mechanisms, and causes. *In* Wright, H. E., Jr., J. E. Kutzbach, T. Webb, III, W. F. Ruddiman, F. A. Street-Perrott, and P. J. Bartlein (eds.), *Global Climates Since the Last Glacial Maximum.* Minneapolis: University of Minnesota Press, pp. 514–535.

Weber, F. R., T. D. Hamilton, D. M. Hopkins, C. A. Repenning, and H. Haas, 1981: Canyon Creek: A late Pleistocene vertebrate locality in interior Alaska. *Quaternary Research,* **16**: 167–180.

Wells, G. L., 1983: Late-glacial circulation over central North America revealed by aeolian features. *In* Street-Perrott, A., M. Beran, and R. Ratcliffe (eds.), *Variations in the Global Water Budget.* Dordrecht: Reidel, pp. 317–330.

Wells, S. G., L. D. McFadden, and J. D. Schultz, 1990: Eolian landscape evolution and soil formation in the Chaco dune field, southern Colorado Plateau, New Mexico. *Geomorphology,* **3**: 517–546.

Wintle, A. G., N. Lancaster, and S. R. Edwards, 1994: Infrared stimulated luminescence dating of late-Holocene aeolian sands in the Mojave Desert, California, U.S.A. *The Holocene,* **4**: 74–78.

Wolfe, S. A., 1997: Impact of increased aridity on sand dune activity in the Canadian Prairies. *Journal of Arid Environments,* **36**: 421–432.

Wolfe, S. A., D. J. Huntley, and J. Ollerhead, 1995: Recent and late Holocene sand dune activity in southwestern Saskatchewan. *Geological Survey of Canada Current Research,* **1995B**: 131–140.

Wolfe, S. A., D. R. Muhs, P. P. David, and J. P. McGeehin, 2000: Chronology and geochemistry of late Holocene eolian deposits in the Brandon Sand Hills, Manitoba, Canada. *Quaternary International,* **67**: 61–74.

Wright, H. E., Jr., 1970: Vegetational history of the Central Plains. *In* Dort, W., Jr., and J. K. Jones, Jr. (eds.), *Pleistocene and Recent Environments of the Central Great Plains.* Lawrence: Univ. of Kansas Department of Geology Special Publication, pp. 157–172.

Zárate, M., and A. Blasi, 1991: Late Pleistocene and Holocene loess deposits of the southeastern Buenos Aires Province, Argentina. *GeoJournal,* **24**: 211–220.

Zárate, M., and A. Blasi, 1993: Late Pleistocene-Holocene eolian deposits of the southern Buenos Aires Province, Argentina: A preliminary model. *Quaternary International,* **17**: 15–20.

Zárate, M., and J. L. Fasano, 1989: The Plio-Pleistocene record of the central eastern Pampas, Buenos Aires Province, Argentina: The Chapadmalal case study. *Palaeogeography, Palaeoclimatology, Palaeoecology,* **72**: 27–52.

Zárate, M., and N. Flegenheimer, 1991: Geoarchaeology of the Cerro La China Locality (Buenos Aires, Argentina): Site 2 and Site 3. *Geoarchaeology,* **6**: 273–294.

Zárate, M., R. Kemp, M. Espinosa, and L. Ferrero, in press: Pedosedimentary and palaeoenvironmental signficance of a Holocene alluvial sequence in the southern Pampas, Argentina. *The Holocene.*

Zimbelman, J. R., S. H. Williams, and V. P. Tchakerian, 1995: Sand transport paths in the Mojave Desert, southwestern United States. *In* Tchakerian, V. P. (ed.), *Desert Aeolian Processes.* London: Chapman & Hall, pp. 101–129.

13

Periods of Wet Climate in Cuba: Evaluation of Expression in Karst of Sierra de San Carlos

JESÚS M. PAJÓN, ISMAEL HERNÁNDEZ,
FERNANDO ORTEGA, AND JORGE MACLE

Abstract

The goal of this chapter is to present a quantitative assessment of the paleoclimatic evolution and paleo-environmental processes in western Cuba during the Pleistocene, focusing on the interval from the Wisconsin to the present. The paleoclimatic data obtained are of regional relevance. Most geomorphologic and geodynamic features observed in Cuba are the result of Pleistocene changes in sea level, temperature, humidity, and rainfall, which played a major role in landscape genesis and evolution. Paleoflow, paleodrainage, and paleorainfall values for the western karst region of Cuba were estimated by using scallop analysis and fluvial load analysis in active and inactive galleries of caves. These galleries represent different levels in the cave system that formed in response to sea-level changes during the Pleistocene. The reconstructed paleorainfall values document pluvial phases during the late Pleistocene. An abrupt climatic warming that occurred at the beginning of the Holocene was dated by ^{14}C to 11,520 ± 50 years B.P. Based on the relationship of ca. 0.21 ‰ per degree Celsius, the oxygen isotope composition of samples from one stalagmite suggests a temperature increase of about 9.5°C since that time. Copyright © 2001 by Academic Press.

Resumen

El principal objetivo de esta investigación es, presentar una valoración cuantitativa de la evolución paleoclimática y los procesos paleoambientales en Cuba Occidental durante el Pleistoceno, especialmente desde la Glaciación Wisconsin hasta el Presente. La data paleoclimatica obtenida es de una relevancia regional. Los rasgos geomorfológicos y geodinámicos observados en Cuba durante el periodo Pleistoceno son el resultado, entre otros factores, de las variaciones del nivel del mar, la temperatura, la humedad, y las precipitaciones, los cuales han desempeado un rol fundamental en la génesis y evolución del relieve. A partir del analisis de simetría de scallops y la carga fluvial en galeras activas e inactivas de cavidades subterráneas en la región del karst occidental de Cuba, se estimaron valores de paleoflujos, paleodrenajes y paleoprecipitaciones. Estas galeras representan varios niveles en el sistema cavernario, y su formación esta relacionada con los cambios glacioeustáticos del nivel del mar durante el Pleistoceno. Los valores obtenidos de paleoprecipitaciones reconstruidas documentan las fases pluviales durante el Pleistoceno Tardío.

Un calentamiento climático abrupto fue detectado a comienzos del Holoceno, cuya fase inicial fue datada por ^{14}C en 11,520 ± 50 años B.P. La composición de isótopos de oxígeno en muestras de una estalagmita parece indicar un calentamiento a través del tiempo hasta el presente, con un cambio de temperaturas de alrededor de 9.5°C, para una relación de 0.21 permil/grado C.

13.1. INTRODUCTION

13.1.1. Location of Study Region

The Cuyaguateje River basin in western Cuba has an area of ca. 842 km². The basin is located between the Sierra de Infierno to the northeast and the Sierra de In-

fierno to the southwest. The karstic massifs of Quemados, Cabezas, Sumidero, Resolladero, Pesquero, Mesa, and Guane are located in the area, including the main study site discussed in this chapter, Sierra de San Carlos (Fig. 1).

13.1.2. Geology

The region is characterized by a pronounced lithological and structural complexity, and several different models have been presented to explain its evolution (Piotrowska, 1978; Pszczolkowski, 1978; Iturralde-Vinent, 1977, 1988).

In general, the stratigraphic units dip gradually to-

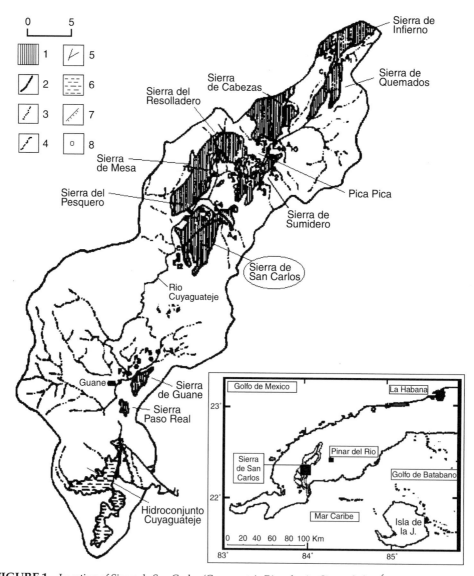

FIGURE 1   Location of Sierra de San Carlos (Cuyaguateje River basin, Sierra de los Órganos, western Cuba). 1, karst massifs; 2, permanent streams; 3, intermittent streams; 4, dry river beds; 5, water conduits in subterranean caves; 6, lagoons and lakes; 7, bank of a dam; 8, points of hydrological and hydrochemical measurements.

ward the southeast. The limestone portion is represented by the Jagua and Guasasa formations (Jurassic), as well as the Pons (Cretaceous) and Ancón (Paleogene) formations (Pszczolkowski, 1978). The terrigenous carbonate Pica-Pica formation is of Paleogene–Neogene age. The units of Alturas de Pizarras del Norte and Sur are composed mainly of terrigenous sequences of the San Cayetano formation of lower and middle Jurassic age.

13.1.3. Geomorphology

In the central portion of the basin, constituted by carbonate massifs, karstic and fluviokarstic processes have been the major geomorphologic agents. From the Pleistocene to the present, karst processes are associated mainly with climatic, physicochemical, biochemical, and hydrological factors, but are also controlled by tectonic and fracturing processes, which created conical and tower karst features (Kegel and Turmkarst).

The karstic phenomena are represented on the surface by the presence of exokarstic morphology, including poljes, mountain pits, crests, and boulder fields. Below the surface, caves have formed. In some cases, extensive cave systems exist, such as the Majaguas-Cantera, Santo Tomás, Palmarito, Fuentes, and Constantino caves, with more than 30 km of explored and mapped underground galleries.

In the lower portion of the basin, Quaternary fluvial plains, fluviomarine sediments, and deltas form the relief, with variable sediment thicknesses.

13.1.4. Hydrogeology

The northeast-to-southwest orientation of the network of surface drainage is related to structural and geomorphologic patterns and, locally, to tributary streams.

The upper and middle portions of the basin are characterized by surface drainage of rivers and streams flowing from the Alturas de Pizarras. After passing through contact valleys, they reach the karst massif through underground galleries. Underground, stream flow increases due to the autochthonous supply of water in the karst. There are also other surface streams originating from springs in the area of the carbonate massif, some with considerable annual flow, which contribute to the water balance of the area.

The hydrogeological system as a whole is formed by both surface and underground drainage and plays an important role in the hydrogeochemical, paleoclimatic, and paleoenvironmental processes. These processes, together with other geodynamic factors, have influenced the genesis and evolution of the karstic phenomena.

13.1.5. Modern Climate

Climate (Díaz, 1984) is characterized by mean annual temperatures between 21.8° and 24.9°C, rainfall of 1600–1800 mm/year (1720 mm average), and mean relative humidity of 77%. In the Sierra de los Órganos, a quantitative relation has been found between elevation and annual rainfall, increasing from ca. 1200 mm at 10–20 m to 2200 mm at 500 m elevation (Gagua et al., 1976). In historical times between 1750 and 1871, rainfall at the Santiago de Cuba meteorological station was low and increased significantly between 1871 and 1984. Temperatures have increased by 0.5°C since the nineteenth century and by 0.2°–0.3°C in the last 40 years (Casablanca meteorological station in Havana City; Celeiro, 1999). Sea level has increased at the Siboney Station (eastern Cuba) by 2.9 mm/year during the last 30 years (Hernández, personal communication).

13.1.6. Previous Work

Based on the presence of carbonates, silicates, and gypsum deposits, Ortega and Arcia (1982) reconstructed a paleorainfall map for Cuba for the last glacial period (Wisconsin) that suggested arid climates. On the basis of the coastal eolian formations of the Pleistocene in the western part of Cuba, Shanzer et al. (1975) suggested that arid climates were synchronous with glaciations at higher latitudes.

Based on the study of the continental Quaternary deposits in central and western Cuba, Kartashov and Mayo (1976) established the existence of oscillations between pluvial and arid phases. The authors correlated the pluvial periods with glacial advances in the Northern Hemisphere.

Based on weathering profiles of the Guane and Guevara Quaternary formations, Kartashov et al. (1981) divided the Cuban Pleistocene into two stages: a lower wet stage and an upper dry stage. Ortega (1983) presented a paleotemperature map for Cuba by estimating the elevation of the periglacial zone from sea surface temperature (SST) reconstructions in the Gates paleoclimate model (Gates, 1976).

From scallop analysis in a fossil fluvial gallery in the Perfecto cave in the Cuyaguateje region, Valdés (1974) calculated a paleorainfall value of 1400 mm. This suggests that the climate in the region at that time was more humid than it is today. Based on geomorphological correlation, an Illinoian age was assigned by Acevedo (1971). Although the study of scalloping features in underground galleries, as well as the investigation of paleohydrological and paleoprecipitation processes, has been the focus of much research (White and White, 1970; Goodchild and Ford, 1971; Wigley, 1972; Curl,

1974; Valdés 1974; Lauritzen, 1981, 1982; Lauritzen et al., 1983; Molerio, 1997), the problems concerning the paleoclimate and paleoenvironment in the tropical karstic areas have received little attention. The most recent paleoclimate research in Cuba is related to the project *Paleoclimate of the Cuban Quaternary: A Quantitative Characterization*, by the Cuban National Program for Global Change and the Cuban Environment. In this chapter, we present data from a multidisciplinary study of the karst region in western Cuba.

13.1.7. Karst of Sierra de San Carlos

The geologic (Fig. 2), geomorphologic, hydrogeologic, and geodynamic characteristics of the Sierra de San Carlos karst and its surroundings are reviewed in detail in several papers (Acevedo, 1971; Acevedo and Valdés, 1974; Acevedo and Gutiérrez, 1976; Fagundo et al., 1978; Molerio, 1978).

The Sierra de San Carlos karst region is characterized by several cave levels related to past hydrological changes. Contemporary and past karstic processes are a result of changes in allochthonous and autochthonous drainage. Fluvial genesis also played an important role during karstification. These caves must have formed during major pluvial periods, probably during interglacial periods when sea level was high. Evidence shows the existence of alternating periods of erosion and infilling. For example, there are sediments of different particle sizes that are representative of complex matrix infilling in seasonally active, fossil, and subfossil galleries and a great number of *scallops* along gallery walls and ceilings, indicating recent and past erosional activities. In addition, large pebbles, composed mainly of sandstone, shale, and limestone, are deposited along several kilometers of active, seasonally active, and fossil underground galleries, originating from the Alturas de Pizarras del Sur. Several generations of speleothems of different mineralogical composition, including goethite, hematite, and ghassoulite, clearly indicate paleoclimatic and paleoenvironmental changes (Pajón et al., 1985; Pajón and Valdés, in press).

13.2. QUATERNARY SEDIMENTS AND CAVE LEVELS

The Majaguas-Cantera cave system in the Sierra de San Carlos in western Cuba has nine underground levels ranging from 50–290 m elevation. There are two active levels at 70 m, one seasonally active level at 80–90 m, one episodically active level at 90–110 m, and five subfossil levels at 120–290 m (Molerio, 1992; Flores,

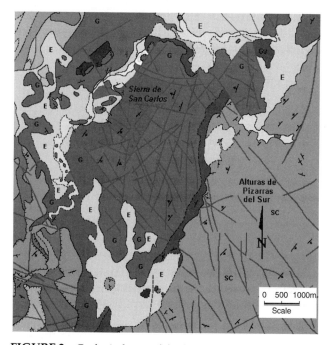

FIGURE 2 Geological map of the Sierra de San Carlos karst and its surroundings (Sierra de los Órganos, Pinar del Río, Cuba). (From Pszczolkowski et al., 1975.)

Symbol	Formation/member (description)
Jurassic deposits	
G	Guasasa formation: thin micritic limestones and some imbedded flint
Gv	San Vicente member: micritic limestones, calcarenites, and dolomitized limestones, thickly layered
Sc	San Cayetano formation: shales, sandstones, and aleurolites
Neogene-Quaternary deposits	
E	Ansenada formation: sandstones, conglomerates, some cemented aleurolites, mainly with materials from the San Cayetano formation

Other symbols: _____ Stratigraphic boundaries. ------Assumed stratigraphic boundaries. ___ Known deep and surface faults. ---Assumed faults. ⊥ Dip of the layers.

1995). These cave levels and the karstification processes are related to sea-level changes resulting from climate changes during the Quaternary.

A good correlation ($r = 0.99$) exists between the altitude of the cave levels in the Sumidero underground system and the kaolinite content of the Quaternary sediments deposited inside the caves (Table 1). The kaolinite content was estimated by using multiple regression analysis of absorption in the infrared spectra for quartz

TABLE 1   Correlation Between the Cave Levels in the Sumidero Underground System of Western Cuba and the Kaolinite Content of the Quaternary Sediments Inside the Caves

Era of speleogenesis	Cave level	Absolute altitude (m)	Kaolinite (K) content (%)	Kaolinite (K)/ Quartz (Q) [K/Q ratio]
Pre-Quaternary (Pliocene)	Cima	195–250		
Nebraska Q_I	Soterráneos cave	170–195	51–53	0.08
Kansas Q_{II}	Pio Domingo cave	130–180	35–43	0.64
Illinois Q_{III}	Perfecto cave	100–138	17–42	0.24
Wisconsin early (Iowa) Q^I_{IV}	Clara cave	85–132	7	0.07
Wisconsin late Q^{II}_{IV}	Sumidero	80–130	6	0.06
Holocene				

Note: The content of kaolinite in the Quaternary sediments deposited in the cave levels was obtained by multiple regression of absorption in the infrared spectrum.

$$K (\%) = 7.7882\, A_3 / A_4 + 46.7491\, A_5 / A_6 - 25.7192,$$ where $A_3 = A\ (925\ cm^{-1})$, $A_4 = A\ (808 \pm 3\ cm^{-1})$, $A_5 = A\ (1040\ cm^{-1})$, $A_6 = A\ (1117 \pm 3\ cm^{-1})$

and kaolinite (Fagundo et al., 1984a,b). The highest caves are the oldest (Acevedo, 1971) and are related to ancient base levels of paleovalleys of erosive and erosive-denudative karstic origin. These caves formed as a consequence of variations in sea level and the associated adjustments of the equilibrium profiles of rivers and streams. Lower sea levels produced hanging caverns representing relics of earlier processes. Each cave level reflects one level of erosion (Acevedo, 1970). The ages of the cave levels were estimated based on the geomorphologic evolution of the region (Acevedo, 1970, 1971).

The parameter kaolinite to quartz (K/Q) has been used to estimate the degree of evolution of Quaternary sediments. The increase in the percentage of kaolinite and the K/Q ratio in the upper cave levels suggest periods of increased humidity or, more likely, long exposure times. Hydrological behavior, which can be derived by analysis of granulometric characteristics as they relate to the mechanism of selective transport, must also be considered.

According to our research in karst areas (representing 77% of the Cuban surface), the late Pleistocene and Holocene seem to have been marked by transport and deposition of large pebbles and also by a well-defined distribution of scallops on the walls, ceilings, and floors of the underground passages, indicating erosive and corrosive processes.

13.2.1. Paleoclimatic Indices

To obtain a quantitative assessment of the paleo-rainfall and paleoflow in the study area, we analyzed

different morphological parameters in the cave systems.

To measure the geometrical parameters on scalloped surfaces in cave passages (length, width, depth, and alpha and beta angles), a SCALLOGRAPHER EPC-01 device was constructed and used in the measurement of thousands of scallops in underground passages.

Random sampling of the fluvial load in representative underground passages was used to determine the length, width, depth, and weight of several generations of pebbles. A detailed description of the studied sites (system, subsystem, sector, place, and point), including geographical coordinates, was made, and topographic maps (surface and profile) of the points of scallop measurement were drawn to estimate the area and perimeter of the transversal section in the gallery.

This information was included in the Quaternary Paleoenvironmental Data Base (see Appendix).

The graphical method of statistical symmetry analysis of scallops proposed by Lauritzen (1981), based on the inflection point angles of scallops (alpha and beta angles), was used to determine the paleohydrological behavior of the main transects in the underground galleries (Fig. 3).

By testing a normal distribution for all the measured scallop formations in the system (number of measurements generally ≥30), the paleoflow and direction of the water flow were determined.

A brief description of the transects (Fig. 3).

• Transect No. 1: The data show a clear paleoflow direction from the basin of the Majaguas stream in the contact valley between the Alturas de Pizarras del Sur

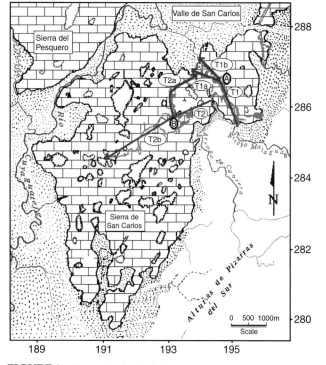

FIGURE 3 General paleohydrological behavior of the main underground transects of the Majaguas-Cantera cavernary system, in the karst of Sierra de San Carlos, obtained by using the scallop symmetry analysis.

and the San Carlos massif (T1) to the Ensenada de Bordallo, through Hoyo de los Helechos-Salón de los Gigantes (T1a). Probably during Wisconsin times (Flores, in preparation) the Hoyo de los Helechos collapsed, causing a change in the local drainage. At the point El Aserradero (point A), phreatic tube passages of the subsystem Aserradero-La Llave-Cueva Fría-Cueva Pajón (T1b) were excavated.

• Transect No. 2: The main direction of the paleoflow is from the basin of the Cantera stream, in the contact valley between Alturas de Pizarras del Sur and the San Carlos massif (T2), to the inner part of the massif, through the Laminador and Submarine galleries. At the confluence point of El Meandro–El Comedor (point B),

there are two different directions of paleoflow patterns, which suggest that the waters first moved (T2a) toward the Ensenada de Bordayo (northwest) through the sectors Dos Anas and Roloff (which are conjugated galleries associated with episodic water levels between 90 and 110 m) and subsequently flowed in the opposite direction, toward the deep sectors (T2b) Ignacio Agramonte, Marcel Loubens, and Alcatraz (seasonal galleries associated with seasonal levels with altitudes between +80 and +90 m).

The change of the hydraulic gradient at the confluence point B of Transect T2 demonstrates that water moved in a direction opposite to that found in the former gallery (sectors Dos Anas and Roloff). This movement could relate to paleohydrological processes associated with the confluence at point A in Transect T1, perhaps in response to a change in the base level due to a decrease in sea level during the Sangamon–Wisconsin (?) interval.

These results document the general paleohydrological behavior of the main underground transects of the cave system, as well as of the paleodrainage basins; the paleodrainage basins correspond to the present basins, although their morphometric parameters have changed since the late Pleistocene.

The estimates of paleoclimatic indices (paleorainfall, paleoflow, annual paleoflow, and paleovelocity) presented in Table 2 are based on measurements and analyses of underground morphologies (scallops) and fluvial load (pebbles) in the resurgence of the Majaguas stream, representing active and seasonally active galleries (in the Majaguas-Cantera cave system). The possible age of the scallop evolution ranges from the late Sangamon to the present (Flores, in preparation). These results show two major pluvial periods, with precipitation substantially above the modern values of ca. 1800 mm and a decreasing trend in paleorainfall from the early Wisconsin to the present.

The results presented are preliminary, and more data will be needed to confirm the paleoclimate interpretations and the chronology of the features.

TABLE 2   Preliminary Results of Paleoclimatic Indices

Scallop Database	Paleorainfall (mm)	Age (scallops, years B.P.)	Paleoflow (m³/sec)	Annual paleoflow (m³/year) × 10⁷	Paleoflow (m/sec)
SRAMAS 54+55	5000–6000	110,000	31.6	100.0	4.2
SRAMAS 56+57	3000–4000	60,000	5.5	57.9	3.3
SRAMAS 58+59	2500–3000	18,000–5,000	3.7	4.9	1.1

Based on the measurement and analysis of scallops and pebbles in active and seasonal galleries in the Majaguas-Cantera cave system (Sierra de San Carlos). The paleoclimatic indices were calculated with the computer program PALEORAINFALL.

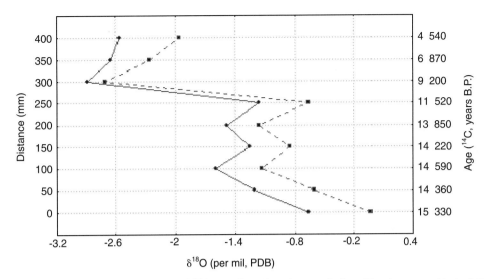

FIGURE 4 Oxygen isotope ratios and ^{14}C age of a stalagmite sample from Dos Anas cave (dashed line: analysis at the University of Florida, Gainesville, by Jason Curtis; continuous line: analysis at the University of Edinburgh by Sandy Tudhope and Colin Chilcott).

13.3. STABLE ISOTOPE ANALYSIS

A 420-mm long stalagmite from the Dos Anas cave was collected for isotopic analysis. Nine samples of 2 mg each were extracted by using a 1-mm drill at intervals of 50 mm along the growth axis. Oxygen and carbon stable isotope analyses were performed on crushed 0.5- to 1.0-mg subsamples. The carbonate powder was treated with 100% orthophosphoric acid at 90°C in an automatic carbonate preparation system, and the resulting CO_2 was analyzed on a VG Isogas Prism mass spectrometer. All analyses of $\delta^{18}O$ and $\delta^{13}C$ are reported in per mil (‰) relative to the Pee Dee Belemnite (PDB) standard.

The oxygen isotope composition of the stalagmite changes to isotopically lighter values from the oldest sample 1 (−0.668‰) to the youngest sample 9 (−2.570‰), with a particularly large change between samples 6 and 7. This shift may indicate a warming trend through time to the present (Fig. 4). If the changes in isotopic composition were due to temperature alone, the magnitude of the temperature change would have been ca. 9.5°C, based on a relationship of ca. 0.21‰ per degree Celsius (Pajón et al., in preparation).

An abrupt warming was detected before the beginning of the Holocene (values of $\delta^{18}O$ shift from −1.168 to −2.892‰), dated by ^{14}C to 11,520 ± 50 years B.P. This shift could be related to the Younger Dryas climate oscillation, but more high-resolution data are needed to confirm this result.

13.4. [14]C AGE AND GROWTH RATE OF STALAGMITE

The radiocarbon age of the stalagmite of Dos Anas cave is 15,330 ± 50 years B.P. Since stalagmites can be formed only in air-filled passages, the age, as well as the isotopic data, suggests that the seasonal cave passage (80–90 m above sea level [asl]) had been inactive with respect to the allochthonous drainage since that time. The age demonstrates that about 200 mm of the speleothem was formed during the Holocene, which represents an interglacial climate.

The mean growth rates of the stalagmite are in the range of 0.03–0.14 mm/year. Before 13,850 ± 50 years B.P., the rates were much faster, related perhaps to the much higher CO_2 levels in the drip waters than in the cave air.

13.4.1. Potential for Paleoclimate Research in Cuba

In studying large-scale linkages between paleoclimate patterns in the Americas, Cuba represents an important region. First, its geographical position represents a transitional area between the Northern and Southern Hemispheres. Second, its location in the tropics close to the North American continent offers an exceptional opportunity to study the influence of continental glaciations on tropical and subtropical environments. Practically all observed geomorphologic features in Cuba were formed during the Quaternary, a time of major changes in sea level and climate. Cuba offers a great range of features for paleoclimate research, such as lakes, hundreds of kilometers of coast with marine limestone terraces, a shallow and extensive submarine platform covered with different generations of fossil and active coral reefs, geological formations of Quaternary limestone, and soil deposits. Unique features are ancient underground fluvial caves at different elevations, which reflect local variations in the water table related to changes in sea level. These caves represent fragments of ancient river courses, with their bed load sediments perfectly preserved in many places, alternating with carbonate sediments of chemical origin (stalagmites, sinters, etc.). These carbonate deposits allow dating and paleotemperature reconstruction using isotopic and paleomagnetic methods. In addition, the study of both the granulometric spectra of the bed load sediments and of the coarse materials (pebbles, etc.), as well as the study of the scalloped cave conduits, allows estimates of past water discharges, runoff, and other parameters. Since the drainage basins are well defined, paleorainfall can be estimated.

Our main interest in the next phase of the research of the Cuban Working Group for Paleoclimate Studies is to study the recent climate history using high-resolution lake sediments for paleolimnological analyses, cave deposits for stable isotope geochemistry and dating, and terrestrial sediments for pollen analysis to reconstruct late Quaternary vegetation changes.

Acknowledgments

We thank Gabriel García, Vladimir Otero, Ernesto Flores, Leonardo Flores, Leslie Molerio, Jorge de Huelves, Zochil González, and Manuel Rivero for discussions and assistance in the field. Tom Guilderson of the University of Florida, Gainesville, and Colin Chilcott of the University of Edinburgh are thanked for providing radiometric dating and stable isotope laboratory analyses.

We thank Sandy Tudhope and Jason Curtis for comments concerning isotopic studies.

We are deeply indebted to Vera Markgraf and Roger Y. Anderson, who helped revise an earlier version of this chapter and deserve special thanks.

APPENDIX: CONTENT OF THE QUATERNARY PALEOENVIRONMENTAL DATA BASE FOR WESTERN CUBA

Quaternary Paleoenvironmentatl Data Base

	Name	Data set name	Information in Data set
	Lithologic and litho-stratigraphic data	LEXICO (IGP[a])	Information on lithology, age, halostratotype, etc. of sediments.
	Paleohydrologic data	SCALLOPS PEBBLES	Geometrical characteristics of erosion morphologies, underground galleries in karstic areas; geometrical mass parameters of fluvial load.
	Climatic data	GeoCLIM	Climatic database of Cuba Historic and present network of Belen School and Casablanca Meteorological stations.
	Oceanographic data	Sea level change	30-year record of sea level changes at the Siboney Station.
Content of CPP data and proxy database	Hydrochemical data	GeoWAT-1	Hydrochemical parameters: T, pH, SPC (25°C), pCO_2; saturation index of calcite, dolomite, dolomite, and gypsum; physicochemical and geochemical indices.
	Speleological data	GeoCLEV-1	Information on the cave system in the Majaguas-Cantera system.
	Analytical methods	GeoIREM	Data on X-ray diffraction, infrared spectroscopy, atomic emission, and particle-induced X-ray emission.
	Paleomagnetics	MAGSUS-1	Data on magnetic susceptibility of Quaternary sediments, bedrock, speleothems, and other deposits.
	Sampling points	TSAMPLE	General information on sampling points of soils, Quaternary sediments, speleothems, and fluvial load (georeference information).

[a]IGP, Instituto de Geología y Paleontología.

QUATERNARY PALEOENVIRONMENTAL DATA BASE SYSTEM

The purpose of the Quaternary Paleoenvironmental Data Base System (DBS) of the Cuban Quaternary is the integration and management of multidisciplinary information. The database includes information on geology, hydrogeology, and paleomagnetic and analytical data, as well as information on present climate variability and contemporary geodynamic processes.

Another important purpose is the inclusion of the DBS in the World Data Center-A for Climatology as part of the PAGES Initiative (Webb et al., 1994; Markgraf, 1996).

The Quaternary Paleoenvironmental Data Base for Western Cuba is the base of the computer program "PaleoSys" (Hernández et al., in preparation) and includes information on paleohydrology, climate, sealevel changes, physicochemical analysis, and magnetic susceptibility and codifications for sampling points in the cave environment and its surroundings. PaleoSys (first version) is written in Visual Basic Version 4.0 for Windows, with access to numeric information, as well as sketches, bar and line diagrams, and a collection of images (maps and slides) for the Majaguas-Cantera cave system experimental area in the western karst of Cuba.

References

Acevedo, G. M., 1970: Contribución al estudio de la evolución geomorfológica de la Sierra de los Órganos, Pinar del Río, Cuba. *Revista Tecnológica,* **8(2):** 15–34.

Acevedo, G. M., 1971: Geomorfología de Sumidero y sus inmediaciones, Sierra de los Órganos, Pinar del Río, Cuba. *Revista Tecnológica,* **9(3–4):** 33–54.

Acevedo, G. M., and R. Gutiérrez, 1976: El Sistema Cavernario Majaguas-Canteras. *Revista Voluntad Hidráulica,* **13:** 18–30.

Acevedo, G. M., and J. J. Valdés, 1974: Introducción de métodos geomorfológicos e hidrogeológicos cuantitativos en la evolución del sistemas cavernarios, aplicación al Sistema Cavernario Majaguas-Canteras. *Revista Tecnológica,* **12(1):** 29–39.

Celeiro, M., 1999: El Método Histórico en las Investigaciones Climáticas en Cuba: Contribución al Transecto PEP-I del PAGES. Libro "El Caribe: Contribución al Conocimiento de su Geografía." Instituto de Geografia Tropical (IGT), Agencia de Medio Ambiente / Ministerio de Ciencia, Tecnologia y Medio Ambiente (AMA / CITMA), pp. 48–55.

Curl, R., 1974: Deducing flow velocity in cave conducts from scallops. *National Speleological Society Bulletin,* **36(2):** 1–5.

Díaz, C. L. R., 1984: Monografía Hidrológica de la cuenca del río Cuyaguateje (clima). Archivo Instituto de Geografía. ACC. Unpublished report 191, 120 pp.

Fagundo, J. R., J. M. Pajón, J. J. Valdés, and J. Rodríguez, 1978: Comportamiento químico-físico de las aguas kársticas de la cuenca del Río Cuyaguateje. *Revista Ingeniería Hidráulica, Ciudad de La Habana,* **11(3):** 251–274.

Fagundo, J. R., J. J. Valdés, and J. M. Pajón, 1984a: Estudio de los Sedimentos Cuaternarios de la Cuenca del río Cuyaguateje, mediante Espectroscopia Infraroja y Difracción de Rayos X. *Revista Voluntad Hidráulica,* **12:** 53–61.

Fagundo, J. R., J. J. Valdés, and J. M. Pajón, 1984b: Determinación del contenido de los minerales cuarzo y caolinita en sedimentos mediante espectroscopia infraroja. *Revista de Ciencias Químicas, Centro Nacional de Investigaciones Científicas,* **15(2):** 327–333.

Flores, V. E., 1995: Niveles de cavernamiento y fluctuaciones glacioeustáticas cuaternarias de Cuba Occidental. Congreso Internacionl LV Aniversario de la Sociedad Espeleológica de Cuba, pp. 90–91.

Flores, V. E., in preparation: Niveles de Cavernamiento en el Sistema Cavernario Majaguas-Cantera, Sierra de San Carlos, Pinar del Río, Cuba.

Gagua, G., S. Zarembo, and A. Izquierdo, 1976: Sobre el nuevo mapa isoyético (3^{era} versión). *Revista Voluntad Hidráulica, La Habana,* **13(37):** 35–41.

Gates, L. W., 1976: Modeling the ice-age climate. *Science,* **191:** 1138–1144.

Goodchild, M. F., and D. C. Ford, 1971: Analysis of scallop patterns by simulation under controlled conditions. *Journal of Geology,* **79:** 52–62.

Hernández, I., A. Quintana, J. M. Pajón, and O. Sánchez, in preparation: PaleoSys: Sistema de Bases de Datos Paleoclimáticos del Occidente de Cuba.

Iturralde-Vinent, M., 1977: Ophiolites and volcanic arcs in eastern Cuba. *Journal of Petroleum Geology,* **20(2):** 250–251.

Iturralde-Vinent, M., 1988: *Naturaleza Geológica de Cuba.* La Habana: Científico-Técnica, 246 pp.

Kartashov, I. P., and N. A. Mayo, 1976: Esquema estratigráfico y división del sistema cuaternario de Cuba. *In Sedimentación y Formación del Relieve de Cuba en el Cuaternario.* Moscow: Russian Academy of Sciences (Nauka), pp. 5–33.

Kartashov, I. P., A. G. Chernyajovski, and L. Peñalver, 1981: *El Antropógeno en Cuba.* Moscow: Russian Academy of Sciences (Nauka), 147 pp.

Lauritzen, S. E., 1981: Statistical symmetry analysis of scallops. *National Speleological Society Bulletin,* **43:** 52–55.

Lauritzen, S. E., 1982: The paleocurrents and morphology of Pikkaggrottene, Svartisen, North Norway. *Norsk Geografiska Tidsskrift,* **4(82):** 183–209.

Lauritzen, S. E., A. Ive, and B. Wilkinson, 1983: Mean annual runoff and the scallop flow regime in a subarctic environment. Preliminary results from Svartisen, North Norway. *Transactions of the British Cave Research Association,* **10:** 97–102.

Markgraf, V., 1996: PANASH: Paleoclimas de los Hemisferios Norte y Sur. El Programa PANASH. Transecta de las Américas / PEP I. PAGES Series 96–6, 97 pp. (Spanish translation.)

Molerio, L. L., 1978: Geomorfología e Hidrogeología de la Sierra del Pesquero, Provincia de Pinar del Río, Cuba. Unpublished report.

Molerio, L. L., 1992: Distribución del cavernamiento en las Sierras del Pesquero, San Carlos, Resolladero y Mesa, Pinar del Río, Cuba. II Congreso Espeleologica Latinoamericano y el Caribe, Pinar del Río, Cuba, pp. 19–20.

Molerio, L. L. F., 1997: Indicadores hidráulicos del Paleoflujo Subterráneo en la Sierra de Quemado, Pinar del Río, Cuba. Unpublished report.

Ortega, F., 1983: Una hipótesis sobre el clima de Cuba durante la Glaciación de Wisconsin. *Ciencias de la Tierra y el Espacio,* **7:** 57–68.

Ortega, F., and M. Arcia, 1982: Determinación de las lluvias en Cuba durante la glaciación de Wisconsin, mediante los relictos edáficos. *Ciencias de la Tierra y el Espacio,* **4:** 85–104.

Pajón, J. M., and J. J. Valdés, in press: Mineralogical characterization of speleothems and sediments of Cuban caves: A paleogeographic approximation. *In El Carso y las Cuevas del Archipielago Cubano.* Instituto Cubano del Libro, Editorial Científico-Técnica.

Pajón, J. M., J. H. Curtis, S. Tudhope, S. E. Metcalfe, C. E. Grimm, and M. Brenner, in preparation: Isotope records of a stalagmite from Dos Anas Cave in Pinar del Río Province, Cuba. Paleoclimatic Implications.

Pajón, J. M., J. R. Fagundo, and J. J. Valdés, 1985: Ocurrencia de goethita y hematita en la Cueva Superior del Arroyo Majaguas. Sistema Cavernario Majaguas-Cantera. Pinar del Río, Cuba. *Revista Voluntad Hidráulica*, **22**: 41–45.

Piotrowska, K., 1978: Nappe structures in the Sierra de los Órganos western Cuba. *Acta Geologica Polonica*, **28(1)**: 97–170.

Pszczolkowski, A., 1978: Geosynclinal sequences of the Cordillera de Guaniguanico in western Cuba: Their lithostratigraphy, facies development and paleogeography. *Acta Geologica Polonica*, **28(1)**: 1–96.

Pszczolkowski, A., Piotrowska, K., Myczynski, R., Piotrowski, J., Grodzicki, A., Skupinski. G., Haczewski, G., y Danilewski, D., 1975: Texto explicativo para el mapa geológico en escala 1:250 000 de la provincia de Pinar del Río. *Archivo Instituto de Geología y Paleontología*. Academia de Ciencias de Cuba, p. 743.

Shanzer, E. V., O. M. Petrov, and G. L. Franco, 1975: Sobre las formaciones costeras del Holoceno en Cuba. Las terrazas pleistocénicas de la Región Habana-Matanzas y los sedimentos asociados a ellas. *Academia de Ciencias de Cuba. Serie Geográfica*, **21**: 3–26.

Valdés, J. J., 1974: Nuevo elemento para el estudio cuantitativo de los carsos obtenidos mediante el análisis dimensional y su utilidad en el cálculo de paleoprocesos geohidrológicos. *Revista Tecnológica*, **11**: 23–32.

Webb, S. R., M. D. Anderson, and J. T. Overpeck, 1994: Editorial: Archiving data at the World Data Center-A for Paleoclimatology. *Paleoceanography,* **9(3)**: 391–393.

White, B. W., and L. E. White, 1970: Channel hydraulics of free-surface streams in caves. Caves and karst. *Research in Speleology*, **12(6)**: 41–48.

Wigley, T. M. L., 1972: Analysis of scallop patterns by simulation under controlled condition. *Journal of Geology,* **79**: 52–62.

Lacustrine Evidence

14

Identifying Paleoenvironmental Change Across South and North America Using High-Resolution Seismic Stratigraphy in Lakes

D. ARIZTEGUI, F. S. ANSELMETTI, K. KELTS,
G. O. SELTZER, AND K. D'AGOSTINO,

Abstract

High-resolution seismic profiling in lakes coupled with targeted coring provides powerful tools for the comparative analysis of environmental change along the Pole-Equator-Pole: Americas (PEP 1) transect. Mapping seismic unconformities in Lake Titicaca (Bolivia), Lake Mascardi, Laguna Cari-Laufquen Grande (Argentina), Laguna Miscanti (Chile) and Great Salt Lake and Yellowstone Lake (USA) builds the acoustic stratigraphic framework to interpret the history of lacustrine sedimentation and choose optimal coring sites for multiproxy sediment studies. Most of these seismic unconformities are associated with past lake-level changes.

Lake Mascardi (41°S, 215 m deep) exemplifies a proglacial lake sequence. Seismic and coring evidence indicates two major phases of environmental change coinciding with the late glacial/Holocene transition and a lower amplitude change within the mid-Holocene sequence. Lake Titicaca (18°S, 285 m) and Laguna Cari-Laufquen Grande (40°S, only 3 m deep) illustrate the acoustic character of relatively closed basins without direct glacial overprints. In these cases, the seismic data yield information on changes in the precipitation-evaporation pattern of the past as reflected by major lake-level changes. Lake Titicaca suggests

drops up to 100 m below present levels. Laguna Cari-Laufquen Grande shows several phases of drastic desiccation recognized by distinctive mound structures. Laguna Miscanti (23°S, 8 m, in the Atacama) illustrates acoustic layering in a small, groundwater-fed saline lake and distinctive evidence for a brief mid-Holocene desiccation phase.

Great Salt Lake (41°N, 8 m) acoustic stratigraphy documents a cold playa to deep Lake Bonneville evolution, followed by drastic drawdown to variable Holocene depths from near desiccation to modern levels. Profiles document neotectonic faulting. Yellowstone Lake (44°N, 90 m) illustrates acoustic signatures in a large caldera where clastics from the rapid retreat of an ice cap are overlain by uniform Holocene diatom oozes.

Results for these lacustrine basins of different origins and latitudes show that paleoclimate patterns also lead to characteristic acoustic packages. Lakes commonly have sediment sequences with high-impedance contrasts, and thus, acoustic boundaries are equivalent to lithological changes related to environmental dynamics. A feature that stands out in all examples is a distinctive shift in seismic facies in each system corresponding to the late glacial/Holocene transition. Copyright © 2001 by Academic Press.

Resumen

El perfilaje sísmico de alta resolución en cuencas lacustres combinado con el estudio de testigos sedimentarios proveen una poderosa herramienta para el análisis comparativo de cambios ambientales a lo largo de la transecta PEP 1. El mapeo de discontinuidades sísmicas en los lagos Titicaca (Bolivia), Mascardi, Cari-Laufquen Grande (Argentina), Miscanti (Chile), Great Salt y Yellowstone (Estados Unidos) permite definir un marco acústico-estratigráfico para reconstruir la sedimentación lacustre y elegir los sitios más adecuados para realizar estudios multidisciplinarios de detalle en los sedimentos. La mayoría de dichas discontinuidades sísmicas están relacionadas con antiguos cambios en el nivel de los lagos.

El Lago Mascardi (41°S, 215 m de profundidad) ejemplifica una secuencia de lago proglacial. Datos sísmicos combinados con el analisis de testigos sedimentarios evidencian dos fases mayores de cambio ambiental. La primera de ellas coincide con la transición Tardío Glacial-Holoceno mientras que la segunda, de menor amplitud, ocurrió durante el Holoceno medio. El Lago Titicaca (18°S, 285 m de profundidad) y la Laguna Cari-Laufquen Grande (40°S y solo 3 m de profundidad) ilustran el comportamiento acústico de cuencas lacustres relativamente cerradas sin influencia directa de actividad glacial. El carácter cerrado de éstas dos cuencas amplifica los cambios en el balance "precipitación-evaporación" ocurridos en el pasado que se ven reflejados como substanciales cambios en el nivel de los lagos. El Lago Titicaca muestra cambios en su nivel de hasta 100 m por debajo del nivel actual. La Laguna Cari-Laufquen Grande por su parte muestra varias fases drásticas de desecación que se observan en los perfiles sísmicos como montículos. La Laguna Miscanti (23°S, 8 m de profundidad) en Atacama (Chile) ejemplifica el ordenado comportamiento acústico de un pequeño lago alimentado por aguas subterráneas con claras evidencias de una breve fase de desecación durante el Holoceno medio.

El Lago Great Salt (41°N, 8 m de profundidad) presenta una estratigrafía acústica que varía desde sedimentos de playa hasta la formación del profundo Lago Bonneville seguido de drásticas caídas en el nivel de las aguas durante el Holoceno, alcanzando profundidades que variaron desde la desecación hasta aquella de los niveles actuales. Los perfiles sísmicos en ésta cuenca documentan la presencia de actividad neotectónica (fallas). El Lago Yellowstone (44°N, 90 m de profundidad) es una gran caldera donde al material clástico depositado durante la rápida retracción de una calota glacial le sigue una uniforme secuencia Holocena compuesta de diatomeas.

Los resultados de los estudios sísmicos en éstas cuencas lacustres de diferentes orígenes y a diferentes latitudes muestran la relación existente entre patrones paleoclimáticos conocidos y un determinado comportamiento acústico. Los sedimentos lacustres presentan a menudo altos contrastes en su impedancia y por lo tanto marcados límites acústicos equivalentes a cambios litológicos. Estos últimos están a su vez relacionados a variaciones en la dinámica del medio ambiente. El claro cambio en las facies sísmicas durante la transición Tardío Glacial-Holoceno observado en todas las cuencas lacustres aquí consideradas es un buen ejemplo de la utilización de ésta metodología para estudios paleoambientales.

14.1. INTRODUCTION: SEISMIC REFLECTION STRATIGRAPHY AND CLIMATE

Paleoclimate studies are commonly confronted with the critical question of whether a single core is representative of a uniform, continuous sediment package in a basin. Profiling with centimeter-scale seismic resolution can provide direct lithological and geometrical information to evaluate this question.

The concepts of seismic sequence stratigraphy are applied widely for marine geophysical surveying, but can be applied to lake systems as well if basic system differences are considered. In marine studies, seismic sequence analyses provide information on factors such as eustatic sea-level fluctuations and tectonic events. In lake studies, seismic stratigraphy can yield valuable information on lake-level fluctuations and rapid changes in the sedimentary system and facies. Such changes are commonly due to climatic factors, and the responses in lakes may be much faster and more drastic than for ocean margins. The variety of lacustrine lithologies, from proglacial muds to soil surfaces, carbonate precipitates, or organic-rich or diatomaceous muds, generally impart high-impedance contrasts to sequences. Thus, strong reflectors are commonly equivalent to distinctive lithological layering. This is a reason why some seismic equipment companies illustrate their sales brochures with examples from lake deposits.

Seismic profiling studies in lakes may be loosely divided between basinal surveys of the thickness, structure, tectonics, and extent of sedimentary patterns and those in direct support of sediment coring programs. Because lacustrine events are associated with environmental and climatic changes, seismic surveying is becoming an essential tool for the selection of optimal sites for coring long sediment sequences with continuous paleoclimate records. Not only do the profiles document the undisturbed nature of a sediment package, but they also can be used to correlate distinctive reflector packages among core networks and map the extent of lake-level changes. This information is important for developing a representative environmental sequence from multiple sites and basins as the framework for broad regional paleoclimate analyses and eventually combining them into a global network of lake studies. Lake sequences commonly have excellent temporal resolution for determining the response rates and vectors of climate changes over the continents—information that is needed to validate and inspire climate models. Without seismic stratigraphy, it may be difficult to recognize unconformities, slumps, or gaps in spatially limited core sections.

The application of high-resolution seismic profiling in lakes proceeded along with general advances in seismic profiling technology. The high-resolution, 3.5-kHz pinger sound source is related to the stroboscope, also invented at the Massachusetts Institute of Technology (MIT) by H. Edgerton. In the 1960s, J. Cousteau and Edgerton recorded profiles from Lake Titicaca as part of a film series and found distinctive *pagoda structures* in the subsurface postglacial layering of Lake Geneva (Serruya et al., 1966). Müller and Gees (1968) used a new *air gun* source to discover a tri-part stratigraphy for 200 m of glacial fill in Lake Constance. Both the high-resolution and air-gun systems were used to define acoustic stratigraphy in Lake Zürich and perialpine lakes in support of mass-flow and heat-flow studies (Hsü and Kelts, 1970; Matter et al., 1971; Finckh and Kelts, 1976; Finckh et al., 1984) and as a foundation for coring operations (e.g., Degens et al., 1984; Giovanoli et al., 1984; Johnson et al., 1987; Niessen et al., 1993; Mullins et al., 1991; Chondrogianni et al., 1996; Mullins et al., 1996). Seismic profiling became an important tool for the exploration of sedimentation in the Great Lakes of North America (e.g., Wold et al., 1981; Johnson, 1980) and the rift lakes of East Africa (e.g., Degens et al., 1971, 1973; Scholz and Rosendahl, 1988; Johnson et al., 1995).

The concept of seismic sequence stratigraphy was developed in marine systems to partition a sedimentary succession into several seismically defined sequences (e.g., Mitchum and Vail, 1977; see review in Cross and Lessenger, 1988). Sequences are defined by unconformable surfaces, commonly interpreted as a result of relative sea-level fluctuations in the case of marine environments. This concept of partitioning a sedimentary record by seismic stratigraphy can also be applied in lacustrine systems if adequate understanding of differences in systems is applied. Unconformities can be used to define lake-level changes, but lakes rarely follow Walter's law of gradually changing later-

al facies leading to construction of a vertical sequence. Lake-level changes may be drastic and abrupt, leaving immature deltas stranded and heavily incised. Lensoid fan packages, or even shoreline sands, may appear to be unmotivated within well-layered deepwater pelagic sequences.

Scholz and Rosendahl (1988) utilized sequence stratigraphic principles such as on-lap and off-lap relationships to propose controversial sudden lake-level drops up to hundreds of meters in Lakes Tanganyika and Malawi. This approach has been successfully applied to other sequences such as those for Lake Huron (Moore et al., 1994) and Lake Titicaca (Seltzer et al., 1998), where a succession of lake-level low stands and a lake-level curve were determined from on-lapping seismic reflections and erosional surfaces combined with the mapping of exposed strandlines.

Mapping of seismic facies is also a tool for separating sedimentary packages and processes due to strong variance in petrophysical character and high-impedance contrast common in lake sediments. A succession of high-amplitude, parallel reflections, for example, defines layering within a distinct unit. Strong petrophysical signatures are useful for correlating reflections among coring sites. Gas-rich zones are commonly transparent and are common in lake sequence zones with rapid burial of terrestrial organic matter. Thick mass flows or turbidites may also have acoustic transparency, but will have distinctive geometries, thickening and thinning according to bathymetry (e.g., Giovanoli et al., 1984). Multiple, thin-bedded turbidites will have parallel layering that ponds, whereas pelagic sedimentation will drape uniformly over rise and basinal areas. Differences in seismic thickness of sedimentary packages can be exploited in a targeted coring campaign. Thinned or eroded sequences on highs can be cored for maximum age-depth recovery, whereas thicker sequences may provide higher resolution, or the timing and history of recurring events.

In this chapter, we briefly review a series of case studies that have combined seismic stratigraphy and seismically targeted coring to provide information essential to the interpretation of paleoenvironmental reconstruction of lake systems along the Pole-Equator-Pole: Americas (PEP 1) transect. These examples are selected to illustrate diverse environments:

• From midlatitude Patagonia, Lake Mascardi is a proglacial, open system, and Laguna Cari-Laufquen Grande is a large, shallow *meseta*, closed-basin system.
• From the Peruvian-Bolivian Altiplano, Lake Titicaca illustrates a large, deep, closed tectonic lake with extensive drainage.
• From the high-altitude Atacama, Laguna Miscanti

illustrates unique aspects of a small, groundwater-fed, shallow saline lake.

• From the Great Basin, Great Salt Lake illustrates seismic stratigraphy covering the evolution of the Bonneville phases.

• From Wyoming, Yellowstone Lake illustrates post-glacial pelagic sedimentation in a large caldera setting.

We pursue three main goals with these examples of seismic stratigraphy: (1) to illustrate and encourage applications of seismic surveying in lakes of different types as tools for selecting optimal coring sites and recognizing past lake-level fluctuations, (2) to highlight new results from lakes of South America, and (3) to document some surprising similarities in the acoustic evidence and timing of abrupt changes in conjunction with major paleoclimate events.

14.2. METHODOLOGY

The case studies for this chapter utilized various higher frequency, high-resolution seismic profiling systems generally with penetration of 20–50 m and decimeter resolution. The records were not subjected to extensive processing, but display high online clarity.

Approximately 100 km of seismic profiles were collected in Lakes Mascardi and Cari-Laufquen Grande in 1996 and 1998 by using an Ocean Research Equipment (ORE)-GEOPULSE 3.5-kHz single-channel pinger system with a vertical seismic resolution of approximately 10–20 cm. In these lakes, the system achieved a maximum of 40–50 m of penetration. A conventional Global Positioning System (GPS) was used for navigation. Approximately 600 km of high-resolution seismic reflection data were collected from Lake Titicaca in April 1996 by utilizing an EdgeTech X-Star subbottom profiler operating at a sweep band width of 2–12 kHz with GPS navigation. Data were collected on a network of intersecting lines (Fig. 1C) to trace seismic-stratigraphic boundaries throughout the basin.

The few cross sections of Laguna Miscanti were made in 1995 with a GEOPULSE boomer (1–12 kHz) and hydrophone system at 170 J. Profiling in Great Salt Lake in 1979 used an EDO-Western 3.5-kHz system, with a shielded hydrophone tuned to a water depth of 8 m, and navigation by sextant. A series of profiles to accompany Kullenberg coring in Yellowstone in 1992 were derived from a vintage 3.5-kHz ORE 154 pinger and GPS navigation.

14.3. LAKE MASCARDI

Lake Mascardi (30 km² and 200 m maximum depth) is a typical proglacial lake located ca. 800 m above sea level (asl) and 15 km from the glacier front of a tongue of the Tronador ice cap (41°10′ S, 71°53′ W; 3554 m asl). Precipitation falls mostly as snow in the winter season, reaching maximum values of ca. 4000 mm/year (Drago, 1973). The western branch of the lake is fed directly by glacial meltwater through the Upper Manso River (Fig. 1B). Previous work showed that the extent of the Tronador ice cap is sensitive to both winter precipitation derived from the southern Pacific westerlies and to mean summer temperatures (Villalba et al., 1990). Thus, the Lake Mascardi sediments record fluctuations in glacial meltwater activity that provide a record of Southern Hemisphere postglacial climate variability (Ariztegui et al., 1997 and references therein).

Approximately 60 km of 3.5-kHz profiles allow a reconstruction of the lake bathymetry and sediment structure, as well as the separation of the effects of climate and neotectonics on lake sedimentation. This is important because the lake is located in an area of significant Holocene volcanic activity. The profiles record up to 50 m below the lake floor, representing approximately the last 15,000 years of infill history. Sedimentation is characterized by a paucity of chaotic debris and a relatively simple infill stratigraphy. The bedrock surface and the thick proglacial sediments reflect glacial erosion and the impact of proglacial meltwater influxes. Although the predominant pattern of sedimentation comprises simple and continuous basin infilling, variable sedimentation rates, as well as hiatuses, were identified in certain areas of the lake.

Comparison of the seismic results with multiproxy analyses from a suite of sediment cores established a well-dated lithostratigraphy (Ariztegui et al., 1997). Figure 2 shows uninterpreted and interpreted seismic sections with seismic units in position A–B at a water depth of 30 m (see Fig. 1B for locations of cross sections). Assuming a p-wave velocity of 1460 m/sec in the sediments, 10 m/sec of two-way travel time in the seismic profiles corresponds to 7.3 m in the sediment cores. Different shades of gray in the interpreted seismic section outline seismic sequences, which were mapped by tracing reflection terminations abutting unconformities. Radiocarbon ages could be assigned to sediment layers equivalent to seismic reflectors. Distinctive seismic facies were used to distinguish Holocene (coherent high-amplitude reflections) and late glacial deposits (low amplitude to transparent). These differences in the seismic facies are caused by differences in their physical properties. The mapping of reflector patterns that define seismic unconformities indicate two major environmental events that probably represent lake-level changes (Fig. 2). The first of these unconformities corresponds to the late glacial/Holocene transition and is marked by a distinct on-lap

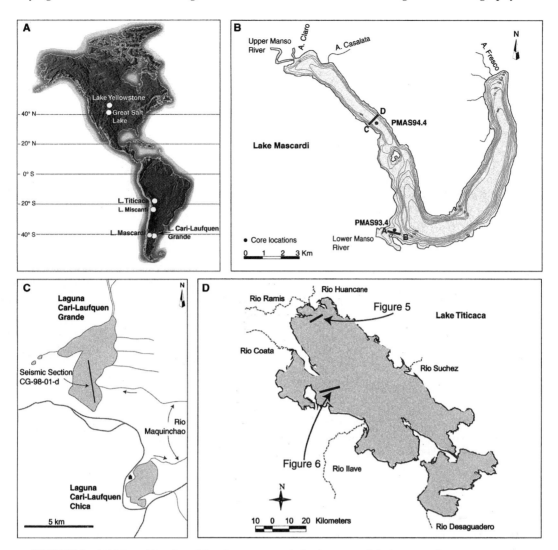

FIGURE 1   (A) Map of North and South America with the locations of the lacustrine basins with seismic profiles (Figs. 2–9) discussed in the text. Maps of (B) Lake Mascardi (Argentina), (C) Laguna Cari-Laufquen Grande (Argentina), and (D) Lake Titicaca (Bolivia-Peru).

surface. The sedimentation at the relatively shallow site of Fig. 2 is very sensitive to changes in water depth. Consequently, this change in lake level was recorded in the sediments not only as a change in lithology, but also as a change in sediment geometry, documented by the on-lapping reflections. Both the acoustic and lithologic features indicate an abrupt rather than a smooth change in the depositional environment. The second unconformity seems to be associated with an event in the mid-Holocene. It shows a less prominent impedance contrast than the Holocene boundary and, thus, may reflect a more gradual change in environmental conditions.

Figure 3 shows the uninterpreted (left) and interpreted (right) seismic section from position C–D at a water depth of 100 m. Compared to the A–B section, this section lies closer to the primary inflow to the lake and, thus, is closer to the source of sediments. The seismic profile in section C–D combined with core data clearly shows that this area of the lake is characterized by much higher sedimentation rates, therefore providing a higher resolution record of the Holocene. Because this section lies in deeper water than section A–B, the sedimentary processes are less influenced by environmental change than those from the shallower areas of the lake. Consequently, the sedimentary geometry is rather conformable. The late glacial/Holocene transition in Lake Mascardi is marked by a change in core lithology from fine-grained uniform muds to muds with more clastic intercalations. This

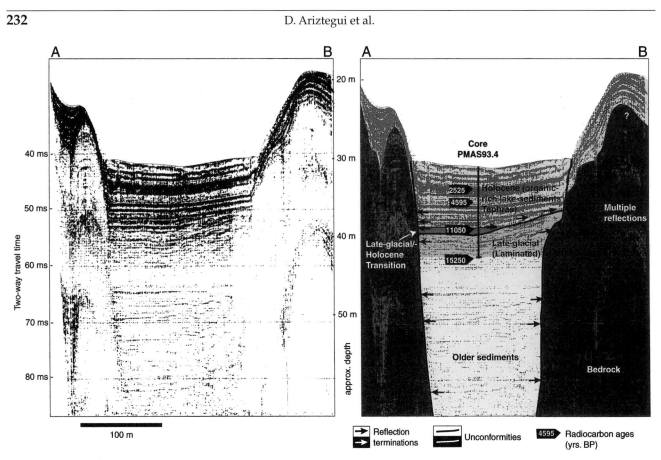

FIGURE 2 Lake Mascardi: uninterpreted (left) and interpreted (right) seismic profile in position A–B (refer to Fig. 1 for location). The late glacial/Holocene transition is seismically defined by a distinct on-lap surface, documenting a significant change in environment.

change is registered by changes in the seismic facies from low-amplitude internal reflections in the late glacial period to a facies characterized by layered internal reflectors of gradually increasing amplitudes for the Holocene.

14.4. LAGUNA CARI-LAUFQUEN GRANDE

The closed basins of Lagunas Cari-Laufquen Grande and Chica (Fig. 1C) lie in a tectonic depression surrounded by basalt plateaus or *mesetas* of the Mesozoic to Tertiary age (Coira, 1979). In contrast to Lake Mascardi, this region was not affected by the last glaciation of the Andes. Laguna Cari-Laufquen Grande is located at an elevation of 800 m and is an ephemeral, brackish water body with an average depth of 3.0 m during the rainy season. Mean annual precipitation is ca. 200 mm, falling primarily in the winter months (May to August), when the southern westerly storm tracks shift toward the equator. Mean annual temperature is 4°C, and the prevailing winds are from the west. Paleo-shorelines

have been observed and mapped at elevations of up to 68 m above the present lake level (Coira, 1979), although older shorelines up to 100 m above the present lake level have apparently been found (Galloway et al., 1988). During these lake-level high stands, the smaller lake, Cari-Laufquen Chica, merged with Cari-Laufquen Grande to form a large paleolake. Today, they are connected only by the Río Maquinchao (Fig. 1C). Results of dating paleo-shorelines indicate that higher lake levels than exist today occurred ca. 19,000 [14]C B.P. (Galloway et al., 1988) and between 14,000 and 10,000–8000 B.P. (Bradbury et al., 2000). Fine-grained lacustrine deposits underlying the upper two shorelines contain diatoms and ostracodes, which suggest deposition in a deeper, saline, and alkaline lake.

A seismic survey using a 3.5-kHz pinger system was undertaken in both lakes. With a maximum water depth of approximately 4 m, both lakes present major challenges to seismic surveying because strong multiple reflections and ringing usually obscure seismic data in such shallow water. Although the seismic record of Cari-Laufquen Chica did not show clear subsurface in-

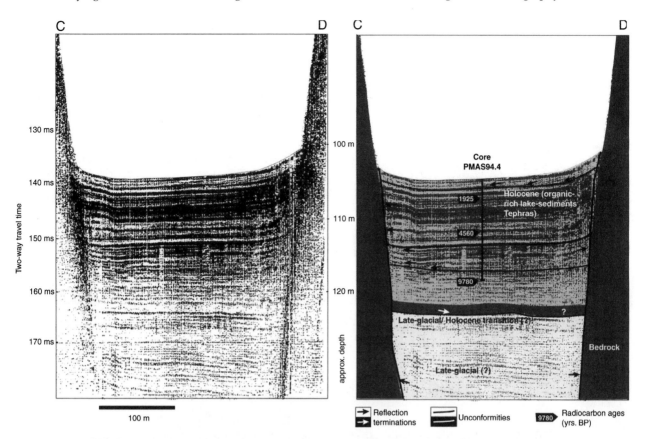

FIGURE 3   Lake Mascardi: uninterpreted (left) and interpreted (right) seismic profile in position C–D (refer to Fig. 1 for location). In contrast to profile A–B (Fig. 2), which lies in shallower water, this section is characterized by only minor unconformities. However, a distinct increase in the amplitude of internal reflections in the Holocene can be recognized.

formation, seismic sections obtained in Laguna Cari-Laufquen Grande yielded good acoustic stratigraphy (Fig. 4) because of the high-amplitude sediment impedance contrasts. The lines are characterized by a very weak water-bottom multiple because receiver gain can be kept low. This allows imaging of subsurface geometries in great detail to a depth of more than 15 m. Such good-quality seismic data, however, were obtained only in parts of the lake. In other areas, the seismic quality was limited, probably due to the patchy distribution of gas-rich sediments in which exolved gas bubbles presumably dissipate the seismic energy (Fig. 4).

In spite of the limitations of acoustic stratigraphy, subsurface geometries imaged by seismic profiling clearly document major structures related to paleolake levels. Several reflectors show strong relief (Fig. 4). These moundlike features could unlikely form by deepwater processes in the middle of a flat, 10-km-wide basin. We interpret such mounds as having formed by either erosional or constructional processes at or near desiccation levels in the lake. Potentially, the

acoustic character could also correspond to barlike features of fluvial gravels or sands. The lowest of these prominent unconformities is labeled seismic horizon A, which defines the acoustic basement. The surface is a barrier to reflected acoustic energy, likely indicating a well-indurated lithology below reflector A. The relief on reflector A is almost 10 m and is subsequently infilled by the lacustrine deposits that eventually form the modern flat-lake bathymetry. This geometry documents a period of very low lake levels or even subaerial exposure during formation of reflector A. Periods of higher lake levels yielded the overlying, relief-infilling lacustrine strata.

These overlying strata display numerous on-lapping geometries, as indicated by arrows in Fig. 4. Within the package, two horizons (B and C) form the base of mounds, approximately 1 m high, that are located on top of flat lacustrine substrates. These moundlike structures are most likely formed in a shallow lake as evidence of paleo-shorelines or sand/silt bars and spit features. The mounds probably comprise mainly silty

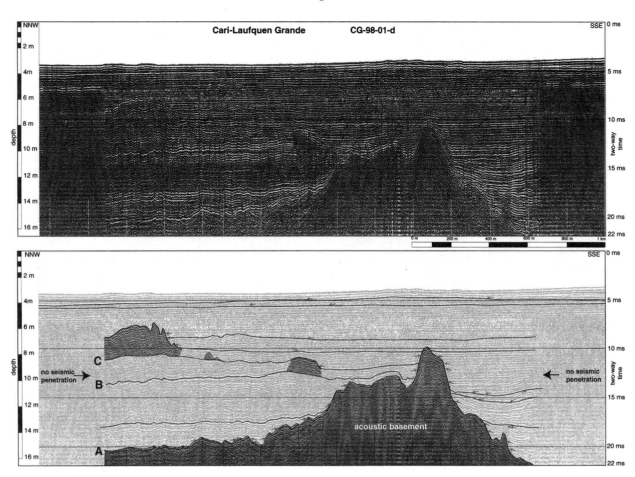

FIGURE 4 Laguna Cari-Laufquen: uninterpreted (top) and interpreted (bottom) seismic profile (refer to Fig. 1 for location). Reflection A defines the acoustic basement, which is not penetrated by the acoustic signal. It documents a paleorelief of up to 10 m in height and is either the result of erosion of a completely desiccated lake or an accumulation of river-transported gravels or sands. Horizons B and C mark the base of positive relief structures, which are likely paleo-shorelines. Thus, all three horizons are a result of periods when lake levels were lower than they are today or when the lake dried out.

sand sediments rather than coarse gravels because seismic signals penetrated these structures and imaged the underlying sediments.

The example of Cari-Laufquen Grande seismic stratigraphy is encouraging because of the useful information gained in spite of what would be considered an environment too shallow for viable seismic profiling. Three seismic horizons (A, B, and C; Fig. 4) represent desiccation surfaces, representing periods during which the lake level was lower than it is today. Thus, these seismic sequences and associated unconformities can be used to support estimates of changing rates of evaporation and/or variations in water input during the late Quaternary period. Using the recent data for optimal coring would suggest that reflector C could most efficiently be cored at the highest elevation of C, where it occurs only ca. 4 m below the lake floor (Fig. 4). A core retrieved farther to the north of this *basement*

high will be useful for a thicker, more continuous temporal reconstruction of the sedimentation history—a project that is in progress.

14.5. LAKE TITICACA

With a maximum water depth of 285 m, Lake Titicaca is located at the northern end of the Peruvian-Bolivian Altiplano extending from 14°–21°S at ca. 3600–4000 m asl (Fig. 1D). The high-resolution seismic reflection data from Lake Titicaca have been used to look at both recent lake-level change (Seltzer et al., 1998) and sedimentation in the lake during the late Quaternary period (D'Agostino, 1998). Seismic data obtained in the regions surrounding the deltas in the lake and in regions shallower than a water depth of ca. 85 m show erosional features that have been interpreted as having

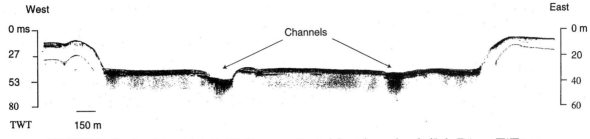

FIGURE 5   Erosional channels in the Río Huancane-Ramis delta at the north end of Lake Titicaca. TWT, two-way travel time (meters per second).

formed during a low stand of the lake ~100 m below present depths during the early and middle Holocene (Seltzer et al., 1998). Channels incised on the tops of the deltas and erosional truncation of seismic reflections appear to have formed relatively recently, as there is little or no sedimentation above these features (Fig. 5). Independent confirmation and dating of the lake-level change are provided by sediment core analyses (Cross et al., 2000). Apparently, the low stand of the lake was produced by more arid conditions during the late glacial and an early Holocene phase, followed by a rapid transgression of the lake at ca. 3600 [14]C B.P. (Abbott et al., 1997; Seltzer et al., 1998; Cross et al., 2000).

Lake Titicaca (18°S) provides an example of a large, deep, closed tectonic basin with large catchment influences. Although the lake is surrounded by mountain ranges that reach over 6000 m and are glaciated today, the lake itself was never glaciated. The result is a relatively continuous sediment stratigraphy that spans the

last glacial and the interglacial transition. The glacial-to-interglacial transition can be readily identified in the seismic stratigraphy as a transition from a package of relatively high-amplitude, evenly spaced reflections below to a 3- to 5-m thick unit of low-amplitude, closely spaced reflections (Fig. 6). This transition can be traced throughout the basin (D'Agostino 1998) and corresponds to a relatively inorganic, gray silt characteristic of glacial sedimentation with an abrupt transition to relatively organic-rich silts that have accumulated during the Holocene. At greater depths in the sediment (~40 m subbottom), an unconformity has been imaged at the limits of the penetration of the high-resolution data. This unconformity appears to represent a major low stand of the lake during the late Pleistocene (Fig. 6) (D'Agostino, 1998). Thus, the seismic data from Lake Titicaca indicate major hydrological changes in the basin and allow researchers to determine the best sites for piston coring and potential drilling for the recovery of long records.

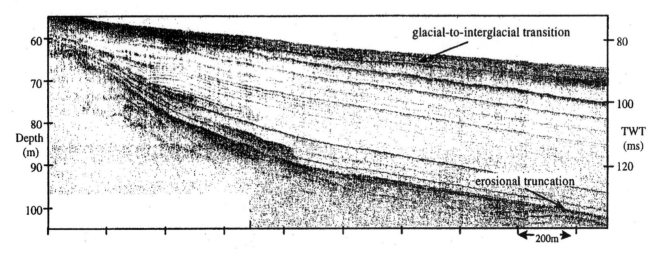

FIGURE 6   The glacial-to-interglacial transition and a deeper unconformity in the seismic stratigraphy as indicated with arrows in a representative cross section from Lake Titicaca. TWT (m/sec), two-way travel time (meters per second).

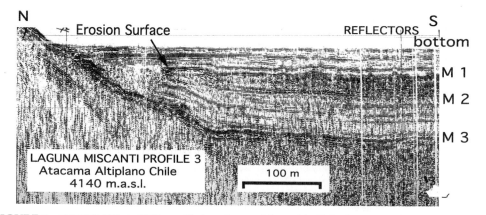

FIGURE 7 GEOPULSE 1- to 7-kHz profile from Laguna Miscanti in Chile showing mid-Holocene desiccation surface separating two parallel-layered lacustrine reflection packages (Valero-Garcés et al. 1996).

14.6. LAGUNA MISCANTI

Several small saline lakes are scattered along the extremely arid Chilean Atacama highlands and are fed by groundwater. Valero-Garcés et al. (1996) investigated the limnogeology of one of the largest lakes, Laguna Miscanti (23°45′ S, 4000 m asl, 15.5 km², ca. 10 m deep), with evidence of ancient, stromatolitic shorelines up to 20 m above the present surface. The modern sediments from a 3-m core comprise mainly magnesium carbonate precipitates with some gypsum; bioclasts of charophytes, ostracodes, and diatoms; and some aquatic organic matter. Siliciclastic input is low without surface inflow, and the residual component is likely eolian with some tephra glass.

Reconnaissance seismic profiles by a cumbersome GEOPULSE boomer provided surprisingly high subsurface resolution, high-impedance contrast, and penetration to 15 m. Four prominent reflections (bottom, M1, M2, M3) separate four acoustic units. A classic offlap and irregular reflection at 3 m subbottom identifies clear evidence of a major mid-Holocene desiccation surface of limited duration (Fig. 7), although dating remains equivocal due to groundwater reservoir effects on radiocarbon ages. Cores that hit this layer are characterized by aragonite crusts and dolomitic playa mud. The basin margin is steep, linking to M3, and is interpreted as a Pliocene volcanic or alluvial basement that outcrops nearby and suggests a relatively young sedimentary basin. Reflection M2 occurs at the top of a series of well-stratified reflections consistent with up to 6–10 m of sediment from late glacial/early Holocene higher lake levels. These sediments appear to correlate with shoreline terrace lake deposits. Between the M1 and M2 reflections are irregular, almost chaotic, suggesting fill and perhaps bioherms in an extremely shallow lake. Reflection M1 documents an erosive surface with a marked angular unconformity with the M2–M3 beds along the margins. Above the desiccation surface (M1), layering is uniform, with on-lapping basin margins that are considered to be evidence of a long lacustrine episode of shallow-water pelagial sedimentation until the present.

14.7. GREAT SALT LAKE

Great Salt Lake (~4800 km², 8 m deep) is a remnant of the 300-m deep Lake Bonneville (70,000 km²). Seismic profiling began with sparker and air-gun surveys in 1968–69 to investigate basinal tectonics (Mikulich and Smith, 1974). These surveys identified major fault traces and Quaternary fill tilted gently eastward. They mapped the extent of a mirabilite bed buried beneath 8 m of soft muds (Fig. 8); this bed was drilled earlier during a causeway construction and dated at >11,600 [14]C B.P. by bulk carbonate radiocarbon. During Amoco oil exploration in the 1970s, the lake was surveyed extensively with deep-penetration seismic sources.

In conjunction with an expedition in 1979, 127 km of 3.5-kHz profiling was used to site and correlate piston cores (Spencer et al., 1984). These profiles (Fig. 8) showed a distinctive late Quaternary seismic stratigraphy with a strong, uniform reflection (LB) found 2–8 m subbottom, and correlated with the Lake Bonneville stage. Post-Bonneville sediment packages show subparallel, and some irregular reflections with cut-outs, lensing, or buried hummocky surfaces (Great Salt Lake series) that are consistent with generally shallow, saline lake sediments identified in cores. The profiles suggest lower to mid-Holocene low stand(s), yet undated, that correlate with bioherms rooted on a planation surface

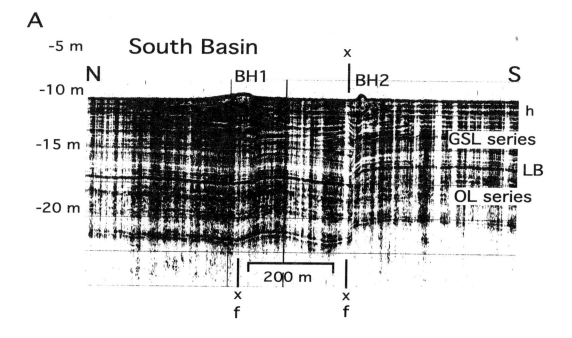

A

South Basin

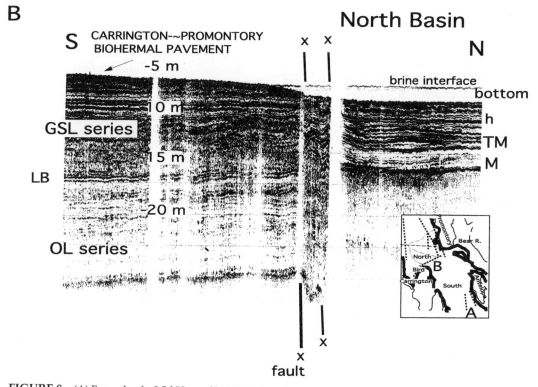

B

North Basin

FIGURE 8 (A) Example of a 3.5-kHz profile (13:04) from Great Salt Lake illustrating midlake section from southern basin (1) neotectonic faulting of older lake beds (OL horizon) beneath a Lake Bonneville maximum (reflection LB); (2) Holocene layered bedding (horizon GSL) above reflection LB; (3) a mid-Holocene, low-stand planation (reflection h) correlates with biohermal mounds (BH1, BH2) overlying fault traces. (B) Profile transition into north basin showing (1) Carrington Island-Promontory ridge with a carbonate-cemented biohermal pavement (mid-Holocene?) over parallel reflections of horizon GSL and OL below reflection LB. (2) A major fault runs along the northern ridge margin with evidence of disrupted bedding from strong recent movements. (3) The northern basin is characterized by a thick layer of (lower Holocene?) mirabilite ($NaSO_4 \cdot 10H_2O$) buried beneath ca. 7–8 m of Holocene sediments (GSL [Great Salt Lake] series). The insert map shows intrabasin profile locations A and B, locations of faults identified by Mikulich and Smith (1974) marked by heavy dotted lines, and an outline of the northern mirabilite basin in light dashes.

overlain by a meter or so of parallel lake beds that mirror the modern, smooth, flat bottom. A link with fault locations suggests the interaction of groundwater inflow and the formation of single constructional mounds, 20 m to ca. 80 m wide.

While not extensive, the profiles in the south basin cross numerous closely spaced faults with 0.5- to 1-m scale displacements. Some scarps break the surface, and others are visible only by the detection of drape folds beneath the LB reflection (>18,000 years B.P.; Fig. 8A). The general fault orientation runs parallel to Antelope Island, and at two sites it conforms to surface traces of deep faults F7 and F8 defined by Mikulich and Smith (1974). A major fault oblique to this trend, the Carrington-Promontory Fault (CPF), shows major recent disruption (Fig. 8B). When lake levels were about 4 m lower than they are today, the north and south basins could have been separated by a nearly emergent Carrington-Promontory Ridge covered in a veneer of a polygonal biohermal pavement (Fig. 8B).

Penetration north of the CPF was stopped ca. 6.75 m subbottom by a high-impedance, horizontal reflection M that is interpreted as a massive mirabilite salt bed described by Eardley (see Spencer et al. 1984). It thins to the north and west where OL reflections again are visible below the salt layer. Above reflection M is a 2-m series and reflection TM, which was hit by core GSL 79-09, confirming mirabilite salt and aragonitic mud interlayers. A controversy continues about whether this salt layer may potentially be seismic evidence of aridity during the Younger Dryas chronozone.

14.8. YELLOWSTONE LAKE

Otis et al. (1977) surveyed the Yellowstone Lake part of the caldera in 1973–74 with a 1-in.³ air gun providing 350 ms/sec penetration and with a 7-kHz system for the surface layer. This procedure generated a bathymetric and a sediment thickness map (190 m maximum), as well as evidence of a deep turbidite basin along paleodrainage. They documented faulting, hydrothermal activity, and a variety of glacial features. In 1992, 3.5-kHz profiling was used to target 10-m-long Kullenberg cores on deep bathymetric highs to collect undisturbed, mainly pelagic sediment sequences without turbidites for a diatom evolution study.

The profile FD between Frank and Dot islands (Fig. 9) crosses the late glacial/Holocene lithology at Core Site FD-2 (Tiller and Kelts, in press). This lithology, with tephra and accelerator mass spectrometer (AMS) radiocarbon age correlations, shows a major change from clastic, bedded mud at ca. 10,600 ¹⁴C B.P. (−8 m) to laminated Holocene diatom muds (−7.5 m) and then burrow mottled to faintly banded, uniform, Holocene diatom ooze from ca. 6500 ¹⁴C B.P. (−5 m) to the present. The parallel-bedded acoustic character of the Pg reflector series is consistent with an interpretation of ponded proglacial turbidites such as were just touched by the corer. The Y reflection at the base of the nearly transparent Holocene package is correlated with two, decimeter-thick, bluish-gray clay sediment layers believed to be ca. 10,000 ¹⁴C years old and derived from the last phreatic eruption into the lake. The Holocene (h series reflections) has a typical pelagic acoustic signature that drapes over the rise. Several weak internal reflectors occur, but do not correlate clearly with lithological features.

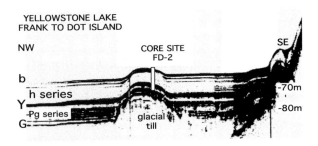

FIGURE 9   3.5-kHz profile section from Yellowstone Lake between Dot and Frank Islands showing core site of 9-m-long YEL-92-FD2 on a rise about 8 m above the 70-m deep plain, where the reflection package thins. Reflection Y divides ponded glacial deposits and postglacial muds from a pelagic Holocene diatom-ooze sequence.

14.9. CONCLUSIONS

The examples presented in this chapter illustrate seismic stratigraphy for different types of lake environments and show how the concepts of seismic sequence stratigraphy can be applied to lacustrine seismic data. Coordinated seismic surveys and sediment coring offer the opportunity for continuing to obtain optimal paleoclimate records from many lakes along the PEP 1 transect. Abbott et al. (2000), for example, recently applied a novel seismic and coring approach to build a lake-level history for Birch Lake, AK. Core analyses are required to calibrate seismic reflection sequences in a given succession as an aid for such lake-level and environmental change reconstructions. Initial survey results illustrate the potential of acoustic techniques to enhance quality, correlation, and comparison of significant environmental events, adding another tool to multiproxy paleoclimate analyses.

Acknowledgments

We are grateful to the collaborators who assisted the seismic and coring campaigns. For Lake Mascardi, we thank in particular Frank Niessen, Andreas Lehmann, Christina Chondrogianni, and Kurt Ghilardi. The study of Laguna Cari-Laufquen was supported by the Patagonian Lake Drilling (PATO) Team headed by Vera Markgraf and Kerry Kelts. Judith McKenzie and Judith Lezaun assisted the field campaign. At both lakes, we benefited from the invaluable help of Arturo Amos and the Programa en Gea Bariloche (PROGEBA) Team from Bariloche, Argentina. At Lake Titicaca, P. Baker, S. Fritz, R. Dumbar, S. Cross, D. Mucciarone, M. Revello, and J. Sanjines contributed substantially to the acquisition and interpretation of seismic data. Seismic profiling at Laguna Miscanti was by Blas Valero-Garcés and Antje Schwalb, assisted by Betina Jenny and Klaus Kammer. Yellowstone profiles are available thanks to Aosheng Wang and with help from M. Abbott, Ed Theriot, and others. We acknowledge financial support from the Swiss National Foundation (NF 21-37689.93 and NF 2100-050862.97/1). For the work at Lake Titicaca, we acknowledge support of the U.S. National Science Foundation (NSF; ATM-9600051). Profiling in Great Salt Lake and Yellowstone Lake was supported by several U.S. NSF grants. The manuscript benefited significantly from thorough reviews by Steve Coleman and Chris Scholz.

References

Abbott, M., M. B. Binford, M. W. Brenner, and K. R. Kelts, 1997: A 3500 ^{14}C yr high resolution record of lake level changes in Lake Titicaca, South America. *Quaternary Research*, **47**: 169–180.

Abbott, M., B. P. Finney, M. Edwards, and K. R. Kelts, 2000: Paleohydrology of Birch Lake, central Alaska: Lake-level reconstructions using seismic profiles and core transect approaches. *Quaternary Research*, **53**: 154–166.

Ariztegui, D., M. M. Bianchi, J. Masaferro, E. Lafargue, and F. Niessen, 1997: Interhemispheric synchrony of late glacial climatic instability as recorded in proglacial Lake Mascardi, Argentina. *Journal of Quaternary Sciences*, **12(4)**: 333–338.

Bradbury, J. P., M. Grosjean, S. Stine, and F. Sylvestre, 2000: Full and late glacial lake records along the PEP 1 transect: Their role in developing interhemispheric paleoclimate interactions. *In* Markgraf, V. (ed.), *Interhemispheric Climate Linkages.* San Diego: Academic Press, Chapter 16.

Chondrogianni, C., D. Ariztegui, F. Niessen, C. Ohlendorf, and G. Lister, 1996: Late Pleistocene and Holocene sedimentation in Lake Albano and Lake Nemi (central Italy). *In* Oldfield, F., and P. Guilizzoni (eds.), Palaeoenvironmental Analysis of Italian Crater Lake and Adriatic Sediments. Memorie dell'Istituto Italiano di Idrobiologia. *International Journal of Limnology*, **55**: 23–38.

Coira, G. L., 1979: Descripción geológica de la Hoja 40D, Ingeniero Jacobacci. *Boletín Servicio Geológico Nacional, Buenos Aires*, **168**: 1–84.

Cross, S. L., P. A. Baker, G. O. Seltzer, S. C. Fritz, and R. B. Dunbar, 2000: A new estimate of the Holocene lowstand level of Lake Titicaca, and implications for tropical paleohydrology. *The Holocene*, **10**: 21–32.

Cross, T. A., and M. A. Lessenger, 1988: Seismic stratigraphy. *Annual Reviews of Earth Planetary Science*, **16**: 319–354.

D'Agostino, P., 1998: High-resolution seismic stratigraphy of Lake Titicaca, Peru/Bolivia. Masters thesis. Syracuse University, New York.

Degens, E. T., R. P. von Herzen, and H. K. Wong, 1971: Lake Tanganyika: Water chemistry, sediments, geological structure. *Naturwissenschaften*, **58**: 229–241.

Degens, E. T., R. von Herzen, and H. K. Wong, 1973: Lake Kivu: Structure, chemistry and biology of an East African rift lake. *Geologische Rundschau*, **62**: 245–277.

Degens, E. T., H. K. Wong, S. Kempe, and F. Kurtman, 1984: A geological study of Lake Van, eastern Turkey. *Geologische Rundschau*, **73(2)**: 701–734.

Drago, E., 1973: Algunas características geomorfológicas de la llanura aluvial del Río Manso superior (prov. de Río Negro). *Revista de la Asosiación de Ciencias Naturales del Litoral*, **4**: 187–200.

Finckh, P., and K. Kelts, 1976: Geophysical investigations into the nature of pre-Holocene sediments of Lake Zürich. *Eclogae Geologicae Helveticae*, **69(1)**: 139–148.

Finckh, P., K. Kelts, and A. Lambert, 1984: Seismic stratigraphy and bedrock forms in perialpine lakes. *Geological Society of America Bulletin*, **95**: 1118–1128.

Galloway, R. W., V. Markgraf, and J. P. Bradbury, 1988: Dating shorelines of lakes in Patagonia, Argentina. *Journal of South American Earth Sciences*, **1**: 195–198.

Giovanoli, F., K. Kelts, P. Finckh, and K. J. Hsü, 1984: Geological framework, site survey and seismic stratigraphy. In Hsü, K. J., and K. Kelts (eds.), Quaternary geology of Lake Zürich: An interdisciplinary investigation by deep-lake drilling. *Contributions to Sedimentology*, **13**: 5–20.

Hsü, K. J., and K. Kelts, 1970: Seismic investigation of Lake Zürich, Part II. *Eclogae Geologicae Helveticae*, **63(2)**: 520–537.

Johnson, T. C., 1980: Late glacial and post glacial sedimentation in Lake Superior based on seismic reflection profiles. *Quaternary Research*, **13**: 380–391.

Johnson, T. C., J. D. Halfman, B. R. Rosendahl, and G. S. Lister, 1987: Climatic and tectonic effects on sedimentation in a rift valley lake: Evidence from Lake Turkana, Kenya. *Geological Society of America Bulletin*, **98**: 439–447.

Johnson, T. C., J. D. Wells, and C. A. Scholz, 1995: Deltaic sedimentation in a modern rift lake. *Geological Society of America Bulletin*, **107(7)**: 812–829.

Matter, A., A. E. Süsstrunk, K. Hinz, and M. Sturm, 1971: Ergebnisse reflexionsseismischer Untersuchungen im Thunersee. *Eclogae Geologicae Helveticae*, **64(3)**: 505–520.

Mikulich, M. J., and R. B. Smith, 1974: Seismic reflection and aeromagnetic surveys of the Great Salt Lake, Utah. *Geological Society of America Bulletin*, **85**: 99–1002.

Mitchum, R. M., Jr., and P. R. Vail, 1977: Seismic stratigraphic interpretation procedure. *In* Payton, C. E. (ed.), Seismic stratigraphy—Applications to hydrocarbon exploration. *American Association of Petroleum Geologists Memoir*, **26**: 117–133.

Moore, T. C., D. K. Rea, L. A. Mayer, D. M. Lewis, and D. M. Dodson, 1994: Seismic stratigraphy of Lake Huron—Georgian Bay and postglacial lake level history. *Canadian Journal of Earth Sciences*, **31**: 1606–1617.

Müller, G., and R. Gees, 1968: Erste Ergebnisse reflexionsseismischer Untersuchungen des Bodensee-Untergrundes. *Neues Jahrbuch für Geologie und Paläontologie Monatsheft*, **6**: 364–369.

Mullins, H. T., N. Eyles, and E. J. Hinchey, 1991: High-resolution seismic stratigraphy of Lake McDonald, Glacier National Park, Montana, U.S.A. *Arctic and Alpine Research*, **23**: 311–319.

Mullins, H. T., E. J. Hinchey, R. W. Wellner, D. B. Stephens, W. T. Anderson, T. R. Dwyer, and A. C. Hine, 1996: Seismic stratigraphy of the Finger Lakes: A continued record of Heinrich event H-1 and Laurentide ice sheet instability. Geological Society of America Special Paper 311, 1–35.

Niessen, F., A. Lami, and P. Guilizzoni, 1993: Climatic and tectonic effects on sedimentation in central Italian volcano lakes (Latium)—Implications from high resolution seismic profiles. *In* Negendank,

J. F. W., and B. Zolitschka (eds.), *Paleolimnology of European Maar Lakes*. Berlin-Heidelberg: Springer-Verlag, pp. 129–148.

Otis, R. M., R. B. Smith, and R. J. Wold, 1977: Geophysical surveys of Yellowstone Lake, Wyoming. *Journal of Geophysical Research,* **82(26)**: 3705–3717.

Scholz, C. A., and B. R. Rosendahl, 1988: Low lake stands in Lakes Malawi and Tanganyika, East Africa, delineated with multifold seismic data. *Science,* **240**: 1645–1648.

Seltzer, G. O., P. Baker, S. Cross, R. Dunbar, and S. Fritz, 1998: High-resolution seismic reflection profiles from Lake Titicaca, Peru-Bolivia: Evidence for Holocene aridity in the tropical Andes. *Geology,* **26(2)**: 167–170.

Serruya, C., O. Leenhardt, and A. Lombard, 1966: Etudes géophysiques dans le Lac Léman. *Archive de Science de Genéve,* **19**: 179–196.

Spencer, R. J., M. J. Baedecker, H. P. Eugster, B. F. Jones, K. Kelts, J.

McKenzie, et al., 1984: Great Salt Lake and precursors, Utah: The last 30,000 years. *Contributions to Mineralogy and Petrology,* **86**: 321–334.

Tiller, C., and K. Kelts, in press: A muddy view of Yellowstone Lake history. Proceedings IV, 125th Yellowstone Park. Geological Society of America Special Paper.

Valero-Garcés, B., M. Grosjean, A. Schwalb, M. Geyh, B. Messerli, and K. Kelts, 1996: Limnogeology of Laguna Miscanti, evidence for mid to late Holocene moisture changes in the Atacama Altiplano (northern Chile). *Journal of Paleolimnology,* **16**: 1–21.

Villalba, R., J. C. Leiva, S. Rubulis, J. Suarez, and L. Lenzano, 1990: Climate, tree-rings and glacial fluctuations in the Río Frías valley, Río Negro, Argentina. *Arctic and Alpine Research,* **22**: 215–232.

Wold, R. J., R. A. Paull, C. A. Wolosin, and R. J. Friedel, 1981: Geology of central Lake Michigan. *Geological Society of America Bulletin,* **65(9)**: 1621–1623.

15

Holocene Climate Patterns in the Americas Inferred from Paleolimnological Records

SHERILYN C. FRITZ, SARAH E. METCALFE, AND WALTER DEAN

Abstract

Holocene paleolimnological records for the Americas that clearly reflect climate are limited in number, particularly for the Southern Hemisphere. Most lacustrine records are of variation in effective moisture (precipitation minus evaporation), except for the Arctic, where temperature is the dominant influence. The millennial-scale patterns of climate variation evident in paleolimnological records are consistent with changes in insolation, modulated in the early Holocene in North America by the influence of the Laurentide ice sheet. In the midlatitudes of both North and South America and in the Southern Hemisphere tropics, the mid-Holocene was drier than it is today, although the onset, magnitude, and duration of aridity vary considerably. In contrast, the Northern Hemisphere tropics were wetter than they are today at times during the mid-Holocene. The broader scale patterns of mid-Holocene moisture variation are consistent with a northerly migration of both the Intertropical Convergence Zone (ITCZ) and the mean position of the jet stream. In North America, the termination of the mid-Holocene dry interval was apparently time transgressive, beginning earlier in the northwest than in north-central North America; in the East, there currently are too few sites to clearly characterize late Holocene moisture patterns from lake records. We speculate that spatial variation in moisture increased in the late Holocene, perhaps as a result of the diminished seasonality of insolation and increased meridional flow. We also suggest that studies of high-frequency climate variation spanning different intervals of the Holocene are needed to investigate the interactions of large-scale boundary conditions, such as insolation and the ice sheet, with other climate forcing mechanisms in controlling spatial and temporal climate patterns. Copyright © 2001 by Academic Press.

Resumen

El número de los registros paleolimnológicos en las Americas que reflejan claramente el clima son muy limitados, particularmente en el Hemisferio Sur. A excepción del Artico, donde la temperatura es la influencia dominante, la mayoría de los registros lacustres presentan una variación en la humedad efectiva (precipitación menos evaporación). Los patrones de variación climática a escala de milenio, evidente en los registros paleolimnológicos, son consistentes con los cambios en la irradiación solar, modulados en el Holoceno temprano en Norteamerica por la influencia de la glaciación continental Lauréntida. En las latitudes intermedias de Norte- y Sudamerica así como en el trópico del Hemisferio Sur, el Holoceno medio fué más seco que en el presente, aunque el inicio, la magnitud y la duración de la aridez variaron considerablemente. En contraste, el trópico del Hemisferio Norte durante el Holoceno medio fué más humedo que en el presente. Estos patrones en la variación de la humedad durante el Holoceno medio son consistentes con una migración hacia el Norte del ITCZ y la posición promedio de la Corriente de Aire troposférica. En la America del Norte, la terminación del período seco en el Holoceno medio fué aparentemente diácrono, comenzando más temprano en el Noroeste que en el norte del Centro Norteamericano. Actualmente, hay todavía muy pocos sitios en el Este Norteamericano que claramente caractericen los patrones de humedad en los registros lacustres durante el Holoceno tardío. Por lo tanto, se especula que la variación espacial de la humedad se incrementó en el Holoceno tardío, posiblemente como resultado de la disminución estacional de la irradiación solar y el incremento del flujo del aire meridional. También se sugiere que los estudios de la variación climática de alta frecuencia que abarquen diferentes intervalos del Holoceno son necesarios para investigar las interacciones de las condiciones de los límites de amplia escala, tales como la irradiación solar y la glaciación continental, con otras fuerzas climáticas que controlan los patrones climatológicos temporales y espaciales.

15.1. INTRODUCTION

Climate affects lakes through its influence on hydrological, thermal, and chemical processes (Hostetler, 1995). Climate directly controls lake levels (Street-Perrott and Harrison, 1985) and the ionic concentrations and composition of lake water via precipitation and evaporation at the lake surface, as well as through climate-induced changes in groundwater and surface-water inflow and outflow. Air temperature and wind determine the degree of thermal stratification and the timing and duration of ice cover. These direct climatic controls, in turn, influence a suite of chemical and biological processes. Although records for individual basins reflect a mixture of regional climate forcing mechanisms and specific catchment characteristics, the value of paleolimnological records in reconstructing past climates is well established (e.g., Eugster and Kelts, 1983; Gasse et al., 1987; Bradbury, 1991; Fritz, 1996). In this chapter, we review Holocene paleolimnological records for the Americas and use these data to suggest continental and interhemispheric patterns of Holocene climate variation. Because the number of high-resolution records is very small, we focus on centennial- and millennial-scale patterns, although we do summarize the few studies of decadal variation.

Paleolimnological records for the Americas that clearly reflect climatic change are not abundant, and spatial coverage is extremely patchy. Most of these records involve proxy variables that are interpreted in terms of changes in the hydrologic budget for lakes in arid to subhumid regions, where evaporation is a large component of the water budget, and we emphasize these records in the following discussion. However, in high-latitude or high-elevation regions, thermally driven shifts in stratification or ice cover can also be derived from lacustrine records. We have restricted our discussion to inferences of climate history from paleolimnological proxies—that is, proxies that are derived from within the lake itself or are used to infer lacustrine processes. Thus, most of these inferences are based on diatoms or ostracodes or on sediment geochemistry, mineralogy, and lithology. We use uncalibrated radiocarbon rather than calendar ages (unless otherwise specified) because these are most often used in the primary literature, which forms the foundation for this chapter.

15.2. NORTH AMERICA

For North America, most Holocene paleolimnological records are for the northern continental interior, with relatively few for western and eastern regions (Fig. 1; Table 1) or for high latitudes (Fig. 2; Table 1). In nearly all North America records, the mid-Holocene is an interval of severe aridity, although the onset, duration, and magnitude of aridity vary considerably among the sites (Table 1).

15.2.1. Western North America

In the Great Basin and elsewhere in the southwestern United States, many lake basins became dry in the early Holocene, following late glacial levels that were higher than they are now. Thus, there are few complete Holocene records. The large lakes in the northern Great Basin, all part of the glacial Lake Lahontan and Lake Bonneville systems, underwent dramatic expansion and contraction during the late Quaternary (Benson et al., 1990, 1996), but there are relatively few studies of Holocene fluctuations. Lithologic data, combined with algal remains, for Ruby Marshes in Nevada (Thompson, 1992) suggest dry to shallow conditions from the early through the mid-Holocene, with deepening after 4700 B.P., a pattern similar to that in Diamond Pond in Oregon (Wigand, 1987). Pore-fluid data for a core from Walker Lake in Nevada, which occupies the southernmost basin of glacial Lake Lahontan, indicate it became a playa at ca. 9000 B.P. (Benson, 1978). Refilling of Walk-

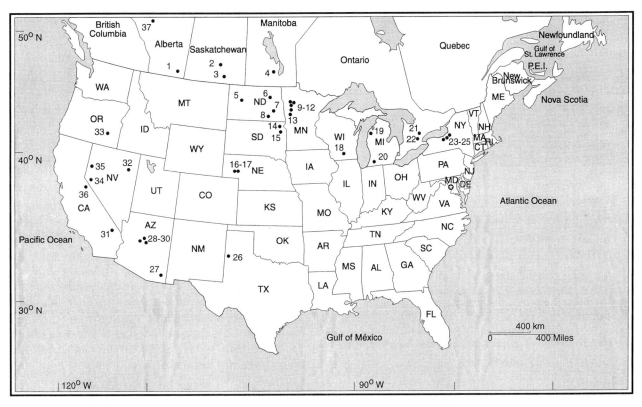

FIGURE 1 Map of the United States and southern Canada showing locations of sites mentioned in the text. (See Table 1 for an explanation of the numbers.)

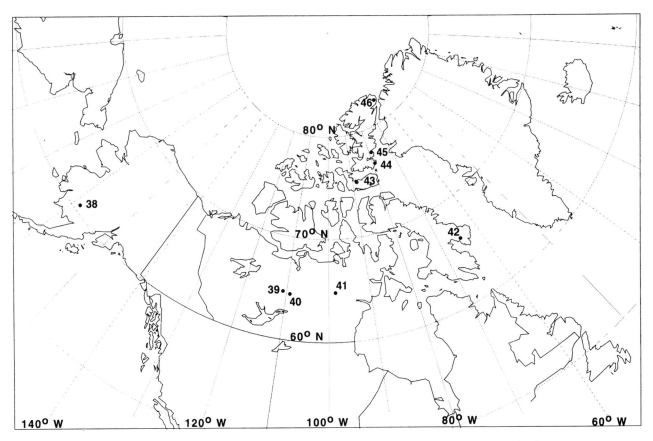

FIGURE 2 Map of boreal and Arctic regions of North America showing locations of sites mentioned in the text. (See Table 1 for an explanation of the numbers.)

TABLE 1   List of North American Lake Sites Referred to in the Text

Site	Location	Ref.
1.[a]	Chappice, Alberta	Vance et al., 1992
2.	Ceylon, Saskatchewan	Last, 1990
3.	Waldsea, Saskatchewan	Last and Schweyen, 1985
4.	Manitoba, Manitoba	Teller and Last, 1981
5.	Rice, North Dakota	Yu and Ito, 1999
6.	Devils, North Dakota	Fritz et al., 1991
7.	Moon, North Dakota	Laird et al., 1996a,b
8.	Coldwater, North Dakota	Xia et al., 1997
9.	Elk, Clearwater Co., Minnesota	Bradbury and Dean, 1993; Forester et al., 1987
10.	Williams, Minnesota	Schwalb et al., 1995
11.	Shingobee, Minnesota	Schwalb et al., 1995
12.	Adley, Minnesota	Digerfeldt et al., 1993
13.	Elk, Grant Co., Minnesota	Smith et al., 1997
14.	Pickerel, South Dakota	Schwalb and Dean, 1998; Smith, 1991
15.	Medicine, South Dakota	Valero-Garcés et al., 1995; Radle et al., 1989
16.	Swan, Nebraska	Loope et al., 1995; Mason et al., 1997
17.	Crescent, Nebraska	Loope et al., 1995; Mason et al., 1997
18.	Mendota, Wisconsin	Winkler et al., 1986
19.	Herring, Michigan	Wolin, 1996
20.	Austin, Michigan	Krishnamurthy et al., 1995
21.	Crawford, Ontario	Yu et al., 1997
22.	Weslemkoon, Ontario	Edwards et al., 1996
23.	Seneca, New York	Anderson et al., 1997
24.	Owasco, New York	Dwyer et al., 1996
25.	Cayuga, New York	Mullins, 1998
26.	Lubbock, Mustang Spring, Texas	Holliday, 1989
27.	Cochise, Arizona	Waters, 1989
28.	Potato, Arizona	Anderson, 1993
29.	Montezuma Well, Arizona	Blinn et al., 1994
30.	Stoneman, Arizona	Hasbargen, 1994
31.	Silver Lake Playa, California	Enzel, 1992
32.	Ruby Marsh, Nevada	Thompson, 1992
33.	Diamond Pond, Oregon	Wigand, 1987
34.	Walker, Nevada	Bradbury, 1987; Benson, 1978
35.	Pyramid, Nevada	Benson, 1978, Benson and Thompson, 1987
36.	Mono, California	Stine, 1994
37.	Pine, Alberta	Campbell, 1998
38.	Farewell, Alaska	Hu et al., 1998
39.	Queens, Northwest Territories	MacDonald et al., 1993; Edwards et al., 1996
40.	Toronto, Northwest Territories	Edwards et al., 1996; Wolfe et al., 1996
41.	Whatever, Northwest Territories	Edwards et al., 1996
42.	Amarok, Baffin Island	Wolfe, 1996; Wolf and Hartling, 1996
43.	Unnamed Lake, Devon Island	Gajewski et al., 1997
44.	Elison, Camp Pond, and Col Pond, Ellesmere Island	Douglas et al., 1994
45.	Rock Basin, Ellesmere Island	Smol, 1983
46.	C2, Ellesmere Island	Lamoureux and Bradley, 1996

[a]The numbers correspond to site locations in Fig. 1 (sites 1–37) and Fig. 2 (sites 38–46).

er Lake at ca. 4700 B.P., probably by geomorphic means (diversion of the Walker River), yielded a late Holocene record of moisture variation, reconstructed primarily from diatom and ostracode data (Bradbury, 1987; Bradbury et al., 1989). The data indicate low lake levels and high salinity between 2400 and 2000 B.P., followed by subsequent refilling. Pore fluids in sediments from Pyramid Lake, another remnant of Lake Lahontan, suggest it remained in existence during the early Holocene when Walker Lake was dry (Benson, 1978). Dating of tufa deposits around Pyramid Lake (Benson and Thompson, 1987) suggests low lake levels between 2500 and 1800 B.P. Mono Lake in California, underwent a series of oscillations during the Holocene and reached its Holocene high stand at ca. 3400 B.P. By 1800 B.P., it had fallen to a low stand and subsequently underwent a series of century- and subcentury-scale oscillations (Stine 1990). These oscillations include two intervals of prolonged drought in the Medieval period (A.D. 900–1300), both of which exceeded historic droughts in magnitude and duration (Stine, 1994).

For Arizona, lacustrine records indicate moist conditions in the early Holocene, followed by dry conditions after 8500 B.P. (Waters, 1989; Anderson, 1993; Blinn et al., 1994; Hasbargen, 1994). Within the dry mid- to late Holocene, two marl units in a core from Lake Cochise (Wilcox playa), dated at ca. 5400 and between 4000 and 3000 B.P., suggest mid-Holocene pluvial intervals when the basin filled (Waters, 1989). A continuous lacustrine diatom sequence for Montezuma Well suggests higher lake levels between 5000 and 3500 B.P., followed by a brief arid period to 3000 B.P. (Blinn et al. 1994). The lithology of a core from Potato Lake indicates that it refilled at ca. 3000 B.P., after mid-Holocene desiccation (Anderson, 1993). However, diatom, pollen, and macrofossil records for nearby Stoneman Lake suggest persistent aridity until 2000 B.P. (Hasbargen, 1994). In southern California, a short episode of increased moisture at ca. 3600 B.P., as well as a more recent event at 390 B.P., is indicated by interbedded lake sediments in a presently dry basin in the Mojave Desert (Enzel, 1992). Thus, there is considerable evidence for aridity throughout the Southwest between ca. 8500 and 5500 B.P. and of episodes of increased moisture between 5400 and 3000 B.P. However, late Holocene patterns of lake-level change are spatially variable, and thus, the nature of climate variation is unclear. Lithologic data for basins in the southern High Plains of Texas show lacustrine deposition prior to 8500 B.P., followed by desiccation in a number of locations and continuous deposition in others until ca. 5500 B.P. (Holliday, 1989). Deposits of eolian origin and widespread desiccation of lake basins between 6500 and 4500 B.P. suggest maximum aridity in this interval.

Although pollen data for the Pacific Northwest are abundant, there are virtually no paleolimnological studies of climate variation. The available lithologic data suggest that lake levels were very low in the early Holocene and rose sometime after 5000 B.P. (Barnosky et al., 1987).

15.2.2. Northern Great Plains and the Midwest

In the northwestern part of the Canadian Plains (western and central Alberta), many lakes were dry at the onset of the Holocene, suggesting arid conditions (Schweger and Hickman, 1989). Flooding of these formerly dry basins began at various times in the mid-Holocene (7500–4000 B.P.), although low lake levels and high salinities persisted at most sites until 4000 B.P. (Schweger and Hickman, 1989; Vance et al., 1992, 1995). In contrast, for southern Saskatchewan (Ceylon Lake) and southern Manitoba (Lake Manitoba), mineralogical and lithologic data indicate that these lakes were fresh; therefore, conditions must have been wet until 8000 B.P., followed by high salinities and low lake levels to 4500 B.P. (Teller and Last, 1981; Last, 1990).

For the northern Great Plains of the United States, evidence from diatoms, ostracodes, and carbonate geochemistry indicates that lakes were fresh and lake levels were high at the onset of the Holocene. However, levels began to decline in the early Holocene, and many lakes became hydrologically closed. Once they were hydrologically closed, the lakes showed an increase in salinity in response to decreases in effective moisture, although there was regional variation in the magnitude and rate of salinity change and, thus, in the pattern of inferred climate (see summary in Laird et al., 1996a). For North Dakota, diatom and ostracode stratigraphies for several lakes indicate that the rate of salinity increase accelerated after 8000 B.P., and maximum salinity was reached at ca. 6500 B.P., although clearly conditions remained quite dry for several thousand years afterward (Laird et al., 1996a; Xia et al., 1997). Similarly, in a suite of lakes in western Minnesota, paleolimnological proxies (diatoms, ostracodes, isotopes, and mineralogy) indicate gradually increasing aridity from the early Holocene onward, with an accelerated rate of increase after 8000 B.P. and maximum aridity in the interval between 7200 and 5200 B.P. (Forester et al., 1987; Bradbury and Dean, 1993; Digerfeldt et al., 1993; Schwalb et al., 1995; Smith et al., 1997). Further east in Wisconsin (Winkler et al., 1986), a sand lens in cores from Lake Mendota suggests lake-level lowering and aridity beginning at ca. 6900 B.P.

For the southern part of the northern Great Plains, poor dating control and the small number of sites preclude a clear understanding of Holocene moisture patterns. In the meromictic Medicine Lake in South Dakota, maximum diatom-inferred salinity was reached by 9000 B.P., and high salinity persisted throughout the mid-Holocene (Radle et al., 1989). However, sedimentological and geochemical evidence suggests that the lowest lake levels may have occurred several thousand years later (Valero-Garcés et al., 1995). Isotopical and geochemical data for Pickerel Lake in South Dakota similarly indicate the onset of dry conditions at ca. 9000 B.P., but a shift in ostracode species dominance at ca. 5500 B.P. suggests increased aridity at this time relative to earlier millennia (Smith, 1991; Schwalb and Dean, 1998). Thus, the timing of maximum aridity in this region is currently uncertain.

Many lakes and wetlands in the Nebraska Sandhills were formed just prior to the onset of the Holocene by blockage of drainage channels by eolian sand during an interval of extreme aridity. In the western Nebraska Sandhills during the middle Holocene (ca. 6000 B.P.), such a blockage by sand caused a 25-m rise in the regional water table (Loope et al., 1995; Mason et al., 1997). As a result, lakes formed in several interdune depressions and accumulated unusually thick sequences of organic sediments. Thin sand layers in several of these lakes represent fluctuations in lake levels during episodes of late Holocene (post-1500 B.P.) dune movement (Mason et al., 1997).

Although evidence for a mid-Holocene dry interval is widespread throughout the northern Great Plains and the Midwest, it is unclear whether the dry climate was cool or warm. For Elk Lake in Minnesota, ostracode and oxygen isotope data suggest that, although conditions were dry beginning at ca. 6800 B.P., temperatures were cold until 5900 B.P. and then became warmer (Forester et al., 1987; Dean and Stuiver, 1993). Isotopic data for Pickerel Lake in South Dakota have been also interpreted as indicative of a cool early to mid-Holocene (Schwalb and Dean, 1998; see also Xia et al., 1997). In contrast, pollen analyses for Elk Lake suggest warmer July temperatures than now from ca. 7000 B.P. through the mid-Holocene (Bartlein and Whitlock, 1993). Unfortunately, there are presently insufficient data for other sites to confirm the nature of early to mid-Holocene regional temperature.

Sites throughout the northern Great Plains show a decrease in salinity and an increase in moisture beginning between 5000 and 4000 B.P. (reviewed in Laird et al., 1996a; see also Bradbury and Dean, 1993; Schwalb et al., 1995; Smith et al., 1997; Xia et al., 1997). In most locations, moist conditions have persisted to modern times, although for many sites, the data suggest large fluctuations between wet and very dry, especially after 2200 B.P. (Laird et al., 1996b; Schwalb and Dean, 1998).

In Wisconsin, a transition from sand to gyttja at 3200 B.P. marks higher lake levels and enhanced moisture at Lake Mendota (Winkler et al., 1986), slightly later than at most sites to the west, a pattern also suggested by vegetation data for Iowa (Baker et al., 1992).

At present, only a few high-resolution lacustrine records of decadal or subdecadal climate variation exist for North America, and these are restricted to north-central North America (Laird et al., 1996b; Dean, 1997; Campbell, 1998). Data for several lakes in North Dakota (Laird et al., 1996b; Yu and Ito, 1999) suggest that the climate of the last 2000 years was hydrologically complex, with frequent alternations between wet and dry intervals. The data indicate that severe drought was characteristic of Medieval times (A.D. 900–1300), but that this interval was not unusual relative to previous centuries. Similarly, intervals of major drought were also typical within the Little Ice Age (LIA) (A.D. 1400–1850), although extremely fresh conditions, suggesting high precipitation, are evident from A.D. 1330–1430 and in the early decades of the 1800s. The data suggest intervals of multiple decades to more than a century when drought was frequent and severe, indicating that the frequency of extreme events may have been greater at times during the last 2000 years than in the twentieth century. A record of eolian activity inferred from sedimentary aluminum and quartz concentrations in Elk Lake in Minnesota also indicates decadal- to century-scale cycles of drought during the last 1500 years (Dean, 1997), with the most distinct drought intervals corresponding in time to the LIA (see also Fritz et al., 1994). Similar cycles occur in magnetic susceptibility and organic carbon in Pickerel Lake in South Dakota (Schwalb and Dean, 1998).

A high-resolution, sediment grain size record for a lake in boreal parkland in central Alberta (Campbell, 1998) suggests several directional late Holocene shifts in streamflow magnitude into the lake. Median sediment grain size in a midlake deepwater core, which is argued to be directly related to stream inflow (coarse grain size indicates high flow), decreases markedly during Medieval times relative to the previous ca. 1000 years, suggesting an interval of very dry conditions. Inferred moisture increased during the last 600 years and was higher than at any other time in the 4000-year record.

15.2.3. Great Lakes Region

Interpretations of lake records for the Great Lakes region of eastern North America disagree about Holocene moisture patterns. Most records suggest that lake levels were highest during the early Holocene and

subsequently fell. Most authors agree that the climate was warm from 8000–2000 B.P. (Stuiver, 1970; Drummond et al., 1995; Krishnamurthy et al., 1995); some suggest that maximum aridity occurred from 8000 to 5500 B.P. (Colman et al., 1990; Edwards et al., 1996; Anderson et al., 1997), whereas others suggest it was driest between 5000 and 2000 B.P. (Dwyer et al., 1996; Yu et al., 1997). In Lake Michigan, increased carbonate deposition, bands of iron sulfide, and ostracode taxa characteristic of higher salinity in core sections dated at between 7000 and 5000 B.P. have been interpreted as evidence for a drier climate, and the data suggest that lake levels in the Great Lakes may have been as much as 50 m lower than they are now (Colman et al., 1990; Rea et al., 1994). Several records for the Finger Lakes region of New York, to the east, show the highest calcium carbonate deposition between 10,000 and 5000 B.P. (Dwyer et al., 1996; Anderson et al., 1997; Mullins, 1998), caused by enhanced carbonate precipitation as a result of higher water temperatures. Some of the climatic interpretations are based on oxygen isotopes in carbonates, which vary with changes in temperature, moisture source, and evaporative concentration. At Crawford Lake in southern Ontario, apparent depositional hiatuses in shallow-water cores between 5000 and 2000 B.P., coupled with lighter $\delta^{18}O$ values in marls, are interpreted as evidence of reduced input of summer Gulf moisture, which is relatively heavy (Yu et al., 1997). In contrast, a nearby site shows heavier values of $\delta^{18}O$ during this interval, attributed to increased flow of Gulf moisture (Edwards et al., 1996). The latter interpretation is broadly consistent with a compilation of lithologic data on lake-level fluctuations in eastern North America, which shows fewer basins with low lake levels beginning at 4000 B.P. (Webb et al., 1993). Intensified water vapor transport from the Gulf of México from 8000 until 2000 B.P. has been suggested from a record of δD in lacustrine organic matter in western Michigan (Krishnamurthy et al., 1995), although here the relative roles of temperature versus moisture source cannot be determined. The inability to disentangle temperature versus precipitation influences from these lacustrine proxy records is a general problem, which may be the cause of some of the discordant interpretations mentioned previously. Clearly, more sites are needed from eastern North America, as well as a more developed understanding of lacustrine isotopic variations.

A single, detailed record of lake-level fluctuations at multidecadal resolution has been inferred from grain size analysis of the sediments of Herring Lake, where water levels are controlled by fluctuations in the adjacent Lake Michigan (Wolin, 1996). An increase in coarser grained sediments is used to infer low lake stages from A.D. 0–150, 450–780, 920–1010, 1220–1450, and after 1670, thus indicating significant decadal- to century-scale variations in atmospheric circulation patterns.

15.2.4. Boreal and Arctic North America

There are relatively few paleolimnological records of Holocene climate for northern latitudes. For northwestern Alaska, trace element analyses of ostracodes suggest that the climate was cold and dry in the early Holocene and remained dry until after 7000 B.P. (Hu et al., 1998). The data suggest that the temperature increased during the early Holocene (see also Bradbury et al., 2000), with peak temperatures between 8500 and 8000 B.P., followed by cooling until 7000 B.P. The climate probably warmed coincident with increasing effective moisture between 7000 and 6000 B.P., but became cooler between 4500 and 1500 B.P.

In central Canada, at sites just north of the modern tree line, a rapid change at 5000 B.P. in $\delta^{18}O$ of aquatic cellulose and diatom species composition, coincident with a shift from tundra to forest tundra, suggests warming and an increase in effective moisture, which persisted for ca. 1000 years (MacDonald et al., 1993; Edwards et al., 1996; Pienitz et al., 1999). Isotopic data also suggest a several-hundred-year interval of particularly wet climate centered at ca. 2700 B.P. (Wolfe et al., 1996).

For the Arctic, diatom studies have been used to reconstruct changes in lacustrine productivity, chemistry, and habitat associated with changes in temperature (Douglas and Smol, 1999). On Ellesmere Island, mid-Holocene increases in diatom species associated with aquatic mosses indicate the expansion of these mosses as a result of decreased ice cover during warm intervals (Smol, 1983). Early to mid-Holocene increases in diatom concentrations also probably reflect the warmer temperatures and longer open-water periods, which allowed enhanced diatom production (Williams, 1994; Wolfe, 1996; Wolfe and Hartling, 1996). For the High Arctic, the data suggest a colder climate after ca. 4000 B.P. (Smol, 1983). For other parts of the Arctic (Wolfe and Hartling, 1996), however, Holocene changes in diatom assemblages and sediment chemistry primarily reflect variations in lake water chemistry, driven by changes in catchment vegetation and alkalinity generation within the lake, rather than the direct influence of temperature.

Patterns of late Holocene climate change have been addressed with sedimentological and microfossil studies of Arctic lakes, including some with varved sediments. Interpretations based on varve thickness for a lake on Ellesmere Island suggest a cool climate from 3000–2400 B.P., warmer conditions at 2400–1200 B.P., subsequent cooling until ca. 700 B.P., and variable conditions during the LIA interval (Lamoureux and Bradley, 1996). Analyses for the last few centuries for several Arctic lakes suggest increased sediment deposition, increased diatom concentrations, and changes in diatom species composition indicative of warming, beginning at some sites in the mid-nineteenth century and at others in the twentieth century (Douglas et al., 1994; Gajewski et al., 1997; Overpeck et al., 1997).

15.2.5. Climate Patterns in North America

A synthesis of lake records for North America (Figs. 3 and 4) suggests millennial-scale patterns of lacustrine and climate change driven by changes in major boundary conditions, particularly insolation and wastage of the Laurentide ice sheet. The evidence for lowered lake levels and severe aridity at the onset of the Holocene (10,000 B.P.) in the Pacific Northwest and northern Great Basin (Fig. 3) is consistent with the higher summer insolation and a resultant intensification of the eastern Pacific subtropical high, which blocked the flow of westerly moisture (Barnosky et al., 1987). These conditions subsequently diminished after 5000 B.P. as insolation decreased. The higher frequency moisture fluctuations apparent in the late Holocene records for the northern Great Basin are likely a result of winter precipitation variability associated with changes in the mean position of the winter jet and a consequent shift in the position of westerly storm tracks and high-pressure cells (Stine, 1994).

In the Southwestern United States, the modern hydrologic regime is influenced by both winter snow and rain, as well as by summer monsoonal precipitation. A climate model for 9000 cal. B.P. suggests that there was an upper level anticyclone over the Colorado Plateau (Kutzbach, 1987), a configuration that is presently associated with wetter summers. Winter precipitation, although reduced from its late Pleistocene maximum, was also probably higher in the early Holocene than it is today (Van Devender et al., 1987) because the winter jet stream was still displaced south of its current position. After 9000 B.P., the northerly migration of the winter jet stream resulted in decreased winter precipitation across the southwestern United States and widespread desiccation of lake basins. Some sites show pluvial phases within the mid- to late Holocene arid interval, suggesting that there were intervals of intensified summer monsoons or of winter floods caused by a southerly displacement of the North Pacific storm tracks (Enzel, 1992). Late Holocene moisture patterns appear to be spatially variable, perhaps as a result of the complex

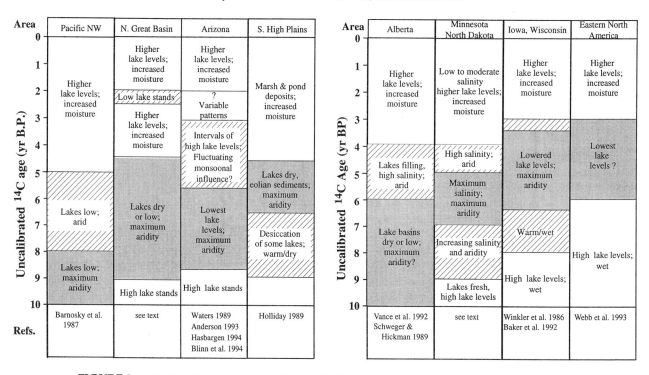

FIGURE 3 Paleoclimatic interpretations of selected paleolimnological records for the United States and southern Canada. Gray areas show the timing of maximum aridity, and other times of relatively low effective moisture are indicated by the hatched areas.

terrain and consequent variability in precipitation patterns (Mock, 1996).

In the northern continental interior of North America, between ca. 40° and 50°N latitude, the onset of arid conditions occurred just prior to the beginning of the Holocene (10,000 B.P.) at western sites (Alberta) and in the early to mid-Holocene (9000–6500 B.P.) eastward (southeastern Saskatchewan, Manitoba, North Dakota, Minnesota, and Wisconsin) (Fig. 3). Climate modeling (Hostetler et al., 2000; COHMAP [Climate of the Holocene—Mapping of Pollen Data] Members, 1988) suggests that this contrast between western and central North America in the onset of aridity is a result of early Holocene modulation of insolation-driven aridity by the persistent Laurentide ice sheet in eastern North America, which created steep climatic gradients at the margin of the ice sheet. For North Dakota and western Minnesota, the data suggest that the driest conditions within the mid-Holocene were between 7200 and 5200 B.P. The timing of maximum aridity within the mid-Holocene interval in Saskatchewan, Manitoba, and Wisconsin is not clear, either because of the small number of sites or because interpretations are based on lithology and mineralogy, which show thresholds rather than continuous responses to limnological and climatic change. In interior regions just to the south

(South Dakota and Nebraska), which were more distant from the ice sheet, severe aridity may have occurred earlier; however, there are currently too few sites to assert this with any confidence.

Moisture in the North America continental interior derives largely from the Gulf of México, with a precipitation maximum in summer. Seasonal dry intervals are the result of strong zonal (westerly) flow and the dominance of dry Pacific or Arctic air masses, which divert Gulf moisture to the east. Extending these observations to Holocene lacustrine records suggests that there was strong zonal flow beginning in the early Holocene, increasing in strength at ca. 7000 B.P., such that areas to the east were more frequently dry. The onset of increased moisture began at varied times in the mid-Holocene in central and western Alberta, between 4700 and 4000 B.P. throughout Saskatchewan, Manitoba, North Dakota, and Minnesota, but not until ca. 3200 B.P. in eastern Wisconsin (Fig. 3). This apparent west-to-east time-transgressive pattern suggests a series of shifts in the location and shape of the planetary Rossby waves, which changed the mean position of the summer jet stream and the position of high- and low-pressure cells across the North American continent.

The overall pattern of mid-Holocene aridity throughout western North America is consistent with a

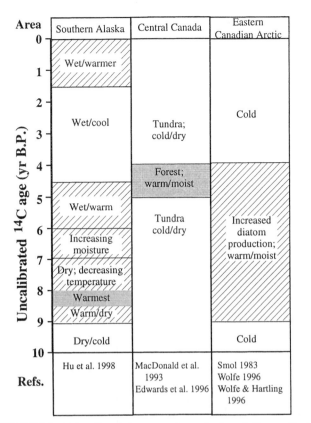

FIGURE 4   Paleoclimatic interpretations of selected sites north of 60°N latitude. Warm intervals are indicated by the hatched areas, and gray areas show the timing of maximum warmth.

northerly migration in the mean position of the jet stream, which shifted the locus of major storm tracks northward. Similarly, the mid-Holocene transition from a drier to wetter climate at 5000 B.P. in southern Alaska and regions above the modern tree line in central Canada (Fig. 4) has been attributed to northerly migration of the Arctic Front, allowing increased penetration of moist air masses.

15.3. CENTRAL AMERICA AND THE CIRCUM-CARIBBEAN REGION

A relatively limited number of paleolimnological records exist for México, Central America, and the circum-Caribbean region, between 30° and 8°N latitude. Records for Central America south of México are predominantly palynological studies (e.g., Horn, 1993; Islebe et al., 1996; Islebe and Hooghiemstra, 1997) and will not be considered in this chapter (see Grimm et al., 2000). The locations of sites reviewed in this section are shown in Fig. 5.

15.3.1. México

The highest density of Holocene lacustrine records comes from the basins of the trans-Mexican volcanic belt, which crosses México from east to west at about 19°N. In spite of the large number of published records, the Holocene dating control for many records is surprisingly poor. In addition, many records show significant human disturbance over the last 3500 years or longer, and, in some cases, deliberate drainage of basins in the post-Hispanic period has resulted in deflation and loss of late Holocene sediments.

The best-dated Holocene sequence comes from the Zacapu basin in Michoacán (Petrequin et al., 1994; Metcalfe, 1995). The magnitude of fluctuations over the Holocene is small relative to those in the late Pleistocene; however, diatom sequences indicate that conditions may have been slightly wetter between ca. 7000 and 6000 B.P., followed by a dry period until ca. 5000 B.P. Water levels then recovered, and there may have been a brief episode of lake-level rise at ca. 2900 B.P. The largest fluctuation in lake level, however, was a distinct drying at ca. 1100 B.P.

Although the basin of México has been the focus of much research (Bradbury, 1989; Lozano-Garcia et al., 1993; Caballero-Miranda and Ortega-Guerrero, 1998), the dating control for Holocene sequences is poor, and catchment disturbance and probable deflation of later sediments compromise good Holocene records in many parts of the basin. Water levels increased during the terminal Pleistocene (14,000–10,000 B.P.), but the early Holocene was marked by low lake levels throughout, followed by generally wetter, although possibly quite variable, conditions. One of the sequences from Chalco, in the southern part of the basin (Caballero-Miranda et al., 1995), records a dry period at ca. 5500 B.P.

Other Holocene sequences are available for Lake Chichonahuapan in the upper Lerma basin, Lake Pátzcuaro in Michoacán, and La Piscina de Yuriria in southern Guanajuato. The upper Lerma diatom record (Metcalfe et al., 1991) indicates a lake transgression after 8000 B.P. which ended by 6000 B.P. (Metcalfe, unpublished data). A strong drying episode occurred prior to 4600 B.P., and a late Holocene transgression peaked at ca. 1600 B.P., which was followed by a marked regression and desiccation. The interpretation of Holocene changes in the Pátzcuaro basin (Watts and Bradbury, 1982; O'Hara et al., 1994) is again constrained by the scarcity of dates. A diatom record (Bradbury, in press) indicates that Holocene lake levels were generally lower than those in the Pleistocene. A possible shift to more arid conditions at ca. 5000 B.P. was reported by Watts and Bradbury (1982). Ostracode data (Bridgwater et al., 1998) indicate three wet phases over the last 4000

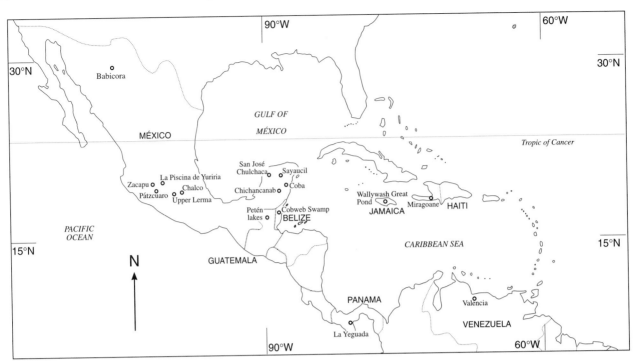

FIGURE 5   Map of Central America and the circum-Caribbean region showing locations of sites mentioned in the text.

years separated by longer dry periods. The most complete Holocene record for La Piscina de Yuriria (Davies, 1995) is based on ostracode isotopic and trace element analyses. The early Holocene was marked by fluctuating conditions, followed by generally wetter conditions between 6700 and 5500 B.P. The diatom record for La Piscina de Yuriria (Metcalfe et al., 1994) covers only the last 4000 years and indicates conditions drier than those now at the base of the sequence and between ca. 3700 and 3300, 2950 and 2700, and 2300 and 800 B.P.

For northern México, the only published lacustrine records come from the Alta Babícora basin in Chihuahua (Ortega Ramirez, 1995; Metcalfe et al., 1997). Early Holocene conditions were clearly very much wetter than those now, whereas the mid-Holocene was dry, possibly at its driest at ca. 6000 B.P. Slightly wetter conditions were then reestablished until ca. 3000 B.P., after which there was basin desiccation, although the record is poorly preserved.

There are a number of records from the Mexican part of the Yucatán Peninsula, which should be considered together with the sequences from adjacent lowland Guatemala and Belize (see Sections 15.3.2 and 15.3.3). In this karstic geology, hard-water errors and inputs of colluvium of unknown isotopic composition have resulted in the rejection of many radiocarbon dates

(Vaughan et al., 1985). Many lake levels are directly, or indirectly, affected by sea level. Records covering most of the Holocene include Lakes Chichancanab, Cobá, and San José Chulchacá. The original work on Chichancanab (Covich and Stuiver, 1974) extended back to the Pleistocene, whereas a more recent paper by Hodell et al. (1995) provides a full Holocene record. Interpretations based on the oxygen isotopic composition of biogenic carbonate indicate that lake levels rose slowly after 8200 B.P. and rapidly from ca. 7200 B.P., and wet conditions prevailed between 7100 and 3000 B.P. (Hodell et al., 1995). The climate then became drier, with the driest conditions of the last 8000 years between 1300 and 1100 B.P. A core from Lake Cobá (Whitmore et al., 1996) has a date near the base of 7410 ± 80 B.P., but only a patchy diatom record. The lake did not fill until ca. 7600 B.P. and was then dry and saline. Lake levels subsequently rose to form a deep freshwater lake prior to 2600 B.P., followed by a series of poorly dated fluctuations. The sequence for San José Chulchacá (Leyden et al., 1996) covers most of the Holocene, with pollen and some isotopic data from a long core and diatom data from a short core. The early Holocene was dry, but moist conditions, possibly wetter than those now, were established by 6800 B.P. There was a dry period between 6000 and 5000 B.P., followed by the reestablishment of wetter conditions until 3500 B.P. After this, there was drying until ca. 1500 B.P. The diatom record indicates lower salinities between 1860 and 1010 B.P., followed by increased salinity, which suggests drier, shallower conditions. The record for Sayaucil (Whitmore et al., 1996) suggests that the modern lake is deeper and fresher than at any time in the past.

High-resolution studies of the last 2000 years are derived from the $\delta^{18}O$ record of ostracode carbonate in two Yucatán lakes (Curtis and Hodell, 1996) and suggest significant decadal to centennial climate variability. The data indicate that the interval from A.D. 310–1050 was frequently dry, with periods of extreme aridity that persisted for multiple decades. The collapse of Mayan culture is dated at between A.D. 800 and 900 and is correlated with an interval of severe drought. Moisture generally increased from A.D. 1050 onward, but within this time are several intervals of moderate aridity lasting one to two decades.

15.3.2. Belize

A record covering the mid-Holocene has been obtained from Cobweb Swamp in Belize (Jacob and Hallmark, 1996). This record provides a detailed sedimentological history of the Holocene, but the site has been influenced by sea-level changes and long-term anthropogenic impacts.

15.3.3. Guatemala

Pioneering tropical paleolimnological work in Guatemala began in the 1960s (Cowgill et al., 1966; Tsukada and Deevey, 1967). Studies in the Petén lake district of northern Guatemala have been frustrated, however, by problems with obtaining a reliable radiocarbon chronology (see Leyden et al., 1994). With the absence of a radiometric chronology, pollen records were divided into zones, which were then correlated with archaeological events of known ages (Vaughan et al., 1985). Conditions appear to have been wetter than those now during the early part of the Holocene, with drier conditions becoming established from ca. 5600 B.P. Later records are significantly affected by human disturbance.

15.3.4. Panama

Paleolimnological data are confined to Lake La Yeguada in central Panama (Piperno et al., 1990). Between 10,000 and 6000 B.P., the diatom record is poor (Bush et al., 1992). Phytoliths are reduced greatly in number between 7000 and 5000 B.P., and sedimentation

rates slow significantly (Piperno et al., 1990). Piperno et al. (1990) suggest that streamflow to the site was greatly reduced between 7000 and 3800 B.P., lowering the water level and indicating drier conditions. The absence of Podostemataceae (river weed) phytoliths from ca. 7650 B.P. points to reduced streamflow into the lake, and lower lake levels are inferred at this time, as deposition of lacustrine sediments ceased on a raised beach deposit at the lake margin. However, the mechanisms that caused loss of the diatom record remain unclear (Bush et al., 1992). The reappearance of diatoms after 5000 B.P., albeit in small quantities, may suggest a slight increase in effective moisture. The vegetation record for this site shows evidence of human disturbance since the beginning of the Holocene.

15.3.5. The Caribbean

Published paleolimnological records are available for only two sites in the Caribbean: Wallywash Great Pond in Jamaica (Street-Perrott et al., 1993) and Lake Miragoane in Haiti (Hodell et al., 1991). A lack of freshwater lakes and potential dating problems in karstic environments account for the scarcity of paleolimnological information for the region. Of the two records, only the record for Lake Miragoane is of sufficient resolution to examine Holocene climatic trends in any detail. Analysis of stable isotopes in the core from Wallywash Great Pond indicates three cycles of wet and dry conditions during the Holocene. However, accurate radiocarbon dating of these events has not been possible because of evidence of significant contamination of sediments with *dead* carbon from springs surrounding the lake (Street-Perrott et al., 1993).

Oxygen isotope data for ostracodes in the Lake Miragoane core show major changes in precipitation and evaporation through the Holocene (Hodell et al., 1991). For the early Holocene, declining $\delta^{18}O$ values indicate a shift from the more arid conditions of the Pleistocene to a wetter climate and higher lake levels. The lowest $\delta^{18}O$ values in the Miragoane record are between 7000 and 5300 B.P. Following a slight increase in evaporation relative to precipitation at ca. 5200 B.P., lake levels remained relatively high until approximately 3200 B.P. (Hodell et al., 1991). The short, dry episode at ca. 5200 B.P. may correspond to other abrupt dry phases evident elsewhere in the region at ca. 5000 B.P. The pollen record indicates that mesic forests developed in the vicinity during the mid-Holocene when lake levels were relatively high. After 3200 B.P., a marked increase in $\delta^{18}O$ points to drier conditions. By 2400 B.P., evaporation rates reached their highest levels, marking the beginning of a late Holocene dry phase.

15.3.6. Venezuela

The only study for northern South America is for Lake Valencia, where diatom, mineralogical, and sedimentological data (Bradbury et al., 1981) show a transition at ca. 8500 B.P. from a shallow, saline, closed-basin lake to an overflowing, low-salinity system, driven largely by rising sea levels. Diatom assemblages and sedimentology suggest an interval of lower lake levels between ca. 7000 and 6000 B.P., followed by generally high levels until at least 3000 B.P. Effective moisture has decreased over the last 2200 years, marked by falling lake levels and less extensive forests (Leyden, 1985).

15.3.7. Regional Climate Patterns

Central America and the circum-Caribbean region today are within a summer precipitation regime, with moisture derived mainly from the Atlantic through the Northeast Trades (McGregor and Nieuwolt, 1998). Locally, Pacific moisture sources may be significant, particularly from tropical cyclones. A monsoon-type circulation occurs over México and into the southwest and central United States in the Northern Hemisphere summer, with moisture from both the Gulf of México/Caribbean and the Pacific Ocean (Douglas et al., 1993).

Figures 6 and 7 summarize available data for the period 7000–4500 B.P. for Central America and the circum-Caribbean region. A combination of poor dating control and variable preservation of certain types of microfossils means that the number of data points for different time periods is variable. It has been clear from review of the literature that there is a need for much better dating control on Holocene sequences. This need is particularly evident for periods when it appears that the climate may have fluctuated quite rapidly (e.g., 6000–5000 B.P.).

Lake levels were generally low during the late glacial period because of southerly displacement of the Intertropical Convergence Zone (ITCZ) and/or because of shifts in the position and strength of the Northeast Trades (see Bradbury et al., 2000). The only exception to this trend is in northern México, which probably came under the influence of the westerlies which were displaced equatorward (see earlier). The picture for 7000 B.P. (Fig. 6) is rather mixed. It is clear that a number of low-lying sites were still affected by sea levels well below those now. Modern summer rainfall systems do not appear to have become fully established. The data for northern México (Babicora), taken with records for other data sources and adjacent areas north of the United States—México border (see earlier), seem

to indicate conditions still wetter than those now (persisting from the late Pleistocene), possibly due to enhanced winter precipitation. It would be valuable to have data for sites to the west of the Sierra Madre Occidental and around the Gulf of California for this time period. How far south any enhanced winter precipitation was effective remains unclear.

The picture for 6000 B.P. (Fig. 6) suggests that wetter conditions had become established over much of the region, although Panama may be an exception. Lake levels in the lowlands of the Yucatán had risen rapidly by this time, whereas conditions in northern México were becoming drier. This pattern may reflect a strong summer precipitation regime, together with warmer temperatures, possibly associated with shifts in the position and strength of the Bermuda High (Hodell et al., 1991; Metcalfe et al., 1997) and changes in the strength of the summer Mexican monsoon. By 5000 B.P. (Fig. 7), a number of sites in central México record increased aridity, although this may have been quite short lived. Whether this was due to a further increase in temperature or a decrease in precipitation cannot be determined. Conditions in the Chihuahuan desert were moving toward those now. Data for 4500 B.P. (Fig. 7) show an intensification of the drying in central México, suggesting a decrease in the westerly penetration of summer storms. The records for many sites are hard to interpret over this interval.

The data for 3000 B.P. show considerable variation between lakes in the same areas. This variation seems to be indicative of another period of fluctuating climate, with short episodes of higher or lower lake levels. An increasing number of lacustrine records are affected by human impact, in some cases apparently completely obscuring the paleoclimatic signal. Indeed, it has been argued that the climate of Central America did not change significantly during the late Holocene (Rue, 1987). This assumption was based entirely on palynological records available at the time, and it is now evident that significant and abrupt climatic changes did occur throughout the Holocene.

After ca. 3000 B.P., some lakes in central México showed rising lake levels, and high lake stands initially persisted in the Yucatán, whereas the Caribbean became dry. Between 2000 and 1000 B.P., however, the Yucatán and central México showed some of the lowest lake levels of the Holocene, whereas moisture appears to have increased in the Caribbean (Hodell et al., 1991, 1995).

Across the region as a whole, it is possible to identify areas that often display coherent patterns of change. Northern México seems to be quite different from the rest; the records from the Mexican Yucatán and Haiti often show the same trends. The lakes in central Méx-

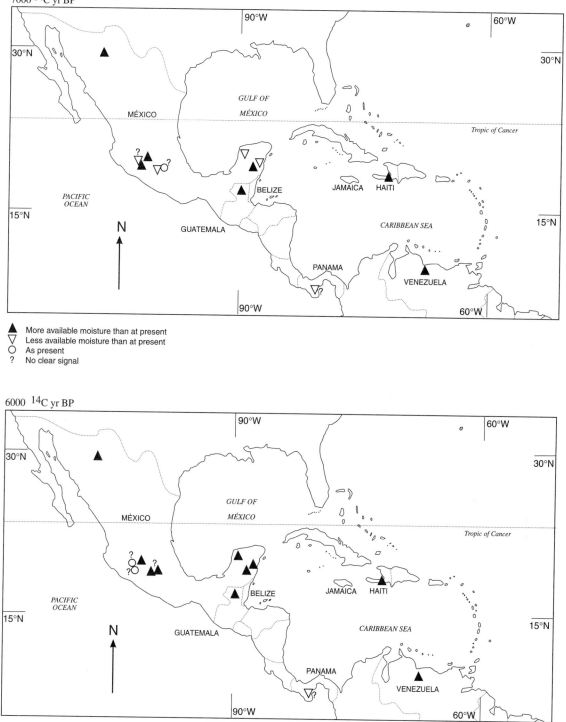

FIGURE 6   Maps of lake-level status of lakes in Central America and the circum-Caribbean at 7000 and 6000 B.P.

ico are very varied—in spite of being within the same summer rainfall regime as the Yucatán and Caribbean today, they show the same signal as these regions only at 6000 B.P. It should be noted perhaps that a number of central Mexican records are affected by volcanic activity over the period 5200–4300 B.P., which complicates paleoclimatic interpretation. The site in Panama is different from most of the others, indicating dry con-

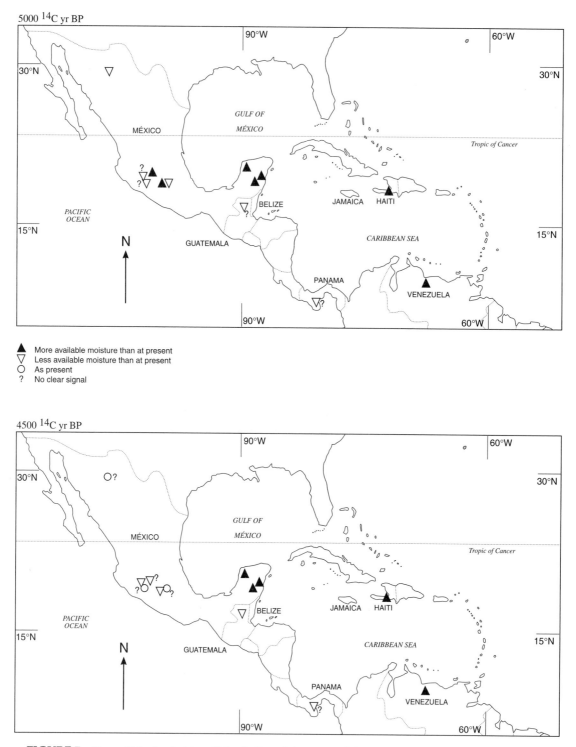

FIGURE 7   Maps of lake-level status of lakes in Central America and the circum-Caribbean at 5000 and 4500 B.P.

ditions over most of the Holocene. Some countries have no published Holocene data (e.g., Cuba), and there may be significant paleoclimatic gradients that cannot be identified from the present coverage of sites.

This may be an area where implementation of a regional-scale climate model would yield particularly interesting results and offer up many testable hypotheses.

15.4. SOUTHERN HEMISPHERE

Few detailed Holocene records exist at present for the Southern Hemisphere (Fig. 8). For the northern Altiplano of Bolivia and Peru, dating of paleolake shorelines (Servant and Fontes, 1978; Servant et al., 1995) and sedimentological and ostracode data from cores in the small basin of Lake Titicaca indicate high lake stands during late glacial times and a subsequent decline (Wirrmann and DeOliveira Almeida, 1987; Wirrmann and Mourguiart, 1995), although the timing of the regression is not very well constrained. Clearly, the interval from 8000–3600 B.P. was one of extreme aridity in

both the northern and southern Altiplano, and many lakes went dry or dropped significantly below modern levels. The small basin of Lake Titicaca desiccated except in a single, small, deep hole (Wirrmann and De-Oliveira Almeida, 1987; Abbott et al., 1997a), and seismic studies in the large basin suggest a lake-level decline of nearly 100 m (Seltzer et al., 1998). Sedimentological and geochemical records for small, glacier-fed lakes in the northern Altiplano suggest the onset of glacial retreat between 10,700 and 9700 B.P., the absence of glaciers in the watershed of the lakes after 8800 B.P., and episodic drought throughout the mid-Holocene (Abbott et al., 1997b). Similarly, in Lake Miscanti in

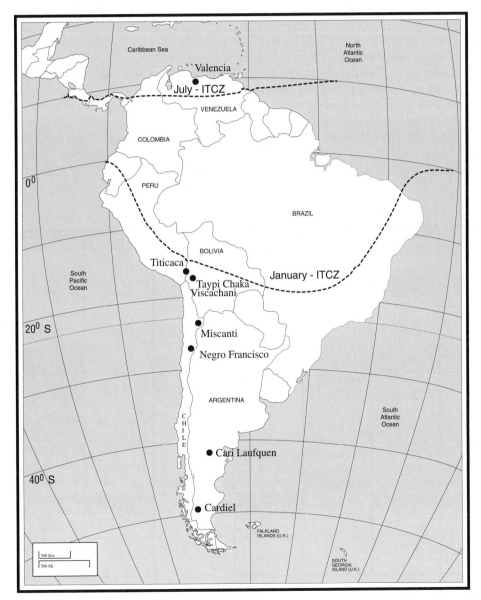

FIGURE 8   Map of South America showing locations of sites mentioned in the text and the positions of the Intertropical Convergence Zone (ITCZ) at present.

northern Chile at 23°S, a transition from finely laminated carbonate mud to banded carbonate sediments with allochthonous debris and sand suggests a decline in lake levels after 8000 B.P., following early Holocene high stands (Grosjean et al., 1995; Valero-Garcés et al., 1996). Sedimentological, mineralogical, and geochemical data for Laguna del Negro Francisco at 27°S in southern Chile show that desiccation occurred prior to 6000 B.P., and frequent subaerial exposure of the lake bottom occurred between 6000 and 3800 B.P. Evidence from the latter site indicates that the dry mid-Holocene was punctuated by humid phases of relatively short duration (Grosjean et al., 1997).

Lake levels began to rise throughout the Altiplano at ca. 3600 B.P. and attained modern levels at ca. 3000 B.P. in the southern Altiplano (Valero-Garcés et al., 1996; Grosjean et al., 1997; Nuñez et al., 2000) and by 2100 B.P. in the northern Altiplano, as evidenced by lithological, biological, and geochemical data for both Titicaca and the smaller lakes (Wirrmann and Mourguiart, 1995; Abbott et al., 1997a,b; Cross et al., 2000). However, clearly the climate of the last 2000 years was variable, punctuated by intervals of both high effective moisture (Grosjean et al., 1997) and drought (Abbott et al., 1997a; Binford et al., 1997).

In northern Patagonia (41°S), lake shorelines show an early Holocene decline in lake levels, following late glacial and glacial high stands (Galloway et al., 1988). In southern Patagonia, geomorphic evidence from Lago Cardiel (Stine and Stine, 1990) similarly suggests a decline in lake levels from an early Holocene high stand after 7700 B.P., followed by persistent low levels until ca. 5100 B.P. A brief lake-level rise at 5100 B.P. is followed by several thousand years of dry conditions, with a series of century-scale episodes of moderate lake-level rise during the last 2000 years (Stine, 1994).

15.4.1. South American Climate Patterns

Most of the precipitation in the South American Altiplano falls during the austral summer and is highly correlated with convection in the Amazon basin. Precipitation variability is linked to shifts in the position and intensity of the Bolivian High, an upper troposphere high-pressure system, which controls the upper atmospheric winds and advection of moisture from the Amazon (Lenters and Cook, 1997). In the contemporary climate system, a northward shift in the position of the Bolivian High is associated with reduced precipitation on the Altiplano, and this scenario may provide an analog for mid-Holocene circulation patterns and the resultant mid-Holocene aridity (Fig. 9b). These pos-

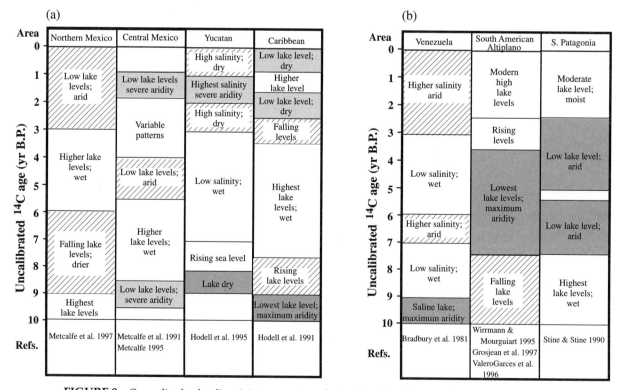

FIGURE 9   Generalized paleoclimatic interpretations of paleolimnological records from (a) Central America and the Caribbean region and (b) South America. Gray areas show the timing of maximum aridity, and other times of relatively low effective moisture are indicated by the hatched areas.

tulated shifts in atmospheric circulation patterns are consistent with the reduced seasonality of insolation at tropical latitudes in the Southern Hemisphere at 9000 cal. B.P. (ca. 8450 ^{14}C B.P.). Climate model simulations for this time (Kutzbach and Guetter 1986) suggest cooler summers, warmer winters, and a decrease in the vapor pressure gradient between land and sea, as well as a decrease in P–E (precipitation minus evaporation) driven largely by a decline in convective precipitation.

The southernmost regions of the Altiplano, such as the Chilean site Laguna del Negro Francisco, also receive winter precipitation controlled by frontal storms and cutoff lows associated with outbreaks of polar air. Here, the long mid-Holocene aridity may result in part from blocking of westerly precipitation by intensification of the Southeast Pacific anticyclone (Markgraf et al., 1994; Grosjean, 1997). The higher frequency periods of enhanced moisture within the mid-Holocene in this region may reflect increased winter precipitation or alternatively intense summer storm activity (Grosjean, 1997).

Patagonia receives moisture from westerly sources driven by the northward movement of storm tracks in winter. Thus, one explanation for the early (northern Patagonia) to mid-Holocene (southern Patagonia) aridity is a southerly shift in the mean position of the winter storm tracks (Markgraf et al. 1994). Alternatively, a decrease in outbreaks of polar air and diminished easterly moisture from the western Atlantic may be involved in causing dry episodes (Stine and Stine, 1990).

15.5. CONCLUSIONS

Throughout the Americas, the broad millennial-scale patterns of climate variation are consistent with insolation-driven forcing. However, at a submillennial scale, the considerable spatial variability in the pattern and rate of climate change points to the influence of other climatic controls, as well as the differing response times of individual systems to climate. In North America in the early Holocene, increased seasonality of insolation resulted in dry conditions in the Northwest, but moist conditions in the Southwest and northern México. The exact timing of the reestablishment of the modern summer rainfall regime in the latter areas has been the subject of considerable debate, as the paleoecological data are not consistent with modeled reconstructions of COHMAP (Spaulding and Graumlich, 1986; Van Devender et al., 1990). In the northern Great Plains, the Midwest, and eastern North America, conditions were still cool and moist in the early Holocene because of modulation of insolation by the persistent Laurentide ice sheet. As the ice sheet waned, the continental

interior between 40° and 50°N latitude saw increasing aridity, and the Southwest dried as the jet stream moved northward. By 7000 B.P., arid conditions were widespread throughout western North America, with maximum Holocene aridity in the northern continental interior between 7000 and 5000 B.P. (Fig. 3). The data suggest strengthened zonal flow across western North America after 7000 B.P., which blocked the penetration of summer monsoonal precipitation into the continental interior, such that regions to the east were more frequently dry.

After 6000 B.P., lakes in the Pacific Northwest and the western portion of the Canadian Great Plains, many of which were dry throughout the early and mid-Holocene, began to refill. In the northern Great Basin and elsewhere in the Southwest, conditions remained dry until ca. 4500 B.P., although some lake records show pluvial intervals between 5000 and 3000 B.P. Dry conditions persisted in Saskatchewan, Manitoba, North and South Dakota, Nebraska, and Minnesota until at least 4500 B.P. and apparently until ca. 3200 B.P. in Wisconsin. Paleolimnological data for eastern North America are presently equivocal with respect to mid- and late Holocene moisture patterns. Thus, the evidence suggests that mesic conditions began earlier in northwestern North America than in the Southwest, northern Plains, and Midwest, all of which receive significant summer moisture. Similarly, for the northern Great Plains and the Midwest, the data suggest that increased summer moisture began earlier in the west (4500 B.P.) than to the east (3000 B.P.). A sequence of changes in the location and configuration of the planetary Rossby waves, which initially resulted in enhanced meridional moisture flow into western areas but not eastward, may explain the west-to-east gradient in moisture patterns within the northern continental interior.

For the tropics of the Northern and Southern Hemispheres, the data also show hydrologic variation in response to millennial-scale forcing. Between 22° and 8°N in Central America and the circum-Caribbean region, lake levels were generally low during the late glacial period (Fig. 9a) because of southerly displacement of the ITCZ and/or weakening of easterly moisture sources. Lake-level patterns at the onset of the Holocene are regionally variable, in part because of the hydrologic influence of rising sea level on low-lying sites. By 7000 years B.P., however, nearly all the sites from central México eastward suggest wetter conditions than exist today (Figs. 6 and 9a) and thus enhanced July precipitation, probably caused by shifts in the position and strength of the Bermuda High (Hodell et al., 1991; Metcalfe et al., 1997) associated with northerly movement of the ITCZ. Between 6000 and 4000 B.P., many sites in central México became drier, whereas high lake levels persisted in the Yucatán and Caribbean, suggesting a decrease in the westerly penetration of summer storms. Dry conditions began at ca. 3000 B.P. in Haiti, but not until 2000 B.P. in the Yucatán.

The data for both North America and the circum-Caribbean regions suggest west-to-east gradients in the establishment of late Holocene moisture regimes. These patterns reflect a sequence of shifts in the loci of high- and low-pressure systems driving moisture transport, probably in response to enhanced meridional circulation patterns as summer insolation diminished.

Lakes in the Southern Hemisphere tropics also show strong moisture fluctuations correlative with changes in insolation forcing, but the large-scale pattern of change is often inverse in sign to that at similar tropical latitudes in the Northern Hemisphere (Fig. 9b). Lake Titicaca and other smaller lakes in the Andean Altiplano showed high stands in the late glacial period and subsequent drops, and there is widespread evidence for extreme mid-Holocene aridity. In contrast, lake levels in Haiti (Fig. 9a) and Venezuela (Fig. 9b) were generally high throughout the mid-Holocene. Lake levels began to rise throughout the Altiplano at ca. 3600 B.P. and attained modern levels at ca. 3000 B.P. in the south and by 2100 B.P. in the north, whereas in the Northern Hemisphere tropics, late Holocene lake levels were frequently low. The inverse behavior of the Northern and Southern Hemisphere tropics is consistent with insolation-driven shifts in the ITCZ, which would differentially affect the two regions. In both regions, most of the precipitation falls in summer months: during May to October from central México east to the Caribbean as the ITCZ shifts northward to its northern limit (at about 10°N), and in November to March in the Altiplano as the ITCZ moves back to the south. Thus, a northward shift in the mean position of the ITCZ during the mid-Holocene (Fig. 8) and a strengthening of monsoon-type circulation would cause increased precipitation in the Northern Hemisphere tropics, while shifting the locus of precipitation north of lakes in the tropics of the Southern Hemisphere.

Interestingly, this pattern of inverse response to millennial-scale climate forcings of the Northern and Southern Hemisphere tropics is also suggested for patterns of centennial-scale climate variation. A comparison of a lake-level curve over the last few thousand years for Lake Titicaca, derived from sedimentological data (Abbott et al., 1997a), with a high-resolution record from Punta Laguna in the Yucatán, based on $\delta^{18}O$ in biogenic calcite (Curtis and Hodell, 1996), suggests that from ca. 3500–2000 B.P., Titicaca underwent several episodes of drying, whereas Punta Laguna was,

on average, wet. Similarly, during the major droughts of the Early and Late Classic periods in the Yucatán (Curtis and Hodell, 1996), lake levels in Titicaca were consistently high. Clearly, these patterns are based on a single pair of sites, and the temporal resolutions of these two data sets are quite different. Nonetheless, the suggestion that the response of the two hemispheres is opposite in sign at centennial timescales deserves further investigation.

Among the interesting issues with respect to Holocene climate are the relative roles of forcing mechanisms other than insolation in affecting climate variation. In northwestern North America, the influence of insolation forcing was clearly quite strong in the early to mid-Holocene, but in the continental interior and eastern North America, climate was modulated by the Laurentide ice sheet until the mid-Holocene. As the ice sheet melted and the seasonality of insolation diminished, other aspects of the Earth–atmosphere–ocean system may have played a bigger role in controlling regional to continental climates. An analysis of the rate of limnological change at decadal to centennial scales for a single site in the northern Great Plains (Laird et al., 1998) suggests that, at this scale, the last 2000 years were more highly variable than the period from 9000–7000 B.P. The inference is that climate change in the early Holocene responded in a nearly linear fashion to strong directional changes in insolation forcing. However, as this influence decreased, higher frequency variation either increased or became more apparent. This increase in variability is consistent with the hypothesis that the onset of El Niño/Southern Oscillation (ENSO) variation occurred in the late Holocene (Shulmeister and Lees, 1995; Sandweiss et al., 1996; Rodbell et al., 1999). At present, there are very few records with decadal to centennial resolution for the early through late Holocene, and, thus, it is unclear whether or not different intervals within the Holocene vary in the degree of centennial, decadal, and annual climate variability.

A synthesis of the data also suggests that there may be more spatial variability in moisture patterns in the late Holocene than in the mid-Holocene, and, thus, that small-scale, local to regional, controls may differ in their influence under different large-scale boundary conditions. The increase in spatial variability in the late Holocene is apparent in Central America and the circum-Caribbean region, but some of the North American data also suggest a similar pattern, such as the records from the southwestern United States. It is not apparent whether the variation is a result of site-specific differences in limnological response to a broadly similar climate or alternatively is caused by spatial variation in climate itself. Nonetheless, as the density of sites increases, attention should be given to spatial patterns of variation. For example, in the modern climate record, widespread continental droughts, such as the Dust Bowl droughts of the mid-twentieth century, are infrequent, but it is possible that such events may have been more common at times in the past, such as during the mid-Holocene when summer insolation was still quite high relative to today.

Although the Holocene limnological and climatic variability is smaller in magnitude than Pleistocene changes, the data indicate extremes more severe and persistent than observation- and instrument-based records of the last 150 years (Laird et al., 1996b; Overpeck, 1996; Dean, 1997). In most extratropical regions, early to mid-Holocene aridity is unparalleled in the modern record, and the few detailed records of the last few thousand years also suggest severe climate, either in magnitude or duration, relative to the twentieth century. In the limited number of tropical sites with data of sufficient resolution, the driest climates of the Holocene appear to have occurred within the last 2000 years. Clearly, a more extensive set of high-resolution paleoecological data are needed to better understand spatial patterns in terms of the timing of extreme climatic intervals and the relative roles of the multiple controls on regional and continental climates.

Acknowledgments

Sarah Davies assisted in the collation and interpretation of data from the circum-Caribbean region.

References

Abbott, M., M. W. Binford, M. Brenner, and K. Kelts, 1997a: A 3500 ^{14}C high-resolution record of water-level changes in Lake Titicaca, Bolivia/Peru. *Quaternary Research*, 47: 169–180.

Abbott, M. B., G. O. Seltzer, K. R. Kelts, and J. Southon, 1997b: Holocene paleohydrology of the tropical Andes from lake records. *Quaternary Research*, 47: 70–80.

Anderson, R. S., 1993: A 35,000 year vegetation and climate history from Potato Lake, Mogollon Rim, Arizona. *Quaternary Research*, 40: 351–359.

Anderson, W. T., H. T. Mullins, and E. Ito, 1997: Stable isotope record from Seneca Lake, New York: Evidence for a cold paleoclimate following the Younger Dryas. *Geology*, 25: 135–138.

Baker, R. G., L. J. Maher, C. A. Chumbley, and K. L. van Zant, 1992: Patterns of Holocene environmental change in the midwestern United States. *Quaternary Research*, 37: 379–389.

Barnosky, C. W., P. M. Anderson, and P. J. Bartlein, 1987: The northwestern U.S. during deglaciation: Vegetational history and paleoclimatic implications. *In* Wright, H. E., and W. Ruddiman (eds.), *North America and Adjacent Oceans During the Last Deglaciation, The Geology of North America*, v. K-3. Boulder, CO: Geological Society of America, pp. 289–321.

Bartlein, P. J., and C. Whitlock, 1993: Palaeoclimatic interpretation of the Elk Lake pollen record. *In* Bradbury, J. P., and W. E. Dean (eds.), *Elk Lake, Minnesota: Evidence for Rapid Climatic Change in the North-Central United States*. Boulder, CO: Geological Society of America, pp. 275–293.

Benson, L. V., 1978: Fluctuations in the level of pluvial Lake Lahontan during the last 40,000 years. *Quaternary Research*, 9: 300–318.

Benson, L., and R. S. Thompson, 1987: The physical record of lakes in the Great Basin. *In* Ruddiman, W., and H.E. Wright (eds.), *North America and Adjacent Oceans During the Last Deglaciation*. Boulder, CO: Geological Society of America, pp. 241–260.

Benson, L., J. W. Burdett, M. Kashgarian, S. P. Lund, F. M. Phillips, and R. O. Rye, 1996: Climatic and hydrologic oscillations in the Owens Lake basin and adjacent Sierra Nevada, California. *Science*, 274: 746–749.

Benson, L. V., D. R. Currey, R. I. Dorn, K. R. Lajoie, C. G. Oviatt, S. W. Robinson, G. I. Smith, and S. Stine, 1990: Chronology of expansion and contraction of four Great Basin lake systems during the past 35,000 years. *Palaeogeography, Palaeoclimatology, Palaeoecology*, 78: 241–286.

Binford, M., A. L. Kolata, M. Brenner, J. W. Janusek, M. T. Seddon, M. Abbott, and J. Curtis, 1997: Climate variation and the rise and fall of an Andean civilization. *Quaternary Research*, 47: 235–248.

Blinn, D. W., R. H. Hevly, and O. K. Davis, 1994: Continuous Holocene record of diatom stratigraphy, paleohydrology, and anthropogenic activity in a spring-mound in southwestern United States. *Quaternary Research*, 42: 197–205.

Bradbury, J. P., 1987: Late Holocene diatom paleolimnology of Walker Lake, Nevada. *Archiv für Hydrobiologia/Supplement*, 79(1): 1–27.

Bradbury, J. P., 1989: Late Quaternary lacustrine paleoenvironments in the Cuenca de México. *Quaternary Science Reviews*, 8: 75–100.

Bradbury, J. P., 1991: Late Cenozoic lacustrine and climatic environments at Tule Lake, northern Great Basin, U.S.A. *Climate Dynamics*, 6: 275–285.

Bradbury, J. P., 1997: Sources of glacial moisture in Mesoamerica. *Quaternary International*, 43/44: 97–110.

Bradbury, J. P., in press: Limnologic history of Lago de Pátzcuaro, Michoacán, México for the past 48,000 years; impacts of climate and man. *Palaeogeography, Palaeoclimatology, Palaeoecology*.

Bradbury, J. P., and W. E. Dean (eds.), 1993: *Elk Lake, Minnesota: Evidence for Rapid Climate Change in the North-Central United States*. Boulder, CO: Geological Society of America.

Bradbury, J. P., R. M. Forester, and R. S. Thompson, 1989: Late Quaternary paleolimnology of Walker Lake, Nevada. *Journal of Paleolimnology*, 1: 249–267.

Bradbury, J. P., M. Grosjean, S. Stine, and F. Sylvestre, 2000: Full and late glacial lake records along the PEP1 transect: Their role in developing interhemispheric paleoclimate interactions. *In* Markgraf, V. (ed.), *Interhemispheric Climate Linkages*. San Diego: Academic Press, Chapter 16.

Bradbury, J. P., B. Leyden, M. Salgado-Labouriau, W. M. Lewis, C. Schubert, M. W. Binford, D. G. Frey, D. R. Whitehead, and F. H. Weibezahn, 1981: Late Quaternary environmental history of Lake Valencia, Venezuela. *Science*, 214: 1299–1305.

Bridgwater, N. D., J. A. Holmes, and S. L. O'Hara, 1998. Complex controls on the trace-element chemistry of non-marine ostracods: An example from Lake Pátzcuaro, central México. *Palaeogeography, Palaeoclimatology, Palaeoecology*, 148: 117–131.

Bush, M. B., D. R. Piperno, P. C. Colinaux, L. A. De Oliveira Krissek, M. C. Miller, and W. E. Rowe, 1992: A 14,300 yr paleoecological profile of a lowland tropical lake in Panama. *Ecological Monographs*, 62: 251–275.

Caballero-Miranda, M., and B. Ortega-Guerrero, 1998: Lake levels since about 40,000 years ago at Lake Chalco, near México City. *Quaternary Research*, 50: 69–79.

Caballero-Miranda, M., S. Lozano-García, B. Ortega-Guerrero, and J. Urrutia-Fucugauchi, 1995: Historia ambiental del sistema lacustre del sureste de la Cuenca de México. *Memoria del Segundo Seminario Internacional de Investigadores de Xochimilco,* **1**: 12–22+figs.

Campbell, C., 1998: Late Holocene lake sedimentology and climate change in Southern Alberta, Canada. *Quaternary Research,* **49**: 96–101.

COHMAP Members, 1988: Climatic changes of the last 18,000 years: Observations and model simulations. *Science,* **241**: 1043–1052.

Colman, S. M., G. A. Jones, R. M. Forester, and D. S. Foster, 1990: Holocene paleoclimatic evidence and sedimentation rates from a core in southwestern Lake Michigan. *Journal of Paleolimnology,* **4**: 269–284.

Covich, A., and M. Stuiver, 1974: Changes in oxygen 18 as a measure of long-term fluctuations in tropical lake levels and molluscan populations. *Limnology and Oceanography,* **19**: 682–691.

Cowgill, U. M., C. Goulden, G. E. Hutchinson, R. Patrick, A. A. Recek, and M. Tsukada, 1966: The history of Laguna de Petenxil, a small lake in northern Guatemala. *Memoirs of the Academy of Arts and Sciences,* **XVII**: 1–126.

Cross, S., P. A. Baker, G. O. Seltzer, S. C. Fritz, and R. Dunbar, 2000: A new estimate of the Holocene low stand level of Lake Titicaca and its implications for regional paleohydrology. *The Holocene,* **10**: 21–32.

Curtis, J. H., and D. A. Hodell, 1996: Climate variability on the Yucatán Peninsula (México) during the past 3500 years, and implications for Maya cultural evolution. *Quaternary Research,* **46**: 37–47.

Davies, H. L., 1995: Quaternary palaeolimnology of a Mexican crater lake. Ph.D. thesis. Kingston University, London.

Dean, W. E., 1997: Rates, timing, and cyclicity of Holocene eolian activity in north-central United States: Evidence from varved lake sediments. *Geology,* **25**: 331–334.

Dean, W. E., and M. Stuiver, 1993: Stable carbon and oxygen isotope studies of the sediments of Elk Lake, Minnesota. *In* Bradbury, J.P., and W. E. Dean (eds.), *Elk Lake, Minnesota: Evidence for Rapid Climate Change in the North-Central U.S.* Boulder, CO: Geological Society of America, pp. 163–180.

Digerfeldt, G., J. E. Almendinger, and S. Björck, 1993: Reconstruction of past lake levels and their relation to groundwater hydrology in the Parkers Prairie sandplain, west-central Minnesota. *Palaeogeography, Palaeoclimatology, Palaeoecology,* **94**: 99–118.

Douglas, M., R. Maddox, and K. Howard, 1993: The Mexican monsoon. *Journal of Climate,* **6**: 1665–1677.

Douglas, M. S. V., and J. P. Smol, 1999: Freshwater diatoms as indicators of environmental change in the High Arctic. *In* Stoermer, E. F., and J. P. Smol (eds.), *The Diatoms: Applications for the Environmental and Earth Sciences.* London: Cambridge University Press, pp. 227–244.

Douglas, M. S. V., J. P. Smol, and W. Blake, 1994: Marked post-18th century environmental change in high Arctic ecosystems. *Science,* **266**: 416–419.

Drummond, C. N., W. P. Patterson, and J. C. G. Walker, 1995: Climatic forcing of carbon-oxygen isotopic covariance in temperate-region marl lakes. *Geology,* **23**: 1031–1034.

Dwyer, T. R., H. T. Mullins, and S. C. Good, 1996: Paleoclimatic implications of Holocene lake-level fluctuations, Owasco Lake, New York. *Geology,* **24**: 519–522.

Edwards, T. W. D., B. B. Wolfe, and G. M. MacDonald, 1996: Influence of changing atmospheric circulation on precipitation δ ^{18}O-temperature relations in Canada during the Holocene. *Quaternary Research,* **46**: 211–218.

Enzel, Y., 1992: Flood frequency of the Mojave River and the formation of late Holocene playa lakes, southern California, U.S.A. *The Holocene,* **2**: 11–18.

Eugster, H. P., and K. Kelts, 1983: Lacustrine chemical sediments. *In* Goudie, A. S., and K. Pye (eds.), *Chemical Sediments and Geomorphology.* San Diego: Academic Press, pp. 321–368.

Forester, R. M., L. D. Delorme, and J. P. Bradbury, 1987: Mid-Holocene climate in northern Minnesota. *Quaternary Research,* **28**: 263–273.

Fritz, S. C., 1996: Paleolimnological records of climate change in North America. *Limnology and Oceanography,* **41**: 882–889.

Fritz, S. C., D. R. Engstrom, and B. J. Haskell, 1994: Little Ice Age aridity in the North American Great Plains: A high-resolution reconstruction of salinity fluctuations from Devils Lake, North Dakota, U.S.A. *The Holocene,* **4**: 69–73.

Fritz, S. C., S. Juggins, R. W. Battarbee, and D. R. Engstrom, 1991: Reconstruction of past changes in salinity and climate using a diatom-based transfer function. *Nature,* **352**: 706–708.

Gajewski, K., P. B. Hamilton, and R. McNeely, 1997: A high resolution proxy-climate record from an Arctic lake with annually laminated sediments on Devon Island, Nunavut, Canada. *Journal of Paleolimnology,* **17**: 215–225.

Galloway, R. W., V. Markgraf, and J. P. Bradbury, 1988: Dating shorelines of lakes in Patagonia, Argentina. *Journal of South American Earth Sciences,* **1**: 195–198.

Gasse, F., J. C. Fontes, J. C. Plaziat, P. Carbonel, I. Kaczmarska, P. De Deckker, I. Soulié-Marsche, Y. Callot, and P. A. Dupeuble, 1987: Biological remains, geochemistry and stable isotopes for the reconstruction of environmental and hydrological changes in the Holocene lakes from North Sahara. *Palaeogeography, Palaeoclimatology, Palaeoecology,* **60**: 1–46.

Grimm, E. C., S. Lozano-García, H. Behling, and V. Markgraf, 2000: Holocene vegetation and climate variability in the Americas. In Markgraf, V. (ed.) *Interhemispheric Climate Linkages,* San Diego, Academic Press, pp. 325–370.

Grosjean, M., 1997: Late Quaternary humidity changes in the Atacama Altiplano: Regional, global climate signals and possible forcing mechanisms. *Zentralblatt für Geologie und Palaontologie Teil I,* **3**: 581–592.

Grosjean, M., M. A. Geyh, B. Messerli, and U. Schotterer, 1995: Late-glacial and early Holocene lake sediments, ground-water formation and climate in the Atacama Altiplano 22–24 degrees South. *Journal of Paleolimnology,* **14**: 241–252.

Grosjean, M., B. L. Valero-Garces, M. A. Geyh, B. Messerli, U. Schotterer, H. Schreier, and K. Kelts, 1997: Mid- and late-Holocene limnogeology of Laguna del Negro Francisco, northern Chile, and its palaeoclimatic implications. *The Holocene,* **7**: 151–159.

Hasbargen, J., 1994: A Holocene paleoclimatic and environmental record from Stoneman Lake, Arizona. *Quaternary Research,* **42**: 188–196.

Hodell, D. A., J. H. Curtis, and M. Brenner, 1995: Possible role of climate in the collapse of classic Maya civilization. *Nature,* **375**: 391–394.

Hodell, D. A., J. H. Curtis, G. A. Jones, A. Higuera-Gundy, M. Brenner, M. W. Binford, and K. T. Dorsey, 1991: Reconstruction of Caribbean climate change over the past 10,500 years. *Nature,* **352**: 790–793.

Holliday, V. T., 1989: Middle Holocene drought on the southern Great Plains. *Quaternary Research,* **31**: 74–82.

Horn, S. P., 1993: Postglacial vegetation and fire history in the Chirripo Paramo of Costa Rica. *Quaternary Research,* **40**: 107–116.

Hostetler, S. W., 1995: Hydrological and thermal response of lakes to climate: Description and modeling. *In* Lerman, A., D. Imboden, and J. Gat (eds.), *Physics and Chemistry of Lakes.* Berlin: Springer-Verlag, pp. 63–82.

Hostetler, S. W., P. J. Bartlein, P. U. Clark, E. E. Small, and A. M. Solomon, 2000: Simulated influence of Lake Agarsizonttu climate of central North America 11,000 years ago. *Nature,* **405**: 334–337.

Hu, F. S., E. Ito, L. B. Brubaker, and P. M. Anderson, 1998: Ostracode geochemical record of Holocene climatic change and implications for vegetational response in the northwestern Alaska Range. *Quaternary Research,* **49**: 86–95.

Islebe, G. A., and H. Hooghiemstra, 1997: Vegetation and climate history of montane Costa Rica since the last glacial. *Quaternary Research Review,* **16**: 589–604.

Islebe, G. A., H. Hooghiemstra, M. Brenner, J. H. Curtis, and D. A. Hodell, 1996: A Holocene vegetation history from lowland Guatemala. *The Holocene,* **6**: 265–271.

Jacob, J. S., and C. T. Hallmark, 1996: Holocene stratigraphy of Cobweb Swamp, a Maya wetland in northern Belize. *Geological Society of America Bulletin,* **108**: 883–891.

Krishnamurthy, R. V., K. A. Syrup, M. Baskaran, and A. Long, 1995: Late glacial climate record of midwestern United States from the hydrogen isotope ratio of lake organic matter. *Science,* **269**: 1565–1567.

Kutzbach, J. E., 1987: Model simulations of the climatic patterns during the deglaciation of North America. *In* Ruddiman, W. F., and H. E. Wright (eds.), *North America and Adjacent Oceans During the Last Deglaciation.* Boulder, CO: Geological Society of America, pp. 425–446.

Kutzbach, J. E., and P. J. Guetter, 1986: The influence of changing orbital parameters and surface boundary conditions on climate simulations for the past 18,000 years. *Journal of Atmospheric Science,* **43**: 1726–1759.

Laird, K. R., S. C. Fritz, E. C. Grimm, and B. F. Cumming, 1998: Early Holocene limnologic and climatic variability in the northern Great Plains. *The Holocene,* **8**: 275–286.

Laird, K. R., S. C. Fritz, E. C. Grimm, and P. G. Mueller, 1996a: Century-scale paleoclimatic reconstruction from Moon Lake, a closed-basin lake in the northern Great Plains. *Limnology and Oceanography,* **41**: 890–902.

Laird, K. R., S. C. Fritz, K. A. Maasch, and B. F. Cumming, 1996b: Greater drought intensity and frequency before A.D. 1200 in the northern Great Plains, U.S.A. *Nature,* **384**: 552–555.

Lamoureux, S. F., and R. S. Bradley, 1996: A late Holocene varved sediment record of environmental change from northern Ellesmere Island, Canada. *Journal of Paleolimnology,* **16**: 239–255.

Last, W. M., 1990: Paleochemistry and paleohydrology of Ceylon Lake, a salt-dominated playa basin in the northern Great Plains, Canada. *Journal of Paleolimnology,* **4**: 219–238.

Last, W. M., and T. H. Schweyen, 1985: Late Holocene history of Waldsea Lake, Saskatchewan, Canada. *Quaternary Research,* **24**: 219–234.

Lenters, J. D., and K. H. Cook, 1997: On the origin of the Bolivian High and related circulation features of the South American climate. *Journal of Atmospheric Science,* **54**: 656–677.

Leyden, B., 1985: Late Quaternary aridity and Holocene moisture fluctuations in the Lake Valencia basin, Venezuela. *Ecology,* **66**: 1279–1295.

Leyden, B. W., M. Brenner, D. A. Hodell, and J. Curtis, 1994: Orbital and internal forcing of climate on the Yucatán Peninsula for the past ca. 36,000 ka. *Palaeogeography, Palaeoclimatology, Palaeoecology,* **109**: 193–210.

Leyden, B. W., M. Brenner, T. Whitmore, J. H. Curtis, D. Piperno, and B. H. Dahlin, 1996: A record of long- and short-term climatic variation from northwest Yucatán: Cenote San José Chulchacá. *In* Fedick, S. L. (ed.), *The Managed Mosaic: Ancient Maya Agriculture and Resource Use.* Salt Lake City: The University of Utah Press, pp. 30–50.

Loope, D. B., J. B. Swinehart, and J. P. Mason, 1995: Dune-dammed paleovalleys of the Nebraska Sand Hills: Intrinsic versus climatic controls on the accumulation of lake and marsh sediments. *Geological Society of America Bulletin,* **107**: 396–406.

Lozano-Garcia, M. S., B. Ortega-Guerrero, M. Caballero-Miranda, and J. Urrutia-Fucugauchi, 1993: Late Pleistocene and Holocene paleoenvironments of Chalco Lake, central Mexico. *Quaternary Research,* **40**: 332–342.

MacDonald, G. M., T. W. D. Edwards, K. A. Moser, R. Pienitz, and J. P. Smol, 1993: Rapid response of treeline vegetation and lakes to past climate warming. *Nature,* **361**: 243–246.

Markgraf, V., J. R. Dodson, P. Kershaw, M. S. McGlone, and N. Nichols, 1994: Evolution of late Pleistocene and Holocene climates in the circum—South Pacific lake areas. *Climate Dynamics,* **6**: 193–211.

Mason, J. P., J. B. Swinehart, and D. B. Loope, 1997: Holocene history of lacustrine and marsh sediments in a dune-blocked drainage, southwestern Nebraska Sand Hills, U.S.A. *Journal of Paleolimnology,* **17**: 67–83.

McGregor, G. R., and S. Nieuwolt, 1998: *Tropical Climatology. An Introduction to the Climates of Low Latitudes.* New York: Wiley.

Metcalfe, S. E., 1995: Holocene environmental change in the Zacapu basin, Mexico: A diatom-based record. *The Holocene,* **5**: 196–208.

Metcalfe, S. E., A. Bimpson, A. J. Courtice, S. L. O'Hara, and D. M. Taylor, 1997: Climate change at the monsoon / westerly boundary in northern Mexico. *Journal of Paleolimnology,* **17**: 155–171.

Metcalfe, S. E., F. A. Street-Perrott, S. L. O'Hara, P. E. Hales, and R. A. Perrott, 1994: The palaeolimnological record of environmental change: Examples from the arid frontier of Mesoamerica. *In* Millington, A. C., and K. Pye (eds.), *Environmental Change in Drylands: Biogeographical and Geomorphological Perspectives.* New York: Wiley & Sons, pp. 131–145.

Metcalfe, S. E., F. A. Street-Perrott, R. A. Perrott, and D. D. Harkness, 1991: Palaeolimnology of the Upper Lerma Basin, central México: A record of climatic change and anthropogenic disturbance since 11,600 yr B.P. *Journal of Paleolimnology,* **5**: 197–218.

Mock, C. J., 1996: Climatic controls and spatial variations of precipitation in the western United States. *Journal of Climate,* **9**: 1111–1125.

Mullins, H. T., 1998: Environmental change controls of lacustrine carbonate, Cayuga Lake, New York. *Geology,* **26**: 443–446.

Núñez, L., M. Grosjean, and I. Cartajena, 2000: Human dimensions of late Pleistocene / Holocene arid events in southern South America. *In* Markgraf, V. (ed.), *Interhemispheric Climate Linkages,* San Diego: Academic Press, Chapter 7.

O'Hara, S. L., S. E. Metcalfe, and F. A. Street-Perrott, 1994: On the arid margin: The relationship between climate, humans and the environment. *Chemosphere,* **29**: 965–981.

Ortega Ramirez, J., 1995: Los paleoambientes Holocenicos de la Laguna de Babicora, Chihuahua, México. *Geofisica Internacional,* **34**: 107–116.

Overpeck, J., 1996: Warm climate surprises. *Science,* **271**: 1820–1821.

Overpeck, J. P., K. Hugen, D. Hardy, R. Bradley, R. Case, M. Douglas, B. Finney, K. Gajewski, G. Jacoby, A. Jennings, S. Lamoureux, A. Lasca, G. MacDonald, J. Moore, M. Retelle, S. Smith, A. Wolfe, and G. Zielinski, 1997: Arctic environmental change of the last four centuries. *Science,* **278**: 1251–1256.

Petrequin, P., C. Arnauld, P. Carot, M.-F. Fauvet-Bertelot, R. Lamy Rousseau, M. Magny, S. E. Metcalfe, O. J. Polaco, H. Richard, M. M. Rios Paredes, and M. S. Xelhuantzi Lopez, 1994: 8,000 años de la Cuenca de Zacapu. Evolución de los paisajes y primeros desmontes. *Cuadernos de Estudios Michoacanos,* **6**: 1–144.

Pienitz, R., J. P. Smol, and G. M. MacDonald, 1999: Paleolimnological reconstruction of Holocene climatic trends from two boreal treeline lakes, Northwest Territories, Canada. *Arctic, Antarctic, and Alpine Research,* **31**: 82–93.

Piperno, D., M. Bush, and P. A. Colinvaux, 1990. Palaeoenvironments and human occupation in late-glacial Panama. *Quaternary Research,* **33**: 108–116.

Radle, N. J., C. M. Keister, and R. W. Battarbee, 1989: Diatom, pollen, and geochemical evidence for the paleosalinity of Medicine Lake, S. Dakota, during the late Wisconsin and early Holocene. *Journal of Paleolimnology*, **2**: 159–172.

Rea, D. K., T. C. Moore, Jr., T. W. Anderson, C. F. Lewis, D. M. Dobson, D. L. Dettman, A. J. Smith, and L. A. Mayer, 1994: Great Lakes paleohydrology: Complex interplay of glacial meltwater, lake levels and sill depths. *Geology*, **22**: 1059–1062.

Rodbell, D., G. O. Seltzer, D. M. Anderson, D. B. Enfield, M. B. Abbott, and J. H. Newman, 1999: A high-resolution ~15,000 years record of El Niño driven alluviation in southwestern Ecuador. *Science*, **183**: 516–520.

Rue, D. J., 1987: Early agriculture and early Postclassic Maya occupation in western Honduras. *Nature*, **326**: 285–286.

Sandweiss, D. H., J. B. Richardson, E. J. Reitz, H. B. Rollins, and K. A. Maasch, 1996: Geoarchaeological evidence from Peru for a 5000 year onset of El Niño. *Science*, **273**: 1531–1533.

Schwalb, A., and W. E. Dean, 1998: Stable isotopes and sediments from Pickerel Lake, South Dakota, U.S.A.: A 12 ky record of environmental changes. *Journal of Paleolimnology*, **20**: 15–30.

Schwalb, A., S. M. Locke, and W. E. Dean, 1995: Ostracode $\delta^{18}O$ and $\delta^{13}C$ evidence of Holocene environmental changes in the sediments of two Minnesota lakes. *Journal of Paleolimnology*, **14**: 281–296.

Schweger, C. E., and M. Hickman, 1989: Holocene paleohydrology of central Alberta: Testing the general-circulation-model climate simulations. *Canadian Journal of Earth Sciences*, **26**: 1826–1833.

Seltzer, G. O., P. A. Baker, S. Cross, R. Dunbar, and S. C. Fritz, 1998: High-resolution seismic reflection profiles from Lake Titicaca, Peru-Bolivia: Evidence for Holocene aridity in the tropical Andes. *Geology*, **26**: 167–170.

Servant, M., and J.-C. Fontes, 1978: Les lacs quaternaires des hauts plateaux des Andes Boliviennes premières interprétations paléoclimatiques. *Cahiers, ORSTOM, Séries Geologie*, **10**: 9–23.

Servant, M., M. Fournier, J. Argollo, S. Servant-Vildary, F. Sylvestre, D. Wirrmann, and Ybert, J.-P., 1995: La dernierè transition glaciaire/interglaciaire des Andes tropicales sud (Bolivie) d'apres l'étude des variations des niveaux lacustres et des fluctuations glaciaires. *Comptes Rendus de l'Academie des Sciences*, Paris, Seriés II, **320**: 729–736.

Shulmeister, J., and B. Lees, 1995: Pollen evidence from tropical Australia for the onset of an ENSO-dominated climate at c. 4000 B.P. *The Holocene*, **5**: 10–18.

Smith, A. J., 1991: Lacustrine ostracodes as paleohydrochemical indicators in Holocene lake records of the north-central United States. Ph.D. thesis, Brown University, Providence, RI.

Smith, A. J., J. J. Donovan, E. Ito, and D. R. Engstrom, 1997: Groundwater processes controlling a prairie lake's response to middle Holocene drought. *Geology*, **25**: 391–394.

Smol, J. P., 1983: Paleophycology of a high Arctic lake near Cape Herschel, Ellesmere Island. *Canadian Journal of Botany*, **61**: 2195–2204.

Spaulding, W. G., and L. Graumlich, 1986: The last pluvial climatic episode in the deserts of southwestern North America. *Nature*, **320**: 441–444.

Stine, S., 1990: Late Holocene fluctuations of Mono Lake, eastern California. *Palaeogeography, Palaeoclimatology, and Palaeoecology*, **78**: 333–381.

Stine, S., 1994: Extreme and persistent drought in California and Patagonia during medieval time. *Nature*, **369**: 546–549.

Stine, S., and M. Stine, 1990: A record from Lake Cardiel of climate change in southern South America. *Nature*, **345**: 705–708.

Street-Perrott, F. A., and S. P. Harrison, 1985: Lake levels and climate reconstruction. *In* Hecht, A. D. (ed.), *Paleoclimate Analysis and Modeling*. New York: John Wiley & Sons, pp. 291–331.

Street-Perrott, F. A., P. E. Hales, R. A. Perrott, J. C. Fontes, V. R. Switsur, and A. Pearson, 1993: Late Quaternary palaeolimnology of a

tropical marl lake: Wallywash Great Pond, Jamaica. *Journal of Paleolimnology*, **9**: 3–22.

Stuiver, M., 1970: Oxygen and carbon isotope ratios of fresh-water carbonates as climatic indicators. *Journal of Geophysical Research*, **75**: 5247–5257.

Teller, J. T., and W. M. Last, 1981: Late Quaternary history of Lake Manitoba, Canada. *Quaternary Research*, **16**: 97–116.

Thompson, R. S., 1992: Late Quaternary environments in Ruby Valley, Nevada. *Quaternary Research*, **37**: 1–15.

Tsukada, M., and E. S. Deevey, 1967: Pollen analysis from four lakes in the southern Maya area of Guatemala and El Salvador. *In* Cushing, E. J., and H. E. Wright (eds.), *Quaternary Palaeoecology*. New Haven, CT: Yale University Press, pp. 303–331.

Valero-Garcés, B. L., K. Kelts, and E. Ito, 1995: Oxygen and carbon isotope trends and sedimentological evolution of a meromictic and saline lacustrine system: The Holocene Medicine Lake basin, North American Great Plains, U.S.A. *Palaeogeography, Palaeoclimatology, Palaeoecology*, **117**: 253–278.

Valero-Garcés, B. L., M. Grosjean, A. Schwalb, M. Geyh, B. Messerli, and K. Kelts, 1996: Limnogeology of Laguna Miscanti: Evidence for mid- to late Holocene moisture changes in the Atacama Altiplano (northern Chile). *Journal of Paleolimnology*, **16**: 1–21.

Van Devender, T. R., T. L. Burgess, R. S. Felger, and R. M. Turner, 1990: Holocene vegetation of the Hornoday Mountains of northwestern Sonora, México. *Proceedings of the San Diego Society of Natural History*, **2**: 1–19.

Van Devender, T. R., R. S. Thompson, and J. L. Betancourt, 1987: Vegetation history of the deserts of southwestern North America: The nature and timing of the Late Wisconsin–Holocene transition. *In* Ruddiman, W., and H. E. Wright (eds.), *North America and Adjacent Oceans During the Last Deglaciation*. Boulder, CO: Geological Society of America, pp. 323–352.

Vance, R. E., A. B. Beaudoin, and B. H. Luckman, 1995: The paleoecological record of 6 ka B.P. climate in the Canadian prairie provinces. *Geographie Physique et Quaternaire*, **49**: 81–98.

Vance, R. E., R. W. Mathewes, and J. J. Clague, 1992: A 7000-year record of lake-level change on the northern Great Plains: A high-resolution proxy of past climate. *Geology*, **20**: 879–882.

Vaughan, H. H., E. S. Deevey, and S. E. Garrett-Jones, 1985: Pollen stratigraphy of two cores from the Petén Lake District. *In* Pohl, M. (ed.), *Prehistoric Lowland Maya Environment and Subsistence Economy*. Cambridge, MA: Harvard University Press, pp. 73–89.

Waters, M. R., 1989: Late Quaternary lacustrine history and paleoclimatic significance of pluvial Lake Cochise, southeastern Arizona. *Quaternary Research*, **32**: 1–11.

Watts, W. A., and J. P. Bradbury, 1982: Paleoecological studies at Lake Patzcuaro on the west-central Mexican plateau and at Chalco in the Basin of México. *Quaternary Research*, **17**: 56–70.

Webb, T., P. J. Bartlein, S. P. Harrison, and K. H. Anderson, 1993: Vegetation, lake levels, and climate in eastern North America for the past 18,000 years. *In* Wright, H.E. (ed.), *Global Climates Since the Last Glacial Maximum*. Minneapolis: University of Minnesota Press, pp. 415–467.

Whitlock, C., P. J. Bartlein, V. Markgraf, and A. C. Ashworth, 2000: The midlatitudes of North and South America during the last glacial maximum and early Holocene: Similar paleoclimatic sequences despite differing large-scale controls. *In* Markgraf, V. (ed.) *Interhemispheric Climate Linkages*. San Diego, Academic Press, pp. 391–416.

Whitmore, T., M. Brenner, J. H. Curtis, B. H. Dahlin, and B. W. Leyden, 1996: Holocene climatic and human influences on lakes of the Yucatán Peninsula, México: An interdisciplinary, paleolimnological approach. *The Holocene*, **6**: 273–287.

Wigand, P. E., 1987: Diamond Pond, Harney County, Oregon: Vegetation history and water table in the eastern Oregon desert. *Great Basin Naturalist,* **47**: 427–458.

Williams, K. M., 1994: Paleolimnology of three Jackman Sound Lakes, southern Baffin Island, based on down-core diatom analyses. *Journal of Paleolimnology,* **10**: 129–139.

Winkler, M., A. Swain, and J. Kutzbach, 1986: Middle Holocene dry period in the northern midwestern United States, lake levels and pollen stratigraphy. *Quaternary Research,* **25**: 235–250.

Wirrmann, D., and L. F. DeOliveira Almeida, 1987: Low Holocene level (7700 to 3650 years ago) of Lake Titicaca (Bolivia). *Palaeogeography, Palaeoclimatology, Palaeoecology,* **59**: 315–323.

Wirrmann, D., and P. Mourguiart, 1995: Late Quaternary spatio-temporal limnological variations in the Altiplano of Bolivia and Peru. *Quaternary Research,* **43**: 344–354.

Wolfe, A. P., 1996: A high-resolution late-glacial and early Holocene diatom record from Baffin Island, eastern Canadian Arctic. *Canadian Journal of Earth Sciences,* **33**: 928–937.

Wolfe, A. P., and J. W. Hartling, 1996: The late Quaternary development of three ancient tarns on southwestern Cumberland Peninsula, Baffin Island, Arctic Canada: Paleolimnological evidence from diatoms and sediment chemistry. *Journal of Paleolimnology,* **15**: 1–18.

Wolfe, B. B., T. W. D. Edwards, R. Aravena, and G. M. MacDonald, 1996: Rapid Holocene hydrologic change along boreal tree-line revealed by δ^{13}C and δ^{18}O in organic lake sediments, Northwest Territories, Canada. *Journal of Paleolimnology,* **15**: 171–181.

Wolin, J. A., 1996: Late Holocene lake-level and lake development signals in Lower Herring Lake, Michigan. *Journal of Paleolimnology,* **15**: 19–45.

Xia, J. J., B. J. Haskell, D. R. Engstrom, and E. Ito, 1997: Holocene climate reconstructions from tandem trace-element and stable-isotope composition of ostracodes from Coldwater Lake, North Dakota, U.S.A. *Journal of Paleolimnology,* **17**: 85–100.

Yu, Z., and E. Ito, 1999: Century-scale drought frequency in the northern Great Plains: Possible solar forcing? *Geology,* **27**: 263–266.

Yu, Z., J. H. McAndrews, and U. Eicher, 1997: Middle Holocene dry climate caused by change in atmospheric circulation patterns: Evidence from lake levels and stable isotopes. *Geology,* **25**: 251–254.

16

Full and Late Glacial Lake Records Along the PEP 1 Transect: Their Role in Developing Interhemispheric Paleoclimate Interactions

J. PLATT BRADBURY, MARTIN GROSJEAN, SCOTT STINE,
AND FLORENCE SYLVESTRE

Abstract

Lacustrine records that encompass the full and late glacial periods and extend into the early Holocene document climate changes through the effects climate has on limnology. Wind intensity, temperature, and water balance are the principal climate effects recorded by lake sediments. A selection of lake records along the Pole-Equator-Pole: Americas (PEP 1) transect from Alaska to southern Patagonia reveals broad zonal regions of climate-controlled paleolimnology that change as the climate systems respond to changing forcing mechanisms operative during the past 20,000 years. In the Arctic, lakes record climate mostly as temperature signals that change the effective moisture balance and their productivity. South of the Laurentide ice sheet, lakes also responded to warming temperatures and reduced anticyclonic wind systems between the full glacial and the early Holocene. Lakes in the Great Basin were high during the full glacial period as a result of southerly storm tracks displaced by the mass of continental ice in central and eastern Canada. This effect can be documented as far south as central México. A related effect, the disruption of Intertropical Convergence Zone (ITCZ) and the moisture-bearing trade winds responsible for maintaining lakes in Central America and northern South America, caused basins in this region to be shallow, saline, or dry during the full glacial period and to become reestablished only by the latest glacial and early Holocene. Easterly moisture, perhaps related to the interaction of the South Atlantic high-pressure system and ocean temperatures, appears to have controlled lake stage elevations and limnology throughout much of central South America. The details of these interactions will require sites that are analyzed more carefully with adequate chronological control. Full and late glacial lake conditions at sites in Patagonia, and perhaps those farther north as well, responded to an increased intensity of cold polar outbreaks to variable degrees. The single site in southern South America that was included in this chapter apparently did not receive moisture in significant amounts until the early Holocene, as westerly precipitation finally became established in the region. Throughout the PEP 1 transect, the effect of temperature in determining lake levels and salinity operated in conjunction with precipitation sources and seasonality. This analysis indicates that lakes recorded generally synchronous climate changes by similar but sometimes opposite responses. This synchroneity does not imply the same causality, even though the dynamics of climate change between the full glacial and the Holocene were clearly linked and mutually interdependent.

Copyright © 2001 by Academic Press.

Resumen

Los registros lacustres que aparecen en la era del maximo glacial y tardi glacial, y que se extienden hasta el Holoceno temprano, documentan cambios climáticos a través de los efectos que el clima tiene en la limnología. Intensidad del viento, temperatura, y balance del agua son los principales factores del clima registrados en los sedimentos de lagos. Una selección de los registros de lagos situados a lo largo del transecto PEP 1, desde Alaska hasta el sur de Patagonia, muestran grandes regiones zonales de la paleolimnología, controladas por el clima, que cambia como respuesta del sistema climático al cambio de los principales mecanismos vigentes durante los últimos 20.000 años. En el Ártico, los lagos registran el clima, principalmente como señales de la temperatura que cambian el balance de humedad y su productividad. Al sur de la capa de hielo de Laurentide, los lagos también respondían al aumento de la temperatura y a la reducción del sistema de vientos anticiclónico, entre la era glacial y los inicios del Holoceno. Los lagos del Great Basin tuvieron nivel alto durante la era del maximo glacial como resultado de las tormentas cuyas trayectorias se hallaban desplazadas al sur debido a la masa de hielo continental del centro y este de Canadá. Este efecto puede observarse incluso hasta en México central. Otro efecto unido a este, la ruptura de la Zona de Convergencia Intertropical y el comportamiento de los vientos cargados de humedad responsables del mantenimiento de los lagos en América Central y el norte de Sudamérica, causó el escaso nivel de agua, salinidad o sequía durante la era glacial, y restablecidos solo en la última glaciación y a inicios del Holoceno. La humedad del este, quizá ligada a la interacción del sistema de altas presiónes del Atlántico sur y las temperaturas del océano, parece haber controlado las elevaciones de nivel de lagos y limnología en gran parte del territorio central de Sudamérica. Los detalles de esta interacción necesitarán lugares que sean analizados cuidadosamente con controles adecuados de cronología. Las condiciones de los lagos en la era del maximo glacial y tardi glacial, en lugares de Patagonia, y quizá más hacia el norte, respondían con distinta intensidad al aumento de incursiones de aire frío polar. El único lugar de Sudamérica que ha sido incluído en este estudio aparentemente no recibió una cantidad de humedad significativa hasta el inicio del Holoceno, cuando la precipitación del oeste se establece en la región. A través del transecto PEP1, el efecto de la temperatura en la determinación del nivel de los lagos y la salinidad, actua de forma conjunta con la precipitación y la estacionalidad. Este análisis indica que los lagos registran generalmente sincronizados los cambios climáticos, con parecidas o a veces opuestas reacciones. Esta sincronización no implica una misma causalidad, aunque la dinámica del cambio climático entre la era glacial y el Holoceno esté claramente conectada, y sea interdependiente mútuamente.

16.1. INTRODUCTION

The climate of the Western Hemisphere is defined by global wind systems set in motion by temperature contrasts between the poles and the equator. These elements of atmospheric dynamics deliver the moisture and, in part, the temperature regimes and wind stresses that determine the limnology of lakes. This chapter explores the relation between climate and lakes along a transect (the Pole-Equator-Pole: Americas [PEP 1] transect) extending from Alaska to southern Patagonia, and it extrapolates past climate states from their sediment records. The probable causes of limnoclimatic changes relative to altered glacial global wind systems remain speculative, but a coherent synthesis demands that all elements of the puzzle fit into a picture that is at least as focused as individual lake chronologies allow.

Limnological histories as revealed by paleontological and sedimentological proxies from lake sediment sequences form complex and detailed, often site-specific, records of limnoclimatic change at potentially high resolution. As such, they do not provide unequivocal evidence for deciphering interhemispheric changes in global climate dynamics. A considerable danger exists of over- or misinterpreting a lake record in terms of climate, particularly global climate (e.g., as forced by insolation), when, in fact, only local circumstances have caused proxy variations. The only suitable approach to answering large-scale questions from such fine-scale records is to look for spatial and temporal consistencies among site records that demand overarching explanations related to climate dynamics at regional to hemispheric scales. This is the approach we attempt to use. It will become evident that the number of suitable sites is still too few to develop particularly coherent interhemispheric linkages of past-climate changes in required detail. Nevertheless, the outlines of the major full and late glacial climatic interactions can be seen from intersite comparisons. It is hoped that such comparisons will stimulate further research into the question of past-climate interactions.

There are basically three ways in which insolation and the atmosphere—generally through the distribution, temporal residence, and interaction of air masses—significantly affect lakes: (1) by governing the amount and seasonality of moisture sources, including loss of moisture through evaporation and lags in moisture delivery as a result of snow accumulation and subsequent melting; (2) by controlling the annual and seasonal variations in temperature as determined by distribution, interaction, and residence times of air masses; and (3) by directing the amount and seasonality of wind stress as related to temperature gradients and the residence or passage of air masses. Interpretation of paleolimnological records should seek such generalized causal (?) climate-based relations if lake records are to contribute significantly to the understanding of climate change.

Lakes respond to climate changes affecting water balance (precipitation and evaporation related to winds and temperature) by rising and falling and (or) by becoming fresher or more saline. These changes can be complicated by the interplay between surface- and groundwater sources of differing chemistry and volumes and by loss of lake water to the groundwater system. Variable lags between snowmelt and runoff also have important consequences to lake biology. Temperature obviously affects the amount and seasonality of ice cover in temperate lakes and, therefore, influences stratification, nutrient dynamics, and the life histories of many organisms. Wind, apart from its role in enhancing evaporation, determines the length and strength (depth) of circulation in combination with temperature effects on water density. Wind and runoff characteristics thereby play critical roles in determining the oxygen content and nutrient dynamics of lakes and also can indirectly impact the light regime of lakes by turbidity. Cloud and snow cover can directly vary the amount of light a lake receives.

Although these climate–lake interactions have certainly operated to form the lake records summarized below, rarely has their impact on the record been understood at more than a superficial level. Their effects vary from lake to lake, depending on lake size, depth, morphometry, degree of protection, and surrounding geological terrain. Climatic, biological, and limnological monitoring are prerequisites to full paleoclimatic interpretation of lake records. Unfortunately, paleoclimate studies without a paleolimnological focus— pollen records, for example—rarely provide even minimal descriptive information about the limnological character of the investigated site. Without such information, dutifully recorded abundances of aquatic microfossils (*Pediastrum, Botryococcus, Isoetes, Myriophyllum,* etc.) become difficult or impossible to properly interpret in either paleolimnological or paleoclimatic terms.

The records summarized in Section 16.2 represent a selection of sites between Alaska and southern Patagonia that lie near the western margin of North and South America (Fig. 1). Site selection aims at highlighting climate changes related to or buffered by the Pacific

FIGURE 1 Locations of sites with paleolimnological records spanning full and (or) late glacial through early Holocene environments. Letter designations indicate lake or site name: Z, Zagoskin Lake; K, Klamath Lake; La, Lake Lahontan; O, Owens Lake; E, Lake Estancia; Ba, Cuenca de Babicora; P, Lago de Pátzcuaro; Q, Laguna Quexil; Y, La Yeguada; Li, Laguna Los Lirios; V, Lago de Valencia; F, Laguna de Fúquene; S, Laguna Surucucho; J, Lago de Junín; T, Lago Titicaca; U, Salar de Uyuni; Le, Laguna Lejía; Be, Salinas del Bebedero; CL, Laguna Cari Laufquen; Ca, Lago Cardiel.

Ocean in an attempt to simplify, if possible, climate–lake relations along the PEP 1 transect. Although the summary is not exhaustive, most major records have been included. Their spotty distribution may guide future research to key areas for more complete coverage. Many records lack sufficient chronological control to build a particularly strong case for climate-mediated limnologic change at high resolutions. Nevertheless, most records document distinct limnological modes associated with full glacial, late glacial, and early Holocene environments. These variable and sometimes detailed records, based on assorted paleolimnological evidence, have been interpreted generally in terms of presumed climate controls on lake character. The age control on all figured lake records is in thousands of radiocarbon years before present.

16.2. SYNOPSIS OF SITE RECORDS

16.2.1. Zagoskin Lake: Norton Sound, Alaska (63.4°N, 162.08°W; 15 m)

Zagoskin Lake occupies a small, circular maar crater (500 m diameter) without significant fluvial input from the surrounding low, peat-mantled, volcanic surface (Ager, 1982). Overall, the geochemical record (e.g., Al) of Zagoskin Lake indicates a complacent lacustrine record with major sediment input from eolian sources (Fig. 2). However, biogenic silica and organic matter that relate to the productivity of the lake system increased after 15,000 ^{14}C B.P., approximately coincident with the pollen transition from herbaceous (tundra) vegetation to *Betula*. Larger increases in organic matter and biogenic silica occurred in the mid- and late Holocene as still warmer climates allowed alder and spruce vegetation to characterize the region (Ager and Brubaker, 1985). Warmer summers and seasonally open lake conditions probably caused this increased productivity. The limnological effects of increasing warmth in Arctic environments are of increasing effective moisture, nutrient fluxes, and more effective illumination as lakes remain free of ice and snow for greater periods of time.

16.2.2. Klamath Lake: Klamath Basin, Oregon (42.311°N, 121.915°W; 1263 m)

The diatom stratigraphy of the Caledonia Marsh core from the margin of Klamath Lake documents the effect of wind on a lacustrine system. During the full glacial period, arid climate and strong easterly wind regimes (Barnosky et al., 1987) allowed the transportation and redeposition of Pliocene diatoms (*Aulacoseira paucistriata*) from outcrops of diatomaceous sediment in the Klamath Basin (e.g., Bradbury, 1991) to the shallow and turbid Klamath Lake (Fig. 3). By 15,000 ^{14}C B.P., the increased effective moisture and reduced wind stress produced an open, but still shallow, eutrophic lake as shown by the stratigraphic distribution of *Stephanodiscus niagarae*. Between 15,000 and 10,000 ^{14}C

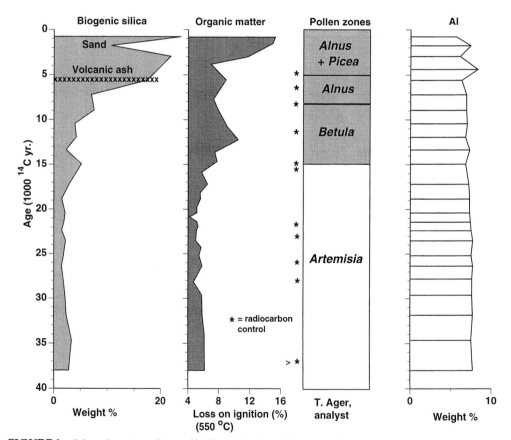

FIGURE 2   Selected stratigraphic profiles from Zagoskin Lake in Alaska. Pollen zonation from T.A. Ager; biogenic silica data courtesy of S.M. Colman; geochemical data from J.P. Bradbury—all of the U.S. Geological Survey.

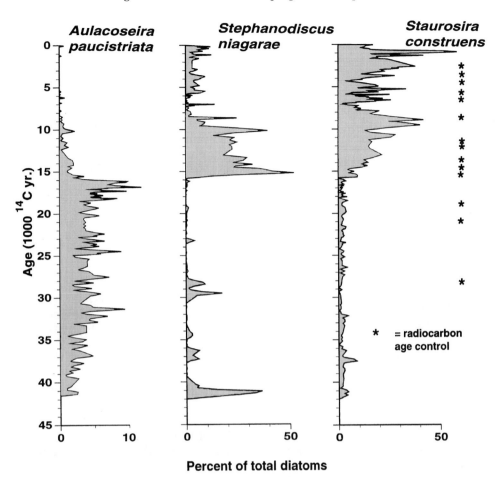

FIGURE 3 Selected diatom profiles from core C-2, Caledonia Marsh, Klamath Lake, Oregon. (Data from J.P. Bradbury.)

B.P., open-water diatoms decreased in favor of marsh epiphytic and benthic species such as *Staurosira construens*, indicating an encroachment of vascular emergent vegetation. Marsh conditions persisted and expanded in the early Holocene, probably as a result of lower water levels and somewhat drier climates than previously existed.

16.2.3. Lake Lahontan: West-Central Great Basin, Nevada (center = 40.0°N, 119.0°W; High Stand = ~1330 m)

The lake level chronology (Fig. 4) of Lake Lahontan is based on >100 ^{14}C dates on shoreline tufa, gastropods, soils, and plant and animal remains (Benson et al., 1990). This large data set indicates intermediate lake levels during the full glacial period (~18,000 ^{14}C B.P.) for Pyramid Lake, the lowest elevation basin in the system. Lake levels fell moderately 17,000–15,000 ^{14}C B.P.

and then rose rapidly to a high stand of 1330 m between 14,000 and 13,000 ^{14}C B.P. Levels fell equally rapidly (<500 years) after 12,500 ^{14}C B.P. to a low stand that separated the major basins in the system (1207 m). Levels again rose minimally (10 m) between 10,500 and 10,000 ^{14}C B.P. and then fell to modern lake elevations by 9500 ^{14}C B.P. (Benson et al., 1992).

The Lahontan and other Great Basin records record high stands that reflect abundant precipitation in the Sierra Nevada to the west from westerly storm tracks displaced to the south (Thompson et al., 1993). Large, shallow, and variably interconnected basins make lake levels of the Lahontan system sensitive to control by interbasin spilling and evaporation that complicate the lake-level history. The highest stand at 14,000–13,000 ^{14}C B.P. may document the passage of the storm track system across the latitude of the Lahontan basin on its way north as the ice sheet became smaller and lower (Thompson et al., 1993), coupled with the rapid melting of Sierra Nevada alpine glaciers.

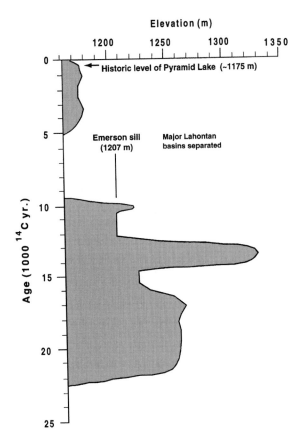

FIGURE 4 Lake stage elevation versus age of the Lahontan basin from data gathered in the area of Pyramid Lake, Nevada. (Compiled from Benson et al., 1990, 1992.)

16.2.4. Owens Lake: Southern Great Basin, California (36.38°N, 118.0°W; 1085 m)

Diatom and ostracode assemblages from cored lake sediments deposited between 25,000 and 14,500 [14]C B.P. document a full, turbid, overflowing lake during and after the full glacial period. The boreal ostracode *Cytherissa lacustris* indicates cold, stable conditions at 24,000–21,000 [14]C B.P., perhaps associated with the permanent residence of the Polar Front at the latitude of Owens Lake at that time. The stratigraphy of *Asterionella formosa*, a planktonic diatom blooming in the summer and fall, after 19,000 [14]C B.P. suggests that Owens Lake became less turbid, but remained fresh following the full glacial period (Fig. 5). Pollen, diatom, and ostracode evidence indicates that precipitation increased at least 1.5–2 times and that summers and winters were at times as much as 10°C colder than they are now (Bradbury and Forester, in press).

A hiatus between 15,500 and 13,500 [14]C B.P.—possibly due to lake shoaling, wave erosion, and redeposition of sediment in deeper parts of the basin—interrupts the record. Between 13,500 and 11,000 [14]C B.P.,

diatoms indicate very wet conditions and high sedimentation rates, perhaps relating to flooding and outwash from receding alpine glaciers. Saline diatoms and ostracodes indicating decreased precipitation and discharge down the Owens River first appeared at ca. 11,000 [14]C B.P. After that time, the record shows alternating freshwater conditions punctuated by short (200-year) saline intervals, which became more saline throughout the early Holocene. By the mid-Holocene, the already shallow Owens Lake desiccated, creating a second hiatus between 6000 and 4000 [14]C B.P. At the end of the nineteenth century and the beginning of the twentieth century, Owens Lake was described as being about 15 m deep and having a salinity of 100–200 g/L (Lee, 1906).

16.2.5. Lake Estancia: Estancia Valley, Central New Mexico (34.50°N, 106.00°W; 1875 m)

A long sequence of late Pleistocene and early Holocene lacustrine deposits exposed by late Holocene deflation outcrop in and around the Lake Estancia playa. The sequence contains ostracodes, diatoms, and pollen, along with geochemical and sedimentological proxies of past lake changes (Bachhuber, 1982; Allen and Anderson, 1993) that document rapid fluctuations between full and intermediate lake-level stages (Fig. 6). A trend toward progressively decreasing lake levels began after 14,000 [14]C B.P. Sediments record strong fluctuations of lake levels, including desiccation and a final brief rise to the approximate early late glacial level between 11,500 and 10,000 [14]C B.P. Lake levels fell thereafter until desiccation at 8000 [14]C B.P., when mid-Holocene deflation began to remove sediments from the basin.

Like the Great Basin records, full glacial moisture was probably associated with more persistent and strengthened westerly storm tracks forced south to the latitude of the Estancia basin. The presence of the ostracode *Cytherissa lacustris* in these deposits at 20,000–19,000 [14]C B.P. suggested exceptionally cold conditions at Lake Estancia at that time (e.g., Bradbury and Forester, in press). Subsequent late glacial moisture pulses lack *C. lacustris*, but contain ostracodes (*Candona rawsoni* and *Limnocythere ceriotuberosa*) that indicate greater seasonal limnoclimatic variability.

16.2.6. Cuenca de Babícora: Chihuahua, México (29.00°N, 108.00°W; 2200 m)

Lacustrine sediments beneath an intermittently dry part of the Alta Babícora basin were exposed by excavation. Apparent, very slow, or variable sediment ac-

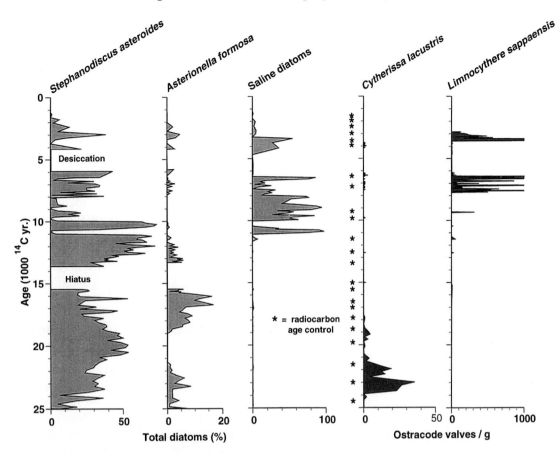

FIGURE 5   Selected diatom and ostracode profiles from Owens Lake, California. (Data from Bradbury and Forester, in press.)

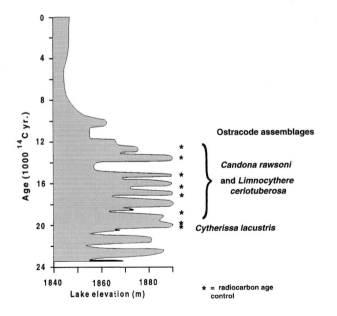

FIGURE 6   Lake stage elevation vs age of the Estancia basin in New Mexico. (Data from Allen and Anderson, 1993.)

cumulation rates make this record difficult to evaluate. The bottom of the section (143 cm) is undated (Metcalfe et al., 1997), but may begin at ca. 13,000 ^{14}C B.P. (estimated by extrapolation), with a diatom assemblage characterized by *Synedra fasciculata, Stephanodiscus asteroides,* and *S. niagarae* that persists with fluctuations into the early Holocene period (Fig. 7). *Synedra fasciculata* is an attached brackish water diatom that often blooms in the summer (Bahls et al. 1984), whereas the *Stephanodiscus* species are planktonic in fresh, open water and bloom in the fall or early winter (*S. niagarae*) and spring (*S. asteroides*) (Bradbury and Forester, in press). The mixed assemblage probably indicates strong seasonality in water depth and hydrochemistry. Winter precipitation could provide fresh, open-water conditions to support first *S. niagarae* and then *S. asteroides* later in the spring. Evaporation during the arid summer would have increased lake salinity and decreased water depth to provide suitable habitats for *Synedra fasciculata.*

The inadequate late glacial chronology does not al-

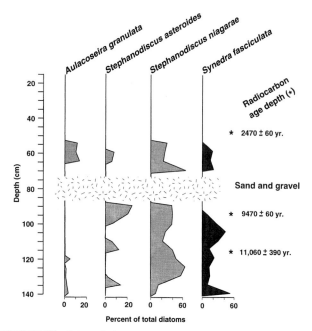

FIGURE 7 Selected diatom profiles from Cuenca de Babícora, Chihuahua, México. (Data from Metcalfe et al., 1997.)

low useful discussion of the relative changes of these diatom taxa, but the assemblage clearly indicates a winter precipitation climate regime compatible with packrat midden evidence (Van Devender, 1990). Somewhat shallower, more saline environments (*S. fasciculata*) may have predominated at ca. 13,000 and 10,000 [14]C B.P. (extrapolated), and the site seems to have desiccated shortly after 9500 [14]C B.P. The *Stephanodiscus / Synedra* diatom assemblage—accompanied by *Aulacoseira granulata*, a summer, freshwater planktonic species—reappeared afterward, apparently documenting an undated Holocene interval of increased effective moisture.

16.2.7. Lago de Pátzcuaro: Michoacán, México (19.5°N, 101.5°W; 2040 m)

The full glacial diatom record from a Lago de Pátzcuaro core (Bradbury, in press) was dominated by a *Stephanodiscus* species presumed to have bloomed in the spring under cold, low-light regimes (Fig. 8). Its presence suggests lake circulation and (or) hydrologic input in the winter and spring. An *Aulacoseira* species, characteristic of taxa that live in large lakes with deep mixing regimes, and *Stephanodiscus niagarae*, a diatom that usually blooms in the fall, probably indicate maximum effective moisture at 19,000–13,000 [14]C B.P. The presence of these late summer and fall diatoms re-

quired sufficient precipitation and cool temperatures to maintain Lago de Pátzcuaro as a fresh and comparatively deep system throughout the year, even after the end of the winter–spring rain season. Perhaps this interval of low temperatures and abundant moisture related to the glacial episodes dated on Iztaccíhuatl at the same time (Vazquez-Selem 1998). Late glacial pollen of *Juniperus, Artemisia,* and *Isoetes* (Watts and Bradbury 1982) also imply cooler and moister conditions in the winter.

After 10,000 [14]C B.P., *Aulacoseira ambigua*, a summer diatom species, dominates with increasing numbers of other shallow-water taxa. These data indicate a general shallowing and warming of the lake after 10,000 [14]C B.P. and a shift of nutrient input to the summer months by lake circulation at that time and (or) by nutrient influx during the summer precipitation season as happens today in central México.

16.2.8. Laguna Quexil: Guatemala (16.93°N, 89.82°W; 110 m)

Sediment mineralogy and pollen investigations on a core from Laguna Quexil in the Petén district of Guatemala allow important late Wisconsin limnological interpretations despite inadequate chronological control (Leyden et al., 1993) (Fig. 9). Gypsiferous clays bearing abundant grass and juniper pollen, but devoid of *Botryococcus* indicate extremely arid conditions during intervals presumed to represent the full and early postglacial periods. Analog vegetation communities live at higher elevations in environments 6.5°–8°C cooler than now at Laguna Quexil. Carbonate sedimentation and Chenopodiineae pollen increased afterwards, peaking by an estimated 12,500 [14]C B.P. These changes indicate that Laguna Quexil was alkaline and slightly saline, with fluctuating levels that responded to modest changes in effective moisture. By the early Holocene, organic sedimentation with abundant lowland rain forest pollen assemblages (Moraceae) characterize the record, indicating substantial increases in moisture and temperature to modern levels.

16.2.9. La Yeguada: Panama (8.45°N, 80.85°W, 650 m)

The 14,000 [14]C B.P. lake core record at La Yeguada (Bush et al., 1992) contains diatom, pollen, geochemical, and mineralogical evidence of a cool and dry (but seasonally wet) climate between 14,300 and 10,800 [14]C B.P. Shallow-water (attached) and saline diatoms characterize the calcium-rich levels between 14,300 and 12,800 [14]C B.P. (Fig. 10). This record corre-

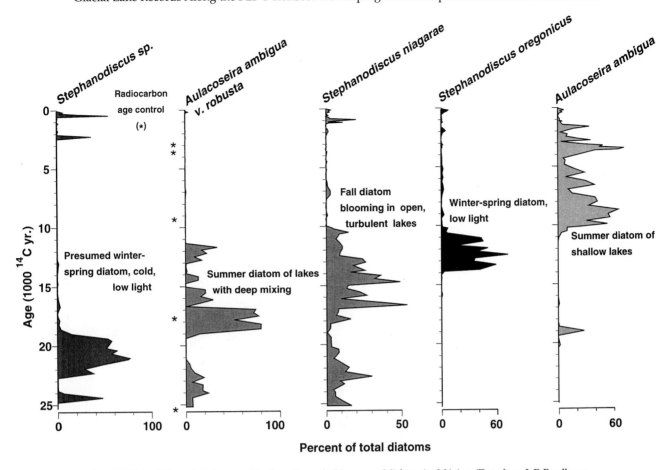

FIGURE 8   Selected diatom profiles from Lago de Pátzcuaro, Michoacán, México. (Data from J. P. Bradbury, in press.)

sponds to similar environments at Lake Valencia and indicates a modest precipitation deficit in the early late glacial period. However, strong fluctuations between *Aulacoseira granulata* early in this interval suggest submillennial-scale climatic fluctuations that require further chronological and paleontological resolution. After 12,800 ^{14}C B.P., *A. granulata* and *Cyclotella stelligera* dominated and replaced indicators of shallow, saline conditions in a productive, seasonally fluctuating lake. The laminated sediments and *C. stelligera* imply seasonal stratification under warm climates. Smectite deposition predominated throughout the late glacial period, indicating moderate chemical weathering and therefore somewhat dry, cool environments, as did the presence of *Quercus* and other montane plants.

Warm and wet conditions indicated by a dominance of kaolinite followed (10,500–8200 ^{14}C B.P.), but an absence of diatoms during this interval remains unexplained. Silica limitation—perhaps related to leached,

silica-poor tropical soils and persistent lake stratification and reduced nutrient influx—may have been relevant in this case. Therefore, the early Holocene paleolimnological record contains little information of value in evaluating paleoclimate. La Yeguada appears to have freshened earlier than Lago Valencia, perhaps reflecting a west-to-east gradient in moisture availability in the early late glacial period. The mid- to late Holocene dominance of *Aulacoseira ambigua* may suggest lower rates of nutrient supply (Cumming et al., 1995) to La Yeguada than before, perhaps as a result of progressive soil weathering.

16.2.10. Laguna Los Lirios: Mérida, Venezuela (8.3°N, 71.8°W; 2300 m)

Geochemical and clay analyses from the 17,500 ^{14}C B.P. Los Lirios core (Weingarten, 1988) broadly divide the record into periods before and after 9300 ^{14}C B.P. Be-

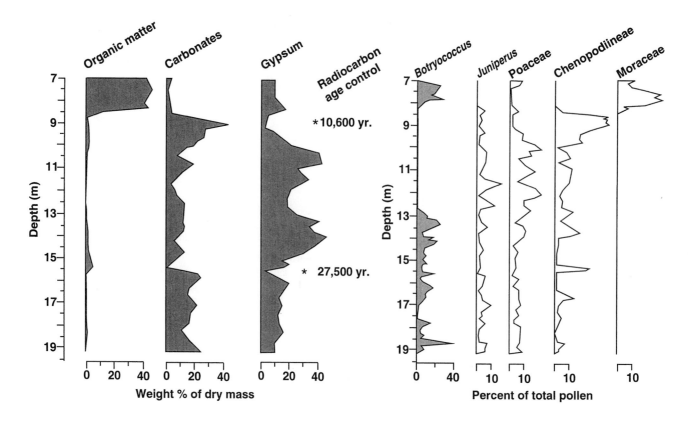

* = radiocarbon date depth**

FIGURE 9 Selected geochemical, algal, and pollen profiles from Laguna Quexil in the Petén district of Guatemala. (Data from Leyden et al., 1993.)

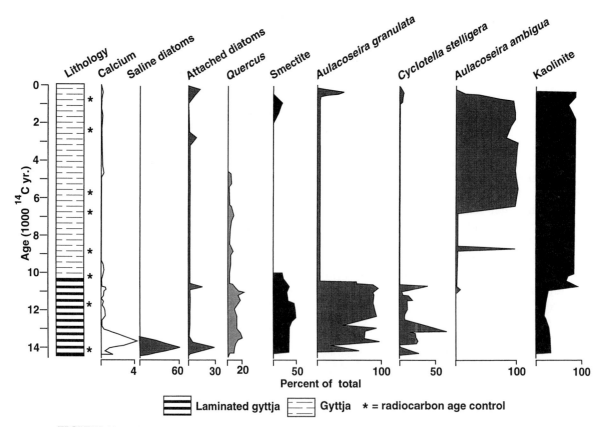

FIGURE 10 Selected diatom, geochemical, and mineralogical profiles from La Yeguada in Panama. (Data from Bush et al., 1992.)

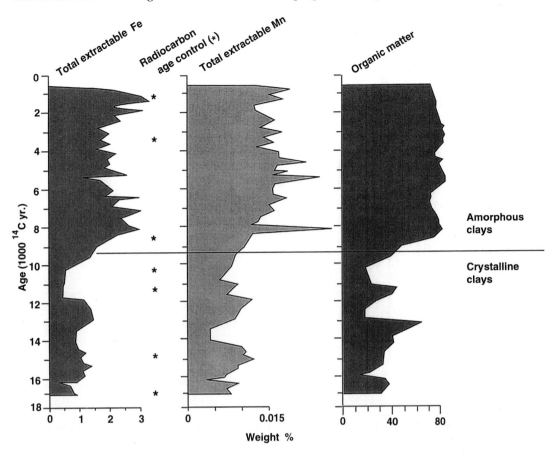

FIGURE 11   Selected geochemical profiles from Laguna Los Lirios in Venezuela. (Data from Weingarten, 1998.)

fore 9300 [14]C B.P., crystalline clays (principally chlorite and illite) and relatively lower amounts of organic matter and of extractable Fe and Mn characterize the late glacial period. After 9300 [14]C B.P., clays are largely amorphous, and organic matter and extractable Fe and Mn increase (Fig. 11). Crystalline clays reflect cold and relatively dry climatic conditions in which physical weathering predominated. The increased abundances of organic matter and of extractable Fe and Mn after 9300 [14]C B.P. indicate increased lake productivity coupled with greater development of surrounding vegetation; the chemical reduction and subsequent deposition of Fe and Mn were thereby increased. Amorphous clays that characterized the Holocene document increased temperatures and moisture and, consequently, increased chemical weathering.

Fluctuations of these parameters in the record, especially during the late glacial period (Fig. 11), almost certainly reflect climatic variations (Weingarten, 1988). The record suggests that cold and dry conditions at 17,000 [14]C B.P., which become cold and wet at 15,000–14,000 [14]C B.P., possibly coincide with maximum

glaciation. Variable, but generally dry climates between 13,400 and 11,300 [14]C B.P. became quite dry between 11,300 and 9800 [14]C B.P. Increasingly warmer and wetter climates followed, to become very warm and wet by 8000 [14]C B.P. Mid-Holocene climates are cooler and wet, reflecting neoglacial climates.

Cold and wet climates in the Venezuelan Andes at 15,000–14,000 [14]C B.P. contrast with presumed dry climates at Lake Valencia at the same time, possibly reflecting the effect of elevation and (or) the more westerly location of the Los Lirios record. Records for both Lake Valencia and Laguna Los Lirios document increased moisture levels by and following 10,000 [14]C B.P., which probably indicated the development of easterly (trade wind) source areas of moisture (e.g., the Gulf of México).

16.2.11. Lago de Valencia: Venezuela (10.18°N, 67.7°W; 404 m)

A 7.3-m core from the deepest part of Lake Valencia has a basal ^{14}C age of 12,900 B.P. (Fig. 12). Stiff clays,

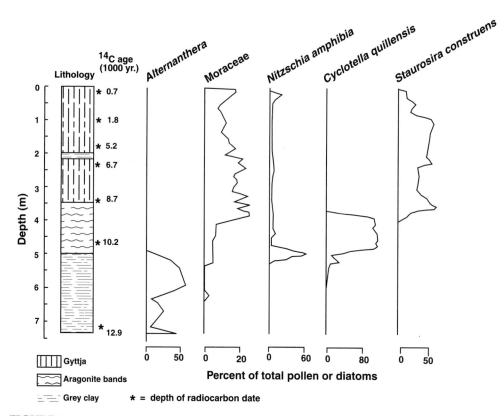

FIGURE 12   Selected pollen and diatom profiles from Lago de Valencia in Venezuela. (From Bradbury et al., 1981.)

mostly barren of diatoms but containing pollen of *Alternanthera,* a herbaceous saline marsh plant (Leyden, 1985), characterized the lower 2.5 m and indicated that Lake Valencia was shallow and variably saline before 10,000 [14]C B.P. (Bradbury et al., 1981). Saline planktonic diatoms (*Cyclotella quillensis*) appeared at 10,000 [14]C B.P. and corresponded with the initial filling of Lago Valencia by 8700 [14]C B.P. Thereafter, the lake periodically drained and became fresh as indicated by the diatom *Staurosira construens.* Abundant Moraceae pollen indicates the replacement of dry scrub vegetation by forests in the early Holocene period (Leyden, 1985). The late glacial and early Holocene paleolimnologic changes at Lago Valencia resemble those at Laguna Quexil.

16.2.12. Laguna de Fúquene: Valle de Ubaté, Colombia (5.45°N, 73.77°W; 2580 m)

The poorly dated Fúquene record from the Colombian Andes shows that dry full glacial climates supported páramo (grass-dominated) vegetation (Van

Geel and Van der Hammen, 1973). *Myriophyllum* and *Isoetes* indicate cold, shallow, moderately alkaline, but oligotrophic, lacustrine conditions during that time (Fig. 13). By 11,000 [14]C B.P., warmer and moister climates promoted a change to *Quercus*-rich Andean forest vegetation in the drainage basin. Aquatic vegetation dominated by *Potamogeton* suggests a possible increase in lake level and productivity.

At the higher elevation site of La Laguna in this area (4.92°N, 74.33°W, 2900 m), palynological studies from a wetland core (Helmens et al., 1996) document cold, oligotrophic conditions (*Isoetes*) between ~15,000 and ~8000 [14]C B.P., with an intermediate, undated interval of less *Isoetes*, but high Poaceae pollen content assumed to correlate with the Younger Dryas episode. Increased *Botryococcus* at 8300–8000 [14]C B.P. may suggest conditions of deeper, more eutrophic, and warmer water. This record resembles the Fúquene core in character, although available chronological control indicates differences in timing of climate-related paleolimnological changes. The higher elevation of La Laguna may explain the more persistent record of *Isoetes* (to 8000 [14]C B.P.) than was observed in the Fúquene core. Increased *Botryococcus* correlates with pollen evidence for a rise in

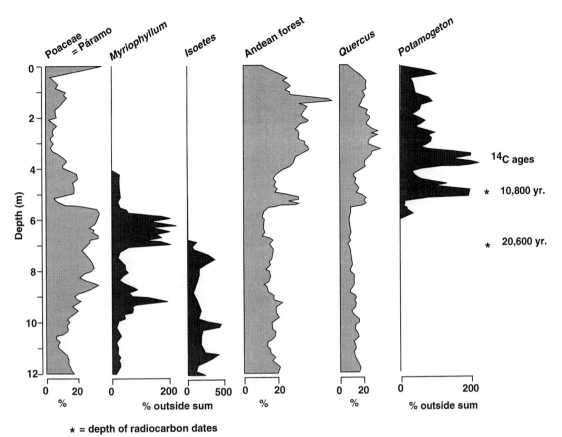

FIGURE 13 Selected pollen profiles from Laguna de Fúquene in Colombia. (Data from Van Geel and Van der Hammen, 1973.)

the tree line of Andean vegetation (warmer climate). It is unlikely that this site suffered significant moisture deficits in the late Pleistocene.

Sparse paleolimnological data from Laguna de Fúquene and other Colombian records suggest a transition from colder and possibly drier early late glacial climates to warmer and wetter conditions afterwards. A reduction or southerly displacement of northeast trade winds associated with the Intertropical Convergence Zone (ITCZ) seems a logical cause for dry, late glacial environments (Bradbury, 1997). An increase in the water balance could reflect a reestablishment of easterly moisture sources associated with latest glacial and early Holocene increased summer insolation coupled with warm sea surface temperatures (SSTs) and increased fluxes of moisture to the atmosphere.

16.2.13. Laguna Surucucho, Ecuador (2.9°S, 78.8°W; 3180 m)

Pollen in a 12-m core from Laguna Surucucho on the east flank of the Andes documents abundant *Isoetes* in probable full and late glacial intervals, although ran-

dom age reversals make the radiocarbon chronology for this period ambiguous (Fig. 14). *Isoetes* in presumed full glacial intervals may reflect shallow-water conditions, which are also suggested by sandy, inorganic, clay-rich sediment. However, *Isoetes* remained abundant during the late glacial period in laminated black gyttja, which probably indicates moderately increased lake productivity and stratification related to warmer summer temperatures. Abundant *Polylepis* pollen in this interval may imply more effective precipitation. By the Holocene, *Isoetes* falls to low concentrations, coincident with pollen evidence of an Andean forest with a warmer climate and increased moisture (Colinvaux et al., 1997).

This record, although essentially undated, appears to correlate with other low-latitude, high-elevation records that show high *Isoetes* values in full or late glacial levels. Inferred shallow-water conditions could suggest modest moisture deficits in large-capacity basins, although the abundance of *Isoetes* may simply relate to cold, oligotrophic lake conditions and the presence of suitable silty substrates near the lake margin regardless of lake depth.

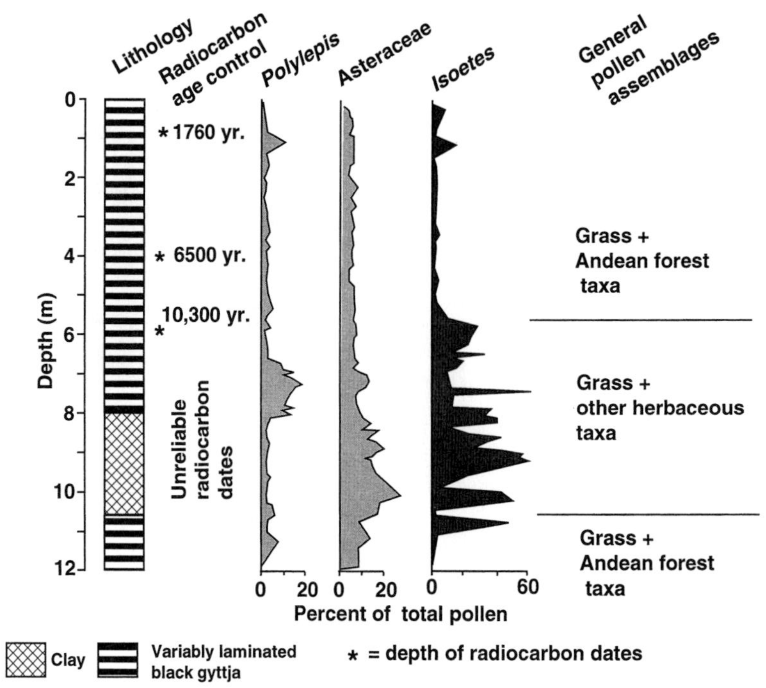

FIGURE 14 Selected pollen profiles from Laguna Surucucho in Ecuador. (Data from Colinvaux et al., 1997.)

16.2.14. Lago de Junín: West-Central Peru (11.0°S, 76.13°W; 4100 m)

Silty clay and abundant *Isoetes* spores characterize the full glacial record of the Junín core (Fig. 15). These indicators are presumed to represent outwash from nearby glaciers entering a cold, shallow, oligotrophic lake (Hansen et al., 1984). Abundant herb (Asteraceae) and *Polylepis/Acaena* pollen indicate relatively dry, puna-like climates in the vicinity, perhaps at lower elevations (Hansen et al., 1984).

By 15,000 ^{14}C B.P., peaty, organic-rich sediments imply greater productivity and decreased influx of glacial silt that lasted to ca. 11,000 ^{14}C B.P. Thereafter, banded marl with organic layers suggests more persistent, carbonate-charged groundwater input with biogenic deposition of calcium carbonate from stems and leaves of submerged aquatic macrophytes. *Isoetes* become rare and pollen (e.g., Moraceae) from east Andean slopes characterize the record throughout the early and middle Holocene (Fig. 15).

The peaty deposits at an extrapolated date of 15,000 ^{14}C B.P. suggest low lake levels and possible moisture deficits during the late glacial period, whereas the transition to marl by 11,000 ^{14}C B.P. signaled the onset of moister and warmer climates with greater weathering of surrounding Paleozoic carbonate rocks. The appearance of extraregional pollen from the east at that time

probably indicates strengthening of the trade winds that serve today as the principal source of moisture in the region. Overall, both the Lago de Junín and the Laguna Surucucho records have similar interpretations.

16.2.15. Lago Titicaca: Bolivian Altiplano (16.0°S, 69.0°W; 3810 m)

In Lake Titicaca (Huiñamarca), ~5 m cores from a water depth of 19 m document lake-level changes reconstructed from sedimentology (Wirrmann, 1987; Wirrmann et al., 1988); *Isoetes*, pollen, and algae (Ybert, 1992); and diatoms (Servant-Vildary, unpublished data). The interpretation of fossil assemblages is based on modern algal assemblages in surface sediments along two transects located in Lake Titicaca (Ybert, 1992). At present, *Botryococcus* and *Pediastrum* are rare (<20%) between 0 and 2 m depth, but increase to 70% between 2 and 4 m and reach 90% between 4 and 200 m. These algae currently associate with the abundant planktonic diatom *Cyclotella andina* in oligosaline Lake Titicaca (Servant-Vildary, 1992). Percentages of *Pediastrum + Botryococcus* in the cores are presumed to indicate similar lake depths in the past and a hydrochemical composition close to that of modern Lake Titicaca. Radiocarbon chronologies for Lake Titicaca require reservoir correction of about −250 years (Abbott et al., 1997).

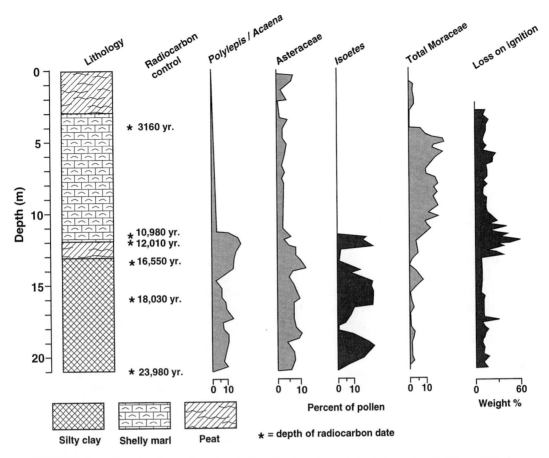

FIGURE 15 Selected pollen and geochemical profiles from Lago de Junín in west-central Peru. (Data from Hansen et al., 1984.)

Moderate percentages of *Botryococcus + Pediastrum* in full glacial levels of the core (19,000 [14]C B.P.) suggest lake elevations similar to those existing now. After 18,200 [14]C B.P., high percentages of *Botryococcus + Pediastrum* in a core labeled TD (Fig. 16) indicate freshwater conditions and modern or higher lake levels than those of today at Lake Titicaca. A high stand (Tauca III at 182 cm in core TD1) occurred at 13,200 [14]C B.P., and an undated lacustrine terrace, located at +5 m above the present level of Lake Titicaca and attributed to the Tauca phase by Servant and Fontes (1978), suggests that Lake Titicaca could have overflowed to the southern basins (Poopó–Coipasa–Uyuni) at this time. A drastic drop of the lake level was recorded by the disappearance of *Botryococcus + Pediastrum* before the 9600 [14]C B.P. date on the correlative algae peak in core TD (Sylvestre et al., 1999). High percentages of fungi after 7000 [14]C B.P. in core TD1 suggests a hiatus at this Huiñamarca locality.

Although a modern ^{14}C reservoir correction is inappropriate for times when the paleolake was lower than it is today, we tentatively correlate this low lake stage

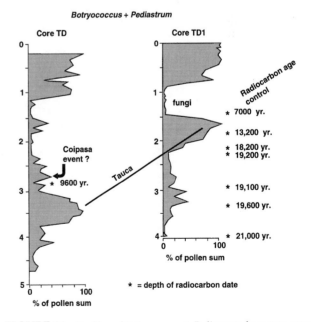

FIGURE 16 Profiles of *Botryococcus + Pediastrum* from two cores in the Huiñamarca basin of Lake Titicaca in Bolivia. (Data from Ybert, 1992.)

with the dry Ticaña event. The reappearance of *Botry-ococcus* + *Pediastrum* (40%) just after 9600 ¹⁴C B.P. in a core (TD; Fig. 16) suggests that Lake Titicaca may have been somewhat deeper for a brief period during the early Holocene, which would correspond with increasing lake levels in the Uyuni–Coipasa basin (Coipasa event). However, this increased lake level is not recorded in core TD1 and must be confirmed by further studies in Lake Titicaca.

16.2.16. Salar de Uyuni–Coipasa: Southern Bolivian Altiplano (20.0°S, 67.0°–68.0°W; 3653 m)

The late glacial and early Holocene lake-level changes on the southern Bolivian Altiplano are reconstructed from outcrops and shorelines on the margin of the Uyuni–Coipasa basin (Servant and Fontes, 1978; Servant et al., 1995; Sylvestre et al., 1996; Sylvestre et al., 1999). The elevation of dated lacustrine deposits above the present bottom of the Salar de Uyuni (3653 m) reflects paleolake levels. The ecology of the fossil diatom assemblages documents the limnological character of

the paleolakes according to the modern distribution of diatom communities from the Altiplano (Servant-Vildary and Roux, 1990; Sylvestre, 1997) and from other regions when no modern analogs were found (Sylvestre et al., 1996). Fossil diatoms from four selected outcrops on the margin of the basin (Sylvestre et al., 1996) provide the data for this analysis. The lake-level chronology is based on 44 radiocarbon dates on inorganic carbonates. The validity of the radiocarbon ages is verified by comparison with six ²³⁰Th/²³⁴U ages. This comparison shows excellent agreement between ¹⁴C calendar ages and ²³⁰Th/²³⁴U ages during lake transgressions and high stands. However, during periods of lake regression, apparent ¹⁴C ages are greater than U/Th ages, suggesting that radiocarbon ages must be corrected for a reservoir effect.

By ca. 15,500 ¹⁴C B.P., lacustrine conditions (Tauca Ia) became established on the southern Bolivian Altiplano. At 3657 m (~4 m above the present Salar de Uyuni), a thin aragonite crust gave a radiocarbon age of 15,430 ± 80 ¹⁴C B.P., calibrated to 18,458 cal. B.P. and Th/U dated at 18,860 ± 390 B.P. Fossil diatom assemblages, dominated by benthic *Denticula subtilis*, indicate shallow, saline water at the lake margin (Fig. 17).

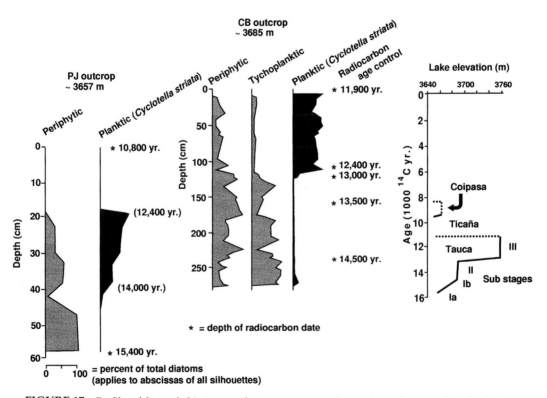

FIGURE 17 Profiles of diatom habitat groups from two outcrops of lacustrine sediment in the Salar de Uyuni basin and the reconstructed lake stage elevation profile for the region. Dashed lines represent lake-level changes of less certainty. (Data from Sylvestre et al., 1999; Sylvestre, 1997.)

After 15,500 [14]C B.P., the level (Tauca Ib) rose rapidly to ~27 m above the present salar (Fig. 17). The dominant planktonic diatom, *Cyclotella striata*, indicated that comparatively deep water covered the 3657-m site. Coeval lacustrine deposits at 3682 m contain the tychoplanktonic diatoms *Fragilaria atomus* and *F. construens subsalina*, which indicated water inputs to the Uyuni–Coipasa basin. Between ca. 14,400 and 13,000 [14]C B.P., the lake level stabilized (Tauca II). Finely laminated sediments at 3685 m contain littoral epiphytic diatoms (*Achnanthes brevipes*, *Cocconeis placentula* var. *euglypta*, *Rhopalodia gibberula*) and tychoplanktonic diatoms (*Staurosirella pinnata*). These assemblages indicate shallow water with episodic freshwater input. The lake level is estimated to have been ~40 m above the present salar (Fig. 17). The lake level increased abruptly at ca. 13,000 [14]C B.P. and reached a hydrological maximum with a depth of ca. 100 m above the present bottom of the Salar de Uyuni (Tauca III). Twenty-one radiocarbon ages place this phase between ca. 12,000 and 13,000 [14]C B.P. An algal bioherm located at 3760 m on the highest shoreline gave a radiocarbon age of 12,930 ± 50, calibrated at 15,330 cal. B.P. and Th/U dated at 15,070 ± 400 B.P. The planktonic diatom *Cyclotella striata* dominates in all outcrops between 3657 and 3685 m and in the bioherms at 3760 m. Most outcrop surfaces are eroded, but two radiocarbon ages suggest that the Tauca phase may have ended after ca. 12,000 [14]C B.P. A sudden drop of the lake level (Ticaña event) is recorded at 3657 m by fluvial sands between <12,000 and >9500 [14]C B.P. These deposits locally contain interbedded clay-silt lenses with mollusk shells and *Denticula subtilis* that imply shallow residual ponds remained around the main basin, which must have been almost dry. A radiocarbon age (11,980 ± 50 B.P.) on mollusk shells may include a reservoir effect related to old (Tauca stage) groundwater. During the early Holocene period, a slight oscillation of the lake level (Coipasa event) is represented by a calcareous crust widely developed around the basin at 3660 m. This crust supports small algal bioherms dominated by *D. subtilis*, indicating shallow, saline water. The Coipasa lake level rose at least ~7 m above the present salar. Nine radiocarbon ages between 11,400 and 10,500 years on the Coipasa event may be more than ~2000 years too old according to comparative radiocarbon and U/Th dates on related samples. These old radiocarbon dates may reflect groundwater discharge of old inorganic carbon from the former lacustrine Tauca transgression (Sylvestre et al., 1999). Consequently, the lacustrine Coipasa event probably occurred between ca. 9500 and 8500 [14]C B.P. Since the early Holocene, the Salar de Uyuni has remained comparatively low (Sylvestre, 1997).

16.2.17. Laguna Lejía: Atacama Desert, Northern Chile (23.50°S, 67.70°W; 4325 m)

Diatoms and sediment studies of high-stand lacustrine outcrops on the south side of Laguna Lejía are radiocarbon dated (with variable and often large reservoir corrections) to 11,500–9700 [14]C B.P. (Geyh et al., 1999). These deposits indicate an initial (11,500 [14]C B.P.) lake-level rise (2–10 m) from previous lake elevations near modern levels (4325 m) and an abrupt rise to +25 m by <10,800 [14]C B.P. (Fig. 18). Laguna Lejía then fell to near modern levels shortly afterwards, but rose rapidly thereafter to another high stand (+25 m) during the early Holocene. After 8800 [14]C B.P., Laguna Lejía progressively fell, reaching modern elevations by ca. 8000 [14]C B.P. and then desiccating in the mid-Holocene (Grosjean et al., 1995; Geyh et al., 1999).

The highest lake levels, 10–24 m above modern, were previously thought to correlate with the Tauca phase of the Uyuni–Coipasa basins to the north (Messerli et al., 1992), but now appear to postdate the maximum lake-level stages recorded farther north (Geyh et al., 1999). Additional chronological control for the general Atacama lake chronology comes from methodologically reliable (i.e., reservoir effect-free) peat intercalated with diatomite and archaeological charcoal associated with the maximum paleobeach ridges (Geyh et al., 1999). The revised chronology of the Lejía/Atacama region would indicate an approximate correlation of high lake stands to the Coipasa lake-level rise in the Titicaca/Uyuni area (Fig. 18).

Annual precipitation is estimated to be as much as 2.8 times modern values at the lake-level maximum (Grosjean et al., 1995). Tropical easterly moisture sources expanded to the south and southwest during these times, and increased cloudiness is thought to have maintained the Atacama lakes (Grosjean and Nuñez, 1994).

16.2.18. Salinas del Bebedero: San Luis Province, Argentina (33.33°S, 66.75°W; 380 m)

Two paleolimnological data sets document past lake-level stages at Salinas del Bebedero. Radiocarbon-dated beach ridges (Fig. 19) indicate high-stand lake elevations of +25 to +20 m above the present-day playa surface between 20,100 and 13,300 [14]C B.P. (González, 1994). Salt-tolerant gastropods associated with the beach ridges reflect dissolution of Mesozoic evaporites in the drainage area of the Río Desaguadero and, as a consequence, the saline composition of the river and the lake basin (Déletang, 1929).

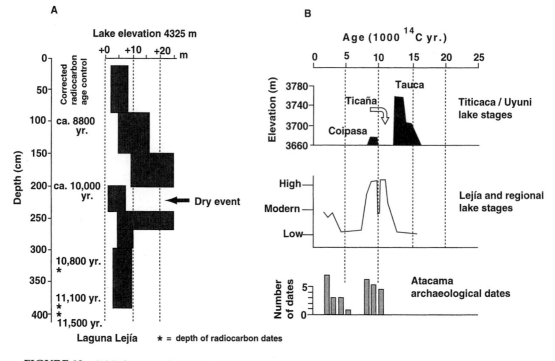

FIGURE 18   (A) Lake stage elevation histogram for Laguna Lejía in Chile. (B) Lake stage elevation curves for the Titicaca-Uyuni region in Bolivia and Laguna Lejía in Chile, and histogram of radiocarbon dates on archaeological material from the Atacama region of Chile. (Data from Geyh et al., 1999.)

By the end of the late glacial period, falling lake levels are recorded by outcrops of diatomaceous silt and clay exposed in the Arroyo Bebedero 5–10 m above the modern playa surface. Freshwater benthic diatoms in deposits dated at between 11,600 and 10,700 [14]C B.P. and between 10,100 and 9600 [14]C B.P. alternate with open- (deep-?) water, saline planktonic diatoms that apparently indicate lake-level fluctuations (Fig. 19). Low, but comparatively fresh lake stages may have reflected less saline base flow conditions in Arroyo Bebedero that supported marsh environments surrounding the basin and drained to a shallow saline playa at the basin center. Open-water, planktonic, saline diatoms (*Cyclotella choctawatcheeana*) characterize the intervening high lake stages (10,700–10,100 and 9600–9100 [14]C B.P.) and imply abundant surface flow from the saline Río Desaguadero that entered Salinas del Bebedero via Arroyo Bebedero (González and Maidana, 1998). A hiatus and archaeological remains on beach ridges within the basin 9100–8600 [14]C B.P. suggest a dry period, although the lake rose afterward and remained open and at generally intermediate stages for an unknown period according to the distribution of planktonic saline diatoms (González and Maidana, 1998).

Today, increased flow in the Río Desaguadero responds to precipitation and snowmelt in the Andes, mostly from westerly storms originating in the Pacific. For example, high snowfall during the 1983 El Niño caused major flooding of the Salinas del Bebedero (M.A. González, personal communication). Summer precipitation from Atlantic sources probably does not play an important role in the hydrologic balance of the playa today (González and Maidana, 1998), although it is possible that easterly moisture played an important role in the early Holocene freshwater marsh environments recorded by *Pseudostaurosira brevistriata* (Fig. 19).

16.2.19. Laguna Cari Laufquen: Río Negro Province, Argentina (41.13°S, 69.42°W; 800 m)

Radiocarbon-dated lake sediments and lithoid tufa on shorelines above the modern floor of Laguna Cari Laufquen describe a preliminary lake-level curve (Fig. 20) that extends and modifies earlier work on the paleolimnology of this basin (Galloway et al., 1988). The largest and most persistent high stage of the lake occurred between 18,400 and ca. 13,000 [14]C B.P., although the gravels representing the termination of this event

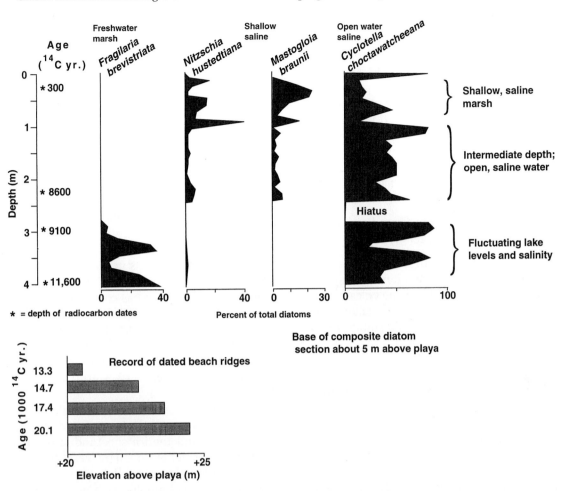

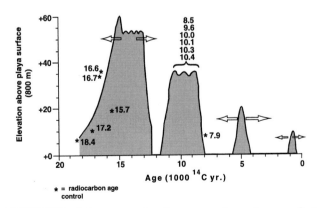

FIGURE 19   Selected diatom profiles from outcrops and elevation of dated beach ridges surrounding Salinas del Bebedero in Argentina. (Data from González, 1994; González and Maidana, 1998.)

are undated (Stine, unpublished). Apparently, the lake did not rise sufficiently at this time to persistently drain. Nevertheless, the 60-m high stand was probably within 10 m of the sill elevation. The calcareous clays

FIGURE 20 Lake stage elevation curve from Laguna Cari Laufquen in Argentina. (Data from S. Stine.)

and silts deposited by the lake at this high stand contain rare diatoms and ostracodes and indicate that the late glacial lacustrine system was turbid and saline, but not especially productive. The clear, productive water required to produce stromatolites was apparently confined to protected bays distant from the main sediment sources. Deposits from the second high stand (+35 m), centered on 10,000 ^{14}C B.P., suggest limnological conditions similar to those during the late glacial period. The abrupt termination of the 35-m stage has good age control and appears to correlate with the lake stage chronology of the Atacama region. Although clear stratigraphic and geomorphic evidence exists for minor Holocene lake rises, at this time their chronology is unknown.

Lake levels at Laguna Cari Laufquen respond to snowmelt and precipitation on mountain drainages east of the Andes. Presumably, the full glacial (?), late glacial, and early Holocene high stands at Cari Laufquen document greater precipitation at this latitude;

this increased precipitation was also expressed by Andean full and late glacial glacier advances recorded near Bariloche (e.g., at Aguado; Markgraf and Bianchi, 1999) and in Lago Mascardi at 15,000–13,000 [14]C B.P. (Ariztegui et al., 1997). The source of the precipitation is unclear. A westerly source conflicts with pollen evidence at this latitude west of the Andes (Markgraf et al., 1992). Southeasterly moisture related to increased activity of mobile polar highs or to cold air outbreaks from Antarctica and southernmost South America remains a possibility (e.g., Leroux, 1993).

16.2.20. Lago Cardiel: Santa Cruz Province, Argentina (49.0°S, 71.25°W; 276 m)

Lago Cardiel is a large (460 km²), turbid, deep (>70 m), endorheic lake in southern Patagonia occupying a volcano-tectonic depression in Neogene volcanic rocks and underlying Cretaceous sediments. Lake hydrochemistry is dominated by sodium and sulfate, but the comparatively fresh water (~3 g/L) suggests lake recharge to the groundwater table.

Shoreline dates from lithoid tufa indicate lake depths up to 55 m above present water levels between 9800 and 7700 [14]C B.P. (Fig. 21). Intermediate to low lake levels existed at 20,000 [14]C B.P., but transgressive deposits with a minimum age of 31,400 [14]C B.P. and possibly related strandlines indicate a very high stage preceding the full glacial period. Mid- to late Holocene lake levels higher than modern levels (10–20 m) imply

a complex and variable lacustrine history (Stine and Stine, 1990). At present, the lake continues to fall below recorded historical levels.

16.3. DISCUSSION

Despite many irregularities in dating, proxy interpretation, and site analysis, a number of records along the PEP 1 transect contain full glacial to early Holocene paleolimnological information that identifies the past distribution and strength of large-scale climate processes affecting lake levels and limnological processes (Fig. 22). The northernmost record, Lake Zagoskin, reflects a warming of global climate related to shrinkage of continental ice sheets in the Northern Hemisphere. The persistence of ice in Canada throughout much of the late glacial period has compromised the search for full and late glacial records at those latitudes. Presumably, such records would also document progressive limnologic change related mostly to ice disappearance and replacement of anticyclonic circulation around the ice sheet by jet stream- (westerly) dominated climate systems present today.

The Klamath Lake record from Caledonia Marsh appears to document the retreat of cold, dry, easterly circulation around the ice sheet-induced, high-pressure system and its replacement by westerly storm tracks. It is difficult to tell from this lake margin site to what extent late glacial limnological fluctuations reflect pro-

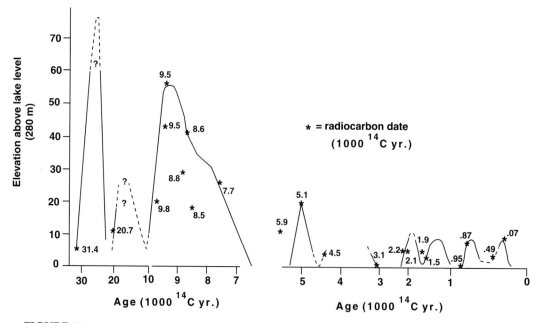

FIGURE 21 Lake stage elevation curves for Lago Cardiel in Argentina. Note variable age scale on abscissa. (Data from Stine and Stine, 1990.)

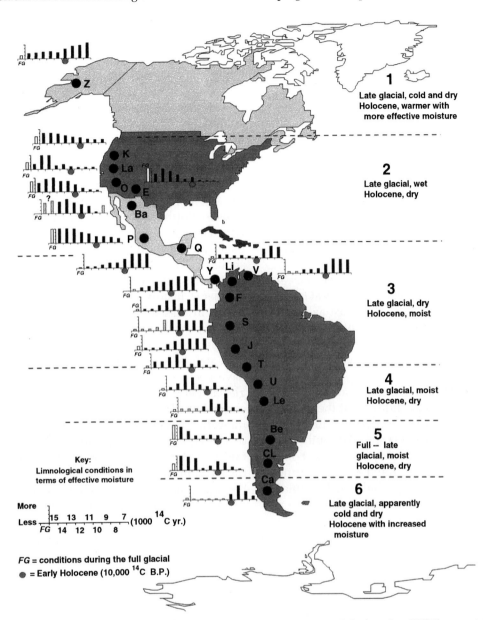

FIGURE 22   Summary of lacustrine changes along the Pole-Equator-Pole: Americas (PEP 1) transect. Black histograms represent generic lacustrine conditions (lake level, salinity, etc.) expressed in dimensionless terms of effective water balance as compared to modern conditions. Unfilled histograms represent full glacial (FG) lake conditions for comparison. Shaded histograms represent questionable interpretations based on data that are lacking. Red dots on histogram abscissas identify the early Holocene period (10,000 [14]C B.P.). Dashed horizontal lines reflect major zones of contrasting full to late glacial and Holocene lake conditions. Locations of sites are given in Fig. 1. (From Markgraf et al., 2000. With permission.)

gressive development of marsh seral stage communities around the lake. A midlake core, however, indicates that climate was the principal driver of limnological change at this site.

To the south, lakes in the Great Basin and southwestern United States were high and fresh during all or part of the full and late glacial periods as a consequence of displaced westerly storm tracks and cooler mean annual temperatures related to the residence of polar and arctic air masses in this region. The early Holocene witnessed desiccation in these basins as the storm tracks and air masses retreated northward following the progressive melting and retreat of the continental ice sheets. Chronologies require further refinement to demonstrate and track this progressive displacement, if indeed it happened. As the records now stand—including those from northern and central México, which also apparently received westerly moisture during the late glacial period—the retreat of this moisture source may have been very rapid. High lake levels at all sites between 13,000 and 12,000 [14]C B.P. fall by 10,000 [14]C B.P., although the southernmost site in this Northern Hemisphere grouping, Lago de Pátzcuaro, seems to have received increasing amounts of easterly moisture as the westerly sources declined. This decline resulted in little change in the lake level, but a clear change in the seasonality of diatom productivity. At Owens Lake, the diatom and ostracode records show that this basin fluctuated between full fresh and shallow saline conditions during the late glacial–Holocene transition (~11,000–9000 [14]C B.P.), with the changes taking place in less than a century. Such high-resolution records illustrate that the *progressive* retreat of climate systems responsible for maintaining large lakes in the Great Basin was not unidirectional, but involved at least as much variability as is seen today in the climate of this region.

Low-elevation Mesoamerican and northern South American lakes were all low or dry during the full glacial period and most of the late glacial period. Lake levels began to rise as easterly precipitation arrived or developed in this region. Dry conditions at these sites during the late glacial period are consistent with other Caribbean records, which suggest a south or southwest displacement of the Azores High and a consequent displacement or reduction of the ITCZ and tropical marine moisture sources (Bradbury, 1997). Hadley cell circulation may have been reduced by tropical cooling and a reduction of the latitudinal temperature gradient between the tropics and subtropics (Rind, 1998).

The filling of the Caribbean lake basins after 10,000 [14]C B.P. may represent a retreat of the Azores High and reestablishment of the northeasterly trade winds, which bring moisture to the region.

At the latitude of Central America and northern South America, there seems to be a west-to-east gradient in the timing of the transition to high lake conditions. La Yeguada appears to have risen, warmed, and freshened earlier than Laguna Los Lirios and Lago de Valencia at approximately the same latitude. The potential significance of this gradient could relate to the way in which the ITCZ was spatially manifested and strengthened by the end of the late glacial period. More well-dated sites in this region will help define the longitudinal gradients of climate change suggested by these data.

High-elevation fens and lakes in equatorial South America probably never suffered severe moisture deficits, but the abundance of *Isoetes* in late glacial and early Holocene records suggests cold temperatures and reduced productivity, perhaps associated with greater

and more persistent lake ice cover. The arrival of warmer temperatures that included easterly precipitation in the latest glacial and early Holocene periods had important limnological effects on lake productivity, sediment chemistry, and mineralogy that reflected increased weathering and nutrient delivery to these lacustrine systems. In general, these systems became more productive in climatic regimes of increased warmth and precipitation that sponsored greater coverage of Andean forests.

If the ITCZ and its associated easterly trade wind precipitation were forced south by full glacial conditions in North America (e.g., the Laurentide ice sheet), which thereby deprived low-latitude lakes of moisture, its residence at higher latitudes in South America should be registered in lake records south of Lago de Junín (11°S); this lake is the southernmost one with a lake-level history similar to that of the low-latitude basins in Mesoamerica and northern South America. Whereas the records from central South America (16°–23.5°S) show variable latest glacial and early Holocene high stands, none of the basins in this zone appear to have been high or particularly fresh during the full glacial period—although according to the available chronology, Lake Titicaca rose to a high stand shortly thereafter. Although high-elevation sites near or north of the equator remained relatively high during the mid-Holocene, high-elevation sites farther south (Titicaca, Uyuni, and Lejía) desiccated or at least shrank by the mid-Holocene. Presumably, the high Holocene levels of lakes north of 16°S latitude reflect an intensification or concentration of easterly moisture sources related to the release of latent heat at high latitudes. There is no convincing evidence that the easterly (trade wind) moisture source existed at higher latitudes in South America during full and late glacial times in anything resembling its present form.

Lake Titicaca and Salar de Uyuni have comparable records of a major late glacial high lake stage, called the Tauca stage, a dry (Ticaña) event, and a minor resurgence in the early Holocene period, the Coipasa event. The major high stand at Laguna Lejía, in contrast, seems to correlate more closely to the Coipasa event according to reliable radiocarbon dates. Its magnitude was considerably greater than the Coipasa event farther north. Although Lake Titicaca subsequently rose to comparatively high levels in the late Holocene, the Salar de Uyuni remained low (Sylvestre, 1997).

The source of moisture for these lake basins appears to have come from the east, probably as a result of moisture advected by the South Atlantic subtropical high-pressure zone. The South Atlantic Ocean was warm by the time of the Tauca high stage event and could serve as a moisture source for lakes at the latitude of Titicaca and Salar de Uyuni. The comparatively equatorial position of the South Atlantic subtropical high may reflect the diminishing effect of the Laurentide ice sheet at that time and the continuing role of Antarctic sea ice in determining gradients of atmospheric circulation. Possibly the large Coipasa-age lake stand at Laguna Lejía (11,000–9000 [14]C B.P.) records the diminishing effect of Antarctica on the latitude of the South Atlantic subtropical high and its subsequent migration farther poleward at the beginning of the Holocene—coincident with the development and strengthening of the easterly trade wind circulation patterns that characterize the region today.

The two Patagonian sites in southern South America (Salinas del Bebedero [33°S] and Laguna Cari Laufquen [41°S]) have similar lake-level histories reminiscent of the North American Great Basin pattern—high levels during the full and late glacial periods and lower levels by the early Holocene. These lakes are fed by rivers from the Andes. Today, this segment of the Andes receives westerly moisture primarily during winter, when storm tracks shift toward the equator (Markgraf et al., 1992). Consequently, higher lake levels at Bebedero in the past could have related to a more persistent location, and (or) intensification, of storm tracks at these latitudes. The extensive evidence of full and late glacial moraines (>13,900 [14]C B.P.) west of the Andes (Lowell et al., 1995) may testify to the importance of westerly precipitation in the Andes at that time. However, pollen records on the west side of the Andes (La Serena: 32°S; Puerto Octay: 40°S) and immediately east of the Andes (Gruta del Indio: 34.75°S; Aguado: 40°S) do not indicate increased moisture during the full and late glacial periods (Villagran and Varela, 1990; Moreno, 1997; D'Antoni, 1983; Markgraf and Bianchi, 1999). If westerly moisture sources were not responsible for high full and late glacial levels at Salinas del Bebedero and Laguna Cari Laufquen, it may become necessary to invoke southerly or easterly moisture associated with polar outbreaks to accomplish this (Leroux, 1993; Marengo and Rogers, 2000). Evidence of lower glacial snow lines and, hence, of more intensive glaciation on the east side of the Andes at latitudes 24°–28°S (Fox and Strecker, 1991) also suggest the possibility of easterly moisture sources at times in the past.

A single date on lithoid tufa at the highest latitude site, Lago Cardiel (49°S), indicates a full glacial lake stage marginally higher than average Holocene levels. No record of late glacial high stands exists for this site, but very high (+55 m) lake levels after 10,000 [14]C B.P. at Lago de Cardiel are consistent with a moisture increase due to lowered seasonality and a wide amplitude of storm tracks that reached to 50°S in response to warmer southern oceans, reduced sea ice, and a shallow thermal gradient between Antarctica and Patagonia.

16.4. CONCLUSIONS

Viewed from this very general, low-resolution perspective, full glacial–early Holocene lake records along the PEP 1 transect (Fig. 22) document only six major interacting climate scenarios (Fig. 23):

1. Unglaciated Arctic lake systems witness progressive warming related to deglaciation by a reduction in the thickness and seasonal extent of ice cover and a concomitant increase in productivity. Additional records could certainly show other climate/lake interactions related to changing moisture sources and winds, changing seasonality of insolation, and varying styles of land/lake interactions mediated by vegetation and erosion changes under climate control.

2. Midlatitude, western North American late glacial lakes responded to an intensification and greater persistence of westerly moisture delivered by storm tracks forced south by climate changes that encouraged the growth of ice sheets and southerly displacement of polar air masses.

3. South of central México and north of Lago Titicaca in southern Bolivia (16.0°S), all lacustrine systems seem to have been low or dry in the full glacial period, apparently as a consequence of substantially reduced ITCZ/ trade wind circulation systems that today provide moisture to these basins. Late glacial moisture increases characterized the western sites, whereas the timing of the arrival of easterly moisture approximately coincided with the beginning of the Holocene at the eastern lake sites (Lago de Valencia and Laguna Los Lirios). The spatial coherence and climatic significance of this apparent gradient must await additional study of well-dated records.

4. Lake records from southern South America at latitudes 16°–23.5°S were also low or dry during the full glacial period, but record important high stands during the late glacial period. The most notable of these (the Tauca lake stage) occurred at Lago Titicaca and Salar de Uyuni, but apparently not at Laguna Lejía (23.5°S), which rose to higher than modern levels only during the latest glacial and early Holocene. Barring unforeseen chronological problems at these sites, the reasons for this rise are not understood at present. They may relate to cool ocean temperatures coupled with the location of the northern, moisture-bearing limbs of high-pressure systems generated by polar outbreaks or other mechanisms. The oceans may have been too cold and

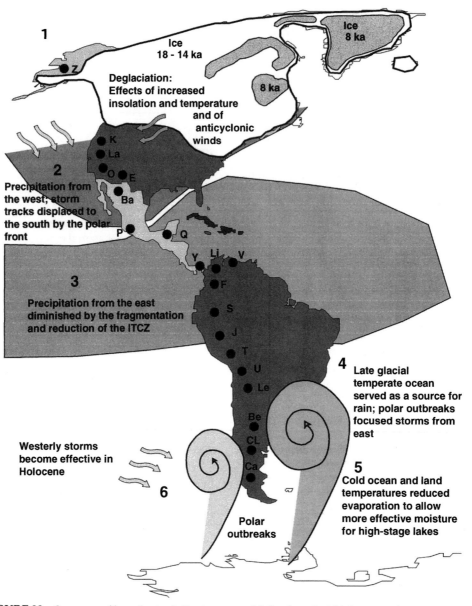

FIGURE 23 Summary of hypothesized climate causes of full to late glacial lake status changes compared to the Holocene along the Pole-Equator-Pole: Americas (PEP 1) transect. The numbers that identify roughly coherent paleoclimatic scenarios correspond to the numbered conclusions in Section 16.4. ka = thousands of radiocarbon years before present. Locations of sites are given in Fig. 1.

(or) the appropriate wind systems too far equatorward to provide moisture to Laguna Lejía until the early Holocene.

5. Between 30° and 41°S latitude, two lake systems, Salinas del Bebedero and Laguna Cari Laufquen, document high lake stages during the full and early late glacial periods and lower levels thereafter. These changes may also reflect moisture sources related to outbreaks of polar air and the interaction with low-pressure systems in the western Atlantic. Such pro-

cesses would have been most active and intense in the winter, and precipitation derived from them could have maintained lakes in otherwise arid environments. Cooler overall temperatures clearly contributed to a positive moisture balance. Low temperatures may have been more important than precipitation per se in reducing evaporation and allowing high lake stages.

6. South of latitude 45°S, Lago Cardiel is the only studied representative of lake systems with sediment and geomorphic evidence of past lake stages higher

than modern levels. Lago Cardiel has no record of full or late glacial stages higher than its present, comparatively low levels. If it was indeed low during the full and late glacial periods, the reason may be ascribed to a cold South Pacific Ocean and a northward displacement of westerly storm tracks beyond the latitude of Lago Cardiel. If so, higher than modern levels in the early Holocene period must reflect warm ocean sources of moisture finally reaching relevant latitudes to the south and the poleward retreat of the westerlies to deliver the moisture to the lake as happens today.

At this scale of resolution, it appears that past interhemispheric climate interactions in the Western Hemisphere are synchronous, but not necessarily of the same sign, and are therefore linked, at least indirectly. The massive Laurentide/Cordilleran ice sheet or those processes that sustained it may have forced climate changes throughout the western Northern Hemisphere and perhaps even as far south as central South America. The southerly displacement of westerly storm tracks in North America has been modeled (Kutzbach, 1987), and western North America lake records register this displacement. Easterly precipitation sources associated with the ITCZ diminished in importance or disappeared altogether, perhaps as a consequence of general cooling in tropical and subtropical regions. At least no records in northern and central South America studied so far appear to document increased easterly (ITCZ) precipitation during full or early late glacial times. To account for dry conditions during the full and late glacial periods, the ITCZ is supposed to have resided farther north by South American workers (e.g., Martin et al., 1997) and farther south by North American workers (e.g., Bradbury, 1997). Perhaps the resolution to this problem lies in the weakening, restriction, and redirection of easterly moisture sources by the effects of mobile polar highs as has been suggested by Leroux (1993).

A late glacial expansion of southern South American lakes as a result of intensified and more northerly displaced westerly moisture sources conflicts with pollen evidence at the same latitude west of the Andes. As a consequence, precipitation of moisture from the southwest Atlantic, related to outbreaks of polar air originating over a very cold Antarctica, may have provided a positive hydrologic balance for these lacustrine systems. The synchroneity of midlatitude lake-level records in both hemispheres probably relates to much colder polar climates ultimately thought to be caused by reduced heat transport to the north and south related to reduced thermohaline circulation controlled by glacial meltwater processes in the North Atlantic Ocean (e.g., Alley, 1995).

To provide relevant input to interhemispheric paleoclimate interactions and histories, future paleolimnological studies must concentrate on high-resolution, multiproxy analyses, including coherent, fine-scale sediment facies analyses with chronological control at millennial scales. Most important, paleolimnological records must be convincingly related to local and regional climate controls through both neolimnological and climate monitoring.

References

Abbott, M. B., M. W. Binford, M. Brenner, and K. R. Kelts, 1997: A 3500 ^{14}C yr high-resolution record of water-level changes in Lake Titicaca, Bolivia/Peru. *Quaternary Research*, **47**: 70–80.

Ager, T. A., 1982: Vegetational history of western Alaska during the Wisconsin glacial interval and the Holocene. *In* Hopkins, D. M., J. V. Mathews, Jr., C. E. Schweger, and S. B. Young (eds.), *Paleoecology of Beringia*. New York: Academic Press, pp. 75–93.

Ager, T. A., and L. Brubaker, 1985: Quaternary palynology and vegetational history of Alaska. *In* Bryant, V., Jr., and R. Holloway (eds.), *Pollen Records of Late Quaternary North American Sediments*. Dallas, TX: American Association of Stratigraphic Palynologists Foundation, pp. 353–384.

Allen, B. D., and R. Y. Anderson, 1993: Evidence from western North America for rapid shifts in climate during the last glacial maximum. *Science*, **260**: 1920–1923.

Alley, R. B., 1995: Resolved, the Arctic controls global climate change. In Smith, W. O., and J. M. Grebmeier (eds.), Arctic oceanography: Marginal ice zones and continental shelves. *Coastal and Estuarine Studies*, **49**: 263–283.

Ariztegui, D., M. M. Bianchi, J. Masafero, E. Lafargue, and F. Niessen, 1997: Interhemispheric synchrony of late glacial climatic instability as recorded in proglacial Lake Mascardi, Argentina. *Journal of Quaternary Sciences*, **12**(4): 333–338.

Bachhuber, F. W., 1982: Quaternary history of the Estancia Valley, central New Mexico. New Mexico Geological Society Guidebook, 33rd Field Conference, pp. 343–346.

Bahls, L. C., E. E. Weber, and J. O. Jarvie, 1984: Ecology and distribution of major diatom ecotypes in the southern Fort Union Coal Region of Montana. U.S. Geological Survey Professional Paper 1289, 151 pp.

Barnosky, C. W., P. M. Anderson, and P. J. Bartlein, 1987: The northwestern U.S. during deglaciation: Vegetational history and paleoclimatic implications. *In* Ruddiman, W. F., and H. E. Wright, Jr. (eds.), *North America and the Adjacent Oceans During the Last Deglaciation*, Vol. K-3, *The Geology of North America*. Boulder, CO: Geological Society of America, pp. 289–321.

Benson, L. V., D. R. Currey, R. I. Dorn, K. R. Lajoie, C. G. Oviatt, S. W. Robinson, G. I. Smith, and S. Stine, 1990: Chronology of expansion and contraction of four Great Basin lake systems during the past 35,000 years. *Palaeogeography, Palaeoclimatology, Palaeoecology*, **78**: 241–276.

Benson, L., D. Currey, Y. Lao, and S. Hostetler, 1992: Lake-size variations in the Lahontan and Bonneville basins between 13,000 and 9000 ^{14}C years B.P. *Palaeogeography, Palaeoclimatology, Palaeoecology*, **95**: 19–32.

Bradbury, J. P., 1991: Late Cenozoic diatom stratigraphy and paleolimnology of Tule Lake, Siskiyou Co. California. *Journal of Paleolimnology*, **6**: 205–255.

Bradbury, J. P., 1997: Sources of glacial moisture in Mesoamerica. *Quaternary International*, **43/44**: 97–110.

Bradbury, J. P., in press: Limnologic history of Lago de Pátzcuaro, Michoacán, México for the past 48,000 years; impacts of climate and man. *Palaeogeography, Palaeoclimatology, and Palaeoecology.*

Bradbury, J. P., and R. M. Forester, in press: Environment and paleolimnology of Owens Lake, California: A record of climate and hydrology for the past 50,000 years. Washington, DC: Smithsonian Institution.

Bradbury, J. P., B. W. Leyden, M. Salgado-Labouriau, W. M. Lewis, Jr., C. Schubert, M. W. Binford, D. G. Frey, D. R. Whitehead, and F. H. Weibezahn, 1981: Late Quaternary environmental history of Lake Valencia, Venezuela. *Science,* **214**: 1299–1305.

Bush, M. B., D. R. Piperno, P. A. Colinvaux, P. E. de Oliveira, L. A. Krissek, M. C. Miller, and W. E. Rowe, 1992: A ^{14}C 300-year paleoecological profile of a lowland tropical lake in Panama. *Ecological Monographs,* **62**: 251–275.

Colinvaux, P. A., M. B. Bush, M. Steinitz-Kannan, and M. C. Miller, 1997: Glacial and postglacial pollen records from the Ecuadorian Andes and Amazon. *Quaternary Research,* **48**: 69–78.

Cumming, B. F., S. E. Wilson, R. J. Hall, and J. P. Smoll, 1995: Diatoms from British Columbia (Canada) lakes and their relationship to salinity, nutrients, and other limnological variables. *Bibliotheca Diatomologica,* **31**: 207.

D'Antoni, H. L., 1983: Pollen analysis of Gruta del Indio. *Quaternary of South America and Antarctic Peninsula,* **1**: 83–104.

Déletang, L. F., 1929: Contribución al estudio de las salinas Argentinas. La Salina del Bebedero y sus relaciones con el sistema hidrográfico Andino o de Desaguadero: Ministerio de Agricultura de la Nación, Division General de Minas, Geología e Hidrología, Publicación 47, pp. 1–69.

Fox, A. N., and M. R. Strecker, 1991: Pleistocene and modern snowlines in the central Andes (24–28°S). *Bamberger Geographische Schriften,* **11**: 169–182.

Galloway, R. W., V. Markgraf, and J. P. Bradbury, 1988: Dating shorelines of lakes in Patagonia, Argentina. *Journal of South American Earth Sciences,* **1**: 195–198.

Geyh, M. A., M. Grosjean, L. Nuñez, and U. Schotterer, 1999: Radiocarbon reservoir effect and the timing of the late-glacial/early Holocene humid phase in the Atacama Desert, northern Chile. *Quaternary Research,* **52**: 143–153.

González, M. A., 1994: Salinas del Bebedero Basin (República Argentina). *In* Kelts, K., and E. Gierlowski-Cordesch (eds.), *Global Inventory of Lake Basins.* Cambridge, U.K.: Cambridge University Press, pp. 381–386.

González, M. A., and N. I. Maidana, 1998: Post Wisconsin paleoenvironments at Salinas del Bebedero basin (33°20′ S, 66°45′ W, 380 m asl; San Luis, Argentina). *Journal of Paleolimnology,* **20**: 353–368.

Grosjean, M., and A. L. Nuñez, 1994: Late glacial, early and middle Holocene environments, human occupation and resource use in the Atacama (Northern Chile). *Geoarchaeology,* **9**: 271–286.

Grosjean, M., M. A. Geyh, B. Messerli, and U. Schotterer, 1995: Late glacial and early Holocene lake sediments, ground-water formation and climate in the Atacama Altiplano 22–24°S. *Journal of Paleolimnology,* **14**: 241–252.

Hansen, B. C. S., H. E. Wright, and J. P. Bradbury, 1984: Pollen studies in the Junín area, central Peruvian Andes. *Geological Society of America Bulletin,* **95**: 1454–1465.

Helmens, K. F., P. Kuhry, N. W. Rutter, K. Van Der Borg, and A. F. De Jong, 1996: Warming at 18,000 yr B.P. in the tropical Andes. *Quaternary Research,* **45**: 289–299.

Kutzbach, J. E., 1987: Model simulations of the climatic patterns during the deglaciation of North America. *In* Ruddiman, W. F., and H. E. Wright (eds.), *North America and Adjacent Oceans During the Last Deglaciation,* Vol. K-3, *The Geology of North America.* Boulder, CO: Geological Society of America, pp. 425–446.

Lee, W. T., 1906: Geology and water resources of Owens Valley, California. U.S. Geological Survey Water Supply and Irrigation Paper 181, 28 pp.

Leroux, M., 1993: The mobile polar high: A new concept explaining present mechanisms of meridional air-mass and energy exchanges and global propagation of paleoclimatic changes. *Global Planetary Change,* **7**: 69–93.

Leyden, B. W., 1985: Late Quaternary aridity and Holocene moisture fluctuations in the Lake Valencia basin, Venezuela. *Ecology,* **66**(4): 1279–1295.

Leyden, B. W., M. Brenner, D. A. Hodell, and J. H. Curtis, 1993: Late Pleistocene climate in the Central American lowlands. *In* Swart, P. K., K. C. Lohmann, J. McKenzie, and S. Savin (eds.), *Climate Change in Continental Isotopic Records,* Washington, DC: American Geophysical Union, Geophysical Monograph 78, pp. 165–178.

Lowell, T. V., C. J. Heusser, B. G. Andersen, P. I. Moreno, A. Hausser, L. E. Heusser, D. R. Schlüchter, D. R. Marchant, and G. H. Denton, 1995: Interhemispheric correlation of late Pleistocene glacial events. *Science,* **269**: 1541–1549.

Marengo, J. A., and J. C. Rogers, 2000: Polar air outbreaks in the Americas: Assessments and impacts during modern and past climates. *In* Markgraf, V. (ed.), *Interhemispheric Climate Linkages.* San Diego: Academic Press, Chapter 3.

Markgraf, V., and M. M. Bianchi, 1999: Paleoenvironmental changes during the last 17,000 years in western Patagonia: Mallín Aguado, Province of Neuquen, Argentina. *Bamberger Geographische Schriften,* **19**: 175–193.

Markgraf, V., Baumgartner, T. R., Bradbury, J. P., Diaz, H. F., Dunbar, R. B., Luckman, B. H., Seltzer, G. O., Swetnam, T. W., Villalba, R., 2000: Paleoclimate reconstruction along the Pole-Equator-Pole transect of the Americas (PEP 1). *Quaternary Science Reviews,* **19**: 125–140.

Markgraf, V., J. R. Dodson, A. P. Kershaw, M. S. McGlone, and N. Nichols, 1992: Evolution of late Pleistocene and Holocene climates in the circum–South Pacific lake areas. *Climate Dynamics,* **6**: 193–211.

Martin, L., J. Bertaux, T. Correge, M.-P. Ledru, P. Mourguiart, A. Sifeddine, F. Soubies, D. Wirrmann, K. Suguio, and B. Turc, 1997: Astronomical forcing of contrasting rainfall changes in tropical South America between 12,400 and 8800 cal yr B.P. *Quaternary Research,* **47**: 117–122.

Messerli, B., M. Grosjean, K. Graf, U. Schotterer, H. Schreier, and M. Vuille, 1992: Die Veränderungen von Klima und Umwelt in der Region Atacama (Nordchile) Seit der letzten Kaltzeit. *Erdkunde,* **46**: 257–272.

Metcalfe, S. E., A. Bimpson, A. J. Courtice, S. L. O'Hara, and D. M. Taylor, 1997: Climate change at the monsoon? Westerly boundary in northern México. *Journal of Paleolimnology,* **17**: 155–171.

Moreno, P. I., 1997: Vegetation and climate near Lago Llanquihue in the Chilean Lake District between 20 200 and 9500 ^{14}C yr BP. *Journal of Quaternary Science,* **12**: 485–500.

Rind, D., 1998: Latitudinal temperature gradients and climate change. *Journal of Geophysical Research,* **103**: 5943–5971.

Servant, M., and J. C. Fontes, 1978: Les lacs quaternaires des hauts plateaux des Andes Boliviennes. Premières interprétations paléoclimatiques. *Cahiers ORSTOM, Série Géologie,* **10**(1): 9–23.

Servant, M., M. Fournier, J. Argollo, S. Servant-Vildary, F. Sylvestre, D. Wirrmann, and J.-P. Ybert, 1995: La denière transition glaciaire/interglaciaire des Andes tropicales sud (Bolivie) d'après l' étude des variations des niveaux lacustres et des fluctuations glaciaires. *Comptes Rendus Académie des Sciences de Paris, ser. II a,* **320**: 729–736.

Servant-Vildary, S., 1992: The diatoms. *In* Dejoux, C., and A. Iltis (eds.), *Lake Titicaca: A Synthesis of Limnological Knowledge.* Dordrecht: Kluwer Academic, pp. 163–176.

Servant-Vildary, S., and M. Roux, 1990: Multivariate analysis of diatoms and water chemistry in Bolivian saline lakes. *Hydrobiologia,* **197**: 267–290.

Stine, S., and M. Stine, 1990: A record from Lake Cardiel of climate change in southern South America. *Nature,* **345**: 705–707.

Sylvestre, F., 1997: Le denier transition glaciaire-interglaciaire (18000–8 000 ^{14}C ans B.P.) des Andes tropicales sud Bolivie) d'après l'étude des diatomées: These doctoral Museum National d'Histoire Naturelle (Paris), 243 pp.

Sylvestre, F., S. Servant-Vildary, M. Fournier, and M. Servant, 1996: Lake-levels in the southern Bolivian Altiplano (19°–21°S) during the late glacial based on diatom studies. *International Journal of Salt Lake Research,* **4**: 281–300.

Sylvestre, F., M. Servant, S. Servant-Vildary, C. Causse, and M. Fournier, 1999: Lake-level chronology in the south Bolivian Altiplano (18°–23°S) during late-glacial time and the early Holocene. *Quaternary Research,* **51**: 54–66.

Thompson, R. S., C. Whitlock, P. J. Bartlein, S. P. Harrison, and W. G. Spaulding, 1993: Climatic changes in the western United States since 18,000 yr B. P. *In* Wright, H. E., Jr., J. E. Kutzbach, T. Webb, III, W. F. Ruddiman, F. A. Street-Perrot, and P. J. Bartlein (eds.), *Global Climates Since the Last Glacial Maximum.* Minneapolis: University of Minnesota Press, pp. 468–513.

Van Devender, T. R., 1990: Late Quaternary vegetation and climate of the Chihuahuan Desert, United States and México. *In* Betancourt, J. L., T. L. Van Devender, and P. S. Martin (eds.), *Packrat Middens.* Tucson: University of Arizona Press, pp. 104–133.

Van Geel, B., and T. Van der Hammen, 1973: Upper Quaternary vegetational and climatic sequence of the Fúquene area (eastern cordillera, Colombia). *Palaeogeography, Palaeoclimatology, Palaeoecology,* **14**: 9–92.

Vázquez-Selem, L., 1998: Glacial chronology of Iztaccíhuatl volcano, Central México, based on cosmogenic [36]Cl exposure ages and tephrochronology [abs.]. American Quaternary Association Program and Abstracts of the 15th Biennial Meeting, Puerto Vallarta, México, p. 174.

Villagran, C., and Varela, J., 1990: Palynological evidence for increased aridity on the central Chilean coast during the Holocene. *Quaternary Research,* **34**: 198–207.

Watts, W. A., and J. P. Bradbury, 1982: Paleoecological studies at Lake Pátzcuaro on the west-central Mexican Plateau and at Chalco in the Basin of México. *Quaternary Research,* **17**: 56–70.

Weingarten, B., 1988: Geochemical and clay-mineral characteristics of lake sediments from the Venezuelan Andes: Modern climatic relations and paleoclimatic interpretation. Ph.D. thesis. University of Massachusetts, Amherst, 213 pp.

Wirrmann, D., 1987: El Lago Titicaca: Sedimentología y paleohidrología durante el Holoceno (10 000 años B.P. actual). Informe, UMSA, ORSTOM, La Paz, 6, 61 pp.

Wirrmann, D., P. Mourguiart, and L. F. de Oliveira Almeida, 1988: Holocene sedimentology and ostracodes repartition in Lake Titicaca. Paleohydrological interpretations. *In* Rabassa, J. (ed.), *Quaternary of South America and Antarctic Peninsula,* **6**: 89–127.

Ybert, J. P., 1992: Ancient lake environments as deduced from pollen analysis. *In* Dejoux, C., and A. Iltis (eds.), *Lake Titicaca: A Synthesis of Limnological Knowledge.* Dordrecht: Kluwer Academic, pp. 49–62.

Vegetation Evidence

CHAPTER

17

Paleotemperature Estimates for the Lowland Americas Between 30°S and 30°N at the Last Glacial Maximum

M. B. BUSH, M. STUTE, M.-P. LEDRU, H. BEHLING, P. A. COLINVAUX,
P. E. DE OLIVEIRA, E. C. GRIMM, H. HOOGHIEMSTRA, S. HABERLE,
B. W. LEYDEN, M.-L. SALGADO-LABOURIAU, AND R. WEBB

Abstract

Paleoecological data for the lowland neotropics and subtropics between 30°N and 30°S are compiled to provide an overview of climatic conditions at the time of the last glacial maximum. A clear consensus emerges from both fossil pollen and noble gas proxies that lowland climates were ca 5°C cooler at 18,000 [14]C B.P. than they are now. In many records, this period appears to have been the coldest portion of the last glacial episode, but in others, periods between 33,000 and 30,000 [14]C B.P. and between 14,000 and 12,000 [14]C B.P. appear to have been somewhat cooler. These data suggest that while there was a general cooling across the lowlands, local climates were subject to periodic warmer and cooler events. More data are needed to determine the effects of such climatic oscillations on biotic assemblages in the lowland neotropics and subtropics.

Within the community of paleoclimatologists, considerable debate continues regarding the amount of precipitation reduction at the last glacial maximum. Some drying is probable, but more data and improved methods to quantify paleoprecipitation are needed to resolve the current debate. Copyright © 2001 by Academic Press.

Resumen

Los datos paleoclimáticos de las zonas bajas neotropicales entre los 30°N y 30°S han sido recogidos para proporcionar una sinopsis de las condiciones climáticas en tiempos del último máximo glacial. Un evidente consenso aparece, con datos de polen fósil y de gases nobles, respecto a que estos climas tenían 5°C menos en 18.000 [14]C B.P. que en la actualidad. En muchos registros, este periodo parece haber sido el momento más frío del último episodio glacial, pero en otros, los periodos entre 33.000 y 30.000 [14]C B.P., y entre 14.000 y 12.000 [14]C B.P. parecen haber sido algo más fríos. Estos datos sugieren que mientras tenía lugar un enfriamiento general en las zonas de estudio, los climas locales estuvieron sujetos a periódicos eventos cálidos y fríos. Serán necesarios más datos para determinar los efectos de las oscilaciones climáticas en los conjuntos bióticos en la zona tropical y subtropical.

En el seno de la comunidad de paleoclimatólogos, continua el debate con respecto a la reducción de la cantidad de precipitación en el último máximo glacial. Es probable alguna sequía, pero más datos y la mejora de los métodos para cuantificar la paleoprecipitación son necesarios para resolver el actual debate.

17.1. INTRODUCTION

An accurate representation of the temperature gradients from the poles to the equator is fundamental to deriving reliable climate models. A persistent problem for modelers is to establish appropriate temperature gradients for the past. It is probably fair to say that both research effort and confidence in paleotemperature reconstructions are positively correlated with increasing latitude. Robust temperature reconstructions for the last 18,000 years at high latitudes are based on core records from deep ocean sediment, ice caps, and lake sediments. In the midlatitudes of the Americas (30°–70°N), an abundance of marine oxygen isotope and foraminiferal records provide data on past sea surface temperature (SST) (e.g., Broecker, 1986; CLIMAP Project Members, 1976, 1981). Terrestrial and nearshore sedimentary records have drawn on a wide array of proxy indicators that provide paleotemperature records for the land (e.g., Davis, 1981; Webb, 1987; Heusser, 1995; Haberle, 1998; Smith and Betancourt, 1998). On the whole, the temperature reconstructions of midlatitude terrestrial and oceanic systems are in close agreement (e.g., CLIMAP Project Members, 1981; Kutzbach and Guetter, 1986). In South America, at equivalent latitudes, a growing body of palynological, glaciological, and entomological data provide convincing paleotemperature reconstructions from the present to 18,000 [14]C B.P. (e.g., Markgraf, 1993). However, comparatively few records from nearshore or terrestrial environments are available for the subtropics to the equator (30°–0°) in either hemisphere.

Paleoclimate data for tropical regions began to appear in the 1960s with fossil pollen records from montane Colombia (van der Hammen and Gonzalez, 1960) and Costa Rica (Martin, 1964). Mercer began investigations of relict moraines in Peru (e.g., Mercer and Palacios, 1977) and spawned an active field of Andean moraine-based reconstructions of high-elevation cooling (e.g., Clapperton, 1987; Seltzer, 1990). In the 1980s, further data were gathered from the Sabana de Bogotá and other sites in Andean Colombia (Hooghiemstra, 1984, 1989) and the Junín Plateau of Peru (Hansen et al., 1984). By the mid-1980s, a consensus emerged of high-elevation cooling during glacial times. Some glaciological data from Ecuador suggested that mid-Pleistocene (ca. 40,000–30,000 [14]C B.P.) ice advances had pushed further downslope than those of 18,000 [14]C B.P. (Clapperton, 1987). However, it was not clear whether these lowermost moraines were the product of lower temperatures or greater moisture availability. Van der Hammen and Hooghiemstra (2000) suggest a cooling of ca. 8°C in the Andes at 18,000 [14]C B.P. and suggest that this was the coolest period documented in the late Pleistocene at the Sabana de Bogotá.

The CLIMAP (1976) SST estimates coincided with conventional biogeographic wisdom (e.g., Haffer, 1969; Vanzolini, 1970), which demanded that lowland tropical temperatures remained constant, or nearly so, during glacial cycles. However, this hypothesis remained untested because of logistic difficulties and technical doubts about the feasibility of lowland tropical palynology (Faegri, 1966). The first data that provided a glimpse of glacial conditions from the lowland neotropics were equivocal.

Van der Hammen and his team pioneered lowland palynology in South America with core sections from Ogle Bridge, Guyana, and the Alliance Borehole, Surinam (van der Hammen, 1963, 1974; Wijmstra, 1969). In both sequences the records show Poaceae-rich pollen spectra replacing mangrove and mesic forest pollen. A reasonable interpretation is that closed forest was replaced by more xeric communities. However, as neither sequence is dated (apart from one date of >45,000 [14]C B.P.), it is impossible to assign these spectra to any particular pre-Holocene interval. Similarly, two undated diagrams from Rondônia on the southern fringe of forested Amazonia apparently depict a northward range expansion of savanna elements. Absy and van der Hammen (1976) attributed the savanna expansion to glacial age aridity. Though tantalizing, these data sets did not provide reliable climatic data for the last glacial period. Better evidence for Pleistocene drying came from the Lake Valencia record, Venezuela, which suggests climates were cooler and drier during glacial times than now (Bradbury et al., 1981).

In this chapter, we provide an overview of a growing body of well-dated paleoecological data for the lowland (<1200 m elevation) neotropics and provide a consensus estimate of lowland neotropical paleotemperature at 18,000 [14]C B.P.

17.2 METHODS

The principal techniques used to date in reconstructions of terrestrial paleotemperatures for the lowland neotropics have been pollen in lake sediments and noble gases dissolved in groundwater. The collection and processing of pollen samples are broadly standardized around the techniques outlined by Faegri and Iversen (1989). The area of methodology that is worthy of review is in the quantification of paleotemperature.

Lowland palynological temperature reconstructions are generally based on the movement of indicator taxa. Apparently, stenothermic pollen taxa, or ones that at least occupy identifiable habitat ranges, are selected as indicator species. The modern range of these taxa is then compared with past ranges, and from this an inference is made about past climate. Van der Hammen and Gonzalez (1960) were the first researchers to correlate the downslope movement of pollen taxa to changes in temperature. They achieved this translation through the application of moist air adiabatic lapse rates. For instance, if in the past a species was documented 1000 m downslope of its present position, and the local moist air adiabatic lapse rate was 5°C of cooling per 1000 m of ascent, the paleotemperature change was 5°C cooler than present values.

Most researchers have since adopted this method of calculating temperature change, but we should examine some of its assumptions.

The first assumption is that there has been no significant evolutionary change in the requirements of the pollen taxa. Given the short time interval (20–50 tree generations since the last glacial interval), it is unlikely that evolution is a major problem, especially when more than one pollen taxon is exhibiting the same trend.

A second assumption is that the moist air adiabatic lapse rate has not changed. Studies of moist air adiabatic lapse rates reveal that they are rigidly constrained by the physical properties of air and are unlikely to have wavered outside of a narrow range (Webster and Streten, 1978; Rind and Peteet, 1985). Moist air associated with cloud forests and the wet lowland forests has a lapse rate of ca. 5°C. At the other extreme, desert dry air can have lapse rates approaching 7°C (Webster and Streten, 1978). Thus, the greatest potential change would be 2°C/1000 m of ascent. For many years, changes in ice age lapse rates were used to explain the anomalously warm oceans compared with the cool Andes (e.g., Haffer, 1991). However, as will be demonstrated later, when there is evidence of cooling, the forests are mesic or humid, indicating the presence of moist air. In other words, the *Alnus, Hedyosmum, Weinmannia, Podocarpus,* and *Drimys* populations that spilled down the flank of the Andes were of species adapted to the moist conditions of the cloud forest in which they now live. With paleoecological evidence to show that elevations as low as 1000 m above sea level (asl) had saturated air, it is not possible to discount evidence of cooling on the basis of steepened lapse rates.

A further criticism of using pollen to describe annual average paleotemperature is that the range of plants (therefore the elevation at which they grow) is determined not by mean annual temperatures, but by absolute minima. The distribution of plants is determined by the coldest night they survive rather than mean temperature. One way to test whether observing minimum temperatures rather than mean temperatures would

yield more information on species ranges is to plot both mean and minima data against elevation, and fit a regression line for each set of values. If the regression lines have a similar slope, then it is legitimate to use plant ranges to derive mean temperature values. Of course, the minima are likely to be more ecologically revealing in terms of determining the cause of the distribution, but that is a separate issue.

Detailed long-term climatic data on temperature maxima and minima are scarce for the neotropics, but a data set that provides a transect of daily minimum temperatures from Manaus, across lowland Ecuador, to the crest of the Andes is shown in Fig. 1. This data set is far from perfect, and some records were kept for only a few years. A relatively short run of data will not affect the mean temperature values, but may underestimate occasional bouts of extreme cooling. However, the lowland records did include an episode of *friagem* cooling, and it is unlikely that much lower temperatures would be experienced under modern conditions. As a first approximation, this data set clearly makes the point that tropical temperature minima are generally closely correlated to mean temperatures.

Climatic requirements of *Araucaria*—e.g., mean winter temperature, number of days of frost, and length of dry season—are used by Ledru (1991, 1992, 1993) and De Oliveira (1992) to infer paleoclimates associated with a Pleistocene range expansion of this genus. If it is assumed that *Araucaria* distributions are bound by these variables, a comparison of climatic data from the modern range with that of the Pleistocene range pro-

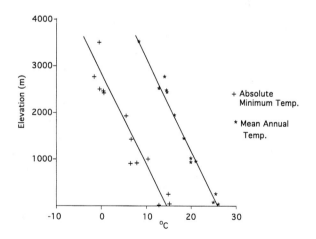

FIGURE 1 Modern mean annual and minimum temperatures for Ecuadorian weather stations (Centro Ecuadoriano de Investigación Geográfica [CEDIG] 1983) plotted against elevation. The regression line through the mean temperature data represents a 5°C/1000 m ascent, representing moist air adiabatic lapse rate. The line through the minima data represents a best-fit regression line. Note how the two lines for minimum and mean temperatures are virtually parallel.

vides estimates of changes in temperature and precipitation. This technique is freed from assumptions about lapse rates and, therefore, provides a valuable alternate means to measure paleotemperature. The above-cited authors used the movement or expansion of *Araucaria* forest from southern into southeastern Brazil (20°–25°S) to infer past-climate change. Behling and Lichte (1997) adopted a similar technique as they documented the movement, or expansion, of subtropical grassland from southern into southeastern Brazil. They found Pleistocene assemblages rich in subtropical grassland species approximately 7° of latitude farther north than their present range. Basing their climatic inference on modern weather data for the two areas, they infer an ice age cooling of between 4° and 8°C.

Another way to assess temperature using whole community values rather than indicator species has formed the basis of conventional transfer functions (e.g., Imbrie and Kipp, 1971; CLIMAP Project Members, 1976; Bonnefille et al., 1990). A number of problems are inherent in this approach, such as the lack of modern analogs for past assemblages and an inherent tendency toward underestimating any change. Many pollen taxa within an assemblage provide no detailed climatic information and can be regarded as catholic. If a full range of analog sites existed, the diluting effect of many catholic species would not matter, but without a full array of analogs, the presence of catholic species inevitably moderates the signal of climate change. The solution is to exclude the catholic species from the analysis and use a selection of stenothermic species. This compromise between using single indicator species and whole communities can be used to estimate response surfaces for precipitation or temperature. This technique could provide a paleothermometer that is independent of lapse rates.

It has been suggested that biome boundaries could be used to model past-climate change. The strength of the biome approach is that it is independent of lapse rates, it does not rely on modern communities being exactly those of the past (though intermediate vegetation types between recognizable modern biome types are a problem), and it should reduce subjectivity in interpretation. However, this technique also has problems that are particularly severe in the tropical lowlands (Marchant et al., in press).

Biome models assume that there will always be a biome to replace the existing one, but in the case of the lowland tropics there is none. Applying such models to lowland tropical paleoecology brings into focus a philosophical problem inherent in the concept of the biome—that modern conditions are normal. But, they are not. Glacial age conditions were the norm of the last 2 million years, and modern times are oppressively hot.

At 0° latitude and at 50 m elevation in the middle of the vast Amazonian plain, there is nowhere to retreat when it gets warmer. Some of the most stenothermic species that flanked the Andes escaped upslope to cooler climates at the beginning of the Holocene and will stay there until normal conditions return. The majority of species stay where they are because there are no *hotter adapted* species to displace them. Thus, the lowland tropics are unique—they really cannot show a warmer than usual signal (remember glacial conditions are the norm), other than an upslope migration of a few species. Biome models may be more appropriate in other settings, but they will fail in the lowland tropics because the lowland tropic biome is an endpoint in the biome continuum.

A second problem with taking the results of biome models at face value is more mechanical. Because the models treat a biome as a uniform climatic mass, the only changes indicated are when one biome replaces another. In other words, two regions that occur within the same biome—say, Atlanta and New York, which both occur in a temperate forest biome—would be accorded the same climate. It is clear that there could be substantial climatic change and yet no change in biome. Where biomes do change, relatively massive changes in climate are inferred. Neighboring areas that experience similar climatic change, but are judged to be biome constant, are suggested to have had a constant climate. Not all climate effects are geographically gradual, but we suggest the biome is too coarse a descriptive unit to elucidate paleoclimatic change in the tropics.

The only possible biome change that could be registered in Amazonia would be a transition from forest to savanna. Clearly, it is unsatisfactory to reduce all possible climatic variants to a simple "either savanna or forest." Under this kind of biome construction, vast areas will show no climatic change, and within the constructs of their model they are precisely correct. During the last glacial period, savanna did not replace large areas of forest, nor did lowland forests give way to Paramo grasslands or even to montane forest. Given the observed vegetation changes documented in the Amazon basin, over the greatest portion of the area there were no changes in biome; but this does not mean that there was not a significant change in temperature or precipitation.

17.3. NOBLE GASES DISSOLVED IN GROUNDWATER

In recent years, a new approach that provides an independent paleotemperature proxy has evolved, i.e., the measurement of atmospheric noble gases dissolved in radiocarbon-dated groundwater. The principle of the "noble gas thermometer" is based on the temperature dependency of the solubility of noble gases (Ne, Ar, Kr, and Xe, and in certain cases also N_2) in water (Mazor, 1972; Andrews and Lee, 1979; Stute and Schlosser, 1993). Groundwater percolating through the unsaturated zone continually equilibrates with soil air until it reaches the water table. While moving deeper into the subsurface, the groundwater becomes isolated from the atmosphere and carries the imprinted climate signal, recorded by its noble gas composition, along. Fluctuations of the water table typically result in the partial or complete dissolution of trapped air bubbles. In most cases, the individual processes can be separated by an iterative procedure (Stute and Schlosser, 1993; Stute et al., 1995). In suitable groundwater flow systems, the resulting *noble gas temperature* closely reflects the mean annual ground (soil) temperature at the water table in the recharge area. This climate signal is best preserved in confined aquifers. However, a groundwater flow system acts as a low-pass filter (Stute and Schlosser, 1993). At 18,000 [14]C B.P., for example, this smoothing effect is equivalent to a moving average of several thousand years. The weaknesses of this technique are the uncertainty in the excess air correction, the radiocarbon dating, and the smoothing of climate information; the advantage is that it is based on a simple physical principle and it is not sensitive to local or short-term climate fluctuations.

Two records have been obtained so far for the low-latitude Americas, i.e., for southern Texas (Stute et al., 1992) and northeastern Brazil (Stute et al., 1995). Both records indicate a 5°C cooler climate during the last glacial maximum.

17.4. A BRIEF REVIEW OF PALEOTEMPERATURE SIGNALS FROM SELECTED LOWLAND RECORDS

The first record of glacial age sediment from Central America was from beneath Lake Gatún (ca. −30 m elevation compared with modern sea level) in Panama (Bartlett and Barghoorn, 1973) (Fig. 2). Lake Gatún was formed when the Panama Canal was constructed and the Chagres River was dammed. Marshes that once flanked the Chagres and Trinidad Rivers are now submerged by Lake Gatún. Cores raised from the flooded marshes provided sediments that were riverine in origin. Because of the potential inclusion of pollen that had been reworked or transported long distances, Flenley (1979) urged caution in their interpretation. However, the source rivers drain low-elevation basins that presently would not support montane forest elements.

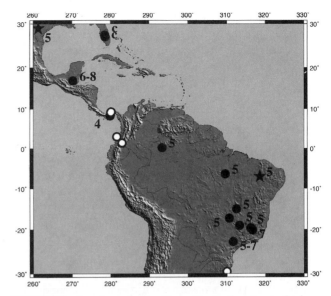

FIGURE 2 Locations and cooling estimates for sites in the lowland neotropics and subtropics at the last glacial maximum. Numbers indicate degrees Celsius (°C) of cooling compared with modern temperatures. Where the cooling is not quantified, it is indicated by "c." Sites marked with a circle are based on pollen records; those marked with a star are based on groundwater isotopic data.

Indeed, the modern pollen rain throughout their catchment areas would be remarkably uniform (Bush, personal observation). The largest changes in pollen abundance in this record reflect the migration of mangrove in response to changes in sea level. However, the presence of *Iriartea*, Ericaceae, *Ilex*, and *Podocarpus* pollen in late glacial assemblages indicates a downslope migration of these taxa of between 500 and 1000 m. Bartlett and Barghoorn suggested that this downslope shift represented a minimum temperature depression of 2.5°C. In an extensive study of modern pollen rain in Panama, Bush et al. (personal observation) did not find *Iriartea* pollen in any assemblage at less than 1000 m. Assuming a lapse rate of 5.5°C/1000 m, it would appear likely that the Gatún record documents at least a 5°C cooling. The complete lack of *Quercus* pollen in this record is striking, as pollen of this genus is abundant at two central Panamanian sites of glacial age (El Valle, 500 m: Bush and Colinvaux, 1990; and La Yeguada, 650 m: Bush et al., 1992). The modern pollen rain study shows *Quercus* to be ubiquitous above 1700 m elevation, and so we may infer that the temperature depression was not sufficient to bring *Quercus* down to sea level. Thus, the cooling would not have exceeded 8°C.

At Lake Quexil in Guatemala (Leyden et al., 1993, 1994), the coolest time on record was between 24,000 and 14,000 [14]C B.P., when temperatures were between 6° and 8°C lower than present values. As there is clear evidence of drying during the glacial maximum at this site, it is possible that there would have been some steepening of the lapse rate. For this reason, Leyden considers the lower of these estimates to be more realistic.

The first compelling evidence that glacial cooling affected the Amazon basin came with the discovery of *Podocarpus* timber at 1100 m elevation at Mera, Ecuador (Liu and Colinvaux, 1985). Equatorial *Podocarpus* species are almost exclusively montane, seldom living at elevations lower than 1800 m. Liu and Colinvaux (1985) inferred a temperature depression of ca. 4°C for the period between 33,000 and 30,000 [14]C B.P. on the basis of this evidence. A more detailed analysis of the sediments and the discovery of a second site at San Juan Bosco (970 m elevation) in Ecuador widened the list of cool indicator taxa to include *Magnolia, Drimys, Alnus, Hedyosmum, Weinmannia,* and grasses of the three-carbon (C3) photosynthetic pathway. With further dating, a cooling of 7.5°C was suggested for the period from 33,000–30,000 [14]C B.P. and 4°C for the period from 30,000–26,000 [14]C B.P. (Bush et al., 1990).

Since 1990, new data sets for different lowland ecosystems have produced further evidence of a substantial temperature depression during the glacial period. Lagoa Crominia in the Cerrado of central Brazil suggested a cooling of 5°C at 18,000 [14]C B.P. (Ferraz-Vicentini and Salgado-Labouriau, 1996). Lagoas dos Olhos and Serra Negra (De Oliveira, 1992) and the swamp of Salitre (Ledru, 1993) all document the Pleistocene expansion of *Araucaria* forests. Records for southern Brazil indicate the expansion of subtropical grasslands (Behling and Lichte, 1997; Behling et al., 1998). In each case, a 5°C lowering of temperature during glacial times is inferred. In the lowland Amazon, lakes perched atop massifs, such as the Serra dos Carajas (Absy et al., 1991) and the Hill of Six Lakes (Colinvaux et al., 1996), have provided long records of the lowland forest environment. The Carajas record is interpreted by Absy et al. (1991) primarily in terms of wet and dry events, but the pollen record is consistent with a glacial cooling (Absy, personal communication). Two records from the Hill of Six Lakes contain significant amounts of *Podocarpus, Hedyosmum,* and *Weinmannia* pollen, leading to the suggestion of a 5°C glacial cooling (Colinvaux et al. 1996).

17.5. THE DATA SET TO DATE: 30°N TO 30°S

Table 1 presents a summary of available data documenting paleotemperature at 18,000 [14]C B.P. for the tropical and extratropical regions of the New World.

17.6. DISCUSSION

It is evident from this data set that few records have sediments explicitly dated to 18,000 [14]C B.P. Ledru (1992) and Ledru et al. (1998) have suggested that the period from ca. 25,000–16,000 [14]C B.P. was a time of regionwide aridity in which most lake basins dried out. A priori arguments, such as a cooler ocean and land surface would have reduced evaporation, and hence cloud formation, and weakened circulations would have brought less moisture onshore, are powerful. Some reduction in precipitation and lake levels during the last glacial is very likely. There is no question that depositional rates slowed at a number of sites from Panama to Brazil during this period. However, more sites and better dating are needed before it can be resolved whether this period was a single phase of major widespread aridity or local asynchronous drying events.

With our paucity of sites, the vastness of the area considered, and current inability to assess paleoprecipitation from pollen records (see Colinvaux et al., 2000, for a review, and a contrasting review by Hooghiemstra and van der Hammen, 1998), it is to be expected that there will be considerable debate within our community over the extent and duration of possible dry events. Divergent views are held within this group of authors; however, none espouse such serious drying as to elevate lapse rates sufficiently to explain away the signature of cooling.

17.6.1. The Potential Influence of CO_2 and UV Radiation on Past Vegetation Assemblages

Street-Perrott et al. (1997) suggested that past concentrations of CO_2 may have been at least as important as cooling in determining the elevation at which species grew during glacial times. They suggest that under glacial conditions, with atmospheric concentrations of CO_2 close to 180 ppm, plants using the four-carbon photosynthetic pathway (C4) or crassulacean acid metabolism (CAM) would be expected to outcompete those using the C3 pathway. It has been argued that, in Africa, the lowering of tree lines apparent in some pollen records was, in part, an artifact of palynologists' inability to resolve the difference between C3 (montane) and C4 (lowland) grasses (Street-Perrott et al., 1997). Street-Perrott et al. suggest that instead of an invasion of midelevations by C3 grasses, high Poaceae pollen percentages reflect C4 grasses outcompeting seedlings of C3 trees during the glacial maximum.

The principal problem encountered by a C3 plant faced with low CO_2 concentrations is drought stress (Woodward 1993). In western Amazonia and on the

TABLE 1 Locations, Ages, and Paleotemperature Data for Lowland Neotropical Sites Used in Reconstructions of Paleotemperature at 18,000 [14]C B.P.[a]

Site	Latitude, longitude	Elevation (m)	Dates	Lab no.	Calibrated years B.P. min. cal. age (cal. age) max. cal. age	Delta T°C at 18,000 (21,500)	Max. cooling	Date of max. cooling	Ref.
Carrizo Aquifer (U.S.A.)	29°N, 98°40'W	−98.75	10,600 ± 3,000 11,700 ± 3,000 20,800 ± 3,000 27,200 ± 3,000		19,560 (12,530) 5,300 17,600 (13,640) 9,650 [24,400] [30,500]	−5	−5	Uniform throughout period considered	Stute et al., 1992 Stute and Clark, unpublished ^{14}C data
Lake Annie (U.S.A.)	27°12'N, 81°25'W	40	4,715 ± 95 13,010 ± 165 37,000 ± 3,200	I-6889 I-6888 I-6025	5,580 (5,350) 5,310 15,730 (15,450) 15,150 [38,900]	Cooler, unquantified			Watts, 1975
Lake Tulane (U.S.A.)	27°35'N, 81°30'W	34	9,810 ± 90 10,940 ± 120 13,730 ± 130 17,170 ± 210 20,380 ± 239 24,240 ± 400 26,120 ± 440 32,300 ± 450 35,700 ± 650 >33,000 35,600 ± 400 39,600 ± 500 >46,000	WIS-1753 WIS-1648 WIS-1649 WIS-1754 WIS-1650 WIS-1755 WIS-1651 QL-4630 QL-4631 WIS-1652 QL-4057 QL-4058 QL-4632	11,000 (10,990) 10,950 12,980 (12,860) 12,740 16,650 (16,460) 16,270 20,720 (20,350) 20,000 [24,000] [27,700] [29,500] [34,900] [37,800] >[35,500] [37,700] [41,100] >[46,600]	Cooler, unquantified			Grimm et al., 1993
Quexil (Guatemala)	16°55'N, 89°49'W	110	10,750 ± 460 10,300 ± 110 10,630 ± 110 12,790 ± 60 27,450 ± 500	SI-5257 AA-3062 AA-3063 b-92902 AA-3064	13,120 (12,680) 12,130 12,330 (12,150) 11,900 12,680 (12,560) 12,420 15,270 (15,100) 14,910 [30,800]	−6 to −8	−6.5 to −8	24,000 (27,500) to 14,000 (16,800)	Leyden et al., (1993)
Lake Gatún (Panama)	9°16'N, 79°52'W	−30	9,600 ± 300 11,300 ± 200 35,500 ± 2,500	UCLA-185 UCLA-186 UCLA-1025	11,000 (10,750) 10,220 13,430 (13,210) 13,010 [37,600]	N/A	−2.5 reinterpreted as −5	11,000 (12,900) to 10,000 (11,200)	Bartlett and Barghoorn, 1973
La Yeguada (Panama)	8°27'N, 80°51'W	650	8,840 ± 130 10,210 ± 130 10,530 ± 100 11,250 ± 140 11,610 ± 180 14,230 ± 370 13,670 ± 210 12,910 ± 140	b-26102 b-25923 b-24739 b-24738 b-25924 b-25925 b-25696 b-24241	9,970 (9,880) 9,650 12,240 (11,980) 11,340 12,580 (12,450) 12,310 13,310 (13,160) 13,010 13,770 (13,160) 13,330 17,490 (17,060) 16,610 16,660 (14,430) 13,830 15,340 (15,290) 15,020	N/A	−6	14,000 (16,800) to 12,000 (14,000)	Bush et al., 1992

(continues)

TABLE 1 (continued)

Site	Latitude, longitude	Elevation (m)	Dates	Lab no.	Calibrated years B.P. min. cal. age (cal. age) max. cal. age	Delta T°C at 18,000 (21,500)	Max. cooling	Date of max. cooling	Ref.
El Valle (Panama)	8°20′N, 80°10′W	500	8,330 ± 150	b-27721	9,450 (9,370) 9,040	−4	−6	14,000 (16,800)	Bush and Colinvaux, 1990
			14,180 ± 250	b-29038	17,300 (17,000) 16,700				
			19,420 ± 330	b-27722	[23,100]				
			31,850 ± 1,800	b-27723	[34,500]				
			>35,000	b-27724	>[37,200]				
			30,720 ± 800	b-25697	[33,600]				
San Juan Bosco (Ecuador)	3°3′N, 78°27′W	970	26,020 ± 300	b-27144	[29,400]	N/A	−7.5	ca. 33,000 (35,500)	Bush et al., 1990
			30,990 ± 350	b-27145	[33,800]				
Mera (Ecuador)	1°29′N, 77°06′W	1100	26,530 ± 270	b-10170	[29,600]	N/A	−7.5	33,000 (35,500)	Bush et al., 1990
			31,870 ± 970	b-27143	[34,500]				
			33,520 ± 1,010	b-9618	[35,900]				
Lagoa Pata (Brazil)	0°16′N, 66°41′W	250–300	5,800 ± 70	b-63417	6,720 (6,630) 6,490	−5	−5	14,000 (16,800)	Colinvaux et al., 1996
			14,230 ± 60	b-91489	17,170 (17,060) 16,950				
			15,560 ± 60	b-90306	18,570 (18,460) 18,360				
			17,840 ± 300	b-75109	21,700 (21,280) 20,840				
			18,020 ± 70	b-90307	21,700 (21,510) 21,320				
			30,830 ± 220	b-91490	[33,600]				
			31,390 ± 540	b-75110	[34,200]				
			32,010 ± 630	b-88941	[34,700]				
			34,650 ± 420	b-89715	[36,900]				
			37,830 ± 1,300	b-88942	[39,600]				
			38,860 ± 920	b-68529	[40,500]				
			42,010 ± 1,240	b-68530	[43,200]				
Lagoa Verde (Brazil)	0°16′N, 66°41′W	250–300	12,050 ± 50	CAMS-47775	14,190 (14,050) 13,920	−5	−5	?	Bush et al., unpublished
			2,790 ± 50	CAMS-47776	2,940 (2,860) 2,790				
			12,480 ± 60	b-95704	14,800 (14,620) 14,450				
			17,100 ± 70	CAMS-47777	20,480 (20,250) 20,050				
			16,410 ± 70	CAMS-47778	19,470 (19,310) 19,170				
			18,430 ± 100	b-95705	[22,200]				
			19,740 ± 70	b-95706	[23,400]				
			19,170 ± 120	CAMS-47779	[22,900]				
			18,680 ± 130	CAMS-47780	[22,400]				
			23,600 ± 450	OS-1320	[27,100]				
			>43,800	OS-1321	>[44,700]				

Site	Lat., Long.	Elevation	^{14}C age (yr B.P.)	Lab number	Calendar age				Reference
Carajas (Brazil)	6°20′S, 50°25′W	700	10,460 ± 150 12,520 ± 120 22,870 ± 500 23,670 ± 500 24,520 ± 820 28,660 ± 1,000 51,200 ± 2,000		12,550 (12,370) 12,150 14,920 (14,680) 14,450 [26,400] [27,200] [28,100] [31,800] [51,200]	−5	−5	9000 (10,000)	Absy et al., 1991
Serra Grande Aq (Brazil)	7°S, 41°30′W	−41.50	13,100 ± 3,000 14,800 ± 3,000 16,000 ± 3,000		18,980 (15,590) 11,680 21,240 (17,700) 13,750 22,700 (18,860) 15,300 [51,200]	−5	−5	Uniform throughout period considered	Stute et al., 1995
Aguas Emendadas	15°S, 47°35′W	1040	7,220 ± 50 21,450 ± 100	OBDY 1152 OBDY 1193	8,060 (7,960) 7,930 [25,100]	−5	−5	14,000 (16,800)	Barberi et al., 1995
Crominia (Brazil)	17°17′S, 49°25′W	710	6,680 ± 90 13,150 ± 50 32,060 ± 520 32,390 ± 680 32,580 ± 1,640	UtC 45715 OBDY 956 UtC 45716 UtC 64283 UtC 45717	7,560 (7,490) 7,400 15,800 (15,670) 15,530 [34,700] [35,000] [35,100]	−5	−5	14,000 (16,800)	Ferraz-Vicentini and Salgado-Labouriau, 1996
Lagoa dos Olhos (Brazil)	19°38′S, 43°54′W	730	6,710 ± 140 15,530 ± 110 19,410 ± 160	b-53327 b-53328 b-35394	7,640 (7,540) 7,400 18,570 (18,440) 18,300 [23,100]	−5	−5		De Oliveira, 1992
Serra Negra (Brazil)	19°S, 46°45′W	1170	14,280 ± 90 39,930 ± 540 >46,180	b-53314 b-53315 b-53321	17,250 (17,120) 16,990 [41,400] >[46,800]	−5	−5		De Oliveira, 1992
Salitre (Brazil)	19°S, 46°46′W	1050	9,150 ± 80 10,440 ± 110 10,350 ± 230 12,890 ± 80 14,230 ± 150 16,800 ± 100 28,740 ± 500 32,030 ± 500 >50,000	OBDY 570 OBDY 495 OBDY 496 OBDY 550 OBDY 571 OBDY 552 OBDY 470 OBDY 471	10,280 (10,040) 10,010 12,490 (12,350) 12,170 12,520 (12,230) 11,720 15,440 (15,260) 15,060 17,250 (17,060) 16,870 20,050 (19,820) 19,610 [31,900] [34,700] >[50,000]	−5	−5	11,000 (12,900)	Ledru, 1993
Catas Altas (Brazil)	20°05′S, 43°22′W	755	8,310 ± 295 20,490 ± 165 19,960 ± 530 21,550 ± 440 22,087 ± 1,580/ −2,190 37,880 ± 930 >47,740	Hv20825 Hv20826 Hv20827 Hv20828 Hv20829 Hv20830 Hv20831	9,520 (9,300) 8,780 [24,100] [23,600] [25,200] [25,700] [39,700] >[48,100]	−7	−7	28,000 (31,300) to 18,000 (21,500)	Behling, 1998
Botucatu (Brazil)	22°48′S, 48°23′W	770	5,678 ± 37 19,180 ± 190 25,750 ± 170 22,900 ± 130 >32,360	UtC-5544 Hv-20824 UtC-5545 UtC-5546 Hv-20824	6,490 (6,450) 6,410 [22,900] [29,200] [26,400] >[34,900]	−5 to −7	−5 to −7		Behling et al., 1998

[a] Calendar ages were calculated with CALIB 3.0.3c (Stuiver and Reimer, 1993) for the interval up to 18,367 ^{14}C B.P. In the interval from 18,367 to 27,120 ^{14}C B.P., a linear conversion is used assuming corresponding calendar ages of 22,115 and 27,120 cal. years B.P., respectively. The calendar ages are based on Bard et al. (1990). For the interval 27,120–50,000 ^{14}C B.P., Mommersteeg (1998) assumes a difference of 3350 cal. years at 27,120 ^{14}C B.P., which gradually diminished to a difference of zero calendar years at an age of 50,000 ^{14}C B.P., based on the work of Mazuad et al. (1991) and Laj et al. (1996). Calibrated ages are shown in parentheses, flanked by a 1 sigma age range.

Andean flanks, where rainfall can exceed 5 m/year, drought is less of a problem than for plants on the slopes of Mt. Kenya. Generally, in the neotropics the plants that are observed invading new areas during glacial times are C3 trees. Indeed, where Poaceae did become more abundant at San Juan Bosco and Mera, it was determined on the basis of their phytoliths that C3 grasses were invading areas presently occupied by C4 grasses (Bush et al., 1990). The replacement of C4 with C3 grasses clearly runs counter to the pattern expected if CO_2 were the factor most limiting to plant growth. Cooling rather than low CO_2 concentrations is a more likely cause of this invasion.

At drier locations, such as the Carajas plateau, which receives ca. 2000 mm of precipitation per annum, or across much of the eastern flank of Amazonia, it is possible that reduced precipitation during glacial periods could have provided C4 and CAM plants with a competitive advantage. Indeed, at Carajas, the facultative CAM plant *Cuphea* increases in abundance during glacial times (Absy et al., 1991). However, plenty of C3 trees remain in the glacial landscape, and the greatest peaks of grasses (potentially a major C4 component of vegetation) occur in the Holocene, not during the Pleistocene. Overall, low concentrations of CO_2 cannot be discounted as an important ecological factor in these drier parts of the neotropics, but they do not appear to be as important as temperature effects.

Reduced CO_2 concentrations and reduced cloud cover could result in increased inputs of ultraviolet (UV) radiation at the Earth's surface (Flenley, 1993). Flenley (1998) suggested that under these circumstances montane plants that are more highly adapted to UV radiation would have a competitive advantage. The highest elevations would have had the most exposure to UV radiation and would have experienced the greatest change in incoming radiation. Flenley suggests that some of the difference observed in paleotemperature between the highest and lowest locations may be attributable to changes in UV radiation. In the Andes, the differential in cooling between 2550 m elevation and the lowland elevations appears to have been ca. 3°C at 18,000 [14]C B.P., reflecting a cooling of 8°C for the Sabana de Bogotá and ca. 5°C at sites below 1000 m elevation. Although some part of this temperature discrepancy between high- and lower elevation sites might be attributable to the effects of UV radiation, a larger variable is likely to be the error of the methods used to estimate paleotemperature.

17.6.2. The Timing of Peak Cooling

From our data set, even though there are 16 records from the lowland neotropics and subtropics that span

the period of 18,000 [14]C B.P., only half have sediments directly dated to between 16,000 and 20,000 [14]C B.P. The available records document a 5°C cooling for this period, which is typical of late Pleistocene times; i.e., it was not especially cold. Several, though certainly not all, records show pronounced cold events at approximately 33,000–30,000 [14]C B.P., when temperatures may have dipped as much as 7.5°C below present levels, and again at ca. 14,000 [14]C B.P., with temperatures between 5° and 6°C cooler than present levels. These episodes of more intense cooling appear to coincide with phases of glacial re-advance in Ecuador (Clapperton, 1987; Seltzer, 1990), whereas no distinct re-advance at 18,000 [14]C B.P. has been documented for any of the tropical Andean glaciers. However, the cooling evident in the records of San Juan Bosco and Mera at 33,000–30,000 [14]C B.P. does not appear to have been influenced by precipitation, and at least two data points suggest this time to have been both wet and very cold (Bush et al., 1990). While the terrestrial pollen data provide no regional consensus for the timing of maximal cooling, they suggest that the period from 33,000–12,000 [14]C B.P. was predominantly cold with marked local climatic events that provided oscillations between warmer and cooler conditions. This is supported by a continuous pollen record from the Amazon deep-sea fan, spanning the last 50,000 years B.P., which shows a maximum contribution from cold-adapted taxa, such as *Podocarpus* and *Hedyosmum*, between 19,800 and 11,000 cal. B.P. (Haberle and Maslin, 1999).

It is also important to note that the terrestrial paleoclimatic records for other tropical regions are in close accord with our estimate for the neotropics. In Southeast Asia, lower montane rain forest taxa invaded the bottomland rain forest, and Flenley (1998) suggests a mean annual temperature depression of 6°–10°C (Flenley, 1998). In Africa, numerous high-elevation pollen records documented a descent of forest that would have required a cooling of ca. 9°C (e.g., Coetzee, 1964; Livingstone, 1967; Taylor, 1990); but in the lowlands, the cooling appears to have been more modest (Giresse et al., 1994). In a study of African lowland sites using transfer functions derived from a suite of modern analogs, Bonnefille et al. (1990) estimated a cooling of 4° ± 2°C. Street-Perrott et al. (1997) suggest that when CO_2 effects are factored in, even at high elevations, the actual cooling may have been closer to 2°C.

Just as some tropical cooling is a consistent pattern, the extent of precipitation change is variable. Indeed, none of the tropical regions show a consistent pattern, and the pattern seen in Amazonia of some sites appearing to dry while others remained humid is repeated in Africa and Southeast Asia (see Flenley, 1998 for a summary).

17.6.3. Implications of Cooled Tropical Lowlands

The descent of upland vegetation elements 1000 m below their modern limits is apparent in both high- and low-elevation pollen records at 18,000 ^{14}C B.P. As humid conditions extended from the foot of the Andes to 2500 m elevation, it is safe to assume a lapse rate that approximated to 5°C/1000 m of ascent, and thereby a 5°C cooling. Another way to view this is that frost may have occurred as low as 800 m elevation at the equator. Other critical temperature thresholds, such as permanent damage to plant lipid membranes that occurs between 8 and 13°C (Graham and Patterson, 1982), are likely to have been experienced in all but the most sheltered locations. Cold stress would have affected populations throughout the study area, causing the reassortment of species into unique communities. Populations at the tips of peninsulas such as Florida, on islands, and close to the equator could not have migrated to escape cold episodes and may have experienced the greatest species losses. Given that this last glacial interval was not significantly colder than those that had preceded it in the Quaternary, a wave of extinctions would not be expected. Any truly cold-intolerant species would have been eliminated by earlier glacial episodes, possibly explaining the greater diversity of Miocene than Holocene palynomorphs in Amazonia (van der Hammen and Absy, 1994). However, as in the midlatitudes, from Florida to the equator and south to the Mato Grosso, the Pleistocene landscape would have been filled with assemblages of plants and animals without modern analogs.

It is apparent that a reconciliation is needed between the marine and terrestrial records. Reconstructions of SSTs suggested that the tropical oceans adjacent to South America had barely cooled at 18,000 ^{14}C B.P. (CLIMAP Project Members, 1981). Temperature depressions of only 1°–2°C were suggested by analysis of foraminiferal communities recovered from deep ocean cores. These reconstructions were based on the comparison of modern analog and fossil communities. At high and midlatitudes, the degree of communality (overlap of modern and fossil assemblages) was strong (approaching 100%), giving confidence to the temperature reconstructions. However, seven of nine cores within 2° latitude of the equator had statistically insignificant communality during the period 20,000–14,000 ^{14}C B.P. (sensu Mix et al., 1986). Weak communality translated to tenuous paleoclimatic reconstructions, yet the CLIMAP (1981) tropical paleotemperatures for 18,000 ^{14}C B.P. were widely accepted. A reevaluation of the CLIMAP data is being conducted (Alan Mix, personal communication), and these results will be eagerly awaited by the terrestrial community. In the meantime, isotopic analyses of Pleistocene Caribbean coral reefs suggest that in these nearshore environments, SSTs were 5°C cooler than present values (Guilderson et al., 1994). A similar cooling of the sea surface was documented by using the relative abundance of double bonds in alkenones from marine sediments (Bard et al., 1997). These investigators inferred a 4°C cooling for the equatorial Atlantic at the time of the last glacial maximum and a cooling of as much as 7°C in some nearshore samples from the Caribbean.

Taking a different approach, Webb et al. (1997) maintained a modern ocean heat transport in their model of glacial SST. After allowing for reduced sea levels, lowered CO_2 concentrations, enlarged ice masses, and energy being channeled through glacial North Atlantic Intermediate Water rather than via the North Atlantic Deep Water, they predicted a 5.5°C cooling of tropical SSTs.

17.7. CONCLUSIONS

Overall, the paleoecologic and isotopic data set from the lowland neotropics reveals a remarkable constancy in the estimate of temperature for the period around 18,000 ^{14}C B.P. In almost all cases, a cooling approximating to 5°C is recorded. It is heartening to find a growing body of independent temperature estimates emerging from the marine and modeling communities that are in close accord with the terrestrial estimates of paleotemperature. The terrestrial data also strongly suggest that brief climatic oscillations lasting a few millennia centered on 30,000 and 14,000 ^{14}C B.P. may have produced even stronger local cooling episodes with temperatures as much as 7.5°C below present levels. Although our community is generally united in its estimate of a lowland neotropical cooling of 5°C for much of the last glacial period, it must be recognized that this is a very crude estimate. We have very few records from an enormous area, and, in many cases, the sedimentary sequences apparently contain gaps or at least rapid changes in depositional rate.

Our community is divided over the important issue of the magnitude and timing of precipitation change during glacial episodes. The forthcoming debate promises to be both vigorous and rewarding, stimulating, it is hoped, the invention of a technique to provide reliable estimates of paleoprecipitation. The only certainty that we have now is that we have too few data to address the issue in any authoritative way.

Although the potential impact of lowered concentrations of CO_2 on the success of individual species is recognized, there is no evidence to suggest that C4 or

CAM plants outcompeted C3 plants in the lowland neotropics. Indeed, the species replacements are almost always one C3 tree replacing another. The influence of changes in CO_2 concentration in this ecosystem is considered to be far smaller than influences of temperature, precipitation, and hydrology. Similarly, while the high elevations of the Andes may have been affected by higher levels of UV radiation during glacial episodes, it seems unlikely that the migration of midelevation or lowland taxa would have been significantly affected.

To unravel the complexities of episodic climatic events within the last glacial cycle, more long records, analyzed in great detail and with good isotopic dating, are clearly needed. In our quest to determine the climatic history of the lowland neotropics, our community has arrived at the starting line, not at the winner's post.

References

Absy, M. L., and T. van der Hammen, 1976: Some palaeoecological data from Rondônia, southern part of the Amazon basin. *Acta Amazonica*, 6: 293–299.

Absy, M. L., A. Cleef, M. Fournier, L. Martin, M. Servant, A. Sifeddine, M. Ferreira da Silva, F. Soubiès, K. Suguio, B. Turcq, and T. van der Hammen, 1991: Mise en évidence de quatre phases d'ouverture de la forêt dense dans le sud-est de L'Amazonie au cours des 60,000 dernières années. Première comparaison avec d'autres regions tropicales. *Comptes Rendus de l' Academie Science Paris, Series II*, 312: 673–678.

Andrews, J. N., and D. J. Lee, 1979: Inert gases in groundwater from the Bunter Sandstone of England as indicators of age and paleoclimate trends. *Journal of Hydrology*, 41: 233–252.

Barberi, M., M.-L. Salgado-Labouriau, K. Suguio, L. Martin, B. Turcq, and J.-M. Flexor, 1995: Analise palinologica da vereda de Aguas Emendadas (DF). *V. Congreso da Associacao Brasileira de Estudos do Quaternario*, Universidad Fedderale Fluminense, Niteroi, Rio de Janeiro, Brazil.

Bard, E., B. Hamelin, R. G. Fairbanks, and A. Zindler, 1990: Calibration of the ^{14}C timescale over the past 30,000 years using mass spectrometric U-Th ages from Barbados. *Nature*, 345: 405–410.

Bard, E., F. Rostek, and C. Sonzogni, 1997: Interhemispheric synchrony of the last deglaciation inferred from alkenone paleothermometry. *Nature*, 385: 707–710.

Bartlett, A. S., and E. S. Barghoorn, 1973: Phytogeographic history of the Isthmus of Panama during the past 12,000 years (a history of vegetation, climate and sea-level change). *In* Graham, A. (ed.), *Vegetation and Vegetational History of Northern Latin America*. Amsterdam: Elsevier, pp. 203–299.

Behling, H., 1998: Late Quaternary vegetational and climatic changes in Brazil. *Review of Palaeobotany and Palynology*, 99: 143–156.

Behling, H., and M. Lichte, 1997: Evidence of dry and cold climatic conditions at glacial times in tropical southeastern Brazil. *Quaternary Research*, 48: 348–358.

Behling, H., M. Lichte, and A. W. Miklos, 1998: Evidence of a forest free landscape under dry and cold climatic conditions during the last glacial maximum in the Botucatú region (São Paulo State), Southeast Brazil. *Quaternary of South America and Antarctic Peninsula*, 11: 99–110.

Bonnefille, R., J. C. Roeland, and J. Guiot, 1990: Temperature and rainfall estimates for the past 40 000 years in equatorial Africa. *Nature*, 346: 347–349.

Bradbury, J. P., B. W. Leyden, M. Salgado-Labouriau, W. M. Lewis, Jr., C. Schubert, M. Binford, D. G. Frey, D. R. Whitehead, and F. H. Weibezahn, 1981: Late Quaternary environmental history of Lake Valencia, Venezuela. *Science*, 214: 1299–1305.

Broecker, W. S., 1986: Oxygen isotope constraints on surface ocean temperatures. *Quaternary Research*, 26: 121–134.

Bush, M. B., and P. A. Colinvaux, 1990: A long climatic and vegetation record from lowland Panama. *Journal of Vegetation Science*, 1: 105–118.

Bush, M. B., P. A. Colinvaux, M. C. Wiemann, D. R. Piperno, and K.-B. Liu, 1990: Pleistocene temperature depression and vegetation change in Ecuadorian Amazonia. *Quaternary Research*, 34: 330–345.

Bush, M. B., D. R. Piperno, P. A. Colinvaux, P. E. De Oliveira, L. A. Krissek, M. C. Miller, and W. E. Rowe, 1992: A 14,300 year paleoecological profile of a lowland tropical lake in Panama. *Ecological Monographs*, 62: 251–275.

Centro Ecuatoriano de Investigación Geográfica (CEDIG), 1983: *Los Climas Del Ecuador. Documentos de Investigación*, No. 4, Quito, Ecuador.

Clapperton, C. M., 1987: Maximal extent of the late Wisconsin glaciation in the Ecuadorian Andes. *Quaternary of South America and Antarctic Peninsula* 5: 165–180.

CLIMAP Project Members, 1976: The surface of the ice-age Earth. *Science*, 191: 1131–1137.

CLIMAP Project Members, 1981: Seasonal reconstruction of the Earth's surface at the last glacial maximum. *Geological Society of America Map and Chart Series*, 36.

Coetzee, J. A., 1964: Evidence for a considerable depression of vegetation belts during the upper Pleistocene on the east African mountains. *Nature*, 204: 564–566.

Colinvaux, P. A., P. E. De Oliveira, and M. B. Bush, 2000: Amazonian and neotropical plant communities on glacial time-scales. *Quaternary Science Reviews*, 19: 141–169.

Colinvaux, P. A., P. E. De Oliveira, E. Moreno, M. C. Miller, and M. B. Bush, 1996: A long pollen record from lowland Amazonia: Forest and cooling in glacial times. *Science*, 274: 85–88.

Davis, M. B., 1981: Quaternary history and the stability of forest communities. *In* West, D. C., H. H. Shugart, and D. B. Botkin (eds.), *Forest Succession: Concepts and Application*. New York: Springer-Verlag, pp. 132–154.

De Oliveira, P. E., 1992: A palynological record of late Quaternary vegetational and climatic change in southeastern Brazil. Ph.D. thesis. The Ohio State University, Columbus.

Faegri, K., 1966: Some problems of representativity in pollen analysis. *Palaeobotanist*, 15: 135–140.

Faegri, K., and J. Iversen, 1989: *Textbook of Pollen Analysis*, 4th ed. Copenhagen: Munksgaard, 328 pp.

Ferraz-Vicentini, K. R., and M. L. Salgado-Labouriau, 1996: Palynological analysis of a palm swamp in central Brazil. *Journal of South American Earth Science*, 9: 207–219.

Flenley, J. R., 1979: *A Geological History of Tropical Rainforest*. London: Butterworth.

Flenley, J. R., 1993: Cloud forest: The Massenerhebung effect and ultraviolet insolation. *In* Hamilton, L. S., J. O. Juvik, and F. M. Scatena (eds.), *Tropical Montane Cloud Forest: Proceedings of an International Symposium*. East-West Center, Honolulu, pp. 94–96.

Flenley, J. R., 1998: Tropical forests under the climates of the last 30,000 years. *Climatic Change*, 39: 177–197.

Giresse, P., J. Maley, and P. Brenac, 1994: Late Quaternary paleoenvironments in the Lake Barombi-Mbo (West Cameroon) deduced from pollen and carbon isotopes of organic matter. *Palaeogeography, Palaeoclimatology, Palaeoecology*, 107: 65–78.

Graham, D., and B. D. Patterson, 1982: Responses of plants to low, nonfreezing temperatures: Proteins, metabolism and acclimation. *Annual Review of Plant Physiology*, 33: 347–372.

Grimm, E. C., G. L. Jacobson, Jr., W. A. Watts, B. C. S. Hansen, and K. A. Maasch, 1993: A 50,000-year record of climate oscillations from Florida and its temporal correlation with the Heinrich events. *Science,* **261**: 198–200.

Guilderson, T. P., R. G. Fairbanks, and J. L. Rubenstone, 1994: Tropical temperature variations since 20,000 years ago: Modulating interhemispheric climate change. *Science,* **263**: 663–665.

Haberle, S. G., 1998: Late Quaternary vegetation change in the Tari basin, Papua New Guinea. *Palaeogeography, Palaeoclimatology, Palaeoecology,* **137**: 1–24.

Haberle, S. G., and M. Maslin, 1999: Late Quaternary vegetation and climate change in the Amazon basin based on a 50,000 year pollen record from the Amazon fan: ODP Site 932. *Quaternary Research,* **51**: 157–183.

Haffer, J., 1969: Speciation in Amazonian forest birds. *Science,* **165**: 131–137.

Haffer, J., 1991: Avian species richness in tropical South America. *Studies in Neotropical Fauna and Environment,* **25**: 157–183.

Hansen, B. C. S., H. E. Wright, and J. P. Bradbury, 1984: Pollen studies in the Junín area, central Peruvian Andes. *Geological Society of America Bulletin,* **95**: 1454–1465.

Heusser, L. E., 1995: Pollen stratigraphy and paleoecologic interpretation of the 160-K.Y. record from Santa Barbara basin, Hole 893A. *In* Kennett, J. P., J. G. Badaulf, and M. Lyle (eds.), *Proceedings of the Ocean Drilling Program, Scientific Results,* Vol. 146. Ocean Drilling Program, College Station, TX, pp. 265–279.

Hooghiemstra, H., 1984: *Vegetational and Climatic History of the High Plain of Bogotá, Colombia: A Continuous Record of the Last 3.5 Million Years. Dissertationes Botanicae,* **79**: 1–368.

Hooghiemstra, H., 1989: Quaternary and upper-Pliocene glaciations and forest development in the tropical Andes: Evidence from a long high-resolution pollen record from the sedimentary basin of Bogotá, Colombia. *Palaeogeography, Palaeoclimatology, Palaeoecology,* **72**: 11–26.

Hooghiemstra, H., and T. van der Hammen, 1998: Neogene and Quaternary development of the neotropical rain forest: The refugia hypothesis, and a literature review. *Earth Science Reviews,* **44**: 147–183.

Imbrie, J., and N. G. Kipp, 1971: A new micropaleontological method for quantitative paleoclimatology: Application to a late Pleistocene Caribbean core. *In* Turekian, K. K. (ed.), *Late Cenozoic Glacial Ages.* New Haven, CT: Yale University Press, pp. 71–181.

Kutzbach, J. E., and P. J. Guetter, 1986: The influence of changing orbital parameters and surface boundary conditions on climate simulations for the past 18,000 yrs. *Journal of Atmospheric Sciences,* **43**: 1726–1759.

Laj, C., A. Mazaud, and J.-C. Duplessy, 1996: Geomagnetic intensity and ^{14}C abundance in the atmosphere and ocean during the past 50 kyr. *Geophysical Research Letters,* **23**: 2045.

Ledru, M.-P., 1991: Etude de la pluie pollinique actuelle des forêts du Brésil central: climat, végétation, application à l'étude de l'évolution paléoclimatique des 30 000 dernières années." Unpublished Ph.D. thesis. Museum National d'Histoire Naturelle, Paris, 169 pp.

Ledru, M.-P., 1992: Modification de la végétation du Brásil central entre la dernière époque glaciaire et l'interglaciaire actuel. *Comptes Rendus de l'Academie des Sciences,* Paris, II, **314**: 117–123.

Ledru, M.-P., 1993: Late Quaternary environmental and climatic changes in central Brazil. *Quaternary Research,* **39**: 90–98.

Ledru, M.-P., J. Bertaux, A. Sifeddine, and K. Suguio, 1998: Absence of last glacial maximum records in lowland tropical forest. *Quaternary Research,* **49**: 233–237.

Leyden, B. W., M. Brenner, D. A. Hodell, and J. A. Curtis, 1993: Late Pleistocene climate in the Central American lowlands. *In* Swart, P. K., K. C. Lohmann, J. McKenzie, and S. Savin (eds.), *Climate*

Change in Continental Records. Washington, DC: American Geophysical Union, pp. 165–178.

Leyden, B. W., M. Brenner, D. A. Hodell, and J. A. Curtis, 1994: Orbital and internal forcing of climate on the Yucatán Peninsula for the past ca. 36 ka. *Palaeogeography, Palaeoclimatology, Palaeoecology,* **109**: 193–210.

Liu, K.-B., and P. A. Colinvaux, 1985: Forest changes in the Amazon basin during the last glacial maximum. *Nature,* **318**: 556–557.

Livingstone, D. A., 1967: Postglacial vegetation of the Ruwenzori Mountains in equatorial Africa. *Ecological Monographs,* **37**: 25–52.

Marchant R., H. Behling, A. Cleef, S. Harrison, H. Hooghiemstra, V. Markgraf, C. Prentice, M. L. Absy, T. Ager, R. Anderson, C. Baied, S. Bjorck, R. Byrne, M. Bush, J. Duivenvoorden, J. Flenley, P. De Oliveira, B. van Geel, K. Graf, S. Haberle, T. van der Hammen, B. Hansen, S. Horn, P. Kuhry, M.-P. Ledru, B. Leyden, S. Garcia-Lozano, A. B. M. Melief, P. Moreno, N. T. Moar, A. R. Prieto, G. van Reenen, M. L. Salgado-Labouriau, F. Schabitz, and E. J. Schreve-Brinkman, in press: Pollen-based biome reconstructions for Latin America at 0, 6000 and 18,000 radiocarbon years. *Journal of Biogeography BIOME 6000 Special Issue.*

Markgraf, V., 1993: Climatic history of Central and South America since 18,000 yr [14]C B.P.: Comparison of pollen records and model simulations. *In* Wright, H. E., Jr., J. E. Kutzbach, T. Webb, III, W. F. Ruddiman, F. A. Street-Perrott, and P. J. Bartlein (eds.), *Global Climates Since the Last Glacial Maximum.* Minneapolis: University of Minnesota Press, pp. 357–386.

Martin, P. S., 1964: Paleoclimatology and a tropical pollen profile. Report of the VIth International Congress on the Quaternary, Warsaw, 1961, Vol. ii, pp. 319–323. Paleoclimatological section, Lodz. (Also Contribution 46, Program in Geochronology, University of Arizona, Tucson.)

Mazor, E., 1972: Paleotemperatures and other hydrological parameters deduced from noble gases dissolved in groundwaters, Jordan Rift Valley, Israel. *Geochimica Cosmochimica Acta,* **36**: 1321–1336.

Mazuad, A., C. Laj, E. Bard, M. Arnold, and E. Tric, 1991: Geomagnetic field control of ^{14}C production over the last 80 kyr: Implications for the radiocarbon time-scale. *Geophysical Research Letters,* **18**: 1885–1888.

Mercer, J. H., and O. Palacios, 1977: Radiocarbon dating of the last glaciation in Peru. *Geology,* **5**: 600–604.

Mix, A. C., W. F. Ruddiman, and A. McIntyre, 1986: Late Quaternary paleoceanography of the tropical Atlantic. 1. Spatial variability of annual mean sea-surface temperatures, 0–20,000 years [14]C B.P. *Paleoceanography,* **1**: 43–66.

Mommersteeg, H., 1998: Vegetation development and cyclic and abrupt climatic changes during the late Quaternary: Palynological evidence from the Colombian Eastern Cordillera. Ph.D. thesis. Hugo de Vries Laboratory, University of Amsterdam.

Rind, D., and D. Peteet, 1985: Terrestrial conditions at the last glacial maximum and CLIMAP sea-surface temperature estimates: Are they consistent? *Quaternary Research,* **24**: 1–22.

Seltzer, G. O., 1990: Recent glacial history and paleoclimate of the Peruvian-Bolivian Andes. *Quaternary Science Reviews,* **9**: 137–152.

Smith, F. A., and J. L. Betancourt, 1998: Response of bushy-tailed woodrats (*Neotoma cinerea*) to late Quaternary climatic change in the Colorado Plateau. *Quaternary Research,* **50**: 1–11.

Street-Perrott, A. F., Y. Huang, R. A. Perrott, G. Eglinton, P. Barker, L. B. Khelifa, D. D. Harkness, and D. O. Olago, 1997: Impact of lower atmospheric carbon dioxide on tropical mountain ecosystems. *Science,* **278**: 1422–1426.

Stuiver, M., and P. J. Reimer, 1993: Extended ^{14}C database and revised CALIB radiocarbon calibration program. *Radiocarbon,* **35**: 215–230.

Stute, M., and P. Schlosser, 1993: Principles and applications of the noble gas paleothermometer. *In* Swart, P. K., K. C. Lohmann, J. McKenzie, and S. Savin (eds.), Climate Change in Continental

Isotopic Records. Washington, DC: American Geophysical Union, *Geophysical Monograph* **78**: 89–100.

Stute, M., M. Forster, H. Frischkorn, A. Serejo, J. F. Clark, P. Schlosser, W. S. Broecker, and G. Bonani, 1995: Cooling of tropical Brazil (5°C) during the last glacial maximum. *Science,* **269**: 379–383.

Stute, M., P. Schlosser, J. F. Clark, and W. S. Broecker, 1992: Paleotemperatures in the southwestern United States derived from noble gas measurements in groundwater. *Science,* **256**: 1000–1003.

Taylor, D. M., 1990: Late Quaternary pollen records from two Ugandan mires: Evidence for environmental change in the Rukiga highlands of southwest Uganda. *Palaeogeography, Palaeoclimatology, Palaeoecology,* **80**: 283–300.

van der Hammen, T., 1963: A palynological study on the Quaternary of British Guiana. *Leidse Geologische Mededelingen,* **29**: 125–180.

van der Hammen, T., 1974: The Pleistocene changes of vegetation and climate in tropical South America. *Journal of Biogeography,* **1**: 3–26.

van der Hammen, T., and M. L. Absy, 1994: Amazonia during the last glacial. *Palaeogeography, Palaeoclimatology, Palaeoecology,* **109**: 247–261.

van der Hammen, T., and E. Gonzalez, 1960: Upper Pleistocene and Holocene climate and vegetation of the Sabana de Bogotá (Colombia, South America). *Leidse Geologische Mededelingen,* **25**: 126–315.

van der Hammen, T., and H. Hooghiemstra, 2000: Neogene and Quaternary history of vegetation, climate and plant diversity in Amazonia. *Quaternary Science Reviews,* **19**: 725–742.

Vanzolini, P. E., 1970: Zoologia sistematica, geografia e a origem das especies. *Instituto Geografico São Paulo. Serie Teses e Monografias,* **3**: 1–56.

Watts, W. A., 1975. A late Quaternary record of vegetation from Lake Annie, south-central Florida. *Geology,* **3**: 344–346.

Webb, R. S., D. H. Rind, S. J. Lehmann, R. J. Healy, and D. Sigman, 1997: Influence of ocean heat transport on the climate of the last glacial maximum. *Nature,* **385**: 695–699.

Webb, T., III, 1987: The appearance and disappearance of major vegetational assemblages: Long term vegetational dynamics in eastern North America. *Vegetatio,* **69**: 177–188.

Webster, P., and N. Streten, 1978: Late Quaternary ice-age climates of tropical Australasia, interpretation and reconstruction. *Quaternary Research,* **10**: 279–309.

Wijmstra, T. A., 1969: Palynology of the Alliance well. *Geologie en Mijnbouw,* **48**: 125–133.

Woodward, F. I., 1993: Plant responses to past concentrations of CO_2. *Vegetatio,* **104/105**: 145–155.

18

Neotropical Savanna Environments in Space and Time: Late Quaternary Interhemispheric Comparisons

HERMANN BEHLING AND HENRY HOOGHIEMSTRA

Abstract

In order to obtain a better insight into past vegetational and climatic changes along the Pole-Equator-Pole: Americas (PEP 1) transect, 32 late Quaternary pollen records from savanna and forest–savanna transition regions of the South American neotropics, north and south of the equator, have been compared. During pre-full glacial times, environmental changes in savannas were spatially complex. Some records show either stable grassland where forest exists today or a repeated alternation between forest and savanna. During the full glacial period, neotropical savannas, both north and south of the equator, expanded due to markedly drier conditions. The Amazon rain forest area must have been reduced during this period. In the southern neotropical regions, savanna area was reduced and replaced by subtropical grassland by cold climatic conditions during glacial periods. During the late glacial period, climate changed to wetter conditions north of the equator at 11,500 ^{14}C B.P. (or later) and south of the equator earlier at ca. 16,000–14,000 ^{14}C B.P. in montane regions. Wetter conditions were not recorded in the high plains or lowlands. During the early Holocene (until ca. 6000–5000 ^{14}C B.P.), the climate was drier in most of the South American savannas than during the late glacial and late Holocene. Early Holocene distribu-

tion of savanna was much larger than during the late Holocene. The general synchrony of paleoenvironmental changes since the full glacial period, from neotropical savanna sites north and south of the equator, suggests that changes in the latitudinal migration of the Intertropical Convergence Zone (ITCZ) may have played an important role. The movement of the high-pressure cell over the South Atlantic and changes in frequency of the tracks of the Antarctic cold fronts were also important. Copyright © 2001 by Academic Press.

Resumen

Con el objeto de obtener un mejor conocimiento de los cambios de vegetación y clima en el pasado a lo largo del transecto Polo-Ecuador-Polo (PEP 1), se compararon 32 registros de polen del Cuaternario Tardío de regiones de transición sabana y bosque-sabana de los neotrópicos suramericanos, al norte y al sur del ecuador. Durante tiempos pre-Pleniglaciales el cambio ambiental en las sabanas era espacialmente complejo. Hay registros que indican la presencia de pastizales estables donde hoy en día existe bosque, o bien se encuentran registros con una alternancia repetida entre bosque y sabana. Durante el periodo Pleniglacial las sabanas neotropicales, tanto al norte como al sur del ecuador, se expandieron en consecuencia de condiciones secas marcadas y el área de bosque Amazónico tropical lluvioso debió ser sido reducido. Durante los periodos glaciales y condiciones climáticas frías, el área de sabanas en las regiones neotropicales del sur habia sido limitado y reemplazado por pastizales subtropicales. Durante el Tardiglacial, alrededor de 11,500 ^{14}C B.P., (o más tarde), el clima cambió por condiciones más húmedas al norte del ecuador. Al sur del ecuador, en regiones montañosas, este cambio se realizó más temprano alrededor de 16,000–14,000 ^{14}C B.P. No se registraron condiciones más húmedas en altillanuras o en las tierras bajas. Durante el Holoceno Temprano (hasta ca. 6000–5000 ^{14}C B.P.) la mayoría de registros de las sabanas Suramericanas sugiere un clima más seco que durante los periodos del Tardiglacial y el Holoceno Tardío. La sincronía general de cambios paleo-ambientales desde el Pleniglacial, de localidades de sabana tropical al norte y al sur del ecuador, sugiere que los cambios en migración latitudinal de la ZCIT (Zona de Convergencia Inter-Tropical) debieron haber jugado un papel importante. También fueron importantes el movimiento de la célula de alta presión sobre al Atlántico sur y los cambios en la frecuencia de curso de los frentes antárticos fríos.

18.1. INTRODUCTION

Different savanna ecosystems cover large areas in the South American neotropics, second in area to the Amazon rain forest. Savanna vegetation is found north and south of the equator and is separated by rain forest. Most of the savanna regions in South America are characterized by a seasonal climate with a rainy period and a marked dry period, related to the annual movement of the Intertropical Convergence Zone (ITCZ). Pollen analysis of late Quaternary sediment deposits indicates past changes in the distribution of savanna vegetation in space and time. Comparisons of glacial and Holocene pollen records north and south of the equator help our understanding of past vegetational and climatic changes along the Pole-Equator-Pole: Americas (PEP 1) transect. Further, the comparison of records from South American savannas contributes to the ongoing discussion of the forest refugia hypothesis. According to this hypothesis, savanna vegetation replaced significant portions of the present-day Amazon rain forest area during glacial times, and the modern rain forest area was reduced to a number of forest refugia, separated from each other by different types of savanna vegetation (e.g., Haffer, 1969; Whitemore and Prance, 1987; Clapperton, 1993; Van der Hammen and Absy, 1994). That the South American rain forest has not been markedly reduced to forest refugia during the last glacial period has also been suggested by several publications (e.g., Colinvaux, 1987, 1993; Bush, 1994; Colinvaux et al., 1996; Behling, 1996).

The first palynological studies in present-day neotropical savannas were carried out by Van der Hammen (1963) in Guyana and by Wijmstra and Van der Hammen (1966) in Colombia and Suriname. Recently, 11 new sites were studied in Brazil and Colombia by Behling (see Table 1). In this chapter, we compare paleorecords for savanna sites in both hemispheres to improve our understanding of the dynamic histories of these ecosystems and the sequence of climatic change along the PEP 1. Earlier overview papers (e.g., Van der Hammen, 1983; Markgraf, 1993) did not compare savanna sites north and south of the equator. We consider only radiocarbon-dated records (except for one), which provide 32 pollen records for this study. Relatively small savanna areas in Central America (e.g., México, Cuba, and Panama) and pollen sites in southern subtropical Brazil are not included in this interhemispheric comparison.

18.2. MODERN NEOTROPICAL SAVANNA ENVIRONMENTS

18.2.1. Climate

Annual precipitation and its temporal distribution, which are controlled by the seasonal migration of the ITCZ and topographic features, cause significant climatic differences among different climate regions in the neotropics. The general atmospheric circulation of the tropical regions east of the Andes is controlled by the position of the ITCZ, which shifts from 7°–9°N in January to 10°–20°S in July. South America north of the equator is influenced by the northeasterly trade winds blowing from the Caribbean, whereas the region south of the equator is influenced by the Atlantic southeasterlies, which originate from the subtropical high-pressure cell over the South Atlantic Ocean. As a result, two different main climate types are represented in the neotropics: a climate without a dry season in the equatorial latitudes, and climates with marked dry seasons north and south of the equatorial latitudes. North of the equator the dry season is from October to March, and south of the equator it is from April to September. For most of the savanna regions with seasonal climates, the annual precipitation ranges from 1200–2000 mm, and the average annual temperature is 20°–26°C. Advections of Antarctic cold fronts produce heavy rainfall when they meet tropical air masses. These weather pat-

TABLE 1   List of 32 Studied Sites from Neotropical Savannas and Site-Specific Information

No.	Site[a]	Coordinates	Elevation (m asl)	Age range (years B.P.)	[14]C No.	Ref.
Northern Hemisphere						
Llanos Orientales: Colombia						
1	Laguna Sardinas	4°58'N, 69°28'W	180	0–11,600	6	Behling and Hooghiemstra, 1998
2	Laguna Angel	4°28'N, 70°34'W	200	2,000–10,030	5	Behling and Hooghiemstra, 1998
3	Laguna El Pinal	4°08'N, 71°23'W	180	0–18,000	6	Behling and Hooghiemstra, 1999
4	Laguna Carimagua	4°04'N, 70°14'W	180	0–8,270	6	Behling and Hooghiemstra, 1999
5	Laguna de Agua Sucia	3°35'N, 73°31'W	300	0–5,500	4	Wijmstra and Van der Hammen, 1966
6	Laguna Loma Linda	3°18'N, 73°23'W	310	0–8,720	8	Behling and Hooghiemstra, in preparation
Gran Sabana:Venezuela						
7	Quebrada Arapán	5°20'N, 61°06'W	900	0–ca. 3,100	1	Rull, 1991
8	Valle de Urué	5°02'N, 61°10'W	940	0–ca. 1,700	1	Rull, 1991
9	Laguna Divina Pastora	4°42'N, 61°04'W	800	0–ca. 5,300	1	Rull, 1991
10	Laguna Santa Teresa	4°43'N, 61°05'W	880	0–ca. 5,100	1	Rull, 1991
11	Santa Cruz de Mapaurí	4°56'N, 61°06'W	940	0–ca. 9,800	1	Rinaldi et al., 1990
Coastal Savanna: Guyana						
12	Ogle Bridge	6°50'N, 58°10'W	0	0–45,000	2	Van der Hammen, 1963
Coastal Savanna: Suriname						
13	Alliance Borehole (T 28)	5°53'N, 54°54'W	0		—	Wijmstra, 1971
Roraima-Rupununi Savanna: Suriname						
14	Lake Moriru	3°54'N, 59°31'W	110	0–(18,000?)	2	Wijmstra and Van der Hammen, 1996
Coastal Savanna: French Guiana						
15	GUY 2	5°44'N, 53°51'W	0	0–9,000	2	Tissot and Marius, 1992
Rupununi–Rio Branco Savanna: Northern Brazil						
16	Lago do Galheiro	3°7'N, 60°41'W	70	0–1,200	1	Absy, 1979
Southern Hemisphere						
Coastal Savanna: Northern Brazil						
17	Lago Arari	0°39'S, 49°70'W	0	0–7,500	2	Absy, 1985
18	Lagoa da Curuça A	0°46'S, 47°51'W	35	9,500–11,700	3	Behling, 1996
19	Lagoa da Curuça B	0°46'S, 47°51'W	35	0–9,500	4	Behling, 1996
20	Lago do Aquiri	3°10'S, 44°59'W	10	6,700–7,450	3	Behling and Costa, 1997
Cerrado: Northern Brazil						
21	Carajas CSS2	6°20'S, 50°25'W	700–800	0–>50,000	13	Absy et al., 1991
22	Katira	9°S, 63°W	80	18,000–49,000?	4	Van der Hammen and Absy, 1994 Absy and Van der Hammen, 1976
Cerrado: Central Brazil						
23	Agua Emendadas	15°S, 47°35'W	1040	0–26,000	6	Barberi 1994; Salgado-Labouriau et al., 1998
24	Cromínia	17°15'S, 49°25'W	730	0–32,000	7	Ferraz-Vicentini and Salgado-Labouriau, 1996; Salgado-Labouriau et al., 1997
Cerrado: Southeastern Brazil						
25	Lago do Pires	17°57'S, 42°13'W	390	0–9,700	6	Behling, 1995b
26	Lagoa Nova	17°58'S, 42°12'W	390	ca. 100–10,020	4	Behling, in preparation
27	Lagoa Serra Negra	19°S, 46°57'W	1170	0–>40,000	15	De Oliveira, 1992
28	Salitre	19°S, 46°46'W	1050	0–50,000	14	Ledru, 1993; Ledru et al., 1994, 1996
29	Lagoa Salvana	19°31'S, 42°25'W	240–300	0–ca. 10,200	3	Rodriges-Filho et al., submitted
30	Lago dos Olhos	19°38'S, 43°54'W	730	0–19,000	7	De Oliveira, 1992
31	Lagoa Santa	19°38'S, 43°54'W	730	0–6,200	2	Parizzi et al., 1998
32	Catas Altas	20°05'S, 43°22'W	755	ca. 18,000– >48,000	6	Behling and Lichte, 1997

[a] Site locations are shown in Fig. 4.

terns occur mainly in the regions of southern and southeastern Brazil. One consequence of these weather patterns is that these regions have little or no marked dry season during the austral winter (Snow, 1976; Nieuwolt, 1977; Nimer, 1989; Weischet, 1996).

18.2.2. Savanna Regions in South America

The term *savanna* is used by several authors in different ways (Eiten, 1992). Classifications of neotropical savannas, for example, were published by Cole (1986) and Eiten (1972, 1982). In this chapter, savanna encompasses vegetation types with a certain floristic composition ranging from tropical, treeless, grassy savanna to savanna woodland with shrubs and trees, including cerrado vegetation. These vegetation types occur under seasonal climatic conditions. Also, edaphic savannas, which occur on the sandstone plateaus in the Amazon

basin, for example, are included. Most savanna regions lie below 500 m elevation. As shown in Fig. 1, major areas of neotropical savanna vegetation are distributed between 9° and 3°N and 6° and 23°S (Sarmiento and Monasterio, 1975; Sarmiento, 1983; Hueck and Seibert, 1972; Seibert, 1996).

North of the equator, the *Llanos Orientales* in Colombia and the *Orinoco Llanos* in Venezuela represent the major neotropical savanna ecosystems. The primary floristic and ecological data for these savannas were published by Cuatrecasas (1989), Gentry (1993), Huber (1987), Hueck (1966), and Vareschi (1980). For the Llanos Orientales, several groups of different savanna communities have been recognized by Blydenstein (1967): (1) savannas with scattered forested areas, (2) savannas with a gradually increasing density of shrubs, and (3) savanna trees following a gradient of increasing moisture from dry through seasonally inundated to hu-

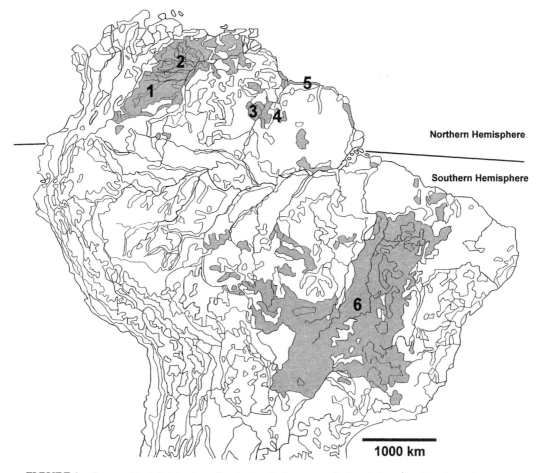

FIGURE 1   Geographical distribution of major tropical savannas in South America: (1) Llanos Orientales, (2) Orinoco Llanos, (3) Gran Sabana, (4) Rupununi–Rio Branco and Roraima-Rupununi savanna, (5) coastal Guianan savanna belt, (6) cerrado. (Adapted from Seibert, 1996.)

FIGURE 2   Photo of savanna vegetation in the Llanos Orientales of Colombia.

mid. According to Cuatrecasas (1989), characteristic taxa for grass savannas are in the family Poaceae, including the genera *Andropogon, Eragrostis, Axonopus, Paspalum, Aristida, Ctenium,* and *Panicum.* Important shrubs belong to the family Melastomataceae (*Miconia, Tibouchina*), Fabaceae (*Clitoria*), Ceasalpiniaceae (*Cassia*), Lamiaceae (*Hyptis*), Sterculiaceae (*Waltheria*), Malvaceae (*Sida, Pavonia*), and other families. Characteristic larger shrubs or trees that can be less than 1 m tall are *Curatella americana* (Dilleniaceae), *Byrsonima crassifolia* (Malpighiaceae), *Bowdichia virgilioides* (Fabaceae), and *Palicourea rigida* (Rubiaceae). The Colombian Llanos Orientales is characterized by the predominance of grass savanna with little presence of shrubs and small trees (Fig. 2). A forest occurs mainly as a gallery forest, but also in patches between the rivers. The distribution of gallery forest in the grassy and relatively flat landscape follows the drainage system of the Andes and is mainly in a west-to-east orientation. Dense stands of the palm *Mauritia* (morichal) occur along waterways and lake borders and in swampy areas. Fires are frequent in this savanna ecosystem.

Other large savanna areas are the *Gran Sabana* in southeastern Venezuela, located on a high plain between ca. 700 and 1000 m elevation; the *Rupununi–Rio Branco* savanna in Suriname; and the *Roraima-Rupununi* savanna in Brazil. Savanna vegetation also occurs along the northeastern coast of South America (Guyana, Suriname, French Guiana, and in some smaller areas of

northern Brazil in Amapá State). The coastal Guianan savanna belt has a yearly precipitation of 2000–3000 mm without a marked dry season; and therefore, it represents an edaphic savanna.

Savanna areas within the Amazon rain forest area north and south of the equator are mostly *savanna islands.* These areas can be divided into edaphic and nonedaphic savannas. Edaphic savannas occur in seasonally flooded areas (e.g., coastal savannas, Marajó savanna, riverine flooded savannas) and on white, sandy soils (e.g., coastal savannas, Amazon terra firme savannas). Nonflooded savannas in Amazonia occur on lateritic and sandy soils—e.g., the savannas of Humaitá (Gottsberger and Morawetz, 1986)—and correspond to the savanna type of the Llanos Orientales. These isolated floras might be relicts of a former, more widespread savanna area (Huber, 1982; Pires and Prance, 1985; Prance, 1996).

In the southern neotropics, the *cerrado* is the major savanna region (covering ca. 1.8 million km^2) (Fig. 1). Cerrado vegetation, which is also a savanna type (Fig. 3), occurs between 4° and 23°S, with the core area on the highland (Planalto) in central Brazil (Ferri, 1976). Other important savanna regions are the *Pantanal* in eastern Brazil and the *Llanos de Mojos* in Bolivia. The cerrado vegetation is sensitive to frost and has its southernmost limit in southeastern Brazil, where a few isolated islands of cerrado occur within the area of semideciduous forest (Hueck, 1956; Silberbauer-

FIGURE 3   Photo of savanna vegetation (cerrado) in southeastern Brazil.

Gottsberger et al., 1977). Cerrado can be classified into five physiognomic vegetation forms (Eiten, 1972, 1982): (1) grasslands (campo limpo), (2) grasslands with small shrubs and occasionally small trees (campo sujo), (3) open or closed low tree and/or scrub woodlands (campo cerrado), (4) tree and scrub woodland with 2- to 5-m-tall trees and an open tree canopy (cerrado, in the strict sense), and (5) arboreal woodlands with 5- to 15-m-tall trees with a semiclosed or closed tree canopy (cerradão). The modern composition and distribution is the result of several environmental constraints, including pedological, physical, biotic, and climatic factors. Goldsmith (1974) showed that variation in cerrado communities is strongly correlated with available soil moisture and water. Soil fertility also controls the savanna types (Goodland and Pollard, 1973; Oliveira-Filho et al., 1989). It is interesting to note that the floral relationships between different savanna types and their geographical distributions are controlled by geomorphological development of the landscape (Cole, 1960, 1982, 1986). Other authors (Ferri, 1973; Furley and Ratter, 1988) suggest that fire frequency is the most important factor in determining the physiognomy and geographical distribution of cerrado types in Brazil. The xeromorphic cerrado vegetation is well adapted to fire, which plays an important role in the savanna ecosystem, under recent conditions of human impact as well as under natural conditions (Coutinho, 1982; Eiten, 1975). The floristic composition of the extensive cerra-

do area is not homogeneous and has contact zones with other vegetation types such as the Amazon rain forest, deciduous and semideciduous forest, and caatinga. Floristic analysis of 98 cerrado sites, including Amazon savannas, showed that cerrado vegetation is very heterogeneous (Ratter et al., 1996).

18.3. POLLEN RECORDS AND PAST NEOTROPICAL SAVANNA ENVIRONMENTS

Table 1 and Fig. 4 show the relevant pollen sites available for this study and their site characteristics, such as location, ^{14}C age range, number of radiocarbon dates, and references. Today, a few sites are within contact zones or within the forested areas of transition zones. The 32 pollen records listed are located between 7°N and 21°S, extending over a distance of more than 4000 km.

18.3.1. Northern Neotropics

18.3.1.1. Llanos Orientales of Colombia

Six radiocarbon-dated pollen records from six lake cores are available for the Llanos Orientales and are located along a ca. 500-km-long transect extending northeast from the Andes to the Venezuelan border (Table 1, (Wijmstra and Van der Hammen, 1966;

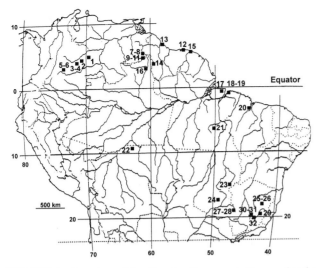

FIGURE 4 Locations of the sites of pollen records discussed in the text. The numbers (1–32) refer to Table 1.

Behling and Hooghiemstra, 1998, 1999, submitted.). No pollen records are available for the Orinoco Llanos in Venezuela, which belongs to the same savanna ecosystem. The Laguna El Pinal record for the central part of the Llanos Orientales extends back to the full glacial period at ca. 18,000 [14]C B.P. Sites Laguna Sardinas and Laguna Angel contain late glacial and Holocene deposits.

During full glacial time, savanna dominated by grasses (Poaceae) and a very few woody taxa, such as *Byrsonima* and *Curatella*, characterized the landscape in the Laguna El Pinal region. The lake was shallow and ephemeral. Gallery forests along the river system must have been rare. Such vegetation reflects the driest climatic conditions of the last 18,000 years; the annual precipitation must have been markedly lower and the annual dry season longer than they are today. During the late glacial period, at ca. 10,690 [14]C B.P., Laguna El Pinal developed into a permanent shallow lake. Permanent lakes also came into existence elsewhere in the Llanos Orientales (Laguna Angel ca. 10,030 [14]C B.P. and Laguna Sardinas ca. 11,570 [14]C B.P.), indicating a change to wetter climatic conditions than before. Increased humidity also is suggested in the region of Laguna El Pinal by greater forest representation, probably reflecting only gallery forests, and by the increased presence of *Mauritia* and *Alchornea* in the region of Laguna Sardinas.

Higher lake levels of Laguna El Pinal and Laguna Carimagua (ca. 8270 [14]C B.P.), point to wetter conditions during the Holocene than during full glacial times, but the late glacial period was apparently wetter than the early Holocene. The increase of the forest-

ed area in the records of Laguna Angel at 5260 [14]C B.P., Laguna Sardinas at 6390 [14]C B.P., and Laguna Carimagua at 5570 [14]C B.P. reflect a change from dry early Holocene conditions to wetter conditions during the mid-Holocene. The period after ca. 3800 [14]C B.P. is characterized by a marked increase in the palm *Mauritia* and / or *Mauritiella* in the Llanos Orientales, which is interpreted as reflecting increased human impact on the savanna vegetation under the wettest climatic conditions of the Holocene. The detailed Holocene pollen record from Laguna Loma Linda (Behling and Hooghiemstra, submitted) comes from the contact zone between the savanna of the Llanos Orientales and the Amazon rain forest (Fig. 5). The grass-dominated savanna in the early Holocene changed at ca. 6060 [14]C B.P. into a grass / shrub savanna with a marked expansion of the Amazon forest / gallery forest. At ca. 3600 [14]C B.P., forests started to dominate the landscape, reflecting a further continuing change to wetter climatic conditions with a shorter annual dry season. During the last 2300 years, the proportion of forest has continuously decreased and stands of *Mauritia* have increased, pointing to an increasing human influence in this region.

18.3.1.2. Gran Sabana of Venezuela

Five Holocene records are available for the Gran Sabana in southeastern Venezuela at elevations between 800 and 950 m (Rinaldi et al., 1990; Rull, 1991 [Table 1]). The longest record is from Santa Cruz de Mapaurí, with an extrapolated basal age of ca. 9800 ^{14}C years. Grass savanna predominates, with changes in the floristic composition that are interpreted as oscillations between drier and wetter periods. The Quebrada Arapán record indicates a predominance of grassland without marked changes during the last ca. 3100 ^{14}C years. At Laguna Divina Pastora, a swamp surrounded by treeless savanna was present between ca. 5300 and 2700 [14]C B.P., but forests must have been nearby. From ca. 2700–1500 [14]C B.P., the swamp changed into a lake, which was larger than it is today and was surrounded by abundant stands of *Mauritia*. Laguna Santa Teresa was a lake larger than it is now from ca. 5100–3900 [14]C B.P. and was surrounded by forest. After 3900 [14]C B.P., the lake became smaller and the forest retreated, but *Mauritia* stands markedly expanded. Since ca. 1700 [14]C B.P., the modern situation has prevailed, i.e., with a smaller lake and denser stands of *Mauritia*. The valley Urué record indicates that between 1700 and 1050 [14]C B.P., dense stands of *Mauritia* existed, apparently surrounded by savanna. The modern environmental setting with stands of *Mauritia* and a treeless savanna developed after 1050 [14]C B.P.

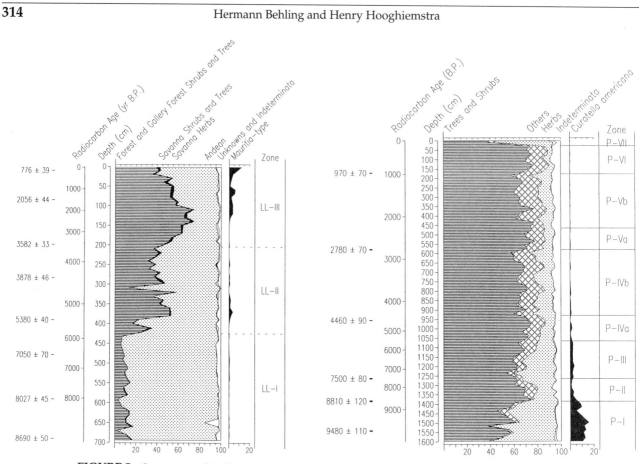

FIGURE 5 Summary pollen diagrams of the sites Laguna Loma Linda (Colombia) and Lago do Pires (Brazil).

18.3.1.4. Rupununi Savanna of Suriname

Of the three published records for the Rupununi savanna, only the pollen record for Lake Moriru has radiocarbon time control (Wijmstra and Van der Hammen 1966). Based on the two radiocarbon dates, the base of the core may extend into the late glacial or full glacial period. Prior to ca. 10,000 (or 9000 [14]C B.P.), the pollen record reflects a closed savanna woodland with abundant *Byrsonima* and some periods with more grass savanna. The authors suggest that a marked extension of savanna could be coeval with the Younger Dryas or Allerod intervals. From 10,000 (or 9000) to 5000 [14]C B.P., the proportions of woodland and grassland savanna changed repeatedly. After 5000 [14]C B.P., and especially during the last 3000 ^{14}C years, grassland savanna became dominant.

18.3.1.5. Roraima-Rupununi Savanna of Northern Brazil

The savanna region around Boa Vista, in the Brazilian state of Roraima, belongs to the same savanna system of Rupununi in Suriname. Only one short record,

for Lago Galheiro, has been published (Absy 1979). The 40- to 50-cm interval of this 260-cm-long core has been dated to 205 ± 75 [14]C B.P. There are two gaps in the sediment record, suggesting dry periods. A grass-dominated savanna with little presence of arboreal taxa and *Mauritia* occurs throughout this late Holocene with no marked changes in the vegetation.

18.3.1.6. Coastal Savanna of Guyana, Suriname, and French Guiana

In the coastal region, palynological studies were undertaken in Guyana (Van der Hammen, 1963), Suriname (Wijmstra, 1971), and French Guiana (Tissot and Marius, 1992). The 120-m-long record from the Alliance Borehole (T28) in Suriname is not dated. The pollen diagram apparently shows several glacial–interglacial cycles with mangrove vegetation (*Avicennia, Rhizophora*) during interglacial periods and savanna vegetation (Poaceae, *Curatella, Byrsonima*) during glacial periods. This sequence of vegetational change is also found in the pollen record of Ogle Bridge in Suriname, with a basal age of >45,000 ^{14}C years. During the full glacial period, savanna vegetation was present. After this period, a sea-level rise changed the environment of these sites, and locally, a mangrove developed. Swamp savanna, with mainly Poaceae and Cyperaceae, has been present during the last 5000 years in the coastal regions of Guyana, Suriname, and French Guiana.

18.3.2. Southern Neotropics

18.3.2.1. Coastal Savannas of Northern Brazil

The pollen record of Lago Arari, on Marajó Island in the mouth of the Amazon River in northeastern Pará State, represents the last 7500 ^{14}C years (Absy, 1985). At ca. 6520 [14]C B.P., there is a marked change from the more or less closed to open swamp savanna and forest. Several smaller fluctuations between forest and savanna vegetation occur during the late Holocene.

Two cores from Lagoa da Curuça (Behling, 1996), an isolated lake about 15 km from the Atlantic Ocean in northeastern Pará State, reflect the last 11,700 ^{14}C years (Profile A) and 9440 ^{14}C years (Profile B). The pollen spectra are indicative of closed and dense Amazon rain forest. The rare occurrence of the conifer *Podocarpus* suggests a marked cooling at low latitudes during late glacial times. The higher representation of Poaceae pollen and charcoal particles during the Holocene suggests the presence of anthropogenic open vegetation.

The pollen record from Lago Aquiri in northern Maranhão State (Behling and Costa, 1997) represents the period from 6700–7450 [14]C B.P. The presence at that time of mangrove vegetation indicates that the Atlantic coast migrated deeply inland within the drainage system of the Rio Mearim. Amazon rain forest occurred behind the mangrove zone, while cerrado vegetation played only a minor role in this area. The uppermost part of the core, probably representing the last 150 years, shows very different environmental conditions, with seasonally inundated swamp savannas and secondary forest. This change reflects disturbance related to human settlement.

18.3.2.2. Cerrado of Northern Brazil

The core from a swamp on the Carajas table mountain, at 700–800 m elevation in southeastern Pará State, has a basal date of >50,000 [14]C B.P. (Absy et al., 1991). The pollen record shows several alternations between arboreal and herbaceous savanna taxa (Poaceae, Asteraceae, *Borreria, Cuphea*), interpreted as alternations between forest and edaphic savanna in the surrounding lowland and on the hilltops. Savanna extension, reflecting dry episodes, occurred at ca. 60,000 and ca.

40,000 [14]C B.P., between ca. 22,900 and 12,500 [14]C B.P., and between ca. 7500 and 3000 [14]C B.P.

The pollen record from the Katira Creek valley in the state of Rondônia, western Brazil, represents the period from 49,000 (41,300) to ca. 18,000 (?) [14]C B.P., but includes only five pollen samples (Absy and Van der Hammen, 1976; Van der Hammen and Absy, 1994). The pollen diagram shows a predominance of *Mauritia* at the base of the core, followed by an interval in which Poaceae pollens dominate. This result suggests that the present-day rain forest at this site was replaced by savanna vegetation during glacial times.

18.3.2.3. Cerrado of Central Brazil

Two pollen records from palm swamps are available: the record from Agua Emendadas represents the period of 26,000–3500 [14]C B.P. (Barberi, 1994; Salgado-Labouriau et al., 1998), and the record from Cromínia reaches back to 32,000 [14]C B.P. (Ferraz-Vicentini and Salgado-Labouriau, 1996). In the Agua Emendadas record, the period from 26,000–21,500 [14]C B.P. is marked by abundant grassy vegetation and the presence of marsh, cerrado, and gallery forest taxa. Pollen grains of the palm *Mauritia* were not found. This period has been interpreted as moist and cold. The core interval representing the period from 21,500–7200 [14]C B.P. contains a thin sand layer without palynomorphs, suggesting that a climate with a long dry season prevailed. This period is followed by a transitional phase until ca. 5000 [14]C B.P.; this transitional phase is characterized by the development of a *Mauritia* palm swamp, suggesting an increase in moisture. During the late Holocene, vegetation and climate were similar to present-day conditions.

The swamp near Cromínia is located about 320 km southwest of Agua Emendadas. In the lower part of the core, three samples from the interval of 281–154 cm depth have a radiocarbon age of ca. 32,000 ^{14}C years. Vegetational and climatic conditions inferred from this core interval are similar to modern conditions: i.e., the presence of cerrado, gallery forest, and *Mauritia* palm swamp under semihumid seasonal climatic conditions. From 32,000 to ca. 20,000 [14]C B.P., the vegetation was treeless grassland with gallery forest, without the presence of a palm swamp. This period is interpreted as having wetter climatic conditions and lower temperatures than today. The period from ca. 18,500 to ca. 10,500 [14]C B.P. apparently had a landscape with reduced vegetation cover under very dry climatic conditions. During the following period, until ca. 6500 [14]C B.P., conditions were dry, and since ca. 5000 [14]C B.P., the modern vegetational composition is found in this region.

18.3.2.4. Cerrado of Southeastern Brazil

Several pollen records are available from sites in the southeastern Brazilian highland between ca. 700 and 1200 m elevation. The pollen records of Salitre and Serra Negra are from inland sites about 750 km from the Atlantic Ocean. The vegetation of this mountain region is complex, including montane vegetation, semideciduous forest, and cerrado. The base of the pollen record from Lagoa Campestre of Salitre has a radiocarbon age of >32,000 ^{14}C years (Ledru, 1993; Ledru et al., 1994, 1996) and is located only a few kilometers from the site Lagoa da Serra Negra with a pollen record dating back to >42,000 ^{14}C B.P. (De Oliveira, 1992).

In the pollen record of Salitre, the period from ca. 50,000–40,000 ^{14}C B.P. is marked by low arboreal pollen percentages reflecting arid climatic conditions. The following period until 27,000 ^{14}C B.P. shows high percentages of tree pollen (mainly Myrtaceae), which are indicative of moist climatic conditions. After a gap in the sedimentary sequence, the period from 16,000–11,000 ^{14}C B.P. reflects a succession of different forest types, and moisture must have increased. After 12,000 ^{14}C B.P., the presence of characteristic Araucaria forest taxa indicates colder and moister climatic conditions. An absence of these taxa during a following short period possibly reflects an equivalent of the Younger Dryas interval. At the beginning of the Holocene, from 10,000–8500 ^{14}C B.P., the Araucaria forest increased again, indicating a cool and moist climate. After 8500 ^{14}C B.P., the Araucaria forest was replaced by a mesophytic semideciduous forest, suggesting a warmer and drier climate. At 5000 ^{14}C B.P., a short interval with high percentages of Poaceae reflects aridity. After 4000 ^{14}C B.P., the climate was moister and comparable to present-day conditions. Higher percentages of Poaceae and Cyperaceae indicate a short, dry period at ca. 1000 ^{14}C B.P.

The pollen record of Serra Negra shows several changes in vegetation and climate before 40,000 ^{14}C B.P. (zones SN 1 to SN 6 in Ledru et al., 1998). Zone SN 7, representing the period of ca. 40,000–14,340 ^{14}C B.P., shows alternating cold and warm conditions, suggesting a time transgressive mosaic of different forest types reflecting a very cold climate with alternating moister and drier phases. A possible hiatus in the record during this period, including the full glacial period, cannot be excluded (Ledru et al., 1998). During most of the Holocene, the regional vegetation was dominated by a mosaic of cerrado and semideciduous forest, indicative of a seasonal climate. Comparisons and correlations between the Serra Negra and the Salitre records have already been presented by Ledru et al. (1996).

The pollen record from Catas Altas reflects the period of >48,000 to ca. 18,000 ^{14}C B.P. (Behling and Lichte, 1997). The landscape was characterized by grasslands with small areas of subtropical gallery forests with Araucaria. Today, this region is mainly covered by semideciduous forest with some cerrado vegetation in drier areas. Subtropical grasslands apparently migrated more than 750 km during glacial times, from southern to southeastern Brazil, reflecting a dry and cold climate with the occurrence of heavy frosts. The average temperature must have been 5°–7°C colder than today. The somewhat stronger presence of gallery forests suggests that the period from >48,000–26,500 ^{14}C B.P. was slightly wetter.

Lagoa dos Olhos (De Oliveira, 1992) and Lagoa Santa (Parizzi et al., 1998) are located 1 km from each other in karst terrain at 730 m elevation near the city of Belo Horizonte, ca. 400 km from the Atlantic Ocean. The Lagoa dos Olhos record is interpreted as an open canopy mosaic forest of Podocarpus, Rapanea, and Caryocar from 19,500–13,700 ^{14}C B.P., suggesting colder and more humid climatic conditions than exist now. During most of the Holocene, the region was apparently occupied by a mosaic of cerrado and semideciduous forest, indicative of strongly seasonal precipitation. The pollen record of Lagoa Santa starts at ca. 5400 ^{14}C ka during a period of increasing moisture. From ca. 5400 to ca. 4600 ^{14}C B.P., the presence of swamp vegetation reflects drier climatic conditions than are present now. After ca. 4600 ^{14}C B.P., a mosaic of cerrado, semideciduous forest, and gallery forest reflects climatic conditions similar to those of today, with a dry season of 3–4 months.

The lakes of Pires (Behling, 1995b), Nova (Behling, in preparation), and Silvana (Rodrigues-Filho et al., submitted) are located ca. 250 km from the Atlantic Ocean in the semideciduous forest region of the Atlantic lowland. The best-studied record is the 16-m-long core from Lago do Pires, representing the Holocene (Fig. 5). During the early Holocene from 9720–5530 ^{14}C B.P., cerrado vegetation with some gallery forest indicates low precipitation and a dry season of ca. 6 months. Within this period, an interval from 8810–7500 ^{14}C B.P. with expanding gallery forests suggests greater precipitation and a shorter dry period (6–5 months). From 5530–2780 ^{14}C B.P., the vegetation was characterized by extensive forest in the valleys and cerrado on hills, indicating higher precipitation and a dry season of ca. 5 months. From 2780–970 ^{14}C B.P., denser cerrado vegetation on the hills and slopes suggests a dry season of 4–5 months. During the latest Holocene, the presence of a closed semideciduous forest is indicative of modern climatic conditions. Inferred environmental conditions for Lago Nova (located near Lago do Pires) and

Lagoa Silvana (located 180 km south of Lago do Pires) are similar. Both records indicate that during the end of the late glacial period, savanna was the main vegetation in this region.

18.4. DISCUSSION

When pollen records from the main neotropical savanna regions are compared, differences in local environmental setting, in dating control, and in temporal resolution have to be taken into account. In the following, we discuss separately the development of coastal savannas, Northern Hemisphere inland savannas, and cerrado.

18.4.1. Coastal Savannas

In both hemispheres, the history of coastal savannas is strongly influenced by marine transgressions and regressions, but climate change and human impact may also be important. During the low sea levels of the full glacial period, coastal savannas were found in Guyana and Suriname, and the mangrove zone migrated seaward along with the shoreline. Savanna was replaced by mangroves as the sea level rose after the last deglaciation. Since mid-Holocene times (ca. 5000 ^{14}C B.P.), or somewhat earlier, savanna swamps have been found in the Guyanas, on Marajó Island in Brazil, and in Maranhão State (at least for the last 150 years), related to the changing sea level and perhaps human influence. Coastal savannas located on poor soils are primarily edaphic savannas, but may also reflect climatically controlled seasonal inundation. Today, these coastal regions have a relatively high annual rainfall of 2000–3000 mm and a relatively short dry season of 2–3 months (Walter and Lieth, 1967).

The pollen data for the coastal region of northeastern Pará (Behling, 1996) show no evidence of savannas during the late glacial period (since 11,700 ^{14}C B.P.), although some believe this region supported savanna vegetation during glacial times (e.g., Clapperton, 1993). The increased presence of charcoal in records from areas that have had open vegetation during the Holocene may reflect human activity in this region.

18.4.2. Inland Savannas (Northern Hemisphere)

Most inland savannas exist as a consequence of strong seasonality of climate. The best-studied savanna ecosystem in the Northern Hemisphere is the Llanos Orientales in Colombia. The pollen record of Laguna El Pinal, in the center of the savanna area, extends to the full glacial period and indicates extensive grass savanna with sparse savanna trees and some forested areas; the latter probably represent a gallery forest. Annual precipitation must have been markedly lower and the dry season must have been longer than it is today (modern conditions are ca. 2000 mm annual rainfall and a dry season of ca. 5 months). The age of the pollen record from Lake Moriru in the Rupununi savanna is uncertain; therefore, it is unclear which of the intervals, if any, showing *closed savanna woodland* and *grass savanna* is of full glacial age.

The formation of permanent shallow lakes and the increase of gallery forests occurred at the end of the late glacial period, indicating a change to wetter conditions. According to Wijmstra and Van der Hammen (1966), the marked expansion of savanna in the Lake Moriru record occurred during the late glacial period. It is unclear whether this expansion is coeval to the increase in gallery forests in the Llanos Orientales at the end of the late glacial period. During the early Holocene, the area of gallery forest in the Llanos Orientales decreased again, indicating relatively dry conditions.

An increase of forest area in the Llanos Orientales during the mid-Holocene reflects a change from dry early Holocene to wetter mid-Holocene climatic conditions. The initial phase of this change in precipitation is probably best documented in the high-resolution pollen record of Laguna Loma Linda (Fig. 5) at 6060 ^{14}C B.P. Laguna Loma Linda is located in the transitional zone of the Llanos Orientales savanna and the Amazon rain forest. Laguna de Agua Sucia in the Llanos Orientales and the lakes of Divina Pastora and Santa Teresa developed after this date, which suggests a synchronous increase in precipitation in the savanna region in northern South America between 6000 and 5000 ^{14}C B.P. The record of lake Moriru in the Rupununi savanna suggests more abundant grass savanna after 5000 ^{14}C B.P. and apparently does not follow this pattern.

A second significant and synchronous climate shift in the Llanos Orientales to even wetter conditions is marked by an increase of forested area and is dated between 3900 and 3600 ^{14}C B.P. At about this time, most of the pollen records for the Llanos Orientales (Behling and Hooghiemstra, 1998, 1999) and from the Gran Sabana (Rull, 1992) document a marked increase of the palm *Mauritia*, which today forms the main floristic aspect in this landscape. An expansion of *Mauritia* palms is considered an effect of greater human activity under the wettest climatic conditions recorded. The first evidence of humans in northern South America dates back to the late glacial period (e.g., Roosevelt et al., 1996; Behling, 1996), but during the late Holocene human impact on the savanna ecosystem may have been especially significant. As a consequence, vegetational change registered in pollen records may be masked by anthropogenic impact (e.g., higher frequency of savanna burning), making it difficult to detect changes forced by climate, either synchronous or opposite in nature.

We also conclude that the physical setting and the distance to the nearest forested areas may have an impact on the sensitivity of a pollen record to register climatic change. We postulate that the record from Mapaurí in the Gran Sabana does not show marked changes during the Holocene for this reason.

18.4.3. Inland Savannas (Southern Hemisphere)

Several pollen records from the southern neotropics extend into pre-full glacial times. The limits of radiocarbon dating lead to a restriction in time control; e.g., three radiocarbon dates in the lower part of the Cromínia record, with core depths between 281 and 154 cm, have the same age of ca. 32,000 ^{14}C years.

However, the pre-full glacial conditions seem to be complex and difficult to understand. Some records show no marked changes—e.g., Katira and Catas Altas—while others show significant alternations between forest and savanna—e.g. Carajas, Salitre, and Serra Negra. Periods with marked dry conditions are inferred from the Katira record (49,000 to ca. 18,000 [14]C B.P.) and also from the Catas Altas record (>47,000 to ca. 18,000 [14]C B.P.). In the latter record, the period before 27,000 [14]C B.P. seems to be slightly wetter. If the extrapolated dates are correct, expansion of the savanna due to aridity occurred in the Carajas region at ca. 60,000, 40,000, and 20,000 [14]C B.P. and in the Salitre region between ca. 50,000 and 40,000 [14]C B.P. Periods with relatively wet conditions are inferred from the pollen records of Carajas (ca. 40,000–22,800 [14]C B.P.), Aguas Emendadas (26,000–21,500 [14]C B.P.), Cromínia (32,000–20,000 [14]C B.P.), and Salitre and Serra Negra (40,000–27,000 [14]C B.P.). Based on the records discussed earlier, it seems that the area of the present cerrado vegetation, except for the Katira and Catas Altas records, experienced markedly wetter climatic conditions between ca. 40,000 and ca. 27,000 B.P. (or 20,000 [14]C B.P.).

However, the following aspects have to be considered. The date for forest expansion in the Carajas pollen record is based on the increase of *Ilex* pollen (the complete pollen record is still not available), and that in the Salitre record is based on an increase in Myrtaceae pollen; but both taxa might represent only local vegetation. We believe that the absence of *Mauritia* palm pollen in the Agua Emendadas record may not relate to cooler (Salgado-Labouriau et al., 1998) but rather to dri-

er conditions. For instance, the pollen record from Cromínia, which is located 320 km southwest of Agua Emendadas, shows high percentages of *Mauritia* at 32,000 [14]C B.P. Furthermore, it has to be considered that the records from Carajas, Salitre, and Serra Negra originate from a mountain area where orographic rains may cause *locally* higher precipitation, whereas the lowlands or the surrounding highlands of lower elevation, such as Catas Altas, may have received much less precipitation.

Climatic conditions during the full glacial period are better understood than are those during pre-full glacial periods. Many of the lake records show gaps in the sediment sequences, a problem that was summarized by Ledru et al. (1998) and suggests markedly dry conditions. Also, pollen samples from full glacial deposits in the Catas Altas record indicate that dry and cold conditions prevailed. The semideciduous forest area must have been reduced, and in the southern regions cerrado vegetation must have been replaced by subtropical grassland (Behling and Lichte, 1997; Behling, 1998).

The late glacial period is characterized by forest expansion in the Carajas mountains since ca. 12,500 [14]C B.P. The pollen records from central Brazil (Agua Emendadas and Cromínia) suggest that dry conditions prevailed (Salgado-Labouriau et al., 1998), but the dry climate during the early Holocene may have caused the erosion or oxidation of deposits of late glacial age. Between 16,000 and 11,000 [14]C B.P., the Salitre record shows a succession of different forest types, which reflects increasing moisture. The absence of *Araucaria* forests after 11,000 [14]C B.P. is interpreted to reflect a Younger Dryas event (Ledru et al., 1996; Ledru, 1993). Pollen data for the lakes Nova and Silvana in the southeastern Brazilian lowlands indicate the occurrence of cerrado and, thus, dry climatic conditions at the end of the late glacial period. The records show consistently that during the late glacial period, mountain regions received more rain than lowland regions or the high plains of central Brazil (see also Behling, 1997a). In the lowland of southeastern Brazil, grasslands of glacial age were replaced by different types of cerrado vegetation in areas with long annual dry periods (5–6 months) and by semideciduous forest in areas with shorter annual dry seasons (3–4 months). Initial forest expansion probably originated from gallery forests and from small forested areas in regions without the influence of strong frosts and sufficient moisture.

The pollen record for Lago do Pires is the most detailed Holocene record for savannas of the southern neotropics (Fig. 5) (Behling, 1995b). Cerrado occurred in the Atlantic lowlands from the early Holocene until 5500 [14]C B.P., and fires apparently were frequent. The distribution of cerrado was much larger than it is today, and it extended in easterly and southerly directions clear to the Atlantic Ocean. The present-day islands of cerrado vegetation in southeastern Brazil probably are remnants of one large continuous area that existed in the past. Annual precipitation must have been lower and the dry season 2 months longer than they are today. During the early Holocene, between 8800 and 7500 [14]C B.P., there is a moist phase in the Lago do Pires record, characterized by an expansion of gallery forests. A moist period is recorded from 10,000–8500 [14]C B.P. in the pollen record of Salitre (Ledru, 1993), but is not evidenced in the other records. Dry conditions in the early Holocene are evident from the Carajas record, which shows such conditions until 3000 [14]C B.P. or somewhat earlier. A change to wetter conditions during the mid-Holocene occurs in central Brazil (Agua Emendadas and Cromínia) and from southeastern Brazil (Salitre and Lagoa Santa) and seems to be synchronous with the increase in precipitation in the savannas of the southern neotropics. During the late Holocene, there are still remnants of cerrado on drier hilltops or slopes in the Lago do Pires region. A development to denser cerrado vegetation types or a reduction in the distribution of cerrado, reflects a second important increase in moisture at ca. 2780 [14]C B.P. Changes to wetter climatic conditions also are evidenced in the records for Serra Negra and Lago dos Olhos. The wettest period during the Holocene occurred after 970 [14]C B.P., when the entire Lago do Pires region was covered by semideciduous forests. Such conditions are not clear in other records for the cerrado, perhaps as a consequence of the low sample resolution. Very wet conditions have been found in the subtropical region of southern Brazil since ca. 1000 [14]C B.P. in Santa Catarina State (Behling, 1995a) and since 1500 [14]C B.P. in Paraná State (Serra Campos Gerais) (Behling, 1997b), where *Araucaria* forests and campos are now common.

18.5. INTERHEMISPHERIC COMPARISONS AND CAUSES OF ENVIRONMENTAL CHANGES

Records of pre-full glacial paleoenvironmental change for the inland savannas in the southern neotropics still have no equivalents for the northern neotropics, which prevents any comparison. Records of full glacial conditions indicate very dry conditions for both hemispheres, and savanna vegetation was markedly expanded at low latitudes. This first impression is based on only a few data points (only one from the northern neotropical inland savanna) within the huge area under discussion. Thus, how far savannas ex-

panded, and if the Amazon rain forest was reduced to several refugia, cannot be answered either by records for these savanna sites or by the few pollen records for the rain forest sites. Several sites in the Amazon basin do show the presence of rain forest during various intervals of the last glacial period (e.g., Colinvaux et al., 1996; Behling, 1996; Behling et al., 1998); but these sites are from regions where modern precipitation is very high, and such regions might represent the glacial forest refugia (Van der Hammen and Absy, 1994; Hooghiemstra, 1997; Hooghiemstra and Van der Hammen, 1998; Van der Hammen and Hooghiemstra, submitted).

The pollen signal from Amazon Fan sediments (Haberle, 1997; Hoorn, 1997) also is not decisive and is interpreted in different ways. Unchanging proportions of grasses and rain forest taxa can be taken as evidence for a stable forest ecosystem (Colinvaux, 1996; Haberle, 1997) and can thus be used as an argument to reject the forest refugia hypothesis; or, the same evidence can be considered suspect (Hooghiemstra and Van der Hammen, 1998) because of the potential for strong mixing of sediments from different age and source areas.

There was a major reduction of cerrado area in the southeast Brazilian regions during the pre-full glacial and the full glacial periods (Catas Altas record). Cerrado was replaced by subtropical grasslands and small areas of gallery forests. The cerrado area may have been reduced by the stronger presence of dry cold fronts, which moved across the Brazilian highlands far to the north during glacial times. Therefore, cerrado may have been limited during the pre-full glacial and full glacial periods to the northern regions of southeastern Brazil, where frost did not occur. Another suggestion is that the South Atlantic high-pressure cell was less stable during the pre-full glacial and full glacial times. Another important factor that may have changed climatic conditions is the ITCZ, which might have migrated seasonally in a more narrow latitudinal band, especially in the southern neotropical savanna regions, thus reducing precipitation and enhancing seasonality.

During the late glacial period, there is a general change to wetter climatic conditions in the neotropics, both north and south of the equator, but wetter conditions are not indicated by the two records for central Brazil or by the lowland records for southeastern Brazil. The change from a dry full glacial period to wetter conditions starts earlier in the higher elevation regions of southeastern Brazil, at 16,000–14,000 [14]C B.P., than in the Llanos Orientales, where it starts at and after ca. 11,500 [14]C B.P. Larger annual excursions of the ITCZ in both hemispheres, as shown by Martin et al. (1997), and / or strong northward shifts of the Antarctic cold front could be the reason for higher precipitation rates and a shorter annual dry season. Northward-moving cold fronts over the Brazilian highlands would also explain the higher precipitation levels at higher elevation mountains that form barriers (e.g., Salitre and Serra Negra) in contrast with drier surrounding lowlands. Further, conditions of less precipitation and a 2 month longer dry season—which are suggested in the pollen records for the lowlands in southeastern and southern Brazil for the early Holocene—can be explained by a stronger influence of the dry tropical continental air masses (Behling, 1993), which would have blocked polar cold fronts.

We note similar changes in vegetation and climate when the two most detailed and best-dated Holocene pollen records are compared (Fig. 5): Laguna Loma Linda from the Llanos Orientales and Lago do Pires from southeastern Brazil. The two sites are 4150 km apart, and today, both are located in the transition zone between forest and savanna. The first change occurs at 6050 ^{14}C B.P. (Loma Linda) and at 5530 ^{14}C B.P. (Pires), and the second occurs at 3590 ^{14}C B.P. (Loma Linda) and at 2780 ^{14}C B.P. (Pires). Both changes are ca. 500 years younger in the Pires than in the Loma Linda record, but when evidence from other sites is considered, changes in both northern and southern neotropics are more or less synchronous. There is a synchronous change to higher precipitation rates and a shorter annual dry season at mid- and late Holocene times in both hemispheres. A plausible explanation may relate to the geographical position of the ITCZ-influenced zone and a greater latitudinal range for the annual migration of the ITCZ. Improved concepts of atmospheric circulation in different time slices, and the forcing mechanisms involved, are badly needed to understand the paleodata in the context of atmospheric circulation patterns.

18.6. CONCLUSIONS

Thirty-two pollen records from different neotropical savanna regions and from forest–savanna contact zones in the neotropics of South America between the latitudes 7°N and 20°S (distance of more than 4000 km) have been compared for paleoenvironmental change. Sites in transitional zones between savanna and forest, such as Laguna Loma Linda and Lago do Pires, were most sensitive to past environmental change. Here, seasonality of climate is strong, and thresholds in the climate system and ecosystem are easily crossed. The number of pollen records is very small with regard to the huge area and the potential impact of this area on the global climate system.

The environmental changes in savannas appear to

have been spatially complex during pre-full glacial times. Records for the pre-full glacial period south of the equator either show stable grassland where today forest exists (e.g., Katira and Catas Altas) or a repeated alternation between forest and savanna (e.g., Carajas, Salitre, and Serra Negra). During the full glacial period, neotropical savannas, both north and south of the equator, expanded due to markedly drier conditions. The rain forest area must have been reduced. Whether these drier conditions led to *forest refugia* cannot be answered from the available evidence. Due to pre-full glacial and full glacial cooling, the area of tropical savanna was markedly reduced in the southern portion of its modern distribution. In both hemispheres, climate started to become wetter during the late glacial period—in the northern neotropics at ca. 11,500 ^{14}C B.P. (or later) and earlier at ca. 16,000–14,000^{14}C B.P. in the southern neotropics—but this was restricted to montane regions and was not found in the high plains or lowlands. The early Holocene (until ca. 6000–5000 ^{14}C B.P.) was drier in most of the South American savannas than during the late glacial and late Holocene. Early Holocene distribution of savanna was much larger than it was during the late Holocene.

Overall, marked vegetational and climatic changes occurred in the savanna regions. Changes since the full glacial period seem to be mostly synchronous along the PEP 1 transect crossing the equator. The general synchrony of changes in both hemispheres suggests that the latitudinal migration of the ITCZ may have played an important role. The migration of the high-pressure cell over the South Atlantic was also important, as were changes in the frequency of the tracks of Antarctic cold fronts.

Acknowledgments

The authors thank the reviewers Mark B. Bush, Barbara M. Leyden, and Vera Markgraf for constructive and valuable comments on the manuscript. The first author thanks The Netherlands Academy of Sciences (KNAW) for financial support in preparing this chapter.

References

Absy, M. L., 1979: A palynological study of Holocene sediments in the Amazon basin. Ph.D. thesis. University of Amsterdam, 86 pp.

Absy, M. L., 1985: Palynology of Amazonia: The history of the forests as revealed by the palynological record. *In* Prance, G. T., and T. E. Lovejoy (eds.), *Key Environments of Amazonia*. Oxford: Pergamon, pp. 72–82.

Absy, M. L., and T. Van der Hammen, 1976: Some palaeoecological data from Rondônia, southern part of the Amazon basin. *Acta Amazonia*, **6**: 293–299.

Absy, M. L., A. M. Cleef, M. Fournier, L. Martin, M. Servant, A. Sifeddine, M. Ferreira da Silva, F. Soubies, K. Suguio, B. Turcq, and T. Van der Hammen, 1991: Mise en évidence de quatre phases d'ouverture de la forêt dense dans le sud-est de l'Amazonie au cours des 60,000 dernières années. Première comparaison avec d'autres régions tropicales. *Comptes Rendus Academie des Sciences, Paris, Série II,* **312**: 673–678.

Barberi, M., 1994: Paleovegetacao e paleoclima no Quaternário tardio da vereda de Águas Emendadas, DF. Mestrado, University of Brasilia, **93**: 1–110.

Behling, H., 1993: Untersuchungen zur spätpleistozänen und holozänen Vegetations—und Klimageschichte der tropischen Küstenwälder und der Araukarienwälder in Santa Catarina (Südbrasilien). *Dissertationes Botanicae,* **206**: 1–149.

Behling, H., 1995a: Investigations into the late Pleistocene and Holocene history of vegetation and climate in Santa Catarina (S Brazil). *Vegetation History and Archaeobotany,* **4**: 127–152.

Behling, H., 1995b: A high resolution Holocene pollen record from Lago do Pires, SE Brazil: Vegetation, climate and fire history. *Journal of Paleolimnology,* **14**: 253–268.

Behling, H., 1996: First report on new evidence for the occurrence of *Podocarpus* and possible human presence at the mouth of the Amazon during the late-glacial. *Vegetation History and Archaeobotany,* **5**: 241–246.

Behling, H., 1997a: Late Quaternary vegetation, climate and fire history from the tropical mountain region of Morro de Itapeva, SE Brazil. *Palaeogeography, Palaeoclimatology, Palaeoecology,* **129**: 407–422.

Behling, H., 1997b: Late Quaternary vegetation, climate and fire history in the *Araucaria* forest and campos region from Serra Campos Gerais (Paraná), S Brazil. *Review of Palaeobotany and Palynology,* **97**: 109–121.

Behling, H., 1998: Late Quaternary vegetational and climatic changes in Brazil. *Review of Palaeobotany and Palynology,* **99**: 143–156.

Behling, H., in preparation: Late Quaternary vegetation, climate and fire history from Lagoa Nova in the southeastern Brazilian lowland.

Behling, H., and M. L. Costa, 1997: Studies on Holocene tropical vegetation, mangrove and coast environments in the state of Maranhão, NE Brazil. *Quaternary of South America and Antarctic Peninsula,* **10**: 93–118.

Behling, H., and H. Hooghiemstra, 1998: Late Quaternary palaeoecology and palaeoclimatology from pollen records of the savannas of the Llanos Orientales in Colombia. *Palaeogeography, Palaeoclimatology, Palaeoecology,* **139**: 251–267.

Behling, H., and H. Hooghiemstra, 1999: Environmental history of the Colombian savannas of the Llanos Orientales since the last glacial maximum from lake records El Pinal and Carimagua. *Journal of Paleolimnology,* **21**: 461–476.

Behling, H., and H. Hooghiemstra, submitted: Holocene Amazon rain forest–savanna dynamics and climatic implications: High resolution pollen record Laguna Loma Linda in eastern Colombia. *Journal of Quaternary Science.*

Behling, H., and M. Lichte, 1997: Evidence of dry and cold climatic conditions at glacial times in tropical SE Brazil. *Quaternary Research,* **48**: 348–358.

Behling, H., J. C. Berrio, and H. Hooghiemstra, 1999: Late Quaternary pollen records from the middle Caquetá river basin in central Colombian Amazon. *Palaeogeography, Palaeoclimatology, Palaeoecology,* **145**: 193–213.

Blydenstein, J., 1967: Tropical savanna vegetation of the Llanos of Colombia. *Ecology,* **48**: 1–15.

Bush, M. B., 1994: Amazonian speciation: A necessarily complex model. *Journal of Biogeography,* **21**: 5–17.

Clapperton, C., 1993: *Quaternary geology and geomorphology of South America.* Amsterdam: Elsevier, 779 pp.

Cole, M. M., 1960: Cerrado, caatinga and pantanal: Distribution and origin of the savanna vegetation of Brazil. *Geographical Journal,* **126**: 168–179.

Cole, M. M., 1982: The influence of soils, geomorphology and geology on the distribution of plant communities in savanna ecosystems. *In* Huntley, B. J., and B. H. Walker (eds.), *Ecology of Tropical Savannas. Ecological Studies,* Berlin: Springer-Verlag, **42**: 145–174.

Cole, M. M., 1986: *The Savannas: Biogeography and Geobotany.* London: Academic Press, 438 pp.

Colinvaux, P. A., 1987: Amazon diversity in the light of the paleoecological record. *Quaternary Science Reviews,* **6**: 93–114.

Colinvaux, P. A., 1993: Pleistocene biogeography and diversity in tropical forests of South America. *In* Goldblatt, P. (ed.), *Biological Relationships Between Africa and South America.* New Haven, CT: Yale University Press, pp. 473–499.

Colinvaux, P. A., 1996: Quaternary environmental history and forest diversity in the neotropica. *In* Jackson, J. B. C., A. F. Budd, and A. G. Coates (eds.), *Evolution and Environment in Tropical America.* Chicago: The University of Chicago Press, pp. 359–405.

Colinvaux, P. A., P. E. De Oliveira, J. E. Moreno, M. C. Miller, and M. B. Bush, 1996: A long pollen record from lowland Amazonia: Forest and cooling in glacial times. *Science,* **274**: 85–87.

Coutinho, L. M., 1982: Ecological effects of fire in Brazilian cerrado. *In* Huntley, B. J., and B. H. Walker (eds.), *Ecology of Tropical Savannas. Ecological Studies.* Berlin: Springer-Verlag, **42**: 273–291.

Cuatrecasas, J., 1989: Aspectos de la vegetación natural en Colombia. *Perez-Arbelaezia,* **II**(8): 155–283. Jardín Botánico de Bogotá "Jose Celestino Mutis."

De Oliveira, P. E., 1992: A palynological record of late Quaternary vegetational and climatic change in southeastern Brazil. Ph.D. thesis. The Ohio State University, Columbus, 238 pp.

Eiten, G., 1972: The cerrado vegetation of Brazil. *Botanical Review,* **38**: 205–341.

Eiten, G., 1975: The vegetation of the Serra do Roncador. *Biotropica,* **7**: 112–135.

Eiten, G., 1982: Brazilian 'savannas'. *In* Huntley, B. J., and B. H. Walker (eds.), *Ecology of Tropical Savannas. Ecological Studies.* Berlin: Springer-Verlag, **42**: 25–47.

Eiten, G., 1992: How names are used for vegetation. *Journal of Vegetation Science,* **3**: 419–424.

Ferraz-Vicentini, K. R., and M. L. Salgado-Labouriau, 1996: Palynological analysis of a palm swamp in central Brazil. *Journal of South American Earth Sciences,* **9**: 207–219.

Ferri, M. G., 1973: Sobre a origem, a manutenção e a transformação dos cerrados, tipos de savana do Brasil. *Revista de Biologia,* **9**: 1–13.

Ferri, M. G., 1976: Ecologia dos cerrados. *In* Ferri M. G. (coord.), IV Simpósio sobre o cerrado. Livraria Itatiaia Editora (LTDA), São Paulo, pp. 15–33.

Furley, P. A., and J. A. Ratter, 1988: Soil resources and plant communities of the central Brazilian cerrado and their development. *Journal of Biogeography,* **15**: 97–108.

Furley, P. A., J. Proctor, and J. A. Ratter (eds.), 1992: *Nature and Dynamics of Forest Savanna Boundaries.* London: Chapman & Hall, 616 pp.

Gentry, A. H., 1993: *Woody Plants of Northwest South America.* Washington, DC: Conservation International, 895 pp.

Goldsmith, F. B., 1974: Multivariate analysis of tropical grassland communities in Mato Grosso, Brazil. *Journal of Biogeography,* **1**: 111–122.

Goodland, R., and R. Pollard, 1973: The Brazilian cerrado vegetation: A fertility gradient. *Journal of Ecology,* **61**: 219–224.

Gottsberger, G., and W. Morawetz, 1986: Floristic, structural and phytogeographical analysis of the savannas of Humaitá (Amazonas). *Flora,* **178**: 41–71.

Haberle, S., 1997: Upper Quaternary vegetation and climate history of the Amazon basin: Correlating marine and terrestrial pollen records. *Proceedings of the Ocean Drilling Program, Scientific Results,* **155**: 381–396.

Haffer, J., 1969: Speciation in Amazonian forest birds. *Science,* **165**: 131–137.

Hooghiemstra, H., 1997: Tropical rain forest versus savanna: Two sides of a precious medal? A comment. *NWO/Huygenslezing 1997.* The Hague: Netherlands Organization for Scientific Research, pp. 31–43.

Hooghiemstra, H., and T. Van der Hammen, 1998: Neogene and Quaternary development of the neotropical rain forest: The forest refugia hypothesis, and a literature overview. *Earth Science Reviews,* **44**: 147–183.

Hoorn, C., 1997: Palynology of the Pleistocene glacial-interglacial cycles of the Amazon Fan (Holes 940A, 944A, and 946A). *Proceedings of the Ocean Drilling Program, Scientific Results,* **155**: 397–409.

Huber, O., 1982: Significance of savanna vegetation in the Amazon territory of Venezuela. *In* Prance, G. T. (ed.), *Biological Diversification in the Tropics.* New York: Columbia University Press, pp. 221–244.

Huber, O., 1987: Neotropical savannas: Their flora and vegetation. *Trends in Ecology and Evolution,* **2**: 67–71.

Hueck, K., 1956: Die Ursprünglichkeit der brasilianischen "Campos cerrados" und neue Beobachtungen an ihrer Südgrenze. *Erdkunde,* **11**: 193–203.

Hueck, K., 1966: *Die Wälder Südamerikas.* Stuttgart: Fischer, 422 pp.

Hueck, K., and P. Seibert, 1972: *Vegetationskarte von Südamerika.* Stuttgart: Fischer, 71 pp. + map.

Ledru, M. P., 1993: Late Quaternary environmental and climatic changes in central Brazil. *Quaternary Research,* **39**: 90–98.

Ledru, M. P., H. Behling, M. Fournier, L. Martin, and M. Servant, 1994: Localisation de la forêt d'*Araucaria* du Brésil au cours de l'Holocène. Implications paléoclimatiques. *Comptes Rendus de l'Academie des Sciences, Paris,* **317**: 517–521.

Ledru, M. P., J. Bertaux, A. Sifeddine, and K. Suguio, 1998: Absence of last glacial maximum records in lowland tropical forests. *Quaternary Research,* **49**: 233–237.

Ledru, M. P., P. I. Soares Braga, F. Soubiés, M. Fournier, L. Martin, K. Suguio, and B. Turcq, 1996: The last 50,000 years in the neotropics (southern Brazil): Evolution of vegetation and climate. *Palaeogeography, Palaeoclimatology, Palaeoecology,* **123**: 239–257.

Markgraf, V., 1993: Climatic history of Central and South America since 18,000 yr. B.P: Comparison of pollen records and model simulations. *In* Wright, H. E., Jr., J. E. Kutzbach, T. Webb, III, W. F. Ruddiman, F. A. Street-Perrott, and P. J. Bartlein (eds.), *Global Climates Since the Last Glacial Maximum.* Minneapolis: University of Minnesota Press, pp. 357–385.

Martin, L., J. Bertaux, T. Corrège, M.-P. Ledru, P. Mourguiart, A. Sifeddine, A. Soubiès, D. Wirrmann, K. Suguio, and B. Turcq, 1997: Astronomical forcing of contrasting rainfall changes in tropical South America between 12,400 and 8800 cal yr. B.P. *Quaternary Research,* **47**: 117–122.

Nieuwolt, S., 1977: *Tropical Climatology.* London: Wiley & Sons, 198 pp.

Nimer, E., 1989: *Climatologia do Brasil.* Rio de Janeiro: IBGE, 421 pp.

Oliveira-Filho, A. T., G. J. Shepherd, F. R. Martins, and W.H. Stubblebine, 1989: Environmental factors affecting physiognomic and floristic variation in an area of cerrado in central Brazil. *Journal of Tropical Ecology,* **5**: 413–431.

Parizzi, M. G., M. L. Salgado-Labouriau, and H. C. Kohler, 1998: Genesis and environmental history of Lagoa Santa, southeastern Brazil. *The Holocene,* **8**: 311–321.

Pires, J. M., and G. T. Prance, 1985: The vegetation types of the Brazilian Amazon. *In* Prance, G. T., and T. E. Lovejoy (eds.), *Key Environments of Amazonia.* Oxford: Pergamon, pp. 109–145.

Prance, G. T., 1996: Islands in Amazonia. *Philosophical Transactions of the Royal Society of London,* **B 351**: 823–833

Ratter, J. A., S. Bridgewater, R. Atkinson, and J. F. Ribeiro, 1996: Analysis of the floristic composition of the Brazilian cerrado vegetation II: Comparison of the woody vegetation of 98 areas. *Edinburgh Journal of Botany,* **53**: 153–180.

Rinaldi, M., V. Rull, and C. Schubert, 1990: Analisis paleoecologico de una turbera en la Gran Sabana (Santa Cruz de Mapauri), Venezuela: Resultados preliminares. *Acta Cientifica Venezolana,* **41**: 66–68.

Rodriges-Filho, S., H. Behling, G. Irion, and G. Müller, submitted: Early to mid-Holocene climatic transition in southeastern Brazil: Evidence from sediments of Lake Silvana. *Geology.*

Roosevelt, A. C., M. Lima da Costa, C. Lopes Machado, M. Michab, N. Mercier, H. Valladas, J. Feathers, W. Barnett, M. Imazio da Silveira, A. Hederson, B. Sliva, B. Chernoff, D. S. Reese, J. A. Holman, N. Toth, and K. Schick, 1996: Paleoindian cave dwellers in the Amazon: The peopling of the Americas. *Science,* **272**: 373–384.

Rull, V., 1991: Contribución a la paleoecologia de Pantepui la Gran Sabana (Guayana Venezolana): Clima, biogeografía, y ecología. *Scientia Guaianae,* **2**: 1–133.

Rull, V., 1992: Successional patterns of the Gran Sabana (southeastern Venezuela) vegetation during the last 5000 years, and its responses to climatic fluctuations and fire. *Journal of Biogeography,* **19**: 329–338.

Salgado-Labouriau, M. L., M. Barberi, K. R. Ferraz-Vicentini, and M. G. Parizzi, 1998: A dry climatic event during the late Quaternary of tropical Brazil. *Review of Palaeobotany and Palynology,* **99**: 115–129.

Salgado-Labouriau, M. L., V. Casseti, K. R. Ferraz-Vicentini, L. Martin, F. Soubiès, K. Suguio, and B. Turcq, 1997: Late Quaternary vegetational and climatic changes in cerrado and palm swamp from central Brazil. *Palaeogeography, Palaeoclimatology, Palaeoecology,* **128**: 215–226.

Sarmiento, G., 1983: The savannas in tropical America. *In* Bourlière, F. (ed.), *Tropical Savannas.* Amsterdam: Elsevier, pp. 245–288.

Sarmiento, G., and M. Monasterio, 1975: A critical consideration of the environmental conditions associated with the occurrence of savanna ecosystems in tropical America. *In* Golley, F. B., and E. Medina (eds.), *Tropical Ecological Systems: Trends in Terrestrial and Aquatic Research.* Berlin: Springer-Verlag, pp. 223–250.

Seibert, P., 1996: *Farbatlas Südamerika: Landschaft und Vegetation.* Stuttgart: Ulmer, 288 pp.

Silberbauer-Gottsberger, I., W. Morawetz, and G. Gottsberger, 1977: Frost damage of cerrado plants in Botucatu. *Biotropica,* **9**: 253–261.

Snow, J. W., 1976: The climate of northern South America. *In* Schwerdtfeger, W. (ed.), *World Survey of Climatology, 12. Climates of Central and South America.* Amsterdam: Elsevier, pp. 295–403.

Tissot, C., and C. Marius, 1992: Holocene evolution of the mangrove ecosystem in French Guiana: A palynological study. *In* Singh, K. P., and J. S. Singh (eds.), *Tropical Ecosystems: Ecology and Management.* New Delhi: Wiley Eastern, pp. 333–347.

Van der Hammen, T., 1963: A palynological study on the Quaternary of British Guyana. *Leidse Geologische Mededelingen,* **29**: 126–168.

Van der Hammen, T., 1983: The palaeoecology and palaeogeography of savannas. *In* Bourlière, F. (ed.), *Tropical Savannas.* Amsterdam: Elsevier, pp. 19–35.

Van der Hammen, T., and M. L. Absy, 1994: Amazonia during the last glacial. *Palaeogeography, Palaeoclimatology, Palaeoecology,* **109**: 247–261.

Van der Hammen, T., and H. Hooghiemstra, in press: Neogene and Quaternary history of vegetation, climate and plant diversity in Amazonia. *Quaternary Science Reviews.*

Vareschi, V., 1980: *Vegetationsökologie der Tropen.* Stuttgart: Ulmer, 293 pp.

Walter, H., and H. Lieth, 1967: *Klimadiagramm-Weltatlas.* Jena: Fischer, 80 pp.

Weischet, W., 1996: *Reogionale Klimatologie, 1. Die Neue Welt.* Stuttgart: Teubner, 468 pp.

Whitemore, T. C., and G. T. Prance, 1987: *Biogeography and Quaternary History in Tropical America.* London: Oxford Univ. Press, 214 pp.

Wijmstra, T. A., 1971: The palynology of the Guiana coastal basin. Ph.D. dissertation. University of Amsterdam, 63 pp.

Wijmstra, T. A., and T. Van der Hammen, 1966: Palynological data on the history of tropical savannas in northern South America. *Leidse Geologische Mededelingen,* **38**: 71–90.

19

Holocene Vegetation and Climate Variability in the Americas

ERIC C. GRIMM, SOCORRO LOZANO-GARCÍA, HERMANN BEHLING,
AND VERA MARKGRAF

Abstract

The striking feature of Holocene climate change along the Pole-Equator-Pole: Americas (PEP1) transect is that both hemispheres were generally warmer and drier in the early and middle Holocene than they are today, but with some exceptions. In North America, the northwestern and southeastern parts of the continent were warmer and drier because of increased insolation and strengthening of both the Bermuda High and the Eastern Pacific Subtropical High. However, the heat low in the midcontinent strengthened the monsoon in the southwestern United States, which was wetter in the early to middle Holocene than it is today, although still warmer. In the southwestern United States and the basin of México, seasonality of precipitation was enhanced in the early Holocene with different results in the mountains and lowlands. Wetter conditions also existed in a band from Texas to the eastern Midwest. Northeast Canada, which was still covered by the Laurentide ice sheet, was the exception to an early Holocene thermal maximum. In South America, the early Holocene was also warmer and drier in most areas both north and south of the equator. In the lowlands, savanna regions were drier over a larger area than they are today, and the Andes were drier from Colombia to Tierra del Fuego, except at high elevations

in northern Argentina and Chile. During the early and middle Holocene, westerly storm tracks were more tightly focused than they are today between 45° and 50°S. Holocene climate change in the Amazon tropical rain forest was relatively minor. A reduced north–south migration of the Intertropical Convergence Zone (ITCZ) during the early to middle Holocene is a possible explanation for the synchrony of maximum aridity in the two hemispheres. The mean position of the ITCZ must also have remained more northerly year-round in the early to middle Holocene, reducing easterly moisture throughout South America. Throughout the Americas, the late Holocene was generally wetter and cooler. Copyright © 2001 by Academic Press.

Resumen

La característica más llamativa del cambio climático en el Holoceno a lo largo del transecto PEP1 es que ambos hemisferios fueron en general más cálidos y secos que hoy, a inicios y mediados del periodo, pero con algunas excepciones. En América del norte, el noroeste y el sudeste del continente fueron más cálidos debido a un incremento de la insolación y reforzamiento de los anticiclones subtropicales de Bermuda y del este del Pacifico. Sin embargo, la baja térmica del centro del continente reforzó el monzón en el suroeste de los Estados Unidos, el cual fue más húmedo que hoy a inicios y mediados del Holoceno, aunque todavía más cálido. En el suroeste de los Estados Unidos y en la cuenca de México, la estacionalidad de la precipitación aumentó a inicios del Holoceno con diferentes resultados en las montañas y en las tierras bajas. Condiciones más húmedas existieron también en el sector comprendido de Texas al este del Medio oeste. El noreste de Canada fue la excepción a un inicio del Holoceno de máximo térmico, el cual estaba aún cubierto por la capa de hielo Laurentide. En Sudamérica, el Holoceno inicial fue también más cálido y seco en la mayoría de áreas, tanto al norte como al sur del Ecuador. En las tierras bajas, las regiones de sabana fueron más secas y extensas que hoy, y los Andes, más secos desde Colombia a Tierra de Fuego, a excepción de las elevadas cotas del norte de Argentina y Chile. Durante inicios y mediados del Holoceno, las trayectorias del oeste de las tormentas fueron aún más fuertes centradas que hoy entre los 45° y 50° sur. El cambio climático del Holoceno en la selva tropical amazónica fue relativamente menor. Una reducida migración de la ITCZ (Zona de Convergenica Intertropical) durante inicios y mediados del Holoceno es una posible explicación a la sincronización del máximo de aridez en ambos hemisferios. La posición media del ITCZ debió haber quedado más hacia el norte durante todo el año, a inicios y mediados del Holoceno, reduciendo la humedad del este a través de Sudamérica. De lado a lado de América, el Holoceno tardio fue en general más húmedo y frío.

19.1. INTRODUCTION

Vegetation is one of the most important terrestrial proxies for evaluating climate change. In this chapter, we document vegetation changes during the Holocene in the Americas and discuss the climate signal that vegetation changes imply. The primary data are pollen and macrofossils from lake and peatland sediments. Pollen is a particularly valuable climate proxy because it typically forms a continuous time series and registers the timing and rates of climate change. Macrofossils do not occur as regularly nor do they sample the vegetation as completely as pollen, but because they are not so widely dispersed, they have major value in that they indicate local occurrence. Macrofossils have been particularly valuable for tracing fluctuations in tree line, which is particularly sensitive to climate.

A great diversity in physical geography and vegetation occurs throughout the Americas from the polar tundras to the tropics. In this chapter, we cannot exhaustively treat every vegetation type and region; that would require an entire book. Several recent volumes have described the major vegetation–climate trends in the Americas (e.g., Wright, 1983; Ruddiman and Wright, 1987; Wright et al., 1993), and our aim is not to duplicate these efforts. We will emphasize sites that have been published since these volumes were published. We consider major geographic regions and describe representative sites (Fig. 1 and Tables 1–3) that illustrate the regional climate trends during the Holocene. For the most part, these sites are in the Global Pollen Database available from the World Data Center-A for Paleoclimatology. We use calendar years throughout this chapter, except where otherwise noted. We consider the beginning of the Holocene to be ca. 11,500 B.P. By this time, the continental glaciers had retreated from most of North America, although ice did not completely disappear from northeastern Canada until ca. 6000 B.P. (Dyke and Prest, 1987).

19.2. NORTH AMERICA NORTH OF MÉXICO

19.2.1. Northern Alaska and Northwest Canada

The vegetation of Alaska and adjacent northwest Canada is boreal forest and tundra today. The dominant trees of the boreal forest are *Picea, Larix, Betula,* and *Populus. Pinus contorta* occurs in central Yukon. Tundra occurs in the far north, in westernmost Alaska, and at higher altitudes in the many mountain ranges. Shrub tundra dominated by *Betula glandulosa/B. nana* and *Alnus viridis* is widespread.

Betula, Alnus, and *Picea* dominate the Holocene of northern Alaska and northwestern Canada. Livingstone (1955) first identified the two-zone sequence typical of the Holocene across the region: an early Holocene *Betula* zone and a middle to late Holocene *Alnus* zone. The pollen in these zones derives from predominantly shrubby *Betula* and *Alnus.* The pollen diagram from Screaming Yellowlegs Pond (Edwards et al., 1985) is representative of northern Alaska (Fig. 2). The radiocarbon dates of the *Betula/Alnus* transition vary from ca. 10,300–8200 B.P., with much geographic variability but perhaps a west-to-east trend toward a younger age (Anderson, 1985). However, the radiocarbon dates are almost all on bulk sediment with low organic matter and typically slow sedimentation rates, and in many cases, the dates are not exactly on the transition from *Betula* to *Alnus.* Therefore, the existence or nonexistence of synchrony in this zone boundary is uncertain (Brubaker et al., 1983).

Picea shows more regional variability. It immigrated very rapidly from the east and south, increasing before *Alnus* in the east but after *Alnus* in the west. At sites

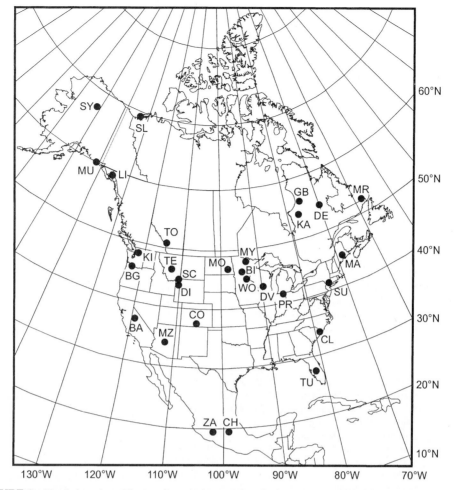

FIGURE 1 North America with sites discussed in text for which summary pollen diagrams are shown. BA, Balsam Meadow; BI, Billys Lake; BG, Battle Ground Lake; CH, Lake Chalco; CL, Clear Pond; CO, Como Lake; DE, Delorme 2; DI, Divide Lake; DV, Devils Lake; GB, Lake GB2; KA, Kanaaupscow; KI, Kirk Lake; LI, Lily Lake; MA, Mansell Pond; MO, Moon Lake; MR, Moraine Lake; MU, Munday Creek; MY, Myrtle Lake; MZ, Montezuma Well; PR, Pretty Lake; SC, Slough Creek Pond; SL, Sleet Lake; SU, Sutherland Pond; SY, Screaming Yellowlegs Pond; TE, Telegraph Creek; TO, Toboggan Lake; TU, Lake Tulane; WO, Wolsfeld Lake; ZA, Lake Zacapu.

where the pollen analyst has separated the two *Picea* species, *P. glauca* increased first, followed by *P. mariana*, after which at many sites *P. glauca* subsequently decreased (Anderson et al., 1988; Brubaker et al., 1983; Cwynar and Spear, 1995; Spear, 1993).

Sites in northern Alaska and the Yukon had a number of taxa that were north of their present range in the early Holocene, including *Populus, Myrica, Typha,* and *Juniperus,* which indicate warmer conditions than today (Ritchie et al., 1983). Total pollen influx of trees and shrubs was also typically highest in the early Holocene (Ritchie, 1984). The pollen data indicate that conditions warmer than today existed from ca. 13,000–8000 B.P., with maximum warmth at ca. 11,500 B.P., and that the warmest conditions existed before *Picea* migrated

into the region. Soon after *Picea* arrived, it became established north of its present range and at higher elevations than today. In the Tuktoyaktuk Peninsula, northern Yukon, *Picea* occurred north of its present range in the early Holocene (Ritchie and Hare, 1971; Spear, 1993), for example, at Sleet Lake (Spear, 1993) (Fig. 2); and in the southern Ogilvie Ranges, central Yukon, *Picea* occurred at higher elevations than today at sites that are now in shrub tundra (Cwynar and Spear, 1995). Cooling is evident in the pollen data by ca. 9000 B.P., but temperatures remained warmer than modern until ca. 5000 B.P. (Ritchie, 1984). The late Holocene expansion of *Alnus* indicates wetter summers (Anderson, 1985). In general, vegetation changes in the Yukon indicate cooler and wetter growing sea-

TABLE 1 Sites in North America North of Mexico

Site name	Region	Coordinates	Elevation (m)	Ref.
Sleet Lake	Northern Alaska	69°17′ N, 133°35′ W	50	Spear, 1993
Screaming Yellowlegs Pond	Northwest Canada	67°35′ N, 151°25′ W	650	Edwards et al.,1985
Lily Lake	Southeast Alaska	59°12′ N, 135°24′ W	230	Cwynar, 1990
Munday Creek	Southeast Alaska	60°02′ N, 141°58′ W	88	Peteet, 1986
Pleasant Island	Southeast Alaska	58°21′ N, 135°40′ W	150	Hansen and Engstrom, 1996
Prince William Sound	Southeast Alaska	60°58′ N, 149°00′ W	6	Heusser, 1983
Pinecrest Lake	Pacific Northwest	50°30′ N, 121°30′ W	320	Mathewes and Rouse, 1975
Marion Lake	Pacific Northwest	49°18′30″ N, 122°32′50″ W	305	Mathewes, 1973
Kirk Lake	Pacific Northwest	48°14′ N, 121°37′ W	190	Cwynar, 1987
Davis Lake	Pacific Northwest	46°35′30″ N, 122°15′00″ W	282	Barnosky, 1981
Battle Ground Lake	Pacific Northwest	45°48′00″ N, 122°29′30″ W	155	Barnosky, 1985
Little Lake	Pacific Northwest	44°10′04″ N, 123°34′56″ W	217	Worona and Whitlock, 1995
Balsam Meadow	Sierra Nevada	37°11′45″ N, 119°15′37″ W	2015	Davis et al., 1985; Anderson et al., 1985
Bonaparte Meadows	Northern Rocky Mountains	48°46′40″ N, 119°03′49″ W	1021	Mack et al., 1979
Waits Lake	Northern Rocky Mountains	48°10′29″ N, 117°47′10″ W	593	Mack et al., 1978d
Simpsons Flats	Northern Rocky Mountains	48°19′08″ N, 118°44′24″ W	535	Mack et al., 1978c
Big Meadow	Northern Rocky Mountains	48°43′39″ N, 117°33′35″ W	1040	Mack et al., 1978a
Hager Pond	Northern Rocky Mountains	48°35′51″ N, 116°58′17″ W	860	Mack et al., 1978b
Tepee Lake	Northern Rocky Mountains	48°09′45″ N, 115°24′28″ W	1270	Mack et al., 1983
Telegraph Creek	Northern Rocky Mountains	46°31′52″ N, 112°23′44″ W	2130	Brant, 1980
Lost Trail Pass Bog	Northern Rocky Mountains	45°41′35″ N, 113°57′09″ W	2152	Mehringer et al., 1977
Divide Lake	Northern Rocky Mountains	43°56′ N, 110°14′ W	2628	Whitlock and Bartlein, 1993
Slough Creek Pond	Northern Rocky Mountains	44°56′ N, 110°21′ W	1884	Whitlock and Bartlein, 1993
Toboggan Lake	Northern Rocky Mountains	50°46′ N, 114°36′ W	1480	MacDonald, 1989
Rapid Lake	Northern Rocky Mountains	42°43′37″ N, 109°11′38″ W	3134	Fall et al., 1995
Como Lake	Southern Rocky Mountains	37°33′ N, 105°30′ W	3523	Jodry et al., 1989; Shafer, 1989
Lake Emma	Southern Rocky Mountains	37°49′ N, 107°35′ W	3740	Carrara et al., 1991; Elias et al., 1991
Crested Butte	Southern Rocky Mountains	38°45′ N, 106°50′ W	2800	Markgraf and Scott, 1981
Keystone Ironbog	Southern Rocky Mountains	38°52′ N, 107°02′30″ W	2920	Fall, 1997b
Potato Lake	Southern Rocky Mountains	34°27′44″ N, 111°20′43″ W	2205	Anderson, 1993
Montezuma Well	Southwest deserts	34°38′57″ N, 111°45′08″ W	1088	Davis and Shafer, 1992
San Augustin Plains	Southwest deserts	33°52′ N, 108°15′ W	2069	Markgraf et al., 1984
Lake Cochise	Southwest deserts	32°08′ N, 109°50′ W	1260	Waters, 1989
Sierra Bacha	Southwest deserts	29°50′ N, 112°29′ W	220	Anderson and Van Devender, 1995
Rosebud	Northern Great Plains	43°00′02″ N, 101°05′12″ W	894	Watts and Wright, 1996
Rice Lake	Northern Great Plains	48°00′36″ N, 101°31′48″ W	620	Grimm, unpublished
Pickerel Lake	Northern Great Plains	45°30′18″ N, 97°16′18″ W	563	Watts and Bright, 1968
Moon Lake	Northern Great Plains	46°51′27″ N, 98°09′32″ W	444	Laird et al., 1996
Myrtle Lake	Midwest	47°58′56″ N, 93°23′09″ W	399	Janssen, 1968
Pogonia Bog Pond	Midwest	45°02′08″ N, 93°37′47″ W	291	Swain, 1979
Wolsfeld Lake	Midwest	45°00′16″ N, 93°34′16″ W	292	Grimm, 1983
Billys Lake	Midwest	46°16′15″ N, 94°33′02″ W	382	Jacobson and Grimm, 1986
Roberts Creek	Midwest	42°59′10″ N, 91°30′01″ W	308	Baker et al., 1992
Devils Lake	Midwest	43°25′05″ N, 89°43′55″ W	294	Baker et al., 1992
Lake Mendota	Midwest	43°06′ N, 89°25′ W	257	Winkler et al., 1986
Pretty Lake	Midwest	41°34′30″ N, 85°15′04″ W	294	Williams, 1974
Crawford Lake	Midwest	43°30′ N, 79°50′ W	102	Yu et al., 1997
Rice Lake	Midwest	44°17′ N, 78°19′ W	187	Yu and McAndrews, 1994
Sutherland Pond	Northeast	41°23′29″ N, 74°02′16″ W	380	Maenza-Gmelch, 1997
Rogers Lake	Northeast	41°22′ N, 72°07′ W	91	Davis, 1969
Mansell Pond	Northeast	45°02′30″ N, 68°44′ W	58	Almquist-Jacobson and Sanger, 1995
Owasco Lake	Northeast	42°44′ N, 76°27′ W	217	Dwyer et al., 1996
Moraine Lake	Northeastern Canada	52°16′ N, 58°03′ W	385	Engstrom and Hansen, 1985
Lake GB2	Northeastern Canada	56°06′ N, 75°17′ W	300	Gajewski et al., 1993
Kanaaupscow	Northeastern Canada	54°01′40″ N, 76°38′38″ W	200	Richard, 1979
Delorme 2	Northeastern Canada	54°25′25″ N, 69°55′47″ W	538	Richard et al., 1982
Lake Tulane	Southeast	27°35′09″ N, 81°30′13″ W	36	Grimm et al., 1993
Lake Annie	Southeast	27°12′24″ N, 81°21′05″ W	34	Watts, 1975
Buck Lake	Southeast	27°14′07″ N, 81°19′57″ W	27	Watts et al., 1996
Barchampe Lake	Southeast	30°37′20″ N, 83°15′14″ W	34	Watts et al., 1996
Langdale Pond	Southeast	30°38′30″ N, 83°11′42″ W	58	Watts et al., 1996
Clear Pond	Southeast	33°47′55″ N, 78°57′13″ W	10	Watts et al., 1996

(continues)

TABLE 1 (*continued*)

Site name	Region	Coordinates	Elevation (m)	Ref.
White Pond	Southeast	34°10′05″ N, 80°46′35″ W	89	Watts, 1980
Camel Lake	Southeast	30°16′33″ N, 84°59′27″ W	20	Watts et al., 1992
Tombigbee	Southeast	33°36′16″ N, 88°13′25″ W	49	Whitehead and Sheehan, 1985
Cupola Lake	Southeast	36°47′57″ N, 91°05′22″ W	259	Smith, 1984
Ferndale Bog	Southeast	34°24′26″ N, 95°48′18″ W	277	Albert and Wyckoff, 1981

TABLE 2 Sites in Mexico, Central America, and the Caribbean

Site name	Country	Area	Coordinates	Elevation (m)	Ref.
Tlaloqua	México	Puebla-Tlaxcala	19°15′ N, 98°15′ W	3100	Ohngemach, 1977; Heine, 1988
Lake Chalco	México	Basin of México	19°15′ N, 98°55′ W	2240	Lozano-García et al., 1993; Lozano-García and Ortega-Guerrero 1994; Caballero-Miranda and Ortega-Guerrero, 1998; Sosa-Najera et al., 1998
Lake Texcoco	México	Basin of México	19°26′ N, 99°01′ W	2200	Lozano-García and Ortega-Guerrero, 1998
Lake Texcoco-Mulco	México	Basin of México	19°53′ N, 98°21–98°25′ W	2500	Caballero-Miranda et al., 1999
Lake Quila	México	Trans-Mexican Volcanic Belt	19°04′ N, 99°19′ W	3010	Almeida-Leñero, 1997
Lake Zempoala	México	Trans-Mexican Volcanic Belt	19°03′ N, 99°18′ W	2800	Almeida-Leñero, 1997
Lake Zacapu	México	Michoacan	19°51′ N, 101°40′ W	1980	Metcalfe, 1995; Xelhuantzi-López, 1994; Arnauld et al., 1997
Lake Cuitzeo	México	Michoacan	19°53′–20°04′ N, 100°50′–101°19′ W	1831	Velázquez-Durán, 1998
Lake Pátzcuaro	México	Michoacan	19°55′ N, 101°39′ W	2035	Watts and Bradbury, 1982; O'Hara et al., 1994
La Hoya de San Nicolas de Parangueo	México	Valle Santiago	20°23′ N, 101°17′ W	1700	Brown, 1984
Laguna San Pedro	México	Nayarit	21°12′30″ N, 104°45′ W	1300	Brown, 1984, 1985
Cenote San José Chulchacá	México	Yucatán	21°38′ N, 87°37′ W	1–3	Leyden et al., 1996
Lake Chichancanab	México	Yucatán	19°50′ N, 88°45′ W	15	Hodell et al., 1995
Lake Cobá	México	Yucatán	20°50′ N, 88° W	<25	Leyden et al., 1998
Laguna Cocos	México	Yucatán	18°10′ N, 88°40′ W	20	Hansen, 1990
Lake Sayaucil	México	Yucatán	20°41′ N, 88°49′ W	<25	Whitmore et al., 1996
Punta Laguna	México	Yucatán	20°38′ N, 87°37′ W	14	Curtis et al., 1996
Quexil	Guatemala	Petén lowlands	16°55′ N, 89°49′ W	110	Vaughan et al., 1985; Leyden, 1984; Leyden et al., 1993
Salpeten	Guatemala	Petén lowlands	16°59′ N, 89°40′ W	104	Leyden, 1984, 1987
Petén Itzá	Guatemala	Petén lowlands	16°55′ N, 89°50′ W	110	Islebe et al., 1996b; Curtis et al., 1998
Lachner Bog	Costa Rica	Sierra de Talamanca	9°43′ N, 83°56′ W	2400	Martin, 1964
Lago de las Morrenas	Costa Rica	Sierra de Talamanca	9°29′ N, 83°29′ W	3480	Horn, 1993
Lago Chirripó	Costa Rica	Sierra de Talamanca	9°29′05″ N, 83°29′52″ W	3520	Horn, 1993; Islebe and Hooghiemstra, 1997
La Chonta	Costa Rica	Sierra de Talamanca	9°41′ N, 83°57′ W	2310	Islebe and Hooghiemstra, 1997; Hoogheimstra et al., 1992
La Trinidad	Costa Rica	Sierra de Talamanca	9°50′ N, 84° W	2700	Islebe et al., 1996a; Islebe and Hoogheimstra, 1997
La Yeguada	Panama	Veraguas Province	8°27′ N, 80°51′ W	650	Piperno er al., 1990; Bush et al., 1992
Lake Miraoane	Haiti		18°24′ N, 73°05′ W	20	Hodell et al., 1991
Church's Blue Hole	Bahamas	Andros Island	24°45′ N, 77°53′ W	0	Kjellmark, 1996
Gatún Basin	Panama	Canal Zone	8°27′ N, 80°51′ W	650	Bartlett and Barghoorn, 1973

TABLE 3 **Sites in South America**

Site name	Country	Area/vegetation	Coordinates	Elevation (m)	Ref.
Laguna Sardinas	Columbia	Llanos Orientales	4°58′N, 69°28′W	180	Behling and Hooghiemstra, 1998
Laguna Angel	Columbia	Llanos Orientales	4°28′N, 70°34′W	200	Behling and Hooghiemstra, 1998
Laguna El Pinal	Columbia	Llanos Orientales	4°08′N, 70°23′W	180	Behling and Hooghiemstra, 1999
Laguna Carimagua	Columbia	Llanos Orientales	4°04′N, 70°14′W	180	Behling and Hooghiemstra, 1999
Laguna Loma Linda	Columbia	Llanos Orientales	3°18′N, 73°23′W	310	Behling and Hooghiemstra, 2000
Laguna de Agua Sucia	Columbia	Llanos Orientales	3°35′N, 73°31′W	300	Wjimstra and van der Hammen, 1966
Laguna Divina Pastora	Venezuela	Gran Sabana	4°42′N, 61°04′W	800	Rull, 1991
Laguna Santa Teressa	Venezuela	Gran Sabana	4°43′N, 61°05′W	880	Rull, 1991
Santa Cruz de Mapaurí	Venezuela	Gran Sabana	4°56′N, 61°06′W	940	Rinaldi et al., 1990
Lake Moriru	Surinam	Rupununi savanna	3°54′N, 59°31′W	110	Wjimstra and van der Hammen, 1966
Laguna Piusbi (Chocó)	Columbia	Chocó rain forest in the Pacific lowland	1°53′N, 77°56′W	80	Behling et al., 1998
Pantano de Monica 1	Columbia	Amazon rain forest	0°42′S, 72°04′W	160	Behling et al., 1998
Pantano de Monica 2	Columbia	Amazon rain forest	0°42′S, 72°04′W	160	Behling et al., 1998
Pantano de Monica 3	Columbia	Amazon rain forest	0°42′S, 72°04′W	160	Behling et al., 1998
Mariname I	Columbia	Amazon rain forest	0°44′S, 72°04′W	160	Urrego, 1994/1997
Mariname II	Columbia	Amazon rain forest	0°45′S, 72°03′W	160	Urrego, 1994/1997
Lago Ayauchi	Ecuador	Amazon rain forest	2°05′S, 78°01′W	500	Bush and Colinvaux, 1988
Lago Kumpaka	Ecuador	Amazon rain forest	3°02′S, 77°49′W	700	Liu and Colinvaux, 1988
Lake Pata	Brazil	Amazon rain forest	0°16′S, 66°41′W	300	Colinvaux et al., 1996; De Oliveira, 1996
Lagoa da Curuça B	Brazil	Amazon rain forest	0°46′S, 47°51′W	35	Behling, 1996
Lago Surara	Brazil	Amazon rain forest	4°09′S, 61°46′W	76	Absy, 1979
Carajas CSS2	Brazil	Amazon rain forest	6°20′S, 50°25′W	700–800	Absy et al., 1991
Agua Emendadas	Central Brazil	Cerrado	15°S, 47°35′W	1040	Barberi, 1994; Salgado-Labouriau et al., 1998
Crominia	Central Brazil	Cerrado	17°15′S, 49°25′W	730	Ferraz-Vincentini and Salgado-Labouriau, 1996; Salgado-Labouriau et al., 1997
Lago do Pires	SE Brazil	Cerrado, semidecidous forest, and mountain vegetation	17°57′S, 42°13′W	390	Behling, 1995a
Lagoa Serra Negra	SE Brazil	″	19°S, 46°57′W	1170	De Oliveira, 1992
Salitre	SE Brazil	″	19°S, 46°46′W	1050	Ledru, 1993; Ledru et al., 1994, 1996
Lago dos Olhos	SE Brazil	″	19°38′S, 43°54′W	730	De Oliveira, 1992
Logoa Santa	SE Brazil	″	19°38′S, 43°54′W	730	Parizzi et al., 1998
Morro de Itapeva	SE Brazil	″	22°47′S, 45°28′W	1850	Behling, 1997a
Serra do Rio Rastro	S Brazil	*Araucaria* forest and campos region on the southern Brazilian highland and tropical Atlantic rain forest in the southern lowland	28°23′S, 49°33′W	1420	Behling, 1995b, 1998
Morro da Igreja	S Brazil	″	28°11′S, 49°52′W	1800	Behling, 1995b, 1998
Serra da Boa Vista	S Brazil	″	27°42′S, 49°09′W	1160	Behling, 1993, 1998
Poço Grande	S Brazil	″	26°25′S, 48°52′W	10	Behling, 1993, 1998
Serra Campos Gerais	S Brazil	″	24°40′S, 50°13′W	1200	Behling, 1997b, 1998
Arroyo Sauce Chico	Argentina	Pampa	38°05′S, 62°16′W		Prieto, 1989
Fortín Necochea	Argentina	Pampa	37°23′S, 61°08′W		Nieto and Prieto, 1987
Empalme Querandíes	Argentina	Pampa	37°00′S, 60°07′W		Prieto, 1989
Cerro La China	Argentina	Pampa	37°57′S, 58°37′W	200	Prieto and Páez, 1998
Arroyo Las Brusquitas	Argentina	Pampa	38°14′S, 58°52′W	5	D'Antoni et al., 1985
La Primavera	Columbia	Paramo	4°N, 74°10′W	3525	Melief, 1985
Laguna Huatacocha	Peru	Superpuna	10°46′S, 76°37′W	4500	Hansen et al., 1984
Laguna Fúquene	Colombia	Subparamo	5°40′N, 73°53′W	2580	van Geel and van der Hammen, 1973

(continues)

TABLE 3 (*continued*)

Site name	Country	Area/vegetation	Coordinates	Elevation (m)	Ref.
Caunahue	Chile	Valdivian rainforest, west slope of the Andes	40°S, 72°W	500	Markgraf, 1991a
Mallín Aguado	Argentina	*Nothofagus dombeyi-Austrocedrus* forest, east slope of the Andes	40°S, 71°29′W	840	Markgraf and Bianchi, 1999
Harberton	Argentina	Subantarctic mixed evergreen-deciduous forest	54°53′S, 67°10′W	20	Markgraf, 1991a

sons after 6800 B.P., when *Picea mariana* and *Alnus* expanded at the expense of *Picea glauca* (Cwynar and Spear, 1995). Therefore, the Holocene trend is toward cooler, wetter summers, with the warmest conditions during the earliest Holocene.

19.2.2. Southeast Alaska

The modern vegetation of southeast Alaska is Pacific coastal forest consisting of *Picea sitchensis, Tsuga heterophylla,* and *T. mertensiana. Pinus contorta* var. *contorta* occurs in the southern part of the area, especially on muskegs, which are widespread. In the early Holocene, *Alnus* dominated, as in the Alaskan interior, although it was probably the subspecies *A. viridis* ssp. *sinuata,* which is present today, instead of *A. viridis* ssp. *crispa,* which occurs in the interior. *Pinus contorta* migrated

very rapidly into the southern part of the area, reaching Lily Lake by at least 12,000 years ago (Fig. 3) (Cwynar, 1990). *Picea sitchensis, Tsuga heterophylla,* and *T. mertensiana* migrated up the coast during the Holocene. *Picea* arrived first, but the order of the two *Tsuga* species varied regionally. *T. mertensiana* preceded *T. heterophylla* at Pleasant Island (Hansen and Engstrom, 1996) at the southern end of the region and at Prince William Sound (Heusser, 1983) at the northern end, but *T. heterophylla* preceded *T. mertensiana* by at least 1000 years at Lily Lake (Cwynar, 1990) and Munday Creek (Peteet, 1986). Pleasant Island, Lily Lake, Munday Creek, and Prince William Sound form a southeast–northwest transect along the Pacific coast. Arrival times along this transect were 10,700, 10,700, 8400, and 2800 B.P. for *Picea sitchensis;* 9800, 7800, 3600, and 2800 B.P. for *Tsuga mertensiana;* and 9500, 8500, 4200, and 2000

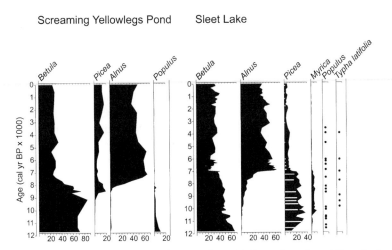

FIGURE 2 Representative pollen diagrams from northern Alaska and northwest Canada: Screaming Yellowlegs Pond (Edwards et al., 1985) and Sleet Lake (Spear, 1993). *Picea glauca* is shown as white bars in the Sleet Lake diagram.

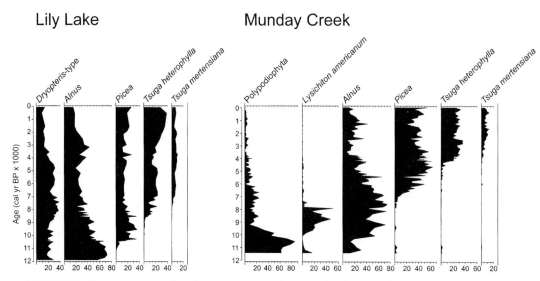

FIGURE 3   Representative pollen diagrams from southeast Alaska: Lily Lake (Cwynar, 1990) and Munday Creek (Peteet, 1986).

B.P. for *T. heterophylla.* In the southern part of Alaska, *Pinus contorta* reappeared in the middle to late Holocene, following *Picea* and the two *Tsuga* species. Increased moisture and paludification probably led to the increase in *Pinus contorta* ssp. *contorta,* which grows on muskegs (Hansen and Engstrom, 1996).

Most investigators (e.g., Peteet, 1986; Hansen and Engstrom, 1996; Heusser, 1983) ascribe the development of coastal forest to increasing precipitation and to increasing temperature in the early Holocene in the southern part of the area (Hansen and Engstrom, 1996). The development of coastal forest at Prince William Sound at the time of neoglacial cooling suggests that precipitation rather than temperature was the critical factor. The progressive cooling and increasing moisture during the Holocene suggest a weakening of the Pacific Subtropical High and an intensification of the Aleutian Low and associated cyclonic storms (Heusser et al., 1985).

Investigators have debated the relative importance of climate change versus migrational lag to explain the migration of conifers along the coast. However, macrofossil evidence indicates the presence of species long before pollen percentages increase beyond background values. A cone of *Picea sitchensis* indicates its presence in the Glacier Bay region 2500 years before pollen indicates certain presence (Ager, 1983). *Pinus contorta* macrofossils occur throughout a core from near Yakutat, Alaska, from which pollen data suggest a recent arrival (Peteet, 1993). Thus, the macrofossil data support the climate change hypothesis. In summary, the pollen and macrofossil data suggest that the early Holocene was warmer

and drier than it is today and that the trend throughout the Holocene has been toward a cooler and moister climate.

19.2.3. Pacific Northwest and the Sierra Nevada

The modern vegetation of the Pacific Northwest is conifer forest with *Tsuga heterophylla, Thuja plicata,* and *Pseudotsuga menziesii. Alnus rubra* is an important postfire species. The lowlands between the Coastal Ranges and the Cascade Range are drier; *Quercus garryana* occurs in the Willamette Valley. A north–south transect of sites have a consistent vegetation history: Pinecrest Lake (Mathewes and Rouse, 1975), Marion Lake (Mathewes, 1973), Kirk Lake (Cwynar, 1987), Davis Lake (Barnosky, 1981), Battle Ground Lake (Barnosky, 1985), and Little Lake (Worona and Whitlock, 1995).

High percentages of *Alnus, Pseudotsuga,* and *Pteridium* characterize the early Holocene, indicating drier conditions and high fire frequencies. Charcoal studies at Kirk Lake (Cwynar, 1987) and Little Lake (Long et al., 1998) confirm higher fire frequencies. The *Alnus-Pseudotsuga-Pteridium* zone began before 12,000 B.P. at most sites and ended between 7400–5000 B.P. The pollen diagram from Kirk Lake (Fig. 4) is representative of the Pacific Northwest. At Battle Ground Lake (Fig. 4) in the southern Puget Trough in southwest Washington, *Pseudotsuga–Quercus* savanna occupied the area from 12,000–5000 B.P. (Barnosky, 1985). Northward advance of *Pseudotsuga* and higher tree lines indicate warmer temperatures. Between 9800 and 7800 B.P., *Pseudotsuga*

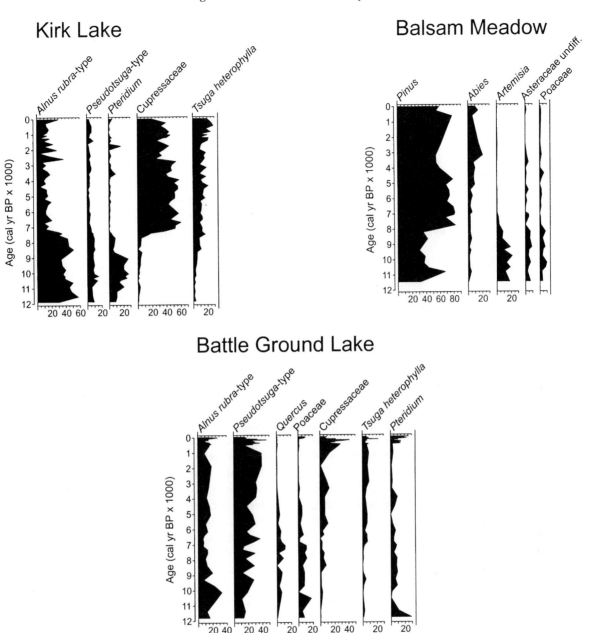

FIGURE 4   Representative pollen diagrams from the Pacific Northwest and the Sierra Nevada: Kirk Lake (Cwynar, 1987), Battle Ground Lake (Barnosky, 1985), and Balsam Meadow (Davis et al., 1985).

occurred north of its present range in northeast Vancouver Island (Hebda, 1983). Fossil wood above modern timberline dated to 10,200–9200 B.P. in the southeastern Coast Mountains of British Columbia show that timberline was at least 60 m higher and perhaps as much as 130 m higher in the early Holocene than it is today (Clague and Mathewes, 1989).

Between 8000 and 5000 B.P., *Thuja plicata* and *Tsuga heterophylla* expanded, indicating cooler, moister conditions; and lower charcoal indicates a reduction in fire

frequency (Cwynar, 1987; Long et al., 1998). The *Thuja–Tsuga* expansion began at ca. 7400 B.P. at northern sites (Pinecrest Lake, Marion Lake, and Kirk Lake) and ca. 1000 years later at southern sites (Davis Lake and Little Lake). At Battle Ground Lake, which is in a somewhat drier area, *Thuja–Tsuga* forest did not develop until ca. 5000 B.P. In the Olympic Peninsula, *Picea sitchensis* and *Tsuga heterophylla* forest developed after ca. 6800 B.P., and *Thuja plicata* expanded after 3200 B.P. (Heusser, 1974, 1977). All of the sites from Pinecrest Lake to Little

Lake show an increase in *Pseudotsuga* and *Alnus* sometime after 2000 B.P., indicating a return to somewhat drier conditions and increased fire frequency (Worona and Whitlock, 1995).

Sites in the Sierra Nevada range in California also show evidence of early Holocene aridity. At high-altitude sites in the central Sierra Nevada, an open *Pinus contorta* forest prevailed. After ca. 6800 B.P., populations of *Tsuga mertensiana* and *Abies magnifica* increased, and after 2500 B.P., these two species retreated downslope (Anderson, 1990). At lower elevations, *Artemisia* dominated from 13,000–7800 B.P., and conditions were much drier than they are today. *Pinus* forest prevailed from 7800–3200 B.P., and *Abies* increased after 3200 B.P., indicating a cooler and wetter climate (Davis et al., 1985). The pollen diagram from Balsam Meadow (Anderson et al., 1985; Davis et al., 1985) illustrates these trends (Fig. 4). Thus, the climate trends in the Sierra Nevada parallel those reconstructed from pollen records farther north, with the warmest and driest conditions during the early Holocene.

19.2.4. Northern Rocky Mountains

The modern vegetation of the northern Rocky Mountains is conifer forest with grassland or steppe in the intermontane valleys. The vegetation history varies geographically and altitudinally. Unfortunately, many of these sites are not in the pollen database.

Similar to the Pacific Northwest, the mountainous regions in northeastern Washington, northern Idaho, and northwestern Montana were warmer and drier in the early Holocene than they are today. Three sites in northeastern Washington, which are now in *Pinus–Pseudotsuga/Larix* forest, were in either *Artemisia* steppe or open *Pinus* savanna. This drier, warmer period was 11,500–5600 B.P. at Bonaparte Meadows (Mack et al., 1979) and Waits Lake (Mack et al., 1978d). The mire at Simpsons Flats (Mack et al., 1978a) was dry from 10,200–7600 B.P., after which open *Pinus* savanna occurred until 4400 B.P. Similar to sites in more mesic areas to the east, Simpsons Flats shows evidence of a cooler period from 4400–2800 B.P., when *Abies* and *Picea* were more abundant. Sites that are now in more mesic *Tsuga heterophylla* forest were open *Pinus* forest or savanna in the early Holocene. This period of more open vegetation was 11,200–3500 B.P. at Big Meadow (Mack et al., 1978b), 9300–3200 B.P. at Hager Pond (Mack et al., 1978c), and 11,500–4400 B.P. at Tepee Lake (Mack et al., 1983). At all three of these sites, a period cooler and moister than today followed with increased *Picea* and *Abies*. *Tsuga heterophylla* increased in the late Holocene: 2800 B.P. at Tepee Lake, 2400 B.P. at Big Meadow, and 1400 B.P. at Hager Pond.

At high-altitude sites in southwestern Montana and in the Yellowstone region, *Pseudotsuga* was present at higher altitudes in the early to middle Holocene than it is today. At Telegraph Creek (Fig. 5), *Pseudotsuga* was present from 11,500–4400 B.P., indicating early Holocene warming and cooling after 4400 B.P. (Brant, 1980). At Lost Trail Pass Bog (Mehringer et al., 1977), *Pseudotsuga* was present from 7800–4400 B.P., and at Divide Lake (Whitlock and Bartlein, 1993) (Fig. 5) it was present from 10,200–5700 B.P. Thus, the signal is somewhat mixed. Telegraph Creek and Divide Lake suggest early to middle Holocene warmth; whereas at Lost Trail Pass Bog, the warmer period is confined to the middle Holocene, more similar to the Great Plains to the east. At Toboggan Lake (Fig. 5) on the east slope of the Rockies in Alberta, Canada, *Picea* was less abundant in the middle Holocene from 8100–7300 B.P. (MacDonald, 1989), similar to the Great Plains records of middle Holocene warmth. On the other hand, the record from Rapid Lake (Fall et al., 1995), a high-altitude lake at tree line in the Wind River Range south of the Yellowstone region, indicates early Holocene warmth. Tree line was 100–150 m higher than it is today from 12,600–3200 B.P., and the alpine trees *Pinus albicaulis*, *Picea*, and *Abies* were more abundant than they are today. By 12,600 B.P., climate was warmer than it is today, and warmer conditions persisted until 3200 B.P. when tree line retreated.

In contrast to the precipitation minimum in the early Holocene at Divide Lake, Whitlock and Bartlein (1993) interpret increased *Pseudotsuga* in the late Holocene at Slough Creek Pond (Fig. 5) in the northern Yellowstone region to indicate a late Holocene precipitation minimum. They relate the temporal variability in the timing of maximum Holocene drought to the spatial variability in precipitation. Sites having an early Holocene warm/dry maximum are in areas having maximum precipitation in winter, whereas sites with a later Holocene warm/dry maximum are in areas having maximum precipitation in summer. Over a larger area, increased spatial and temporal variability in the timing of maximum drought may be owing to the precipitation source—either the Gulf of México or the Gulf of California. Based on patterns in the Midwest and Southwest, the Gulf of México source was minimal during the early and middle Holocene and more important during the late Holocene, whereas the Gulf of California source was the reverse. The possibility of both sources further complicates the pattern. A greater spatial array of multiproxy climate records from various elevations will be necessary to better understand the complicated spatial and temporal patterns of climate history in the northern Rocky Mountains of the United States.

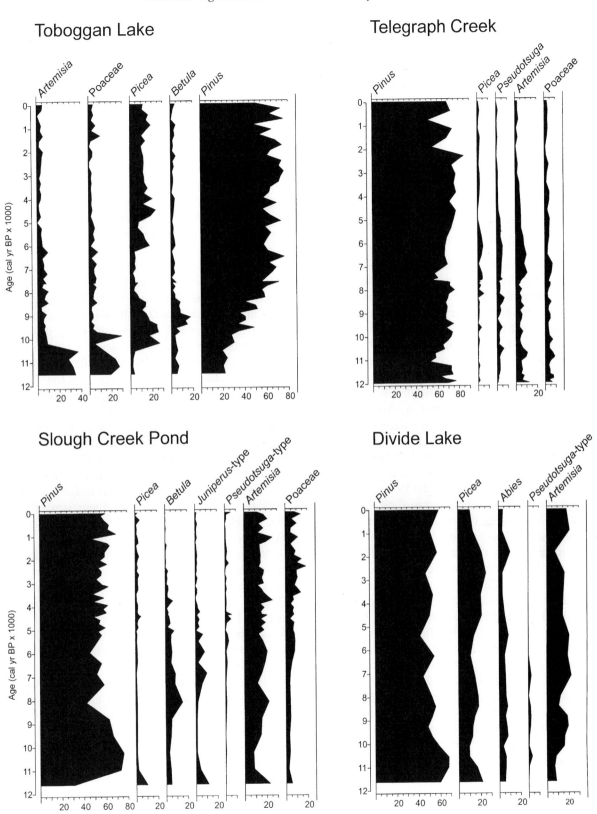

FIGURE 5 Representative pollen diagrams from the northern Rocky Mountains: Toboggan Lake (Mac-Donald, 1989), Telegraph Creek (Brant, 1980), Slough Creek Pond (Whitlock and Bartlein, 1993), and Divide Lake (Whitlock and Bartlein, 1993).

In summary, most sites in the northern Rocky Mountains were warmer in the early Holocene than they are now, and these conditions persisted until the late Holocene (see also Chapter 21). Many sites show evidence of cooling sometime after 4400 B.P., and some sites were cooler after this time then they are today, with the modern climate and vegetation regime becoming established sometime after 2800 B.P. At a few sites, the palynological evidence suggests that warmer conditions were delayed until the middle Holocene, more similar to the pattern on the Great Plains.

19.2.5. Southern Rocky Mountains and Southwest Deserts

In the southern Rocky Mountains and Southwest deserts today, conifer forests dominate at higher altitudes in the mountains, whereas desert scrub occurs at low altitudes. Maximum precipitation is in July and August, in contrast to May and June in the northern Rocky Mountains (Thompson et al., 1993). Factors complicating proxy climate data are altitudinal variability and the importance of summer versus winter precipitation. In general, the southern Rockies and Southwest were wetter and warmer in the early Holocene than they are today and became drier and cooler after 6500 B.P.

A number of studies have focused on tree line as a climate indicator. The upper tree line is particularly sensitive to temperature, whereas the lower tree line is responsive primarily to precipitation (Fall, 1997b). In the San Juan Mountains of southwest Colorado, evidence from pollen, fossil insects, and fossil wood at Lake Emma indicates that tree line was 80–140 m higher than modern levels in the early Holocene from ca. 11,000–4000 B.P. (Carrara et al., 1991; Elias et al., 1991). Isotopic analysis of fossil wood also indicates increased summer precipitation (Friedman et al., 1988). At Como Lake (Fig. 6), which lies 100 m below tree line in the Sangre de Cristo Mountains of southern Colorado, the upper tree line was highest in the earliest Holocene, 11,500–10,200 B.P., as much as 200 m higher than it is today (Jodry et al., 1989; Shafer, 1989).

In the Crested Butte area of central Colorado, the upper tree line was higher and the lower tree line was lower from 11,500 to 4400 B.P., also indicating a warmer and wetter climate during the early to middle Holocene than today (Markgraf and Scott, 1981). Fall (1997b) studied a number of sites at various altitudes in the Colorado Front Range. From 10,200–4400 B.P., the upper timberline was as much as 270 m higher than it is today, whereas the lower tree line descended, indicating a warmer and wetter climate than today. At Keystone Ironbog, fire-sensitive *Picea–Abies* forest predominated

from 9000–2800 B.P.; climate then became drier, fires became more frequent, and *Pinus contorta* forest has prevailed since (Fall, 1997a).

In contrast to these reconstructions of warmer climate in the early Holocene, Thompson (1990) argued for a cooler climate in the Great Basin from 11,300–7800 B.P., when upper montane plants occurred 400–500 m below their modern limits. However, if moisture controls these lower limits instead of temperature, the climate signal may be of increased precipitation rather than decreased temperature, an interpretation more in agreement with the Southwest. During the middle Holocene, 7800–4400 B.P., the upper tree line was 100 m higher than it is today, indicating warmer temperatures.

Packrat midden data from the deserts indicate that the early Holocene was wetter than today, although the details vary. Van Devender (1990a,b) argued for a wetter climate from 13,000–4400 B.P. in the Sonoran and Chihuahuan Deserts. In the Grand Canyon, Cole (1990) interpreted the data to indicate a warm, wet climate from 13,000–9500 B.P., becoming drier after 9500 B.P. Packrat midden and pollen data from the Sierra Bacha in Sonora, México, indicate greater precipitation in the early Holocene from 11,300–10,400 B.P. (Anderson and Van Devender, 1995). A gap in the record exists between 10,400 B.P. and 6100 B.P., by which time essentially modern conditions existed. The pollen data from Montezuma Well (Fig. 6) on the northern edge of the Sonoran Desert indicate *Juniperus* woodland from 13,000–9400 B.P., with a long-term decline in precipitation and the development of desert scrub from 9400–5500 B.P. (Davis and Shafer, 1992).

Water levels in desert lakes were higher in the early to middle Holocene. Lake Cochise in southeastern Arizona had deep phases between 10,000 and 3500 B.P. (Waters, 1989). A permanent lake existed on the San Augustin Plains in western New Mexico in the early Holocene from 11,300–5700 B.P. After 5700 B.P., the modern playa and salt flats developed (Markgraf et al., 1984). However, shallow, high-altitude lakes on the Mogollon Rim in northern Arizona and in the Chuska Mountains in northern New Mexico were frequently dry during the early and middle Holocene (Anderson, 1993; Jacobs, 1983, 1985; Wright et al., 1973), resulting in truncated Holocene records. Potato Lake on the Mogollon Rim was dry from 11,200–3000 B.P., and lake levels rose after 3000 B.P. (Anderson, 1993). Thus, these lake-level records conflict with the evidence from Colorado and the low-altitude deserts for a wetter early to middle Holocene. Seasonality of precipitation may be the explanation, inasmuch as the vegetation and water levels of desert lakes respond primarily to summer monsoonal precipitation, whereas water levels in the

Como Lake

Montezuma Well

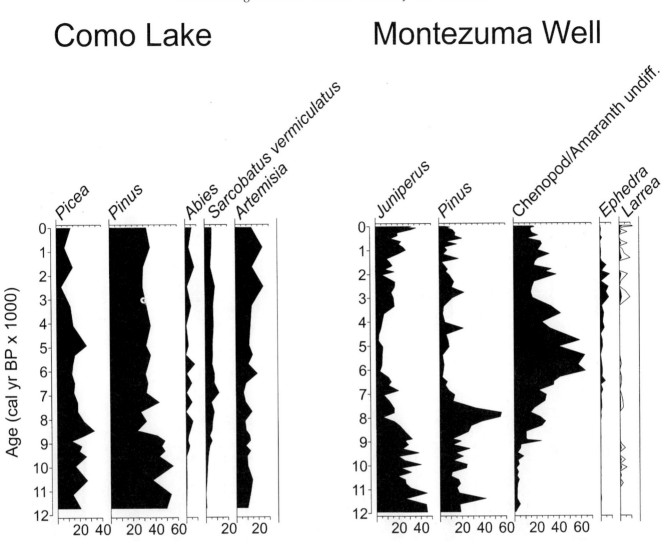

FIGURE 6 Representative pollen diagrams from the southern Rocky Mountains and Southwest deserts: Como Lake (Jodry et al., 1989; Shafer, 1989) and Montezuma Well (Davis and Shafer, 1992).

high-altitude lakes of northern Arizona and New Mexico are dependent on winter snow for replenishment. If this interpretation is correct, then the data indicate an early Holocene maximum in summer monsoonal precipitation, but with decreased winter precipitation. In the late Holocene, summer precipitation was less, but winter precipitation, especially at high altitudes, was greater.

19.2.6. Midwest and Northern Great Plains

The Midwest and northern Great Plains region, extending from the Dakotas to Ohio, includes the northern Great Plains, the western Great Lakes region, and the Prairie Peninsula. The vegetation varies from mixed-grass prairie in the northern Great Plains to tall-grass prairie in the Prairie Peninsula, a wedge of grassland

projecting eastward from the Great Plains into the Midwest from western Iowa to Indiana. Before European settlement, deciduous forest covered most of Indiana and Ohio, although islands of prairie extended into Ohio. Bordering the northern side of the Prairie Peninsula were fire-tolerant oak savannas and woodlands. More fire-sensitive deciduous forest (*Big Woods*) dominated by *Ulmus americana*, *Acer saccharum*, and *Tilia americana*, together with *Quercus*, occurred behind effective firebreaks. Northwards, these forests graded into northern hardwood–pine forests with *Betula*, *Populus*, and *Pinus* in addition to mesic hardwoods. *Picea*, *Abies*, *Larix*, and *Thuja* are important in northern Minnesota, Wisconsin, and Michigan where the Great Lakes forest grades into the boreal forest.

Picea forest covered the area at the end of the Pleistocene and declined rapidly ca. 11,500 years ago. The

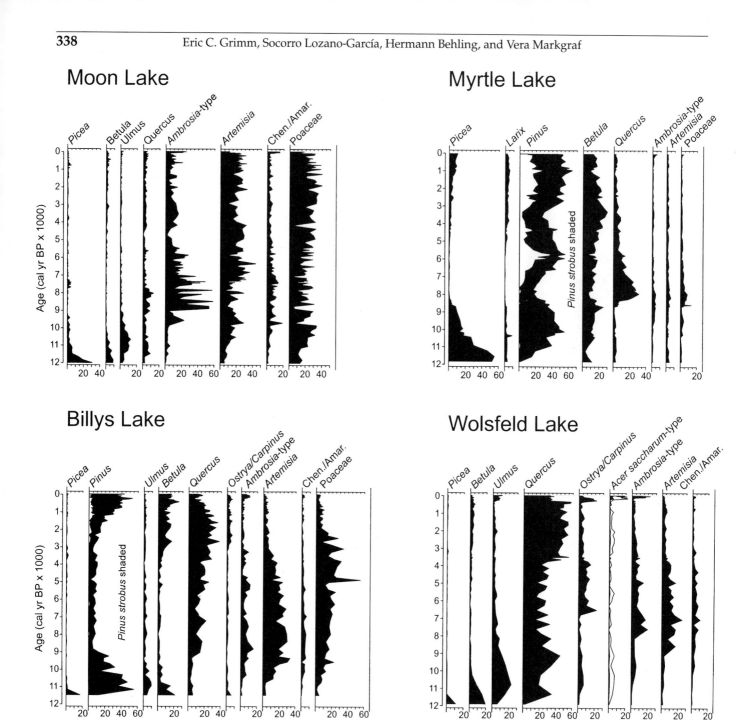

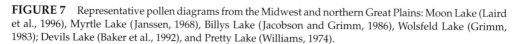

FIGURE 7　Representative pollen diagrams from the Midwest and northern Great Plains: Moon Lake (Laird et al., 1996), Myrtle Lake (Janssen, 1968), Billys Lake (Jacobson and Grimm, 1986), Wolsfeld Lake (Grimm, 1983); Devils Lake (Baker et al., 1992), and Pretty Lake (Williams, 1974).

Picea decline is precipitous at almost all sites and, within the dating errors, is synchronous across the region, except in northernmost Minnesota and southern Ontario, where it is at least 1000 years later and where *Picea* continued to persist throughout the Holocene. The regional synchrony of the *Picea* decline is somewhat problematic because it occurred during a well-known radiocarbon plateau at ca. 10,000 ^{14}C years ago and because many of the bulk-sediment radiocarbon dates are

on sediment with low organic matter and are too old because of contamination by older carbon.

The east-to-west precipitation gradient existing today is evident in the vegetation succession that occurred following the *Picea* decline. In the western Dakotas—for example, at Rosebud (Watts and Wright, 1966) and Rice Lake (Grimm, unpublished data)— prairie succeeded *Picea* forest directly. In the eastern Dakotas—for example, at Moon Lake (Fig. 7) (Laird et

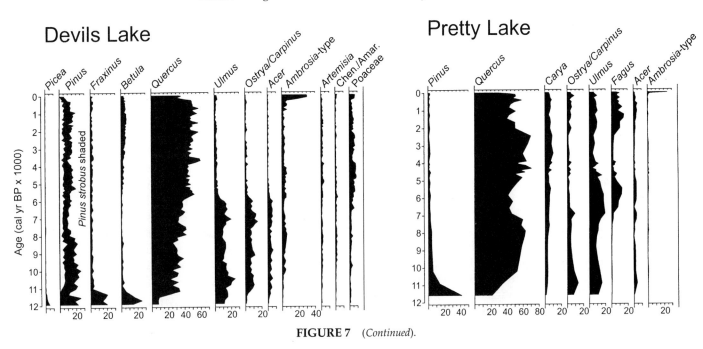

FIGURE 7 (*Continued*).

al., 1996) and Pickerel Lake (Watts and Bright, 1968)—a short phase of *Ulmus–Quercus* forest occurred between *Picea* forest and full prairie development, although prairie began expanding as *Picea* declined. Farther east in the Prairie Peninsula and in southern Minnesota and Wisconsin, the *Ulmus–Quercus* phase lasted longer, and prairie did not begin to develop until after this phase. In Minnesota and western Iowa, prairie began to develop ca. 9000 years ago. Prairie development was much delayed, however, in eastern Iowa, southern Wisconsin, and Illinois, where *Ulmus–Quercus* forest persisted until 6800–6300 years ago. North of the early Holocene *Ulmus–Quercus* forest in Minnesota, *Pinus* forest developed after the *Picea* decline. During the *Ulmus–Quercus* phase, *Ulmus* percentages commonly exceeded 40%, greater than anywhere in the modern vegetation. *Ulmus* prefers poorly drained soils, suggesting wet conditions in the early Holocene. The occurrence of other mexic taxa, such as *Acer saccharum, Fraxinus,* and *Tilia,* during this phase and very low percentages of herb pollen all point to wet conditions. However, the climate gradually became drier, and in the western part of the region, the maximum of Holocene drought was 8000–7000 years ago. Prairie advanced northeast of its modern range, and *Quercus* forest developed in northeastern Minnesota, for example, at Myrtle Lake (Fig. 7) (Janssen, 1968). By ca. 6000 years ago, the *Quercus* phase ended in northeastern Minnesota, and forest began reinvading prairie. In northwestern Minnesota, *Pinus* forest reoccupied prairie following a short *Quercus–Populus* phase, while farther south *Quercus* invaded prairie. Be-

cause of the propensity of prairie to burn, forest could not invade prairie easily, and the reinvasion occurred in a steplike manner over several thousand years (Grimm, 1983; Almendinger, 1992). Some of the most recent reinvasions coincided with the Little Ice Age (LIA) (Almendinger, 1992). Clearly, a wetter climate must have been responsible for the forest reinvasion of prairie, but because of the resistance of prairie to forest invasion, the exact relationship between climate and the position of the prairie–forest border at any one time is difficult to evaluate. Today, the prairie–forest border is stabilized along firebreaks, and no climate difference exists between one side of the firebreak and the other (Grimm, 1983, 1984).

A high-resolution study at Moon Lake in North Dakota shows that in the early to middle Holocene from ca. ~9000–5500 B.P., climate was more variable than during the later Holocene (Laird et al., 1996). Some indication of this greater variability also exists in central Minnesota, for example, at Pogonia Bog Pond (Swain, 1979) and Wolsfeld Lake (Fig. 7) (Grimm, 1983), but the herbaceous grassland community in North Dakota is probably more sensitive to short-term climate changes than communities with long-lived trees.

In Minnesota and westward, the driest time of the Holocene was the early to middle Holocene from 9000–6000 B.P. However, *Ulmus* remained prominent in the eastern Prairie Peninsula until ca. 6000 B.P., where the driest part of the Holocene was from 6300–3200 B.P. This trend is most clearly evident at Roberts Creek in northeastern Iowa and Devils Lake (Fig. 7) in south-central Wisconsin (Baker et al., 1992). At Roberts

Creek, *Ulmus* forest gave way to prairie at 6300 B.P. *Quercus* increased at 3800 B.P., and the modern mosaic of prairie and oak groves became established. At Devils Lake, mesic forest with *Ulmus, Ostrya, Acer, Carya,* and *Tilia* persisted until ca. 6300 B.P., after which more xeric *Quercus* forest with many fewer mesic trees dominated (Fig. 7). At Lake Mendota in south-central Wisconsin, lake levels were lower from 7400–3700 B.P. At 7400 B.P., a shift occurred from mesic forest to *Quercus* savanna with increased fire frequency; after 3700 B.P., fires decreased and the *Quercus* forest closed (Winkler et al., 1986).

In the middle Holocene, a steep climate gradient existed just west of the Iowa–Wisconsin border, with a drier than modern climate west of this line and a wetter than modern climate east of this line. The timing of maximum dryness was before 6300 B.P. to the west and after 6300 B.P. to the east, although the maximum severity of drought was certainly greater to the west. After 4000 years ago, sites throughout the Prairie Peninsula region showed a tendency toward a wetter climate and the development of modern vegetation. The late period of maximum dryness extends east of the Prairie Peninsula into Ohio and Ontario and is particularly evident in the pollen profiles for *Fagus*. The dating is somewhat variable, and significant hard-water error in bulk-sediment dates may be partly responsible. Nevertheless, a number of sites show a decline in *Fagus* sometime after 6000 B.P. and an increase again sometime after 3000 B.P. The pollen diagram from Pretty Lake (Williams, 1974) (Fig. 7) in northeastern Indiana clearly shows this trend. The region of late-middle Holocene dryness extends into southern Ontario. Lake-level and δ^{18}O studies at Crawford Lake indicate low water levels from 5600–2000 B.P. (Yu et al., 1997), and water levels were also low at Rice Lake in Ontario during this time (Yu and McAndrews, 1994; Yu et al., 1996).

By ca. 3000 B.P., the climate dichotomy between the western and eastern Midwest disappeared, and sites throughout the region showed a trend toward a wetter and cooler climate. *Betula* increased at many sites in the northern part of the area—for example, Billys Lake (Fig. 7) (Jacobson and Grimm, 1986)—and *Picea* increased at sites in northern Minnesota—for example, Myrtle Lake (Fig. 7) (Janssen, 1968). LIA cooling is evident at sites in Wisconsin (Gajewski, 1987, 1988; Gajewski et al., 1985) and is particularly striking in south-central Minnesota, where the Big Woods rapidly developed from oak woodland, for example, at Wolsfeld Lake (Fig. 7) (Grimm, 1983).

19.2.7. Southern High Plains

Paleobotanical evidence for climate and vegetation change on the southern plains is sparse, and tapho-

nomic problems have severely biased the record (Holliday, 1987). However, stratigraphic, geomorphic, pedologic, and stable-isotope data from Texas and New Mexico provide a paleoclimatic record. According to these data, the climate during the early Holocene (11,800–9000 B.P.) was cool with fluctuating moisture, and the most arid part of the Holocene was 9000–5000 B.P. (Holliday, 1989, 1995, 2000). During this time, wetlands dried up and dune fields became enlarged and more active. Faunal and other evidence from the Edwards Plateau in central Texas shows a trend toward a drier climate throughout the Holocene, culminating between 5800–2400 B.P. (Toomey et al., 1993), somewhat later than in west Texas. This pattern is reminiscent of that in the Midwest, where the driest part of the Holocene was later in the eastern sector.

19.2.8. Northeast

The northeastern United States is topgraphically diverse, and vegetation varies with latitude, altitude, and distance from the Atlantic Ocean. Today the vegetation is *Tsuga–Pinus*–northern hardwoods forest at lower altitudes with *Picea–Abies* forest at high altitudes. More xeric and fire-adapted *Quercus–Castanea–Carya* forest occurs in southern New England. *Pinus strobus* and *Quercus* dominated the early Holocene vegetation of the Northeast. *Tsuga* appeared in the southern part of the region from 11,000–10,000 B.P. and gradually increased, eventually replacing *Pinus strobus* as the dominant taxon in northern New England. *Fagus* increased after the decline of *Pinus strobus*. *Betula* was abundant during the *Pinus* zone, but increased following the decline of *Pinus* in the middle Holocene. *Betula* includes several species, and more fire-tolerant species may have predominated during the *Pinus* phase. *Tsuga* declined precipitously at ca. 5600 B.P., probably because of a pathogen or insect outbreak (Davis, 1981; Allison et al., 1986). *Fagus* and other species replaced *Tsuga*, which recovered several thousand years later in some areas, but never did regain its former importance in others. In more northern parts of the region and at higher altitudes, *Picea* has increased during the last 2000 years. In southern New England, *Fagus* and *Tsuga* have been present during the Holocene, but have not been particularly abundant. *Carya* and *Castanea* expanded after *Fagus* increased and were codominants with *Quercus*. Sutherland Pond (Maenza-Gmelch, 1997) (Fig. 8) in southeastern New York and Rogers Lake in Connecticut (Davis, 1969) are typical records for southern New England.

Davis et al. (1980) conducted pollen and macrofossil studies at sites near the altitudinal limits of various conifers in the White Mountains of New Hampshire. *Pinus strobus* grew as much as 350 m above its present

Sutherland Pond   Mansell Pond

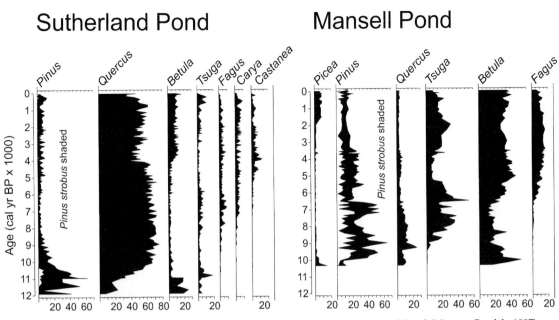

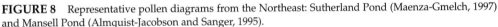

FIGURE 8   Representative pollen diagrams from the Northeast: Sutherland Pond (Maenza-Gmelch, 1997) and Mansell Pond (Almquist-Jacobson and Sanger, 1995).

limit soon after arriving at ca. 10,200 B.P., indicating that the climate was already warmer by that time than it is now (Davis et al., 1980). *Picea* had almost reached its modern altitudinal limit by 11,500 B.P. *Tsuga* grew 300–400 m above its modern limit soon after arriving at 7800 B.P. *Tsuga* began to retreat downslope by ca. 5700 B.P., but occurred above its present limit until the LIA. After *Tsuga* declined at 5600 B.P., it reestablished at lower altitudes by ca. 3200 B.P., but not at higher altitudes. *Picea* was rare throughout most of the Holocene until it expanded ca. 2000 years ago. These data indicate a warmer and drier early to middle Holocene, with the warmest period from 10,200 to at least 5600 B.P., when *Tsuga* disappeared from the higher sites. The last 2000 years, corresponding with the expansion of *Picea*, were the coldest part of the Holocene (Gajewski, 1987, 1988; Russell et al., 1993).

Pinus strobus not only reached its maximal altitudinal limits in the early Holocene, it also reached its maximum latitudinal limits (Jacobson and Dieffenbacher-Krall, 1995), essentially attaining its modern limits by 10,200 B.P. From 7800–4400 B.P., it had large populations north of its present limit. The timing of the northward expansion of *P. strobus* suggests that the period of maximal warmth occurred later than in the White Mountains.

Webb et al. (1993) used pollen response surface techniques to reconstruct soil moisture and mean annual precipitation at 3000 radiocarbon-year intervals for the Northeast. This study indicates that the climate was drier than it is today during times when *Pinus* domi-

nated at 10,200 B.P. (9000 [14]C B.P.) with a return to moist conditions at 6800 B.P. (6000 [14]C B.P.). Lithostratigraphic studies by Webb (1990) indicate low water levels during the early Holocene *Pinus* maximum.

Pollen, lake-level, and charcoal studies at Mansell Pond in Maine (Fig. 8) provide a multiproxy record of climate and vegetation change (Almquist-Jacobson and Sanger, 1995). *Pinus strobus* was high between 10,200 and 6500 B.P. However, in this record, this *P. strobus* interval is different from many other records in New England in that a *Tsuga* phase is intercalated from 8200–7300 B.P. The lake level was at least 5.9 m below the modern level at 8000 B.P. Increased charcoal during the *P. strobus* phase indicates greater burning. Other studies also indicate a greater frequency of forest fires in the early Holocene *P. strobus* forest than in the hardwood forests that succeeded it (Anderson et al., 1992; Clark et al., 1996). At Mansell Pond, the lake level rose at 6700 B.P., and *Tsuga* increased again at 6500 B.P. The lake level increased again at 3400 B.P., but fell slightly with the increase of *Picea* 2000 years ago, indicating perhaps a cooler, but drier, late Holocene climate.

Water-level studies at Owasco Lake in New York indicate a low stand and dry climate at 10,200 B.P., in agreement with other studies in New England (Dwyer et al., 1996). A high stand occurred at 7700 B.P., during the time of the early *Tsuga* phase at Mansell Pond, providing additional evidence of a dry-wet-dry climate cycle during the early Holocene *Pinus strobus* phase.

In summary, the climate in the Northeast was warmest and driest during the early Holocene from

10,200 to ca. 6000 B.P., with the most extreme conditions early at ca. 10,200 B.P. The climate became moister and cooler after the *Pinus strobus* phase, with the establishment of hemlock–northern hardwoods forest, with abundant *Tsuga, Fagus,* and *Betula.* Distinctly cooler, but perhaps somewhat drier, conditions commenced ca. 2000 years ago with the expansion of *Picea.*

19.2.9. Northeastern Canada

In the northeastern part of Canada, which lies north of the *Tsuga–Pinus*–northern hardwoods forest, the vegetation grades northward from boreal forest to forest-tundra to tundra. Dominant trees in the boreal forest are *Picea glauca, P. mariana, Pinus banksiana, Betula papyrifera,* and in the southeastern part of the region *Abies balsamea. B. papyrifera* is an important postfire species. This part of North America was the last to be deglaciated, with much of the area still ice covered in the early Holocene.

In southeastern Labrador, ice had retreated by 12,500 B.P., and tundra persisted until ca. 10,200 B.P. Lamb (1980, 1984) recognized four regional zones: a *Betula–Salix*–Cyperaceae zone representing tundra from 12,500–10,200 B.P., an *Alnus–Betula* zone representing shrub-tundra from 10,200–6800 B.P., an *Abies–Picea glauca* zone from 6800–5700 B.P., and a *Picea mariana* zone from 5700 B.P. to the present. Moraine Lake (Fig. 9) (Engstrom and Hansen, 1985) is representative of northeast Canada. The development of shrub-tundra at 10,200 B.P. indicates a warming trend, but in contrast to the northwestern Arctic, the early Holocene was cooler than the middle and late Holocene. The *Abies* peak from 6800–5700 B.P. may represent the warmest part of the Holocene based on the greater abundance of *Abies* farther south. Widespread paludification after 5700 B.P. favored *Picea mariana,* which almost completely replaced *P. glauca* by 4400 B.P. A cooler, wetter climate may have favored the development of peat, but pollen-influx data suggest that the warmest time was at ca. 4400 B.P., with cooling after 2700 B.P. (Lamb, 1980, 1984).

Sites in northern Quebec were not deglaciated until 7400–6400 B.P. (Richard, 1979; Richard et al., 1982; Gajewski et al., 1993). A short-lived herb zone, possibly representing successional vegetation, followed ice retreat, but all the important boreal trees and shrubs were present almost immediately. At some sites, for example, Lake GB2 (Gajewski et al., 1993) and Kanaaupscow (Richard, 1979) (Fig. 9), a peak of *Populus,* which may also have been successional, followed the herb zone. At Lake GB2, a peak of *Larix* accompanied the *Populus* peak. *Picea, Alnus,* and *Betula* predominated thereafter. At Kanaaupscow, forest-tundra with *Picea mariana, Alnus viridis,* and *Betula glandulosa* dominated the rest of

the Holocene. Forest density was highest from 6800–2800 B.P., after which *Pinus banksiana* became more prominent. At Lake GB2 in the northern boreal forest, *Alnus* was most abundant until 4200 B.P., after which *Pinea* dominated. In forest-tundra sites, the tundra component increased after 3000 B.P., indicating progressive cooling. At Delorme 2 (Fig. 9), dense *Picea mariana* forest existed from 6300–5000 B.P., with forest opening after 5000 B.P. (Richard et al., 1982). *Betula papyrifera* was more abundant from 6600–5500 B.P. The greater prevalence of *Betula* in the boreal forest in the middle Holocene is evident in sites throughout the eastern boreal forest and may indicate a greater incidence of fire owing to a warmer, drier climate (Jacobson et al., 1987).

Although sites in the northern boreal forest and forest-tundra show evidence of decreasing cover and expansion of tundra in the late Holocene from 5000–3000 B.P., no evidence exists that *Picea* ever occurred north of its present limit or that any latitudinal retreat of the forest limit occurred (Gajewski et al., 1993; Gajewski and Garralla, 1992; Lamb, 1985). However, in the mountainous terrain of northern Labrador, tree line fell ca. 40 m between 3200 and 1000 B.P. and another 30 m from 1000–250 B.P., although a latitudinal shift did not occur.

Whereas much of the continent was warmest in the early Holocene, including Alaska and northwest Canada, northeastern Canada was still ice covered. The warmest temperatures followed deglaciation from ca. 6000–3000 B.P., and progressive cooling has occurred since 3000 B.P. or earlier. Considerable vegetation change is associated with extensive paludification, which a cooler, wetter climate may have initiated.

19.2.10. Southeast

With the exception of Florida, sites with good Holocene records are few in the southeastern United States, and the vegetation history is not as well known as those in the Northeast and Midwest. Most sites are small sinkholes, and the Holocene sedimentary record is often confined to the top 1 or 2 m. At many sites, a significant local swamp or wetland component consisting of taxa such as *Nyssa, Cephalanthus,* and *Alnus* obscures the upland signal. The Holocene sediments in many of these sinkholes are low in organic matter, and radiocarbon dates of bulk sediment are suspect. Florida, however, has a good Holocene record. The lakes are sinkholes, but are much larger and deeper than elsewhere in the Southeast. Many of these lakes were dry in the late glacial period and did not fill with water until sea level rose in the early Holocene.

Lake Tulane (Grimm et al., 1993) (Fig. 10), Lake An-

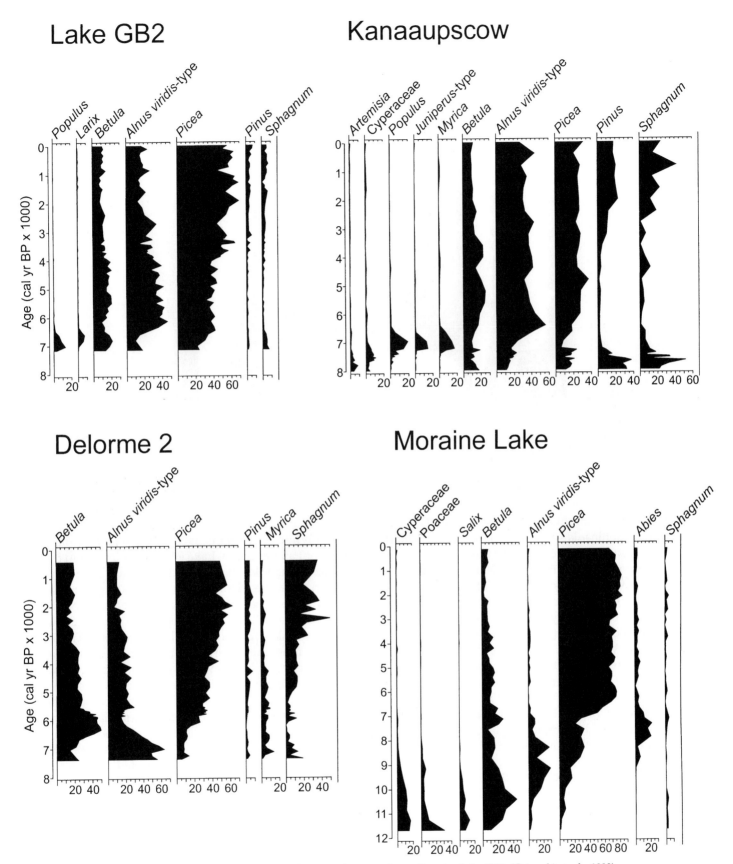

FIGURE 9 Representative pollen diagrams from northeast Canada: Lake GB2 (Gajewski et al., 1993), Kanaaupscow (Richard, 1979), Delorme 2 (Richard et al., 1982), and Moraine Lake (Engstrom and Hansen, 1985).

Clear Pond

Lake Tulane

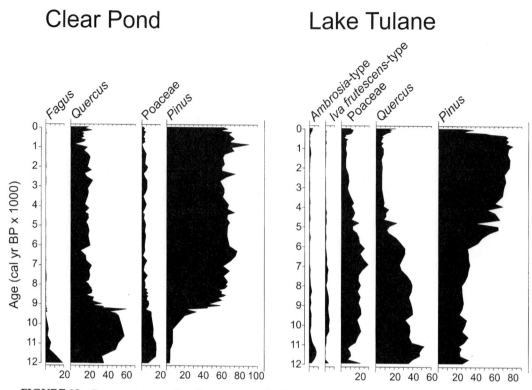

FIGURE 10 Representative pollen diagrams from the Southeast: Clear Pond (Watts et al., 1996) and Lake Tulane (Grimm et al., 1993).

nie (Watts, 1975), and Buck Lake (Watts et al., 1996) in the Florida peninsula are deep lakes with excellent Holocene records. At Buck Lake, which is now over 17 m deep and was dry until after 10,000 B.P., a shallow-water assemblage occurs at the base of the sediment at over 20 m depth (Watts et al., 1996). The dominant pollen types of the early Holocene are *Quercus*, Poaceae, and moderate amounts of *Pinus*. The vegetation was sclerophyllous oak scrub with scattered *Pinus*, with either an open understory or with prairie openings, similar to that occurring on xeric dune tops in Florida today. After ca. 5500 B.P., *Pinus* became much more abundant, prairie herbs declined, and extensive swamps developed, all indicating increased precipitation (Watts, 1980; Watts and Hansen, 1988, 1994; Watts et al., 1996).

Farther north on the Coastal Plain, *Pinus* increased earlier: between 8500 and 7500 B.P. at Barchampe Lake and Langdale Pond just south of the Georgia border and between 10,000 and 9000 B.P. at Clear Pond (Fig. 10) in northeastern South Carolina (Watts et al., 1996). Thus, the Florida peninsula appears to have been drier in the early Holocene than sites farther north. *Fagus* was abundant in the late glacial and earliest Holocene at Clear Pond, and its presence indicates very humid conditions on the sandy soils of the Coastal Plain. *Fa-*

gus and *Carya* were also abundant in the earliest Holocene at White Pond (Watts, 1980) on the South Carolina piedmont and at Camel Lake (Watts et al., 1992) in the Florida panhandle. These humid conditions are not evident in the earliest Holocene in the Florida peninsula.

Thirteen species of *Pinus* occur in eastern North America (Kral, 1993). The three northern pines occur in the Great Lakes region and in boreal regions. The 10 southern pines are widely distributed in the southeastern United States, but they do have different ranges and climatic requirements. Some species occur as far north as New England (e.g., *Pinus rigida* to southern Maine), whereas *P. elliottii* and *P. clausa* are essentially restricted to Florida. *P. palustris*, *P. echinata*, and *P. taeda* are important pines on the Coastal Plain, and these three are probably the main species that expanded there in the early Holocene. *P. palustris* occurs in central Florida today, but the most important species there are the two pines with the most southerly distributions, *P. elliottii* and *P. clausa*.

Pinus spread westward from its early Holocene stronghold on the Coastal Plain of northern Florida to North Carolina. In central and southern Florida, *Pinus* populations expanded ca. 5500 years ago, although the taxon was present in the early Holocene. *P. elliottii* and *P. clausa*, which expanded in central and southern Florida, today are essentially restricted to Florida and, therefore, are not the species that spread northwards from the Coastal Plain. The southward expansion of pine in Florida may result from different climatic causes from the taxon's westward expansion, which probably was related to migration rather than to expansion of a population already in place. *P. echinata* and *P. taeda* have migrated westward rapidly during the last 3000 years, reaching the Tombigbee site in northeastern Mississippi (Whitehead and Sheehan, 1985), Cupola Pond in southeastern Missouri (Smith, 1984), and Ferndale Bog in eastern Oklahoma (Albert and Wyckoff, 1981), all within the last 3000–2000 years. Webb et al. (1987) argued that increasing winter temperatures were responsible for the expansion of the southeastern *Pinus* forest.

19.3. MÉXICO, CENTRAL AMERICA, AND THE CARIBBEAN

Most of the Holocene pollen records from México are from two main areas: the central highlands or Trans-Mexican Volcanic Belt (TMVB) at 22°–18°N and the northern and southern Yucatán Peninsula (Fig. 11).

In the tropical regions, the trades and the subtropical high-pressure belt are the major influences on climate. Precipitation occurs over most of México during summer when the Intertropical Convergence Zone (ITCZ) is at its northernmost position and the easterly trade winds collect humidity from the warm waters of the Gulf of México and the Caribbean Sea. Precipitation along the Gulf coast is higher than along the Pacific coast. During winter, the ITCZ moves southward, and the central and southern regions in México are dry. Occasionally, winter precipitation occurs related to arctic air outbreaks colliding with humid air from the Gulf of México (see also Chapter 3). Hurricanes or tropical cyclones, which develop in the Gulf of México and Pacific Ocean during summer and early autumn, are another significant source of precipitation.

Topography greatly influences precipitation and temperature patterns; closed basins in the central highlands receive 600–800 mm/year, and the Sierras receive up to 1600 mm/year. Temperatures are relatively low with annual averages between 12° and 18°C. In the low-lying Yucatán Peninsula, with a maximum elevation of 300 m, precipitation is less than along the Gulf coast. The northern part of the peninsula receives 800

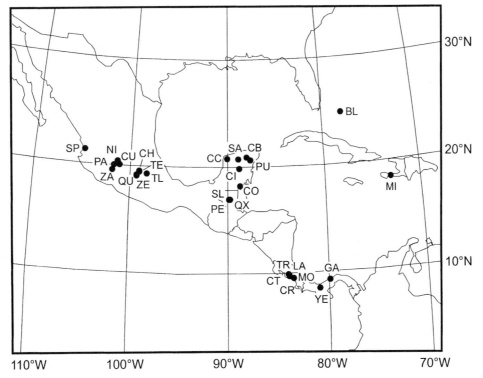

FIGURE 11 México and Central America with sites discussed in the text. BL, Church's Blue Hole; CB, Lake Cobá; CC, Cenote San José Chulchacá; CH, Lake Chalco; CI, Lake Chichancanab; CO, Laguna Cocos; CR, Lago Chirripó; CT, La Chonta; CU, Lake Cuitzeo; GA, Gatún Basin; LA, Lachner Bog; MI, Lake Miragoane; MO, Lago de las Morrenas; NI, La Hoya de San Nicolas de Parangueo; PA, Lake Pátzcuaro; PE, Lago Petén Itzá; PU, Punta Laguna; QU, Lake Quila; QX, Quexil; SA, Lake Sayaucil; SL, Salpeten; SP, Laguna San Pedro; TE, Lake Texcoco; TL, Tlaloqua; TR, La Trinidad; YE, La Yeguada; ZA, Lake Zacapu; ZE, Lake Zempoala.

Lake Chalco, Core E, Basin of México
Analyst: Susana Sosa•Najera

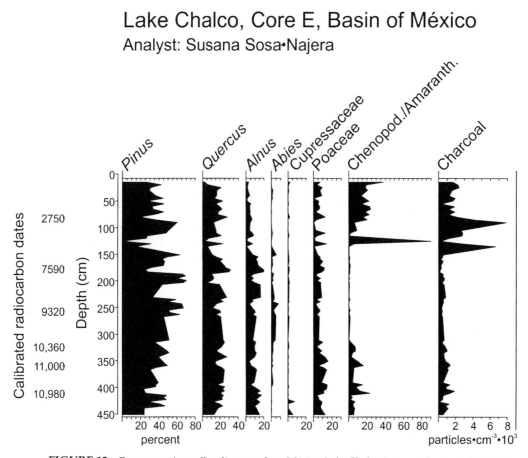

FIGURE 12   Representative pollen diagrams from México: Lake Chalco (Lozano-García et al., 1993; Lozano-García and Ortega-Guerrero, 1994; Sosa-Najera et al., 1998), and Lake Zacapu (Lozano-García and Xelhuanzti-López, 1997).

mm/year precipitation, and the south-central region receives up to 1600 mm/year. Temperatures have an annual mean above 22°C.

Major centers of pre-Hispanic culture developed in both areas, the Maya in the Yucatán Peninsula and several groups in central México, and environmental degradation with deforestation and agricultural activities has affected the landscape for the last 3500 years (Metcalfe et al., 1989; Leyden et al., 1998; Curtis et al., 1996).

19.3.1. Central México

In the mountainous region of central México, numerous internally drained basins formed related to volcanic activity during the late Neogene and Quaternary, associated with the oblique subduction of the Cocos Plate under southwestern México (Morán-Zenteno, 1994). Several closed basins have deep lakes, such as Lakes Cuitzeo and Pátzcuaro, but in general the lakes

are relatively shallow or seasonally dry. Physiographically the area is a volcanic plateau with several volcanic peaks with altitudes up to 5000 m.

The dominant climate in this region is temperate with mild winters and summer rains. Tropical climates occur in the eastern part, and hot, arid climates occur in areas where mountain ranges block the moisture-bringing easterly trade winds. The vegetation in the mountainous areas is primarily mixed *Pinus–Quercus* forest, although several different species of *Pinus* also occur in other ecological communities. Repeated volcanic disturbances have created microhabitats conducive to species evolution. Thus México, with half of the world's *Pinus* species, has the greatest diversity of this taxon on Earth (Styles, 1993). Similarly, the central mountainous region of México has the highest diversity of *Quercus* in the Western Hemisphere, and approximately one-third of the world's species of *Quercus* occur in México (Nixon, 1993).

Lake Zacapu, Central México
Analyst: Susana Xelhuantzi-López

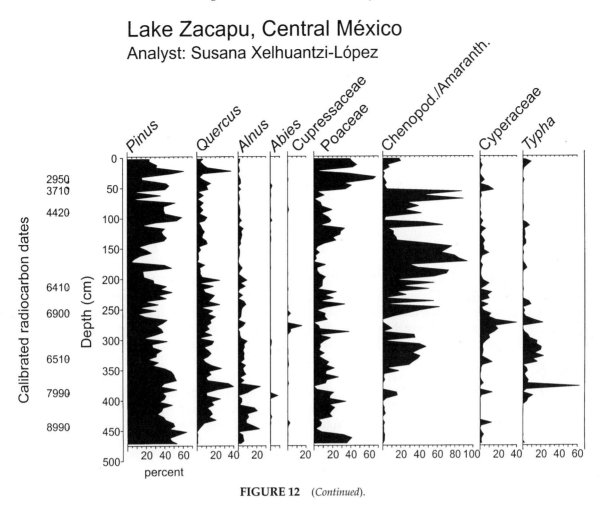

FIGURE 12   (*Continued*).

19.3.1.1. Puebla/Tlaxcala Area

Palynological studies from Tlaloqua Crater Maar at 3100 m elevation (Ohngemach, 1977), together with the regional glacial chronology, provide a climate history of the eastern highlands (Heine, 1988; Straka and Ohngemach, 1989). Conditions in the early Holocene were mesic; alpine meadows occurred from 11,000–10,000 B.P. and were replaced by mixed *Pinus* forest at 9000 B.P. Above a thick pumice layer dated at 2760 B.P., open *Pinus* forest replaced the mixed-pine forest. High amounts of *Abies* pollen indicate cooler-moister conditions in the late Holocene from 1290–1440 B.P. Farming in the area began 4000–3000 years ago (Ohngemach, 1977).

19.3.1.2. Basin of México

Pollen records from Lake Chalco (Fig. 12) in the southern basin of México show a temperate mesic early Holocene with cold winters based on peaks of *Quercus* and the occurrence of *Abies* from 11,000–8000 B.P.

(Lozano-García et al., 1993; Lozano-García and Ortega-Guerrero, 1994; Sosa-Najera et al., 1998). In central México, at this time, conditions were cold and dry, as indicated by relatively high amounts of Cupressaceae pollen in the records (Watts and Bradbury, 1982; Bradbury, 1997). The subsequent decrease of *Abies* and Cupressaceae pollen indicates the end of glacial conditions in the basin of México. Pollen records from the early Holocene are interpreted as indicating an increase in moisture, which is also suggested by glacier advances at 12,000–10,500, 10,000, and 8800–7400 B.P. in the high elevations around the basin (Vázquez-Selem and Phillips, 1998). However, low lake levels in the basin of México and halophytic diatom assemblages suggest high evaporation rates from 10,000–6000 B.P. (Bradbury, 1989; Caballero-Miranda and Ortega-Guerrero, 1998). In Lake Texcoco, the shallowest lake in the central basin of México, a hiatus in sedimentation occurred from 13,000–6500 B.P. (Lozano-García and Ortega-Guerrero, 1998). The conflicting evidence for wetter conditions in the mountains and drier conditions in the basin of México in the early Holocene may be due to several factors. Higher summer temperatures may have promoted high evaporation and high E–P (evaporation–precipitation) ratios from the shallow lakes in the basin of México. A difference in seasonality of precipitation may also be involved, with increased winter snowfall promoting glacial advance at high altitudes, coupled with decreased summer precipitation resulting in lower lake levels at lower altitudes. This situation is opposite to that in the southwestern United States, where the low-elevation deserts were wetter and the higher elevation mountains were drier in the early Holocene. At Lake Quila, a high-elevation site (3010 m) 65 km southwest of the basin of México, a humid phase with *Alnus* and *Pinus* forests occurred at 10,950 B.P., followed by a *Pinus* forest expansion that probably indicates cooler conditions from 9000–6400 B.P. (Almeida-Leñero, 1997).

During the middle Holocene, a short period (7000–6000 B.P.) of drier conditions was established in the southern sector of the basin of México. Lower values of arboreal pollen, especially *Abies* and Cupressaceae, mark this period, as does an increase in fire frequency as evidenced from the charcoal record (Fig. 12). At Lake Texcoco, lake sedimentation resumed at 7300 B.P., and arboreal pollen declined somewhat (Lozano-García and Ortega-Guerrero, 1998). However, at the high-elevation Lake Quila site, no major change in vegetation occurred in the pollen record from this time (Almeida-Leñero, 1997).

Increases in moisture are evident in late Holocene limnological records from the southern, central, and northeastern regions of the basin of México (Bradbury, 1989; Lozano-García et al., 1993; Caballero-Miranda and Ortega-Guerrero, 1998; Caballero-Miranda et al., 1999). Higher lake levels at all sites indicate wetter conditions, but the vegetation apparently did not respond to this wetter climate. Instead, the pollen data indicate a significant increase in herbaceous taxa related to human activities. A late Holocene pollen record from Lake Zempoala (2800 m) near Lake Quila indicates a temperate, semidry climate with abundant *Quercus*, *Salix*, and *Alnus* at the end of the middle Holocene, and this trend continues until 2000 years ago. After this time, the *Abies* forests expanded at Lakes Quila and Zempoala, indicating cooler and more humid conditions, with a short dry period at 1500 B.P. The human impact in this high-altitude region is less important than in other parts of central México (Almeida-Leñero, 1997).

19.3.1.3. Western-Central Region

Toward the eastern section of the TMVB, records from three lakes, Pátzcuaro, Cuitzeo, and Zacapu, in the Michoacan area provide a history of Holocene climate changes. Tephra layers, changes in sedimentation rates, and magnetic susceptibility data provide evidence for the recent volcanic and tectonic history of these three sites. According to Israde Alcantara (1997), Lake Cuitzeo evolved under intense faulting activity, which controlled sedimentation. Some of these fault systems are currently active, and Holocene lava flows are present along the western margin of Lake Zacapu. This volcanic activity complicates paleoclimate interpretations (Metcalfe, 1997; Lozano-García and Xelhuantzi-López, 1997). Pollen and diatom records from Lake Pátzcuaro indicate the late Pleistocene/Holocene environmental history (Watts and Bradbury, 1982). Lake Cuitzeo is the second largest lake in central México, and recent paleolimnological and palynological data provide more information on late Quaternary environmental changes (Velázquez-Durán, 1998). One of the better dated Holocene sequences from central México is from Lake Zacapu, where multiproxy studies of pollen, diatoms, and sediment stratigraphy provide information on environmental history (Metcalfe, 1992; Xelhuantzi-López, 1994; Arnauld et al., 1997).

The Holocene record of Lake Cuitzeo begins with evidence for a dense forest after a hiatus in sedimentation from 15,000–9000 B.P. Pine forest dominated the early Holocene, and the diatom *Cyclotella meneghiniana*, which grows in swamps, indicates low lake levels (Velázquez-Durán, 1998).

The Lake Zacapu record (Fig. 12) shows evidence for intermittent volcanic activity, and the pollen record begins in the later part of the early Holocene. From 9000–8000 B.P., *Pinus* forest with *Quercus*, *Alnus*, and Poaceae dominated. The marsh was shallow with circumneutral pH and abundant macrophytes. A major volcanic and tectonic event sometime during 8000–7000 B.P. compromised the record, and increases in erosion rates and fires around the basin are evident (Xelhuantzi-López, 1994; Arnauld et al., 1997).

At Lake Pátzcuaro, pollen and diatom data show that the early Holocene was moister than the late Pleistocene, based on a decrease in Cupressaceae pollen and an increase in planktonic diatoms (Watts and Bradbury, 1982; Bradbury, 1997).

The record from La Hoya de San Nicolás de Parangueo, a caldera lake located between central México and the Chihuahua Desert, shows the development of *Pinus* forest from 13,000–3400 B.P., with a peak of Chenopodiaceae-Amaranthaceae at the Pleistocene/Holocene transition (Brown, 1984; Metcalfe et al., 1989). The establishment of grassland at lower elevations and *Pinus* forest at higher elevations continued during the early and middle Holocene.

At Lake Cuitzeo, mixed *Pinus–Quercus* forest developed and herbaceous pollen became more abundant in the middle Holocene. The diatom data show a trend to drier conditions, culminating with the deposition of a sand layer (Velázquez-Durán, 1998). At Lake Zacapu, conditions began to stabilize at the earliest part of the middle Holocene. The marsh area became successively smaller, suggesting a trend to less humid climate. Increases in Chenopodiaceae-Amaranthaceae and Poaceae, coupled with low values of boreal taxa and diatoms, indicate a change to a semiarid climate. Water levels reached a minimum between 6800 and 4400 B.P. By the end of the middle Holocene, wetter conditions returned, the marsh expanded, and *Pinus* forests reexpanded. The reduction of *Alnus* and an increase in *Pinus* pollen mark the beginning of the middle Holocene at Lake Pátzcuaro; shallow-water littoral diatoms became abundant during this time. The record from Laguna San Pedro, located in the westernmost part of the TMVB, shows moderate mid-Holocene aridity, with a change from mixed *Pinus–Quercus* parkland to a *Quercus* parkland occurring at 5000 B.P. (Brown, 1985). Comparable to other central México records, human presence is documented in the pollen diagrams, with *Zea* pollen in the Lake Pátzcuaro record and intense deforestation at Lake Zacapu, La Hoya de San Nicolás de Parangueo, and Lake Cuitzeo.

19.3.2. Yucatán Peninsula

The Yucatán Peninsula is one of the northernmost tropical areas of the Americas. It lies between 17°50′–21°30′ N and 87°00′–91°00′ W and includes the states of Quintana Roo, Yucatán, and Campeche in México, as well as Belize and northern Guatemala (Petén lowlands). The peninsula is a flat platform composed mainly of tertiary marine limestone with thin soils, with elevations mostly below 200 m. Climate in the Yucatán Peninsula is tropical subhumid, with a summer wet season. Precipitation has an east-to-west gradient, becoming drier westward. The northwestern region is relatively dry, with 500–1000 mm/year precipitation. Dominant vegetation is low deciduous forest with trees 6–8 m tall and thorn-scrub forest with trees 3–8 m tall. Dominants of the low deciduous forest are *Jatropha goumeri, Metopium brownei, Alvaradoa, Bursera simaruba, Maclura tinctoria, Bumelia, Mimosa bahamensis,* and *Caesalpinia vesicaria.* Dominants in thorn-scrub forest include *Acacia, Mimosa bahamensis, Pithecolobium, Leucaena, Senna, Piscidia piscipula,* and *Bursera simaruba.* Precipitation in the eastern region is >1500 mm/year, and a semi-evergreen forest dominates. Important trees are *Brosimum alicastrum, Manilkara zapota, Vitex gaumeri, Lysiloma latisiliquum,* and *Swietenia macrophylla* (Flores and Carbajal, 1994).

In this karst landscape of the Yucatán Peninsula, the presence of numerous lakes and water-filled sinkholes (cenotes) offers the possibility for paleoecological studies. At sites in the north-central Yucatán Peninsula, sedimentation began after 8000 B.P., when groundwater rose probably because of both rising sea level and increased precipitation (Whitmore et al., 1996). In the Petén lowlands in the southernmost portion of the peninsula, deeper lakes also preserved a late Pleistocene record as well as the Holocene record (Islebe et al., 1996b).

Multiproxy paleoecological studies of several lakes in the Yucatán Peninsula provide information on environmental and climatic change during the Holocene and on human impact related to the rise and collapse of the Maya civilization. Climate proxies include geochemistry, stable isotopes, diatoms, phytoliths, and pollen (Curtis et al., 1996; Islebe et al., 1996b; Hodell et al., 1995; Leyden, 1984, 1987; Leyden et al., 1994, 1996, 1998; Vaughan et al., 1985; Whitmore et al., 1996). We present here palynological data from Lake Cobá and Cenote San José Chulchacá.

Two cores from Cenote San José Chulchacá (Fig. 13), located in the driest part of the Yucatán Peninsula, cover the last 8120 years B.P. (Leyden et al., 1996). Limnological, oxygen isotope, and palynological data suggest a dry early Holocene. Cyperaceae pollen is absent, suggesting low water levels. Significant values of pollen of *Brosimum,* a canopy tree of the medium-high semievergreen forest, suggest an increase in precipitation at 7600 B.P., although mesic forest probably did not grow around the site (Leyden et al., 1996). Isotopic data and the absence of *Brosimum* indicate another dry episode from 7000–5280 B.P. The return of *Brosimum* indicates a return to wetter conditions from 5280–3500 B.P. The presence of Poaceae and *Trema* pollen in the regional vegetation suggests a trend to drier environments with open, dry forests from 3000–1350 B.P. According to the pollen data, the changes in vegetation at Cenote San José Chulchacá indicate that dry forest has existed since the middle Holocene, and only variations between different types of dry communities have occurred.

The paleoecological record at Lake Cobá in the east-central part of the peninsula starts at 8370 B.P. with a wooded swamp represented by *Eleocharis, Typha,* and *Dalbergia* from 8000–6600 B.P. (Leyden et al., 1998). Stable isotopes from ostracodes show that the lake was shallow and saline during this time (Whitmore et al., 1996). Significant amounts of *Piscidia* pollen and other taxa such as *Metopium* indicate semideciduous dry forest in the region. During the middle Holocene, 6600–4000 B.P., a more diverse plant community expanded, with *Brosimum,* Moraceae, and *Bursera* indicating a dense forest. In the late Holocene, between 2700 and 1250 B.P., herbaceous taxa (Poaceae and Chenopodi-

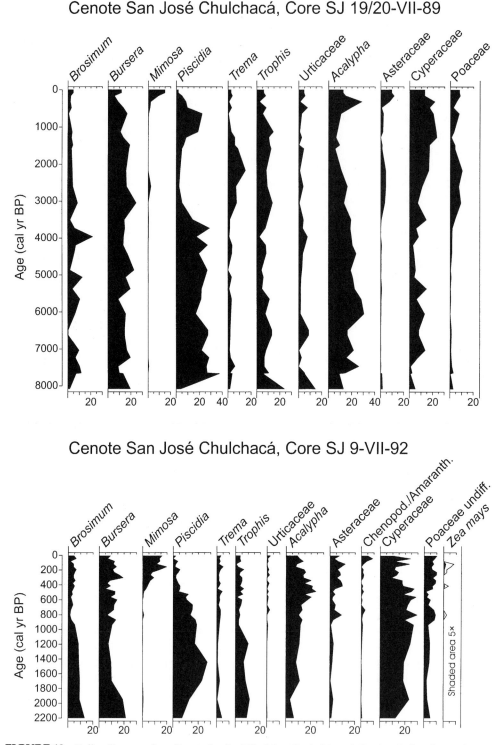

FIGURE 13   Pollen diagrams from Cenote San José Chulchacá in the Yucatán Peninsula (Leyden et al., 1996).

aceae) dominate the pollen spectra, indicating forest clearance. *Zea* pollen provides additional evidence of human activities. During this period, pollen and limnological records indicate intense environmental dis-

turbance; forest cover diminished drastically, and *Zea* pollen is abundant. The diatom data indicate fresher and deeper lake conditions, with the deepest episode at 1300 B.P. From 1250–700 B.P., the Late Classic period,

Zea, Chenopodiaceae-Amaranthaceae, and Caryophyllaceae pollen decreased. This change may have been associated with the growth of the city of Cobá and the shift of the milpa fields farther away from the city. The Maya collapse, which may have resulted from drier conditions in the peninsula (Curtis et al., 1996; Hodell et al., 1995), is documented in the pollen diagram from Cobá by a decline in Chenopodiaceae-Amaranthaceae, an absence of *Zea,* and an increase in tree pollen, represented by *Brosimum, Trema,* Moraceae, and *Piscidia.*

At Laguna Cocos in northern Belize, a pollen sequence that covers the last 6300 years shows a sequence of changes from predominantly arboreal vegetation to nonarboreal vegetation and a return to arboreal vegetation. The presence of *Zea* and other herbs related to agricultural activities indicates a strong human influence (Hansen, 1990).

Stable isotopes, diatoms, and geochemical data from Lake Chichancanab and Lake Sayaucil provide additional evidence for climate change. Lake Chichancanab refilled after 8200 B.P., similar to other water bodies in the Yucatán region (Hodell et al., 1995). Conditions continued to become wetter, and lake levels rapidly increased after 8000 B.P. A change to drier conditions and lower lake levels occurred between 1260 and 980 B.P. A record from Lake Sayaucil covers the last 3000 years. After the initial lake-level rise, a trend to drier conditions occurred from 3200–1940 B.P. Subsequently, lake levels rose at the same time as human settlement activity increased, and wet conditions continued until recent times (Whitmore et al., 1996). Paleoclimatic reconstruction based on vegetation history is consistent with the data obtained from other climate proxies. At Punta Laguna, the late Holocene record indicates climatic variability based on $\delta^{18}O$ data. This record documents a dry period from 1700–800 B.P., contemporaneous with the Maya collapse (Curtis et al., 1996).

19.3.2.1. Petén Lowlands (Guatemala)

The Holocene sediment sequences from the Petén lowlands have dating problems owing to the hardwater effect (Deevey and Stuiver, 1964) and to the scarcity of datable carbon. In the pioneering work of Vaughan et al. (1985), only 5 of 23 dates from several cores were reliable. Vaughan et al. (1985) described six pollen zones, P-1 to P-6, which, because of chronological problems, they correlated with dated archeological events. Other investigators have used this zonation in the Petén Peninsula. The recent work of Islebe et al. (1996b) at Lake Petén Itzá improved the chronology for this tropical region.

According to Leyden et al. (1994), zones P-1 and P-2 from Quexil and Salpeten show a moist early Holocene

based on the replacement of *Pinus, Quercus,* and mesic temperate hardwoods by rain forest taxa such as Moraceae (zone P-1). Higher lake levels also indicate increased precipitation (Leyden, 1987; Brenner, 1994a). The recent work of Islebe et al. (1996b) at Lake Petén Itzá, the largest and deepest lake in the region, reconstructed Holocene paleoclimatic changes with more precision. Abundant Moraceae/Urticaceae pollen are the signal of lowland, semi-evergreen forest and mesic conditions. The pollen sequences of Quexil, Salpeten, and Petén Itzá suggest that the tropical mesic forest was fully developed after 10,000 B.P. and that the environments in the Petén lowlands at that time were moister than they are today. Inferences from pollen data agree with $\delta^{18}O$, magnetic susceptibility, and geochemical climate-proxy data from Salpeten and Quexil. All these data suggest that conditions in the early Holocene were mesic. However, a contradiction exists at Petén Itzá, where the presence of Moraceae-Urticaceae pollen suggests humid conditions, whereas high $\delta^{18}O$ values from shell material suggest increased E–P ratios during this period (Curtis et al., 1998).

Middle Holocene pollen records from Salpeten and Quexil (zone P-2) show the development of more open vegetation with *Quercus, Cecropia, Byrsonima,* and Melastomataceae. At Petén Itzá, a decline in precipitation occurred at 6500 B.P. with a drastic reduction of Moraceae/Urticaceae at the end of zone P-2. At the same time, elements indicative of more open vegetation increased, such as Poaceae, Melastomataceae, and *Byrsonima,* suggesting drier climates. Increased *Pinus* and *Quercus* also indicate more open vegetation and a drier climate (Islebe et al., 1996a). In the upper part of zone P-2 at Petén Itzá, correlated to zone P-3 of Vaughan et al. (1985), human impact is documented as associated with the Maya population increase.

All the late Holocene palynological records from the Petén lowlands present signs of human impact. The agricultural activities of the Maya altered the vegetation substantially, and the pollen records from Salpeten and Quexil show more open vegetation with abundant herbaceous taxa for the last 4000 years. The human impact significantly increased erosion in the basins, and a layer of clay (the Maya Clay) with thicknesses of up to 7 m was deposited in some lakes from 3200 to 470 B.P. (Curtis et al., 1998). During this period, a reduction in arboreal cover is evident with concomitant expansions of herbaceous taxa (zones P-3 to P-5). Although this anthropogenic unit is not present in all lakes (e.g., Petén Itzá), changes in sediment composition and magnetic susceptibility occur in zone P-3.

After two millennia of strong human disruption to the vegetation, a recovery of tropical forest is evident in the pollen records. In the records from Salpeten, Quexil (P-6), and Petén Itzá (P-4), tropical trees (*Brosimum*, *Trema*, and Moraceae-Urticaceae) increase. Forest regeneration postdates the Maya collapse, which is archeologically dated to A.D. 800–900. The precise time of regeneration is difficult to determine because of problems in radiocarbon dating, but at Petén Itzá forest recovery began ca. 1000 years ago (Islebe et al., 1996b).

19.3.3. Central America

19.3.3.1. Costa Rica

A number of high-elevation sites in the Sierra de Talamanca document vegetation history and shifts in tree line: Lachner Bog (2400 m) (Martin, 1964), Lago de la Morrenas (3480 m) in the Chirripó Paramo (Horn, 1993), and La Chonta (2310 m) and La Trinidad Bogs (2700 m) (Hooghiemstra et al., 1992; Islebe et al., 1996a; Islebe and Hooghiemstra, 1997).

After the La Chonta stadial (13,000–12,250 B.P.), a cool episode at the end of the Pleistocene, climate warmed to modern conditions by 11,000 B.P. (Islebe et al., 1996a). At La Chonta, *Alnus* forests expanded upwards to the modern tree line at 2800–3000 m. The tree line remained stable near Chirripó Paramo throughout the Holocene (Horn, 1993). By the middle Holocene, *Podocarpus–Quercus* forests had developed at La Chonta and La Trinidad, indicating humid conditions with tree line at its present altitude (Islebe et al., 1996a). Over the last 4000 years, *Podocarpus* has declined at La Chonta and La Trinidad, but the only charcoal evidence for natural or anthropogenic fires is at Chirripó Paramo. Thus, the early to middle Holocene (10,800–5000 B.P.) was wetter than today, with a drier period from 5000–1500 B.P. A lack of shifts in the Holocene tree line suggests stable temperatures (Islebe et al., 1996a).

19.3.3.2. Panama

In the Gatún basin in the Panama Canal Zone, mangrove vegetation prevailed during the early Holocene (Bartlett and Barghoorn, 1973). The decline of *Iriartea* and the absence of *Symplocos* and Ericaceae at the end of the early Holocene suggest that temperatures attained their present values at this time. At La Yeguada, in central Panama (Bust et al., 1992; Piperno et al., 1990), the pollen record indicates the presence of moist tropical lowland forest since 12,700 B.P. with *Pilea*, *Trema*, and *Cecropia*. Drier conditions developed during 8000–4700 B.P. at the Gatún basin; pollen of *Ilex* and *Myrica* also suggest more seasonal climates. At La Yeguada, diatoms, phytoliths, and lower lake levels indicate a relatively dry phase from 8200–5500 B.P. (Piperno et al., 1990). A decline in Poaceae and an increase in forest

taxa (*Pilea*, *Cecropia*, and Myrtaceae) may indicate human activity (Bush et al., 1992). During the late Holocene, important changes in vegetation occurred in the Gatún basin, but their climatic significance is unclear because of human impacts documented by *Zea* pollen in the cores.

19.3.4. Caribbean

High-resolution analyses of $\delta^{18}O$ from ostracode shells and geochemistry, pollen, and charcoal data provide evidence indicating the climatic and environmental history for the region of Lake Miragoane, Haiti, for the last 12,400 years (Hodell et al., 1991; Brenner, 1994b). The period 12,400–9300 B.P. was drier than today with low lake levels; montane shrubs and xeric palms dominated the vegetation. A sudden change occurred at 9200 B.P., when Chenopodiaceae-Amaranthaceae increased and $\delta^{18}O$ decreased, suggesting that the climate became increasingly wetter from 9200–2500 B.P. Open forest with shrub elements dominated from 9200–6250 B.P.; it was replaced by more mesic forest with *Ambrosia* and Poaceae in the understory from 6250–4000 B.P., and then by Moraceae forest with *Pseudolmedia*, *Trophis*, *Chlorophora* (*Maclura*), and *Cecropia* from 4000–2500 B.P. It was during the latter interval that the lake reached its maximum depth, and the climate was warm-wet and seasonal. In the late Holocene, at Lake Miragoane, the climate became drier as lake levels declined and open, dry forests expanded, although some relicts of the previous mesic forest elements persisted. The pollen record from Church's Blue Hole on Andros Island in the Bahamas also indicates this dry period with increases of *Pilea*, *Dodonaea*, and dry shrubs (Kjellmark, 1996).

19.3.5. Summary for México, Central America, and the Caribbean

In spite of the limited number of records, dating problems, and marked human influence especially during the late Holocene, climatic trends can be interpreted from pollen data. Evidence for increased moisture is documented for the early Holocene in central México, the Petén lowlands, Central America, and the Caribbean. At Lakes Chalco and Quila in México's east-central sector, montane *Pinus–Quercus–Alnus* forests developed from 12,000–8000 B.P. A dense *Abies* forest existed from 11,000–8000 B.P. (Sosa-Najera et al., 1998; Lozano-García and Ortega-Guerrero, 1994; Almeida-Leñero, 1997). Glacier advances occurred on Iztaccíhuatl Volcano at 12,000–10,500 and 8800–7400 B.P. (Vázquez-Selem and Phillips, 1998). Pollen records from the west-central sector indicate warmer/wetter conditions with a reduction of Cupressaceae pollen at Lake Pátzcuaro (Watts and Bradbury, 1982; Bradbury, 1997) and an expansion of mixed *Pinus–Quercus–Alnus* forest at the shallow marsh of Lake Zacapu (Xelhuantzi-López, 1994). *Quercus* increased at several sites during the early Holocene: at Puebla at 9500 B.P., Lake Chalco at 11,400–8300 B.P., Lake Quila at 9400–7000 B.P., and Lake Zacapu at 9000–6100 B.P. High *Quercus* pollen levels may indicate increased summer temperatures. Although summer rain probably was intensified, it was apparently insufficient to increase lake levels, as is suggested by the diatom records from Lakes Chalco, Pátzcuaro, and Zacapu (Caballero-Miranda, 1995; Bradbury, 1997; Metcalfe, 1995; Fritz et al., 2000). Higher temperatures apparently canceled the effect of higher precipitation, and lower lake levels resulted. Seasonality of precipitation with increased monsoonal (summer) precipitation resulted in more mesic vegetation, but low lake levels may relate to low winter moisture instead.

In the Yucatán Peninsula, the records start at ca. 9000 B.P., when lakes filled as a consequence of rising of sea levels (Hodell et al., 1995), and vegetation records are available from the early Holocene. In the Petén lowlands, the existence of tropical forests composed of Moraceae-Urticaceae after 12,300 B.P. suggests a moist early Holocene (Islebe et al., 1996b). At Salpeten, pollen data indicate that a cool climate continued during the earliest part of the Holocene, and at Lake Quexil temperate mesic forest prevailed. The tropical mesic forest was completely established by 9500 B.P. (Leyden et al., 1994).

The early Holocene was also wetter in Central America. At La Chonta in Costa Rica, upper montane rain forest was established by 10,500 B.P. At La Yeguada in Panama, lowland moist tropical vegetation dominates the entire Holocene pollen record. In the Caribbean, the high-resolution record from Lake Miragoane indicates increasing precipitation during the early Holocene.

In the mid-Holocene, a trend toward a drier climate is evident in central México, although the beginning, duration, and culmination differ among sites. In records from the basin of México, a reduction in *Abies* and *Alnus* and abundant charcoal occurred from 7380–5740 B.P. (Sosa-Najera et al., 1998; Lozano-García and Ortega-Guerrero, 1994). Relatively low lake levels continued, although a trend to higher levels did occur (Caballero-Miranda et al., 1999), suggesting higher evaporation rates. At Lake Quila, less humid conditions led to a reduction of *Alnus* forest and the development of a *Pinus hartwegii* forest from 7800–6300 B.P. (Almeida-Leñero, 1997). In west-central México, a re-

duction of *Alnus* occurred at Lake Cuitzeo at 6820 B.P., while Chenopodiaceae-Amaranthaceae peaked (Velázquez-Durán, 1998), and a reduction of planktonic diatoms occurred at Lake Pátzcuaro at 5740 B.P. (Bradbury, 1997). At Lake Zacapu, *Alnus* decreased after 7160 B.P., while Chenopodiaceae-Amaranthaceae increased from 7380–4500 B.P. (Xelhuantzi-López, 1994). Diatom data indicate a dramatic reduction in water levels.

At San José Chulchacá in the northern Yucatán Peninsula, the record begins at 8120 B.P., showing dry vegetation and low lake levels. After a dry early Holocene, conditions became somewhat wetter, with the presence of *Brosimum*, which indicates mesic forest, by 7600 B.P. Episodes of dry conditions between 7000 and 5280 B.P. and between 3000 and 1350 B.P. bracketed a wetter episode. At Lake Cobá, pollen diversity was low from 8370–6480 B.P. After that time, an increase and diversification of dry semideciduous forest occurred that lasted until 4400 B.P. At Salpeten in the Petén lowlands, a trend to a moister climate with *Brosimum* occurred in the early Holocene (Leyden, 1987; Leyden et al., 1994). At Petén Itzá, high forest dominated, but began to decline at ca. 6360 B.P., probably because of human activities; however, the decrease in $\delta^{18}O$ values provides a clear signal of a moist mid-Holocene climate (Curtis et al., 1998).

In Panama, the records indicate dry conditions at the Gatún basin from 7790–4500 B.P. (Bartlett and Barghoorn, 1973), while at La Yeguada a relatively dry phase occurred from 9000–5740 B.P. with a reduction of lake levels. Human activities confuse the climate signal in the pollen records (Bush et al., 1992). The early Holocene wet conditions continued into the middle Holocene at La Chonta (Islebe et al., 1996a), and at Lago Chirripó no evidence exists for a mid-Holocene vegetation change (Horn, 1993). In the Caribbean, the wettest conditions occurred from 7790–4500 B.P. with the development of mesic forest (Hodell et al., 1991).

Thus, for the middle Holocene, the climate was more arid in central México from 6820–5740 B.P., while in the Yucatán Peninsula and Petén lowlands, somewhat moister but fluctuating conditions occurred. In Central America and the Caribbean, the data indicate humid conditions during this time.

Most of the pollen spectra that cover the late Holocene show a significant human influence for the last thousand years, including the presence of *Zea* pollen, reductions in arboreal cover, changes in vegetation composition, and increases in charcoal particles. The human effect on vegetation varies in timing and intensity and masks the climatic signal in some of the Mesoamerican sequences. In central México, a pattern of human disturbance is evident in several basins for the last 3500 years (Metcalfe et al., 1989). Nevertheless, a trend to moister conditions is suggested from synchronous increases in lake levels at Lakes Texcoco, Chalco, and Tecocomulco in the basin of México at 3500 B.P. (Caballero-Miranda et al., 1999). However, episodes of less humid climates are recorded at Lakes Pátzcuaro and Zacapu for 1200–900 B.P. (Curtis et al., 1996; Metcalfe, 1995).

In the northern part of the Yucatán Peninsula, the wettest period occurred between 6000 and 3000 B.P., interrupted by several dry episodes. A period of humid conditions occurred from 5280–3500 B.P., but drier conditions returned from 3500–1340 B.P. (Leyden et al., 1996). In the central part of the peninsula, the Lake Cobá record shows herbaceous and successional associations at 3500 B.P. related to human activities (Leyden et al., 1998). Drier conditions during the late Holocene are evident at Lakes Chichancanab, Punta Laguna, and Miragoane in Haiti (Hodell et al., 1995; Curtis et al., 1996). Oxygen isotope records document a period of aridity from 1790–930 B.P. in two of the Yucatán lakes (Punta Laguna and Chichancanab), which coincides with the collapse of the Maya civilization and other Mesoamerican cultures (Curtis et al., 1996). Nevertheless, this signal of dryness is not evident in the palynological records, probably because of predominant human influences (Bradbury, 1982). The openness in vegetation in the Petén lowlands during the late Holocene also indicates human disturbance (Islebe et al., 1996b; Curtis et al., 1998).

Panamanian palynological records indicate strong human influences on the vegetation during the late Holocene. Bush et al. (1992) suggested that significant human disturbance in La Yeguada began as early as 11,000 years ago. In the Lago Chirripó record in Costa Rica, high concentrations of charcoal during the late Holocene in subalpine and lower montane forests are indicative of human influences (Horn, 1993). The La Chonta pollen record also points to a drier climate for this period, with an episode of more arid conditions at ca. 1200 B.P. (Islebe et al., 1996a).

19.4. SOUTH AMERICAN LOWLANDS

For South America, records are discussed from the lowlands and the Andes (Fig. 14).

19.4.1. Llanos Orientales (Colombia)

Six lake records from the Llanos Orientales (Wijmstra and van der Hammen, 1966; Behling and Hooghiemstra, 1998, 1999, 2000) indicate that the early Holocene land-

scape was extensive grass savanna with few woody taxa and small areas of gallery forest along and between the rivers. Annual precipitation must have been lower and the annual dry season longer than they are today. In the middle Holocene, conditions became wetter and the forested area increased: at 6000 B.P. at Laguna Angel, at 7300 B.P. at Laguna Sardinas (Fig. 15), and at 6300 B.P. at Laguna Carimagua. After ca. 4200 B.P., stands of palms (*Mauritia* or *Mauritiella*) increased markedly in area in the Llanos. This change in vegetation may have resulted from increased human impact on the savanna under the wettest climatic conditions of the Holocene. The new high-resolution record from Laguna Loma Linda (Table 3), located in the transition area between the savanna and the Amazon rain forest, shows that the grass-dominated savanna in the early Holocene changed at ca. 6900 B.P. into savanna vegetation with a greater presence of woody taxa and with a marked expansion of Amazon and gallery forest. At ca. 3900 B.P., forests began to dominate the landscape, indicating a change to wetter conditions with a shorter annual dry season.

19.4.2. Gran Sabana (Venezuela)

Three pollen records from the Gran Sabana in the southeastern Venezuelan savanna include the middle Holocene (Table 3). The pollen record from Santa Cruz de Mapaurí shows a predominance of grass savanna with changes in floristic composition indicating oscillations between drier and wetter periods (Rinaldi et al., 1990). The pollen record from Laguna Divina Pastora indicates the presence of a swamp surrounded by treeless savanna between ca. 6100 and 2800 B.P., but forest must have been close by (Rull, 1991). From ca. 2800–1500 B.P., the swamp changed into a lake, which was larger than it is today and was surrounded by abundant stands of *Mauritia*. Laguna Santa Teresa also was larger than it is today from ca. 5900 to 4400 B.P. and was surrounded by forest. After 4400 B.P., the lake became smaller and forest retreated, but *Mauritia* stands expanded markedly (Rull, 1991). In contrast to the lakes in the Llanos Orientales, those in the Gran Sabana are not much older than 5900 B.P., suggesting drier conditions before this age.

19.4.3. Rupununi Savanna (Surinam)

Wijmstra and van der Hammen (1966) studied three sites in the Rupununi savanna in Surinam. Dating control is minimal, but the 650-cm-long core from Lake Moriru located in the northernmost part of this savan-

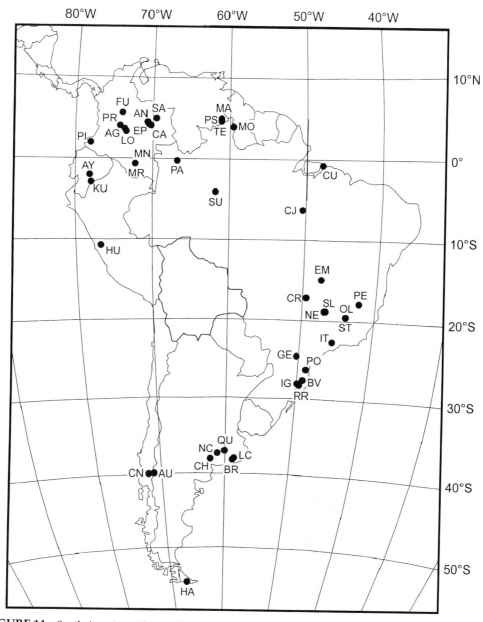

FIGURE 14 South America with sites discussed in the text. AG, Laguna de Agua Sucia; AN, Laguna Angel; AU, Mallín Aguado; AY, Lago Ayauchi; BR, Arroyo Las Brusquitas; BV, Serra de Boa Vista; CA, Laguna Carimagua; CH, Arroyo Sauce Chico; CJ, Carajas; CN, Caunahue; CR, Crominia; CU, Lagoa da Curuça B; EM, Agua Emendadas; EP, Laguna El Pinal; FU, Laguna Fúquene; GE, Serra Campos Gerais; HA, Harberton; HU, Laguna Huatacocha; IG, Morro da Igreja; IT, Morro de Itapeva; KU, Lago Kumpaka; LC, Cerro La China; LO, Laguna Loma Linda; MA, Santa Cruz de Mapaurí; MN, Pantano de Monica; MO, Lake Moriru; MR, Mariname; NC, Fortín Necochea; NE, Lagoa Serra Negra; OL, Lago dos Olhos; PA, Lake Pata; PE, Lago do Pires; PI, Laguna Piusbi (Chocó); PO, Poço Grande; PR, La Primavera; PS, Laguna Divina Pastora; QU, Empalme Querandíes; RR, Serra do Rio Rastro; SA, Laguna Sardinas; SL, Salitre; ST, Lagoa Santa; SU, Lago Surara; TE, Laguna Santa Teressa.

na region has two dates: 7585 ± 75 ^{14}C yr B.P. for 165–195 cm and 6065 ± 95 ^{14}C yr B.P. for 105–115 cm. Between ca. 11,500–10,000 and 5600 B.P., the proportions of woodland and grassland savanna changed repeat-edly. After 5600 B.P., and especially during the last 3000 years, grassland savanna became dominant, suggesting drier conditions or stronger human influences on the savanna.

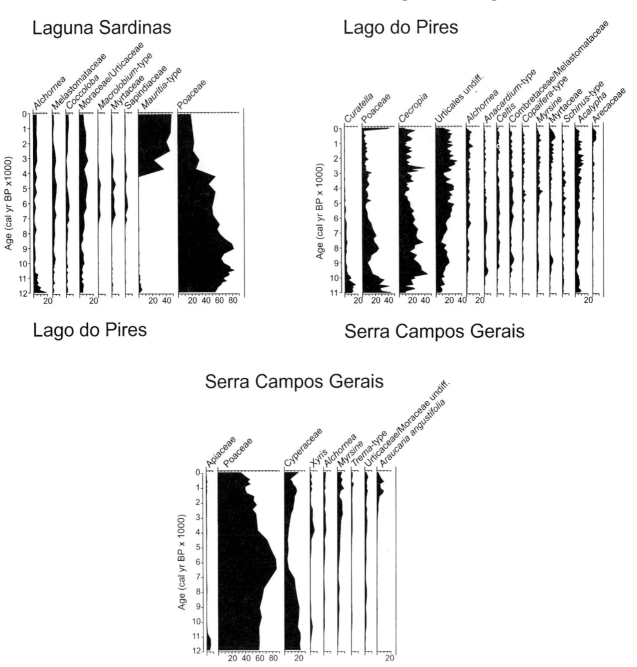

FIGURE 15   Representative pollen diagrams from the South American lowlands: Laguna Sardinas (Behling and Hooghiemstra, 1998), Lago do Pires (Behling, 1995a), and Serra Campos Gerais (Behling, 1997b, 1998).

19.4.4. Chocó Rain Forest in the Pacific Lowland (Colombia)

The pollen record from Laguna Piusbi in the south Colombian Pacific lowlands is the first studied site providing a Holocene history of the Colombian Chocó rain forest area (Behling et al., 1998). An earthquake or

other catastrophic neotectonic event may have formed the closed lake basin probably ca. 4900 B.P. Pollen and spore data illustrate a very diverse floral composition of the tropical rain forest. The Chocó rain forest ecosystem has been very stable, and no marked changes have occurred since the beginning of the record in the middle Holocene. A very humid climate (annual precipitation is >7000 mm) has maintained the high biodiversity and stability of the Chocó rain forest.

19.4.5. Amazon Rain Forest

19.4.5.1. Colombia

The Pantano de Monica 1 site is a swamp on the lower terrace of the Rio Caquetá in the central Colombian Amazon rain forest. During the early Holocene, the swamp was smaller and shallower than it is today with abundant *Mauritia* palm trees. The lower terrace of the Caquetá River must have been better drained than it is today, which may have resulted from changes in the drainage system or from drier conditions than exist today. The records from Pantano de Monica 2 and 3 indicate a drier climate after ca. 3200 B.P. The records from the lower terrace of Rio Caquetá show different forest compositions in the past and indicate that the rain forest environments were not constant during Holocene times (Behling et al., 1999).

Two cores from Mariñame Island, located in the Caquetá River bed (5–6 km south of Pantano de Monica), represent the last 11,500 years (Urrego, 1994/1997). Between 11,500 and 10,200 B.P., open-water conditions prevailed, and the Caquetá River valley was permanently inundated with rapid sedimentation of clay, suggesting relatively wet climatic conditions. At ca. 10,200 B.P., lower water levels with a decreased sedimentation rate probably resulted from a drier climate. *Cecropia* forest developed on the open floodplains. Seasonally inundated várzea forest, indicating relatively dry conditions, followed this period until 7400 B.P. at one site and until 5100 B.P. at a second site on Mariñame Island. After 5100 B.P., the sites on the island became less subject to direct river flooding, and a *Mauritia* swamp developed. Urrego (1994/1997) concluded that climate change at least partly influenced the river dynamics in the study area. In general, lower water levels in the Caquetá River during the early Holocene agree quite well with the paleoenvironmental conclusions from Pantano de Monica 1.

19.4.5.2. Ecuador

Most of the pollen records from western Amazonia are from Ecuadorian lakes along the foothills of the eastern Andes and represent only the late Holocene. The pollen record from Lake Ayauchi at ~500 m elevation provides a Holocene history of the Amazon rain forest since ca. 7800 B.P. (Bush and Colinvaux, 1988). A dry interval occurred between ca. 4800 and 3400 B.P. The record from Lago Kumpaka (700 m elevation) represents the environmental history of a perturbed rain forest for the last 6000 years. According to Liu and Colinvaux (1988), the periods of 4800–4200 and 1400–700 B.P. were drier with more seasonal precipitation.

19.4.5.3. Brazil

The Lake Pata pollen record from Hill of Six Lakes in northwestern Amazonia shows dense Amazon rain forest with several marked vegetation changes during the middle Holocene (Colinvaux et al., 1996; De Oliveira, 1996). Further study is required to determine whether these changes (e.g., the increase of the palm *Mauritia* or *Copaifera*) are due to climate, human disturbance, or other environmental factors. Similarly, the vegetational changes during the middle Holocene (e.g., an increase in *Alchornea*) at Lago Surara in the Amazon basin are still not understood (Absy, 1979).

Lagoa da Curuça B (35 m elevation) lies about 15 km from the Atlantic Ocean in northeastern Pará State. Pollen data show successions of different palm trees in the diverse Amazon rain forest during the middle Holocene (Behling, 1996). Whether these changes are related to climate change or to human disturbance is unclear. A greater representation of Poaceae pollen and carbonized particles suggest clearance and use of fire by humans.

The >50,000 ^{14}C year record from a lake on the small table mountain in Carajas (700–800 m elevation) in southeastern Pará State shows an expansion of savanna and a drier climate between ca. 8300 and 3200 B.P. After this period, the Amazon rain forest expanded, indicating wetter climatic conditions during the late Holocene (Absy et al., 1991).

19.4.6. Cerrado (Central Brazil)

The sites Agua Emendadas (Barberi, 1994; Salgado-Labouriau et al., 1998) and Cromínia (Ferraz-Vicentini and Salgado-Labouriau, 1996; Salgado-Labouriau et al., 1997) are in the central Brazilian cerrado region. The record from the Agua Emendadas palm swamp is barren of pollen before 8000 B.P., suggesting a dry climate with a long dry season in the early Holocene. During a long transitional phase following this period until ca. 5700 B.P., a *Mauritia* palm swamp developed, suggesting an increase in moisture. During the late Holocene, vegetation and climate were similar to those of today with widespread grass savanna and the presence of a marsh surrounded by cerrado and gallery forest. Pollen data from the swamp near Cromínia also show dry conditions in the early Holocene until ca. 7400 B.P., with essentially modern vegetation after ca. 5700 B.P.

19.4.7. Cerrado, Semideciduous Forest, and Mountain Vegetation in Southeastern Brazil

The high-resolution pollen record from Lago do Pires (Fig. 15), located 250 km from the Atlantic Ocean, shows evidence of extended cerrado in this lowland region during the early Holocene until ca. 6300 B.P., indicating a dry season of 5–6 months (Behling, 1995a). From 6300 to 2900 B.P., extensive semideciduous forest predominated in the valleys with cerrado on the hills, indicating higher precipitation and a dry season of about 5 months. From 2900–970 B.P., the pollen record shows more closed cerrado vegetation on the hills and slopes, suggesting a dry season of some 4–5 months. During the latest Holocene, the presence of closed semideciduous forest is indicative of modern climatic conditions.

Lagoa dos Olhos and Lagoa Santa are ~1 km apart and ~400 km from the Atlantic Ocean. The pollen record from Lagoa dos Olhos (De Oliveira, 1992) shows that a mosaic of cerrado and semideciduous forest occupied the region for most of the Holocene, indicating strong seasonality in precipitation. The pollen record from Lagoa Santa (Parizzi et al., 1998) begins at ca. 6200 B.P. during a period of increasing moisture. From ca. 6200–5300 B.P., the record shows the presence of swamp vegetation, indicating drier climatic conditions than those now. After ca. 5300 B.P., the pollen records show a mosaic pattern of cerrado, semideciduous forest, and gallery forest, indicating climatic conditions similar to those of today with a dry season of 3–4 months. Salitre and Serra Negra, which are only a few kilometers from each other but are at different elevations, are further inland, ~750 km from the Atlantic Ocean. Modern vegetation of this mountain region is complex, including mountain vegetation, semideciduous forest, and cerrado. The record from Salitre shows mesophytic semideciduous forest after ca. 9500 B.P., suggesting warmer and drier conditions than previously. At 5700 B.P., a short interval with high percentages of Poaceae indicates an arid interval. After 4400 B.P., the climate became moister, similar to today (Ledru, 1993; Ledru et al., 1994, 1996). The record from Serra Negra indicates that a mosaic of cerrado and semideciduous forest dominated the regional vegetation for most of the Holocene, indicative of a seasonal climate (De Oliveira, 1992).

Morro de Itapeva is located at 1850 m elevation in the Serra da Mantiqueira, which runs parallel to the Atlantic coast west of the Serra do Mar (Behling, 1997a). High-elevation grassland (campos de altitude), Araucaria forest, and cloud forest cover the higher moun-tains. During the early Holocene, cloud forest developed close to the study site, indicating a warm-moist climate on the east-facing slopes. However, rare occurrences of Araucaria and Podocarpus suggest a drier climate on the highland itself. Araucaria and Podocarpus increased during the last 3000 years, indicating an increase in moisture.

19.4.8. Araucaria Forest and Campos Region on the Southern Brazilian Highland and Tropical Atlantic Rain Forest in the Southern Lowland

According to the pollen records from Serra do Rio Rastro, Morro da Igreja, and Serra Campos Gerais (Fig. 15), huge areas of campos vegetation covered the southern Brazilian highland during the early and middle Holocene, suggesting a warm-dry climate with an annual dry season of ca. 3 months (Behling, 1995a,b, 1997a,b, 1998). The initial expansion of Araucaria forest, probably from small areas of gallery forest along the rivers, began at ca. 3200 B.P., suggesting a cooler and somewhat wetter climate than before. A pronounced expansion of Araucaria forest on the highlands, replacing the campos vegetation, occurred in Santa Catarina State (Serra do Rio Rastro and Morro da Igreja) during the last 1000 years and in Paraná State (Serra Campos Gerais) during the last 1400 years. Thus, a very humid climate with no significant dry period characterizes the late Holocene.

Tropical rain forest covers the Atlantic lowland in southern Brazil, and pollen records from this region show no marked changes in climate. The Atlantic lowland and the slopes facing the Atlantic Ocean were apparently wetter than the highlands during the middle Holocene. The record from the Poço Grande lowland indicates a drier period from 5600–5300 B.P., a slightly moister period from 5300–4800 B.P., and a period moister than today from 4800–3800 B.P.

19.4.9. Pampas (Argentina)

Prieto (1996) summarized the late Quaternary history of vegetation and climate from five sites in the Pampas (see Table 3). During the early Holocene, grassland communities existed until 7800 BP in the central Pampas and until ca. 5700 BP in the southwestern portion. For the late Holocene, the pollen data indicate vegetation of psammophytic and halophytic communities similar to those of today's southern Pampas, associated with communities of moister edaphic conditions. A shift to subhumid-dry climate occurred after 5700 BP at Arroyo Sauce Chico and Cerro La China, at ca. 4400 B.P.

at Fortín Necochea, and before 3200 B.P. at Empalme Querandíes.

19.5. THE ANDES

The Andes span a wide latitudinal range from 5°N to 55°S, and the modern vegetation types and altitudinal zonation are quite diverse along the length of the cordillera. In the northern Andes of Colombia, Peru, and Bolivia, the upper limit of the Andean forest lies at ca. 3500 m elevation. The tree line decreases in elevation in the southern Andes to ca. 1500 m at latitude 40°S and to 500 m at latitude 55°S. The northern Andes receive moisture from convection related to the easterly trade winds, whereas the southern Andes are under the influence of the westerlies. Because of regional topographic differences and availability of suitable sites, the pollen records from the northern Andes are primarily from high altitudes above tree line in the paramo and subparamo, whereas records from the southern Andes are from forest zones at lower elevations. Consequently, the Holocene vegetation changes recorded at northern sites primarily involve proportional shifts between Andean forest and paramo elements and, therefore, record treeline changes and temperature shifts. At southern sites, however, the pollen data indicate changes between different forest types, and the inferred climate signal is primarily precipitation and seasonality.

To characterize the Holocene climate changes in the Andes, we discuss representative pollen records from Colombia (La Primavera [Melief, 1985], Fúquene [Van Geel and van der Hammen, 1973]), Peru (Laguna Huatacocha [Hansen et al., 1984]), Chile (Caunahue [Markgraf, 1991a]), and Argentina (Mallín Aguado [Markgraf and Bianchi, 1999] Harberton [Markgraf, 1991a]).

19.5.1. Early Holocene

At high-elevation paramo sites from Colombia to Peru, the dominants of the early Holocene (11,000–7800 B.P.) are Poaceae, subparamo elements (Asteraceae and *Alnus*), and Andean forest elements (Melief, 1985; Hansen et al., 1984, 1994). The pollen diagrams from La Primavera (3525 m) in Columbia and Laguna Huatacocha (4500 m) in Peru illustrate this trend (Fig. 16). Locally, subparamo vegetation replaced the paramo. Compared to the preceding late glacial pollen spectra, which have low species diversity and a low influx of herbaceous taxa, the early Holocene spectra indicate an upslope shift of vegetation zones in response to increased temperatures and moisture.

Subparamo records obtained at elevations below the paramo zone in the northern Andes, for example, Laguna Fúquene, Colombia (2580 m), show a comparable upslope shift of vegetation zones (Fig. 16). Andean forest elements moved upward, especially *Quercus* in Colombia (van Geel and van der Hammen, 1973) and Urticales, *Hedyosmum,* and *Podocarpaceae* in Peru (Hansen et al., 1984, 1994; Hansen and Rodbell, 1995). Sediment records from glacial cirque lakes in the Bolivian Andes suggest increased temperatures in the early Holocene. These sites show greatly increased productivity compared to the preceding late glacial times (Abbott et al., 1997), and receding glaciers in the region suggest greatly decreased precipitation as well as higher temperatures (Abbott et al., 1997). In contrast, the early Holocene was apparently wet and cool in northern Chile and Argentina. A pollen record from the Altiplano in northwestern Argentina shows species-rich herbaceous grassland (Markgraf, 1985; Fernandez et al., 1991), and lake levels were high in northern Chile from 12,800–9500 B.P. (Valero-Garcés et al., 1996; Grosjean et al., 1995; Sylvestre et al., 1999; Bradbury et al., 2000).

Temperatures were apparently higher than they are today throughout the northern Andes in the early Holocene. However, moisture patterns varied latitudinally, with drier conditions than today in Colombia, Peru, and Bolivia, but seasonally more uniform wet conditions in the Andes of northern Chile and Argentina. This difference suggests that during the early Holocene the seasonal shift in latitude of the ITCZ was not as great as it is today. A possible explanation for this behavior may be linked to a different insolation pattern in the early Holocene. Seasonally, insolation was greater in June–July–August (southern winter, northern summer) and lower in December–January–February (southern summer, northern winter). Thus, seasonality was greater than it is today north of the equator, but the opposite was true south of the equator. As a consequence, precipitation was higher north of the equator and lower south of the equator (Abbott et al., 1997; Kutzbach et al., 1993). However, the paleoprecipitation patterns show greater latitudinal differentiation than insolation alone could cause, and other factors may be involved.

In the temperate forest region of the southern Andes, climate patterns were also latitudinally different throughout the Holocene. The sites Caunahue (Markgraf, 1991a) on the west slope of the Andes and Mallín Aguado (Markgraf and Bianchi, 1999) on the east slope of the Andes, both at 40°S, illustrate the middle latitude pattern (Fig. 17). In the early Holocene, forests at this latitude on both sides of the Andes contained a higher

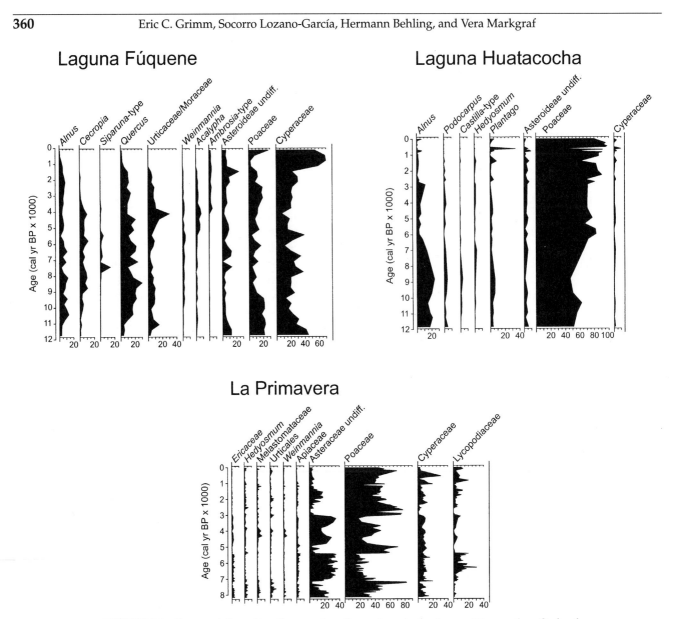

FIGURE 16   Representative pollen diagrams from the northern Andes: Laguna Fúquene (van Geel and van der Hammen, 1973), La Primavera (Melief, 1985), and Laguna Huatacocha (Hansen et al., 1984).

proportion of drought- and disturbance-adapted taxa than are present today. In the forests west of the Andes, *Nothofagus* parkland was more extensive than it is today, with high proportions of *Weinmannia* and other open-forest and disturbance taxa; whereas east of the Andes, substantial numbers of steppe taxa codominated with *Nothofagus dombeyi*-type, and charcoal particles are abundant in sediments from this period. The climate at the middle latitudes must have been warmer and seasonally drier than it is today (Villagrán, 1995; Markgraf, 1991b). At latitudes 45°–50°S, maximum development of Patagonian rain forest and Magellanic moorland occurred west of the Andes in the early

Holocene (Lumley and Switsur, 1993; Ashworth et al., 1991), and lake levels were high east of the Andes (Stine and Stine, 1990). At high southern latitudes south of 50°S, *Nothofagus* woodland and a high fire frequency characterized the early Holocene (Markgraf and Anderson, 1994; Huber and Markgraf, in press; Heusser, 1999). The Harberton record (Fig. 17) illustrates this trend, which suggests that climate must have been seasonally drier and more variable than it is today. During the early Holocene, the westerly storm tracks that determine precipitation patterns in the temperate region must have been more narrowly focused between 45° and 50°S, leaving the regions both to the north and

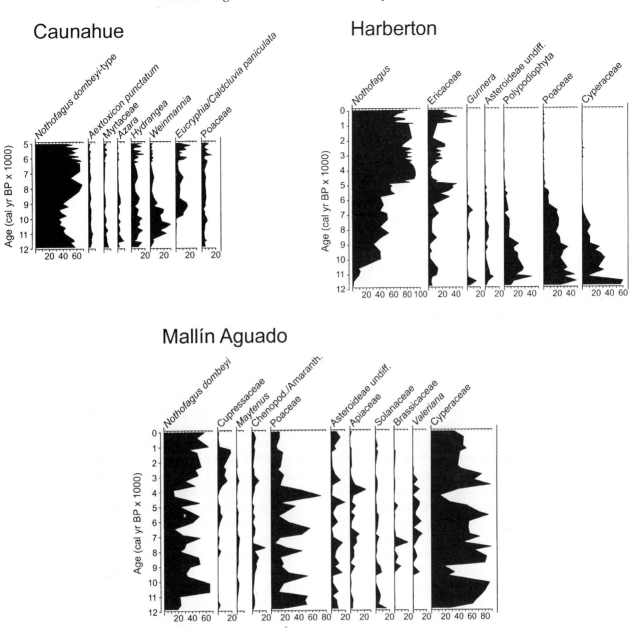

FIGURE 17   Representative pollen diagrams from the southern Andes: Caunahue (Markgraf, 1991a), Mallín Aguado (Markgraf and Bianchi, 1999), and Harberton (Markgraf, 1991a).

south at least seasonally drier than they are today (Markgraf, 1991b, 1993).

19.5.2. Middle Holocene

Except for minor changes in the temperate middle-latitude records from the southern Andes, all other records from South America show increasing temperatures and aridity for the interval between 7800 and 4300 B.P., reaching a maximum between 6500 and 4500 B.P. Water levels in bogs and lakes from Colombia to Tierra

del Fuego decreased, resulting in desiccation intervals at several sites. At Paramo sites in the northern Andes, disturbance taxa as well as forest taxa characteristic of lower elevations increased. Although the increase in pollen from lower Andean forest taxa could have resulted from upslope forest expansion, a more probable explanation is that vegetation was more open at higher elevations and local pollen productivity was lower, resulting in a proportionate increase in pollen from long-distance sources (Hansen and Rodbell, 1995; Markgraf, 1993). Records from the southernmost Andes show an increase in xeric herbs and shrubs, suggesting extreme aridity and windiness. No explanation is obvious for such a widespread, interhemispheric climate anomaly, which simultaneously entailed reduced tropical convection and reduced westerly storm track activity both north and south of the equator.

19.5.3. Late Holocene

After ca. 5500 B.P., human activities, primarily animal husbandry at high elevations and agriculture at lower elevations, began to heavily influence vegetation in the northern Andes. This influence greatly complicates the interpretation of climate change from vegetation records. However, rises in lake levels and renewed glacier growth in Peru and Bolivia suggest that precipitation must have increased (Abbott et al., 1997; Mourguiart et al., 1995).

Taxa composition and latitudinal distribution of forest types began to resemble that of the present after ca. 5500 B.P. in the southern temperate rain forests at latitudes of 40°–55°S. West of the Andes, mesic rain forest expanded at latitude 40°S and extended farther northwards (Markgraf, 1987). East of the Andes, mixed *Austrocedrus/Nothofagus dombeyi*-type forests became established. Enhanced precipitation coupled with greater interannual variability may have resulted from the onset of increased El Niño/Southern Oscillation (ENSO) variability (McGlone et al., 1992; Rodbell et al., 1999; Markgraf and Diaz, 2000). South of latitude 45°–55°S, forests developed to their maximum density of the Holocene, indicating increased westerly precipitation. Thus, in the temperate latitudes, precipitation was more evenly distributed over a wider latitudinal stretch than prior to 5500 B.P. The cause of this pattern may have been increased seasonal variability in insolation, which led to greater latitudinal shifts in the southern westerlies.

19.5.4. Summary for the Andes

Evidence exists for increased temperatures during the early Holocene throughout the Andes. Moisture in the northern Andes was lower than it is today, except at high elevations in northern Chile and Argentina. Because of a steeper temperature gradient in North America owing to the continued presence of the retreating Laurentide ice sheet in Canada, the ITCZ must have remained in a more northerly position year-round, thus reducing the moisture from an easterly source in northern South America. In the southern Andes, temperatures also increased throughout; however, moisture was latitudinally uneven. Both the northern and southern extensions of the temperate rain forests were drier

than they are today, suggesting that the westerly storm tracks were much more focused between latitudes 45°–50°S. In the middle Holocene, moisture decreased further throughout the entire Andes, especially in the northernmost Andes and the southernmost temperate forest region. During this time, both the ITCZ and the southern westerlies must have been less effective sources of moisture than they are today. Vegetation similar to that of today developed at ca. 5500 B.P., but significant human impact complicates climate interpretation from vegetation records, especially in the northern Andes. However, other climate-proxy data suggest cooler and moisture conditions than before, possibly because of greater seasonal shifts of the ITCZ and the southern westerlies, as well as increased ENSO activity.

19.6. SYNTHESIS

The most striking feature of Holocene climate change along the Pole-Equator-Pole: Americas (PEP 1) transect through the Americas is that, overall, the early and middle Holocene were warmer and drier than today in both hemispheres, but with some exceptions. In the Northern Hemisphere, the early to middle Holocene was warmer and drier than today in Alaska, the Pacific Northwest, the Sierra Nevada, the Northeast, and the Southeast. The pattern is mixed in the northern Rocky Mountains, where the warmest and driest phase at some sites was in the early Holocene, similar to the Pacific Northwest; whereas at other sites the warm-dry maximum was delayed until the middle Holocene, more similar to the pattern in the Midwest. In the Southwest and southern Rocky Mountains, the early Holocene was warmer, but wetter than today. In the Southwest, enhanced seasonality is evident, with increased summer and decreased winter precipitation, which resulted in more lush plant communities in the deserts, which rely on summer precipitation, but dry lakes in the mountains, which depend on winter snows. In the basin of México, however, the situation was reversed, with evidence for glacial advances dependent on winter snows in the mountains, but drier conditions in the basin.

In the Midwest and northern Great Plains, the climate during the earliest Holocene was wet, but shifted to a climate drier than that of today by 9000 B.P. in the western part of the region. The middle Holocene maximum of aridity extended from the northern Great Plains, through the southern High Plains, and into the basin of México. The time of maximum drought was delayed in the eastern Midwest and in central Texas until ca. 6500 years ago. Not enough data exist to determine whether this pattern was geographically connected between these two areas. The most notable exception to an early Holocene thermal maximum is in northeast Canada, which remained covered by the receding Laurentide ice sheet during this time. There, the warmest temperatures of the Holocene occurred between 6000–3000 B.P., after final retreat of the ice.

The generally warmer early Holocene, except over the remnant ice sheet, is in agreement with the early Holocene insolation maximum (Kutzbach et al., 1998). The data suggest strengthening of the Eastern Pacific Subtropical High and the Bermuda High in the early Holocene, causing dry climates in the northwestern and southeastern parts of the continent. The heat low in the central part of the continent strengthened the monsoon in the southwestern United States, which was consequently wetter than it is today in the early to middle Holocene. However, the midcontinent was drier in the middle Holocene. In the Midwest, the earliest Holocene was wetter than today, perhaps because anticyclonic motion of the Bermuda High transported moist air from the Gulf of México that collided with the Arctic Front, which was fixed south of its modern position by the still-existing ice sheet to the north. As the ice sheet retreated, the Midwest became drier, and as the subtropical highs weakened in the middle Holocene, the maximum time of aridity shifted eastward.

The climate signal is mixed in the Yucatán Peninsula and Central America. The northern Yucatán was dry in the early Holocene and wetter in the middle Holocene. However, the pattern is reversed in Guatemala, Costa Rica, and Panama, where maximum aridity was in the middle Holocene. Most of the savanna regions of South America both north and south of the equator were drier and more extensive in the early to middle Holocene (see Behling and Hooghiemstra, 2000). The Andes were also warmer and drier in the early and middle Holocene, except at high elevations in northern Argentina and Chile, with the warmest and driest period during the middle Holocene. During the early Holocene and especially during the middle Holocene, westerly storm tracks were more tightly focused than they are today between 45° and 50°S. Hence, north and south of this latitudinal zone, precipitation was greatly reduced, especially east of the Andes. Holocene climate change in the Amazon tropical rain forest was relatively minor. Some areas show some evidence of drier phases during the early or middle Holocene, but others, such as the Chocó rain forest, experienced very little change with a continuously wet climate and no dry season.

The synchrony of maximum aridity in the savanna regions of the two hemispheres and along the entire spine of the Andes in spite of opposite insolation maxima seems paradoxical, but a plausible explanation might be owing to a reduced north–south migration of the ITCZ during the early to middle Holocene. The central equatorial region remained wet throughout the Holocene, but reduced annual migration of the ITCZ would have caused savanna regions to expand and would have exacerbated aridity farther away from the equator. The mean position of the ITCZ must also have remained more northerly year-round in the early to middle Holocene, reducing easterly moisture throughout South America. Throughout Central and South America, the late Holocene was wetter and cooler, as was generally the case in North America. The teleconnections between orbital forcing, the retreating Laurentide ice sheet, the heat low in central North America, and the ITCZ undoubtedly account for general synchrony between the hemispheres, but the mechanisms of these teleconnections remain to be discovered.

References

Abbott, M. V., G. O. Seltzer, K. R. Kelts, and J. Southon, 1997: Holocene paleohydrology of the tropical Andes from lake records. *Quaternary Research*, **47**: 70–80.

Absy, M. L., 1979: A palynological study of Holocene sediments in the Amazon basin. Dissertation. University of Amsterdam, The Netherlands.

Absy, M. L., A. M. Cleef, M. Fournier, L. Martin, M. Servant, A. Sifeddine, M. Ferreira da Silva, F. Soubies, K. Suguio, B. Turcq, and T. van der Hammen, 1991: Mise en évidence de quatre phases d'ouverture de la forêt dense dans le sud-est de l'Amazonie au cours des 60,000 dernières années. Première comparaison avec d'autres régions tropicales. *Comptes Rendus de l'Académie des Sciences, Série II*, **312**: 673–678.

Ager, T. A., 1983: Holocene vegetational history of Alaska. *In* Wright, H. E., Jr. (ed.), *Late Quaternary Environments of the United States. Vol. 2. The Holocene*. Minneapolis: University of Minnesota Press, pp. 128–141.

Albert, L. E., and D. G. Wyckoff, 1981: Ferndale Bog and Natural Lake: Five thousand years of environmental change in southeastern Oklahoma. Oklahoma Archaeological Survey, Studies in Oklahoma's Past, 7.

Allison, T. D., R. E. Moeller, and M. B. Davis, 1986: Pollen in laminated sediments provides evidence for a mid-Holocene forest pathogen outbreak. *Ecology*, **67**: 1101–1105.

Almeida-Leñero, L., 1997: Vegetación, Fitogeografía y Paleoecología del Zacatonal Alpino y Bosques Montañosos de la Región Central de México. Dissertation. University of Amsterdam, The Netherlands.

Almendinger, J. C., 1992: The late Holocene history of prairie, brush-prairie, and jack pine (*Pinus banksiana*) forest on outwash plains, north-central Minnesota, U.S.A. *The Holocene*, **2**: 37–50.

Almquist-Jacobson, H., and D. Sanger, 1995: Holocene climate and vegetation in the Milford drainage basin, Maine, U.S.A., and their implications for human history. *Vegetation History and Archaeobotany*, **4**: 211–222.

Anderson, P. M., 1985: Late Quaternary vegetational change in the Kotzebue Sound area, northwestern Alaska. *Quaternary Research* **24**: 307–321.

Anderson, P. M., R. E. Reanier, and L. B. Brubaker, 1988: Lake Quaternary vegetational history of the Black River region in northeastern Alaska. *Canadian Journal of Earth Sciences*, **25**: 84–94.

Anderson, R. S., 1990: Holocene forest development and paleoclimates within the central Sierra Nevada, California. *Journal of Ecology*, **78**: 470–489.

Anderson, R. S., 1993: A 35,000 year vegetation and climate history from Potato Lake, Mogollon Rim, Arizona. *Quaternary Research*, **40**: 351–359.

Anderson, R. S., and T. R. Van Devender, 1995: Vegetation history and paleoclimates of the coastal lowlands of Sonora, Mexico—Pollen records from packrat middens. *Journal of Arid Environments*, **30**: 295–306.

Anderson, R. S., O. K. Davis, and P. L. Fall, 1985: Late glacial and Holocene vegetation and climate in the Sierra Nevada of California, with particular reference to the Balsam Meadow site. American Association of Stratigraphic Palynologists Contribution Series 16, 127–140.

Anderson, R. S., G. L. Jacobson, Jr., R. B. Davis, and R. Stuckenrath, 1992: Gould Pond, Maine: Late-glacial transition from marine to upland environments. *Boreas*, **21**: 359–371.

Arnauld, C., S. Metcalfe, and P. Petrequin, 1997: Holocene climatic change in the Zacapu basin, Michoacan: Synthesis of results. *Quaternary International*, **43/44**: 173–179.

Ashworth, A. C., V. Markgraf, and C. Villagrán, 1991: Late Quaternary climatic history of the Chilean Channels based on fossil pollen and beetle analyses, with an analysis of the modern vegetation and pollen rain. *Journal of Quaternary Science*, **6**: 279–291.

Baker, R. G., L. J. Maher, Jr., C. A. Chumbley, and K. L. Van Zant, 1992: Patterns of Holocene environmental change in the midwestern United States. *Quaternary Research*, **37**: 379–389.

Barberi, M., 1994: Paleovegetação e paleoclima no Quaternário tardio da vereda de Águas Emendadas. Master's thesis. Universidade de Brasília, Brasil.

Barnosky, C. W., 1981: A record of late Quaternary vegetation from Davis Lake, southern Puget lowland, Washington. *Quaternary Research*, **16**: 221–239.

Barnosky, C. W., 1985: Late Quaternary vegetation near Battle Ground Lake, southern Puget Trough, Washington. *Geological Society of America Bulletin*, **96**: 263–271.

Bartlett, A. S., and E. S. Barghoorn, 1973: Phytogeographic history of the Isthmus of Panama during the past 12,000 years (a history of vegetation, climate and sea level). *In* Graham, A. (ed.), *Vegetation and Vegetational History of Northern Latin America*. New York: Elsevier, pp. 203–299.

Behling, H., 1993: Untersuchungen zur spätpleistozänen und holozänen Vegetations- und Klimageschichte der tropischen Küstenwälder und der Araukarienwälder in Santa Catarina (Südbrasilien). *Dissertationes Botanicae*, **206**. Berlin, J. Cramer.

Behling, H., 1995a: A high resolution Holocene pollen record from Lago do Pires, SE Brazil: Vegetation, climate and fire history. *Journal of Paleolimnology*, **14**: 253–268.

Behling, H., 1995b: Investigations into the late Pleistocene and Holocene history of vegetation and climate in Santa Catarina (S Brazil). *Vegetation History and Archaeobotany*, **4**: 127–152.

Behling, H., 1996: First report on new evidence for the occurrence of *Podocarpus* and possible human presence at the mouth of the Amazon during the late-glacial. *Vegetation History and Archaeobotany*, **5**: 241–246.

Behling, H., 1997a: Late Quaternary vegetation, climate and fire history from the tropical mountain region of Morro de Itapeva, SE Brazil. *Palaeogeography, Palaeoclimatology, Palaeoecology*, **129**: 407–422.

Behling, H., 1997b: Late Quaternary vegetation, climate and fire history in the *Araucaria* forest and campos region from Serra Campos Gerais (Paraná), S Brazil. *Review of Palaeobotany and Palynology*, **97**: 109–121.

Behling, H., 1998: Late Quaternary vegetational and climatic changes in Brazil. *Review of Palaeobotany and Palynology*, **99**: 143–156.

Behling, H., and H. Hooghiemstra, 1998: Late Quaternary palaeoecology and palaeoclimatology from pollen records of the savannas of the Llanos Orientales in Colombia. *Palaeogeography, Palaeoclimatology, Palaeoecology*, **139**: 251–267.

Behling, H., and H. Hooghiemstra, 1999: Environmental history of the Colombian savannas of the Llanos Orientales since the Last Glacial Maximum from lake records El Pinal and Carimagua. *Journal of Paleolimnology*, **21**: 461–476.

Behling, H., and H. Hooghiemstra, 2000: Neotropical savanna environments in space and time: Late Quaternary interhemispheric comparisons. *In* Markgraf, V. (ed.), *Interhemispheric Climate Linkages*. San Diego: Academic Press, Chapter 18.

Behling, H., J. C. Berrio, and H. Hooghiemstra, 1999: Late Quaternary pollen records from the middle Caquetá River basin in central Colombian Amazon. *Palaeogeography, Palaeoclimatology, Palaeoecology*, **145**: 193–213.

Behling, H., H. Hooghiemstra, and A. J. Negret, 1998: Holocene history of the Chocó rain forest from Laguna Piusbi, southern Pacific lowlands of Columbia. *Quaternary Research*, **50**: 300–308.

Bradbury, J. P., 1982: Holocene chronostratigraphy of Mexico and Central America. *Striae*, **16**: 46–48.

Bradbury, J. P., 1989: Late Quaternary lacustrine paleoenvironments in the Cuenca de México. *Quaternary Science Reviews*, **8**: 75–100.

Bradbury, J. P., 1997: Sources of glacial moisture in Mesoamerica. *Quaternary International*, **43/44**: 97–116.

Bradbury, J. P., M. Grosjean, S. Stine, and F. Sylvestre, 2000: Full and late glacial lake records along the PEP 1 transect: Their role in developing interhemispheric paleoclimate interactions. *In* Markgraf, V. (ed.), *Interhemispheric Climate Linkages*. San Diego: Academic Press, Chapter 16.

Brant, L. A., 1980: A palynological investigation of postglacial sediments at two locations along the Continental Divide near Helena, Montana. Dissertation. Pennsylvania State University, State College, Pennsylvania.

Brenner, M., 1994a: Lake Salpeten and Quexil, Petén, Guatemala, Central America. *In* Kelts, K., and E. Gierlowski-Kordesch (eds.), *Global Geological Record of Lake Basins*. Cambridge, United Kingdom: Cambridge University Press, pp. 377–380.

Brenner, M., 1994b: Lake Miragoane, Haiti (Caribbean). *In* Kelts, K., and E. Gierlowski-Kordesch (eds.), *Global Geological Record of Lake Basins*. Cambridge, United Kingdom: Cambridge University Press, pp. 401–403.

Brown, R. B., 1984: The paleoecology of northern frontier of Mesoamerica. Dissertation. University of Arizona, Tucson.

Brown, R. B., 1985: A summary of late-Quaternary pollen records from Mexico west of the Isthmus of Tehuantepec. *In* Bryant, V. M., and R. G. Holloway (eds.), *Pollen Records of Late Quaternary North America Sediments*. Dallas, TX: American Association of Stratigraphic Palynologists Foundation, pp. 71–94.

Brubaker, L. B., H. L. Garfinkel, and M. E. Edwards, 1983: A late Wisconsin and Holocene vegetation history from the central Brooks Range: Implications for Alaskan paleoecology. *Quaternary Research*, **20**: 194–214.

Bush, M. B., and P. A. Colinvaux, 1988: A 7000 year vegetational history from lowland Amazon, Ecuador. *Vegetatio*, **76**: 141–154.

Bush, M. B., D. R. Piperno, P. A. Colinvaux, P. E. De Olivera, L. A. Krissek, M. C. Miller, and W. E. Rowe, 1992: A 14,300 yr paleoecological profile of a lowland tropical lake in Panama. *Ecological Monographs*, **62**: 251–275.

Caballero-Miranda, M., 1995: Late Quaternary paleoenvironments of

Lake Chalco, the Basin of Mexico. Dissertation. University of Hull, Hull, United Kingdom.

Caballero-Miranda, M., and B. Ortega-Guerrero, 1998: Lake levels since about 40,000 years at Lake Chalco, near Mexico City. *Quaternary Research*, **50**: 69–79.

Caballero-Miranda, M., S. Lozano-García, B. Ortega-Guerrero, J. Urrutia, and J. L. Macías, 1999: Environmental characteristics of Lake Tecocomulco, northern Basin of Mexico, for the last 50,000 years. *Journal of Paleolimnology*, **22**: 399–411.

Carrara, P. E., D. A. Trimble, and M. Rubin, 1991: Holocene treeline fluctuations in the northern San Juan Mountains, Colorado, U.S.A., as indicated by radiocarbon-dated conifer wood. *Arctic and Alpine Research*, **23**: 233–246.

Clague, J. J., and R. W. Mathewes, 1989: Early Holocene thermal maximum in western North America: New evidence from Castle Peak, British Columbia. *Geology*, **17**: 277–280.

Clark, J. S., P. D. Royall, and C. Chumbley, 1996: The role of fire during climate change in an eastern deciduous forest at Devil's Bathtub, New York. *Ecology*, **77**: 2148–2166.

Cole, K. L., 1990: Late Quaternary vegetation gradients through the Grand Canyon. *In* Betancourt, J. L., T. R. Van Devender, and P. S. Martin (eds.), *Packrat Middens: The Last 40,000 Years of Biotic Change*. Tucson: University of Arizona Press, pp. 240–258.

Colinvaux, P. A., P. E. De Oliveira, J. E. Moreno, M. C. Miller, and M. B. Bush, 1996: A long pollen record from lowland Amazonia: Forest and cooling in glacial times. *Science*, **274**: 85–87.

Curtis, J. H., M. Brenner, D. A. Hodell, R. A. Balser, G. A. Islebe, and H. Hooghiemstra, 1998: A multi-proxy study of Holocene environmental change in the Maya Lowlands of Petén, Guatemala. *Journal of Paleolimnology*, **19**: 139–159.

Curtis, J. H., D. A. Hodell, and M. Brenner, 1996: Climate variability on the Yucatán Peninsula (Mexico) during the past 3500 years, and implications for Maya cultural evolution. *Quaternary Research*, **46**: 37–47.

Cwynar, L. C., 1987: Fire and the forest history of the North Cascade Range. *Ecology*, **68**: 791–802.

Cwynar, L. C., 1990: A late Quaternary vegetation history from Lily Lake, Chilkat Peninsula, southeast Alaska. *Canadian Journal of Botany*, **68**: 1106–1112.

Cwynar, L. C., and R. W. Spear, 1995: Paleovegetation and paleoclimatic changes in the Yukon at 6 ka B.P. *Géographie Physique et Quaternaire*, **49**: 29–35.

D'Antoni, H. L., M. A. Nieto, and M. V. Mancini, 1985: Pollenanalytic stratigraphy of Arroyo Las Brusquitas profile (Buenos Aires Province, Argentina). *Zentralblatt für Geologie und Paläontologie*, **11/12**: 1721–1729.

Davis, M. B., 1969: Climatic changes in southern Connecticut recorded by pollen deposition at Rogers Lake. *Ecology*, **50**: 409–422.

Davis, M. B., 1981: Outbreaks of forest pathogens in Quaternary history. *Proceedings of the IV International Palynological Conference Lucknow (1976–77)*, **3**: 216–227.

Davis, M. B., R. W. Spear, and L. C. K. Shane, 1980: Holocene climate of New England. *Quaternary Research*, **14**: 240–250.

Davis, O. K., and D. S. Shafer, 1992: A Holocene climatic record for the Sonoran Desert from pollen analysis of Montezuma Well, Arizona, U.S.A. *Palaeogeography, Palaeoclimatology, Palaeoecology*, **92**: 107–119.

Davis, O. K., R. S. Anderson, P. L. Fall, M. K. O'Rourke, and R. S. Thompson, 1985: Palynological evidence for early Holocene aridity in the southern Sierra Nevada, California. *Quaternary Research*, **24**: 322–332.

Deevey, E. S., and M. Stuiver, 1964: Distribution of natural isotopes of carbon in Linsley Pond and other New England lakes. *Limnology and Oceanography*, **9**: 1–11.

De Oliveira, P. E., 1992: A palynological record of late Quaternary vegetational and climatic change in southeastern Brazil. Dissertation. Ohio State University, Columbus.

De Oliveira, P. E., 1996: Glacial cooling and forest disequilibrium in western Amazonia. *Anais da Academia Brasileira de Ciências*, **68**: 129–138.

Dwyer, T. R., H. T. Mullins, and S. C. Good, 1996: Paleoclimatic implications of Holocene lake-level fluctuations, Owasco Lake, New York. *Geology*, **24**: 519–522.

Dyke, A. S., and V. K. Prest, 1987: Late Wisconsinan and Holocene history of the Laurentide ice sheet. *Geographie Physique et Quaternaire*, **41**: 237–264.

Edwards, M. E., P. M. Anderson, H. L. Garfinkel, and L. B. Brubaker, 1985: Late Wisconsin and Holocene vegetational history of the Upper Koyukuk region, Brooks Range, AK. *Canadian Journal of Botany*, **63**: 616–626.

Elias, S. A., P. E. Carrara, L. J. Toolin, and A. J. T. Jull, 1991: Revised age of deglaciation of Lake Emma based on new radiocarbon and macrofossil ages. *Quaternary Research*, **36**: 307–321.

Engstrom, D. R., and B. C. S. Hansen, 1985: Postglacial vegetational change and soil development in southeastern Labrador as inferred from pollen and chemical stratigraphy. *Canadian Journal of Botany*, **63**: 543–561.

Fall, P. L., 1997a: Fire history and composition of the subalpine forest of western Colorado during the Holocene. *Journal of Biogeography*, **24**: 309–325.

Fall, P. L., 1997b: Timberline fluctuations and late Quaternary paleoclimates in the southern Rocky Mountains, Colorado. *Geological Society of America Bulletin*, **109**: 1306–1320.

Fall, P. L., P. T. Davis, and G. A. Zielinski, 1995: Late Quaternary vegetation and climate of the Wind River Range, Wyoming. *Quaternary Research*, **43**: 393–404.

Fernandez, J., V. Markgraf, H. O. Panarello, M. Albero, F. A. Angiolini, S. Valencio, and M. Arriaga, 1991: Late Pleistocene/early Holocene environments and climates, fauna, and human occupation in the Argentine Altiplano. *Geoarchaeology*, **6**: 251–272.

Ferraz-Vicentini, K. R., and M. L. Salgado-Labouriau, 1996: Palynological analysis of a palm swamp in Central Brazil. *Journal of South American Earth Sciences*, **9**: 207–219.

Flores, J. S., and I. E. Carbajal, 1994: Tipos de Vegetación de la Península de Yucatán. *Etnoflora Yucatanense*, **3**: 1–135.

Friedman, I., P. Carrara, and J. Gleason, 1988: Isotopic evidence of Holocene climatic change in the San Juan Mountains, Colorado. *Quaternary Research*, **30**: 350–353.

Fritz, S. C., S. E. Metcalfe, and W. E. Dean, 2000: Holocene climate patterns in the Americas inferred from paleolimnological records. *In* Markgraf, V. (ed.), *Interhemispheric Climate Linkages*. San Diego: Academic Press, Chapter 15.

Gajewski, K., 1987: Climatic impacts on the vegetation of eastern North America during the past 2000 years. *Vegetatio*, **68**: 179–190.

Gajewski, K., 1988: Late Holocene climatic changes in eastern North America estimated from pollen data. *Quaternary Research*, **29**: 255–262.

Gajewski, K., and S. Garralla, 1992: Holocene vegetation histories from three sites in the tundra of northwestern Quebec, Canada. *Arctic and Alpine Research*, **24**: 329–336.

Gajewski, K., S. Payette, and J. C. Ritchie, 1993: Holocene vegetation history at the boreal-forest–shrub-tundra transition in northwestern Quebec. *Journal of Ecology*, **81**: 433–443.

Gajewski, K., M. G. Winkler, and A. M. Swain, 1985: Vegetation and fire history from three lakes with varved sediments in northwestern Wisconsin (U.S.A.). *Review of Palaeobotany and Palynology*, **44**: 277–292.

Grimm, E. C., 1983: Chronology and dynamics of vegetation change in the prairie-woodland region of southern Minnesota, U.S.A. *New Phytologist*, **93**: 311–350.

Grimm, E. C., 1984: Fire and other factors controlling the Big Woods vegetation of Minnesota in the mid-nineteenth century. *Ecological Monographs*, **54**: 291–311.

Grimm, E. C., G. L. Jacobson, Jr., W. A. Watts, B. C. S. Hansen, and K. A. Maasch, 1993: A 50,000-year record of climate oscillations from Florida and its temporal correlation with the Heinrich events. *Science*, **261**: 198–200.

Grosjean, M., M. Geyh, B. Messerli, and U. Schotterer, 1995: Lateglacial and early Holocene lake sediment, groundwater formation and climate in the Atacama Altiplano. *Journal of Paleolimnology*, **14**: 241–252.

Hansen, B. C. S., 1990: Pollen stratigraphy of Laguna de Cocos. *In* Pohl, M. D. (ed.), *Ancient Maya Wetland Agriculture. Excavations on Albion Island, Northern Belize*. Boulder, CO: Westview Press, pp. 155–186.

Hansen, B. C. S., and D. R. Engstrom, 1996: Vegetation history of Pleasant Island, southeastern Alaska, since 13,000 yr B.P. *Quaternary Research*, **46**: 161–175.

Hansen, B. C. S., and D. T. Rodbell, 1995: A late-glacial/Holocene pollen record from the eastern Andes of northern Peru. *Quaternary Research*, **44**: 216–227.

Hansen, B. C. S., G. O. Seltzer, and H. E. Wright, Jr., 1994: Late Quaternary vegetational change in the central Peruvian Andes. *Palaeogeography, Palaeoclimatology, Palaeoecology*, **109**: 263–285.

Hansen, B. C. S., H. E. Wright, Jr., and J. P. Bradbury, 1984: Pollen studies in the Junín area, central Peruvian Andes. *Bulletin of the Geological Society of America*, **95**: 1454–1465.

Hebda, R. J., 1983: Late-glacial and postglacial vegetation history at Bear Cove Bog, northeast Vancouver Island, British Columbia. *Canadian Journal of Botany*, **61**: 3172–3192.

Heine, K., 1988: Late Quaternary glacial chronology of Mexican volcanoes. *Die Geowissenchaften*, **6**(7): 197–205.

Heusser, C. J., 1974: Quaternary vegetation, climate, and glaciation of the Hoh River Valley, Washington. *Geological Society of America Bulletin*, **85**: 1547–1560.

Heusser, C. J., 1977: Quaternary palynology of the Pacific slope of Washington. *Quaternary Research*, **8**: 282–306.

Heusser, C. J., 1983: Holocene vegetation history of the Prince William Sound Region, south-central Alaska. *Quaternary Research*, **19**: 337–355.

Heusser, C. J., 1999: Human forcing of vegetation change since the last ice age in southern Chile and Argentina. *Bamberger Geographische Schriften*, **19**: 211–232.

Heusser, C. J., L. E. Heusser, and D. M. Peteet, 1985: Late-Quaternary climatic change on the American North Pacific coast. *Nature*, **315**: 485–487.

Hodell, D. A., J. H. Curtis, and M. Brenner, 1995: Possible role of climate in the collapse of Classic Maya civilization. *Nature*, **375**: 391–394.

Hodell, D. A., J. H. Curtis, G. A. Jones, A. Higuera-Gundy, M. Brenner, M. W. Binford, and K. Dorsey, 1991: Reconstruction of Caribbean climate change over the past 10,500 years. *Nature*, **352**: 790–793.

Holliday, V. T., 1987: A reexamination of late-Pleistocene boreal forest reconstruction for the southern High Plains. *Quaternary Research*, **28**: 238–244.

Holliday, V. T., 1989: Middle Holocene drought on the southern High Plains. *Quaternary Research*, **31**: 74–82.

Holliday, V. T., 1995: Stratigraphy and paleoenvironments of late Quaternary valley fills on the southern High Plains. *Geological Society of America Memoir*, **186.**

Holliday, V. T., 2000: Folsom drought and episodic drying on the southern High Plains from 10,900–10,200 ^{14}C yr B.P. *Quaternary Research*, **53**: 1–12.

Hooghiemstra, H., A. M. Cleef, G. Noldus, and M. Kappelle, 1992: Upper Quaternary vegetation dynamics and paleoclimatology of La Chonta Bog area (Cordillera de Talamanca, Costa Rica). *Journal of Quaternary Science*, **7**: 205–225.

Horn, S. P., 1993: Postglacial vegetation and fire history in the Chirripó paramo of Costa Rica. *Quaternary Research*, **40**: 107–116.

Huber, U. M., and V. Markgraf, in press: Holocene moisture regime changes and fire frequency at Rio Rubens bog, southern Patagonia.

Islebe, G. A., and H. Hooghiemstra, 1997: Vegetation and climate history of montane Costa Rica since the last glacial. *Quaternary Science Reviews*, **16**: 589–604.

Islebe, G. A., H. Hooghiemstra, and R. van't Veer, 1996a: Holocene vegetation and water level history in two bogs of the Cordillera de Talamanca, Costa Rica. *Vegetatio*, **124**: 155–171.

Islebe, G. A., H. Hooghiemstra, M. Brenner, J. H. Curtis, and D. A. Hodell, 1996b: A Holocene vegetation history from lowland Guatemala. *The Holocene*, **6**: 265–271.

Israde Alcantara, I., 1997: Neogene diatoms of Cuitzeo Lake, central sector of the Trans-Mexican Volcanic Belt and their relationships with the volcano-tectonic evolution. *Quaternary International*, **43/44**: 137–144.

Jacobs, B. F., 1983: Past vegetation and climate of the Mogollon Rim area, Arizona. Dissertation. University of Arizona, Tucson.

Jacobs, B. F., 1985: A middle Wisconsin pollen record from Hay Lake, Arizona. *Quaternary Research*, **24**: 121–130.

Jacobson, G. L., Jr., and A. Dieffenbacher-Krall, 1995: White pine and climate change: Insights from the past. *Journal of Forestry*, **93**(7): 39–42.

Jacobson, G. L., Jr., and E. C. Grimm, 1986: A numerical analysis of Holocene forest and prairie vegetation in central Minnesota. *Ecology*, **67**: 958–966.

Jacobson, G. L., Jr., T. Webb, III, and E. C. Grimm, 1987: Patterns and rates of vegetation change in eastern North America from full-glacial to mid-Holocene time. *In* Ruddiman, W. F., and H. E. Wright, Jr. (eds.), *The Geology of North America, Vol. K-3, North America and Adjacent Oceans During the Last Deglaciation*. Boulder, CO: The Geological Society of America, pp. 277–288.

Janssen, C. R., 1968: Myrtle Lake: A late- and post-glacial pollen diagram from northern Minnesota. *Canadian Journal of Botany*, **46**: 1397–1410.

Jodry, M. A., D. S. Shafer, D. J. Stanford, and O. K. Davis, 1989: Late Quaternary environments and human adaptation in the San Luis Valley, south-central Colorado. *In Water in the Valley: A 1989 Perspective on Water Supplies, Issues & Solutions in the San Luis Valley, Colorado. Eighth Annual Field Trip, 19–20 August 1989. Colorado Ground-Water Association Guidebook*, pp. 189–208.

Kjellmark, E., 1996: Late Holocene climate change and human disturbance on Andros Island, Bahamas. *Journal of Paleolimnology*, **15**: 133–145.

Kral, R., 1993: *Pinus. In* Flora of North America Editorial Committee, *Flora of North America North of Mexico, Vol. 2*. New York: Oxford University Press, pp. 373–398.

Kutzbach, J., R. Gallimore, S. Harrison, P. Behling, R. Selin, and F. Laarif, 1998: Climate and biome simulations for the past 21,000 years. *Quaternary Science Reviews*, **17**: 473–506.

Kutzbach, J. E., P. J. Guetter, P. J. Behling, and R. Selin, 1993: Simulated climatic changes: Results of the COHMAP climate-model experiments. *In* Wright, H. E., Jr., J. E. Kutzbach, T. Webb, III, W. F. Ruddiman, F. A. Street-Perrott, and P. J. Bartlein (eds.), *Global Climates Since the Last Glacial Maximum*. Minneapolis: University of Minnesota Press, pp. 24–93.

Laird, K. R., S. C. Fritz, E. C. Grimm, and P. G. Mueller, 1996: Century-scale paleoclimatic reconstruction from Moon Lake, a closed-basin lake in the northern Great Plains. *Limnology and Oceanography*, **41**: 890–902.

Lamb, H. F., 1980: Late Quaternary vegetational history of southeastern Labrador. *Arctic and Alpine Research*, **12**: 117–135.

Lamb, H. F., 1984: Modern pollen spectra from Labrador and their use in reconstructing Holocene vegetational history. *Journal of Ecology*, **72**: 37–59.

Lamb, H. F., 1985: Palynological evidence for postglacial change in the position of tree limit in Labrador. *Ecological Monographs*, **55**: 241–258.

Ledru, M. P., 1993: Late Quaternary environmental and climatic changes in central Brazil. *Quaternary Research*, **39**: 90–98.

Ledru, M. P., H. Behling, M. Fournier, L. Martin, and M. Servant, 1994: Localisation de la forêt d'*Araucaria* du Brésil au cours de l'Holocène. Implications paléoclimatiques. *Comptes Rendus de l'Académie des Sciences. Séreie III*, **317**: 517–521.

Ledru, M. P., P. I. Soares Braga, F. Soubiés, M. Fournier, L. Martin, K. Suguio, and B. Turcq, 1996: The last 50,000 years in the Neotropics (southern Brazil): Evolution of vegetation and climate. *Palaeogeography, Palaeoclimatology, Palaeoecology*, **123**: 239–257.

Leyden, B., 1984: Guatemalan forest synthesis after Pleistocene aridity. *Proceedings of the National Academy of Sciences*, **81**: 4856–4859.

Leyden, B., 1987: Man and climate in the Maya lowlands. *Quaternary Research*, **28**: 407–414.

Leyden, B. W., M. Brenner, and B. H. Dahlin, 1998: Cultural and climatic history of Cobá, a lowland Maya city in Quintana Roo, Mexico. *Quaternary Research*, **49**: 111–122.

Leyden, B., M. Brenner, D. A. Hodell, and J. Curtis, 1993: Late Pleistocene climate in Central America lowlands. *In* Swart, P. K., K. C. Lohmann, J. McKenzie, and S. Savin (eds.), *Climate Change in Continental Isotopic Records*. Geophysical Monograph 78, Washington, DC: American Geophysical Union, pp. 255–355.

Leyden, B. W., M. Brenner, D. A. Hodell, and J. Curtis, 1994: Orbital and internal forcing of climate on the Yucatán Peninsula for the past ca. 36,000 ka. *Palaeogeography, Palaeoclimatology, Palaeoecology*, **109**: 193–210.

Leyden, B. W., M. Brenner, T. Whitmore, J. H. Curtis, D. R. Piperno, and B. H. Dahlin, 1996: A record of long- and short-term climatic variation from northwest Yucatán: Cenote San José Chulchacá. *In* Fedick, S. L. (ed.), *The Managed Mosaic: Ancient Maya Agriculture and Resource Use*. Salt Lake City: University of Utah Press, pp. 30–50.

Liu, K., and P. A. Colinvaux, 1988: A 5200-year history of Amazon rain forest. *Journal of Biogeography*, **15**: 231–248.

Livingstone, D. A., 1955: Some pollen profiles from Arctic Alaska. *Ecology*, **36**: 587–600.

Long, C. J., C. Whitlock, P. J. Bartlein, and S. H. Millspaugh, 1998: A 9000-year fire history from the Oregon Coast Range, based on a high-resolution charcoal study. *Canadian Journal of Forest Research*, **28**: 774–787.

Lozano-García, M. S., and B. Ortega-Guerrero, 1994: Palynological and magnetic susceptibility records of Lake Chalco, Central Mexico. *Palaeogeography, Palaeoclimatology, Palaeoecology*, **109**: 177–191.

Lozano-García, M. S., and B. Ortega-Guerrero, 1998: Evidences of environmental changes during the late Quaternary, correlation between Texcoco and Chalco sub-basins central Mexico. *Review of Palaeobotony and Palynology*, **99**(2): 77–94.

Lozano-García, M. S., and S. Xelhuantzi-López, 1997: Some problems in the late Quaternary pollen records of Central Mexico: Basin of Mexico and Zacapu. *Quaternary International*, **43**: 117–123.

Lozano-García, M. S., B. Ortega-Guerrero, M. Caballero-Miranda, and J. Urrutia-Fucugauchi, 1993: Late Pleistocene and Holocene palaeoenvironments of Chalco Lake, central Mexico. *Quaternary Research*, **40**: 332–342.

Lumley, S. H., and R. Switsur, 1993: Late Quaternary of the Taitao Peninsula, southern Chile. *Journal of Quaternary Science*, **8**: 161–165.

MacDonald, G. M., 1989: Postglacial palaeoecology of the subalpine forest-grassland ecotone of southwestern Alberta: New insights on vegetation and climate change in the Canadian Rocky Mountains and adjacent foothills. *Palaeogeography, Palaeoclimatology, Palaeoecology*, **73**: 155–173.

Mack, R. N., N. W. Rutter, and S. Valastro, 1978a: Late Quaternary pollen record from the Sanpoil River Valley, Washington. *Canadian Journal of Botany*, **56**: 1642–1650.

Mack, R. N., N. W. Rutter, and S. Valastro, 1979: Holocene vegetation history of the Okanogan Valley, Washington. *Quaternary Research*, **12**: 212–225.

Mack, R. N., N. W. Rutter, and S. Valastro, 1983: Holocene vegetational history of the Kootenai River Valley, Montana. *Quaternary Research*, **20**: 177–193.

Mack, R. N., N. W. Rutter, V. M. Bryant, Jr., and S. Valastro, 1978b: Late Quaternary pollen record from Big Meadow, Pend Oreille County, Washington. *Ecology*, **59**: 956–966.

Mack, R. N., N. W. Rutter, V. M. Bryant, Jr., and S. Valastro, 1978c: Re-examination of postglacial vegetation history in northern Idaho: Hager Pond, Bonner Co. *Quaternary Research*, **10**: 241–255.

Mack, R. N., N. W. Rutter, S. Valastro, and V. M. Bryant, Jr., 1978d: Late Quaternary vegetation history at Waits Lake, Colville River Valley, Washington. *Botanical Gazette*, **139**: 499–506.

Maenza-Gmelch, T. E., 1997: Holocene vegetation, climate, and fire history of the Hudson Highlands, southeastern New York, U.S.A. *The Holocene*, **7**: 25–37.

Marengo, J. A., and J. C. Rogers, 2000: Polar air outbreaks in the Americas: Assessments and impacts during recent and past climates. *In* Markgraf, V. (ed.), *Interhemispheric Climate Linkages*. San Diego: Academic Press, Chapter 3.

Markgraf, V., 1985: Paleoenvironmental history of the last 10,000 years of northwestern Argentina. *Zentralblatt für Geologie und Paläontologie*, **11/12**: 1739–1749.

Markgraf, V., 1987: Paleoenvironmental changes at the northern limit of the subantarctic *Nothofagus* forest, lat. 37°S, Argentina. *Quaternary Research*, **28**: 119–129.

Markgraf, V., 1991a: Younger Dryas in southern South America? *Boreas*, **20**: 63–69.

Markgraf, V., 1991b: Late Pleistocene environmental and climatic evolution in southern South America. *Bamberger Geographische Shriften*, **11**: 271–281.

Markgraf, V., 1993: Climatic history of Central and South America since 18,000 yr B.P.: Comparisons of pollen records and model simulations. *In* Wright, H. E., Jr., J. E. Kutzbach, T. Webb, III, W. F. Ruddiman, F. A. Street-Perrott, and P. J. Bartlein (eds.), *Global Climates Since the Last Glacial Maximum*. Minneapolis: University of Minnesota Press, pp. 357–385.

Markgraf, V., and L. Anderson, 1994: Fire history of Patagonia: Climate versus human cause. *Revista Instituto Geológico, Sao Paulo, Brazil*, **15**: 35–47.

Markgraf, V., and M. M. Bianchi, 1999: Paleoenvironmental changes during the last 17,000 years in western Patagonia: Mallín Aguado, Province of Neuquén, Argentina. *Bamberger Geographische Schriften*, **19**: 175–193.

Markgraf, V., and H. F. Diaz, 2000: The past ENSO record: A synthesis. *In* Diaz, H. F., and V. Markgraf (eds.), *El Niño and the Southern Oscillation. Multiscale Variability and Global and Regional Impacts*. Cambridge, United Kingdom: Cambridge University Press, pp. 465–488.

Markgraf, V., and L. Scott, 1981: Lower timberline in central Colorado during the past 15,000 yr. *Geology,* **9**: 231–234.

Markgraf, V., J. P. Bradbury, R. M. Forester, G. Singh, and R. S. Sternberg, 1984: San Augustin Plains, New Mexico: Age and paleoenvironmental potential reassessed. *Quaternary Research,* **22**: 336–343.

Martin, P. S., 1964: Paleoclimatology and tropical pollen profile. *Report on the VI International Congress on the Quaternary,* Warsaw, 1961, **2**: 319–323.

Mathewes, R. W., 1973: A palynological study of postglacial vegetation changes in the University Research Forest, southeastern British Columbia. *Canadian Journal of Botany,* **51**: 2085–2103.

Mathewes, R. W., and G. E. Rouse, 1975: Palynology and paleoecology of postglacial sediments from the Lower Fraser River Canyon of British Columbia. *Canadian Journal of Earth Sciences,* **12**: 745–756.

McGlone, M. S., A. P. Kershaw, and V. Markgraf, 1992: El Nino/Southern Oscillation climatic variability in Australasian and South American paleoenvironmental records. *In* Diaz, H. F., and V. Markgraf (eds.), *El Niño: Historical and Paleoclimatic Aspects of the Southern Oscillation.* Cambridge, United Kingdom: Cambridge University Press, pp. 435–462.

Mehringer, P. J., Jr., S. F. Arno, and K. L. Petersen, 1977: Postglacial history of Lost Trail Pass Bog, Bitterroot Mountains, Montana. *Arctic and Alpine Research,* **9**: 345–368.

Melief, A. B. M., 1985: Late Quaternary paleoecology of the Parque Nacional los Nevados, Cordillera Central, and Sumapaz Cordillera Oriental areas, Colombia. Dissertation. University of Amsterdam, The Netherlands.

Metcalfe, S. E., 1992: Changing environments of the Zacapu Basin, Central Mexico: A diatom based history spanning the last 30,000 years. Research Paper 38, School of Geography, Oxford, England.

Metcalfe, S. E., 1995: Holocene environmental change in Zacapu basin, Mexico: A diatom based record. *The Holocene,* **5**: 196–208.

Metcalf, S. E., 1997: Paleolimnological records of climate change in México—Frustrating past, promising future? *Quaternary International,* **43/44**: 111–116.

Metcalfe, S. E., F. A. Street-Perrott, R. B. Brown, P. E. Hales, R. A. Perrott, and F. M. Steininger, 1989: Late Holocene human impact on lake basins in central Mexico. *Geoarchaeology,* **4**: 119–141.

Morán-Zenteno, D., 1994: Geology of the Mexican Republic. Studies in Geology 39, American Association of Petroleum Geologists, Tulsa, OK.

Mourguiart, P., J. Argollo, and D. Wirrmann, 1995: Evolución paleohidrológica de la cuenca del Lago Titicaca durante el Holoceno. *Bulletin de l'Institut Français d'Etudes Andines,* **24**: 573–583.

Nieto, M. A., and A. R. Prieto, 1987: Análisis palinològico del Holocene tardío del sitio Fortín Necochea (Partido de Gral. Lamadrid, Provincia de Buenos Aries, Argentina). *Ameghiniana,* **24**: 271–276.

Nixon, K. C., 1993: The genus *Quercus* in Mexico. *In* Ramamoorthy, T. P., R. Bye, A. Lot, and J. Fa (eds.), *Biological Diversity in Mexico: Origins and Distribution.* New York: Oxford University Press, pp. 447–458.

O'Hara, S. L., S. E. Metcalfe, and F. A. Street-Perrott, 1994: On the arid margin: The relationship between climate, humans and the environment. *Chemosphere,* **29**: 965–981.

Ohngemach, D., 1977: Pollen sequence of the Tlaloqua Crater (La Malinche Volcano, Tlaxcala, México). *Boletin de la Sociedad Botanica de México,* **36**: 33–40.

Parizzi, M. G., M. L. Salgado-Labouriau, and H. C. Kohler, 1998: Genesis and environmental history of Lagoa Santa, southeastern Brazil. *The Holocene,* **8**: 311–321.

Peteet, D. M., 1986: Modern pollen rain and vegetational history of the Malaspina Glacier District, Alaska. *Quaternary Research,* **25**: 100–120.

Peteet, D. M., 1993: Postglacial migration history of lodgepole pine near Yakutat, Alaska. *Canadian Journal of Botany,* **69**: 786–796.

Piperno, D. R., M. B. Bush, and P. A. Colinvaux, 1990: Paleoenvironments and human occupation in late-glacial Panama. *Quaternary Research,* **33**: 108–116.

Prieto, A. R., 1989: Palinología de Empalme Querandíes (Provincia de Buenos Aires). Un modelo paleoambiental para el Pleistocene tardío—Holoceno. Dissertation. Universidad Nacional de Mar del Plata, Argentina.

Prieto, A. R., 1996: Late Quaternary vegetational and climatic changes in the Pampa grassland of Argentina. *Quaternary Research,* **45**: 73–88.

Prieto, A. R., and M. M. Páez, 1998: Pollen analysis of discontinuous stratigraphical sequences: Holocene at Cerro La China locality (Buenos Aires, Argentina). *Quaternary of South America and Antartic Peninsula,* **7**: 219–236.

Richard, P., 1979: Contribution à l'histoire postglaciaire de la végétation au nord-est de la Jamésie, Nouveau-Québec. *Géographie Physique et Quaternaire,* **33**: 93–112.

Richard, P. J. H., A. C. Larouche, and M. A. Bouchard, 1982: Age de la déglaciation finale et histoire postglaciaire de la végétation dans la partie centrale du Nouveau-Québec. *Géographie Physique et Quaternaire,* **36**: 63–90.

Rinaldi, M., V. Rull, and C. Schubert, 1990: Analisis paleoecologico de una turbera en la Gran Sabana (Santa Cruz de Mapauri), Venezuela: Resultados preliminares. *Acta Cientifica Venezolana,* **41**: 66–68.

Ritchie, J. C., 1984: *Past and Present Vegetation in the Far Northwest of Canada.* Toronto: University of Toronto Press.

Ritchie, J. C., and F. K. Hare, 1971: Late-Quaternary vegetation and climate near the arctic tree line of northwestern North America. *Quaternary Research,* **1**: 331–342.

Ritchie, J. C., L. C. Cwynar, and R. W. Spear, 1983: Evidence from north-west Canada for an early Holocene Milankovitch thermal maximum. *Nature,* **305**: 126–128.

Rodbell, D. T., G. O. Seltzer, D. M. Anderson, M. B. Abbott, D. B. Enfield, and J. H. Newman, 1999: An ~15,000-year record of El Niño–driven alluviation in southwestern Ecuador. *Science,* **283**: 516–520.

Ruddiman, W. F., and H. E. Wright, Jr. (eds.), 1987: *North America and Adjacent Oceans During the Last Deglaciation, Vol. K-3, Geology of North America.* Boulder, CO: Geological Society of America.

Rull, V., 1991: Contribución a la paleoecologia de Pantepui la Gran Sabana (Guayana Venezolana): Clima, biogeografía, y ecología. *Scientia Guaianae,* **2**: 1–133.

Russell, E. W. B., R. B. Davis, R. S. Anderson, T. E. Rhodes, and D. S. Anderson, 1993: Recent centuries of vegetational change in the glaciated north-eastern United States. *Journal of Ecology,* **81**: 647–664.

Salgado-Labouriau, M. L., M. Barberi, K. R. Ferraz-Vicentini, and M. G. Parizzi, 1998: A dry climatic event during the late Quaternary of tropical Brazil. *Review of Palaeobotony and Palynology,* **99**: 115–129.

Salgado-Labouriau, M. L., V. Casseti, K. R. Ferraz-Vicentini, L. Martin, F. Soubiès, K. Suguio, and B. Turcq, 1997: Late Quaternary vegetational and climatic changes in cerrado and palm swamp from central Brazil. *Palaeogeography, Palaeoclimatology, Palaeoecology,* **128**: 215–226.

Shafer, D. S., 1989: The timing of late Quaternary monsoon precipitation maxima in the southwest United States. Dissertation. University of Arizona, Tucson.

Smith, E. N., Jr., 1984: Late-Quaternary vegetational history at Cupola Pond, Ozark National Scenic Riverways, southeastern Missouri. Master's thesis. University of Tennessee, Knoxville.

Sosa-Najera, S., M. S. Lozano-García, and B. Ortega-Guerrero, 1998: Pleistocene/Holocene transition in Lake Chalco, central Mexico. *American Quaternary Association Program and Abstracts, 15th Biennial Meeting, Puerto Vallarta, México, 5–7 September 1998* p. 161.

Spear, R. W., 1993: The palynological record of late-Quaternary arctic tree-line in northwest Canada. *Review of Palaeobotony and Palynology,* **79**: 99–111.

Stine, S., and M. Stine, 1990: A record from Lake Cardiel of climate change in southern South America. *Nature,* **345**: 705–707.

Straka, H., and D. Ohngemach, 1989: Late Quaternary vegetation history of the Mexican highland. *Plant Systematics and Evolution,* **162**: 115–132.

Styles, B. T., 1993: Genus *Pinus:* A Mexican purview. *In* Ramamoorthy, T. P., R. Bye, A. Lot, and J. Fa (eds.), *Biological Diversity in Mexico: Origins and Distribution.* New York: Oxford University Press, pp. 397–420.

Swain, P. C., 1979: The development of some bogs in eastern Minnesota. Dissertation. University of Minnesota, Minneapolis.

Sylvestre, F., M. Servant, S. Servant-Vildary, C. Causse, M. Fournier, and J.-P. Ybert, 1999: Lake-level chronology on the southern Bolivian Altiplano (18°–23°S) during late-glacial time and the early Holocene. *Quaternary Research,* **51**: 54–66.

Thompson, R. S., 1990: Late Quaternary vegetation and climate in the Great Basin. *In* Betancourt, J. L., T. R. Van Devender, and P. S. Martin (eds.), *Packrat Middens: The Last 40,000 Years of Biotic Change.* Tucson: University of Arizona Press, pp. 200–239.

Thompson, R. S., C. Whitlock, P. J. Bartlein, S. P. Harrison, and W. G. Spaulding, 1993: Climatic changes in the western United States since 18,000 yr B.P. *In* Wright, H. E., Jr., J. E. Kutzbach, T. Webb, III, W. F. Ruddiman, F. A. Street-Perrott, and P. J. Bartlein (eds.), *Global Climates Since the Last Glacial Maximum.* Minneapolis: University of Minnesota Press, pp. 468–513.

Toomey, R. S., III, M. D. Blum, and S. Valastro, Jr., 1993: Late Quaternary climates and environments of the Edwards Plateau, Texas. *Palaeogeography, Palaeoclimatology, Palaeoecology (Global and Planetary Change Section),* **7**: 1–22.

Urrego, L. E., 1994/1997: Los bosques inundables del medio Caquetá: Caracterización y sucesión. Dissertation. University of Amsterdam. Published as: *Estudios en la Amazonia Colombiana/Studies on the Colombian Amazonia,* **XIV.** Tropenbos Colombia, Bototá.

Valero-Garcés, B. L., M. Grosjean, A. Schwalb, M. Geyh, B. Messerli, and K. Kelts, 1996: Limnogeology of Laguna Miscanti: Evidence for mid to late Holocene moisture changes in the Atacama Altiplano (northern Chile). *Journal of Paleolimnology,* **16**: 1–21.

Van Devender, T. R., 1990a: Late Quaternary vegetation and climate of the Chihuahuan Desert, United States and Mexico. *In* Betancourt, J. L., T. R. Van Devender, and P. S. Martin (eds.), *Packrat Middens: The Last 40,000 Years of Biotic Change.* Tucson: University of Arizona Press, pp. 104–133.

Van Devender, T. R., 1990b: Late Quaternary vegetation and climate of the Sonoran Desert, United States and Mexico. *In* Betancourt, J. L., T. R. Van Devender, and P. S. Martin (eds.), *Packrat Middens: The Last 40,000 Years of Biotic Change.* Tucson: University of Arizona Press, pp. 134–165.

van Geel, B., and T. van der Hammen, 1973: Upper Quaternary vegetational and climatic sequence of the Fúquene area (Eastern Cordillera, Colombia). *Palaeogeography, Palaeoclimatology, Palaeoecology,* **14**: 9–92.

Vaughan, H. H., E. S. Deevey, and S. E. Garret-Jones, 1985: Pollen stratigraphy of two cores from the Petén Lake District. *In* Pohl, M. (ed.), *Prehistoric Lowland Maya Environment and Subsistence Econo-*

my. Peabody Museum Papers 77. Cambridge, MA: Harvard University Press, pp. 73–89.

Vázquez-Selem, L., and F. M. Phillips, 1998: Glacial chronology of Iztaccíhuatl Volcano, Central Mexico, based on cosmogenic ^{36}Cl exposure ages and tephrochronology. *America Quaternary Association Program and Abstracts, 15th Biennial Meeting, Puerto Vallarta, México, 5–7 September 1998,* p. 174.

Velázquez-Durán, R., 1998: Palinologia en relación a paleoambientes de los últimos 35,000 años en la Cuenca del lago de Cuitzeo, Mich., México. Dissertation. University of Morelia, Morelia, México.

Villagrán, C. M., 1995: Quaternary history of Mediterranean vegetation in Chile. *In* Arroyo, M. T. K., P. H. Zedler, and M. D. Fox (eds.), *Ecology and Biogeography of Mediterranean Ecosystems in Chile, California and Australia. Ecological Studies,* **108**: 3–20. Springer-Verlag, Stuttgart, Germany.

Waters, M. R., 1989: Late Quaternary lacustrine history and paleoclimatic significance of pluvial Lake Cochise, southeastern Arizona. *Quaternary Research,* **32**: 1–11.

Watts, W. A., 1975: A late Quaternary record of vegetation from Lake Annie, south-central Florida. *Geology,* **3**: 344–346.

Watts, W. A., 1980: The late Quaternary vegetation history of the southeastern United States. *Annual Review of Ecology and Systematics,* **11**: 387–409.

Watts, W. A., and J. P. Bradbury, 1982: Paleoecological studies at Lake Pátzcuaro on the west-central Mexican plateau and at Chalco in the Basin of Mexico. *Quaternary Research,* **17**: 56–70.

Watts, W. A., and R. C. Bright, 1968: Pollen, seed, and mollusk analysis of a sediment core from Pickerel Lake, northeastern South Dakota. *Geological Society of America Bulletin,* **79**: 855–876.

Watts, W. A., and B. C. S. Hansen, 1988: Environments of Florida in the late Wisconsin and Holocene. *In* Purdy, B. (ed.), *Wet Site Archaeology.* West Caldwell, NJ: Telford Press, pp. 307–323.

Watts, W. A., and B. C. S. Hansen, 1994: Pre-Holocene and Holocene pollen records of vegetation history from the Florida peninsula and their climatic implications. *Palaeogeography, Palaeoclimatology, Palaeoecology,* **109**: 163–176.

Watts, W. A., and H. E. Wright, Jr., 1966: Late-Wisconsin pollen and seed analysis from the Nebraska Sandhills. *Ecology,* **47**: 202–210.

Watts, W. A., E. C. Grimm, and T. C. Hussey, 1996: Mid-Holocene forest history of Florida and the coastal plain of Georgia and South Carolina. *In* Sassaman, K. E., and D. G. Anderson (eds.), *Archaeology of the Mid-Holocene Southeast.* Gainesville: University Press of Florida, pp. 28–40.

Watts, W. A., B. C. S. Hansen, and E. C. Grimm, 1992: Camel Lake: A 40,000-yr record of vegetational and forest history from northwest Florida. *Ecology,* **73**: 1056–1066.

Webb, R. S., 1990: Late Quaternary water-level fluctuations in the northeastern United States. Dissertation. Brown University, Providence, RI.

Webb, R. S., K. H. Anderson, and T. Webb, III, 1993: Pollen response-surface estimates of late-Quaternary changes in the moisture balance of the northeastern United States. *Quaternary Research,* **40**: 213–227.

Webb, T., III, P. J. Bartlein, and J. E. Kutzbach, 1987: Climatic change in eastern North America during the past 18,000 years: Comparisons of pollen data with model results. *In* Ruddiman, W. F., and H. E. Wright, Jr. (eds.), *North America and Adjacent Oceans During the Last Deglaciation, Vol. K-3, Geology of North America.* Boulder, CO: Geological Society of America, pp. 447–462.

Whitehead, D. R., and M. C. Sheehan, 1985: Holocene vegetational changes in the Tombigbee River Valley, eastern Mississippi. *American Midland Naturalist,* **113**: 122–137.

Whitlock, C., and P. J. Bartlein, 1993: Spatial variations of Holocene climatic change in the Yellowstone region. *Quaternary Research,* **39**: 231–238.

Whitmore, T. J., M. Brenner, J. H. Curtis, B. H. Dahlin, and B. W. Leyden, 1996: Holocene climatic and human influences in lakes of the Yucatán Peninsula, Mexico: An interdisciplinary, paleolimnological approach. *The Holocene*, 6: 273–287.

Wijmstra, T. A., and T. van der Hammen, 1966: Palynological data on the history of tropical savannas in northern South America. *Leidse Geologische Mededelingen*, 38: 71–90.

Williams, A. S., 1974: Late-glacial–postglacial vegetational history of the Pretty Lake region, northeastern Indiana. U.S. Geological Survey Professional Paper 686-B, U.S. Government Printing Office, Washington, DC.

Winkler, M. G., A. M. Swain, and J. E. Kutzbach, 1986: Middle Holocene dry period in the northern midwestern United States: Lake levels and pollen stratigraphy. *Quaternary Research*, **25**: 235–250.

Worona, M. A., and C. Whitlock, 1995: Late Quaternary vegetation and climate history near Little Lake, central Coast Range, Oregon. *Geological Society of America Bulletin*, **107**: 867–876.

Wright, H. E., Jr. (ed.), 1983: *Late-Quaternary Environments of the United States. Volume 2. The Holocene*. Minneapolis: University of Minnesota Press.

Wright, H. E., Jr., A. M. Bent, B. C. S. Hansen, and L. J. Maher, Jr., 1973: Present and past vegetation of the Chuska Mountains, northwestern New Mexico. *Geological Society of America Bulletin*, **84**: 1155–1180.

Wright, H. E., Jr., J. E. Kutzbach, T. Webb, III, W. F. Ruddiman, F. A. Street-Perrott, and P. J. Bartlein (eds.), 1993: *Global Climates Since the Last Glacial Maximum*. Minneapolis: University of Minnesota Press.

Xelhuantzi-López, S., 1994: Estudio Palinológico y Reconstrucción Paleoambiental del ex -lago de Zacapu, Michoacán. Master's thesis. National University of México, México City.

Yu, Z., and J. H. McAndrews, 1994: Holocene water levels at Rice Lake, Ontario, Canada: Sediment, pollen, and plant-macrofossil evidence. *The Holocene*, **4**: 141–152.

Yu, Z., J. H. McAndrews, and U. Eicher, 1997: Middle Holocene dry climate caused by change in atmospheric circulation patterns: Evidence from lake levels and stable isotopes. *Geology*, **25**: 251–254.

Yu, Z., J. H. McAndrews, and D. Siddiqi, 1996: Influences of Holocene climate and water levels on vegetation dynamics of a lakeside wetland. *Canadian Journal of Botany*, **74**: 1602–1615.

20

Late Glacial Vegetation Records in the Americas and Climatic Implications

MARIE-PIERRE LEDRU AND PHILIPPE MOURGUIART

Abstract

The late glacial, the transition from the last glacial maximum (LGM) to the Holocene, is a key period for understanding the mechanisms of abrupt climatic change. In terrestrial records there is considerable uncertainty about the exact timing, rate, and magnitude of these changes, and the globality of the changes is still debated. To document past environments in the Americas, we reviewed 28 published pollen records for locations between Alaska and Tierra del Fuego and between the Pacific and Atlantic coasts. In this chapter, we consider three time intervals, corresponding to the Oldest Dryas (15,500–14,500 cal. B.P.), the Bølling-Allerød (14,500–12,700 cal. B.P.), and the Younger Dryas (12,700–11,000 cal. B.P.). Each interval has a duration of more than 1000 years, and their chronologies are comparable with those obtained from laminated sediments. In spite of differences in resolution and chronological control, comparison of the data suggests that there is no synchroneity of these late glacial climate oscillations in the Americas. Some vegetation records show no late glacial vegetation changes at all, while others show several oscillations. Three main regions of comparable climate change can be distinguished in these records: (1) high-altitude sites and high northern latitude sites showing synchronous oscillations, but with different paleoenvironmental expressions; (2) midlatitude and tropical-latitude sites showing the same sequence of changes on both sides of the equator; and (3) high southern latitude sites. Because these fluctuations have abrupt onsets and terminations and are assumed to be of global distribution, they cannot be attributed to insolation forcing. Instead, changes in the Arctic heat budget through the thermohaline (North Atlantic Deep Water, NADW) circulation were proposed as the mechanism. We suggest that these changes affected climates as far as the southern extension of the Intertropical Convergence Zone (ITCZ). The mechanism explaining the observed climate fluctuation on land would relate to increased (during cold oscillations) or decreased (during the warm Bølling-Allerød oscillation) penetration of polar air. Under this scenario, the climatic fluctuations documented in records from southernmost South America could be out of phase with those in the Northern Hemisphere. Copyright © 2001 by Academic Press.

Resumen

El Tardi Glacial (TG), la transición del Último Máximo Glacial (LGM) al Holoceno es un período clave para entender los mecanismos que causan los cambios climáticos abruptos. En los registros terrestres hay una

considerable incertidumbre acerca de la duración, frecuencia y magnitud exacta de estos fluctuaciones, y su
globalidad sigue siendo discutida. Para documentar
ambientes pasados en las Americas, hemos revisado 28
registros de polen publicados, situados entre Alaska y
la Tierra del Fuego y entre las costas pacíficas y atlánti-
cas. En este capítulo consideramos tres intervalos de
tiempo, correspondientes al Oldest Dryas (15.500–
14.500 cal años A.P.), Bølling-Allerød (14,500–12,700 cal
años A.P.) y Younger Dryas (12.700–11.000 cal años
A.P.). Estos intervalos tienen más de 1000 años de duración que es comparable con cronologías obtenidas de
sedimentos laminados. No obstante las diferéncias en
resolución y control cronológico, los datos demuestran
que no tuvo sincronía entre Norte y Sur del los Americas durante el intervalo del Tardiglacial. Algunos registros de vegetación no evidencian cambios durante los
intervalos considerados, mientras otros presentan oscilaciones repetidas. Tres areas principales se distin-
guieron de la comparación de los registros: (1) localidades en altura y en altas latitudes en Norte America
muestran oscilaciones sincronizadas, pero con una
señal ambiental diferente en diferentes sitios; (2) altas
latitudes en America del Sur; (3) latitudes medias y latitudes tropicales, que muestran la misma secuencia de
cambios a ambos lados del ecuador. Debido a la alta frecuencia de las fluctuaciones y a su probable carácter
global, eventos como el Younger Dryas o el Bølling-
Allerød no pueden ser atribuidos a un forzamiento de
la insolación. Entonces se propuso un cambio en el balance térmico en el area polar debido a cambios de la circulación "thermohaline" (North Atlantic Deep Water
formation, NADW). Sugerimos que estos cambios en el
Atlantico Norte pudieron haber afectado el clima hasta la extensión austral de la Zona de Convergéncia Inter Tropical (ITCZ). El mecanismo que explica las fluctuaciones observadas en los registros terrestres podría
ser relacionado con el aumento (oscillacion fría), o la
disminución (oscillacion cálida) de la penetración de
massas polares. En este caso, las fluctuaciones climáticas del Sur de Argentina y Chile no estarían en fase con
los del hemisferio Norte.

20.1. INTRODUCTION

The late glacial, the transition from the last glacial
maximum (LGM) to the Holocene, is a key period for
understanding the mechanisms of abrupt climate
change. The late glacial is characterized by three short-
term oscillations, called the Oldest Dryas, Older Dryas,
and Younger Dryas, respectively. They were defined at
the beginning of the century in Denmark at a time
when no radiometric dating was available. The distinc-

tion between the Older Dryas and the Younger Dryas
interval was first established at the Allerød site in Denmark (Hartz and Milthers, 1901) based on sediment
stratigraphy. Both phases, separated by a layer of gyttja that corresponds to the Allerød interstadial, are characterized by the presence of macrofossils of subarctic/
alpine flora, including *Dryas octopetala*, and the absence
of significant amounts of tree birch (*Betula* sp.). The
Bølling interstadial in turn was identified at Bølling Sø,
Denmark, from a layer of sand and silty gyttja containing a high percentage of tree birch pollen (Iversen, 1942,
1954). The underlying inorganic layer was named the
Oldest Dryas interval (in Sanchez-Goñi, 1995). In subsequent years, these intervals were dated in many sites
in Europe, yielding the following chronology: (1) Oldest Dryas, 15,500–14,500 cal. B.P.; (2) Bølling/Allerød,
14,500–12,500 cal. B.P., interrupted by three-decade- to
century-long events of the Inter-Bølling Cold Period
(IBCP), the Older Dryas, and the Inter-Allerød Cold Period (IACP); and (3) the Younger Dryas, 12,500–11,000
cal. B.P. (Fig. 1).

Recent studies on marine cores from the North Atlantic region (Bard et al., 1987, 1990, 1997; Bond et al.
1993; Bond and Lotti, 1995) and on ice cores from
Greenland (Jouzel et al., 1987; Johnsen et al., 1992;
Dansgaard et al., 1993) have highlighted drastic climat-

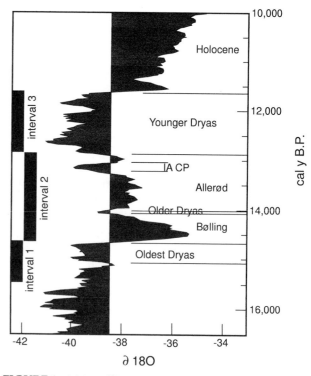

FIGURE 1 Major $\delta^{18}O$ variations and European pollen zone
boundaries during late glacial times. (From Stuiver et al., 1995. With
permission.)

ic changes of very short duration during the late glacial (Fig. 1). It appears that the last deglaciation was a two-step process with large and rapid fluctuations between warmer and colder conditions. Mounting evidence from marine records elsewhere has documented that these abrupt changes were not restricted to the North Atlantic region. In particular, studies from the Cariaco basin (Venezuela), close to the equator, have revealed the same sequence of short-term warmer and cooler periods identified in terrestrial records (Hughen et al., 1996). For the Arabian Sea, Schulz et al. (1998) and Overpeck et al. (1996) showed that following the last glaciation, the monsoon pattern strengthened in a series of steps coinciding with major climate shifts in the polar regions of the Northern Hemisphere. The similarities of events in records from the Cariaco basin, the Arabian Sea, and the high-latitude North Atlantic suggest a common forcing mechanism. For many years, it has been demonstrated that insolation, especially precession, was a dominant factor regulating tropical climate changes during the Pleistocene (Street and Grove, 1979; Prell and van Campo, 1986; Prell and Kutzbach, 1987; Partridge et al., 1997). However, the abrupt climatic changes evidenced during the last deglaciation cannot be accounted for by orbital forcing. Instead, these changes have been linked to rapid reorganizations of the North Atlantic thermohaline circulation (Broecker et al., 1985, 1988; Lehman and Keigwin, 1992; Broecker 1994; Zaucker et al., 1994).

Although in Antarctica the glacial–interglacial warming also occurred in two steps, interrupted by a slight cooling period, the Antarctic Cold Reversal (Jouzel et al., 1995; Sowers and Bender, 1995), it is out of phase with the sequence of climate shifts in the North Atlantic (Broecker, 1998; Blunier et al., 1998).

There is considerable uncertainty in terrestrial records about the exact timing, rate, and magnitude of these changes. The response of the land–atmosphere system to these climatic oscillations is apparently not regionally uniform. Therefore, the global extent of the changes remains in debate.

When the sequences of late glacial fluctuations in palynological continental records along the Pole-Equator-Pole: Americas (PEP 1) transect are compared, the following questions arise:

1. What is the timing of the onset of the late glacial in different types of environments? Given differences in basal ages of the records, this question is often problematic. Records located in the tropical lowlands, for example, start right after the LGM; in high-latitude, high-altitude, and midlatitude records, sedimentation in lakes or bogs generally started only after the glaciers had retreated locally at ca. 15,000 cal. B.P.

2. How are late glacial changes expressed in different vegetation types?

3. Is the sequence of cold (stadial) and warm (interstadial) fluctuations during the time interval between 17,000 and 11,000 cal. B.P. consistent and synchronous over large distances?

In the following, we will discuss three time intervals corresponding to the European late glacial stratigraphy; these intervals are the Oldest Dryas, 15,500–14,500 cal. B.P.; the Bølling-Allerød, 14,500–12,700 cal. B.P.; and the Younger Dryas, 12,700–11,000 cal. B.P.

20.2. PAST ENVIRONMENTS

To document past environments in the Americas, we reviewed 28 published pollen records located between Alaska and Tierra del Fuego and between the Pacific and Atlantic coasts. Table 1 shows geographical information (latitude, longitude, and elevation), radiocarbon dates, calibrated ages, sample resolution (when possible), and references for the 28 sites. Numbers in front of the sites refer to their locations in Fig. 2.

Our purpose was not to review all the available sites from the North American or Latin American pollen databases. Instead, sites were selected to include a wide range of analytical methods that allowed the characterization of late glacial environmental changes, according to the records' temporal resolution, dating control, type of surrounding vegetation, etc. This approach allows us to discuss many of the problems encountered when late glacial terrestrial sites are analyzed, and it enhances our ability to detect and define the character of these changes. Finally, we will refer to published regional compilations to discuss regional or local changes.

20.3. RADIOCARBON DATING CONTROL

One of the most difficult problems in reconstructing paleoclimates is generating a reliable chronology for the records. For the last 30,000 years, radiocarbon dates can be obtained on organic matter or carbonates by using conventional methods or accelerator mass spectrometry. Depending on the material dated, problems can arise. Duplicate dates on different fractions (e.g., bulk sediment, mollusks, plant or wood remains) from the same level can vary considerably. Figure 3 illustrates this problem and demonstrates that only a few records in the Americas present reliable chronologies for the late glacial. Most of the sites have poor chronologic control based on just a few radiocarbon dates. Interpolation between a few dates has to be considered

TABLE 1 Geographical Information on Pollen Records Discussed in the Text, Including Radiocarbon Dates, Calendar Ages, Temporal Resolution, and References

No.	Site	State/country	Latitude	Longitude	Elevation (m)	Resolution (years)	¹⁴C dates	Ages (cal. years B.P.)	Ref.
1	Ongivinuk Grandfather	Alaska, U.S.A.	59°31' N	159°22' W	163	300	13,890 ± 390 13,150 ± 330 9,920 ± 160	16,659 (17,575–15,590) 15,669 (16,580–14,585) 11,006 (12,092–10,616)	Hu et al., 1995
2	Pleasant Island	Alaska, U.S.A.	58°21' N	135°40' W	150	120	13,760 ± 120 12,280 ± 120 11,950 ± 120 11,620 ± 120 11,430 ± 120 11,040 ± 100 10,880 ± 95 10,530 ± 110 10,110 ± 100	15,290 (15,570–15,000) 13,340 (13,540–13,170) 13,010 (13,180–12,850) 12,700 (12,860–12,520) 13,340 (13,480–13,220) 12,950 (13,050–12,860) 12,800 (12,890–12,710) 10,930 (11,010–10,520) 10,200 (10,380–10,040)	Hansen and Engstrom, 1996
3	Olympic Peninsula	Washington, U.S.A.	48°N	123°W	165	150	11,000 ± 150 11,560 ± 160 12,100 ± 310 8,920 ± 100	12,917 (13,230–12,610) 13,482 (13,909–13,132) 14,114 (15,031–13,380) 9,922 (10,040–9,649)	Peterson et al., 1983
4	Little Lake	Oregon, U.S.A.	44°10' N	123°35' W	217		15,920 ± 230 13,640 ± 390 11,990 ± 120 12,720 ± 140 10,940 ± 180 11,320 ± 190 10,400 ± 70	18,800 (19,340–18,430) 16,350 (17,300–15,180) 13,980 (14,370–13,640) 14,980 (15,490–14,490) 12,860 (13,230–12,490) 13,230 (13,680–12,850) 12,300 (12,510–11,990)	Grigg and Whitlock, 1998
5	Gordon Lake	Oregon, U.S.A.	44°21' N	122°15' W	1177		10,790 ± 80 12,640 ± 150 11,520 ± 80 10,400 ± 75 10,010 ± 70	12,720 (12,900–12,520) 14,860 (15,390–14,360) 13,430 (13,700–13,230) 12,300 (12,530–11,960) 11,210; 11,350; 11,110 (11,840–11,000)	Grigg and Whitlock, 1998
6	Killarney Lake	Canada	46° N	66°37'40" W	75		10,770 ± 120 11,180 ± 120	12,699 (12,948–12,420) 13,087 (13,369–12,838)	Levesque et al., 1993
7	Allamuchy Pond	New Jersey, U.S.A.	40°55' N	74°50' W	218	200	9,230 ± 160 10,740 ± 420 12,260 ± 220	10,355 (10,797–9,914) 12,670 (13,521–11,004) 14,317 (15,017–13,738)	Peteet et al., 1993
	Linsley Pond	Connecticut, U.S.A.	41°18' N	72°45' W	65	150	9,920 ± 230 10,440 ± 230 11,500 ± 300	11,006 (12,295–10,424) 12,349 (12,836–11,083) 13,430 (13,700–13,230)	Peteet et al., 1993
8	Jackson Pond	Kentucky, U.S.A.	37°27' N	85°43' W	220	700	12,590 ± 430 17,750 ± 630 11,860 ± 250	13,416 (14,154–12,814) 20,274 (21,160–19,389) 13,825 (14,511–13,255)	Wilkins et al., 1991
9	Mary Jane	Colorado, U.S.A.	39°53' N	105°49' W	2882	100	10,040 ± 190 13,740 ± 160	11,625 (12,350–10,900) 16,473 (16,902–16,014)	Short and Elias, 1987
10	Yosemite NP	California, U.S.A.	38° N	120° W	1554	590	12,380 ± 180 10,420 ± 100 13,690 ± 340	14,478 (15,090–13,961) 12,323 (12,584–11,928) 16,409 (17,242–15,426)	Smith and Anderson, 1992

#	Site	Location	Latitude	Longitude	Elevation		Radiocarbon age (yr B.P.)	Calibrated age (range)	Reference
11	Hendrick Pond	Wyoming, U.S.A.	43°45' N	110°36' W	2073		11,340 ± 100	13,248 (13,518–13,026)	Whitlock, 1993
							14,580 ± 150	17,459 (17,833–17,081)	
							17,160 ± 210	20,340 (21,041–19,657)	
	Lily Lake Fen	Wyoming, U.S.A.	43°45' N	110°19' W	2500–2650		10,170 ± 170	12,699 (12,931–12,440)	Whitlock, 1993
							11,130 ± 110	13,040 (13,292–12,808)	
							12,370 ± 120	14,464 (14,925–14,069)	
							16,040 ± 220	18,915 (19,464–18,468)	
	Fallback Lake	Wyoming, U.S.A.	43°58' N	110°26' W	2597		12,070 ± 120	14,077 (14,493–13,722)	Whitlock, 1993
							12,130 ± 150	14,152 (14,650–13,734)	
							15,640 ± 160	18,539 (18,890–18,196)	
	Cygnet Lake Fen	Wyoming, U.S.A.	44°39' N	110°36' W	2530		11,800 ± 190	13,755 (14,279–13,310)	Whitlock, 1993
12	Browns Pond	Virginia, U.S.A.	38°09' N	79°37' W	620	70	14,490 ± 70	17,358 (17,587–17,128)	Kneller and Peteet, 1993
							13,790 ± 130	16,535 (16,898–16,155)	
							14,300 ± 110	17,143 (17,455–16,836)	
							14,310 ± 110	17,155 (17,456–16,848)	
							13,010 ± 95	15,452 (15,813–15,019)	
							12,910 ± 100	15,292 (15,676–14,831)	
							12,940 ± 110	15,341 (15,745–14,858)	
							13,025 ± 230	15,476 (16,172–14,671)	
13	Camel Lake	Florida, U.S.A.	30°16' N	85°1' W	20		12,610 ± 135	14,812 (15,320–14,333)	Watts et al., 1992
							14,330 ± 275	17,177 (17,814–16,501)	
							10,980 ± 100	12,898 (13,112–12,680)	
							10,020 ± 110	11,200 (12,110–10,900)	
14	Chalco Lake	Mexico	19°30' N	99° W	2240		14,610 ± 470	17,492 (18,473–16,371)	Lozano-García and Ortego-Guerrero, 1994
							12,520 ± 135	14,677 (15,182–14,221)	
							9,395 ± 255	10,369 (11,007–9,902)	
15	La Yeguada	Panama	8°27' N	80°51' W	650		10,210 ± 130	11,984 (12,419–11,009)	Bush et al., 1992
							10,530 ± 100	12,453 (12,691–12,141)	
							11,250 ± 140	13,157 (13,491–12,870)	
							11,610 ± 180	13,538 (14,014–13,143)	
							14,230 ± 370	17,062 (17,904–16,132)	
							13,670 ± 210	16,384 (16,929–15,784)	
							12,910 ± 140	15,292 (15,779–14,731)	
16	Pedro Palo III	Colombia	4°30' N	74°23' W	2000		11,380 ± 130	13,289 (13,629–13,012)	Hooghiemstra and van der Hammen, 1993
							11,950 ± 100	13,931 (14,284–13,629)	
	Pedro Palo V						10,280 ± 90	12,118 (12,420–11,346)	
							10,380 ± 90	12,271 (12,526–11,871)	
17	Laguna Ciega I	Colombia	6°30' N	72°18' W	3500		14,140 ± 120	16,957 (17,283–16,621)	van der Hammen et al., 1980/1981
	Laguna Ciega II						12,830 ± 80	15,162 (15,513–14,739)	Kuhry, 1988
	Laguna Ciega III								
18	Laguna Baja	Peru	7°42' S	77°32' W	3575		12,100 ± 190	14,114 (14,706–13,624)	Hansen and Rodbell, 1995
							10,865 ± 690	12,790 (14,318–10,477)	
19	Mucubají	Venezuela	8°48' N	70°50' W	3650		12,650 ± 130	14,874 (15,367–14,395)	Salgado-Labouriau, 1984
							12,570 ± 130	14,751 (15,245–14,293)	Salgado-Labouriau, 1989
							12,390 ± 250	14,492 (15,304–13,823)	Salgado-Labouriau et al., 1977
							12,250 ± 150	14,304 (14,821–13,870)	
							11,960 ± 100	13,943 (14,298–13,639)	

(continues)

TABLE 1 (*continued*)

No.	Site	State/country	Latitude	Longitude	Elevation (m)	Resolution (years)	^{14}C dates	Ages (cal. years B.P.)	Ref.
20	Lake Valencia	Venezuela	10°16′ N	67°45′ W	403		12,930 ± 500 10,200 ± 350	15,325 (16,721–13,888) 11,962 (12,832–10,481)	Bradbury et al., 1981
21	Pata	Brazil	0°16′ N	66°4′ W	400		14,230 ± 60 15,580 ± 60 17,840 ± 300	17,062 (17,281–18,842) 18,483 (18,692–18,281) 21,277 (22,100–20,375)	Colinvaux et al., 1996
22	Carajás	Brazil	6°20′ S	50°25′ W	700		10,460 ± 200 12,520 ± 130	12,373 (12,800–11,228) 14,677 (15,170–14,231)	Absy et al., 1991
23	Salitre	Brazil	19° S	46°46′ W	950	200–300	14,230 ± 100 12,890 ± 80 10,350 ± 200 10,440 ± 150	17,062 (17,346–16,771) 15,259 (16,719–15,066) 12,228 (12,696–11,010) 12,349 (12,691–11,736)	Ledru, 1993
24	Pichihué II	Chile	42°23′ S	74°00′ W	700		9,150 ± 80 12,760 ± 120 12,405 ± 330 10,220 ± 310	10,042 (10,306–9,973) 15,049 (15,504–14,560) 14,513 (15,564–13,667) 12,005 (10,826–8,669)	Villagrán, 1991
25	Puerto Octay PM13 core PM12 core	Chile	40°56′ S	72°54′ W		45–95	10,140 ± 100 15,270 ± 165 15,290 ± 120 13,940 ± 155 13,790 ± 100 11,840 ± 140 14,300 ± 102 13,980 ± 95 10,560 ± 100 10,770 ± 60 10,330 ± 70	11,808 (12,269–11,007) 18,191 (18,540–17,823) 18,211 (18,497–17,916) 16,719 (17,115–16,304) 16,535 (16,838–16,220) 13,801 (14,220–13,445) 17,212 (17,511–16,907) 16,767 (17,043–16,486) 12,450 (12,690–12,140) 12,699 (12,854–12,529) 12,199 (12,440–11,821)	Moreno, 1997
26	Puerto del Hambre	Chile	53°36′ S	70°55′ W	3	150–200	15,800 ± 200 13,190 ± 80 12,740 ± 260	18,687 (19,132–18,280) 15,728 (16,033–15,382) 15,017 (15,859–14,220)	Heusser, 1995
27	Punta Arenas	Chile	53°09′ S	70°57′ W	75	150	10,940 ± 70 13,400 ± 140 11,960 ± 170 10,840 ± 70	12,860 (13,025–12,689) 16,027 (16,442–15,573) 13,943 (14,451–13,514) 12,766 (12,932–12,588)	Heusser, 1995
28	Harberton	Argentina	54°53′ S	67°10′ W	50	10–50	13,360 ± 280 11,300 ± 200 10,950 ± 170 10,490 ± 475 10,670 ± 150	15,971 (16,719–15,066) 13,207 (13,682–12,809) 12,950 (13,136–12,737) 12,450 (12,665–12,235) 12,650 (12,866–12,434)	Markgraf and Kenny, 1997

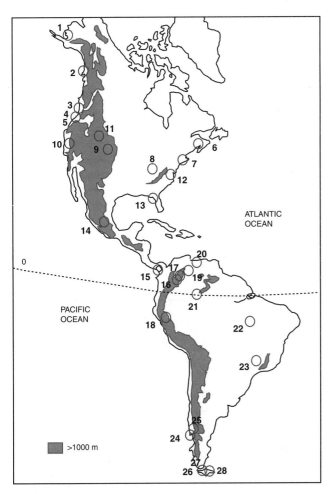

FIGURE 2 Site locations of pollen records along the PEP 1 transect. Locations are given in Table 1.

suspect when we refer to events with a duration of <1000 years. Another problem is related to the dating of carbonates. Radiocarbon dates from lacustrine sediments, at times, can be affected by old carbon (hardwater effect) which produces dates that are too old. Furthermore, the hard-water effect can vary through time in lakes or wetlands. At several coring sites, the dates obtained for different lake levels should be interpreted with great caution, especially in case of isolated ages. In the case of the late glacial, the brief nature of the intervals such as the Older Dryas, IACP, or IBCP requires high temporal resolution, taking into account changes in sedimentation processes that seem to occur in most of the records.

In this chapter, we consider three time intervals corresponding to the Oldest Dryas (15,500–14,500 cal. B.P.), the Bølling-Allerød interval (14,500–12,700 cal. B.P.), and the Younger Dryas (12,700–11,000 cal. B.P.); all intervals have a duration of ≥ 1000 years, which compares with chronologies obtained from revised

sites and laminated sediments (e.g., Cariaco basin). Establishing the presence of the Younger Dryas interval has been complicated by the difficulty of precisely dating its termination with the radiocarbon method. Due to the decreasing concentration of ^{14}C in the atmosphere at that time, radiocarbon dates are nearly the same over an 800-year period (Stuiver and Reimer, 1993).

All radiocarbon dates that cover the late glacial were calibrated (Table 1) by using the CALIB 3.0 program (Stuiver and Reimer, 1993). Late glacial dates older than 10,100 cal. B.P. were calibrated with U/Th and ^{14}C coral sets (Bard et al., 1990, 1993). The 2 sigma standard deviation will be discussed. Only the mean value of the given interval is presented in the text; the whole calibrated interval is given in Table 1. However, we need to keep in mind the probability that a specific age could be the mean value or any other value within this interval. This is an important point to consider when short-term climatic changes such as the late glacial oscillations are discussed. In comparing different vegetation records, therefore, we should consider the entire sequence in order to detect whether there is synchroneity for a specific climatic event or if a temporal lag of climatic events between records is not due to chronological problems. This might be the reason why, in some cases, the interpolated ages do not exactly fit the specific time zone discussed.

We also need to define if in different records there is a difference in the timing of the signal with respect to the beginning or end of the change and if there is a difference in the climatic signal itself. The presence/decrease/increase of the frequency of indicator taxa will be discussed for each record. The presence of the considered indicator taxa will then be translated in terms of warmer/drier/wetter/cooler climatic conditions and compared with the preceding time interval and not with modern data (Figs. 4–6). This climatic reconstruction then will be compared among the records.

20.4. PALEOENVIRONMENTAL RECONSTRUCTIONS

20.4.1. Time Interval Between 15,500 and 14,500 cal. B.P.

Figure 4 shows the qualitative climate reconstructions interpreted from indicator taxa and the interpolated time interval of the occurrence of these taxa. These intervals rarely fit with the intervals defined for the Northern Hemisphere (Fig. 1), but when the standard deviations of the calibrated intervals are considered, it is still possible to compare the records.

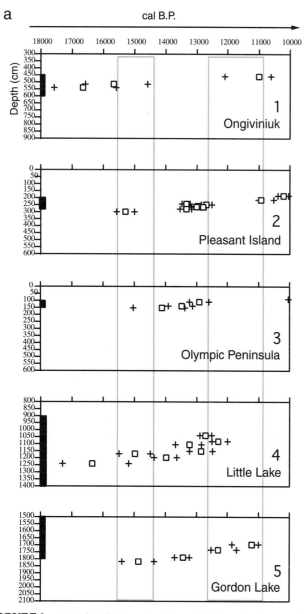

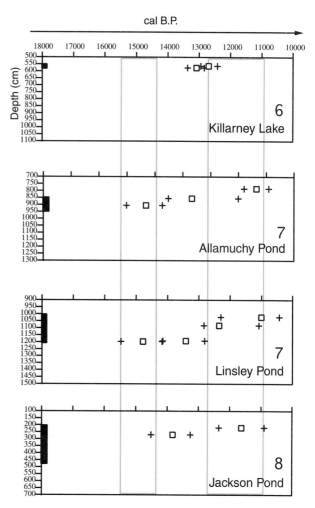

FIGURE 3 (*continued*) A2.

FIGURE 3 Late glacial chronological control for each discussed site. A, sediment thickness that covers the late glacial; B, maximum deviation of the calibrated age; C, mean value of calibrated age; D, minimum deviation of the calibrated age; and E, name of the vegetation record. Light gray lines show the discussed time intervals.

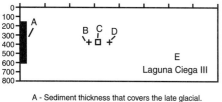

A - Sediment thickness that covers the late glacial.
B - Maximum deviation of the calibrated age.
C - Mean value of calibrated age.
D - Minimum deviation of the calibrated age.
E - Name of the vegetation record.
▨ - Light gray lines show the discussed time intervals.

For Alaska, herb tundra is dominant in the Ongivinuk and Grandfather Lake records (1).* Both the absence of *Betula* and the abundance of Cyperaceae suggest low summer temperatures and abundant soil moisture. This period lasted between ca. 15,600 and 13,000 cal. B.P. In the Pleasant Island record (2), this interval represents the beginning of the record and is characterized by a *Salix* shrub tundra with Poaceae and Cyperaceae. No forest taxa were recorded. This is interpreted as a cold and dry period.

In records from the Olympic Peninsula in Washington (3), the interval between 15,000 and 14,000 cal. B.P. represents the end of a cold interval, with the persistence of open herb- and shrub-dominated communities that contained cactus. The climate is defined as dry with warm summers. For Little Lake in Oregon, record

*The numbers in parentheses pertain to the information in Table 1.

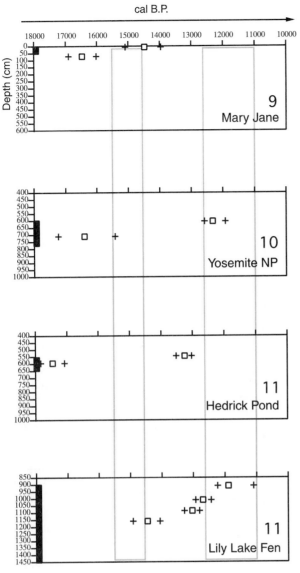

FIGURE 3   (*continued*) A3.

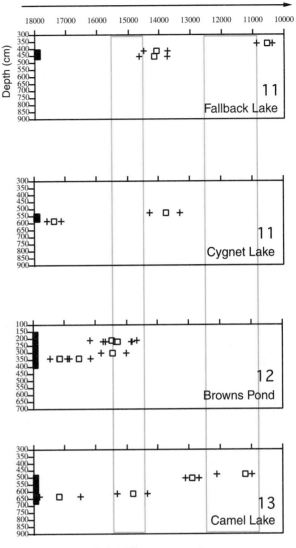

FIGURE 3   (*continued*) A4.

(4), the ability to date macrofossils allowed a high-resolution late glacial record. The record starts at 15,700 cal. B.P. with an open subalpine forest mainly represented by *Tsuga mertensiana*. Between 15,250 and 14,850 cal. B.P., a closed montane forest with primarily *Abies amabilis* is recorded. This record is interpreted to represent a warming compared to the previous interval. Between 14,850 and 14,500 cal. B.P., *Pseudotsuga* dominated in the forest, and an increase of *Tsuga heterophylla* indicates warmer climatic conditions. Between 14,500 and 14,250 cal. B.P., a return to subalpine forests indicates a reversal to a colder climate. At Gordon Lake in Oregon (5), between 15,500 and 14,500 cal. B.P., *T. mertensiana* is recorded together with *Picea* and *Pinus*, suggesting a cooler climate.

In the northeastern Atlantic region, the Allamuchy, New Jersey, and Linsley Ponds, Connecticut, records (7) show the first organic sedimentation at ca. 15,000 cal. B.P. A *Picea-Quercus* assemblage is recorded, attesting to a cool and humid climate. At Jackson Pond in Kentucky (8), with a lower age of older than 16,000 cal. B.P., a *Pinus-Picea* zone is recorded with an increase of deciduous taxa. The environment is defined as an open, taiga-like woodland. At Browns Pond in Virginia (12), *Alnus*, *Pinus*, *Picea*, and *Abies* are recorded prior to 15,000 cal. B.P. At ca. 13,000 cal. B.P., deciduous tree taxa increase—mainly *Ostrya*, *Carpinus*, and *Quercus*. This change indicates a rapid warming and increase in moisture. At Camel Lake in Florida (13), the co-occurrence of *Carya* and *Picea* between 17,000 and 14,800 cal. B.P.

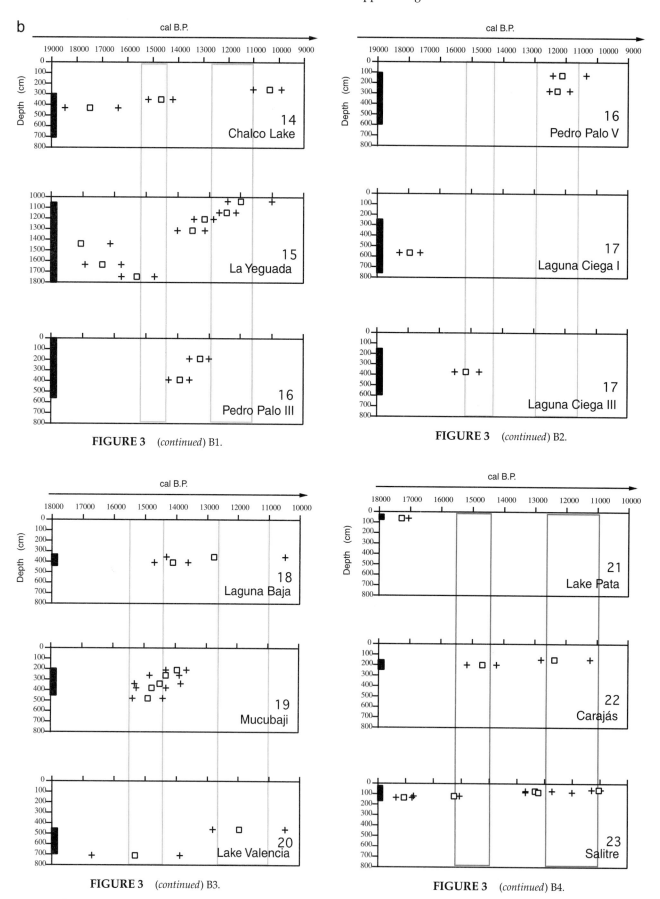

FIGURE 3   (*continued*) B1.

FIGURE 3   (*continued*) B2.

FIGURE 3   (*continued*) B3.

FIGURE 3   (*continued*) B4.

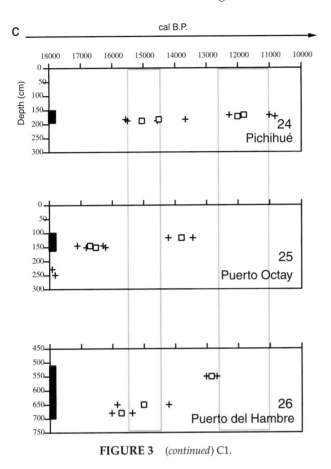

FIGURE 3 (*continued*) C1.

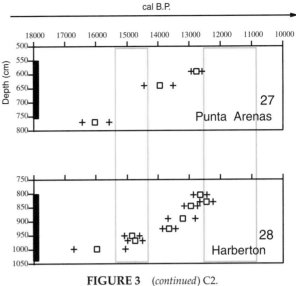

FIGURE 3 (*continued*) C2.

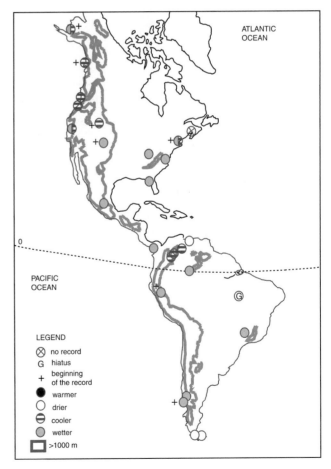

FIGURE 4 Main vegetation types and climatological information for the interval 15,500–14,500 cal. B.P.

lacks a modern vegetation analog. It is interpreted as indicating high precipitation and frost-free climatic conditions. The end of this interval is uncertain.

In the Rocky Mountains, at the site of Mary Jane in Colorado (9), deglaciation began after 17,000 cal. B.P. A mixture of cold-adapted tundra beetles with northern boreal/subalpine vegetation elements is recorded, suggesting a lowering of the tree line by ca. 500 m and representing a cooler climate. In records from Yellowstone and Grand Teton National Parks (11), an increase of *Picea-Pinus* assemblages is recorded between 15,300 and 14,800 cal. B.P., attesting to wetter climatic conditions. In Yosemite National Park, the Swamp Lake (10) record shows vegetation composed of *Pinus*, *Tsuga*, and *Abies* between 16,400 and 12,300 cal. B.P. This assemblage has no modern analog and is interpreted as an enriched mixed forest, attesting to cooler temperatures and an increase in precipitation. Wetter climatic conditions are also interpreted from paleolake levels (Thompson et al., 1993; Webb et al., 1993) and packrat middens, indicating greater seasonality (Betancourt et al., 1990).

At Chalco Lake in Mexico (14), between 17,500 and

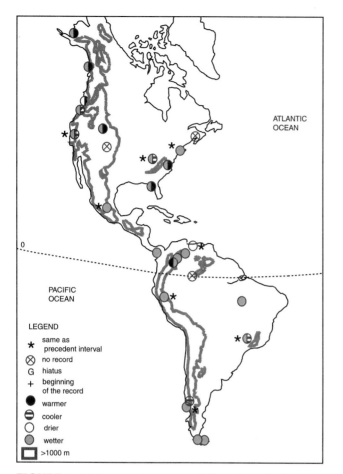

FIGURE 5 Main vegetation types and climatological information for the interval 14,500–12,700 cal. B.P.

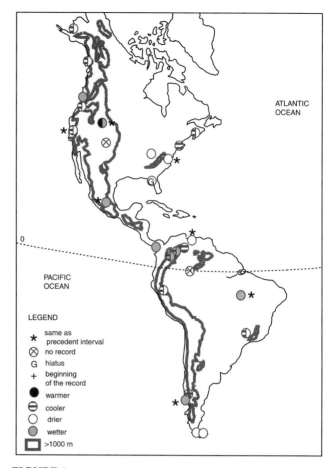

FIGURE 6 Main vegetation types and climatological information for the interval 12,700–11,000 cal. B.P.

11,500 cal. B.P., *Pinus* dominated together with meso-phytic forest, including mainly *Abies, Liquidambar, Carpinus,* and *Corylus.* This assemblage reflects high-moisture conditions. The 17,500 cal. B.P. date is, in fact, the upper limit of an interval between 17,500 and 14,500 cal. B.P. At La Yeguada in Panama (15), between 17,000 and 14,700 cal. B.P., moist montane forest is recorded, suggesting humid climatic conditions.

In the high Andean cordillera, four sites were examined: two in Colombia, one in Peru, and one in Venezuela. In the Pedro Palo, Colombia, record (16), between 15,500 and 14,000 cal. B.P., the dominance of nonarboreal pollen suggests dry climatic conditions. Laguna Ciega, Colombia (17), records an increase in Asteraceae between 15,000 and 14,500 cal. B.P. that indicates cold climatic conditions. This interval is called the La Ciega interval and has also been recognized in other records in this area. The record for Laguna Baja, Peru (18), starts at ca. 16,000 cal. B.P.; moist and wet montane elements are well represented for that time.

Climatic conditions are defined as having been wet, a condition that lasted until ca. 13,500 cal. B.P. In Mucubaji, Venezuela (19), deglaciation started ca. 14,800 cal. B.P. Pollen content at the onset attests to the presence of a desert paramo, suggesting cold conditions between 14,800 and 14,200 cal. B.P.

For the tropical lowlands, we discuss four sites. Lake Valencia, Venezuela (20), in the tropical lowlands, records dominance of *Alternanthera,* Cyperaceae, and Poaceae between 16,700 and 10,500 cal. B.P., suggesting generally arid conditions. For that time, only a minor increase in moisture is indicated by an increase of savanna tree taxa such as *Bursera* and *Spondias.* During the late glacial, Lake Pata, Brazil (21), recorded an assemblage of rain forest taxa with *Podocarpus* for which no modern analogs exist in present-day vegetation. This record was interpreted to indicate cold temperatures and humid climatic conditions. At Carajás, Brazil (22), a gap in sedimentation is recorded until ca. 14,500 cal. B.P. At Salitre, Brazil (23), the presence of *Araucaria* forest elements between 15,300 and 12,300 cal. B.P. attests to cold and humid climatic conditions.

Five sites are discussed for temperate South America. At Puerto Octay, Chile (25), a gap occurs between ca. 15,500 and 14,500 cal. B.P. At Pichihué, Chile (24), the presence of *Nothofagus*, a pioneer tree taxon, and moorland taxa between 15,000 and 14,500 cal. B.P. attests to humid climatic conditions. At Puerto del Hambre (26) and Punta Arenas (27), Chile, the presence of an *Empetrum* steppe-tundra (heath) is recorded between 16,000 and 14,000 cal. B.P., attesting to drier climatic conditions. At Harberton (28), Argentina, *Empetrum* heathland is also recorded between 16,000 and 15,000 cal. B.P., indicating an increase in temperature and a decrease in precipitation.

20.4.2. Time Interval Between 14,500 and 12,700 cal. B.P.

A summary of the results for the period 14,500–12,700 cal. B.P. is reported in Fig. 5 (the asterisk (*) indicates that no change in vegetation or climate occurred compared to the previous interval).

In Alaska, the Ongivinuk record (1) shows an increase of *Betula* shrubs in the tundra landscape between ca. 13,000 and 12,700 cal. B.P. This phase is related to a warmer and moister climate compared to the previous interval. The Pleasant Island (2) record shows the presence of a *Pinus* parkland with *Alnus* and *Salix*, characteristic of a wetter and warmer climate.

At the Olympic Peninsula (3), the presence of seeds of *Opuntia* (Cactaceae) suggests dry and warm summers between 15,000 and 13,000 cal. B.P. At Gordon Lake (5), a closed forest with *Abies amabilis* and *Tsuga heterophylla* is recorded between 14,500 and 12,800 cal. B.P. At Little Lake (4), an increase in the frequency of *Pseudotsuga* is recorded between 14,250 and 12,400 cal. B.P.

Allamuchy and Linsley Pond records (7) show no vegetation change since the previous interval. The same mixed thermophilous deciduous boreal forest with *Picea*, *Larix*, *Abies*, and *Quercus* is recorded, which suggests a cool and humid climate. In the Jackson Pond record (8), there is also no difference compared to the previous interval, although sampling resolution of one sample for 700 years is too low to detect any oscillation. The climate is defined as having been cool and moist. In the Browns Pond record (12), between ca. 13,000 and 10,000 cal. B.P., a *Quercus-Ostrya-Carpinus* association is recorded, attesting to warmer and wetter climatic conditions. At Camel Lake (13), *Picea* disappeared and deciduous tree taxa, *Fagus* and *Quercus*, began to dominate between 14,800 and 11,500 cal. B.P. The climate was warmer and wetter than during the previous interval.

In the Rocky Mountains, the Yellowstone and Grand Teton National Park sites (11) record an increase of *Picea* between 13,500 and 11,500 cal. B.P., attesting to wetter and warmer climatic conditions. In Mary Jane (9), sediment deposition stopped.

On the Pacific coast, pollen assemblages show no change compared to the previous interval and, although no modern analogs exist for this type of rich mixed forest, the climate is defined as cool and wet.

In Central America, no change in vegetation composition occurred at Chalco Lake (14) and La Yeguada (15). In the high Andean cordillera, the Pedro Palo record (16) shows an expansion of open scrub and grass paramo characteristic of a climatic warming between 14,000 and 13,000 cal. B.P. An increase of arboreal pollen is recorded at Laguna Ciega (17) between 14,500 and 13,000 (12,700) cal. B.P., attesting to wetter climatic conditions, particularly between 14,200 and 14,000 cal. B.P. The Laguna Baja record (18) shows no difference in vegetation composition compared to the previous interval.

In the tropical lowland records, at Lake Pata (21), no sediment deposition occurred during this time interval. At Carajás (22), when sedimentation restarts at ca. 14,600 cal. B.P., an increase of arboreal pollen is recorded, attesting to a moist climate until ca. 10,000 cal. B.P. The Salitre record (23) shows no change compared to the previous interval; the climate was still cool and wet.

In southern South America, at Puerto Octay (25), the expansion of North Patagonian rain forest species is indicative of a warming. At Pichihué (24), the development of peat and a succession of different Magellanic tundra taxa characterize a gradual evolution of the edaphic vegetation, which may indicate equally cool but wetter conditions compared to the previous interval.

In the Punta Arenas (27) and Puerto del Hambre (26) records, an increase in *Nothofagus* is recorded between 14,500 and 12,700 cal. B.P., suggesting a wet interval. In Harberton (28), the increase of Poaceae together with mesic taxa attests to an increase in moisture under continuing cooler temperatures.

20.4.3. Time Interval Between 12,700 and 11,000 cal. B.P., Younger Dryas Chronozone

The data for the time interval 12,700–11,000 cal. B.P. are compiled in Fig. 6. In Alaska, at Ongivinuk and Grandfather Lakes (1), the earlier *Betula* shrub tundra was replaced by herb tundra with Poaceae and *Artemisia*, suggesting a climate that was drier and cooler than before.

In Alaska, the Pleasant Island record (2) shows *Pinus* parkland taxa decreased and instead herb tundra returned between 12,300 and 11,400 cal. B.P. This change is interpreted as indicating a drier and cooler climate. In the Olympic Peninsula record (3), the increase in *Pinus*, *Picea*, and *Tsuga* recorded between 13,000 and 11,000 cal. B.P. suggests an increase of precipitation. At Gordon Lake (5), the presence of *T. mertensiana* with *Pinus* and *Abies* characterizes a cooler and drier climate between 12,800 and 11,000 cal. B.P. In the Little Lake record (4), between 12,400 and 11,000 cal. B.P., the increase of a mixed coniferous forest with *Pinus* reversed the expansion of the *Pseudotsuga* forest, attesting to a cooler climate, although there is a lack of clear modern analogs.

On the northeastern Atlantic coast, at the Killarney site (6), an increase of boreal forest taxa is recorded. The northward migration of deciduous tree species was stopped, and a two-step climate reversal was recorded. The first one, called the Killarney oscillation, occurred between 13,000 and 12,800 cal. B.P.; the second one occurred between 12,700 and 11,000 cal. B.P. In the Allamuchy and Linsley Pond records (7), there is no change in vegetation composition compared to the previous interval. The climate was still moist and warm. Records from the Yellowstone area do not show any climatic reversal. The climate was wet and cool as in the previous interval. At Browns Pond (12), no reversal is recorded, although the vegetation assemblage of *Quercus-Ostrya-Carpinus* indicates a drier climate than in the previous interval (new data do show a reversal, D. Peteet, personal communication). At Camel Lake (13), a gap in sedimentation occurred between 12,000 and 11,000 cal. B.P.

In Central America, the Chalco Lake record (14) does not show any change since the previous interval and moist montane forest elements continue. In Costa Rica, at the La Chonta 2 site (not shown), the forest line dropped from 2800–2400 m elevation; this event is interpreted as indicating a 2°–2.5°C cooling (Hooghiemstra et al., 1992). In the La Yeguada record (15), an increase of tropical lowland taxa attests to disturbances of the rain forest and was interpreted to reflect a wetter climate between 13,000 (13,300–12,700) and 12,300 (12,700–12,000) cal. B.P.

In the high Andean cordillera, the Pedro Palo record (16) shows a two-step reversal. The first step, between 13,000 and 12,300 cal. B.P., is characterized by a decrease of Andean and sub-Andean elements, which is interpreted as indicating a cooling. The second step, between 12,300 and 12,000 cal. B.P., shows an increase in paramo elements, which is interpreted as reflecting drier climatic conditions. At Laguna Ciega (17), an increase in Poaceae, together with a decrease of arboreal pollen taxa, is defined as representing a cold and dry interval called *El Abra*. In the Mucubaji record (19),

superparamo plant assemblages increased between 13,500 and 10,500 cal. B.P., attesting to cooler climatic conditions. At Laguna Baja (18), between 13,500 and 11,000 cal. B.P., moist montane forest elements were replaced by Poaceae, and the charcoal concentration also increased. Three environmental and climatic scenarios could explain this observation: (1) the tree line shifted to lower elevations in response to a cooling; (2) paramo vegetation expanded, implying increased aridity, an explanation that is supported by evidence for increased fires; and (3) arboreal pollen are allochthonous and represent locally an increase in long-distance transport of montane forest pollen.

In the tropical lowlands, there is no sediment deposition during this time interval at Lake Pata (21). At Carajás (22), the vegetation composition did not change since the previous interval. At Salitre (23), a brief increase of Apiaceae and Poaceae between 12,300 and 10,000 cal. B.P. a decrease of the *Araucaria* forest elements, is interpreted as indicating a shift to drier climatic conditions.

In the Puerto Octay (25) record, a cooling is indicated by a decrease of rain forest taxa and an increase of *Nothofagus dombeyi*–type, *Maytenus disticha*, and *Podocarpus nubigena* spp. However, at the same time, charcoal particles increased, suggesting that forest fires and not climate could have produced this change in tree taxa. In the Pichihué (24) record, there is no vegetation change. In the Punta Arenas (27), Puerto del Hambre (26), and Harberton (28) records, a Poaceae-*Empetrum* assemblage suggests drier climatic conditions between 12,900 (12,500) and 11,500 cal. B.P.

20.5. GENERAL DISCUSSION: CHARACTERIZATION OF THE LATE GLACIAL

The establishment of modern forests occurred at different times and followed different patterns, probably related as much to the records' locations as to regional climate change. In spite of differences in resolution and unequal chronological control, the data from the Americas show no clear evidence for synchroneity of climatic oscillations during the late glacial interval. We will point out some of the reasons for this lack of synchroneity.

Several vegetation records show no vegetation changes during the intervals considered. In the Pacific Northwest, for instance, the conifer forest with *Pinus*, *Tsuga*, and *Abies* dominated between 16,000 and 14,000 cal. B.P. This forest was replaced by subalpine forest taxa between 14,000 and 11,500 cal. B.P. In eastern North America, a mixed *Picea* forest existed between

17,000 and 13,500 cal. B.P., after which it was replaced by a deciduous *Quercus* forest. In southern Central America, the semi-evergreen forest with *Abies, Liquidambar, Carpinus, Alnus, Corylus, Ulmus, Juglans, Fagus,* and *Populus* was fully developed between 18,500 and 11,000 cal. B.P. In tropical lowlands, rain forest taxa are mixed with mountain forest taxa, such as *Podocarpus* and *Hedyosmum*. This mixed forest became fully developed after 14,000 cal. B.P.

However, other vegetation records show several oscillations that are not coeval with circum-North Atlantic oscillations. For example, in the records from the southernmost latitudes, several oscillations between *Empetrum* heathland and Poaceae steppe-tundra occurred between 16,500 and 13,500 cal. B.P. and 13,500 and 11,000 cal. B.P. Repeated changes in moisture are also recorded for midlatitude records in temperate southern South America, including the onset of peat growth, fluctuations of Magellanic moorland elements, and *Nothofagus* forest taxa. Although the timing of these late glacial fluctuations is not strictly coeval, interhemispheric synchroneity is concluded by some authors (Denton et al., 1999; Moreno et al., 1999; but see the discussion by Markgraf and Bianchi, 1999).

Based on the characteristic patterns of late glacial oscillations, the records discussed previously can be divided into three groups:

1. High-altitude sites and high northern and mid-northern latitude sites showing synchronized oscillations, but with a paleoenvironmental expression or climatic signal different from the circum-North Atlantic oscillations.
2. High southern latitude sites showing oscillations diachronous from those characteristic for the circum–North Atlantic region.
3. Mid-southern latitude and tropical sites showing no oscillations or stepwise changes diachronous from the circum–North Atlantic oscillations.

In general, the late glacial climate oscillations, clearly expressed in polar and tropical ice cores (Thompson et al., 1998; Jouzel et al., 1987) and in marine cores (Bard et al., 1987, 1997; Bond et al., 1993), are rarely recorded as clearly in terrestrial records from the Americas. Instead, the character and amplitude of these oscillations in terrestrial records appear to be greatly affected by site specifics.

During the time interval from 15,500–14,500 cal. B.P., vegetation records show that temperatures were low in the high latitudes of the Northern Hemisphere, in Alaska, in the northwest Pacific, in the northeast Atlantic, and at high-elevation sites in the Colombian and Venezuelan Andes. Conditions were wet in the midlatitudes and in tropical and subtropical forest regions, al-

though there is a severe lack of modern analogs for this period. Conditions were dry in southern South America and along the Venezuelan/southeast Caribbean coast.

During the Bølling-Allerød interval (14,500–12,700 cal. B.P.), conditions became wetter in the Americas, except at sites on the Olympic Peninsula and at Lake Valencia.

The case of the next interval (12,700–11,000 cal. B.P.), which corresponds to the Younger Dryas chronozone, is different and better documented because it has been subject to more detailed analysis and even has led to major controversies (Ashworth and Markgraf, 1989; Peteet et al., 1990, 1993; Curtis and Hodell, 1993; Kuhry et al., 1993; Markgraf, 1993; Mathewes, 1993; Francou et al., 1995; Gasse et al., 1995; Hansen, 1995; Heine, 1993, 1995; Islebe et al., 1995; Leyden, 1995; Mayle and Cwynar, 1995; Osborn et al., 1995; Peteet, 1995; Thompson et al., 1995; van der Hammen and Hooghiemstra, 1995; Clapperton et al., 1997; Menounos and Reasoner, 1997; Yu and Eicher, 1998). Changes in vegetation type and composition shown during this interval that can be attributed to temperature reversals are not clearly observed in Central America (except in Costa Rica; Hooghiemstra et al., 1992), tropical lowlands, and the Southern Hemisphere midlatitudes. Moist forests are well developed in these regions, showing few signs of short-term environmental fluctuations that might correlate with the well-documented climatic cooling in the circum–North Atlantic region. On the other hand, records from the high northern Andean and Central American cordilleras show multiple stepwise vegetation changes, suggesting repeated temperature reversals. The same patterns are also recorded in ice core records from Huascarán, Peru, and Sajama, Bolivia (Thompson et al., 1995, 1998), and by evidence of past glacier fluctuations in the Andean cordilleras (Seltzer, 1994; Francou et al., 1995; Clapperton et al., 1997). Abrupt oscillations in forest development are observed in both the North American Northwest and Southwest. In southern South America, repeated high-amplitude paleoenvironmental changes are also recorded, although the timing is apparently out of phase with the circum–North Atlantic oscillations (Markgraf, 1993; Heusser, 1995; Ariztegui et al., 1997; Moreno, 1997; Markgraf and Bianchi, 1999; Denton et al., 1999; Moreno et al., 1999).

Are all the changes evidenced in the PEP 1 transect synchronous and, therefore, can they be attributed to the same climatic forcing? This unanswered question has recently received new interest. Because of the fluctuations' high frequency and their perhaps global expression (Peteet, 1995; Broecker et al., 1998; Lowell et al., 1995), events like the Younger Dryas or Bølling-Allerød interval cannot be attributed to insolation forcing (Milankovitch cycles). It is now generally thought that the reorganization of the ocean–atmosphere interaction in the Atlantic Ocean (changes in the intensity of the thermohaline circulation) must have played a major role in synchronizing climate oscillations during full glacial and late glacial times (Broecker et al., 1985; Broecker and Denton, 1989). It appears that synchroneity of late glacial oscillations is well established among Greenland, the Cariaco trench, and the Andean cordilleras (in Costa Rica, Colombia, and Peru). We suggest that these oscillations affected climates even as far equatorward as the southern extension of the Intertropical Convergence Zone (ITCZ), with a speculative southernmost limit of the ITCZ during the Northern Hemisphere winter months (Fig. 7). The mechanism explaining the observed climate changes on land would relate to increased (or decreased during the warm

Bølling-Allerød interval) penetration of polar air masses (Latrubesse and Ramonell, 1994; Bradbury, 1997; Bradbury et al., 2000; Lézine and Denèfle, 1997), as presented by the mobile polar high concept of Leroux (1993, 1996). This atmospheric phenomenon today plays an important role in subtropical precipitation patterns, especially during the winter months (Vuille and Ammann, 1997; Marengo and Rogers, 2000).

The major unsolved problem concerns the differences in climatic signal of apparently synchronous events between the North Atlantic (Greenland) and Antarctica, especially during the Younger Dryas chronozone. While several of the Antarctic records (Byrd, Vostok) suggest an opposition of the climate signal, i.e., a Younger Dryas cooling in Greenland opposed to a warming in Antarctica (Blunier et al., 1997; Broecker, 1998), a new Antarctic record (Taylor Dome) is thought to be in phase with the Greenland temperature reversal (Steig et al., 1998). The climatic changes in southern Argentina and Chile (Tierra del Fuego) appear to be supporting the opposition of climate signal; hence, it is likely that the Taylor Dome record is problematic (Markgraf, 1993; Pendall et al., submitted).

20.6. CONCLUSION AND RECOMMENDATIONS

From this overview of late glacial vegetation and inferred climate changes along the PEP 1 transect, several recommendations can be made to advance the issues related to interhemispheric correlation of intervals of rapid change.

20.6.1. Precipitation Versus Temperature Sensitivity

When sites are located in areas where the vegetation is not highly sensitive to temperature change, e.g., sites where precipitation is permanently high (parts of the tropics and temperate rain forest areas), short-term climatic reversals, known to reflect primarily temperature changes elsewhere, are unlikely to be detected. Temperature-sensitive vegetation would be found at high elevations (e.g., tree line) or at least in areas where precipitation is seasonally low and where temperature changes could be perceived by vegetation as moisture changes. In every case, regional aspects greatly affect the character of these oscillations.

20.6.2. Modern Analogs

Another problem in dealing with late glacial climate oscillations is the lack of modern analogs for the vege-

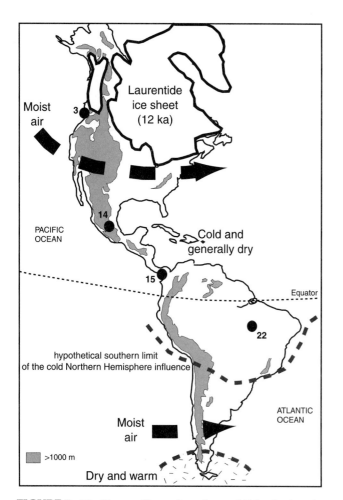

FIGURE 7 The Younger Dryas chron (interval 3) for the Americas deduced from paleovegetation reconstructions, showing the hypothetical southern limit of the influence of cold Arctic air during the Southern Hemisphere summer months. Full circles indicate records where no change is detected.

tation compositions at that time. Assemblages of species that do not occur today complicate interpretations in terms of temperature and precipitation. In eastern North America, pollen records at 14,000 cal. B.P. have the highest number of nonanalog samples (48%), compared to 40% at 17,000 cal. B.P. and 26% during full glacial times (Webb et al., 1993; Thompson et al., 1993; Overpeck et al., 1992).

20.6.3. Radiocarbon Dates

In addition to the problems with radiocarbon plateaus occurring at the time of the short-term climate oscillations, most records discussed have insufficient radiocarbon dates to establish a high-resolution chronology (Fig. 3). This complicates comparison of late glacial fluctuations as well as detection of hiatuses and / or changes in sedimentation rates.

20.6.4. Temporal Resolution

Temporal resolution is often too low in the analyzed records (i.e., sampling intervals are too large) to allow to distinguish short-term oscillations. When a 1- to 2-cm thin sand layer is present in a sediment that is analyzed in a 5- or 10-cm sampling interval, changes in vegetation or climate will go unnoticed. In some cases, when sites show no change compared to the previous interval, this may be due to a low-resolution sampling. This problem might also explain why short-term climatic events such as the Older Dryas are rarely recorded in terrestrial records.

20.6.5. Synchroneity

When dates are calibrated in terms of calendar years, which may produce several ages, synchroneity cannot be proved. For example, in the Gordon and Little Lake records, the Younger Dryas is estimated to date either between 12,400 and 11,000 cal. B.P. or between 12,800 and 11,000 cal. B.P., respectively. For the North Atlantic marine cores, an age of 12,900–11,600 cal. years B.P. is calculated. These calibration ages represent the mean value of a 2 sigma calibration and may represent only one time period. In the Allamuchy and Linsley Pond records, a change is detected between 13,000 and 11,500 cal. B.P. Calibration estimates suggest that this change occurred between 14,000 and 11,000 cal. B.P. or between 12,300 and 10,400 cal. B.P. In La Yeguada, a change in vegetation, characterized by an increase in lowland rain forest taxa, is dated at between 13,000 and 12,300 cal. B.P. Calibration suggests that it happened between 13,300 and 12,700 or 12,700 cal. B.P. and 12,000 cal. B.P. This finding implies that it does not fall into the

Younger Dryas time period as defined in the circum–North Atlantic region.

Acknowledgments

We thank Vera Markgraf for inviting us to participate in the PEP 1 meeting in Mérida (Venezuela) and for asking us to organize this interhemispheric review. Pollen records were selected from lists provided by Vera Markgraf, Cathy Whitlock, and Eric Grimm. A constructive review from Vera Markgraf, Dorothy Peteet, and Cathy Whitlock considerably improved this manuscript.

References

Absy, M. L., A. Cleef, M. Fournier, L. Martin, M. Servant, A. Sifeddine, M. Ferreira da Silva, F. Soubiès, K. Suguio, B. Turcq, and T. van der Hammen, 1991: Mise en évidence de quatre phases d'ouverture de la forêt amazonienne dans le sud-est de l'Amazonie au cours des 60000 dernières années. Première comparaison avec d'autres régions tropicales. *Comptes Rendus Académie des Sciences de Paris*, **312**: 673–678.

Ariztegui, D., M. M. Bianchi, J. Masaferro, E. Lafargue, and F. Niessen, 1997: Interhemispheric synchrony of late-glacial climatic instability as recorded in proglacial Lake Mascardi, Argentina. *Journal of Quaternary Science*, **12**(4): 333–338.

Ashworth, A. C., and V. Markgraf, 1989: Climate of the Chilean channels between 11,000 to 10,000 yr B.P. based on fossil beetle and pollen analyses. *Revista Chilena de Historia Natural*, **62**: 61–74.

Bard, E., M. Arnold, R. G. Fairbanks, and B. Hamelin, 1993: ^{230}Th-^{234}U and ^{14}C ages obtained by mass spectrometry on corals. *Radiocarbon*, **35**: 191–199.

Bard, E., M. Arnold, P. Maurice, J. Duprat, J. Moyes, and J.-C. Duplessy, 1987: Retreat velocity of the North Atlantic polar front during the last deglaciation determined by ^{14}C accelerator mass spectrometer ages from Barbados corals. *Nature*, **328**: 791–794.

Bard, E., B. Hamelin, R. G. Fairbanks, and A. Zindler, 1990: Calibration of ^{14}C timescale over the past 30,000 years using mass spectrometric U-Th ages from Barbados corals. *Nature*, **345**: 405–410.

Bard, E., F. Rostek, and C. Sonzogni, 1997: Interhemispheric synchrony of the last deglaciation inferred from alkenone palaeothermometry. *Nature*, **385**: 707–710.

Betancourt, J. L., T. R. van Devender, and P. S. Martin, 1990: *Packrat Middens: The Last 40,000 Years of Biotic Change.* Tucson: University of Arizona Press.

Blunier, T., J. Chappellaz, J. Schwander, A. Dällenbach, B. Stauffer, T. F. Stocker, D. Raynaud, J. Jouzel, H. B. Clausen, C. U. Hammer, and S. J. Johnsen, 1998: Asynchrony of Antarctic and Greenland climate change during the last glacial period. *Nature*, **394**: 739–743.

Blunier, T., J. Schwander, B. Stauffer, T. Stocker, A. Dällenbach, A. Indermühle, J. Tschumi, J. Chappellaz, D. Raynaud, and J.-M. Barnola, 1997: Timing of the Antarctic Cold Reversal and the atmospheric CO_2 increase with respect to the Younger Dryas event. *Geophysical Research Letters*, **24**(21): 2683–2686.

Bond, G., and R. Lotti, 1995: Iceberg discharges into the North Atlantic on millenia timescales during the last deglaciation. *Science*, **267**: 1005–1010.

Bond, G., W. S. Broecker, S. Johnsen, J. McManus, L. Labeyrie, J. Jouzel, and G. Bonani, 1993: Correlation between climate records from North Atlantic sediments and Greenland ice. *Nature*, **365**: 143–147.

Bradbury, J. P., 1997: Sources of glacial moisture in Mesoamerica. *Quaternary International*, **43/44**: 97–110.

Bradbury, J. P., M. Grosjean, S. Stine, and F. Sylvestre, 2000: Full and late glacial lake records along the PEP 1 transect: Their role in developing interhemispheric paleoclimate interactions. *In* Markgraf, V. (ed.), *Interhemispheric Climate Linkages*. San Diego: Academic Press, Chapter 16.

Bradbury, J. P., B. Leyden, M. L. Salgado-Labouriau, W. M. Lewis, Jr., C. Schubert, M. W. Benford, D. G. Frey, D. R. Whitehead, and F. H. Weibezahn, 1981: Late Quaternary environmental history of Lake Valencia, Venezuela. *Science*, **214**: 1299–1305.

Broecker, W. S., 1994: Massive iceberg discharges as triggers for global climate change. *Nature*, **372**: 421–424.

Broecker, W. S., 1998: Paleocean circulation during the last deglaciation: A bipolar seesaw? *Paleoceanography*, **13**(2): 119–121.

Broecker, W. S., and G. H. Denton, 1989: The role of ocean-atmosphere reorganizations in glacial cycles. *Geochimica Cosmochimica Acta*, **53**: 2465–2501.

Broecker, W. S., M. Andree, W. Wolfli, H. Oeschger, G. Bonani, J. P. Kennett, and D. Peteet, 1988: The chronology of the last deglaciation: Implications to the cause of the Younger Dryas event. *Paleoceanography*, **3**: 1–19.

Broecker, W. S., D. M. Peteet, and D. Rind, 1985: Does the ocean-atmosphere system have more than one stable mode of operation? *Nature*, **315**: 21–25.

Broecker, W. S., D. M. Peteet, I. Hajdas, J. Lin, and E. Clark, 1998: Antiphasing between rainfall in Africa's Rift Valley and North America's Great Basin. *Quaternary Research*, **50**: 12–20.

Bush, M. B., D. R. Piperno, P. A. Colinvaux, P. E. De Oliveira, L. A. Krissek, M. C. Miller, and W. E. Rowe, 1992: A 14,300-yr paleoecological profile of a lowland tropical lake in Panama. *Ecological Monographs*, **62**(2): 251–275.

Clapperton, C. M., M. Hall, P. Mothes, M. J. Hole, J. Still, K. F. Helmens, P. Kuhry, and A. M. D. Gemmell, 1997: A Younger Dryas icecap in the equatorial Andes. *Quaternary Research*, **47**: 13–28.

Colinvaux, P. A., P. E. De Oliveira, J. E. Moreno, M. C. Miller, and M. B. Bush, 1996: A long pollen record from lowland Amazonia: Forest and cooling in glacial times. *Science*, **274**: 85–88.

Curtis, J. H., and D. A. Hodell, 1993: An isotopic and trace element study of ostracods from Lake Miragoane, Haiti: A 10,500 year record of paleosalinity and paleotemperature changes in the Caribbean. *In* Swart, P. K., K. C. Lohmann, J. McKenzie, and S. Savin (eds.), *Climate Change in Continental Isotopic Records:* Washington, DC: American Geophys. Union, Geophysical Monograph 78, pp. 135–152.

Dansgaard, W., S. J. Johnsen, H. B. Clausen, D. S. Dahljensen, N. Gundestrup, C. U. Hammer, C. S. Hvidberg, J. R. Steffensen, A. E. Sveinbjörnsdottir, J. Jouzel, and G. Bond, 1993: Evidence for general instability of past climate from a 250-kyr ice-core record. *Nature*, **364**: 218–220.

Denton, G. H., C. J. Heusser, T. V. Lowell, P. I. Moreno, B. G. Andersen, L. E. Heusser, C. Schüchter, and D. R. Marchant, 1999: Interhemispheric linkage of paleoclimate during the last glaciation. *Geografiska Annaler*, **81A**: 107–154.

Francou, B., Ph. Mourguiart, and M. Fournier, 1995: Phase d'avancée des glaciers au Dryas récent dans les Andes du Pérou. *Comptes Rendus de l' Académie des Sciences de Paris*, **320**: 593–599.

Gasse, J. C., E. B. Evenson, J. Klein, B. Lawn, and R. Middleton, 1995: Precise cosmogenic ^{10}Be measurements in western North America: Support for a global Younger Dryas cooling event. *Geology*, **23**(10): 877–880.

Grigg, L. D., and C. Whitlock, 1998: Late-glacial vegetation and climate change in western Oregon. *Quaternary Research*, **49**: 287–298.

Hansen, B. C. S., 1995: A review of late glacial pollen records from Ecuador and Peru with reference to the Younger Dryas event. *Quaternary Science Reviews*, **14**: 853–865.

Hansen, B. C. S., and D. R. Engstrom, 1996: Vegetation history of

Pleasant Island, southeastern Alaska since 13,000 yr B.P. *Quaternary Research*, **46**: 161–175.

Hansen, B. C. S., and D. T. Rodbell, 1995: A late-glacial/Holocene pollen record from the eastern Andes of northern Peru. *Quaternary Research*, **44**: 216–227.

Hartz, N., and V. Milthers, 1901: Det senglaciale Ler i Allerød Teglvaerksgrav. *Mededelingen Dansk Geoliska Foreningen*, **1**: 31–60.

Heine, J. T., 1993: A reevaluation of the evidence for a Younger Dryas climatic reversal in the tropical Andes. *Quaternary Science Reviews*, **12**: 769–779.

Heine, J. T., 1995: Comments on C. M. Clapperton's "Glacier readvance in the Andes at 12,500–10,000 yr B.P.: Implications for mechanism of late-glacial climatic change." *Journal of Quaternary Science*, **10**(1): 87–88.

Heusser, C. J., 1995: Three late Quaternary pollen diagrams from southern Patagonia and their palaeoecological implications. *Palaeogeography, Palaeoclimatology, Palaeoecology*, **118**: 1–24.

Hooghiemstra, H., and T. van der Hammen, 1993: Late Quaternary vegetation history and paleoecology of Laguna Pedro Palo (sub-Andean forest belt, Eastern Cordillera, Colombia). *Review of Palaeobotany and Palynology*, **77**: 235–262.

Hooghiemstra, H., A. M. Cleef, G. W. Noldus, and M. Kappelle, 1992: Upper Quaternary vegetation dynamics and palaeoclimatology of the La Chonta bog area (Cordillera de Talamanca, Costa Rica). *Journal of Quaternary Science*, **7**(3): 205–225.

Hu, F. S., L. B. Brubaker, and P. A. Anderson, 1995: Postglacial vegetation and climate change in the north Bristol Bay region, southwestern Alaska. *Quaternary Research*, **43**: 382–392.

Hughen, K. A., J. T. Overpeck, L. C. Petersen, and S. Trumbore, 1996: Rapid climate changes in the tropical Atlantic region during the last deglaciation. *Nature*, **380**: 51–54.

Islebe, G. A., H. Hooghiemstra, and K. van der Borg, 1995: A cooling event during the Younger Dryas chron in Costa Rica. *Palaeogeography, Palaeoclimatology, Palaeoecology*, **117**: 73–80.

Iversen, J., 1942: En pollenanalytisk Tidsfaestelse af Ferskvandslagene ved Norre Lyngby. *Mededelingen Dansk Geologiska Foreningen*, **10**: 130–151.

Iversen, J., 1954: The late glacial flora of Denmark and its relation to climate and soil. *Danmarks Geologiska Undersogelse*, **80**(2): 87–119.

Johnsen, S. J., H. B. Clausen, W. Dansgaard, K. Fuhrer, N. Gunderstrup, C. U. Hammer, P. Iversen, J. Jouzel, B. Stauffer, and J. P. Steffensen, 1992: Irregular glacial interstadials recorded in a new Greenland ice core. *Nature*, **359**: 311–313.

Jouzel, J., C. Lorius, L. Merlivat, and J.-R. Petit, 1987: The Antarctic ice record during the late Pleistocene. *In* Berger, W.H., and L.D. Labeyrie (eds.), *Abrupt Climatic Change*. Dordrecht: Reidel, pp. 235–245.

Jouzel, J., R. Vaikmae, J. R. Petit, M. Martin, Y. Duclos, M. Stievenard, C. Lorius, M. Toots, M. A. Mélières, L. H. Burckle, N. I. Barkov, and V. M. Kotlyakov, 1995: The two-step shape and timing of the last deglaciation in Antarctica. *Climate Dynamics*, **11**: 151–161.

Kneller, M., and D. M. Peteet, 1993: Late-Quaternary climate in the ridge and valley of Virginia, U.S.A.: Changes in vegetation and depositional environment. *Quaternary Science Reviews*, **12**: 613–628.

Kuhry, P., 1988: Palaeobotanical-palaeoecological studies of tropical high Andean peatbog sections (Cordillera Oriental, Colombia). *In* van der Hammen, T. (ed.), *The Quaternary of Colombia*. Vaduz: Cramer, 241 pp.

Kuhry, P., H. Hooghiemstra, B. van Geel, and T. van der Hammen, 1993: The El Abra stadial in the Eastern Cordillera of Colombia (South America). *Quaternary Science Reviews*, **12**: 333–343.

Latrubesse, E. M., and C. G. Ramonell, 1994: A climatic model for southwestern Amazonia in last glacial times. *Quaternary International*, **21**: 163–169.

Ledru, M.-P., 1993: Late Quaternary environmental and climatic changes in central Brazil. *Quaternary Research,* **39**: 90–98.

Lehman, S. J., and L. D. Keigwin, 1992: Sudden changes in North Atlantic circulation during the last deglaciation. *Nature,* **56**: 757–762.

Leroux, M., 1993: The mobile polar high: A new concept explaining present mechanisms of meridional air-mass and energy exchanges and global propagation of palaeoclimatic changes. *Global and Planetary Changes,* **7**: 69–93.

Leroux, M., 1996: *La dynamique du temps et du climat.* Paris: Masson.

Levesque, A. J., F. E. Mayle, I. R. Walker, and L. C. Cwynar, 1993: A previously unrecognized late-glacial cold event in eastern North America. *Nature,* **361**: 623—626.

Leyden, B. W., 1995: Evidence of the Younger Dryas in Central America. *Quaternary Science Reviews,* **14**: 833–839.

Lézine, A.-M., and M. Denèfle, 1997: Enhanced anticyclonic circulation in the eastern North Atlantic during cold intervals of the last deglaciation inferred from deep-sea pollen records. *Geology,* **25**(2): 119–122.

Lowell, T. V., C. J. Heusser, B. G. Andersen, P. I. Moreno, A. Hauser, L. E. Heusser, C. Schlüchter, D. R. Marchant, and G. H. Denton, 1995: Interhemispheric correlation of late Pleistocene glacial events. *Science,* **269**: 1541–1549.

Lozano-Garcia, M. S., and B. Ortega-Guerrero, 1994: Palynological and magnetic susceptibility records of Lake Chalco, central México. *Palaeogeography, Palaeoclimatology, Palaeoecology,* **109**: 177–191.

Marengo, J. A., and J. C. Rogers, 2000: Polar air outbreaks in the Americas: Assessments and impacts during recent and past climates. *In* Markgraf, V. (ed.), *Interhemispheric Climate Linkages.* San Diego: Academic Press, Chapter 3.

Markgraf, V., 1993: Younger Dryas in southernmost South America—An update. *Quaternary Science Reviews,* **12**: 351–355.

Markgraf, V., and M. M. Bianchi, 1999: Paleoenvironmental changes during the last 17,000 years in western Patagonia: Mallín Aguado, Province of Neuquen, Argentina. *Bamberger Geographische Schriften,* **19**: 175–193.

Markgraf, V., and R. Kenny, 1997: Character of rapid vegetation and climate change during the late-glacial in southernmost South America in past and future rapid environmental changes. *In* Huntley, B., et al. (eds.), *The Spatial and Evolutionary Responses of Terrestrial Biota,* Vol. 47 of NATO ASI Series. New York: Springer-Verlag, 1, pp. 81–90.

Mathewes, R. W., 1993: Evidence for Younger Dryas-age cooling on the North Pacific coast of America. *Quaternary Science Reviews,* **12**: 321–331.

Mayle, F. E., and L. C. Cwynar, 1995: A review of multi-proxy data for the Younger Dryas in Atlantic Canada. *Quaternary Science Reviews,* **14**: 813–821.

Menounos, B., and M. A. Reasoner, 1997: Evidence for cirque glaciation in the Colorado Front Range during the Younger Dryas chronozone. *Quaternary Research,* **48**: 38–47.

Moreno, P. I., 1997: Vegetation and climate near Lago Llanquihue in the Chilean Lake District between 20,200 and 9500 ^{14}C yr B.P. *Journal of Quaternary Science,* **12**(6): 485–500.

Moreno, P. I., T. V. Lowell, G. L. Jacobson, Jr., and G. H. Denton, 1999: Abrupt vegetation and climate changes during the last glacial maximum and last termination in the Chilean Lake District: A case study from Canal de la Puntilla. *Geografiska Annaler,* **81A**: 107–154.

Osborn, G., C. M. Clapperton, P. T. Davis, M. Reasoner, D. T. Rodbell, G. O. Seltzer, and G. Zielinski, 1995: Potential glacial evidence for the Younger Dryas event in the cordillera of North and South America. *Quaternary Science Reviews,* **14**: 823–832.

Overpeck, J., D. Anderson, S. Trumbore, and W. Prell, 1996: The southwest Indian monsoon over the last 18,000 years. *Climate Dynamics,* **12**: 213–225.

Overpeck, J. T., R. S. Webb, and T. Webb, III, 1992: Mapping eastern North American vegetation change over the past 18,000 years: No-analogs and the future. *Geology,* **20**: 1071–1074.

Partridge, T. C., P. B. Demenocal, S. A. Lorentz, M. J. Paiker, and J. C. Vogel, 1997: Orbital forcing of climate over South Africa: A 200,000-year rainfall record from the Pretoria saltpan. *Quaternary Science Reviews,* **16**: 1125–1133.

Pendall, E., V. Markgraf, U. Huber, J. W. C. White, M. Dreier, and R. Kenny, submitted: A multi-proxy record of Pleistocene-Holocene climate and vegetation changes from a peat bog in Patagonia. *Quaternary Research.*

Peteet, D., 1995: Global Younger Dryas? *Quaternary International,* **28**: 93–104.

Peteet, D. M., R. A. Daniels, L. E. Heusser, J. R. Vogel, J. Southon, and D. E. Nelson, 1993: Late-glacial pollen, macrofossils and fish remains in northeastern United States—the Younger Dryas Oscillation. *Quaternary Science Reviews,* **12**: 597–612.

Peteet, D. M., J. S. Vogel, J. R. Nelson, J. R. Southon, R. J. Nickman, and L. E. Heusser, 1990: Younger Dryas climatic reversal in northeastern United States? AMS ages for an old problem. *Quaternary Research,* **33**: 219–230.

Peterson, K. L., P. J. Mehringer, Jr., and C. E. Gustafson, 1983: Late-glacial vegetation and climate at the Manis Mastadodon site, Olympic Peninsula, Washington. *Quaternary Research,* **20**: 215–231.

Prell, W. L., and J. E. Kutzbach, 1987: Monsoonal variability over the past 150,000 years. *Journal of Geophysical Research,* **92**: 8411–8425.

Prell, W. L., and E. van Campo, 1986: Coherent response of Arabian Sea upwelling and pollen transport to late Quaternary monsoonal winds. *Nature,* **323**: 526–528.

Salgado-Labouriau, M. L., 1984: Late Quaternary palynological studies in the Venezuelan Andes. *Erdwissenschaftliche Forschung,* Bd., **XVIII**: 279–293.

Salgado-Labouriau, M. L., 1989: Vegetation and climatic changes in the Mérida Andes during the last 13,000 years. International Symposium on Global Changes in South America during the Quaternary: Past-Present-Future, São Paulo, Brazil, 8–12 May 1989, Abstracts, pp. 49–55.

Salgado-Labouriau, M. L., C. Schubert, and S. Valastro, Jr., 1977: Paleoecologic analysis of a late Quaternary terrace from Mucubají, Venezuelan Andes. *Journal of Biogeography,* **4**: 313–325.

Sanchez-Goñi, M. F., 1995: The Older Dryas of northern France in a West European context. *Revue de Paléobiologie,* **15**(2): 519–531.

Schulz, H., U. von Rad, and H. Erlenkeuser, 1998: Correlation between Arabian Sea and Greenland climate oscillations of the past 110,000 years. *Nature,* **393**: 54–57.

Seltzer, G. O., 1994: A lacustrine record of late Pleistocene climatic change in the subtropical Andes. *Boreas,* **23**: 105–111.

Short, S. K., and S. A. Elias, 1987: New pollen and beetle analysis at the Mary Jane site, Colorado: Evidence for late-glacial tundra conditions. *Geological Society of America Bulletin,* **98**: 540–548.

Smith, S. J., and R. S. Anderson 1992: Late Wisconsin paleoecologic record from Swamp Lake, Yosemite National Park, California. *Quaternary Research,* **38**: 91–102.

Sowers, T., and M. Bender, 1995: Climate records covering the last deglaciation. *Science,* **269**: 210–214.

Steig, E. J., E. J. Brook, J. W. C. White, C. M. Sucher, M. L. Bender, S. J. Lehman, D. L. Morse, E. D. Waddington, and G. D. Clow, 1998: Synchronous climate changes in Antarctica and the North Atlantic. *Science,* **282**: 92–95.

Street, A. E., and A. T. Grove, 1979: Global maps of lake-level fluctuations since 30,000 years B.P. *Quaternary Research,* **12**: 83–118.

Stuiver, M., and P. J. Reimer, 1993: Extended ^{14}C data base and revised CALIB 3.0 ^{14}C age calibration program. *Radiocarbon,* **35**: 215–230.

Stuiver, M., P. M. Grootes, and T. F. Braziunas, 1995: The GISP2 (^{18}O climate record of the past 16,500 years and the role of the Sun, ocean, and volcanoes. *Quaternary Research,* **44**: 341–354.

Thompson, L. G., M. E. Davis, E. Mosley-Thompson, T. A. Sowers, K. A. Henderson, V. S. Zagorodnov, P.-N. Lin, V. N. Mikhalenko, R. K. Campen, J. F. Bolzan, J. Cole-Dai, and B. Francou, 1998: A 25,000-year tropical climate history from Bolivian ice cores. *Science,* **282**: 1858–1863.

Thompson, L., E. Mosley-Thompson, M. E. Davis, P.-N. Lin, K. A. Henderson, J. Cole-Dai, J. F. Bolzan, and K.-B. Liu, 1995: Late glacial stage and Holocene tropical ice core records from Huascarán, Peru. *Science,* **269**: 46–50.

Thompson, R. S., C. Whitlock, P. J. Bartlein, S. P. Harrison, and G. Spaulding, 1993: Climatic changes in the western United States since 18,000 yr B.P. *In* Wright, H. E., Jr., J. E. Kutzbach, T. Webb, III, W. F. Ruddiman, F. A. Street-Perrott, and P. J. Bartlein (eds.), *Global Climates Since the Last Glacial Maximum,* Minneapolis: University of Minnesota Press, pp. 468–513.

van der Hammen, T., and H. Hooghiemstra, 1995: The El Abra stadial, a Younger Dryas equivalent in Colombia. *Quaternary Science Reviews,* **14**: 841–851.

van der Hammen, T., J. Barelds, H. De Jong, and A. A. De Veer, 1980/1981: Glacial sequence and environmental history in the Sierra Navada del Cocuy (Colombia). *Palaeogeography, Palaeoclimatology, Palaeoecology,* **32**: 247–340.

Villagrán, C., 1991: Desarollo de tundras magallanicas durante la transicion glacial-postglacial en la cordillera de la Costa de Chile,

Chiloé: Evidencias de un evento equivalente al "Younger Dryas"? *Bamberger Geographische Schriften,* Bd., **11**: 245–256.

Vuille, M., and C. Ammann, 1997: Regional snowfall patterns in the high, arid Andes. *Climatic Change,* **36**: 413–423.

Watts, W. A., B. C. S. Hansen, and E. C. Grimm, 1992: Camel Lake: A 40,000-yr record of vegetational and forest history from northwest Florida. *Ecology,* **73**: 1056–1066.

Webb, T., III, P. J. Bartlein, S. P. Harrison, and K. H. Anderson, 1993: Vegetation, lake levels, and climate in eastern North America for the past 18,000 years. *In* Wright, H. E., Jr., J. E. Kutzbach, T. Webb, III, W. F. Ruddiman, F. A. Street-Perrott, and P. J. Bartlein (eds.), *Global Climates Since the Last Glacial Maximum.* Minneapolis: University of Minnesota Press, pp. 415–467.

Whitlock, C., 1993: Postglacial vegetation and climate of Grand Teton and southern Yellowstone National Parks. *Ecological Monographs,* **63**: 173–198.

Wilkins, G. R., P. A. Delcourt, H. R. Delcourt, F. W. Harrison, and M. R. Turner, 1991: Paleoecology of central Kentucky since the last glacial maximum. *Quaternary Research,* **36**: 224–239.

Yu, Z. C., and U. Eicher, 1998: Abrupt climate oscillations during the last deglaciation in central North America. *Science,* **282**: 2235–2238.

Zaucker, F., T. F. Stocker, and W. S. Broecker, 1994: Atmospheric freshwater fluxes and their effect on the global thermohaline circulation. *Journal of Geophysical Research,* **99**(C6): 12443–12457.

CHAPTER

21

The Midlatitudes of North and South America During the Last Glacial Maximum and Early Holocene: Similar Paleoclimatic Sequences Despite Differing Large-Scale Controls

CATHY WHITLOCK, PATRICK J. BARTLEIN, VERA MARKGRAF,
AND ALLAN C. ASHWORTH

Abstract

Paleoenvironmental records from North and South America since the last glacial maximum (LGM) show environmental changes that are similar in timing and direction despite the differing extents of ice cover and opposing seasonal variations of insolation in the two hemispheres. The particular mechanisms that link the variations in the two regions are not well understood. Our understanding of the controls of global and regional climate change during the last 21,000 years is based on information on insolation, ice volume, and atmospheric and ocean conditions and on the spatial and temporal patterns of simulated paleoclimates. These data sets are compared with biotic evidence from temperate western regions of both continents for two periods when the climate controls were extreme: the glacial maximum (ca. 21,000 cal. B.P.), when northern ice sheets were at their greatest size, and the early Holocene (14,000–6000 cal. B.P.), when the amplification of the seasonal cycle of insolation in the Northern Hemisphere was greatest. The full-glacial environment of western North America was characterized by widespread subalpine and tundra taxa in the north and woodland and

montane taxa in the south. This pattern is consistent with a southward displacement of the jet stream, steepened temperature gradients, and a stronger glacial anticyclone produced by the Laurentide ice sheet. In South America, pollen and fossil beetle data indicate that open vegetation was widespread and woodland vegetation was confined to latitudes 36°–43°S. Stronger westerlies and a latitudinal compression of the jet stream provide a plausible explanation. The early Holocene of North America featured an expansion of xerophytic taxa in the northwest and species indicative of increased summer precipitation in the southwest. These data are consistent with the direct effects of greater summer insolation on net radiation, along with the indirect effects of insolation on the strength of the subtropical high-pressure system and the summer monsoon. In South America, warm temperate forests were present at latitudes equatorward of 45°S, rain forest and moorland were optimally developed from latitude 45° to 50°S, and beech woodland was established south of latitude 50°S. Such a gradient may be explained by a weakened subtropical high and summer monsoon and a weakly positive winter insolation anomaly lagged into spring by sea surface temperatures (SSTs) and sea ice extents. We propose that carryover of anomalies in the controls into subsequent months, combined with interactions among surface water and energy balance components, best explain the synchroneity in the paleoclimatic records of the Northern and Southern Hemispheres.

Resumen

Los registros paleoambientales del norte al sur de América desde el Último Máximo Glacial muestran cambios ambientales que son similares en el momento y dirección, a pesar de la diferente extensión de la cobertura de hielo y la opuesta variación estacional de la insolación entre los dos hemisferios. El mecanismo particular que conecta las variaciones entre las dos regiones no se comprende aún bien. Nuestro conocimiento de los parámetros de control del cambio climático global y regional durante los últimos 21,000 años está basado en la información sobre la insolación, del volumen de hielo, de las condiciones atmosféricas y oceánicas, y en los patrones espaciales y temporales de los paleoclimas simulados. Estos conjuntos de datos se han comparado con las evidencias bióticas de las regiones templadas del oeste, en ambos continentes para dos periodos donde los parámetros de control del clima fueron extremos: el máximo glacial (aprox. 21 cal ka), cuando la capa helada del norte llegó a su máxima extensión, y los inicios del Holoceno (14 a 16 cal ka), cuando la amplificación del ciclo estacional de la insolación en el hemisferio norte fue máximo. El ambiente glacial en el oeste de América del norte se caracterizó por una predominante vegetación subalpina y de tundra en el norte y en las montañas, y de vegetación de montaña en el sur. Este patrón es consecuente con el desplazamiento hacia el sur de la corriente en chorro, el elevado gradiente de temperatura, y un más potente anticiclón glacial producido por la capa de hielo Laurentide . En Sudamérica, los datos de polen y escarabajos fósiles indican que predominó la vegetación abierta y que la vegetación árborea se confinó entre lo 36° y 43° de latitud sur. La mayor fuerza de los vientos de componente oeste y la compresión latitudinal de la corriente en chorro proporcionan una explicación verosímil. El inicio del Holoceno en Norte América presentó la expansión de la vegetación xerófita en el noroeste, y de especies que indican el aumento de la precipitación estival en el sureste. Estos datos son consecuentes con los efectos directos del aumento de la insolación de verano en la radiación neta, junto con efectos indirectos de la insolación en la fuerza del sistema de altas presiones subtropical y del monzón de verano. En Sudamérica, bosques lluviosos templados aparecieron desde latitudes inferiores a los 45° sur, bosques lluviosos fríos y moorlands se desarrollaron de forma óptima entre los 45° a 50° latitud sur, y bosques abiertos de *Nothofagus* se establecieron en latitudes superiores a los 50° sur. Tal gradiente puede ser explicado por un debilitamiento del anticiclón subtropical y del monzón de verano, y una pequeña anomalía positiva de la insolación en invierno, retrasada hacia la primavera ena la temperatura superficial del mar y las extensiones de hielo marino. Pensamos que tranferencias de anomalías en los parámetros de control hacia meses posteriores, combinados con las interacciones entre los componentes del balance del agua superficial y del balance energético explicarían mejor la sincronización entre los registros paleoclimáticos del hemisferio norte y sur.

21.1. INTRODUCTION

The temperate regions of western North and South America share similarities in their landforms, climate, and biotic gradients that invite comparison of their environmental histories. Both regions feature rain forest communities that extend from warm temperate forest to seasonal rain forest, to perhumid temperate rain forest, and at high-latitude subpolar rain forest. Tree and shrub richness on both continents declines with increasing latitude. Along latitudinal precipitation and temperature gradients, both regions shift from evergreen rain forest to temperate forest to chaparral, then to subtropical desert at successively lower latitudes, and to tundra and heath at higher latitudes. Moreover, there is a striking west-to-east gradient from species-rich rain forest to species-poor xerophytic forest and woodland, and finally to steppe in both regions, as a result of steep orographic precipitation gradients posed by the western cordillera.

The resemblance in climate and biota is apparently not a recent phenomenon. Paleoecologic records reveal that the temperate latitudes in both hemispheres have experienced changes in climate and biota that have been broadly synchronous and similar in magnitude and direction throughout the late Quaternary period. In general, the causes of regional climate change are described as hierarchical, reflecting a combination of mechanisms that range from those operating at a global scale over several millennia to those occurring at the local scale over the short term (Delcourt et al., 1983; Bartlein, 1988; Prentice, 1992). The large-scale mechanisms of climate change include variations in the seasonal cycle of insolation (incoming solar radiation), the size of continental ice sheets, and the composition of the atmosphere. These controls are responsible for the glacial–interglacial oscillations of the Quaternary period, as well as the reorganizations of biota that accompany each oscillation (Bartlein, 1997).

The comparability of climate and biota of western North and South America is problematic because one of the large-scale controls, the variation in the seasonal cycle of insolation that occurs on glacial–interglacial timescales, is opposite in phase between the two hemispheres. Times of high summer insolation in the Northern Hemisphere occur when summer insolation is low in the Southern Hemisphere and vice versa. Explaining the synchroneity of glacial–interglacial variations between the hemispheres (Broecker and Denton, 1989), as well as the spatial variations of the particular phasing of the regional changes (i.e., Imbrie et al., 1984), has been a long-standing problem in Quaternary paleoclimatology. In particular, the similarity of climatic and biotic response at temperate latitudes in the two hemispheres begs the question: What mechanisms of climate synchronize the regional response of the temperate latitudes when the large-scale controls are different?

In this chapter, we consider the global and regional record of climate change during the last 21,000 years, based on time series of insolation and glacial, ice core, and ocean records and on the spatial and temporal patterns of simulated paleoclimates. We then focus on the biotic response during two periods when the large-scale climate controls were extreme: the last glacial maximum (LGM), ca. 21 cal ka (= 18 ^{14}C ka)*, when northern ice sheets were at their largest size and the seasonal cycle of insolation in the Northern Hemisphere was attenuated, and the early Holocene, ca. 14–6 cal. ka (12–6 ^{14}C ka), when the amplification of the seasonal cycle of insolation was greatest in the Northern Hemisphere and lowest in the Southern Hemisphere. We present hypotheses to explain why the temperate latitudes of western North and South America both show the coldest, driest conditions during the LGM and warm, dry conditions during the early Holocene.

21.2. MODERN SETTING

21.2.1. Climate

The distribution of temperate ecosystems in western North and South America is related primarily to the seasonal patterns of precipitation along latitudinal, longitudinal, and altitudinal gradients (Lawford, 1996; Aceituno et al., 1993; Veblen and Alaback, 1996). Both

*In this chapter, we use cal ka (or kiloannum) to mean 1000s of calendar years before present and ^{14}C ka to mean 1000s of radiocarbon years before present. The radiocarbon ages have been converted to calendar years based on data that accompany the program of Stuiver and Reimer (1993).

western North and South America are affected by conditions in the Pacific Ocean and the interaction between the tropics and extratropics. Key elements of the climate patterns include the strength and position of the Intertropical Convergence Zone (ITCZ); the subtropical high-pressure systems and associated trade wind belts; and in the midlatitudes, upper atmospheric features (i.e., trough and ridge patterns and the locations of polar and subtropical jet streams, and of regions of large-scale subsidence or uplift). These circulation systems show latitudinal shifts related to the seasonal changes in the temperature gradient between the equator and poles. On average, this gradient is steeper in the Southern Hemisphere than in the Northern Hemisphere, primarily because of the extremely cold Antarctic continent, while the greater land/ocean temperature contrast in the Northern Hemisphere leads to greater amplitude of the waves of the westerlies in that hemisphere. As a result, circulation patterns are asymmetric between the two hemispheres, as are their seasonal variations.

During the northern winter/southern summer (December–January–February [DJF]), the northern midlatitude westerlies are displaced equatorward and are accelerated relative to the northern summer (June–July–August [JJA]) (Fig. 1A), while at the surface, the northern subtropical high-pressure systems (STHs) are reduced in size while the Aleutian and Icelandic low-pressure systems are expanded (Figs. 1B and 1C) (Kousky and Ropelewski, 1997). Northwestern North America is dominated by large-scale rising motions at the 500-mbar level (negative values in Fig. 1D), while the region from northern México and the southwestern United States across the cordillera to the Great Plains is dominated by subsidence. Precipitation reaches its maximum during the year along the northwestern coast of North America (Fig. 1E). In the tropics, the ITCZ and the band of precipitation along it are shifted southward (Figs. 1C and 1E).

In the Southern Hemisphere during DJF, the westerlies are slower relative to those during the southern winter (JJA) and are shifted poleward (Fig. 1A), while at the surface, the southern STHs are expanded (Figs. 1B and 1C). Onshore flow from the South Atlantic Ocean into South America is at its seasonal maximum (Fig. 1C), and most of the continent is dominated by large-scale rising motions (Fig. 1D), convergence at the surface, and divergence aloft (Figs. 1A and 1C); as a consequence, precipitation is relatively high over the continent, except for its farthest southern tip. Together, these surface and upper level winds, and the related precipitation patterns, show the wet phase of the "southern monsoon."

During the northern summer/southern winter

FIGURE 1 Features of the present climate of the Western Hemisphere: (A) upper level winds (200-hPa streamlines showing upper level circulation), (B) sea level pressure (SLP), (C) surface winds (925-hPa stream lines), (D) large-scale uplift or subsidence (500-hPa vertical velocity), and (E) precipitation for the northern winter / southern summer (December–January–February, DJF). (F–J) are the same as in A–E for the northern summer / southern winter (June–July–August, JJA).

(JJA), the westerlies are weaker and shifted poleward in the Northern Hemisphere (Fig. 1F), while at the surface, the STHs are expanded (Fig. 1G), while coastal western North America is dominated by large-scale subsidence and dry conditions (Figs. 1H and 1I). Onshore flow and large-scale uplift to the east and south of this region mark the wet phase of the northern monsoon. In the tropics, the ITCZ and its accompanying precipitation maximum are shifted northward. In the Southern Hemisphere, the midlatitude westerlies are stronger than in summer (Fig. 1F), and precipitation reaches its seasonal maximum along the southern west coast of South America. Over most of the continent, subsidence prevails (Fig. 1I), and precipitation is consequently high only in the tropics (Fig. 1J).

There are several distinctive asymmetries in the climates of the two hemispheres. The ITCZ-related precipitation maximum is located north of the equator throughout the year, and the seasonal variability of midlatitude precipitation maxima is stronger in the Northern Hemisphere than in the Southern Hemisphere (Figs 1E and 1J). The amplitude of the waves in the westerlies is greater in the Northern Hemisphere than in the Southern Hemisphere (Figs. 1A and 1F), reflecting the influence of the Andes as the single topographic barrier that influences the large-scale wave pattern in the Southern Hemisphere. Other differences in climates are related to the fact that the land area in the temperate latitudes is large in North America, while South America is relatively narrow. Despite these interhemispheric differences in circulation and continental outline, the climates of the temperate ecosystem regions are generally similar, with cool, wet winters and warm, dry summers.

The regions discussed here are bordered to the east by the complex of climates determined by elevation developed along the cordillera and toward the equator by the Mediterranean-climate regions of California and central Chile. Toward the east and toward the equator, the monsoonal-climate regions of both continents are encountered (Kousky and Ropelewski, 1997).

21.2.2. Vegetation

21.2.2.1. Midlatitudes of Western North America

The predominance of conifers in western North American forests arose in the middle and late Tertiary period as a result of long-term global cooling, tectonism, and changes in atmospheric composition and weathering rates (Barron, 1985; Raymo et al., 1988; Ruddiman and Kutzbach, 1989, 1991). In western North America, summer-dry conditions were intensified with regional tectonism and rain shadow develop-

ment in the late Tertiary period (Barnosky, 1987). Precipitation gradients today are latitudinal, longitudinal (reflecting north-south trending mountain ranges), and altitudinal. Winter precipitation from storms that develop in the vicinity of the Aleutian Low is greatest in the Coast Range, Cascade Range, and Sierra Nevada, and levels decrease eastward and southward. The East Pacific subtropical high-pressure system determines the intensity of summer drought along a latitudinal gradient that also shifts seasonally.

Rain forest extends from northern California to British Columbia and as far north as Kodiak Island in Alaska (Alaback and Pojar, 1997). At low latitudes, warm temperate rain forest occurs in a region of mild, wet winters and cool, dry summers. From southern Oregon to central Vancouver Island, the climate is cool to mild, and snow is uncommon at low elevations and abundant at high elevations. Most rainfall occurs in fall and winter, and less than 10% falls in summer. The seasonal rain forest of this region supports *Pseudotsuga*, *Tsuga heterophylla*, *Abies grandis*, and *Thuja plicata*, and deciduous broad-leaved species are confined to riparian settings and areas of disturbance (Alaback and Pojar, 1997). The driest areas of rain forest, including southeast Vancouver Island, the San Juan Islands, southern Puget Lowland, and Willamette Valley, support oak (*Quercus garryana*) and isolated grassland. At high elevations in the south, coastal ranges support *Tsuga mertensiana*, *Chamaecyparis nootkatensis*, and *Abies* spp. The perhumid temperate forest from central Vancouver Island to southeastern Alaska is a diverse landscape of wetlands, forest, and subalpine meadows as a result of wet, cool conditions throughout the year. *Picea sitchensis*, *T. heterophylla*, *A. amabilis*, *Thuja plicata*, and *C. nootkatensis* are the dominant trees, with a rich understory of ferns and ericaceous taxa. At high elevations in the north, *Tsuga heterophylla* and other cold-intolerant species are replaced by *T. mertensiana* and *C. nootkatensis*, and finally by tundra and heath. Fires are frequent in the southern summer-dry ecosystems, but are rare in the more humid zones in the north.

East of the Cascade Range and Coast Range, forests contain more xerophytic elements as a result of greater continentality and summer drought. The interior Coast Range of British Columbia is covered by forests of *Pinus ponderosa*, *P. contorta*, *Pseudotsuga*, *A. grandis*, and *Larix occidentalis*. *Picea engelmannii*, *Pinus albicaulis*, and *A. lasiocarpa* are present at high elevations. The northern Rocky Mountains support forests of *Pinus ponderosa*, *P. contorta*, and *P. flexilis* at low and middle elevations and forests with *P. albicaulis*, *A. lasiocarpa*, and *Picea engelmannii* at high elevations. Where maritime air penetrates inland, coastal species extend eastward to the Continental Divide in northwestern Montana and southwestern British Columbia and occur together with interior, xerophytic species. In the Sierra Nevada of California, *Pinus contorta, P. jeffreyi, Pseudotsuga, A. magnifica, Pinus longaeva,* and *P. flexilis* are the dominant conifers. The lowlands support grassland and oak savanna in mesic settings, and farther east sagebrush steppe is dominated by *Artemisia tridentata,* other composites, and chenopods. Alpine tundra is present on high peaks throughout the West.

21.2.2.2. Midlatitudes of South America

South America's temperate forests are dominated by evergreen, small-leaved hardwoods, while conifers play only a minor role (Kalin Arroyo et al., 1993). Deciduous forests, dominated by different species of *Nothofagus,* are restricted to high altitudes and high latitudes. This southern temperate flora has evolved from Gondwanan stock, which developed when Southern Hemisphere land areas were connected (Markgraf et al., 1995, 1996). During the middle Tertiary period, neotropical elements were added to the flora, and Andean tectonism in the late Tertiary period resulted in migration of new taxa from the Northern Hemisphere primarily to alpine and subalpine elevations (Moore, 1983; Simpson, 1983). While North American herbaceous taxa migrated to the southernmost tip of South America, North American tree taxa, such as *Alnus* and *Juglans,* reached only latitude 27°S. Thus, the southern temperate arboreal flora never received an influx of the northern temperate flora. The high species diversity and endemism that characterize the South American forests reflect their long history of isolation and the relatively small changes in climate during glacial–interglacial oscillations in mountainous regions (Markgraf et al., 1995; Villagrán, 1988; Villagrán and Armesto, 1993; Kalin Arroyo et al., 1993).

Present-day vegetation is distributed along two precipitation gradients associated with latitude and orography (Lawford, 1996). The latitudinal gradient is related to the seasonal shift of the southern westerly storm tracks. Precipitation from latitude 30° to 40°S falls primarily in winter and ranges from 350 mm/year in Santiago to 2300 mm/year in Puerto Montt. South of latitude 40°S, precipitation reaches a maximum of over 5000 mm/year between latitudes 45° and 50°S, where the southern westerlies dominate year-round. The second, west–east precipitation gradient across the Andes is extremely steep, from 2000 mm/year on the Andean divide in, e.g., Futaleufu, to 800 mm/year in Bariloche—a distance of less than 50 km.

Vegetation types west of the Andes from north to south are (Donoso 1993, 1996): (1) sclerophyllous forests and woodlands between latitudes 25° and 30°S with *Lithraea caustica, Quillaja saponaria, Cryptocarpa alba,* and *Peumus boldus;* (2) semideciduous forests between latitude 30° and 40°S with *Nothofagus obliqua, Aextoxicon punctatum, Laurelia* spp., and *Persea lingue;* (3) species-rich, warm-temperate Valdivian rain forest between latitudes 40° and 45°S with *N. dombeyi, L. sempervirens, L. philippiana, Eucryphia cordifolia, Caldcluvia paniculata, Lomatia ferruginea,* and *Weinmannia trichosperma;* and (4) cool-temperate Patagonian and Magellanic rain forests south of latitude 45°S with *N. betuloides,* several genera and species of Myrtaceae, *Fitzroya cupressoides, Podocarpus nubigena,* and *Drimys winteri.* Nonforested vegetation types include the Magellanic moorland in regions of poor drainage, high precipitation, and high winds along the southwest coast of Chile, with *Astelia pumila, Donatia fascicularis,* and other bog taxa restricted to the Southern Hemisphere; Andean high-elevation, cushion-forming vegetation with *Bolax gummifera, Abrotanella emarginata, Drapetes muscosus,* and *Azorella* spp.; and steppe and steppe-scrub east of the Andes, with bunchgrasses and many primarily shrubby species of Asteraceae.

21.2.3. Beetle Fauna

The temperate coastal rain forests bordering the Pacific Ocean in both North and South America produce deep litter and stable habitats rich in microorganisms and invertebrates. Hatch (1971), in the series *The Beetles of the Pacific Northwest,* laid the foundation for faunistic studies in the coastal regions of western North America. Parsons et al. (1991) and Kavanaugh (1992) are among the recent faunistic studies that provide detailed ecological information from specific geographic regions in Oregon and British Columbia. The insect fauna of the coastal forests is diverse and rich in endemic taxa at the subspecies, species, generic, and even tribal and subfamilial levels. Kavanaugh (1988) proposed that species survived the last glaciation in a series of disjunct refugia, the most significant of which lay along the coast of southern Washington, Oregon, and northern California. Separate refugia also may have existed in the Aleutian Islands, Kodiak Island, coastal southeastern Alaska, and the Queen Charlotte Islands, as well as a much larger Beringian refugium in interior Alaska, Yukon Territory, and northern British Columbia.

The taxonomy and ecology of the southern South America beetle fauna are less well known. There are monographic treatments for some taxa (e.g., Jeannel 1962 for the trechine carabids) and biogeographic analyses (e.g., Kuschel 1969 for the Coleoptera of southern South America), but few faunistic and ecological studies are available. Ashworth and Hoganson (1987) and Ashworth et al. (1991) provided analyses of the beetle faunas along montane transects in the Chilean Lake and Chilean Channels, and Niemela (1990) reported on the distribution of Carabidae along a vegetational transect in Tierra del Fuego.

The southern South American fauna is especially distinctive because of its high degree of endemism. Darlington (1965), in his classic the *Biogeography of the Southern End of the World*, showed that ground beetles displayed endemism to the level of tribe. For example, species in the tribe Migadopini do not occur in the Northern Hemisphere and have a typical Gondwanan distribution in southern South America, including the Falkland Islands, Tasmania and southeastern Australia, and New Zealand and the Auckland Islands. A similar distribution is observed in the most abundant of southern South American weevils (subfamilies Cylindrorhininae and Rhytirhininae [Kuschel, 1969] and the families Protocujidae, Chalcodryidae, Nemonychidae, and Belidae). These groups are either monotypic or are represented by a small number of genera that are inhabitants of *Nothofagus* forest. High endemism within the Southern Hemisphere beetles is attributed to their greater isolation in a smaller, more fragmented land area.

Although the southern and northern faunas are distinctive, there are a number of similarities between them. In both faunas, species richness declines polewards, and the faunas west and east of the mountains are very different. The dominant scavengers of the western rain forest are carabids, but in the eastern xeric forests and steppes, they are replaced by tenebrionids. The Patagonian tenebrionids are a particularly distinctive group of beetles with high endemism, especially in the tribes Nycteliini, Scotobiini, and Praocini. The Carabidae serve to illustrate some faunal similarities between the wet forests of the different hemispheres. In both forests, the dominant predators are relatively large, flightless ground beetles of the supertribe Carabitae; they include the brightly colored Ceroglossini in Chile and the snail-eating Cychrini in the Pacific Northwest. Carabids in the tribes Trechini and Broscini are also well represented in these forests compared with other environments. In the tribe Broscini, the ground beetle *Creobius eydouxi* Guerín of Chile is superficially similar to *Zacotus matthewsi* LeConte of the Pacific Northwest; both are moderately sized flightless beetles that have a metallic luster and fused elytra.

Most of the morphologic similarities are at a superficial level and can perhaps be best explained by evolutionary convergence. However, there are also probably ancient evolutionary links between some of the taxa. Erwin (1985) discussed the biogeographic rela-

tionships within the Trachypachidae, a beetle family that is considered ancestral to the Carabidae because it shares characteristics with both the Hydradephaga and the Geadephaga. In forests from British Columbia to California, the family is represented by the genus *Trachypachus*, and in the forests of southern Chile, it is represented by its sister genus *Systolosoma*. Erwin considered these genera to be ancient relicts that have been disjunct in their existing geographic ranges since the early Tertiary period.

Thus, despite the long distances separating the regions and the differences in the particular controls of regional climate, the western midlatitudes in both hemispheres are remarkably similar in terms of their seasonal variations of climate, the structure of the vegetation, and their general biogeographic characteristics. As we discuss in Section 21.3, this interhemispheric similarity is also a feature of the late Quaternary climatic histories of the two regions.

21.3. CONTROLS AND RESPONSES OF HEMISPHERIC AND REGIONAL CLIMATIC VARIATIONS SINCE THE LAST GLACIAL MAXIMUM

The principal controls of global, hemispheric, and regional climatic variations underwent substantial changes between the LGM and the present (Wright et al., 1993; Webb and Kutzbach, 1998). Over this interval and at these particular spatial scales, the controls include:

- The size and topography of the ice sheets, which produce substantial hemisphere-wide cooling resulting from the high albedo and elevation of the ice sheets, as well as extensive reconfiguration of atmospheric circulation in regions adjacent to the ice sheets.
- The seasonal cycle of insolation, which is reflected by the seasonal cycles of net radiation, air temperature, and evaporative demand.
- The composition of the atmosphere, including the relative abundances of water vapor and trace gases that control the strength of the greenhouse effect, dust, and mineral aerosols.
- Attendant changes in ocean circulation, sea surface temperatures (SSTs), and land-surface cover, which generally amplify the effects of the external controls.

Over longer timescales, the size of the ice sheets is a dependent variable in the climate system, but its changes over the past 21,000 years were gradual enough to regard it as one of the external controls of the regional climatic variations that are the focus here.

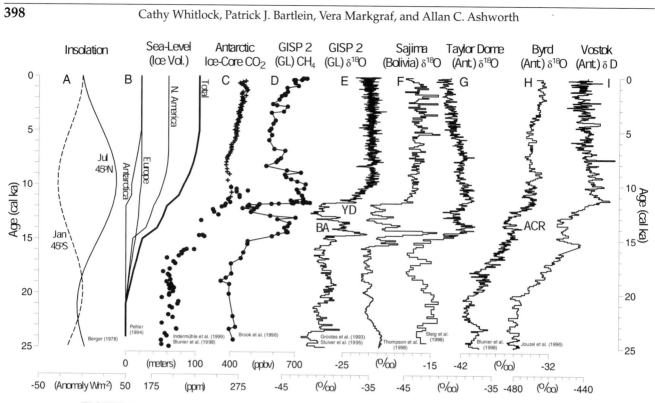

FIGURE 2 Variations over the past 25 ka of (A) insolation (Berger, 1978), (B) ice volume (Peltier, 1994), (C) carbon dioxoide, CO$_2$ (Indermühle et al., 1998; Blunier et al., 1998), (D) methane, CH$_4$ (Brook et al., 1996), along with ice core oxygen isotopic records from (E) Greenland (Grootes et al., 1993; Stuiver et al., 1995), (F) Bolivia (Thompson et al., 1998), and (G–I) Antarctica (Steig et al., 1998; Blunier et al., 1998; Jouzel et al., 1996).

Variations in the composition of the atmosphere are hemispherically symmetrical at the timescales considered here, but, for other controls, hemispheric asymmetries exist in the amplitude (in the case of the ice sheet effects) and sign (in the case of insolation) of their variation (Fig. 2). These patterns, at first glance, present some problems for inferring the causes of the regional environmental responses summarized later. For example, the anomalies of the seasonal cycles of insolation are out of phase between the hemispheres (Fig. 2A). At the LGM, as at present, perihelion occurred during the northern winter, and the tilt of the Earth's axis was relatively low then as now; while at about 10 ka, perihelion occurred in the northern summer, and the tilt was greater than it is now. These orbital variations resulted in summer (July) insolation in the Northern Hemisphere that was greater in the early Holocene than it is now, whereas in the Southern Hemisphere, summer (January) insolation was less than it is now. The volume of ice (expressed in Fig. 2B as the equivalent change in sea level; after Peltier, 1994) was also hemispherically asymmetric, as was its change through time. Most of the ice was in the Northern Hemisphere, and its reduction in size occurred in two major steps

between 15 and 10 ka. In contrast, the smaller ice volume changes of the Southern Hemisphere occurred slowly over this interval.

The greenhouse gases (carbon dioxide and methane, Figs. 2C and 2D) are rapidly dispersed through the atmosphere and have hemispherically symmetric distributions. However, the impact of variations in the concentrations of greenhouse gases is somewhat greater in the north than in the south, owing to the greater land area and, hence, larger impact of the snow/ice-albedo feedback in the north. (This response is also noted in global warming simulations.) Consequently, in general, the major controls have hemispherically asymmetric effects, and in the case of insolation, the effects are opposite.

The opposition of summer insolation in the two hemispheres (Fig. 2A) directs attention toward the potential impact that variations in insolation may have on the seasonal cycle of climate in both hemispheres (Kutzbach and Gallimore, 1988). In particular, the thermal inertia of the mixed layer of the ocean apparently allows the ocean to *carry over* the insolation anomaly from one season into the next. In the Northern Hemisphere for the early Holocene, this process results in the simulation of autumn (September–October–November [SON]) temperatures that are higher than they are now and, consequently, the delayed formation of sea ice (relative to the present) (see also Mitchell et al., 1988), whereas in the Southern Hemisphere the result is a warmer-than-present spring (again, SON) and an earlier seasonal reduction of sea ice extent and thickness.

Two sources of information (independent of the paleoecological records themselves) shed light on the source of the apparent synchroneity of the terrestrial biotic response: ice core records and paleoclimatic simulations with general circulation models (GCMs). (We do not consider in detail a potential third source of information, networks of paleoceanographic records, because at present they are too sparse to shed much light on potential causes of apparent synchroneity; but see Mix et al. [1999] for a recent discussion of potential interhemispheric linkages in paleoceanographic conditions.)

Ice cores provide records of atmospheric composition (including greenhouse gases, dust, and aerosols) that have very high temporal resolution. They also are paleoclimatic records, but the spatial representativeness of the paleoclimatic record is incompletely known and is probably less than hemispheric. Thus, ice core records provide a general indication of climatic variations in the two hemispheres. Paleoclimatic simulations provide spatially explicit depictions (within the limitations of the coarse resolution of the models used) of the responses to changes in boundary conditions set to represent different times. However, the interpretation of these simulations is constrained by (1) their poor or absent representation of important elements of the climate system (e.g., ocean circulation in the case of the simulations considered here), (2) incomplete experimental designs (e.g., specifying present-day vegetation distributions in paleoclimatic simulations, implicitly assuming that there is no potential feedback to climate from changes in vegetation distributions), and (3) fundamental inadequacies in the models. Sequences of simulations can reveal temporal trends in the likely responses to the changing large-scale controls, even though a specific simulation may not accurately portray the response to a set of changes in boundary conditions from their present values. Used together, the ice core results and model simulations help formulate hypotheses about the sources of the synchronous paleoecological responses between the hemispheres.

21.3.1. Ice Core Records

Ice core records show two end-member sequences of variations: a "northern" or "Greenland" sequence as illustrated by the $\delta^{18}O$ data from the Greenland ice core (GISP2) record (Fig. 2E; Grootes et al., 1993; Stuiver et al., 1995) and a southern sequence as illustrated by the $\delta^{18}O$ data from the Byrd record (Fig. 2H; Blunier et al., 1998) or the δD data from the Vostok core (Fig. 2I; Jouzel et al., 1996; Petit et al., 1999). The northern sequence is characterized by little temperature change from the LGM to 15 cal ka, after which a sudden warming occurs, referred to the Bølling/Allerød period in Europe (BA on Fig. 2E). This period is followed by a cool interval (the Younger Dryas climate reversal), abrupt warming again after 12 cal ka, and then gradual warming into the Holocene. In contrast, the southern sequence shows warming commencing just after the LGM (21 cal ka) interrupted by the "Antarctic Cold Reversal" after 15 cal ka (about the time that the northern record shows abrupt warming). After 13 cal ka, warming resumes. (See Alley and Clark [1999] for further discussion of north-south contrasts in paleoclimatic records.)

Were it only for the ice core records, it might be possible to argue that the similarity in the paleoclimatic records from the two hemispheres is related to some intrinsic mechanism in the climate system—probably ocean heat transport—that allowed the Southern Hemisphere to warm after the LGM despite the decrease in summer insolation. However, the ice core record is not that simple. Two other records provide a perspective on the spatial variability of the climate recorded by ice cores. The Taylor Dome $\delta^{18}O$ data from Antarctica (Fig. 2G; Steig et al., 1998) show variations that resemble the northern sequence more than the southern one, as does the tropical montane Sajama (Bolivia) record (Fig. 2F; Thompson et al., 1998). These variations suggest that there may be considerable heterogeneity at the global scale in the climate recorded by ice cores. The variations seen in the ice core records also imply an active link between the ocean and atmosphere and require a mechanism for the transmission of millennial timescale changes around the globe (Clark et al., 1999).

In addition to their low spatial specificity, ice core records also do not provide direct information on the changes in seasonality that might be expected from variations in the seasonal cycle of insolation. In any case, the ice core records generally indicate cold conditions at the LGM, and most show that nearly all of the late glacial warming was completed by 11 cal ka. The exception (in Fig. 2) is the GISP record from Greenland, where warming continues until 10 cal ka, probably due to the lingering effects of remnant ice sheets over North America.

21.3.2. Paleoclimatic Simulations

The paleoclimatic simulations we use here are the set (21, 16, 14, 11, and 6 cal ka and present) created with the

National Center for Atmospheric Research, Community Climate Model version 1 (NCAR CCM1), as described by Kutzbach et al. (1998) and used in Bartlein et al. (1998). This model has relatively coarse resolution and lacks interactive coupling with either the ocean or biosphere, which makes it obsolete by today's standards. Nonetheless, these paleoclimatic simulations remain the only complete sequence for the past 21,000 years that was performed using a single model and a coherent experimental design. Although CCM1 includes a mixed-layer ocean, which allows for simulation of the exchange of heat between the ocean and atmosphere, the absence of an interactive ocean component in the model makes the implicit assumption that ocean circulation changes were not important over the past 21,000 years. This assumption seems unrealistic from the perspective of marine and ice core records, and hence, this sequence of simulations should not be expected to include features that reflect the reorganization of oceanic and atmospheric circulation.

Global and hemispheric average results over the sequence of simulations for some key variables are summarized in Kutzbach et al. (1998). Their analysis suggests that warming in the Southern Hemisphere, between 21 and 11 cal ka, occurs earlier and in a more linear fashion than that in the Northern Hemisphere, where warming is concentrated between 16 and 11 cal ka (Kutzbach et al., 1998). Nevertheless, these differences, while probably not statistically significant, are consistent with ice core records that indicate an earlier start to postglacial warming in the Southern Hemisphere than in the Northern Hemisphere. The fact that both model simulations and ice cores yield the same picture implies that part of the hemispheric asymmetry in warming did *not* involve a reorganization of ocean circulation.

The variation through time of the spatial and seasonal variations of climate produced by the changes in large-scale controls can be evaluated by using the climate model simulations. Figs. 3–8 (see color inserts for these figures) focus on the seasonal cycles of several climate variables and their changes over time. They employ what Tufte (1997) calls "small multiples"–the repeated plotting of small objects (in this case, maps of the Western Hemisphere) with the aim of showing broad-scale patterns (both temporal and spatial in this case). Figures 3–8 are composed of monthly maps of a particular climate variable, plotted with the month of year running down the figure and time period of the simulation running across the figure. The simulated values are plotted on the left-hand side of each illustration, while anomaly values (*paleo* experiments minus the modern *control* simulation) are plotted on the right.

In examining these figures, the important features are those that show patterns across the rows or columns of the small panels, although individual panels can be inspected for details if necessary. Four kinds of patterns may emerge: (1) horizontally oriented patterns that denote changes across millennia but not in the seasonal cycle of a variable; (2) diagonally oriented patterns that mark changes across millennia, and in the seasonal cycle of a variable; (3) features on individual panels that represent local departures from zonal averages; and (4) chaotic patterns that suggest the absence of systematic temporal, spatial, or seasonal trends. Visible in the background of Figs. 3–8 is one of the major controls, the size of the ice sheet. The effects of changing ice sheet size will be most evident as horizontal, as opposed to diagonal, changes in map patterns and as local (over the ice sheets) anomalies that depart from zonally averaged ones.

Figure 3 shows monthly values of insolation (at the top of the atmosphere) and its variations over time (note that the insolation values are longitudinally identical). The panels in the column labeled *Present* show the present-day seasonal variations of insolation and, indirectly, the impact of current values of precession (and the time of year of perihelion) and the tilt of the Earth's axis. The highest values of insolation at present occur in the Southern Hemisphere in December, reflecting the occurrence of perihelion in early January now. In contrast, higher values than these occurred in June and July at 11 cal ka, reflecting the occurrence of perihelion during the northern summer then, as well as the greater tilt than at present. The anomalies show a striking diagonal pattern reflecting the shift in the season of perihelion from the northern spring/southern autumn at 21 cal ka, to the northern summer/southern winter at 11 cal ka, to the northern winter/southern summer at present. At 21 cal ka, insolation had nearly its present distribution through the year, but by 16 cal ka, values during the northern spring (May) at 45°N were 6% greater than they are now. By 6 cal ka, the maximum (positive) insolation anomaly shifted into the later part of the northern summer/southern winter. Maximum negative anomalies show a corresponding shift in the time of occurrence during the year, with the largest anomalies occurring during the northern winter/southern summer at 11 cal ka, when aphelion occurred in December.

Net radiation (Fig. 4) and surface air temperature (Fig. 5) illustrate the effect of the changing seasonal cycle of insolation. The anomalies of net radiation (on the right-hand side of Fig. 4) clearly show the influence of the external insolation forcing, but reveal the modification of this basic pattern by the ice sheets and the mixed-layer ocean in the model. The impact of the Laurentide ice sheet is clearly evident in the negative net radiation anomalies during the Northern Hemisphere summer, and these anomalies remain large so long as the ice sheet is present. The high albedo and elevation of the ice sheets combine to reduce short-wave radiation absorbed at the surface as well as the downward flux of long-wave radiation. Overall, however, the anomalies of net radiation strongly resemble those of insolation.

In contrast, simulated surface air temperature and its anomalies, while expressing some of the characteristic diagonal pattern of insolation and net radiation, mainly show a generally horizontal pattern, reflecting the diminishing influence of the ice sheets and associated *glacial* boundary conditions (e.g., carbon dioxide). The temperature anomalies show gradual increases from 21 to 14 cal ka, with the slightly greater zonally averaged increase in the Southern Hemisphere, noted by Kutzbach et al. (1998), and the slightly greater increase in Northern Hemisphere summer and autumn (southern winter and spring). Locally positive temperature anomalies adjacent to the ice sheet in North America during this interval are related to atmospheric circulation around the Laurentide ice sheet at 21 and 16 cal ka and to the increasing summer insolation at 16 and 14 cal ka (Bartlein et al., 1998). Other isolated anomalies are related to switches of individual grid cells from land at the LGM to water at present.

The simulated temperature anomalies at 11 cal ka show a sharp break from earlier ones. The negative anomalies are generally smaller, and positive temperature anomalies prevail in the interior of both continents in August, despite the continuing presence of an ice sheet in North America and the generally colder-than-present oceans in the Southern Hemisphere. The temperature anomalies at 11 and 6 cal ka illustrate the role of the mixed-layer ocean in *lagging* the net radiation anomaly maxima into subsequent months and seasons. At 6 cal ka, for example, positive temperature anomalies appear in the high latitudes of the Northern Hemisphere despite the occurrence of weakly positive or negative insolation and net radiation anomalies. Overall, the anomalies of temperature reveal the strong influence of the diminishing ice sheets and related large-scale controls and a decreased influence of insolation, in which shifts in the timing of the temperature maxima relative to the insolation maxima result from internal adjustments of the simulated climate.

The patterns of the response of upper-level wind speed (Fig. 6) and sea-level pressure (SLP) (Fig. 7) also show the combination of ice sheet, insolation, and internal controls, with horizontal, diagonal, and local map patterns evident to greater or lesser extents in these atmospheric circulation variables. The anomalies of 500-mbar wind speeds (Fig. 6) are characterized by alternating positive and negative centers that are elongated in an east-to-west direction. The alternation in sign reflects meridional or latitudinal shifts of the zones of fastest winds. For example, the southward shift in the latitude of the Northern Hemisphere polar jet stream, which is a robust feature of paleoclimatic simulations at the LGM, is indicated by the band of positive (blue) wind speed anomalies located equatorward of a band of negative (purple) wind speed anomalies in the northern midlatitudes at 21 cal ka. Similar latitudinal displacements can be noted for the southern westerlies. At 21 and 16 cal ka, positive anomalies predominate, reflecting the steeper latitudinal temperature (and hence pressure) gradient evident in Fig. 5. The wind speed anomalies gradually weaken over the series of simulations and remain strongest along the margin of the northern ice sheet.

In contrast to the primarily horizontal gradation of pattern in the upper level wind speeds, the SLP anomalies (Fig. 7) show a mix of all three patterns. The glacial anticyclone over the Laurentide ice sheet, another robust feature of paleoclimatic simulations, is clearly evident in the simulations for 21 and 16 cal ka and in the summer simulation at 11 cal ka. Its expression at 14 cal ka is somewhat transitional in nature. At 11 and 6 cal ka, lower pressure than at present prevails over North America from June through September (JJAS), while higher pressure than present prevails over adjacent oceans, leading to the simulation of an amplified northern (summer) monsoon. During the southern summer (DJF), weak positive SLP anomalies prevail over South America and lead to less onshore flow than at present. When the 500-mbar wind speed and SLP anomalies are compared, the former gives the impression of being governed mainly by the diminution of the glacial controls (ice sheets and SSTs), while the latter show the additional influence of insolation, net radiation, and temperature changes.

The simulated precipitation anomalies are much noisier than the other variables, but still reveal some general patterns (Fig. 8). The noisiness results from the frequent alternation of large positive and large negative anomalies caused by simple latitudinal shifts in the zones of high precipitation. For example, the generally southward shift in the ITCZ at 21 and 16 cal ka, particularly in the Northern Hemisphere summer, shows up as horizontal bands of contrasting anomalies near the equator—positive (blue) to the south and negative (orange) to the north. Despite the noise, a striking overall trend in the precipitation anomalies is the gradual replacement of negative anomalies with positive ones over the sequence of experiments. Simulated precipitation over the Western Hemisphere is generally lower than at present during glacial times (with locally greater-than-present anomalies related to changes in atmospheric circulation patterns). As hemispheric average temperatures increased over time, so did monthly and annual precipitation.

The potential impact on the regional paleoclimates recorded by the paleoecological data can be further evaluated by transforming the variables produced by the climate model (e.g., temperature and precipitation) into bioclimatic variables (Prentice et al. 1992) that have a direct mechanistic influence on vegetation and, indirectly, on beetles. Growing degree-days provide an index of the energy available for growth and include the effects of the length and intensity of the growing season. The mean temperature of the coldest month provides an index of potential limitations by frost or freezing. The ratio of actual to potential evapotranspiration (AE/PE), commonly called the moisture index, provides a gauge of effective moisture and drought stress upon plants. In humid regions, AE/PE is at or near 1.0, indicating that evapotranspiration is not limited by moisture. In drier regions, moisture limits AE, and the ratio is less than one. The temporal trends in growing degree-days and their anomalies reflect the temporal trends in (summer) insolation, net radiation, and temperature.

Simulated growing degree-days are lower than at present in both hemispheres at 21, 16, and (except in regions distant from the ice sheet in the Northern Hemisphere) 14 cal ka (Fig. 9 [see color insert]) (see Bartlein et al. 1998 for an explanation of positive summer temperature anomalies in Beringia at 14 cal ka). In the Northern Hemisphere, growing degree-days exceed present values at 11 and 6 cal ka, but remain less than at present in the Southern Hemisphere. The Southern Hemisphere trends appear to be a straightforward response to the summer insolation variations (Figs. 2–4). In contrast, in the Northern Hemisphere, the temporal trends reflect the trade-off between the positive summer insolation anomaly and the direct thermal effects of the remnant ice sheets. Once the influence of the ice sheets on circulation (Figs. 6 and 7) and temperature (Fig. 5) is diminished, as at 11 cal ka, growing degree-days increase to greater-than-present values.

Mean temperature of the coldest month (Fig. 9) generally remains lower than at present throughout the sequence of simulations in the Northern Hemisphere and likely reflects the effects of lower-than-present winter insolation in the Northern Hemisphere. In northern North America at 6 cal ka, however, it exceeds present values as a result of the *carryover* of the (positive) summer insolation anomaly into subsequent seasons (see Fig. 5). In the Southern Hemisphere, where winter (JJA) insolation anomalies were positive, the

anomalies of mean temperature of the coldest month are generally negative, again reflecting the *carryover* of the preceding season's insolation, net radiation, and temperature anomalies. Taken together, the temporal variations in growing degree-days and mean temperature of the coldest month indicate that shorter-than-present growing seasons, with cooler-than-present winter conditions, were a feature of both hemispheres until after 14 cal ka. During the early and middle Holocene (11 and 6 cal ka), growing degree-days were greater than present values in the Northern Hemisphere, which reflects the large positive anomalies in summer insolation, net radiation, and temperature despite the shrinking or absent Laurentide ice sheet. In the Southern Hemisphere, these variables show a steady march toward the highest values over the past 21,000 years observed at present. Two factors—the interactions between controls (e.g., positive summer insolation anomalies in the Northern Hemisphere at a time when the Laurentide ice sheet was still extensive) and the phase lag in the temperature response to the insolation forcing introduced by the mixed layer of the ocean—apparently kept the biologically important temperature variables below their present values until 11 cal ka or after. This lag occurred in both hemispheres despite early warming, especially in the Southern Hemisphere.

The ratio of AE/PE integrates effective moisture variations over the year. The temporal changes in this variable reflect the generally drier-than-present conditions that prevailed during the glacial period, particularly in the tropics (see also Farrera et al. 1999). Superimposed on increased global aridity are regional variations created by circulation anomalies, including the glacial anticyclone and colder-than-present conditions around the Laurentide ice sheet (see also Bartlein et al. [1998] for further discussion of the regional anomalies of surface energy and water balance components). Both evapotranspiration variables, AE and PE, attain seasonal maxima during the growing season in summer-wet climate regions. Positive AE/PE anomalies in the summer-wet regions of northern South America throughout the simulations are consistent with the generally lower-than-present growing degree-day anomalies that are observed there (because a cooler atmosphere, reduced vapor-pressure gradient, and less energy available for evapotranspiration all tend to reduce potential evapotranspiration). In contrast, in the southernmost and winter-wet regions of South America, the seasonal cycles of the evapotranspiration variables are out of phase, as they are in northwestern North America. The persistent negative anomalies simulated in winter-wet regions are consistent with the same mechanism—warmer winters than at present would be expected to result in greater-than-present potential evapotranspiration.

In summary, the variations in the anomalies of insolation are out of phase between the hemispheres, and, at first glance, they might be expected to yield opposing or asynchronous regional climatic responses. However, modulation of the seasonal cycles of insolation-related variables by the ocean, the effects of the remnant ice sheets on adjacent regions in the Northern Hemisphere, and interactions among moisture balance variables jointly provide several mechanisms that could synchronize the paleoenvironmental records of the western midlatitude regions of North and South America. After reviewing the paleoenvironmental records of the two regions, we will consider some potential explanations for the common features of those records.

21.3.2. Last Glacial Maximum

21.3.2.1. Paleoclimate Conditions During the Last Glacial Maximum

The potential regional influences of the large-scale controls, as expressed by the climate model simulations, can be summarized by examining simulated upper level and surface winds, SLP, and temperature and precipitation for January and July (Fig. 10 [see color insert]), as well as the simulated bioclimatic variables (Fig. 11 [see color insert]). These simulations should be viewed as illustrations of the potential regional responses to particular combinations of controls and not as reconstructions of the climate.

The CCM1 simulation for 21 cal ka indicates that the Laurentide ice sheet and the related glacial boundary conditions had several general consequences:

- Global cooling (relative to at present), with large regions experiencing conditions 4°–8°C colder than at present.
- Somewhat greater cooling over the continents as opposed to the oceans, at mid- and high latitudes relative to the tropics and in the northern summer / southern winter relative to the northern winter / southern summer.
- A large impact on the upper level winds in both hemispheres, with the southward-shifted jet stream over North America particularly evident in January.
- The development of a glacial anticyclone over the Laurentide ice sheet, and deeper (relative to at present) Aleutian and Icelandic low-pressure systems.

More specifically for western North America (Thompson et al., 1993; Bartlein et al., 1998), the simulations include:

- A steepened latitudinal temperature gradient (relative to at present), with temperatures that were 10°–15°C lower than they are today south of the ice sheet in the northern Rocky Mountains, 5°–10°C lower in the Pacific Northwest and Great Basin, and as little as 5°C lower than they are today in the southwestern United States.
- The southward shift of the North American jet stream to latitude 35°N; such a shift would have robbed the Pacific Northwest of its source of winter precipitation and brought pluvial conditions to the semiarid southwestern United States.
- The development of anticyclonic circulation over the Laurentide ice sheet, which further diminished westerly flow in the Pacific Northwest, bringing cold, dry conditions to the region.
- Large negative anomalies in growing degree-days and the mean temperature of the coldest month.

For South America, the simulations show:

- Less extreme zonally averaged cooling than in the Northern Hemisphere.
- A general southward shift, and latitudinal contraction, of the band of fastest upper level flow, evident in the alternating bands of east-to-west trending wind speed anomalies in Figs. 6 and 10 (with this pattern being somewhat better expressed during the month of June on Fig. 6).
- Generally drier-than-present conditions in the tropics.
- A very slight, and probably not statistically significant, amplification of the southern monsoon, with lower-than-present SLPs over South America in summer (January) and higher-than-present SLPs in winter (July).
- Large negative anomalies in growing degree-days and the mean temperature of the coldest month.

In the model output, the precipitation variations are not particularly strong, and in some cases, the precipitation and AE/PE anomalies may seem counterintuitive compared with the simulated changes in circulation. As is discussed by Bartlein et al. (1998), the low resolution of the model, combined with its crude depiction of topography, tends to move precipitation anomalies from the locations at which they may be expected to be found in the real world and also to limit the response of precipitation. The smoothed depiction of the individual ranges of the Cordillera means that the large precipitation anomalies that could be expected to occur "in the real world" with the replacement of prevailing westerlies with prevailing easterlies is muted or absent and ultimately results in the simulation of positive AE/PE anomalies. Similarly, in South America, the latitudinal shifts and focusing of the westerlies produce only modest simulated precipitation responses, owing to the small number of grid points that represent the region under consideration here and the very smooth topography.

The apparent southward shift in the westerlies in the South American sector in the CCM 1 simulations is consistent with those in a LGM simulation with the higher-resolution UGAMP GCM by Wyrwoll et al. (2000), but not with a latitudinal widening of the zonal flow noted in the latter simulations for the seasonal extremes of DJF and JJA. Fig. 6 illustrates that there is considerable month-to-month variation in the details of the circulation patterns, and it probably is the case that the similarities or differences between the models should not be overemphasized, given the different resolutions of the models and averaging intervals used in presenting the results (i.e., seasonal as opposed to monthly).

The LGM simulation portrays a paleoclimate that in both hemispheres features colder and likely drier (in the real world) conditions than at present. In both hemispheres, the circulation is also quite different. Examination of the full seasonal cycle of the upper-level wind speeds (Fig. 6) and wind directions (Fig. 10) suggests that the very cold conditions year-round over the Laurentide ice sheet act to put the circulation into a year-round "northern winter" mode, displacing all of the pressure systems toward the south. In the Southern Hemisphere, this shift is resisted by the steep temperature gradient along the northern margin of the southern sea ice.

Absent from this discussion is the potentially large role that accompanying reorganizations of the ocean circulation, and consequently heat transport from the tropics to higher latitudes, may have played in governing the regional climate changes. Two observations mitigate this omission. First, as was discussed earlier, at least one of the lines of evidence that suggests a role for ocean heat transport changes, i.e., the earlier deglacial response in the Southern Hemisphere, is apparent in a model without a circulating ocean. Second, climate variations in the North Atlantic region transmitted only through the atmosphere (Hostetler et al., 1999) can apparently produce large changes in remote regions of the kind that are also ascribed to changes in ocean heat transporting mechanisms.

21.3.3. Environmental Response in North America

21.3.3.1. *Vegetation History*

Records of full-glacial vegetation in western North America are available only from south of the cordilleran ice margin. The vegetation patterns are consistent with increasing cold and dry conditions going from the coast to the interior in the Pacific Northwest and with a shift in moisture from the Pacific Northwest to the

American Southwest (Fig. 12 [see color insert]). The Coast Ranges of British Columbia, Washington, and Oregon featured a mixture of tundra and subalpine vegetation during the full-glacial period. *Picea, Tsuga mertensiana,* and *Pinus contorta* formed communities that resembled high-elevation forests in the Coast Range and Cascade Range (Heusser, 1977, 1983; Worona and Whitlock, 1995; Hebda and Whitlock, 1997). Coastal areas close to glacier margins supported alpine tundra with *T. mertensiana* and *P. contorta* (Heusser, 1983). The Puget Lowland, located between the Coast Range and Cascade Range, was covered by subalpine parkland and tundra with species such as *Picea engelmannii* that today are found at high elevations farther east in the Cascade Range and northern Rocky Mountains (Barnosky, 1981, 1985; Worona and Whitlock, 1995). East of the Cascade Range, lowlands were covered by cold steppe vegetation and subalpine parkland (Whitlock and Bartlein, 1997). Thus, the present west-to-east moisture gradient existed during the LGM, but, unlike today, little moisture reached interior lowlands east of the Coast Range. Communities with steppe and xerophytic elements grew within 100 km of the coast on the west side of the Cascade Range. Weakening of the westerlies is also suggested by the occurrence of diatoms originating from Tertiary deposits in eastern Washington in areas west of the Cascade Range (Barnosky, 1985) and in marine deposits off the coast of Washington (Sanchetta et al., 1992). These diatoms are presumed to have been transported by strong easterly surface winds from a cold, sparsely vegetated interior.

In the American Southwest, glacial-age packrat middens from the Great Basin and Mojave Deserts record a downslope expansion of woodland and montane conifers (e.g., *Pinus longaeva, P. flexilis,* and *Abies*) into areas that now support desert and steppe shrubland. The vegetation associations often have no modern analog because they contain mixtures of high- and low-elevation species, but these communities accord well with cold, humid conditions in full-glacial time (Barnosky et al., 1987; Thompson et al., 1993; Spaulding, 1990; Thompson, 1990). The vegetation and lake-level high stands at this time are consistent with more winter precipitation in the Southwest due to the southward shift of the jet stream (e.g., Benson and Thompson, 1987).

21.3.3.2. *Fossil Beetle Data*

Relatively few fossil beetle assemblages have been examined from western North America. In the far northwest, Elias et al. (1996) report fossils from submerged peat beds in the Bering and Chukchi Seas, indicating that a terrestrial environment existed in the Bering Strait until about 13 cal ka. Full-glacial beetles dating from 23.7 cal ka are those of dry tundra heaths and meadows. From 16.8 cal ka until the time of submergence, the beetles are those of more mesic habitats. Mean summer temperature at the time of submergence, based on a mutual climatic range (MCR of Atkinson et al., 1987), is estimated to have been warmer than it is today, whereas mean winter temperature is estimated to have been colder. In southwestern Alaska, the beetle fauna of the last glaciation was depauperate, consisting of a few tundra species (Elias, 1992a). Tundra species persisted through the late glacial and Holocene, but by about 14.7 cal ka, the climate had warmed sufficiently for taxa of open boreal habitats to have colonized the region.

No fossil beetle assemblages have been studied from the islands or the coast of British Columbia. In the Pacific Northwest, only a few beetle assemblages have been examined from deposits of the last glaciation. Nelson and Coope (1982) identified several species from the Kitsap Formation in Seattle, dated at 19.8 cal ka. The species were typical of an open woodland and not the closed canopy forest that exists in the region today. Nelson and Coope (1982) interpreted mean summer temperature to have been similar to that of today. They attributed the open environment to a climate with drier rather than cooler summers. A small fossil assemblage from peat beds near the top of a sea cliff at Kalaloch, dated at between 19.9 and 21.4 cal ka, also indicated an open landscape on the Olympic Peninsula during the last glaciation. However, the occurrence of a species of a chrysomelid beetle from the alpine zone in the Olympic Mountains, together with a few boreal forest species, suggests that the mean July temperature was lower than that of today (Cong and Ashworth, 1996).

In the Rocky Mountains to the east, ice caps were present during the last glaciation, and the alpine fauna was displaced to lower elevations. At Marias Pass in Montana, a few alpine species were reported from an elevation 450 m lower than now at 13.4 cal ka (Elias, 1987). A paleoclimatic interpretation based on 10 fossil beetle assemblages from Utah to Montana, and from elevations ranging from 1548–3325 m above sea level, was summarized in an MCR analysis (Elias, 1996). Mean July temperature at 17.4 cal ka was estimated to have been lower than it is now. Temperatures then increased rapidly so that by 11.4 cal ka mean July temperature was warmer than at present. This interpretation is based on modern comparisons from a large region and a wide range of elevations and needs to be tested further with more robust data.

In the deserts of the southwestern United States, fossil beetle assemblages from the last glaciation have been examined from packrat middens in the Chihuahuan Desert (Elias and Van Devender, 1990; Elias,

1992b), the Sonoran Desert (Hall et al., 1988, 1989), and the Colorado Plateau (Elias et al., 1992). The results of these studies are difficult to interpret in terms of climate change. There is some evidence that beetle species, like plant species, moved to lower elevations during the last glaciation in both the Sonoran Desert (Hall et al., 1989) and the Colorado Plateau (Elias et al., 1992). The distributions of beetle species in this region, however, are not as well known as those of plants, and, consequently, their response to climate change is more difficult to assess. In the Chihuahuan Desert, fossils of beetle species from grasslands indicate that cooler and wetter conditions existed until about 13.95 cal ka (Elias and Van Devender, 1990). In the lowlands of the Sonoran Desert, however, the packrat midden beetle fauna is very similar to that of today, and the conservative response suggests only minimal response to climatic change (Hall et al., 1988).

The response of the western North American beetle fauna to the climate of the LGM was greater toward the interior of the continent, but nowhere as pronounced as at the margins of the Laurentide ice sheet to the north and east. For the LGM, climate reconstructions based on beetle assemblages indicate that mean July temperatures were warmer on the Pacific coast than in the interior of the continent. In eastern North America, including the continental interior, arctic beetle species replaced boreal forest species along the southern margin of the Laurentide ice sheet at about 25.5 cal ka. Colonization was probably from populations that dispersed southward in front of the growing Laurentide ice sheet and westward and eastward from montane refugia in the Appalachian and Rocky Mountains, respectively. This glacial fauna, dominated by tundra species, inhabited the ice margin until about 17.4 cal ka (Morgan, 1987; Schwert, 1992; Ashworth, 1996). Mean July temperature along the ice margin is estimated to have been between 10° and 12°C lower than it is now. Warming, in combination with the ice sheet acting as a barrier to northward dispersal, caused the extirpation of tundra species in the midlatitudes at about 17.4 cal ka (Schwert and Ashworth, 1988). The ice marginal fauna until 14.7 cal ka was characteristic of open terrain but warmer climatic conditions than had existed earlier. By 14.7 cal ka, this fauna was replaced by a boreal forest fauna.

21.3.4. Environmental Response in South America

21.3.4.1. Vegetation History

Andean glaciers at mid- and high latitudes advanced several times between 30 and 14 cal ka, and the maximum advance occurred before 20 cal ka (Clapperton, 1992; Lowell et al., 1995). The biota showed little change from 30 to 14 cal ka (Markgraf, 1991; Villagrán and Armesto, 1993; Ashworth and Hoganson, 1993; Heusser, 1984). Last glacial maximum records from the midlatitude region indicate widespread treeless vegetation east and west of the Andes, except between latitudes 36° and 43°S west of the Andes, where *Nothofagus* woodland was present (Fig. 3). The presence of conifers, such as *Prumnopitys andina* at lower elevations, suggests cooler temperatures (Heusser, 1990). The same climate is suggested by the presence of Magellanic moorland taxa in glacial times in lowland sites that today are *Sphagnum* bogs. These characteristic moorland taxa (*Astelia pumila, Gaimardia australis, Donatia fascicularis,* etc.) grow today at these midlatitudes only in bogs above the tree line in the coastal cordillera and on Isla Chiloe (Villagrán 1990; Ruthsatz and Villagrán 1990). East of the Andes, treeless vegetation was widespread (Fig. 13 [see color insert]). At latitudes south of 50°S, *Empetrum* heath was the dominant vegetation, whereas north of 50°S, Asteraceae-rich shrub-grasslands prevailed (Markgraf 1993a).

The evidence of *Nothofagus* woodland over 7° of latitude as opposed to rain forests over 20° of latitude as occurs today suggests that westerly storm tracks were greatly reduced and focused year-round on a relatively narrow zone. The zone lay north of its present summer location and south of its winter location (Markgraf et al., 1992), which is not inconsistent with the kinds of circulation changes implied by the simulations in Fig. 6. A possible explanation for lower-than-present precipitation in this region dominated by the westerlies is the observation that ocean temperatures in the South Pacific were 2°–4°C lower than they are today.

21.3.4.2. Fossil Beetle Data

During the last glaciation in southern Chile, the North and South Patagonian ice caps coalesced with the montane ice caps in the Cordillera de los Andes to the north to form a Patagonian ice sheet. Reconstructions of this ice sheet show it covering the islands of the southern South American archipelago with the western margin extending well out into the Pacific Ocean (Hollin and Schilling, 1981). The occurrence of several species of flightless beetles in the basal layer of a peat bog at Puerto Edén, Isla Wellington, poses a problem to this interpretation (Ashworth et al., 1991). The peat bog is in a location that is shown in the reconstruction to have been covered by the ice sheet. The position of the bog in a low area between two moraines suggests that it would have started to grow as soon as the area was deglaciated. The occurrence of the flightless carabid beetles *Ceroglossus suturalis, Creobius eydouxi,* and *Cascellius gravesi* on the margins of the bog with the ice

margin nearby suggests that they had colonized the area from a nearby refugium. This interpretation indicates that the archipelago was not entirely glaciated and that biogeographic hypotheses that require this region to have been recolonized during the Holocene from either Tierra del Fuego in the south or from La Isla Grande de Chiloé in the north need to be reexamined. The beetle species that colonized the margins of the bog at Puerto Edén at 15.6 cal ka inhabit the margins of the bog today. Between 11.6 and 13.0 cal ka, during the Younger Dryas Stade, these species had colonized the margins of the western part of the South Patagonian ice cap where they occur now (Ashworth and Markgraf, 1989).

Farther to the north in the Chilean Lake region, several fossil beetle assemblages from lowland sites have been studied (Hoganson and Ashworth, 1992; Hoganson et al., 1989; Ashworth and Hoganson, 1993). From 28.0 cal ka until between 18.0 and 16.5 cal ka, the beetle fauna consisted of only a small number of species (Fig. 14). Several fossil assemblages have been examined from this interval, and all of them are dominated by the weevil *Germainiellus dentipennis* (name revised by Morrone, 1993; formerly *Listroderes dentipennis*). The other species are mostly species of open, wet habitats. The *residuals* were joined by species from habitats at or above the tree line in the Cordillera de la Costa and the Cordillera de los Andes. These species were able to survive in the lowlands because of colder climatic conditions. No species from the south have been discovered in this fauna. Presumably, montane and marine barriers blocked their dispersal.

Based on the occurrence of the high-elevation species at low elevations, the mean January temperature is estimated to have been 4°–5°C lower than it is now. The glacial fauna is very different from the species-rich forest fauna that started to replace it at ca. 17.0 cal ka (Ashworth and Hoganson, 1993). The change in the beetle fauna is first marked by an increase in the ratio of forest to open-ground species and later by an increase in the total number of species (Fig. 14). The last fossil occurrence of the most characteristic beetle of the glacial fauna, the weevil *G. dentipennis*, occurs at ca. 15.45 cal ka. By 15.1 cal ka, several species of the forest fauna had recolonized the deglaciated terrain, and the fauna is inferred to have had a diversity similar to that of the present day.

21.4. EARLY HOLOCENE

21.4.1. Paleoclimate Conditions During the Early Holocene

Between 14 and 6 cal ka, the amplitude of the seasonal cycle of insolation was greater than it is today in

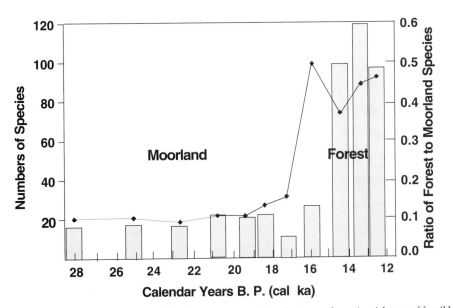

FIGURE 14   Variations in the ratio of forest to open-ground beetle taxa and species richness of fossil beetle assemblages in southern Chile since the last glaciation.

the Northern Hemisphere because of both the greater tilt of the Earth's axis then and the occurrence of perihelion in the Northern Hemisphere summer (Kutzbach et al. 1993). In the Pacific Northwest, insolation was 8.5% greater than it is today in summer and 10% less in winter in the early Holocene, with greatest seasonality occurring between 11 and 10 cal ka. In the Southern Hemisphere, the amplitude of the seasonal cycle and summer insolation were less than they are now. In southern midlatitudes, early Holocene insolation was about 6% less in summer and 4% greater in winter than it is today. These variations likely had a direct effect on temperature and effective precipitation and an indirect effect in terms of changes in the strength of the subtropical high and monsoonal circulation.

The CCM1 simulation for 11 cal ka suggests a hierarchy of responses to these controls:

• In the Northern Hemisphere, as well as globally, the impact of the ice sheet on temperature was much diminished relative to the LGM (Fig. 15 [see color insert]).

• The large impact of the Laurentide ice sheet on circulation was also greatly diminished, the anomalies of upper-level wind speeds were reduced in magnitude, and the areas of the largest anomalies were reduced in size (see also Fig. 6).

• The greater-than-present insolation in the northern summer (Figs. 3 and 4) led to warmer-than-present temperatures over the Northern Hemisphere continents, which increased the land–ocean temperature contrast in the Northern Hemisphere, producing lower

pressure over the continents, higher pressure over the oceans, and, consequently, greater onshore flow into summer-wet regions.

• In both hemispheres, owing to the mechanisms described earlier, growing degree days increased sharply relative to those earlier (Fig. 11), and while they remained lower than they are now in the Southern Hemisphere, most of the LGM-to-present increase in both hemispheres occurred between the 14 and 11 cal ka simulations.

In individual regions,

• Greater summer insolation also strengthened the east Pacific subtropical high-pressure system (Fig. 15), which further intensified summer drought (Thompson et al., 1993; Heusser et al., 1985) and produced lower-than-present AE/PE ratios (Fig. 11).

• In subtropical and tropical South America, there is again a small indication of weaker summer monsoons (slightly higher-than-present SLP over the continent in January accompanied by lower-than-present SLP over the oceans, Fig. 15), with drier-than present conditions resulting (Fig. 15).

21.4.2. Environmental Response in North America

The number of paleoecologic records from the early Holocene is quite extensive compared to the LGM. The effects of higher-than-present summer insolation are well registered along the northwestern coast of North

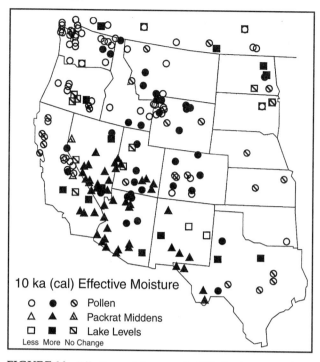

FIGURE 16 Effective moisture map for 10 ka for the western United States (Mock and Brunelle-Daines, 1999. With permission.).

America by a shift toward more xerothermic communities (Fig. 16) (Mock and Brunelle-Daines, 1999). The vegetational changes are a response to the direct effects of greater summer insolation—namely, increased temperatures and decreased effective moisture—as well as to the indirect effects of a strengthening of the northeast Pacific subtropical high. In central and coastal Alaska, greater summer insolation is registered by warmer-than-present temperatures with little change in precipitation (Hansen and Engstrom, 1996; Peteet, 1986; Heusser, 1985). Tundra was replaced by alder thickets in south-central Alaska (Peteet, 1986) and by *Pinus contorta, T. mertensiana,* and *Alnus* in southwestern Alaska (Hansen and Engstrom, 1996) by 10.5 cal ka. Forests of *Picea sitchensis, Alnus,* and *T. heterophylla* developed in southwestern Alaska and the Queen Charlotte Islands between 10.5 and 7.8 cal ka, and the tree line lay above its present elevation (Pellatt and Mathewes, 1997). *Pseudotsuga, A. rubra,* and *Pteridium* were widespread in the seasonal rain forests of southern British Columbia, Washington, and Oregon (Hebda, 1995; Cwynar, 1987; Worona and Whitlock, 1995; Heusser, 1985), and *Quercus garryana,* an indicator of xeric conditions, shifted its range both northward and to higher elevations (Barnosky, 1985; Sea and Whitlock, 1995). Sites where charcoal analysis has been undertaken suggest that fires were more frequent than they are today (see Long et al., 1998; Cwynar, 1987; Mohr et al., 2000). Similarly,

on the east side of the Cascades, the forest-steppe border shifted northward by 100 km and to higher elevations (Mack et al., 1978; Whitlock and Bartlein, 1997). In the northern Rocky Mountains, grasslands were more widespread at low elevations, and *Pinus contorta* was a forest dominant in response to dry conditions and more fires (Whitlock, 1993; Millspaugh et al., 2000). The tree line lay above its present elevation in both the Coast Range and the northern Rockies (Kearney and Luckman, 1983; Clague and Mathewes, 1989).

In the American Southwest, greater-than-present summer insolation apparently increased the onshore flow of moisture from the Gulf of California and Gulf of México, increasing the strength of summer monsoons. The abundance of succulents, which depend on summer precipitation, has been attributed to increased precipitation in the Mojave Desert at this time (Spaulding and Graumlich, 1986), and pollen records from the Colorado Plateau indicate a rise in the upper tree line in the early Holocene that is attributed to increased summer precipitation (Weng and Jackson, 1999). At the subcontinental scale, the trade-off in precipitation regimes is evident between the Pacific Northwest affected by the stronger-than-present subtropical high-pressure system and the American Southwest with stronger monsoons. The pattern of these two precipitation regimes becomes spatially more complex, however, at smaller spatial scales in the Rocky Mountain region, where topography strongly influences air mass patterns. Records from the Yellowstone region indicate that stronger monsoonal circulation in the summer-wet regions led to wetter conditions in the early Holocene, while an expansion of the subtropical high in the adjacent regions caused drier conditions there (Whitlock and Bartlein, 1993; Whitlock et al., 1995). The boundary between summer-wet and summer-dry areas did not shift from its present location, however, because it was and is constrained by topography.

Across most of the Pacific Northwest and into coastal Alaska, the onset of increased moisture preceded the cooling, leading to the definition of a middle Holocene warm, moist interval (the mesothermic period) between 7.7 and 5 cal ka (Hebda, 1995). This period probably represents the changing importance of the direct and indirect effects of greater-than-present insolation through the Holocene. Summer insolation was greater than present levels in the middle Holocene, but lower than that of the early Holocene; this intermediate condition would have maintained high summer temperatures (the direct effects), but perhaps weakened the east Pacific subtropical high. This mesothermal period is characterized by the northward expansion of *Picea* and *Alnus* forest along coastal Alaska (Peteet, 1986) and the inception of peatland development in many regions. Subalpine parkland expanded at high elevations as the upper tree line retreated downslope. In southern British Columbia and northern Washington, *Thuja plicata, Tsuga heterophylla,* and *Picea* replaced *Pseudotsuga* and *Alnus,* and fires became less frequent (Hebda and Whitlock, 1997). Sequoia forests were established in the warm, temperate rain forest region of northern California, and areas of grassland and oak woodland shrank on both sides of the Cascade Range. Modern forest associations and the position of the forest-steppe border were established in the last 3000–5000 years as summer insolation approached present values and the climate became cool and humid. In the interior region of British Columbia, *Pseudotsuga, Betula,* and *T. heterophylla* extended their range at the expense of grassland and pine forest. In the northern Rocky Mountains, mesophytic taxa, such as *Picea, Abies,* and *Pinus albicaulis,* became major constituents of high- and middle-elevation forests after 4.5 cal ka, and *P. contorta* was confined to xeric and disturbed habitats (Thompson et al., 1993).

21.4.3. Environmental Response in South America

Throughout the high southern latitudes, the early Holocene was characterized by *Nothofagus* woodland instead of the present-day diverse vegetation that includes rain forest west of the Andes and woodland and steppe east of the Andes (Fig. 14) (Markgraf, 1993b). The highest prehistoric fire frequencies occurred during the early Holocene south of latitude 50°S (Markgraf and Anderson, 1994). These fires are attributed to climatic instability resulting in repeated intervals of aridity (Markgraf and Anderson 1994; Huber 1998).

Equatorwards, at latitudes 45°–50°S, rain forest and moorland were optimally developed, comparable to present-day conditions (Fig. 14), and lake levels east of the Andes were at their highest (Ashworth et al., 1991; Lumley and Switsur, 1993; Stine and Stine, 1990). Warm-temperate forests north of latitude 45°S and west of the Andes were dominated by drought- and disturbance-adapted taxa (e.g., *N. obliqua, Eucryphia-Caldcluvia* type, and *Weinmannia trichosperma*), which actually expanded south of their present-day distribution (Villagrán, 1991) and were also more abundant east of the Andes (Markgraf, 1984). All this suggests that the westerly storm tracks did not experience the seasonal shifts characteristic of modern times, but instead focused their full strength between latitudes 45° and 50°S (Markgraf et al., 1992). As a result, higher as well as lower latitudes received substantially less precipitation than they do today.

Between 5 and 3 cal ka, moisture in the subtropical latitudes as well as in the high latitudes increased to the highest levels since the LGM. The development of present communities at this time was apparently associated with the establishment of the modern seasonal shift of westerly storm tracks and the increasing influence of El Niño/Southern Oscillation (ENSO), which brought additional moisture to the lower latitudes (McGlone et al., 1992). Increased moisture apparently affected latitudes equatorwards to 32°S, as is recorded by expansion of the relict *Aextoxicon* forests islands (Villagrán and Varela, 1990).

21.5. INTERHEMISPHERIC LINKAGES

Glacial/interglacial oscillations in the Quaternary period affected both North and South America synchronously, but the extent of ice cover was vastly different and the particular mechanisms that link glacial cycles in the two hemispheres are not well understood. Both regions experienced warm, dry conditions in the early Holocene when summer insolation was at its maximum in the Northern Hemisphere and at its lowest in the Southern Hemisphere. In the north, drought developed from the direct and indirect effects of increased summer insolation; in the south, it was caused by weakening monsoons as a result of decreased summer insolation. The fact that contrasting large-scale controls (i.e., differences in the timing of the seasonal cycle of insolation in the two hemispheres) produced drought at the regional scale is remarkable.

We have discussed two hypotheses to explain the synchroneity in the paleoclimatic records of the Northern and Southern Hemispheres:

- Some intrinsic mechanism in the climate system, such as ocean heat transport, that ties the hemispheres together.
- The carryover (or phase lag) of anomalies in the controls into subsequent months or seasons, primarily by the ocean mixed layer, but possibly also by the land surface, combined with interactions among surface water- and energy balance components.

An ocean heat transport mechanism, specifically the thermohaline circulation or "conveyor belt," has been proposed as a way to link the hemispheres on a number of timescales ranging from the glacial–interglacial (Broecker and Denton, 1989; Imbrie et al., 1984) to the millennial (Clark et al., 1999; Alley et al., 1999). In its simplest form, this hypothesis argues that both hemispheres should respond to changes in controls (such as insolation or atmospheric composition) in a fashion consistent with the effect of those controls on the strength of the thermohaline circulation. Ice core and

TABLE 1 Large-Scale Climatic Controls and Regional Climatic and Biotic Responses

	Last glacial maximum (21 cal ka)		Early Holocene period (11 cal ka)	
	Northern hemisphere	Southern hemisphere	Northern hemisphere	Southern hemisphere
Large-scale controls				
Insolation (45° N and 45° S) (relative to present) (3)	Near-present levels	Near-present levels	+8% (July), −10% (Jan.)	+3% (July), −6% (Jan.)
Ice sheet volume (relative to LGM) (2)	Maximum	Maximum	20% LGM	60% LGM
Sea ice thickness (relative to present) (n.s.)	20× (July) 10.4× (Jan.)	3.5× (July) 3.3× (Jan.)	0.34× (July) 0.28 × (Jan.)	1.8× (July) 1.7× (Jan.)
SSTs (relative to present) (5)	4°–8°C colder (July) 4°–8°C colder (Jan.)	4°–8°C colder (July) 4°–8°C colder (Jan.)	1°C colder (July) 1° to 2°C colder (Jan.)	1°C warmer (July) 1–2°C warmer (Jan.)
Regional controls				
Upper-level winds (relative to present) (6, 10, 15)	Southward displacement and acceleration of jet stream (Jan. and July)	Latitudinal compression of jet stream (Jan. and July)	Near-present configuration	Near-present configuration
Surface winds relative to present (7, 10, 15)	Weaker westerlies, winter and summer	Stronger westerlies, especially winter	Stronger STH (PNW), stronger onshore flow (SW)	Stronger STH (winter), weaker STH (summer)
Other aspects of regional circulation systems	Glacial anticyclone	Fast westerlies along sea-ice margin	Stronger summer monsoon	Weaker summer monsoon
Indirect effects of insolation anomalies on seasonal cycles (3, 4, 5)	Absent	Absent	Strong positive summer anomaly lagged into fall and winter	Weak positive winter anomaly lagged into spring
	North America	South America	North America	South America
Regional responses				
Net radiation (relative to present) (4)	Less than present (July), near present (Jan.)	Near present (Jan. and July)	Greater than present (July), less than present (Jan.)	Near present (July), less than present (Jan.)
Temperature (relative to present) (10, 15)	Colder year-round	Colder year-round	Warmer summers, cooler winters, warmer annually	Cooler winters, cooler summers, cooler or near-present annually
Precipitation (relative to present) (10, 15)	Drier winters	Spatially variable	Drier summers, near-present winters	Wetter summers, spatially variable winters

Bioclimatic variables				
Growing degree days *(11)*	Much lower than present	Much lower than present	Greater than present	Less than present
Temperature of the coldest month *(11)*	Much lower than present	Much lower than present	Lower than present	Lower than present
AE/PE moisture index *(11)*	Greater than present	Less than present	Less than present	Spatially variable
Biotic responses				
Vegetation	NW: spread of subalpine, tundra vegetation to low elevations; expansion of xerophytic taxa from east to west; lower-than-present lake levels SW: spread of woodland and montane taxa to low elevations; higher-than-present lake levels	Widespread treeless vegetation. Lat 36°–43°: *Nothofagus* woodland; bogs with moorland taxa > Lat. 45°: expansion of *Empetrum* heath	NW: spread of xerophytic taxa; higher-than-present tree line; expansion of fire-adapted species relative to present SW: spread of species sensitive to monsoonal precipitation (Mojave); higher-than-present tree line; more fire-adapted species than at present	< Lat 45°S: expansion of warm-temperate forests Lat 45°–50°S: rain forest and moorland optimally developed, comparable to present-day conditions; higher-than-present lake levels east of the Andes > Lat 50°: less diverse *Nothofagus* woodland; high fire frequency
Beetles	NW: spread of alpine species to low elevations; depauperate fauna relative to present SW: variable response: expansion of grassland species (Chihuahuan); mixtures of modern and montane species (Sonoran); little faunal change compared to present (Sonoran) Midcontinent: tundra species south to lat. 40°	Spread of high-elevation species to sea level; depauperate fauna	NW: no data SW: similar to present Midcontinent: open boreal forest species; no modern analog	Lat 45°–50°S: similar species to present

Notes: "Northern hemisphere" and "Southern hemisphere" refer to the midlatitude regions discussed in this chapter. Relevant figures are indicated by figure numbers in *italics* in the first column. Sea ice anomalies are for those high-latitude grid points with sea ice at present, extending 30° (or four model grid points) to the west of each region.

marine records suggest that there is more than one mode of variation in the thermohaline circulation (Alley et al., 1999; Stocker, 2000), and so there are likely several scenarios in which the Northern and Southern Hemispheres would vary in a synchronous fashion (and several in which they would vary in opposition [Broecker, 1998; Blunier et al., 1998]). Discriminating among these different scenarios is not feasible using the terrestrial paleoecological evidence alone. Moreover, simulations with models that explicitly do not allow for changes in ocean heat transport (i.e., the ones by Kutzbach et al. [1998] used here, along with those by Hostetler et al. [1999]) suggest that ocean heat transport is not necessary to explain all the regional variations of climate (e.g., the early response in the south).

The second, or carryover, hypothesis seems more plausible. The main mechanism involves the propagation of anomalies that occur in one season to following ones through the thermal inertia of the oceans (Kutzbach and Gallimore, 1988; Mitchell et al., 1988). This carryover is easy to see in the model simulations, and it is also evident in the modern climate as the seasonally varying contrast in isotherm latitude over the continents as compared with over the ocean. It also does not preclude a role for ocean heat transport in the synchroneity of midlatitudes in the two hemispheres. Further testing of this hypothesis is necessary and might best be done in a *forward modeling* approach in which the paleoclimatic data are used to test more formal hypotheses generated by a hierarchy of models.

A combination of mechanisms, including the influence of the oceans on the seasonal cycle of insolation-related variables and the interaction among energy and water balance variables, apparently acted to produce similar paleoenvironmental records in the western temperate regions in both hemispheres (Table 1). This synchroneity, despite the difference in hemispheric forcing, raises some interesting questions about the regional response to potential future global-scale climatic changes. Although the potential forcing of the global climate system from anthropogenic changes in greenhouse gases is not likely to have a strongly seasonally or hemispherically varying pattern, the same cannot be said for aerosols and land-cover changes. Moreover, the potential responses have strong seasonal and regional patterns, with greater anthropogenic warming expected in the winter season than in summer and at higher latitudes than at lower latitudes. Consequently, the potential exists for future regional environmental changes, like those of the past, to be complex, departing in both sign and magnitude from those expected for the hemisphere or globe. At the same time, distant regions may experience similar variations in response to apparently disparate changes in their controls.

Acknowledgments

The research was supported by grants from the National Science Foundation (NSF; ATM-9615822 to CW, ATM-9526139 to VM, ATM-9523584 to PB).

References

Aceituno, P., H. Fuenzalida, and B. Rosenbluth, 1993: Climate along the extratropical west coast of South America. *In* Mooney, H. A., E. R. Fuentes, and B. I. Kronberg (eds.), *Earth Systems Responses to Global Change.* New York: Academic Press, pp. 61–72.

Alaback, P., and J. Pojar, 1997: Vegetation from ridgetop to seashore. *In* Schoonmaker, P. K., B. von Hagen, and E. C. Wolf (eds.), *The Rain Forests of Home: Profile of a North American Bioregion.* Washington, DC: Island Press, pp. 69–87.

Alley, R. B., and P. U. Clark, 1999: The deglaciation of the Northern Hemisphere: A global perspective. *Annual Reviews of Earth and Planetary Sciences,* **27**: 149–182.

Alley, R. B., P. U. Clark, L. D. Keigwin, R. S. Webb, 1999: Making sense of millennial-scale climate change. *In* Clark, P. U., R. S. Webb, and L. D. Keigwin (eds.), *Mechanisms of Global Climate Change at Millennial Time Scales.* Washington, DC: American Geophysical Union, Geophysical Monograph 112, pp. 385–394.

Ashworth, A. C., 1996: The response of arctic Carabidae (Coleoptera) to climate change based on the fossil record of the Quaternary period. *Annales Zoologici Fennici,* **33**: 125–131.

Ashworth, A. C., and J. W. Hoganson, 1987: Coleoptera bioassociations along an elevational gradient in the Lake Region of southern Chile, and comments on the postglacial development of the fauna. *Annals of the Entomological Society of America,* **80**: 865–895.

Ashworth, A. C., and J. W. Hoganson, 1993: The magnitude and rapidity of the climate change marking the end of the Pleistocene in the mid-latitudes of South America. *Palaeogeography, Palaeoclimatology, Palaeoecology,* **101**: 263–270.

Ashworth, A. C., and V. Markgraf, 1989: Climate of the Chilean channels between 11,000 to 10,000 B.P. based on fossil beetle and pollen analysis. *Revista Chilena Historia Natural,* **62**: 61–74.

Ashworth, A. C., V. Markgraf, and C. Villagrán, 1991: Late Quaternary climatic history of the Chilean Channels based on fossil pollen and beetle analyses, with an analysis of the modern vegetation and pollen rain. *Journal of Quaternary Science,* **6**: 279–291.

Atkinson, T. C., K. R. Briffa, and G. R. Coope, 1987: Seasonal temperatures in Britain during the last 22,000 years, reconstructed using beetle remains. *Nature,* **325**: 587–592.

Barnosky, C. W., 1981: A record of late Quaternary vegetation from Davis Lake, southern Puget Lowland, Washington. *Quaternary Research,* **16**: 221–239.

Barnosky, C. W., 1985: Late Quaternary vegetation near Battle Ground Lake, southern Puget Trough, Washington. *Geological Society of America Bulletin,* **96**: 263–271.

Barnosky, C. W., 1987: Response of vegetation to climatic changes of different duration in the late Neogene. *Trends in Ecology and Evolution,* **2**: 247–250.

Barnosky, C. W., P. M. Anderson, and P. J. Bartlein, 1987: The northwestern U.S. during deglaciation: Vegetational history and paleoclimatic implications. *In* Ruddiman, W. F., and H. E. Wright, Jr. (eds.), *North America and Adjacent Oceans During the Last Deglaciation.* Vol. K-3, *The Geology of North America,* Boulder, CO: Geological Society of America, pp. 289–321.

Barron, E. J., 1985: Explanations of the Tertiary global cooling trend. *Palaeogeography, Palaeoclimatology, Palaeoecology,* **50**: 45–61.

Bartlein, P. J., 1988: Late-Tertiary and Quaternary palaeoenvironments. *In* Huntley, B., and T. Webb, III (eds), *Vegetation History.* Dordrecht: Kluwer Academic, pp. 113–152.

Bartlein, P. J., 1997: Past environmental changes: Characteristic features of Quaternary climate variations. *In* Huntley, B., W. Cramer, A. V. Morgan, H. C. Prentice, and J. R. M. Allen (eds.), *Past and Future Rapid Environmental Changes: The Spatial and Evolutionary Response of Terrestrial Biota.* Berlin: Springer-Verlag, pp. 11–29.

Bartlein, P. J., K. H. Anderson, P. M. Anderson, M. E. Edwards, C. J. Mock, R. S. Thompson, R. S. Webb, T. Webb, III, and C. Whitlock, 1998: Paleoclimate simulations for North America over the past 21,000 years: Features of the simulation climate and comparisons with paleoenvironmental data. *Quaternary Science Reviews,* **17**: 549–585.

Benson, L., and R. S. Thompson, 1987: The physical record of lakes in the Great Basin. *In* Ruddiman, W. F., and H. E. Wright, Jr. (eds.), *North America and Adjacent Oceans During the Last Deglaciation. Vol. K-3, The Geology of North America.* Geological Society of America, Boulder, CO: pp. 241–260.

Berger, A., 1978: Long-term variations of caloric insolation resulting from the Earth's orbital elements. *Quaternary Research,* **9**: 139–167.

Betancourt, J. L., T. R. Van Devender, and P. S. Martin, 1990: *Packrat Middens: The Last 40,000 Years of Biotic Change.* Tucson: University of Arizona Press.

Blunier, T., J. Chappellaz, J. Schwander, A. Dällenbach, B. Stauffer, T. F. Stocker, D. Raynaud, J. Jouzel, H. B. Clausen, C. W. Hammer, and S. J. Johnsen, 1998: Asynchrony of Antarctic and Greenland climate change during the last glacial period. *Nature,* **394**: 739–743.

Broecker, W. S., 1998: Paleocean circulation during the last deglaciation: A bipolar seesaw? *Paleoceanography,* **13**: 199–121.

Broecker, W. S., and G. H. Denton, 1989: The role of ocean-atmosphere reorganizations in glacial cycles. *Geochimica Cosmochimica Acta,* **53**: 2465–2501.

Brook, E. J., Sowers, T., and Orchardo, J. 1996. Rapid variations in atmospheric methane concentration during the past 110,000 years. *Science,* **273**: 1087–1091.

Clague, J. J., and R. W. Mathewes, 1989: Early Holocene thermal maximum in western North America: New evidence from Castle peak, British Columbia. *Geology,* **17**: 277–280.

Clapperton, C. M., 1992: La última glaciación y deglaciación en el Estrecho de Magallanes: Implicaciones para el poblamiento de Tierra del Fuego. *Anales Instituto Patagonico,* **21**: 113–128.

Clark, P. U., R. S. Webb, and L. D. Keigwin, 1999: *Mechanisms of Global Climate Change at Millennial Time Scales.* Washingto, DC: American Geophysical Union, Geophysical Monograph 112, 394 pp.

Cong, S., and A. C. Ashworth, 1996: Palaeoenvironmental interpretation of middle and late Wisconsinan fossil coleopteran assemblages from western Olympic Peninsula, Washington. *Journal of Quaternary Science,* **11**: 345–356.

Cwynar, L. C., 1987: Fire and the forest history of the north Cascade Range. *Ecology,* **68**: 791–802.

Darlington, P. J., Jr., 1965: *Biogeography of the Southern End of the World.* Cambridge, MA: Harvard University Press.

Delcourt, H. R., P. A. Delcourt, and T. Webb, III, 1983: Dynamic plant ecology: The spectrum of vegetational change in time and space. *Quaternary Science Reviews,* **1**: 153–175.

Donoso, C., 1993: *Bosques Templados de Chile y Argentina.* Editorial Universitaria, Santiago, Chile, 484 pp.

Donoso, C., 1996: Ecology of *Nothofagus* forests in central Chile. *In* Veblen, T. T., R. S. Hill, and J. Read (eds.), *The Ecology and Biogeography of Nothofagus Forests.* New Haven, CT: Yale University Press, pp. 271–292.

Elias, S. A., 1987: Climatic significance of late Pleistocene insect fossils from Marias Pass, Montana. *Canadian Journal of Earth Sciences,* **25**: 922–926.

Elias, S. A., 1992a: Late Quaternary beetle fauna of southwestern Alaska: Evidence of a refugium for mesic and hygrophilous species. *Arctic and Alpine Research,* **24**: 133–144.

Elias, S. A., 1992b: Late Quaternary zoogeography of the Chihauhuan Desert insect fauna, based on fossil records from packrat middens. *Journal of Biogeography,* **19**: 285–297.

Elias, S. A., 1996: Late Pleistocene and Holocene seasonal temperatures reconstructed from fossil beetle assemblages in the Rocky Mountains. *Quaternary Research,* **46**: 311–318.

Elias, S. A., and T. R. Van Devender, 1990: Fossil insect evidence for late Quaternary climatic change in the Big Bend region, Chihuahuan Desert, Texas. *Quaternary Research,* **34**: 249–261.

Elias, S. A., J. L. Mead, and L. D. Agenbroad, 1992: Late Quaternary arthropods from the Colorado Plateau, Arizona and Utah. *Great Basin Naturalist,* **52**: 59–67.

Elias, S. A., S. K. Short, C. H. Nelson, and H. H. Birks, 1996: Life and times of the Bering land bridge. *Nature,* **382**: 60–63.

Erwin, T. L., 1985: The taxon pulse: A general pattern of lineage radiation and extinction among carabid beetles, *In* Ball, G. E. (ed.), *Taxonomy, Phylogeny and Zoogeography of Beetles and Ants.* Dordrecht: Dr. W. Junk Publishers, pp. 437–472.

Farrera, I., S. P. Harrison, I. C. Prentice, G. Ramstein, J. Guiot, P. J. Bartlein, R. Bonnefille, M. Bush, W. Cramer, U. von Grafenstein, K. Holmgren, H. Hooghiemstra, G. Hope, D. Jolly, S.-E. Lauritzen, Y. Ono, S. Pinot, M. Stute, and G. Yu, 1999: Tropical climates at the last glacial maximum: A new synthesis of terrestrial palaeoclimate data. I. Vegetation, lake-levels and geochemistry. *Climate Dynamics,* **15**: 823–856.

Grootes, P. M., M. Stuiver, J. W. C. White, S. Johnsen, and J. Jouzel, 1993: Comparison of oxygen isotope records from the GISP2 and GRIP Greenland ice cores. *Nature,* **366**: 552–555.

Hall, W. E., C. A. Olson, and T. R. Van Devender, 1989: Late Quaternary and modern arthropods from the Ajo Mountains of southwestern Arizona. *Pan-Pacific Entomologist,* **65**: 322–347.

Hall, W. E., T. R. Van Devender, and C. A. Olson, 1988: Late Quaternary arthropod remains from Sonoran Desert middens, southwestern Arizona and northwestern Sonora. *Quaternary Research,* **29**: 277–293.

Hansen, B. C. S., and D. R. Engstrom, 1996: Vegetation history of Pleasant Island, southeastern Alaska. *Quaternary Research,* **46**: 161–175.

Hatch, M. H., 1971: *The Beetles of the Pacific Northwest.* Part V. Seattle: University of Washington Press.

Hebda, R. J., 1995: British Columbia vegetation and climate history with focus on 6 ka B.P. *Geographie Physique et Quaternaire,* **49**: 55–79.

Hebda, R. J., and C. Whitlock, 1997: Environmental history of the coastal temperate rainforest of northwest North America. *In* Schoonmaker, P. K., B. von Hagen, and E. C. Wolf (eds), *The Rain Forests of Home: Profile of a North American Bioregion.* Washington, DC: Island Press, pp. 227–251.

Heusser, C. J., 1977: Quaternary palynology of the Pacific slope of Washington. *Quaternary Research,* **8**: 282–306.

Heusser, C. J., 1983: Vegetation history of the northwestern United States, including Alaska. *In* Porter, S. C. (ed.), *Late Quaternary Environments of the United States.* Minneapolis: University of Minnesota Press, pp. 239–258.

Heusser, C. J., 1984: Late-glacial-Holocene pollen sequence in the lake district of Chile. *Quaternary Research,* **22**: 77–90.

Heusser, C. J., 1985: Quaternary pollen records from the Pacific Northwest coast: Aleutians to the Oregon-California boundary. *In* Bryant, V. M., Jr., and R. G. Holloway (eds.), *Pollen Records of Late-Quaternary North American Sediments.* Dallas, TX: American Association of Stratigraphic Palynologists Foundation, pp. 141–166.

Heusser, C. J., 1990: Ice age vegetation and climate of subtropical

Chile. *Palaeogeography, Palaeoclimatology, Palaeoecology,* **80**: 107–127.

Heusser, C. J., L. E. Heusser, and D. M. Peteet, 1985: Late-Quaternary climate change on the American North Pacific Coast. *Nature,* **315**: 485–487.

Hoganson, J. W., and A. C. Ashworth, 1992: Fossil beetle evidence for climatic change 18,000–10,000 years B.P. in south-central Chile. *Quaternary Research,* **37**: 101–116.

Hoganson, J. W., M. Gunderson, and A. C. Ashworth, 1989: Fossil beetle analysis. *In* Dillehay, T. D. (ed.), *Monte Verde, A Late Pleistocene Settlement in Chile, 1. Palaeoenvironment and Site Context.* Washington, DC: Smithsonian Institution Press, pp. 215–226.

Hollin, J. T., and D. H. Schilling, 1981: Late Wisconsin-Weichselian mountain glaciers and small ice caps. *In* Denton, G. H., and T. J. Hughes (eds.), *The Last Great Ice Sheets.* New York: Wiley, pp. 179–206.

Hostetler, S. W., P. U. Clark, P. J. Bartlein, A. C. Mix, and N. G. Pisias, 1999: Mechanisms for the global transmission and registration of North Atlantic Heinrich Events. *Journal of Geophysical Research,* **104**(D4): 3947–3953.

Huber, U., 1998: Linkages between Holocene climate, vegetation, and fire regimes in Patagonia. American Quaternary Association 15th Biennial Meeting, Abstracts and Program. 5–7 September 1998, Puerto Vallarta, México.

Imbrie, J., J. D. Hays, D. G. Martinson, A. McIntyre, A. C. Mix, J. J. Morley, N. G. Pisias, W. L. Prell, and N. J. Shackleton, 1984: The orbital theory of Pleistocene climate: Support from a revised chronology of the marine (^{18}O record. *In* Berger, A., J. Imbrie, J. Hays, G. Kukla, and B. Saltzman (eds.), *Milankovitch and Climate.* Dordrecht: Reidel, pp. 269–305.

Indermühle, A., T. F. Stocker, F. Joos, H. Fischer, H. J. Smith, M. Wahlen, B. Deck, D. Mastroianni, J. Tschumi, T. Blunier, R. Meyer, and B. Stauffer, 1999: Holocene carbon-cycle dynamics based on CO_2 trapped in ice at Taylor Dome, Antarctica. *Nature,* **398**: 121–126.

Jeannel, R., 1962: Les trechides de la Paléantarctide Occidentale. Biologie de l'Amerique Australe, vol. 1, *In Etudes sur la Faune du Sol,.* Paris: Presses du CNRS, pp. 527–655.

Jouzel, J., C. Waelbroeck, B. Malaize, M. Bender, J. R. Petit, M. Stievenard, N. I. Barkov, J. M. Barnola, T. King, V. M. Kotyakov, V. Lipenkov, C. Lorius, D. Raynaud, C. Ritz, and T. Sowers, 1996: Climatic interpretation of the recently extended Vostock ice records. *Climate Dynamics,* **12**: 513–521.

Kalin Arroyo, M., J. J. Armesto, F. Squeo, J. Gutierrez, 1993: Global change: Flora and vegetation of Chile. *In* Mooney, H.A., E.R. Fuentes, and B.I. Kronberg (eds.), *Earth Systems Responses to Global Change.* New York: Academic Press, pp. 239–263.

Kavanaugh, D. H., 1988: The insect fauna of the Pacific Northwest coast of North America: Present patterns and affinities and their origins. *Memoir of the Entomological Society of Canada,* **144**: 125–149.

Kavanaugh, D. H., 1992: Carabid beetles (Insecta:Coleoptera:Carabidae) of the Queen Charlotte Islands, British Columbia. *Memoirs of the California Academy of Sciences,* **16**: 1–113.

Kearney, M. S., and B. H. Luckman, 1983: Postglacial history of Tonquin Pass, British Columbia. *Canadian Journal of Earth Sciences,* **20**: 776–786.

Kousky, V. E., and C. F. Ropelewski, 1997: The tropospheric seasonally varying mean climate over the Western Hemisphere (1979–1995). NCEP/Climate Prediction Center ATLAS No. 3, U.S. Dept. of Commerce, National Oceanic and Atmospheric Adminstration, National Weather Service, 135 pp.

Kuschel, G., 1969: Biogeography and ecology of South American Coleoptera. *In* Fittkau, E. J., J. Illies, H. Klinge, G. H. Schwabe, and H. Sioli (eds.), *Biogeography and Ecology in South America.* The Hague: Dr. W Junk Publishers, pp. 709–722.

Kutzbach, J. E., and R. G. Gallimore, 1988: Sensitivity of a coupled atmposphere/mixed layer ocean model to changes in orbital forcing at 9000 years B.P. *Journal of Geophysical Research,* **93**(D1): 803–821.

Kutzbach, J., R. Gallimore, S. Harrison, P. Behling, R. Selin, and R. Laarif, 1998: Climate and biome simulations for the past 21,000 years. *Quaternary Science Reviews,* **17**: 473–506.

Kutzbach, J. E., P. J. Guetter, P. J. Behling, and R. Selin, 1993: Simulated climatic changes: Results of the COHMAP climate-model experiments. *In* Wright, H. E., Jr., J. E. Kutzbach, T. Webb, III, W. F. Ruddiman, F. A. Street-Perrott, and P. J. Bartlein (eds.), *Global Climates Since the Last Glacial Maximum.* Minneapolis: Univer. of Minnesota Press, pp. 5–11.

Lawford, R. G., 1996: North-south variations in west coast hydrometeorological parameters and their significance for earth systems. *In* Lawford, R. G., P. B. Alaback, and E. Fuentes (eds.), *High Latitude Rainforests and Associated Ecosystems of the West Coast of the Americas. Ecological Studies,* **116**: 3–26.

Long, C. J., C. Whitlock, P. J. Bartlein, and S. H. Millspaugh, 1998: A 9000-year fire history from the Oregon Coast Range, based on a high-resolution charcoal study. *Canadian Journal of Forest Research,* **28**: 774–787.

Lowell, T. V., C. J. Heusser, B. G. Andersen, P. I. Moreno, A. Hauser, L. E. Heusser, C. Schluchter, D. R. Marchant, G. H. Denton, 1995: Interhemispheric correlation of late Pleistocene glacial events. *Science,* **269**: 1541–1549.

Lumley, S. H., and R. Switsur, 1993: Late Quaternary chronology of the Taitao Peninsula, southern Chile. *Journal of Quaternary Science,* **8**: 161–165.

Mack, R. N., N. W. Rutter, and S. Valastro, 1978: Late Quaternary pollen record from the Sanpoil River valley, Washington. *Canadian Journal of Botany,* **56**: 1642–1650.

Markgraf, V., 1984: Late Pleistocene and Holocene vegetation history of temperate Argentina: Lago Morenito, Bariloche. *Dissertationes Botanicae,* **72**: 235–254.

Markgraf, V., 1991: Late Pleistocene environmental and climate evolution in southern South America. *Bamberger Geographische Schriften,* **11**: 271–281.

Markgraf, V., 1993a: Climatic history of Central and South America since 18,000 years B.P.: Comparison of pollen records and model simulations. *In* Wright, H. E., Jr., J. E. Kutzbach, T. Webb, III, W. F. Ruddiman, F. A. Street-Perrott, and P. J. Bartlein (eds.), *Global Climates Since the Last Glacial Maximum.* Minneapolis: University of Minnesota Press, pp.357–385.

Markgraf, V., 1993b: Paleoenvironments and paleoclimates in Tierra del Fuego and southernmost Patagonia, South America. *Palaeogeography, Palaeoclimatology, Palaeoecology,* **102**: 53–68.

Markgraf, V., and L. Anderson, 1994: Fire history of Patagonia: Climate versus human cause. *Revista do Instituto Geológico, Sao Paulo, Brazil,* **15**: 35–47.

Markgraf, V., J. R. Dodson, P. A. Kershaw, M. McGlone, and N. Nicholls, 1992: Evolution of late Pleistocene and Holocene climates in circum South Pacific land areas. *Climate Dynamics,* **6**: 193–211.

Markgraf, V., M. McGlone, and G. Hope, 1995: Neogene paleoenvironmental and paleoclimatic change in southern temperate ecosystems–A southern perspective. *Trends in Ecology and Evolution,* **10**: 143–147.

Markgraf, V., E. Romero, and C. Villagrán, 1996: History and paleoecology of South American *Nothofagus* forests. *In* Veblen, T. T., R. S. Hill, and J. Read (eds.), *The Ecology and Biogeography of* Nothofagus *Forests.* New Haven, CT: Yale University Press, pp. 354–386.

McGlone, M., P. A. Kershaw, and V. Markgraf, 1992: El Niño/Southern Oscillation climatic variability in Australasian and South American paleoenvironmental records. *In* Diaz, H. F., and V. Markgraf (eds.), *El Nino, Historical and Paleoclimatic Aspects of the*

Southern Oscillation. Cambridge, UK: Cambridge University Press, pp. 435–462.

Millspaugh, S. H., C. Whitlock, and P. J. Bartlein, 2000: Variations in fire frequency and climate over the last 17,000 years in central Yellowstone National Park. *Geology,* **28**: 211–214.

Mitchell, J. F. B., N. S. Graham, and K. J. Needham, 1988: Climate simulations for 9000 years before present: Seasonal variations and effect of the Laurentide ice sheet. *Journal of Geophysical Research,* **93**: 8283–8303.

Mix, A. C., D. C. Lund, N. G. Pisias, P. Bodén, L. Bornmalm, M. Lyle, and J. Pike, 1999: Rapid climate oscillations in the Northeast Pacific during the last deglaciation reflect Northern and Southern Hemisphere sources. *In* Clark, P. U., R. S. Webb, and L. D. Keigwin (eds.), *Mechanisms of Global Climate Change at Millennial Time Scales.* Washington, DC: American Geophysical Union, Geophysical Monograph 112, pp. 127–148.

Mock, C. J., and A. R. Brunelle-Daines, 1999: A modern analogue of western United States summer paleoclimate at 6000 years before present. *The Holocene,* **9**: 541–545.

Mohr, J. A., C. Whitlock, and C. J. Skinner, 2000: Postglacial vegetation and fire history, eastern Klamath Mountains, California. *The Holocene.*

Moore, D. M., 1983: The flora of the Fuego-Patagonian Cordilleras: Its origins and affinities. *Revista Chilena Historia Natural,* **56**: 123–136.

Morgan, A. V., 1987: Late Wisconsin and early Holocene paleoenvironments of east-central North America based on assemblages of fossil Coleoptera. *In* Ruddiman, W. F., and H. E. Wright, Jr. (eds.), *North America and Adjacent Oceans During the Last Deglaciation. Vol. K-3, The Geology of North America.* Boulder, CO: The Geological Society of America, pp. 353–370.

Morrone, J. J., 1993: Revision sistematica de un nuevo genero de Rhytirrhinini (Coleoptera, Curculionidae), con un analisis biogeographic del dominio subantartico. *Boletin Socedad Biologica Concéption, Chile,* **64**: 121–145.

Nelson, R. E., and G. R. Coope, 1982: A late-Pleistocene insect fauna from Seattle, Washington. *American Quaternary Association 7th Biennial Meeting, Abstracts and Program,* **7**: 146.

Niemela, J., 1990: Habitat distribution of carabid beetles in Tierra del Fuego, South America. *Entomologica Fennica,* **1**: 3–16.

Parsons, G. L., G. Cassis, A. R. Moldenke, J. D. Lattin, N. H. Anderson, J. C. Miller, P. Hammond, and T. D. Schowalter, 1991: Invertebrates of the H. J. Andrews Experimental Forest, Western Cascade Range, Oregon. V. An Annotated List of Insects and Other Arthropods. USDA Technical Report PNW-GTR-290, pp. 1–168.

Pellatt, M. G., and R. W. Mathewes, 1997: Holocene tree line and climate change on the Queen Charlotte Islands, Canada. *Quaternary Research,* **48**: 88–99.

Peltier, R., 1994: Ice Age paleotopography. *Science,* **265**: 195–201.

Peteet, D. M., 1986: Modern pollen rain and vegetational history of the Malaspina Glacier District, Alaska. *Quaternary Research,* **25**: 100–120.

Petit, J. R., J. Jouzel, D. Raynaud, N. I. Barkov, J.-M. Barnola, I. Basile, M. Bender, J. Cahppellaz, M. Davis, G. Dalygue, M. Delmotte, V. M. Kotlyakov, M. Legrand, V. Y. Lipenkov, C. Lorius, L. Pépin, C. Ritz, E. Saltzman, and M. Stievenard, 1999: Climate and atmospheric history of the past 420,000 years from the Vostok ice core, Antarctica. *Nature,* **399**: 429–436.

Prentice, I. C., 1992: Climate change and long-term vegetation dynamics. *In* Glenn-Lewin, D. C., R. K. Peet, and T. T. Veblen (eds.), *Plant Succession: Theory and Prediction.* London: Chapman & Hall, pp. 293–339.

Prentice, I. C., W. Cramer, S. P. Harrison, R. Leemans, R. A. Monserud, and A. M. Solomon, 1992: A global biome model based on plant physiology and dominance, soil properties and climate. *Journal of Biogeography,* **19**: 117–134.

Raymo, M. E., W. F. Ruddiman, and P. N. Froelich, 1988: The influence of late Cenozoic mountain building on ocean geochemical cycles. *Geology,* **16**: 649–653.

Ruddiman, W. F., and J. E. Kutzbach, 1989: Forcing of late Cenozoic Northern Hemisphere climate by plateau uplift in southern Asia and the American West. *Journal of Geophysical Research,* **94**: 18409–18427.

Ruddiman, W. F., and J. E. Kutzbach, 1991: Plateau uplift and climatic change. *Scientific American,* **264**: 66–75.

Ruthsatz, B., and C. Villagrán, 1990: Vegetation pattern and nutrient ecology of a Magellanic moorland on the Cordillera de Piuchué, Chiloé Island, Chile. *Revista Chilena de Historia Natural,* **64**: 461–478.

Sanchetta, C., M. Lyle, L. Heusser, R. Zahn, and J. P. Bradbury, 1992: Late-glacial to Holocene changes in winds, upwelling, and seasonal production of the Northern California Current System. *Quaternary Research,* **38**: 359–370.

Schwert, D. P., 1992: Faunal transitions in response to an ice age: The late Wisconsinan record of Coleoptera in the north-central United States. *Coleopterists Bulletin,* **46**: 68–94.

Schwert, D. P., and A. C. Ashworth, 1988: Late Quaternary history of the northern beetle fauna of North America: A synthesis of fossil and distributional evidence. *Memoirs of the Entomological Society of Canada,* **144**: 93–107.

Sea, D. S., and C. Whitlock, 1995: Postglacial vegetation and climate of the Cascade Range, central Oregon. *Quaternary Research,* **43**: 370–381.

Simpson, B. B., 1983: An historical phytogeography of the high Andean flora. *Revista Chilena de Historia Natural,* **56**: 109–122.

Spaulding, W. G., 1990: Vegetational and climatic development of the Mojave Desert: The last glacial maximum to the present. *In* Betancourt, J. L., T. R. Van Devender, and P. S. Martin (eds), *Packrat Middens—The Last 40,000 Years of Biotic Change.* Tucson: University of Arizona Press, pp.166–199.

Spaulding, W. G., and L. J. Graumlich, 1986: The last pluvial climatic episodes in the deserts of the southwestern North America. *Nature,* **320**: 441–444.

Steig, E. J., E. J. Brook, J. W. C. White, C. M. Sucher, M. L. Bender, S. J. Lehman, D. L. Morse, E. D. Waddington, and G. D. Clow, 1998: Synchronous climate changes in Antarctica and the North Atlantic. *Science,* **282**: 92–95.

Stine, S., and M. Stine, 1990: A record from Lake Cardiel of climate change in southern South America. *Nature,* **345**: 705–707.

Stocker, T. F., 2000: Past and future reorganizations in the climate system. *Quaternary Science Reviews,* **19**: 201–319.

Stuiver, M., and P. J. Reimer, 1993: Extended ^{14}C data base and revised CALIB 3.0 age calibration program. *Radiocarbon,* **35**: 215–230.

Stuiver, M., P. M. Grootes, and R. F. Braziunas, 1995: The GISP2 (^{18}O) climate record of the past 16,500 years and the role of the Sun, ocean and volcanoes. *Quaternary Research,* **44**: 341–354.

Thompson, L. G., M. E. Davis, E. Mosley-Thompson, T. A. Sowers, K. A. Henderson, V. S. Zagorodnov, P.-N. Lin, V. N. Mikhalenko, R. K. Campen, J. F. Bolzan, J. Cole-Dai, and B. Francou, 1998: A 25,000-year tropical climate history from Bolivian ice cores. *Science,* **282**: 1858–1864

Thompson, R. S., 1990: Late Quaternary vegetation and climate in the Great Basin. *In* Betancourt, J. L., T. R. Van Devender, and P. S. Martin (eds), *Packrat Middens—The Last 40,000 Years of Biotic Change,* Tucson: University of Arizona Press, pp. 200–239.

Thompson, R. S., and Anderson, K. H., in press: Biomes of western North America at 18,000, 6,000, and 0 ^{14}C yr B.P. reconstructed from pollen and packrat midden data. *Journal of Biogeography.*

Thompson, R. S., C. Whitlock, P. J. Bartlein, S. P. Harrison, and W. G. Spaulding, 1993: Climate changes in the western United States since 18,000 yr B.P. *In* Wright, H. E., Jr., J. E. Kutzbach, T. Webb,

III, W. F. Ruddiman, F. A. Street-Perrott, and P. J. Bartlein (eds.), *Global Climates Since the Last Glacial Maximum.* Minneapolis: University of Minnesota Press, pp. 468–513.

Tufte, E. R., 1997: *Visual Explanations: Images and Quantities, Evidence and Narrative,* Cheshire, CT: Graphics Press, 156 pp.

Veblen, T. T., and P. B. Alaback, 1996: A comparative review of forest dynamics and disturbance in the temperate rainforests of North and South America. *In* Lawford, R. G., P. B. Alaback, and E. Fuentes (eds.), High latitude rainforests and associated ecosystems of the west coast of the Americas. *Ecological Studies,* **116**: 173–213.

Villagrán, C., 1988: Late Quaternary vegetation of southern Isla Grande de Chiloé. *Quaternary Research,* **29**: 294–306.

Villagrán, C., 1990: Glacial climates and their effects on the history of the vegetation of Chile: A synthesis based on palynological evidence from Isla de Chiloé. *Review of Palaeobotany and Palynology,* **65**: 17–24.

Villagrán, C., 1991: Historia de los bosques templados del sur de Chile durante el Tardiglacial y Postglacial. *Revista Chilena Historia Natural,* **64**: 447–460.

Villagrán, C., and J. J. Armesto, 1993: Full and late glacial paleoenvironmental scenarios for the west coast of southern South America. *In* Mooney, H. A., E. R. Fuentes, and B. I. Kronberg (eds.), *Earth Systems Responses to Global Change.* New York: Academic Press, pp. 195–208.

Villagrán, C., and J. Varela, 1990: Palynological evidence for increased aridity on the central Chilean coast during the Holocene. *Quaternary Research,* **34**: 198–207.

Webb, T., III, and J. E. Kutzbach, 1998: An introduction to "Late Quaternary climates: Data syntheses and model experiments." *Quaternary Science Reviews,* **17**: 465–471.

Weng, C., and S. T. Jackson, 1999: Late-glacial and Holocene vegetation history and paleoclimate of the Kaibab Plateau. *Palaeogeography, Palaeoclimatology, Palaeoecology,* **153**: 179–201.

Whitlock, C., 1993: Postglacial vegetation and climate of Grand Teton and southern Yellowstone National Parks. *Ecological Monographs,* **63**: 173–198.

Whitlock, C., and P. J. Bartlein, 1993: Spatial variations of Holocene climatic change in the Yellowstone region. *Quaternary Research,* **39**: 231–238.

Whitlock, C., and P. J. Bartlein, 1997: Vegetation and climate change in northwest America during the past 125 kyr. *Nature,* **388**: 57–61.

Whitlock, C., P. J. Bartlein, and K. J. Van Norman, 1995: Stability of Holocene climate regimes in the Yellowstone region. *Quaternary Research,* **43**: 433–436.

Worona, M. A., and C. Whitlock, 1995: Late-Quaternary vegetation and climate history near Little Lake, central Coast Range, Oregon. *Geological Society of America Bulletin,* **107**: 867–876.

Wright, H. E., Jr., J. E. Kutzbach, T. Webb, III, W. F. Ruddiman, F. A. Street-Perrott, and P. J. Bartlein, 1993: *Global Climates Since the Last Glacial Maximum.* Minneapolis: University of Minnesota Press.

Wyrwoll, K.-H., B. Dong, and P. Valdes, 2000: On the position of southern hemisphere westerlies at the Last Glacial Maximum: An outline of AGCM simulation results and evaluation of their implications. *Quaternary Science Reviews,* **19**: 881–898.

22

Late Glacial Climate Variability and General Circulation Model (GCM) Experiments: An Overview

DOROTHY M. PETEET

Abstract

This chapter attempts to provide an overview of general circulation model (GCM) experiments for the late glacial period, a time of rapid climate variability. Two basic approaches to modeling the late glacial climate involve the time slice approach and the sensitivity test approach. The time slice approach attempts to model a particular past climate interval using a set of multiple boundary conditions (forcing parameters), while the sensitivity test approach isolates the effect of one boundary condition at a time for a particular time interval. Examples are given for each approach, including studies of various forcing parameters such as changes in insolation, ice sheets, sea surface temperature (SST), ocean transports, and CO_2. Despite the use of different atmospheric general circulation models (AGCMs), including the National Center for Atmospheric Research (NCAR), Geophysical Fluid Dynamics Laboratory (GFDL), Goddard Institute for Space Studies (GISS), and European Community Hamburg Model Version 3 (ECHAM3), there are similarities in the various model responses to forcing parameters, such as insolation and land ice at 12,000 B.P. (all dates are given as calendar years, except when stated differently). With other forcing parameters, i.e., SST and sea ice, the results are not comparable from model to model because duplicate experiments have not been made with different models. Recent focus on coupled ocean–atmosphere GCMs provides new tools for exploring ocean–atmosphere links. A key question emerges as to just how important ocean circulation changes are in affecting global late glacial climate changes. Just how important is lowered CO_2? How are the Northern and Southern Hemisphere climate responses linked? The iterative process of data–model comparison and the development of new questions are essential to this pursuit. Copyright © 2001 by Academic Press.

Resumen

Este capítulo intenta ofrecer un resumen general a cerca de los experimentos realizados con Modelos de Circulación General (GCM) para el periodo tardío-glacial, momento de gran variabilidad climática. Las dos metodologías básicas para modelizar el clima tardío-glacial implican el método de segmentos temporales, y de la prueba de sensibilidad. El método de segmentos temporales intenta modelizar un intervalo climático particular, utilizando un conjunto de multiples condiciones de contorno (parámetros de forzamiento), mientras que el método la prueba de sensibilidad aisla cada vez el efecto de una condición de contorno durante un intervalo de tiempo particular. Se muestran ejemplos de ambos métodos, incluyendo estudios de diferentes parámetros de forzamiento como cambios en la insolación, capas de hielo, temperatura superficial del agua (SST), transporte del océano, y CO_2. A pesar del uso de diferentes Modelos de Circulación General Atmosférica (AGCM), como los de National Center for Atmospheric Research (NCAR), Geophysical Fluids Dynamics Laboratory (GFDL), Goddard Institute for Space Studies (GISS), European Community Hamburg Model Version 3 (ECHAM3), hay semejanza en las respuestas de los distintos modelos respecto a los parámetros de forzamiento como es el caso de la insolación y la superficie helada hace 12,000 años (todas las fechas se muestran en años civiles, salvo si se indica lo contrario). Con otros parámetros de forzamiento, como la temperatura superficial del mar (SST) y el hielo marino, los resultados entre modelos no son comparables puesto que dobles experimentos no han sido realizados con diferentes modelos. Recientes enfoques con modelos de circulación general atmosférica (GCMs) que incluyen la interacción océano-atmósfera proporcionan nuevas herramientas para explorar las conexiones entre océano y atmósfera. Surge entonces la importante cuestión de qué importancia tienen los cambios en la circulación oceánica afectando a los cambios climáticos globales del tardí-glacial; cuánto importante es un bajo nivel de CO_2; cuál es la relación de las respuestas climáticas entre los hemisferios Norte y Sur. El iterativo proceso de comparación entre datos y modelo, a la vez que el desarrollo de nuevas preguntas son esenciales para esta búsqueda.

22.1. INTRODUCTION

The motivation to model past climates began in the 1980s as climate modelers working with paleoecologists, geologists, and paleoceanographers began to focus on the role of climatic forcing parameters, especially to explain glacial climates. Research on modeling full glacial climates continues to the present, but this chapter is restricted to the climate interval between the full glacial and the Holocene, i.e., the late glacial. An overview is presented of the late glacial experiments using atmospheric general circulation models (AGCMs). Although other types of models (energy balance, ocean circulation–biogeochemistry models, etc.) have been used to target specific climate aspects of the Younger Dryas (YD) interval (i.e., Harvey, 1989; Marchal et al., 1999), GCMs are targeted here because AGCMs produce (1) full, three-dimensional atmospheric dynamics based upon the conservation of mass, energy, momentum, and moisture and (2) climate simulations at regional to continental scales that can be compared with

paleoclimate data, thereby testing the model results. These AGCMs are also used in the attempt to reproduce current and future climates.

Two modeling groups, the National Center for Atmospheric Research (NCAR) and the Goddard Institute for Space Studies (GISS), began a series of experiments in the mid-1980s with very different motivations (Kutzbach and Guetter, 1986; Rind et al., 1986). The NCAR group was interested in how the gradual changes of insolation and ice sheet demise since the full glacial affected the Earth as the climate warmed throughout the late glacial into the Holocene. To answer these questions, a series of time slices was selected beginning with 18,000 B.P. The GISS group was interested in modeling a particular interval of rapid climate change (or anomalous climate) that could not be explained by the general pattern of gradual warming following the full glacial interval. Selected for this modeling study were the Allerød warm period and the Younger Dryas cool period. In both approaches, the model results are presented as averages of several years of model simulation. These two very different approaches resulted in a series of experiments with different model results, which then were compared with paleoclimate data, leading to conclusions about how well we understand past climate forcing parameters and climate change.

The interest in the late glacial interval has increased with the recognition of the rapidity of climatic events during this interval and the possibility that future anthropogenic warming could include unexpected and rapid climatic surprises. One of the hypotheses proposed to explain such rapid climate changes are changes in the intensity of the North Atlantic Deep Water (NADW) formation, also called the thermohaline circulation (Broecker, 1997). Experiments with coupled ocean–atmosphere (AGCM) models have resulted in the opportunity to investigate the role of thermohaline changes on global climate. Thus, in a series of coupled atmospheric–ocean general circulation models (AOGCMs), late glacial experiments were made in the last decade by two different groups—the Geophysical Fluid Dynamics Laboratory (GFDL) and the European Community Hamburg Model Version 3 (ECHAM3)—with emphasis on transient forcing parameters, specifically of changes in meltwater input into the North Atlantic and resulting atmospheric effects.

22.2. METHODOLOGY

The late glacial modeling experiments fall into two basic methodological categories: the time slice approach and the sensitivity test approach. This chapter

TABLE 1 General Circulation Model Experiments: Time Slices Approach (A–D) and Sensitivity Test Approach (E–Q)

	Model	Boundary conditions	Sensitivity test	Key results	Conclusions	Further questions
A.	CCM0[a] (AGCM) (Kutzbach and Guetter, 1986; Kutzbach et al., 1987; Kutzbach et al., 1993)	15,000 B.P. orbital perpetual January, July LGM ice sheet[b] LGM SST[c] LGM sea ice		Split jet; July warmer than 18,000 B.P. by 2K at 30°–90° N, Precipitation increase in July; Precipitation decrease in January	Monsoon response larger than response to glacial boundary conditions	Uniform values of global and annual average temperatures fixed 9000 to 0 B.P.? Bias of NCAR CCM to colder winters?
B.		12,000 B.P. orbital 1/2 LGM ice sheet 1/2 LGM SST Cold North Atlantic LGM sea ice		No split jet; Maximal flow over ice sheet; Precipitation increase in northern tropics	Orbital forcing shows warming in summer; Net warming in southeast United States; Monsoon strong in Asia and Africa	
C.	CCM1[a] (AGCM) (Webb et al., 1998; Webb and Kutzbach, 1998)	16,000 B.P. (13,500 ^{14}C B.P.) orbital 16,000 B.P. (13,500 ^{14}C) ice sheets CO_2 210 ppm Fixed-ocean transports		Orbital forcing shows warming greater in summer; Net warming in southeast United States, Beringia	Fundamental climate forcing mechanisms are understood	Climate drift in CCM1? Dynamical ocean? Interactive vegetation? Terrestrial hydrology?
D.		14,000 B.P. (12,000 ^{14}C B.P.) orbital 14,000 B.P. (12,000 ^{14}C) ice sheets CO_2 230 ppm Fixed-ocean transports		Warming south of ice sheets; Tropics stronger monsoon than 16,000 B.P.		
	CCM1 (AGCM) (Bartlein et al., 1998)	Same as above		CCM1 shows less change from LGM than CCM0; Glacial anticyclone not as large in CCM1; July temperature anomalies greater in CCM1 than in CCM0 away from ice sheets; Precipitation gradient across Beringia better developed in CCM1; Mismatches in simulated spread of *Picea, Pseudotsuga, Artemisia*	Model resolution may cause model-data discrepancies; Mismatches in southeast United States and Beringia due to particular atmospheric circulation anomalies, amplified by surface energy and water balances	Analysis of modern climate in key regions? Use of *bioclimatic* models?
	CCM1 (AGCM) (Webb et al., 1998)	Same as above		CCM1 July temperature larger than CCM0 or paleodata south of ice sheets	Glacial anticyclone led to overestimation of high summer temperatures	Model resolution? Soil moisture?

(continues)

TABLE 1 *(continued)*

	Model	Boundary conditions	Sensitivity test	Key results	Conclusions	Further questions
E.	GISS (AGCM) (Rind et al. 1986)	0 and 11,000 B.P. orbital 2/3 LGM ice sheet SST—today and North Atlantic change	11,000 B.P. orbital	Summer temperature increase at midlatitudes Northern Hemisphere	Seasonality and subsidence important	Agreement with data?
			North Atlantic SST Warm/cold	Europe temperatures change with cooling	Agrees with Younger Dryas evidence for European temperature change, not much U.S. temperature change	Gobal distribution of Younger Dryas?
F.	GISS (AGCM) (deMenocal and Rind 1993)	Same as above, lowered Asian orography	Same as above, lowered Asian orography	Increased Asian summer monsoon due to orbital change; North Atlantic SSTs produce cool North African climate Increased winter trade winds with lower orography in Asia	Northwest African climate most sensitive to North Atlantic SST; Arabian and northeast Africa climate sensitive to ice sheets and albedo	
G.	CCM1 (AGCM) (Oglesby et al. 1989)	Today Gulf SST	Gulf—perpetual January SST −3°C −6°C −12°C	Reduction of North Atlantic storm track intensity	Storm track intensity is dependent on Gulf temperature	Other boundary condition changes?
H.	CCM1 (AGCM) (Maasch and Oglesby 1989)	0 and 12,000 B.P. orbital Ice sheet SST	Gulf—perpetual January and July SST −6°C	January temperature decline offsets 12,000 B.P. conditions; July temperature decline strengthens trade winds	Gulf SST important	Other sensitive SST areas?
I.	GISS (AGCM) (Overpeck et al. 1989)	11,000 B.P. orbital Ice sheet SST	Gulf SST −6°C	More intense trade winds	Agrees with some paleodata	Causes for African aridity?
J.	GISS (AGCM) (Peteet et al. 1997)	11,000 B.P. orbital Ice sheet SST	North Pacific SST −2°C	Cooling throughout Northern Hemisphere; increased midlatitude snow	Water vapor important; implications for ice age	Tropics SST cooling?
K.	HAMBURG (AGCM) (Renssen et al. 1996; Renssen 1997; Renssen and Isarin 1998)	0 and 12,000 B.P. orbital Ice sheet SST CO_2 230 ppm	North Atlantic SST cooling North Pacific SST −2°C cooling	Younger Dryas Europe cooling with strong westerly circulation, strong jet; no Southern Hemisphere temperature change	North Atlantic sea ice cover important for result; SSTs important	Unexplained Younger Dryas forcings if Younger Dryas is global, methane?

HAMBURG (AGCM) (Isarin et al. 1997)	Same as above	Same as above	Winter winds from south to southwest in Netherlands, west to southwest in Poland	Winter wind speeds higher closer to ocean, support observations	What are small nonlinear responses?
L. CCM1 (AGCM) (Felzer et al. 1998)	16,000 B.P. (13,500 ^{14}C B.P.) orbital; 16,000 B.P. ice sheets; CO_2 210 ppm; Fixed ocean transports	16,000 B.P. orbital; 16,000 B.P. ice sheets; CO_2 210 ppm	Ice sheets cool Northern Hemisphere; CO_2 cools Southern Ocean	Responses to forcings are linear	
M.	14,000 B.P. (12,000 ^{14}C B.P.) orbital; 14,000 B.P. (12,000 ^{14}C B.P.) ice sheets; CO_2 230 ppm; Fixed-ocean transports	14,000 B.P. orbital; 14,000 B.P. ice sheets; CO_2 230 ppm	14,000 B.P. orbital warming is larger than CO_2 cooling; Asian monsoon is strengthened		
N. CCM1 GENESIS (AGCM) (Fawcett et al. 1997)	0 and 12,000 B.P. orbital Ice sheet; Ocean heat transports; CO_2 250 ppm	Nordic Sea Winter heat transport: -off -on	Heat transport on leads to 2.8°C warming in Greenland; shift of storm tracks toward Greenland	GISP2[d] temperature and precipitation not fully explained by model experiments	Role of CH_4 ocean circulation?
O. HAMBURG (OAGCM) (Mikolajewicz et al. 1997)	Today	500-year meltwater spike to North Atlantic	NADW shutdown leads to 4K North Atlantic cooling, 1–2K cooling in North Pacific	Teleconnections important	? if other boundary conditions changed
P. GFDL[e] (OAGCM) (Manabe and Stouffer 1997)	Today	500-year meltwater spike: -to North Atlantic -to subtropical Atlantic	Placement of meltwater spike important; century-scale recovery	Thermohaline circulation important in Younger Dryas	? reasons for rapid thermohaline recovery
Q. GFDL (OAGCM) (Fanning and Weaver 1997)	Today	Regional meltwater discharge rates	Mississippi meltwater pushes NADW; St. Lawrence inhibits NADW; 2°–3°C global cooling	Global thermohaline circulation provides interhemispheric teleconnection with Southern Ocean	Role of clouds, CO_2? Rapidity of resumption of NADW?

[a]CCMO, Community Climate Model O; CCMI, Community Climate Model I. [b]LGM, last glacial maximum. [c]SSt, sea surface temperature. [d]GISP2, Greenland Ice Sheet Project 2. [e]GFDL, Geophysical Fluid Dynamics Laboratory.

is a review of several of these experiments and includes a summary of some key insights, both from the model results and from the model–paleoclimate data comparisons. In the comparison of model experiments, Table 1 lists the type of model, the references for the experiments listed, the boundary conditions for the experiments, the sensitivity tests (if applicable), the key results of the experiments, the conclusions, and further questions asked by the authors of those experiments. These results are then discussed in more detail, and some differences in the model runs and the conclusions of the different modeling groups are highlighted. CO_2 is not listed as a boundary condition in these experiments unless levels are different from modern values.

22.3. RESULTS

22.3.1. Time Slice Approach (A–D, Table 1)

In the time slice approach, a series of past time intervals are selected, and multiple boundary conditions are chosen to represent each time interval. The GCM is then run using those boundary conditions, and the results are discussed primarily in terms of a comparison between these results and the paleoclimate data for that time interval. Thus, a series of experiments were made by Kutzbach and Guetter (1986) using the NCAR Community Climate Model 0 (CCM0) at 3000-year intervals from 18,000 B.P. (last glacial maximum, LGM) through the present (18,000, 15,000, 12,000, 9000, 6000, 0 B.P.). Boundary conditions for the two late glacial time intervals (15,000 and 12,000 B.P.) are given in Table 1, and the model was run with perpetual January and perpetual July simulations. The results of these experiments were initially given in general terms, as global and hemispheric averages, zonal averages, global maps, and selected regional averages (Kutzbach and Guetter, 1986). One primary conclusion from the set of experiments was that the tropical precipitation and monsoon response could be ascribed more to insolation changes than to glacial boundary conditions.

A more detailed analysis of the individual results at different time intervals was produced by Kutzbach (1987), Kutzbach and Ruddiman (1993), and Kutzbach et al. (1993) within the frame of the Cooperative Holocene Mapping (COHMAP) project. The time slice model results were compared with a set of paleoclimatic summaries (Wright et al., 1993); since the offset between calendar and radiocarbon years (Bard et al., 1990) was not yet known at that time, the model results (in calendar years) were compared with paleoclimate results (in radiocarbon years). The comparison between the model output for the respective time interval focused primarily on the apparent agreement with the

paleoclimate data at regional scales. Details of the 15,000 and 12,000 B.P. runs are discussed in the following sections (see also Table 1A,B).

22.3.1.1. A. AGCM 15,000 B.P. "Snapshot" COHMAP Experiment—Forcing Parameters: 15,000 B.P. Orbital, LGM Ice Sheets and Sea Ice, and CLIMAP LGM SST

In this 15,000 B.P. simulation, while the orbital (insolation) values were changed to those of 15,000 B.P., the other boundary conditions were the same as for 18,000 B.P.—i.e., full glacial ice sheets, sea surface temperature (SST), and sea ice extent. Thus, this snapshot provides a useful comparison with the 18,000 B.P. snapshot in understanding the effects of the insolation change on global climates. The results show that in North America the jet stream continues to split around the Laurentide ice sheet (just as in 18,000 B.P.). In contrast to the 18,000 B.P. model outputs, however, July temperatures are 2 K warmer in the Northern Hemisphere, July precipitation between 30° and 90°N is slightly higher, and January precipitation is slightly lower than before.

22.3.1.2. B. AGCM 12,000 B.P. "Snapshot" COHMAP Experiment—Forcing Parameters: 12,000 B.P. Orbital, Ice Sheets at 50% of LGM, LGM Sea Ice, ½ LGM CLIMAP SST, but North Atlantic Cold

The extent of ice at 12,000 ^{14}C B.P. followed Denton and Hughes (1981). Laurentide ice sheet height was 1650 m, leaving an ice volume of about 40% of that during the glacial maximum. Sea ice was set the same as for the LGM; Climate: Long-Range Investigation, Mapping, and Production (CLIMAP Project Members 1981) SSTs were 50% warmer than those for the LGM, except in the western North Atlantic, where SSTs poleward of 46°N were set at LGM values. Albedo for bare land was set halfway between the value for the LGM and modern values. Comparisons were made with modern climates.

Model outputs for 12,000 B.P. showed that the split jet in North America had disappeared due to lower ice sheets. Instead, the maximum air flow was over the ice sheet, and precipitation had increased in the northern tropics. Net warming occurred in the southeastern United States, and the monsoon was strong in Asia and Africa. Comparisons between the 15,000 and 12,000 B.P. model results and paleoclimate data were made by individual groups of researchers working at regional scales (Wright et al., 1993). These compilations of paleodata are a valuable contribution. Most of the researchers working with paleoclimate data accepted the model results at the regional scales and focused upon general agreements with the model results. One example is Webb et al.'s (1993) observation that lake-level data in eastern North America agreed with the model P–E (precipitation minus evaporation) estimates. By comparing model results with paleoclimate data from China, Winkler and Wang (1993) noted that the dynamic nature of China's late glacial vegetation could not be explained by orbital changes or orography and that insolation seasonality changes had a far greater impact on the environment than Northern Hemisphere ice sheet changes.

Differences between model outputs and paleoclimate data were found for Alaska, the northwestern United States, and the southeastern United States (Wright et al., 1993), where the models simulated conditions far warmer than the paleoclimate reconstructions, which showed substantial cooling at 12,000 B.P. in these areas.

As a time slice update, a new set of model experiments and data synthesis were recently presented by the COHMAP group using the NCAR CCM1 for time slices of 21,000, 16,000, 14,000, 11,000, and 6000 B.P. (Webb and Kutzbach, 1998; Kutzbach et al., 1998; Webb et al., 1998). As these authors note, there were several reasons for the updated comparisons, including the following:

1. The NCAR CCM1 is an improvement over CCM0, with respect to the annual cycle resolution, and improved land surface and upper ocean properties.

2. Given the offset between radiocarbon and calendar years, there is a difference in the relative timing between boundary condition changes and orbital changes (Bard et al. 1990). Thus, for example, 16,000 cal. B.P. relates to 13,500 ^{14}C B.P. years and 14,000 cal. B.P. relates to 12,000 ^{14}C B.P.

3. New data suggest that the heights of the full and late glacial Laurentide and Scandinavian ice sheets were less, while the height of the Greenland ice sheet was greater than assumed previously (after Peltier 1994).

4. Instead of specifying SSTs for each specific time slice, ocean heat transports are fixed, and SSTs are allowed to adjust. Of major significance in this new approach are sensitivity experiments with the CCM1 version of the model, which examine the linearity of the model's response to different forcing factors (see Table 1L,M) (Felzer et al., 1998).

22.3.1.3. C. AGCM 16,000 B.P. "Snapshot" COHMAP Experiment—Forcing Parameters: 16,000 B.P. Orbital, 16,000 B.P. Ice Sheet, Fixed-Ocean Heat Transports, Lower CO$_2$

Orbital values for 16,000 B.P. (13,500 ^{14}C B.P.) are used, along with Peltier's (1994) ice reconstruction,

fixed-ocean transports, and a CO$_2$ value of 210 ppm. The results include global warming in summer due to orbital forcing, including a net warming in the southeastern United States (just as with the CCM0 version), which disagrees with paleoclimatic data.

22.3.1.4. D. AGCM 14,000 B.P. "Snapshot" COHMAP Experiment—Forcing Parameters: 14,000 B.P. Orbital and Ice Sheet, Fixed-Ocean Transports, Lower CO$_2$

Orbital values are used for 14,000 B.P. (12,000 ^{14}C B.P.), along with Peltier's (1994) ice reconstruction, fixed-ocean transports, and a CO$_2$ value of 230 ppm. At 14,000 B.P., warming is present just south of the ice sheets, in contrast to the 12,000 (^{14}C) B.P. CCM0 results (Fig. 19.2 of Webb et al., 1993). The tropics show a pattern similar to that for 16,000 B.P., but a stronger monsoon (Kutzbach et al., 1998). In comparison with CCM0 model results, some improvements of the CCM1 results with paleoclimatic data are noted, such as Alaska's lower temperatures (compared to higher temperatures in the CCM0 experiment), due to the reduction of the split jet.

Bartlein et al. (1998) also reviewed the large-scale features of the simulated climates over North America, using a response surface technique to transform the model-simulated climate into plant taxa distributions, and examined mismatches between the model simulations and data.

For the late glacial interval, they compared the 12,000 ^{14}C B.P. CCM0 run with the 14,000 cal. B.P. CCM1 run and found that (1) the atmospheric circulation change between the LGM and the late glacial is not as large in CCM1 as in CCM0; (2) the glacial anticyclone is not as strong in CCM1 as in CCM0; (3) July temperature anomalies in regions distant from the ice sheet are larger in CCM1 than in CCM0; and (4) the precipitation gradient across Beringia is better developed in CCM1 than in CCM0. Highlighted are mismatches such as between the simulated and observed patterns of *Picea* (spruce) in Beringia. While the model simulation suggests that *Picea* would have expanded in Beringia by 16,000 B.P., in response to the model simulation of conditions that are warmer and wetter than now as early as 16,000 B.P., the data suggest that *Picea* did not expand until the early Holocene. Differences in the spread of *Pseudotsuga* (Douglas fir) and *Artemisia* (sagebrush) between the model and the data are attributed to probable shortcomings in the model's experimental design, specifically perhaps the model's resolution. The mismatches between model outputs and data in the southeastern United States and Beringia are attributed to particular atmospheric circulation anomalies, amplified by surface energy and water balance properties specified in the model.

Webb et al. (1998) and Kutzbach et al. (1998) discuss the model simulations compared with pollen-derived model estimates in eastern North America. They provide difference maps for CCM0 and CCM1 for mean July (Fig. 1 [see color insert]) and January that are very instructive. These maps show that CCM1 results show markedly higher temperatures than CCM0 results for eastern North America south of the ice sheets at 16,000 and 14,000 B.P., which is much warmer than indicated by pollen data. They conclude that the glacial anticyclone simulated over the ice sheet led to an overestimation of summer temperatures south of the ice sheet.

22.3.2. Sensitivity Test Approach (E–Q, Table 1)

The sensitivity approach tests the sensitivity of the GCM to a single boundary condition or set of boundary conditions, such that the result can be compared with a control run and the influence of the changed boundary condition(s) can be identified. Emphasis is placed upon analysis of the model's physics that produced the result. For example, an experiment testing the sensitivity of the Earth's atmosphere to a colder than present North Atlantic Ocean relies on the expertise of the atmospheric physicist to analyze the results of the experiment, including air temperature, precipitation, evaporation, winds, sea level pressure (SLP), heat flux, snow cover, high and low clouds, sensible heat flux, etc. At the same time, paleoclimate data are examined for a simpler set of results (temperature, precipitation, and wind) and compared with results of the model run for the time interval of the sensitivity test. Discrepancies between the model result and the paleoclimate data are emphasized. New data are then targeted, model construction is improved, and new questions are asked.

22.3.2.1. E. AGCM Younger Dryas Experiments— Forcing Parameters: 11,000 B.P. Orbital, ⅔ of Full Glacial Ice Sheet, North Atlantic SST Colder Than Today

The primary focus of the Rind et al. (1986) experiment was to investigate the sensitivity of global climates to colder North Atlantic SSTs during the Younger Dryas (11,000–10,000 ^{14}C B.P.) interval using the GISS atmospheric GCM (Hansen et al., 1983). The general approach was to compare climate model runs with 11,000 ^{14}C B.P. ice sheet and orbital values to similar runs with a colder North Atlantic, using the Ruddiman and McIntyre (1981) SST data north of 40°N. The results showed that with both 11,000 ^{14}C B.P. and present con-

ditions, the colder North Atlantic SSTs produced cooling over western and central Europe, in agreement with paleoclimatic data for the Younger Dryas interval in Europe. Despite the presence of increased land ice and lower ocean temperatures, however, the Younger Dryas summer air temperatures at midlatitudes in the model were warmer than they are today due to changes in the orbital parameters, chiefly precession, and atmospheric subsidence at the perimeter of the ice sheets. Rind et al. (1986) also compiled literature for Younger Dryas climates in North America, Europe, and globally to assess the existence of cooling worldwide. They found that while the model results agreed with paleoclimate data in Europe, they did not show significant cooling in North America. Subsequent research in North America strongly indicated the existence of cooler temperatures during the Younger Dryas interval in records from eastern North America and the Pacific Northwest coast (see summary in Peteet, 1995), which then led to further data acquisition and modeling (J, Table 1).

22.3.2.2. F. AGCM 11,000 B.P. Experiment— Focus on Africa and Asia.

deMenocal and Rind (1993) built on the sensitivity tests of Rind et al. (1986) to explore the sensitivity of the African and Asian monsoons to the cold North Atlantic SST forcing at 11,000 ^{14}C B.P. They found, in agreement with the experiments of Kutzbach and Guetter (1986) and Rind et al. (1986), that the Asian summer monsoon climate is responsive to changes in insolation, but the Asian winter monsoon is only mildly affected. In contrast to the results of Prell and Kutzbach (1987), however, they found that the response to North Atlantic SSTs is larger for the African monsoon than for the Asian monsoon. Cooler North Atlantic SSTs produced cooler conditions over northwestern Africa, but the climate of southern Asia was relatively unaffected. While reducing elevations in Asia produced increased trade winds there, African climate was relatively unaffected by this change. They concluded that while Arabian and northeastern African climate was primarily sensitive to changes in Fennoscandian ice elevation and albedo, northwestern African climate is most sensitive to changes in North Atlantic SSTs. These experiments support the reconstructions based on paleoceanographic evidence. deMenocal and Rind (1993) suggested that climates in northwestern Africa have been strongly regulated by North Atlantic SST variations over the last 2.4 million years. In contrast, Arabian and northeastern African climates were more affected by changes in monsoon intensity related to changes in the extent of high-latitude ice sheets.

22.3.2.3. G–H. AGCM 12,000 B.P. Experiments—Focus on the Gulf of México

Oglesby et al. (1989) produced a series of AGCM experiments with the NCAR CCM that imposed perpetual January temperatures of 3°, 6°, and 12°C lower in the Gulf of México. In all three experiments, North Atlantic storm track intensity was reduced, along with a strong decrease in transient eddy water vapor transport out of the Gulf of México. In the experiments, the region was generally cooler and drier, with a reduction in precipitation. However, warmer, wetter conditions were found over Europe for both the 6° and 12°C SST reductions, but cooler conditions were found for the 3°C reduction. Thus, the magnitude of the imposed SST anomaly affects both the sign and the magnitude of the air temperature response. The authors concluded that Gulf of México meltwater cooling strongly influences the paths of Atlantic storm tracks and may have been responsible for late glacial climatic oscillations seen in paleoclimatic records.

Maasch and Oglesby (1989) expanded on the previous series of experiments to include 12,000 B.P. orbital conditions and perpetual July as well as perpetual January. They found that in the perpetual July simulations with a cool Gulf of México, the Atlantic subtropical high is strengthened and expanded to the west, resulting in stronger trade winds across the Caribbean and southwest Atlantic Oceans. The summer effects of a cooler Gulf of México tended to enhance changes otherwise induced by the 12,000 B.P. boundary conditions.

22.3.2.4. I. AGCM 11,000 B.P. Experiment—Focus on the Gulf of México.

Overpeck et al. (1989), using the GISS AGCM, showed in an experiment with Gulf of México SSTs 6°C cooler than in the experiments of Rind et al. (1986) and 11,000 ^{14}C B.P. boundary conditions (⅔ LGM ice sheet, 11,000 B.P. insolation) that significant temperature changes were produced outside the Gulf of México region. In this model run, lower surface air temperatures, higher SLP, a large negative precipitation anomaly, and more intense trade winds occurred in the western North Atlantic. They concluded that the Gulf of México SSTs must have played a large role affecting deglaciation in the North Atlantic region.

22.3.2.5. J. AGCM Younger Dryas Experiments—Focus on the North Pacific and Snow Cover Over the Continents

Circum-Pacific marine and terrestrial records indicate a series of temperature-inferred oscillations during the late glacial period (Peteet et al., 1997). The purpose of this model run was to test the sensitivity of the Northern Hemisphere air temperatures to North Pacific SST oscillation in the GISS AGCM model. The effects of colder North Pacific SSTs are cooler air temperatures over North America, as well as in parts of Europe and Asia. The colder SSTs result in a large hemispheric response due to the loss of water vapor as a greenhouse gas. The model also shows a significant increase in snow cover at midlatitudes (Fig. 2). The marked sensitivity of Northern Hemisphere climates to a North Pacific SST change has implications for ice age and late glacial climates. The results of this experiment provide a rapid mechanism for widespread cooling that has not been previously addressed.

22.3.2.6. K. AGCM Younger Dryas Experiments Revisited—Focus on Orbital, Ice Sheet, North Atlantic Sea Ice, North Atlantic and North Pacific SSTs

Using the Hamburg AGCM (ECHAM3), Renssen et al. (1996) and Renssen (1997) performed four numerical experiments on the Younger Dryas climate. They altered insolation, ice sheets, atmospheric CO_2 concentration, SSTs in both the North Atlantic and the North Pacific north of 40°N, and North Atlantic sea ice cover. Extended sea ice cover in the North Atlantic proved to be very important in cooling the atmosphere, causing a weakening of the westerlies and a decrease in precipitation. Similar to the results of Rind et al. (1986), they found that paleoclimate evidence for the Younger Dryas cooling agrees with the model output for land areas surrounding the North Atlantic and coastal North Pacific, but does not agree with evidence from tropical or South American data for a Younger Dryas oscillation. Thus, they concluded that either an atmospheric model is unsuitable for modeling Younger Dryas climate or the set of boundary conditions used was insufficient to reconstruct the Younger Dryas climate.

Isarin et al. (1997) used the same Hamburg model results to examine the wind fields both in the Netherlands and in Poland. They found that winter winds were stronger and from the south to southwest in the Netherlands, but from the west to southwest in Poland, supporting the observations of wind direction in the geological data.

22.3.2.7. L–M. AGCM 16,000 and 14,000 B.P. Experiments—Focus on the Orbital, Ice Sheet, and Lowered CO_2 Changes.

Using the NCAR CCM1, sensitivity experiments were run to determine if the climate responses to orbital, ice sheet, and CO_2 change were additive (Felzer et al., 1998). The authors found that the surface tem-

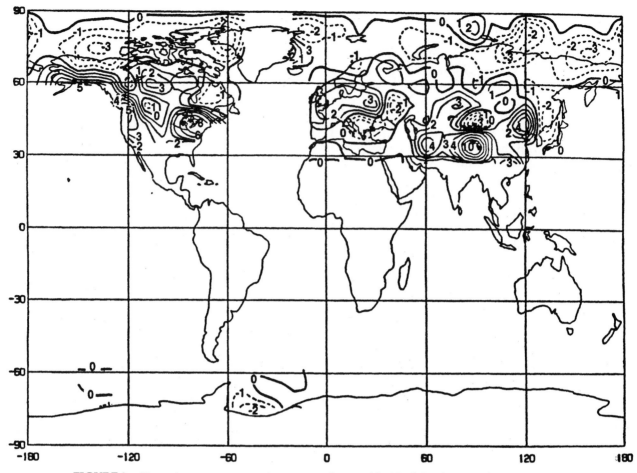

FIGURE 2 Change in percentage annual snow cover due to colder North Pacific sea surface temperatures (SST −2°C) at 11,000 B.P. compared to a model run with today's SST. Grid-point standard deviations are about 0.3%. (From Peteet et al., 1997. With permission.)

perature response is additive for the three parameters. At 16,000 B.P. (13,500 ^{14}C B.P.) and 14,000 B.P. (12,000 ^{14}C B.P.), the simulation differences showed that lower CO_2 (210 ppm at 16,000 B.P., 230 ppm at 14,000 B.P.) contributed to maximum cooling in the oceans, while enhanced orbital insolation during the summer contributed primarily to warming over land. The lower CO_2 concentrations caused the most cooling in regions of sea ice growth in the winter hemisphere, which added to the regional ice sheet cooling during boreal winter, and created large cold anomalies in the Southern Ocean during boreal summer (Fig. 3).

22.3.2.8. N. AGCM GENESIS Younger Dryas Termination Experiment

Two GCM experiments using GENESIS (modified CCM1) were completed for the Younger Dryas time (Fawcett et al. 1997). This model used the boundary conditions of 12,000 B.P. orbital conditions, Peltier's (1994) land ice extent, and 250 ppm CO_2. In the first experiment, the North Atlantic Ocean heat flux was reduced to 10 W/m^2 (watts per square meter heat off), simulating a shutdown in the NADW formation. For the second experiment, the North Atlantic Ocean heat flux was turned on (300 W/m^2, active deepwater formation). Model results over Greenland showed an annual average temperature increase from heat off to heat on of 2.8°C, which is about half of the temperature indicated by the Greenland ice core oxygen isotope profile (e.g., Alley, 2000). The warmer climate experiment showed a shift in storm tracks toward Greenland, but the annual average precipitation rate increased by a few tens of percent, as compared with the 100% accumulation increase documented in the ice core. Thus, the authors concluded that either the model is not fully sensitive to the ocean heat flux forcing or some other factor(s) are active.

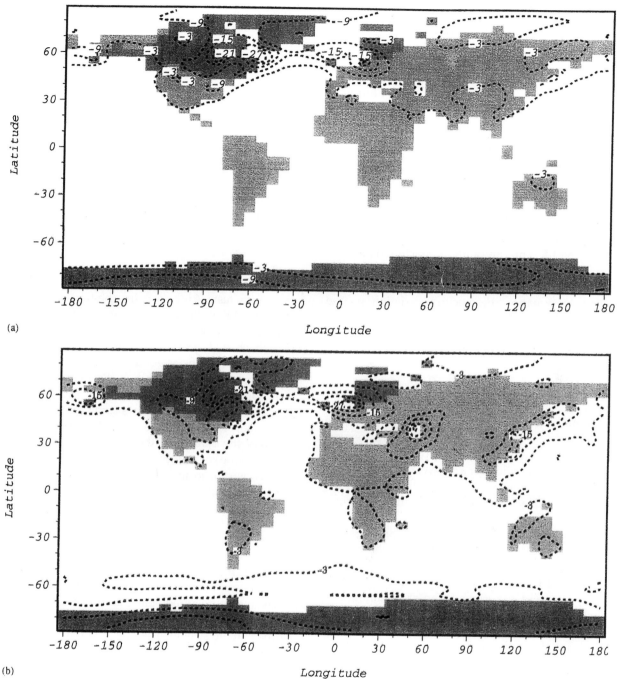

(a)

(b)

FIGURE 3 Surface temperature difference for winter (December–January–February, DJF) between (a) the 14,000 cal. B.P. (14 ka) sensitivity experiment (330 ppm) and (b) the 14 ka simulation experiment (230 ppm) and the pre-industrial control (267 ppm) (Kutzbach et al., 1998). Contour interval is 6°C. The DJF lower orbital insolation in the Northern Hemisphere contributes to the ice sheet cooling effects over the Northern Hemisphere landmasses, while the decreased CO_2 of 37 ppm contributes to the ice sheet cooling of the high-latitude oceans in the Southern Hemisphere. (From Felzer et al., 1998. With permission.)

22.3.3. Coupled Ocean-Atmosphere GCM Experiments

22.3.3.1. O. OAGCM Thermohaline Shutdown Experiment

Mikolajewicz et al. (1997), using the European Community Hamburg Model Version 3/Large Scale Geostrophic (ECHAM3/LSG) coupled ocean–atmosphere model, found a primary atmospheric forcing in the North Pacific due to shutdown of the thermohaline circulation in the North Atlantic. The changes in wind strongly affected Pacific Northwest coastal upwelling, and the atmospheric surface cooling caused increased northeastern Pacific thermocline ventilation. The cooling in the model along the northwestern American coast is about 1 K. Thus, the authors concluded that due to climate teleconnections, a cooling and increased sea ice cover in the North Atlantic Ocean alone seem to be sufficient to enhance thermocline variation in the northeastern Pacific Ocean.

22.3.3.2. P. OAGCM Meltwater Placement Experiment

Manabe and Stouffer (1997) used the GFDL atmospheric GCM coupled with an ocean GCM to explore the response of the coupled model to the discharge of fresh water into the North Atlantic Ocean. In the first experiment, they assumed a freshwater discharge into the northern North Atlantic over a 500-year period. The results showed that the thermohaline circulation in the North Atlantic Ocean weakened, reducing the surface air temperature around the Atlantic land areas and, to a lesser degree, in the circumpolar ocean of the Southern Hemisphere. When the freshwater discharge was stopped, the thermohaline circulation rapidly intensified, regaining its original intensity in a few hundred years. In a second experiment, the same amount of meltwater was discharged into the subtropical North Atlantic again over a 500-year period, and although the thermohaline circulation and climate evolved in a manner similar to that for the first experiment, the magnitude of the thermohaline response was much weaker. Thus, the authors concluded that the freshwater input is much more effective in weakening the thermohaline circulation if it is located in the northern rather than in the subtropical North Atlantic.

22.3.3.3. Q. OAGCM Meltwater Placement and Timing Experiment

Fanning and Weaver (1997) used a coupled atmosphere–ocean–ice model to test the sensitivity of the North Atlantic thermohaline circulation to regional meltwater discharge rates and their possible implica-

tions for Younger Dryas climates. Interestingly, they found that Laurentide freshwater runoff into the North Atlantic prior to the Younger Dryas interval pushes NADW formation beyond the limit of its sustainability. The diversion of freshwater input during Younger Dryas times through the St. Lawrence into the North Atlantic then led to an NADW shutdown altogether. However, the timescale for subsequent resumption of NADW that terminated the Younger Dryas event took ca. 1000 years, while ice core records indicate that the termination occurred in a decade or less (Alley et al., 1993).

22.4. DISCUSSION AND CONCLUSIONS

Two major differences appear when the conclusions from the modeling groups using the time slice approach are compared with the conclusions of those performing sensitivity tests. Authors using the time slice approach suggest (1) that the fundamental climate forcing mechanisms for late glacial climate changes are understood (Kutzbach et al., 1998) and (2) that regional model versus data discrepancies are due to internal model problems, such as model resolution (Bartlein et al.. 1998) or atmospheric circulation over an ice sheet (Webb and Kutzbach et al., 1998). Authors using the sensitivity test approach, in general, question their results more in terms of the validity of the imposed boundary conditions and the sensitivity of their models. Part of this difference is due to the different approach. Most of the time slice experiments focus on gradual forcing of changes in ice sheet and insolation (with fixed SSTs or fixed-ocean heat transports), whereas the sensitivity experiments focus primarily on changes in ocean temperatures (North Atlantic, Gulf of México, and North Pacific) and thermohaline changes. These sensitivity tests have shown the large effects that SSTs have on all aspects of climate at regional and global scales; thus, the sensitivity test community emphasizes this component of the climate system. The question that emerges from these two different experimental approaches is just how important SST and/or heat transport parameters were in forcing late glacial climates.

The COHMAP time slice approach has provided the community with regional syntheses of paleoclimate data that will be valuable for future model experiments (Wright et al., 1993). However, in many cases, the model results for the perpetual January and July experiments using the CCM0 were accepted uncritically by the authors of regional data–model comparisons. Thus, the deficiencies within the model and the limits imposed by possibly erroneous boundary conditions during the late glacial (18,000 B.P. sea ice extent, CLIMAP SST) escaped notice. The severe limitations of all GCMs in accurately simulating regional climates as noted in the Intergovernmental Panel for Climate Change (IPCC, 1996) reports were not recognized. For example, in simulating modern regional climates, all the models produce errors of 2°–3°C in many regions, and rainfall errors are even greater, deviating by as much as 20–50% from present-day values (Houghton et al., 1990, 1992). Although the importance of SST in some model experiments is mentioned, authors attribute the primary late glacial forcing to the extent of ice sheets and insolation parameters.

It is important to note that Kutzbach et al. (1998) have performed comparisons between CCM0 and CCM1 experiments and found better agreement with paleoclimate data in some areas (Beringia) and less agreement in others (land areas adjacent to ice sheets). It is interesting that higher-than-today summer temperatures south of the ice sheets, shown by CCM1 experiments for 12,000 B.P. (Bartlein et al., 1998), are also shown by the GISS model for 11,000 B.P. (Rind et al., 1986) and the Hamburg model (Renssen 1997).

Despite the use of different AGCMs (NCAR, GFDL, GISS, and Hamburg), there is a marked similarity in the various model responses to many of the forcing parameters, such as to the 11,000 or 12,000 B.P. insolation and land ice. With other forcing parameters, including SST, sea ice, and lower CO_2 concentration, the results are not comparable from model to model, and duplicate experiments need to be made with different models. In this respect, the sensitivity approach is preferable to the time slice approach because a single forcing parameter can be identified and its role in the model result can be analyzed, which then aids in understanding the model physics and improving the model itself.

An interesting feature of the late glacial sensitivity and time slice model experiments is the comparison of results for the Southern Hemisphere. Most of the model experiments show no cooling in the Southern Hemisphere during the late glacial in response to the effects of extended ice sheets and insolation (Rind et al., 1986; Renssen, 1997; Kutzbach and Guetter, 1986). However, models with fixed-ocean heat transport instead of fixed SSTs show significant Southern Hemispheric cooling in response to lowered CO_2. For example, the CCM1 model does show substantial cooling in the Southern Hemisphere oceans both at 16,000 and at 14,000 cal. B.P., due to lower CO_2 concentration (Felzer et al., 1998) as a feedback from sea ice changes (Fig. 3). This feature is also a characteristic for the GFDL model in ice age experiments (Broccoli and Manabe, 1987). These results suggest that modelers should increase their efforts to understand the large magnitude of this effect. One relevant question to ask concerning this large cooling is whether the cloud response to the sea ice change is re-

alistic in the models, because it should be a negative feedback.

The recent focus on ocean temperature change is derived from evidence of rapid climate change defining late glacial climate events such as the Allerød warming, the Younger Dryas cooling, and the termination of the Younger Dryas interval. In coupled GCMs, lowering of surface salinity in the North Atlantic and its effect on thermohaline changes were considered as a mechanism explaining these rapid changes. The recent emergence of coupled AOGCM experiments provides new tools for testing ocean–atmospheric links that were previously not possible. This development suggests that modeling climate teleconnections is now feasible, and thus, ocean–atmospheric forcing may be addressed.

In examining late glacial paleoclimatic data, at least four major questions concerning late glacial oscillations remain. The first is whether the Bølling warming was a response to insolation forcing or a rapid *flip* that occurred as a result of some internal oscillation of the climate system. The second is the question of the importance of SST in explaining both the rapidity of the changes and the origin of the changes. The third question that someday may be solved is whether the cooling of the Younger Dryas in the North Pacific is indeed synchronous with that in the North Atlantic, or whether the origin was in the North Atlantic and then spread to the North Pacific by atmospheric climate teleconnections. The fourth question, which is still a controversial topic, concerns the Southern Hemisphere—whether the late glacial changes seen there in paleoclimate records are in phase with, lead, or lag the Northern Hemisphere records.

The most critical step in better defining the late glacial climates is to improve our understanding of the paleoclimate data. This includes improved spatial and temporal coverage, as well as a more precise understanding of the paleoclimate record in all aspects. Only then can the paleoclimate information be better modeled and physically understood. Improvements in modeling, such as more realistic cloud coverage and ocean circulation features, challenge the modeling community. It is the iterative process of data generation and modeling that will lead to better, more informed hypotheses about many puzzles in late glacial climate change.

Acknowledgments

Thanks to B. Liepert, V. Markgraf, and an anonymous reviewer for helpful comments. Funding provided by NASA/GISS and U.S. National Science Foundation grant ATM 94-17160. This is Lamont Doherty Earth Observatory Contribution No. 6102.

References

Alley, R., 2000: The Younger Dryas cold interval as viewed from central Greenland. *Quaternary Science Reviews, 19:* 213–226.

Alley, R., D. A. Meese, C. A. Shuman, A. J. Gow, K. C. Taylor, P. M. Grootes, J. W. C. White, M. Ram, E. D. Waddington, P. A. Mayewski, and G. A. Zielinski, 1993: Abrupt increase in Greenland snow accumulation at the end of the Younger Dryas event. *Nature, 362:* 527–529.

Bard, E., B. Hamelin, and R. G. Fairbanks, 1990: U/Th ages obtained by mass spectrometry in corals from Barbados: Sea level during the past 130,000 years. *Nature, 346:* 456–458.

Bartlein, P. J., K. H. Anderson, P. M. Anderson, M. E. Edwards, C. J. Mock, R. S. Thompson, R. S. Webb, T. Webb, III, and C. Whitlock, 1998: Paleoclimate simulations for North America over the past 21,000 years: Features of the simulated climate and comparisons with paleoenvironmental data. *Quaternary Science Reviews, 17*(6–7): 549–586.

Broccoli, A. J., and S. Manabe, 1987: The influence of continental ice, atmospheric CO_2, and land albedo on the climate of the last glacial maximum. *Climate Dynamics, 1:* 87–99.

Broecker, W. S., 1997: Thermohaline circulation, the Achilles heel of our climate system: Will man-made CO_2 upset the current balance? *Science, 278:* 1582–1588.

CLIMAP Project Members, 1981: Seasonal reconstruction of the Earth's surface at the last glacial maximum. Geological Society of America, Map Chart Series MC, 36.

deMenocal, P. B., and D. Rind, 1993: Sensitivity of Asian and African climate to variations in seasonal insolation, glacial ice cover, sea surface temperature, and Asian orography. *Journal of Geophysical Research, 98*(D4): 7265–7287.

Denton, G., and T. Hughes (eds.), 1981: *The Last Great Ice Sheets.* New York: Wiley.

Fanning, A. F., and A. J. Weaver, 1997: Temporal-geographical meltwater influences on the North Atlantic Conveyor: Implications for the Younger Dryas. *Paleoceanography, 12*(2): 307–320.

Fawcett, P. J., A. M. Agustsdottir, R. B. Alley, and C. A. Shuman, 1997: The Younger Dryas termination and North Atlantic Deep Water formation: Insights from climate model simulations and Greenland ice cores. *Paleoceanography, 12*(1): 23–38.

Felzer, B., R. J. Oglesby, T. Webb, III, and D. E. Hyman, 1996: Sensitivity of a general circulation model to changes in Northern Hemisphere ice sheets. *Journal of Geophysical Research, 101*(D14): 19077–19092.

Felzer, B., T. Webb, III, and R. J. Oglesby, 1998: The impact of ice sheets, CO_2, and orbital insolation on late Quaternary climates: Sensitivity experiments with a general circulation model. *Quaternary Science Reviews, 17*(6–7): 507–534.

Hansen, J., G. Russell, D. Rind, P. Stone, A. Lacis, S. Lebedeff, R. Ruedy, and L. Travis, 1983: Efficient three-dimensional global models for climate studies, models I and II. *Monthly Weather Review, 3:* 609–662.

Harvey, L. D., 1989: Modeling the Younger Dryas. *Quaternary Science Reviews, 8:* 137–149.

Houghton, J. T., B. A. Callender, and S. K. Varney (eds.), 1992: *Climate Change 1992: The Supplementary Report to the IPCC Scientific Assessment.* New York: Cambridge University Press.

Houghton, J. T., G. J. Jenkings, and J. J. Ephraums (eds.), 1990: *Climate Change, the IPCC Assessment.* New York: Cambridge University Press, 364 pp.

IPCC, 1996: *Climate Change 1995: The Science of Climate Change.* Cambridge, UK: Cambridge University Press.

Isarin, R. F. B., H. Renssen, and E. A. Koster, 1997: Surface wind climate during the Younger Dryas in Europe as inferred from aeolian records and model simulations. *Palaeogeography, Palaeoclimatology, Palaeoecology, 134:* 127–148.

Kutzbach, J. E., 1987: Model simulations of the climatic patterns during the deglaciation of North America. *In* Ruddiman, W. F., and H. E. Wright, Jr. (eds.), *North America and Adjacent Oceans During the Last Deglaciation. Vol. K-3, The Geology of North America.* Boulder, CO: Geologial Society of America, pp. 425–446.

Kutzbach, J. E., and P. J. Guetter, 1986: The influence of changing orbital parameters and surface boundary conditions on climate simulations for the past 18,000 years. *Journal of the Atmospheric Sciences, 43*(16): 1726–1759.

Kutzbach, J. E., and W. F. Ruddiman, 1993: Model description, external forcing, and surface boundary conditions. *In* Wright, H. E., J. E. Kutzbach, T. Webb, III, W. F. Ruddiman, F. A. Street-Perrott, and P. J. Bartlein (eds.), *Global Climates Since the Last Glacial Maximum.* Minneapolis: University of Minnesota Press, pp. 12–23.

Kutzbach, J., R. Gallimore, S. Harrison, R. Behling, R. Selin, and F. Laarif, 1998: Climate and biome simulations for the past 21,000 years. *Quaternary Science Reviews, 17*(6–7): 473–506.

Kutzbach, J. E., P. J. Guetter, P. J. Behling, and R. Selin, 1993: Simulated climatic changes: Results of the COHMAP Climate Model Experiments. *In* Wright, H. E., J. E. Kutzbach, T. Webb, III, W. F. Ruddiman, F. A. Street-Perrott, and P. J. Bartlein (eds.), *Global Climates Since the Last Glacial Maximum.* Minneapolis: University of Minnesota Press, pp. 24–93.

Maasch, K. A., and R. J. Oglesby, 1989: Meltwater cooling of the Gulf of México: A GCM simulation of climatic conditions at 12 ka. *Paleoceanography, 5*(6): 977–996.

Manabe, S., and R. J. Stouffer, 1997: Coupled ocean-atmosphere model response to freshwater input: Comparison to the Younger Dryas event. *Paleoceanography, 12*(2): 321–336.

Marchal, O., T. F. Stocker, F. Joos, A. Indermuhle, T. Blunier, and J. Tschumi, 1999: Modelling the concentration of atmospheric CO_2 during the Younger Dryas climate event. *Climate Dynamics, 15:* 341–354.

Mikolajewicz, U., T. J. Crowley, A. Schiller, and R. Voss, 1997: Modelling teleconnections between North Atlantic and North Pacific during the Younger Dryas. *Nature, 387:* 384–387.

Oglesby, R. J., K. A. Maasch, and B. Saltzman, 1989: Glacial meltwater cooling of the Gulf of México: GCM implications for Holocene and present-day climates. *Climate Dynamics, 3:* 115–133.

Overpeck, J. T., L. C. Peterson, N. Kipp, J. Imbrie, and D. Rind, 1989: Climate change in the circum–North Atlantic region during the last deglaciation. *Nature, 338:* 553–557.

Peltier, W. R., 1994: Ice age paleotopography. *Science, 265:* 195–201.

Peteet, D., 1995: Global Younger Dryas? *Quaternary International, 28:* 93–104.

Peteet, D., A. Del Genio, and K. K.-W. Lo, 1997: Sensitivity of Northern Hemisphere air temperatures and snow expansion to North Pacific sea surface temperatures in the Goddard Institute for Space Studies general circulation model. *Journal of Geophysical Research, 102*(D20): 23781–23791.

Prell, W. L., and J. E. Kutzbach, 1987: Monsoon variability over the past 150,000 years. *Journal of Geophysical Research, 92:* 8411–8425.

Renssen, H., 1997: The global response to Younger Dryas boundary conditions in an AGCM simulation. *Climate Dynamics, 13:* 587–599.

Renssen, H., and R. F. B. Isarin, 1998: Surface temperature in NW Europe during the Younger Dryas: AGCM simulation compared with temperature reconstructions. *Climate Dynamics, 14:* 33–44.

Renssen, H., M. Lautenschlager, and C. J. E. Schuurmans, 1996: The atmospheric winter circulation during the Younger Dryas stadial in the Atlantic/European sector. *Climate Dynamics, 12:* 813–824.

Rind, D., D. Peteet, W. Broecker, A. McIntyre, and W. Ruddiman, 1986: The impact of cold North Atlantic sea surface temperatures on climate: Implications for the Younger Dryas cooling (11–10k). *Climate Dynamics,* **1**: 3–33.

Ruddiman, W. F., and A. McIntyre, 1981: The North Atlantic Ocean during the last deglaciation. *Palaeogeography, Palaeoclimatology, Palaeoecology,* **35**: 145–214.

Webb, T., III, and J. E. Kutzbach, 1998: An introduction to "Late Quaternary Climates: Data Synthesis and Model Experiments." *Quaternary Science Reviews,* **17**(6–7): 465–472.

Webb, T., III, K. H. Anderson, P. J. Bartlein, and R. S. Webb, 1998: Late Quaternary climate change in eastern North America: A comparison of pollen-derived estimates with climate model results. *Quaternary Science Reviews,* **17**(6–7): 587-606.

Webb, T., III, W. F. Ruddiman, F. A. Street-Perrott, V. Markgraf, J. E. Kutzbach, P. J. Bartlein, H. E. Wright, Jr., and W. L. Prell, 1993: Climatic changes during the past 18,000 years: Regional syntheses, mechanisms, and causes. *In* Wright, H. E., Jr., J. E. Kutzbach, T. Webb, III, W. F. Ruddiman, F. A. Street-Perrott, and P. J. Bartlein (eds.), *Global Climates Since the Last Glacial Maximum.* Minneapolis: University of Minnesota Press, pp. 524–535.

Winkler, M. G., and P. K. Wang, 1993: The late Quaternary vegetation and climate of China. *In* Wright, H. E., Jr., J. E. Kutzbach, T. Webb, III, W. F. Ruddiman, F. A. Street-Perrott, and P. J. Bartlein (eds.), *Global Climates Since the Last Glacial Maximum.* Minneapolis: University of Minnesota Press, pp. 221–264.

Wright, H. E., Jr., J. E. Kutzbach, T. Webb, III, W. F. Ruddiman, F. A. Street-Perrott, and P. J. Bartlein (eds.), 1993: *Global Climates Since the Last Glacial Maximum.* Minneapolis: University of Minnesota Press.

23

Pole-Equator-Pole Paleoclimates of the Americas Integration: Toward the Big Picture

VERA MARKGRAF and GEOFFREY O. SELTZER

Abstract

Comparison of present- and past-climate records from the Americas along the Pole-Equator-Pole: Americas (PEP 1) transect provided the incentive to search for interhemispheric linkages of climate patterns and their cause(s). Climate and climate proxy records are reviewed from a range of disciplines, with different temporal resolutions and for different time intervals during the last 20,000 years, including the last glacial maximum (LGM), the late glacial interval, and different time windows during the Holocene. Synchroneity of climate change and comparability of climate signals, both temporal and spatial, were the principal parameters for evaluating interhemispheric linkages. The comparisons showed that, depending on the respective temporal and spatial scales represented by the records, major phase changes in climate are synchronous in the Northern and Southern Hemispheres. However, such synchroneity does not necessarily imply the same forcing, as has been shown by comparison of early Holocene vegetation records from the temperate latitudes in North and South America. On the other hand, lack of synchroneity, e.g., as for the late glacial and Little Ice Age (LIA) oscillations, may imply the same overall forcing, but a different response. Thus, there is no simple and universal solution to how climate change is linked interhemispherically. Instead, the character and cause of climate linkages appear to depend on the specific time interval and its regional and hemispheric boundary conditions, on the temporal and spatial resolutions of the climate parameter discussed, and on an understanding of the specific regional response to change. et al. Copyright © 2001 by Academic Press.

Resumen

La comparación entre registros climáticos actuales y del pasado en América, a lo largo del transecto Polo-Ecuador-Polo (PEP 1) proporciona el incentivo necesario para buscar los vínculos interhemisféricos de los patrones climáticos y su causa o causas. El clima y los registros proxy del clima han sido examinados, por una serie de disciplinas, con diferente resolución temporal y para distintos intervalos de tiempo, para los últimos 20000 años, incluyendo el último máximo glacial, el tardi-glacial, y diferentes ventanas temporales del Holoceno. La sincronización en los cambios climáticos y la comparación de la señal climática, tanto espacial como temporal, ha constituido los principales parámetros para evaluar las conexiones interhemisféricas. Las comparaciones han mostrado, que en función de las respectivas escalas espacial y temporal representadas por los registros, los principales cambios de fases en el clima se producen de forma sincrónica en el hemisferio norte y sur. Sin embargo, tal sincronización no significa necesariamente el mismo forzamiento, como muestra la comparación de los registros de vegetación del inicio del Holoceno en latitudes templadas del norte y de sur de América. Por otra parte, la falta de sincronización, como por ejemplo las oscilaciones en el tardi-glacial y en la Pequeña Edad de Hielo (LIA) puede significar el mismo forzamiento total, pero distinta respuesta. De este modo, no hay una solución simple y universal en cómo el cambio de clima está conectado entre hemisferios. En cambio, el carácter y la causa de las conexiones climáticas aparecen como dependientes de un intervalo temporal específico y de su regional y hemisférica condición de contorno, de la resolución temporal y espacial de los parámetros climáticos tratados y del entendimiento de la respuesta regional al cambio.

23.1. INTRODUCTION

The objectives of many global change initiatives are to document the character of past change and to understand the relative roles of anthropogenic versus natural forcing in causing these changes. Because most instrumental records are too short to encompass the entire range of natural variability, paleoclimatic and paleoenvironmental records need to be developed. No single record can adequately represent past global change; records must be developed from the oceans and continents and then be viewed as an assemblage of regional records from which we can understand large-scale forcing. Integration of past evidence of global change must be based on an understanding of the specific response of the environmental systems to regional climate change and also from the perspective of linkages between different regions and systems on hemispheric and global scales.

The Pole-Equator-Pole (PEP) initiatives represent such an international effort under the larger organization of the International Geosphere Biosphere Program, Past Global Change Project (IGBP-PAGES) (PAGES, 1995; IGBP, 1998). The goal of this program is to integrate paleoclimate and paleoenvironmental records along several pole-to-pole transects in order to gain better insight into interhemispheric linkages, which ultimately should lead to an understanding of global change. Because of the symmetry of the North and South American landmasses along the eastern margin of the Pacific Ocean, the Pole-Equator-Pole Americas (PEP 1) transect is especially well suited for such an interhemispheric comparison (Markgraf, 1995; Markgraf et al., 2000). Along this transect, the land areas have a

similar latitudinal distribution of mountain ranges and lowlands, extend continuously to high polar latitudes, and show symmetric relationships to atmospheric and oceanic circulation systems in the Northern and Southern Hemispheres. We might expect that paleoclimate and paleoenvironmental records, especially along the western margin of the Americas, from the continents as well as the oceans, would show comparable histories of climate change, reflecting changes in the interactions between the tropics and extratropics. These interactions occur on daily to millennial timescales. For all of these timescales, records exist along the transect of the Americas, derived from tree rings; corals; and ice cores; and sediment cores from lakes, bogs, and marine environments. These records can be analyzed by different techniques to reconstruct past environmental and climatic conditions. A precondition for a meaningful correlation is adequate chronological control. Only then can synchroneity or relative phasing of change, an essential prerequisite for testing the existence of large-scale interhemispheric and global linkages, be assessed.

Apart from many physiographically related similarities between North and South America, there are also obvious differences that serve to test the interhemispheric extent of specific climate forcings. Such asymmetries include the extent of land area in temperate latitudes—over 4000 km wide in North America compared to 800 km in South America. This difference produces more continental climates in temperate North America compared to temperate South America and causes the tropical precipitation maximum, linked to the Intertropical Convergence Zone (ITCZ), to be located north of the equator, even in its southernmost seasonal position (Cerveny, 1998). Another asymmetry between North and South America characterizes maximum glacial conditions. Whereas the North American landmass was covered by the largest continental ice sheet on the globe, in temperate South America only mountain glaciers and small ice caps existed (Broecker and Denton, 1990). Defining the equatorward extent of the North American ice sheet's climatic influence during the time of maximum glaciation is one of the issues that an interhemispheric comparison of records can address. Another asymmetry of climatic relevance is related to the fact that the Antarctic polar region is markedly colder than the Arctic polar region. Winter air temperatures over the Arctic Ocean are about −35°C compared to −70°C over the center of the Antarctic continent (Hobbs et al., 1998). Today, this results in a pole-equator temperature gradient far steeper in the Southern Hemisphere than in the Northern Hemisphere, producing much stronger winds especially in the southern midlatitudes. Determining the location of the

ITCZ and its related precipitation patterns during times when both hemispheres' pole-equator temperature gradients were steeper than they are today, such as during glacial times, is another question that can be addressed by a transect of records. Finally, past interhemispheric climate asymmetries are also related to differences in solar activity and in seasonal receipt of solar energy (insolation) because of changes in the Earth's orbit and its axis of rotation (Berger and Loutre, 1991). Depending on which of the two principal astronomical parameters of relevance in interhemispheric climate linkages, obliquity or precession, is the critical one affecting the thermal regime at any given place and time in the past, seasonal insolation variation will be either symmetrical (for obliquity) or asymmetrical (for precession). As a consequence, climate change at comparable latitudes in both hemispheres could be either in phase or out of phase. Hence, one of the primary tests for defining interhemispheric environmental and climate linkages is to determine synchroneity of change, including questions on leads and lags.

Interhemispheric comparison of paleoenvironmental records is not a new research focus. By the end of the nineteenth century, projects on Quaternary topics in the Southern Hemisphere, especially in southern South America, were driven by the quest to confirm the notion that there was global synchroneity of climate change (e.g., Markgraf, 1993). Given that at that time no radiometric dating techniques existed for assessing the ages of the records, the primary conclusion was that environmental processes and the magnitude of effects they produced were comparable in both hemispheres at least at the scale of glacial–interglacial cycles. The advent of isotopic techniques has provided the necessary precision to correlate records, thereby avoiding circular reasoning or curve matching. In the following discussion, we will present some of the most intriguing aspects of interhemispheric correlation of records of past climate from the Americas, based on chapters presented in this book and other sources.

23.2. INSTRUMENTAL AND HIGH-RESOLUTION RECORDS OF RECENT CLIMATE CHANGE

Modern climate variability and climate patterns in much of the Americas on interannual timescales are strongly influenced by variations in the El Niño/Southern Oscillation (ENSO) climate anomaly. Much progress has been made in recent years toward understanding the processes of this tropical Pacific ocean–atmosphere anomaly, its global climate teleconnections, and longer term history (e.g., Diaz and Markgraf, 1992,

2000). In addition to documenting interannual variability, paleoclimate records also suggest the existence of decadal- (Cole et al., 1992; Evans et al., 2000; Villalba et al., 2000) and even century-scale ENSO modes (Cole et al., 1992; McGlone et al., 1992; Markgraf and Diaz, 2000). Enfield and Mestas-Nuñez (2000) and Dettinger et al. (2000) analyzed decadal ENSO-like variability in instrumental records. Compared to the interannual ENSO variability, this decadal variability produces qualitatively similar climate patterns in the Americas and sea surface temperature (SST) variability patterns in the Pacific Ocean. There are several hypotheses for identifying the mechanisms that contribute to this ENSO-like variability at different timescales, and it appears that they are linked to the Pacific Decadal Oscillation (PDO) as well as to the North Atlantic Oscillation (NAO). Such an Atlantic–Pacific climate linkage, also proposed by Huang et al. (1998), is of particular interest in the context of paleoclimate discussions. Such a linkage might explain similarities in the patterns of late glacial variability in records from the North Atlantic and North Pacific regions (Kennett and Ingram 1995). However, there remain many unresolved questions, including why the system switches between temporal modes—for example, in A.D. 1850, when tree-ring records from Patagonia and Alaska show a predominantly decadal ENSO-like mode switching to the current interannual mode (Villalba et al. 2000). It is very likely that the mode changes have to do with oceanic reorganization in the tropical Pacific. But, is it the ocean that forces the atmosphere to adapt to the different mode, or is it an atmospheric reorganization that forces the ocean to change? In terms of the atmospheric forcing, changes have been proposed in the interaction between the Hadley cell and the Walker circulation (Kumar et al., 1999; Liu et al., 1999) or changes in the North Atlantic ocean–atmosphere interaction, which could also influence the state of the Pacific Ocean (Huang et al., 1998).

Another climate feature responsible for climate variability in the Americas, as far equatorwards as the subtropical latitudes, is polar outbreaks (Marengo and Rogers, 2000). Although there are differences in the expression of polar outbreaks between North and South America, the meteorological aspects (both near-surface and upper level circulation) are very similar. Only recently has the importance of polar outbreaks been recognized as a major influence on past-climate change. During glacial times when the pole-equator temperature gradients were steeper, a higher frequency of polar outbreaks might have been the cause for much lower temperatures in subtropical and perhaps even tropical latitudes, which are documented in records from both North and South America (Bradbury, 1997; Ledru,

1993). It is possible that at the maximum extent of the Laurentide ice sheet over northeastern North America, Arctic outbreaks could have reached the tropical regions of South America and the Southern Hemisphere (Bush et al., 2000).

23.3. CLIMATE CHANGE AND HUMAN SOCIETIES

Despite advances in technology, climate extremes can still impact human societies today. This is well illustrated by statistics on the impact of frosts related to polar air outbreaks on crops in Florida and Brazil (Marengo and Rogers, 2000). Climate extremes also affected pre-Columbian societies. Examples from the Maya in Central America and the Tiwanaku in the Andes illustrate that population declined following extended droughts (Brenner et al. 2000). Also, pastoral societies were affected by climate change, especially those populations living in semiarid regions, where food resources were restricted to areas around permanent water bodies (Nuñez et al., 2000). According to these authors, after climate changes led to reductions in water supplies, which occurred at different times in the Americas, human populations disappeared from the affected regions. Either these groups relocated elsewhere, or the centers of cultural activity shifted to more suitable regions. An example is the shift of cultures from the Andean highlands in Peru and Ecuador to the coast, which occurred several times during the last 1500 years, apparently synchronously with regional hydrologic changes (Thompson et al., 1992). However, during times of extreme and perhaps especially rapid climate shifts, such as occurred at the end of the Pleistocene when most members of the megafauna became extinct in the Americas, disappearance of entire cultures was the consequence (Nuñez et al., 2000).

Volcanic eruptions also affected human societies in Central America and the southern Andes. Sheets (2000) has advanced an intriguing hypothesis: that more complex societies have greater difficulties coping with massive disruptions caused by volcanic eruptions than simple, egalitarian societies. This hypothesis implies that the vulnerability of human societies to climate change and climate extremes cannot be reduced by technological advances alone. On the contrary, technology, which produces a shift from a nature-dominated to a human-dominated environment, actually renders the societies more vulnerable to natural disasters (Messerli et al., 2000). Clearly there are lessons to be learned from the past, despite the fact that the interactions between environmental change and human societies are highly complex and regionally differentiated.

In addition to immediate local effects, volcanic eruptions also have regional and even hemispheric effects, primarily on temperature. Several empirical and modeling studies have documented a correlation between volcanic eruptions and decreased temperatures on a regional and even hemispheric scale (Bradley and Jones, 1992; Mann et al., 1998; Bryson and Goodman, 1980). Interhemispheric comparison of tree-ring records from North and South America, however, suggests that for eruptions of the last 400 years, volcanic eruptions did not produce global climate effects (Boninsegna and Hughes, 2000). The reason for this may relate to the fact that the last 400 years was not a period of high volcanic activity. During periods of increased frequency and magnitude of volcanic eruptions, such as during the final stages of deglaciation at the end of the Pleistocene, global climate effects of volcanism are well documented (Zielinski et al., 1997).

23.4. GEOMORPHIC RECORDS OF PALEOCLIMATE

One of the major features that defines the Quaternary period is the appearance of continental ice sheets in the Northern Hemisphere. The earlier assumption was that by documenting global synchroneity of continental ice sheet and mountain glacier movements, one could prove the existence of a single large-scale, global climate forcing. Although historical records document that even on a regional scale glaciers do not respond similarly to climate change, this has been generally ignored on longer timescales for which dating control is less exact. The inter-American network on the behavior of glaciers during the Little Ice Age (LIA) discussed by Luckman and Villalba (2000) represents a fundamental advance in this context. The authors combined dated glacier advances during the last 800 years with temperature and precipitation reconstructions from tree-ring records in the American cordillera. They found that, depending on the location along the inter-American transect, glaciers advanced in response to either a precipitation increase, a temperature decrease, or a combination of both. Thus, it is not surprising that at an interhemispheric or even global scale, these climate changes are not synchronous and related to the same forcing. This probably explains the lack of synchroneity of LIA glacier advances in the Americas on decadal scales, although on century-long timescales the temporal asynchronies are less obvious. Essentially, the same problem affects the timing of glacier behavior in the more distant past, be it during the Younger Dryas interval, for which major glacier advances are documented in the circum–North Atlantic land areas, or at times of maximum glacier extent thought to relate to the last glacial maximum (LGM). Clapperton and Seltzer (2000) and Clapperton, (2000) address the synchroneity of full and late glacial glaciation in the American cordillera. The comparison showed that the maximum ice extent of most mountain glaciers occurred between 30,000 and 20,000 B.P.,* clearly earlier than the full glacial maximum of the continental ice sheets of North America and Europe as revealed by the marine isotope stratigraphy. On the other hand, glaciers along the northwest coast of North America and the southwest coast of South America extended to their maxima only at about 15,000 B.P. Because of a positive climate feedback, glaciers affect regional climate (e.g., temperature by changes in albedo and precipitation by changes in the locations of storm tracks). As a consequence, the diachronous behavior of glaciers also affects the timing of changes in local environmental records.

Eolian deposits are also associated with glacial conditions, either derived from glacial outwash sources or deposited during the cold, dry, and windy conditions of the LGM. Hence, despite differences in North and South America in terms of extent and timing of major glacier activity, comparable full glacial climates produced synchroneity of high eolian activity in the Americas (Muhs and Zárate, 2000). Eolian particles in marine sediments and ice cores, including polar ice cores, show global similarities in dust accumulation for the LGM despite different dust-source regions (Thompson, 1977; Petit et al., 1981, 1999; Ram and Koenig, 1997; Kohfeld and Harrison, 2000).

23.5. VEGETATION AND LAKE RECORDS

Terrestrial records, including lake-level and vegetation records, confirm the notion of LGM and late glacial aridity coupled with markedly lower temperatures in the mid- to high latitudes of the Americas (Bradbury et al., 2000; Whitlock et al., 2000; Ledru and Mourguiart, 2000; Kohfeld and Harrison, 2000). Lower LGM temperatures are also confirmed for the Amazon lowlands, based on groundwater temperature reconstructions (Stute et al., 1995) and expansion of Andean forest taxa into the lowlands (Haberle and Maslin, 1999; Colinvaux et al., 2000; van der Hammen and Hooghiemstra, 2000). Controversy, however, continues about a concomitant reduction of precipitation in the tropics, especially in the Amazon lowlands (Bush et al., 2000; Behling and Hooghiemstra, 2000; Ledru et al., 1998; Colinvaux et al., 2000). Only in southwestern North America, including México, was precipitation during

*All dates are in ^{14}C years unless otherwise noted.

the LGM actually higher than it is today (Bradbury et al., 2000; Whitlock et al., 2000). Although reduced rates in sedimentation in lacustrine records from Amazonia between 20,000 and 14,000 B.P. would suggest desiccation at that time (Ledru et al., 1998), different vegetation records show either forests or savanna, interpreted to imply either no precipitation change or markedly less precipitation than today. Other climate proxy data are needed that respond more specifically to precipitation. Another approach to address the question would involve climate modeling experiments, especially in terms of sensitivity testing (Webb et al., 1993; Peteet, 2000; Whitlock et al., 2000). Experiments with different general circulation models (GCMs) (Paleoclimate Modeling Intercomparison Project [PMIP]: Pinot et al., 1999) and model/paleoclimate data comparisons (Farrera et al., 1999; Hostetler and Mix, 1999) suggest that tropical latitudes were both cooler and drier, although on a global scale there were substantial spatial differences.

Lake and vegetation records also provide data indicating the Holocene climate history of the Americas (Fritz et al., 2000; Grimm et al., 2000; Whitlock et al., 2000; Ariztegui et al., 2000). In contrast to the late Pleistocene, the Holocene—and especially the late Holocene—is characterized by a far greater degree of regionalization of climate patterns. The improved regional coverage of paleoenvironmental records and comparable temporal resolution for all these time intervals indicate that the enhanced regionalization of late Holocene climate patterns reflects different climate modes. The most intriguing result from comparison of records in the Americas is that certain regions show the same pattern of climate change despite different forcings, whereas other regions show a different pattern as a result of the same forcing (Whitlock et al., 2000). This result implies that great care must be taken before ascribing the same climate forcing to different regions simply because a climate change is synchronous.

A very interesting period of climate pattern is the mid-Holocene, between ca. 7000 and 5000 B.P. Vegetation, lake-level, and eolian records document that conditions were drier along the whole transect of the Americas, with the exception of the North American Midwest and the northern tropics (Grimm et al., 2000; Fritz et al., 2000; Muhs et al., 2000). Temperature proxy data suggest that in the affected regions aridity was coupled with lower temperatures and that both temperature and precipitation variability overall increased (Markgraf, 1989; Yu et al., 1997; Steig, 1999). The switch at the end of the mid-Holocene (at ca. 4000 B.P.) was quite abrupt, happening within decades to a century. This behavior is also seen in records from the tropical and northern Pacific, as well as Africa (Gagan et al., 1998; Mix et al., 1999; deMenocal et al., 2000). In terms of climate forcing, this widespread and perhaps even global shift to more arid, cold, and variable conditions in the mid-Holocene poses a problem, because at that time boundary conditions, especially insolation, had gradually approached present-day levels. A possible explanation for the abruptness of response in both terrestrial and marine records might be related to nonlinear feedback mechanisms enhancing the otherwise slow response to orbital changes. Paleoclimate models suggest positive feedback mechanisms: in the case of the African records, these would involve a vegetation-albedo feedback, and in the case of the oceanic records a temperature-moisture transport feedback possibly related to a decline in intensity of the thermohaline circulation (Steig, 1999; deMenocal et al., 2000; but see Boyle, 2000).

23.6. INTERHEMISPHERIC CLIMATE LINKAGES

Most published studies of interhemispheric climate linkages involve marine and polar ice core records (Charles et al., 1996; Bard et al., 1997; Broecker, 1998; Ninnemann et al., 1999; Jouzel et al., 1994; Kreutz et al., 1997; Domack and Mayewski, 1999; Bender et al., 1994, 1999; Alley et al., 1999; Petit et al., 1999). In the case of polar ice cores, new techniques, such as correlation of spikes in [10]Be, of atmospheric $\delta^{18}O$ ratio and methane, contained in closed air bubbles in the ice, have facilitated determining interhemispheric linkages by providing exact temporal correlation between Greenland and Antarctica (Bender et al., 1999; Blunier et al., 1998). Interhemispheric comparisons of the ice core records confirm similarity of temperature trends between 35,000 and 75,000 cal. B.P. for the longer term paleotemperature cycles, such as the Dansgaard–Oeschger cycles, as well as for the higher frequency oscillations, such as the Heinrich events (Bender et al., 1999). Even during that interval, however, Antarctic climate change leads that of Greenland by 1000–2500 years (Blunier et al., 1998). On the other hand, the termination of Marine Isotope Stage 2, which marks the onset of the last phase of deglaciation after 20,000 [14]C B.P. and late glacial oscillations between 14,000 and 10,000 [14]C B.P., are out of phase (Sowers and Bender, 1995; Blunier et al., 1998; Bender et al., 1999). An exception to this late glacial out-of-phase behavior comes from the coastal Antarctic Taylor Dome ice core, which shows the termination of the LGM to be in phase with Greenland records (Steig et al., 1998). Local effects have been considered responsible for this difference in Antarctic ice core records, although more likely the chronology of the Taylor Dome record may be problematic (Pendall et al., submitted).

The mechanism proposed to explain interhemispheric synchroneity in marine records was thought to relate to changes in the intensity of the thermohaline circulation. These changes were hypothesized to relate to changes in the intensity of the formation of North Atlantic Deep Water (NADW) in response to changes in freshwater influx into the North Atlantic (Broecker and Denton, 1990; Broecker et al., 1990). Although this mechanism could probably produce synchroneity of change, it would probably result in opposite climate signals, with cooling of the North Atlantic synchronous with warming of the Southern Ocean (Crowley, 1999; Broecker, 1998; Lehman, 1998; Stocker, 1998, 2000). More recent work on marine sediments has shown that the signal of millennial-scale climate variability between 35,000 and 70,000 B.P. was, in fact, synchronous, but of opposite climate sign in both hemispheres, although the data could also be interpreted to reflect phase leads of the Southern Hemisphere by ca. 1500 years (Ninnemann et al., 1999; Stocker, 1998). While this would support the thermohaline circulation hypothesis, the marine data indicate that NADW formation experienced abrupt changes also during interglacial times—times when there was no major change in meltwater input (Ninnemann et al., 1999), implying that the salt oscillator (Broecker et al., 1990) is not the only mechanism that induces thermohaline instability.

Among the mechanisms proposed to explain synchronous but opposite climate changes at high latitudes are changing proportions of greenhouse gases, primarily water vapor, CO_2, and methane. Recent studies, however, document that CO_2 follows rather than leads temperature increases (Sowers and Bender, 1995; Blunier et al., 1998). Water vapor changes, in conjunction with wind-driven mixing processes in the tropical Pacific, continue to be called upon as a possible mechanism for interhemispheric climate synchronization (Kennett and Ingram, 1995; Behl and Kennett, 1996; Mix et al., 1999; Ninnemann et al., 1999). Modeling experiments show that changes in the tropical oceanic regions, especially the tropical Pacific region, could produce changes in the intensity of convection and water vapor (Cane, 1998; Cane and Clement, 1999; Rind, 1998, 2000). Because the tropical oceans are highly responsive to the seasonal cycle, they probably also respond to the longer term precessional (Milankovitch) insolation changes, which are thought to be responsible for past major climate changes (Hays et al., 1976; Imbrie, 1985). Because tropical conditions markedly influence extratropical climates, global changes in temperature and precipitation patterns would be a result, analogous perhaps to the ENSO convection anomalies in the tropical Pacific (e.g., Broecker, 1998; White and Peterson, 1996; Cane, 1998; Cane and Clement, 1999).

Although terrestrial records have been much less the focus of interhemispheric correlation, they are of preeminent importance in view of our need to understand the character and causes of changes that affect human societies. Apart from specific time periods, such as the global search for the Younger Dryas interval (e.g., Denton et al., 1999), the collection of chapters in this book is the first attempt to address interhemispheric climate linkages from terrestrial records in the Americas from a wide range of temporal and spatial scales. A number of the specific research questions listed in the PAGES (1995) program on PEP transects have been directly addressed and answered. These include, for example, the interhemispheric link of climatic effects related to volcanic eruptions, the phase relationship between climate evolution in the Northern and Southern Hemispheres, and the hydrological changes in tropical areas and their relation to changes in ice sheets at higher latitudes. As is also stated in the PAGES (1995) program, several prerequisites need to be fulfilled for a meaningful interhemispheric comparison. These include understanding of present-day climate patterns and anomalies and their expression in climate proxy records, development of detailed and comparable chronologies to avoid curve-matching correlation (Crowley, 1999), multiproxy data analysis to address discrepancies and ambiguities in the interpretation of paleoenvironmental records, and testing of paleoclimate interpretations by paleoclimate modeling experiments to understand forcing mechanisms.

Some exciting new insights illustrate the success of this interhemispheric comparison in the Americas. For instance, data and modeling results for the LGM show that the Northern Hemisphere westerly storm tracks were shifted farther equatorwards than the Southern Hemisphere westerlies. Apparently, the pole-equator temperature gradients, although clearly steeper than they are today, were not symmetrical in both hemispheres. Could it be that the influence of the Laurentide ice sheet, considered responsible for the Northern Hemisphere's equatorward shift of the westerlies, extended even into subtropical latitudes south of the equator? LGM paleoclimate data from the tropics of northern South America (Hughen et al., 1998; Seltzer, 1994) document lower temperatures, a southward shift in the location, and/or a decreased strength of the ITCZ (based on paleoprecipitation patterns in the tropics), which are interpreted to reflect the influence of the marked temperature changes in the North Atlantic Ocean. These climate scenarios are also supported by modeling experiments (Rind, 1998, 2000). When steeper pole-equator temperature gradients coupled with lower tropical SSTs than today are simulated, a stronger Hadley circulation and more zonal winds are the result.

According to Rind (2000), those conditions would produce more precipitation in the tropics because of decreased evaporation. If, however, the tropical SSTs were as warm as they are today and were coupled with the steeper temperature gradient, then the increased tropical precipitation would be primarily over the ocean and not over land. Thus, from a modeling standpoint, the controversy on LGM tropical precipitation cannot be solved unless tropical SSTs are known. However, it is certain that the influence of the extensive ice sheets in the Northern Hemisphere did extend as far south as the tropics and therefore must have altered the precipitation regime over the Amazon.

From the global extent of LGM climate changes, it also follows that the Pacific had to have been a major player, because of its much greater atmospheric water vapor exchange capacity than any other ocean (Rind, 2000). Specifically, for the climate of the Americas, the critical factor is the interaction between both the Pacific and the Atlantic gradients in SST. This might explain the fact that changes in the intensity of NADW circulation did not affect Southern Hemisphere climates at all times (see Ninnemann et al., 1999; Alley et al., 1999).

Another area of particular climatic relevance for the Americas is ENSO variability. Analysis and comparison of instrumental and climate proxy data provide evidence for the existence of decadal-scale modes that are similar to the interannual modes and are apparently stable over extended periods of time. These decadal modes are linked with higher latitude circulation features in both the Pacific and Atlantic regions, explaining the longer term linkages between tropical and extratropical climates. Even when these ENSO-like modes switch, the spatial coherence between the Pacific SST patterns and the climate teleconnection patterns remains. This may explain the fact that ENSO-like climate teleconnection patterns can be identified back in time also in low-resolution records. Given the new understanding of the existence of these decadal- to century-scale ENSO-like modes, the question of how to explain the apparent absence of ENSO-like climate patterns in records prior to ca. 5000 B.P. has now become even more challenging (Sandweiss et al., 1996; Rodbell et al., 1999).

Perhaps the most important outcome of this project on interhemispheric climate linkages in the Americas has been the ability to bring together the entire science community from a wide range of disciplines interested in present- and past-climate change. Although spatially uneven, global change affects, and will continue to affect, all parts of the intricately linked Earth system. Analyzing linkages of present and past environmental and climate variability in the Americas may be the first step in understanding key issues of global change. It is hoped that ultimately this understanding will set in motion mechanisms for incorporating climate change in policy decisions.

Acknowledgments

VM expresses her thanks to F. Oldfield and K. Alverson for helpful comments and discussions on the manuscript. The U.S. National Science Foundation (NSF) is acknowledged for supporting all efforts related to coordinating PEP 1 activities, including the Mérida (Venezuela),1998, meeting that resulted in this book.

References

Alley, R. B., P. U. Clark, L. D. Keigwin, and R. S. Webb, 1999: Making sense of millennial-scale climate change. *In* Clark, P. U., R. S. Webb, and L. D. Keigwin (eds.), *Mechanisms of Global Climate Change.* Washington, DC: American Geophysical Union, Geophysical Monograph 111, pp. 385–394.

Ariztegui, D., K. Kelts, F. S. Anselmetti, G. O. Seltzer, and K. D'Agostino, 2000: Identifying paleoenvironmental change across South America and North America using high-resolution seismic stratigraphy in lakes. *In* Markgraf, V. (ed.), *Interhemispheric Climate Linkages.* San Diego: Academic Press, Chapter 14.

Bard, E., F. Rostek, and C. Sonsogni, 1997: Interhemispheric synchrony of the last deglaciation inferred from alkenone palaeothermometry. *Nature, 385*: 707–710.

Behl, R. J., and J. P. Kennett, 1996: Brief interstadial events in the Santa Barbara basin, NE Pacific, during the past 60 kyr. *Nature, 307*: 515–517.

Behling, H., and H. Hooghiemstra, 2000: Neotropical savanna environments in space and time: Late Quaternary interhemispheric comparisons. *In* Markgraf, V. (ed.), *Interhemispheric Climate Linkages.* San Diego: Academic Press, Chapter 18.

Bender, M., T. Sowers, M.-L. Dickson, J. Orchardo, P. Grootes, P. A. Mayewski, and D. A. Meese, 1994: Climate correlations between Greenland and Antarctica during the past 100,000 years. *Nature, 372*: 663–666.

Bender, M., B. Malaize, J. Orchardo, T. Sowers, and J. Jouzel, 1999: High precision correlations of Greenland and Antarctic ice core records over the last 100 kyr. *In* Clark, P. U., R. S. Webb, and L. D. Keigwin (eds.), *Mechanisms of Global Climate Change.* Washington, DC: American Geophyscai Union, Geophysical Monograph 111, pp. 149–164.

Berger, A., and M.-F. Loutre, 1991: Insolation values for the climate of the last 10 million years. *Quaternary Science Reviews, 10*: 571–581.

Blunier, T., J. Chappellaz, J. Schwander, A. Dallenbach, B. Stauffer, T. F. Stocker, D. Raynaud, J. Jouzel, H. B. Clausen, C. U. Hammer, and S. J. Johnsen, 1998: Asynchrony of Antarctic and Greenland climate change during the last glacial period. *Nature, 394*: 739–743.

Boninsegna, J. A., and M. K. Hughes, 2000: Volcanic signals in temperature reconstructions based on tree-ring records for North and South America. *In* Markgraf, V. (ed.), *Interhemispheric Climate Linkages.* San Diego: Academic Press, Chapter 9.

Boyle, E. A., 2000: Is ocean thermohaline circulation linked to abrupt stadial/interstadial transitions? *Quaternary Science Reviews, 19*: 255–272.

Bradbury, J. P., 1997: Sources of glacial moisture in Mesoamerica. *Quaternary International, 43/44*: 97–110.

Bradbury, J. P., M. Grosjean, S. Stine, and F. Sylvestre, 2000: Full and late glacial lake records along the PEP 1 transect: Their role in developing interhemispheric paleoclimate interactions. *In* Markgraf, V. (ed.), *Interhemispheric Climate Linkages.* San Diego: Academic Press, Chapter 16.

Bradley, R. S., and P. D. Jones, 1992: Records of explosive volcanic eruptions over the last 500 years. *In* Bradley, R. S., and P. D. Jones (eds.), *Climate Since A.D. 1500.* London: Routledge, pp. 606–622.

Brenner, M., D. A. Hodell, J. H. Curtis, M. F. Rosenmeier, M. W. Binford, and M. B. Abbott, 2000: Abrupt climate change and pre-Columbian cultural collapse. *In* Markgraf, V. (ed.), *Interhemispheric Climate Linkages.* San Diego: Academic Press, Chapter 6.

Broecker, W.S., 1998: Paleocean circulation during the last deglaciation: A bipolar seesaw. *Paleoceanography, 13*: 119–121.

Broecker, W. S., and G. H. Denton, 1990: The role of ocean-atmosphere reorganizations in glacial cycles. *Quaternary Science Reviews, 9*: 305–341.

Broecker, W. S., G. Bond, M. Klas, G. Bonani, and W. Wolfli, 1990: A salt oscillator in the glacial North Atlantic. I. The concept. *Paleoceanography, 5*: 469–477.

Bryson, R. A., and B. M. Goodman, 1980: Volcanic activity and climatic changes. *Science, 207*: 1041–1044.

Bush, M. B., M. Stute, M.-P. Ledru, H. Behling, P. A. Colinvaux, P. E. De Oliveira, E. C. Grimm, H. Hooghiemstra, S. Haberle, B. W. Leyden, M.-L. Salgado-Labouriau, and R. Webb, 2000: Paleotemperature estimates for the lowland Americas between 30°S and 30°N at the last glacial maximum. *In* Markgraf, V. (ed.), *Interhemispheric Climate Linkages.* San Diego: Academic Press, Chapter 17.

Cane, M. A., 1998: A role for the tropical Pacific. *Science, 282*: 59–62.

Cane, M. A., and A. C. Clement, 1999: A role for the tropical Pacific coupled ocean-atmosphere system on Milankovitch and millennial timescales. Part II. Global impacts. *In* Clark, P. U., R. S. Webb, and L. D. Keigwin (eds.), *Mechanisms of Global Climate Change.* Geophysical Monograph 111, pp. 373–384.

Cerveny, R., 1998: Present climates of South America. *In* Hobbs, J. E., J. A. Lindesay, and H. A. Bridgman (eds.), *Climates of the Southern Continents: Present, Past and Future.* New York: John Wiley & Sons, pp. 63–106.

Charles, C. D., J. Lynch-Stieglitz, U.S. Ninnemann, and R.G. Fairbanks, 1996: Climate connections between the hemisphere revealed by deep sea sediment core/ice core correlations. *Earth and Planetary Science Letters, 142*: 19–27.

Clapperton, C. M., 2000: Interhemispheric synchroneity of Marine Oxygen Isotope Stage 2 glacier fluctuations along the American cordilleras transect. *Journal of Quaternary Science, 15*: 435–468.

Clapperton, C., and G. O. Seltzer, 2000: Glaciation during Marine Isotope Stage 2 in the American cordillera. *In* Markgraf, V. (ed.), *Interhemispheric Climate Linkages.* San Diego: Academic Press, Chapter 11.

Cole, J. E., G. T. Shen, R. G. Fairbanks, and M. Moore, 1992: Coral monitors of El Niño/Southern Oscillation dynamics across the equatorial Pacific. *In* Diaz, H. F., and V. Markgraf (eds.), *El Niño: The Historical and Paleoclimatic Record of the Southern Oscillation.* Cambridge, UK: Cambridge University Press, pp. 349–376.

Colinvaux, P., P. E. De Oliveira, and M. B. Bush, 2000: Amazonian and neotropical plant communities on glacial time-scales. The failure of the aridity and refuge hypotheses. *Quaternary Science Reviews, 19*: 141–170.

Crowley, T. J., 1999: Correlating high-frequency climate variation. *Paleoceanography, 14*: 271–272.

deMenocal, P., J. Ortiz, T. Guilderson, J. Adkins, M. Sarnthein, L. Baker, and M. Yarusinsky, 2000: Abrupt onset and termination of the African Humid Period: Rapid climate responses to gradual insolation forcing. *Quaternary Science Reviews, 19*: 347–362.

Denton, G. H., C. J. Heusser, T. V. Lowell, P. I. Moreno, B. G. Andersen, L. E. Heusser, C. Schlüchter, and D. R. Marchant, 1999: Interhemispheric linkage of paleoclimate during the last glaciation. *Geografiska Annaler,* **81**: 107–153.

Dettinger, M. D., D. S. Battiti, G. J. McCabe, Jr., and R. D. Garreaud, 2000: Interhemispheric effects of interannual and decadal ENSO-like climate variations on the Americas. *In* Markgraf, V. (ed.), *Interhemispheric Climate Linkages.* San Diego: Academic Press, Chapter 1.

Diaz, H. F., and V. Markgraf (eds.), 1992: *El Niño: The Historical and Paleoclimatic Record of the Southern Oscillation.* Cambridge, UK: Cambridge University Press, 476 pp.

Diaz, H. F., and V. Markgraf (eds.), 2000: *El Niño and the Southern Oscillation: Multiscale Variability and Global and Regional Impacts.* Cambridge, UK: Cambridge University Press, pp. 465–488.

Domack, E. W., and P. A. Mayewski, 1999: Bi-polar ocean linkages: Evidence from late-Holocene Antarctic marine and Greenland ice-core records. *The Holocene,* **9**: 247–251.

Enfield, D. B., and A. M. Mestas-Nuñez, 2000: Interannual to multi-decadal climate variability and its relationship to global sea surface temperatures. *In* Markgraf, V. (ed.), *Interhemispheric Climate Linkages.* San Diego: Academic Press, Chapter 2.

Evans, M. N., A. Kaplan, R. Villalba, and M. A. Cane, 2000: Globality and optimality in climate field reconstructions from proxy data. *In* Markgraf, V. (ed.), *Interhemispheric Climate Linkages.* San Diego: Academic Press, Chapter 4.

Farrera, I., S. P. Harrison, I. C. Prentice, G. Ramstein, J. Guiot, P. J. Bartlein, R. Bonnefille, M. Bush, W. Cramer, U. von Grafenstein, K. Holmgren, H. Hooghiemstra, G. Hope, D. Jolly, S.-E. Lauritzen, Y. Ono, S. Pinot, M. Stute, and G. Yu, 1999: Tropical climates at the last glacial maximum: A new synthesis of terrestrial palaeoclimate data. I. Vegetation, lake-levels and geochemistry. *Climate Dynamics,* **15**: 823–856.

Fritz, S. C., S. E. Metcalfe, and W. E. Dean, 2000: Holocene climate patterns in the Americas inferred from paleolimnological records. *In* Markgraf, V. (ed.), *Interhemispheric Climate Linkages.* San Diego: Academic Press, Chapter 15.

Gagan, M. K., L. K. Ayliffe, D. Hopley, J. A. Cali, G. E. Mortimer, J. Chappell, M. T. McCulloch, and M.J. Head, 1998: Temperature and surface-ocean water balance of the mid-Holocene tropical western Pacific. *Science,* **279**: 1014–1018.

Grimm, E. C., S. Lozano García, H. Behling, and V. Markgraf, 2000: Holocene vegetation and climate variability in the Americas. *In* Markgraf, V. (ed.), *Interhemispheric Climate Linkages.* San Diego: Academic Press, Chapter 19.

Haberle, S. G., and M. A. Maslin, 1999: Late Quaternary vegetation and climate change in the Amazon basin based on a 50,000 year pollen from the Amazon Fan ODP site 932. *Quaternary Research,* **51**: 27–38.

Hays, J. D., I. Imbrie, and N. J. Shackleton, 1976: Variations in the Earth's orbit: Pacemaker of the ice ages. *Science,* **194**: 1121–1132.

Hobbs, J. E., J. A. Lindesay, and H. A. Bridgman (eds.), 1998: *Climates of the Southern Continents: Present, Past and Future.* New York: John Wiley & Sons.

Hostetler, S. W., and A. C. Mix, 1999: Reassessment of ice-age cooling of the tropical ocean and atmosphere. *Nature,* **399**: 673–676.

Huang, J., K. Higuchi, and A. Shabbar, 1998: The relationship between the North Atlantic Oscillation and El Niño–Southern Oscillation. *Geophysical Research Letters,* **25**: 2707–2710.

Hughen, K. A., J. T. Overpeck, S. J. Lehman, M. Kashgarian, J. Southon, L. C. Peterson, R. Alley, and D. M. Sigman, 1998: Deglacial changes in ocean circulation from an extended radiocarbon calibration. *Nature,* **391**: 65–68.

IGBP (International Geosphere Biosphere Program), 1998: Status report and implementation plan. IGBP Report 45, Royal Swedish Academy of Science, Stockholm, Sweden, 236 pp.

Imbrie, J. A., 1985: A theoretical framework for the Pleistocene ice ages. *Journal of the Geological Society,* London, **142**: 417–432.

Jouzel, J., R. D. Koster, R. J. Suozzo, and G. L. Russell, 1994: Stable water isotope behaviour during the LGM: A GCM analysis. *Journal of Geophysical Research,* **99**: 25791–25801.

Kennett, J. P., and G. L. Ingram, 1995: A 20,000 yr record of ocean circulation and climate change from the Santa Barbara basin. *Nature,* **377**: 510–514.

Kohfeld, K. E., and S. P. Harrison, 2000: How well can we simulate past climate? Evaluating the models using global palaeoenvironmental datasets. *Quaternary Science Reviews,* **19**: 321–346.

Kreutz, K. J., P. A. Mayewski, L. D. Meeker, M. S. Twickler, S. I. Whitlow, and I. I. Pittlwala, 1997: Bipolar changes in atmospheric circulation during the Little Ice Age. *Science,* **277**: 1294–1296.

Kumar, K. K., B. Rajagopalan, and M. A. Cane, 1999: On the weakening relationship between the Indian monsoon and ENSO. *Science,* **284**: 2158–2159.

Ledru, M.-P., 1993: Late Quaternary environmental and climatic changes in central Brazil. *Quaternary Research,* **39**: 90–98.

Ledru, M.-P., and P. Mourguiart, 2000: Late glacial vegetation records in the Americas and climatic implications. *In* Markgraf, V. (ed.), *Interhemispheric Climate Linkages.* San Diego: Academic Press, Chapter 20.

Ledru, M.-P., J. Bertaux, A. Sifeddine, and K. Suguio, 1998: Absence of last glacial maximum records in lowland tropical forests. *Quaternary Research,* **49**: 233–237.

Lehman, S. J., 1998: Deglacial abyssal circulation change: Bi-polar seesaw of (almost) global relay? *In* Markgraf, V. (ed.), Abstracts *PEP 1, Pole-Equator-Pole, Paleoclimate of the Americas,* Mérida, Venezuela, 16–20 March 1998.

Liu, Z., R. Jacos, J. E. Kutzbach, S. P. Harrison, and J. Anderson, 1999: Monsoon impact on El Niño in the early Holocene. *PAGES Newsletter,* **7**: 16–17.

Luckman, B. H., and R. Villalba, 2000: Assessing the synchroneity of glacier fluctuations in the western cordillera of the Americas during the last millennium. *In* Markgraf, V. (ed.), *Interhemispheric Climate Linkages.* San Diego: Academic Press, Chapter 8.

Mann, M. E., R. S. Bradley, and M. K. Hughes, 1998: Global-scale temperature patterns and climate forcing over the past six centuries. *Nature,* **392**: 779–787.

Marengo, J. A., and J. C. Rogers, 2000: Polar air outbreaks in the Americas: Assessments and impacts during recent and past climates. *In* Markgraf, V. (ed.), *Interhemispheric Climate Linkages.* San Diego: Academic Press, Chapter 3.

Markgraf, V., 1989: Palaeoclimates of Central and South America since 18,000 B.P. based on pollen and lake-level records. *Quaternary Science Reviews,* **8**: 1–24.

Markgraf, V., 1993: Paleoenvironments and paleoclimates in Tierra del Fuego and southernmost Patagonia, South America. *Palaeogeography, Palaeoclimatology, Palaeoecology,* **102**: 53–68.

Markgraf, V., 1995: PEP I, The Americas Transect. *In* Paleoclimates of the Northern and Southern Hemispheres. *PAGES Series,* **95-1**, 23–42.

Markgraf, V., and H. F. Diaz, 2000: The ENSO record: A synthesis. *In* Diaz, H. F., and V. Markgraf (eds.), *El Niño and the Southern Oscillation: Multiscale Variability and Global and Regional Impacts.* Cambridge, UK: Cambridge University Press, pp. 465–488.

Markgraf, V., T. R. Baumgartner, J. P. Bradbury, H. F. Diaz, R. B. Dunbar, B. H. Luckman, G. O. Seltzer, T. W. Swetnam, and R. Villalba, 2000: Paleoclimate reconstruction along the Pole-Equator-Pole transect of the Americas (PEP 1). *Quaternary Science Reviews,* **19**: 125–140.

McGlone, M. S., A. P. Kershaw, and V. Markgraf, 1992: El Niño/Southern Oscillation climatic variability in Australasian and

South American paleoenvironmental records. *In* Diaz, H. F., and V. Markgraf (eds.), *El Niño: The Historical and Paleoclimatic Record of the Southern Oscillation*. Cambridge, UK: Cambridge University Press, pp. 435–462.

Messerli, B., M. Grosjean, T. Hofer, L. Nuñez, and C. Pfister, 2000: From nature-dominated to human-dominated environmental changes. *Quaternary Science Reviews*, **19**: 459–479.

Mix, A. C., A. E. Morey, N. G. Pisias, and S. W. Hostetler, 1999: Foraminiferal faunal estimates of paleotemperature: Circumventing the no-analog problem yields cool ice age tropics. *Paleoceanography*, **14**: 350–359.

Muhs, D., and M. Zárate, 2000: Late Quaternary eolian records of the Americas and their paleoclimatic significance. *In* Markgraf, V. (ed.), *Interhemispheric Climate Linkages*. San Diego: Academic Press, Chapter 12.

Ninnemann, U. S., C. D. Charles, and D. A. Hodell, 1999: Origin of global millennial scale climate events: Constraints from the Southern Ocean deep sea sedimentary record. *In* Clark, P. U., R. S. Webb, and L. D. Keigwin (eds.), *Mechanisms of Global Climate Change*. Washington, DC: American Geophysical Union, Geophysical Monograph 111, pp. 99–112.

Nuñez, L., M. Grosjean, and I. Cartajena, 2000: Human dimensions of late Pleistocene/Holocene arid events in southern South America. *In* Markgraf, V. (ed.), *Interhemispheric Climate Linkages*. San Diego: Academic Press, Chapter 7.

PAGES, 1995: Panash: Paleoclimates of the Northern and Southern Hemispheres. *PAGES Series*, **95–1**, 92 pp.

Pendall, E., V. Markgraf, J. W. C. White, M. Dreier, and R. Kenny, submitted: A multi-proxy record of Pleistocene-Holocene climate and vegetation changes from a peat bog in Patagonia. *Quaternary Research*.

Peteet, D. M., 2000: Late glacial climate variability in general circulation model (GCM) experiments: An overview. *In* Markgraf, V. (ed.), *Interhemispheric Climate Linkages*. San Diego: Academic Press, Chapter 22.

Petit, J.-R., M. Briat, and A. Royer, 1981: Ice age aerosol content from East Antarctic ice core samples and past wind strength. *Nature*, **293**: 391–394.

Petit, J. R., J. Jouzel, D. Raynaud, N. I. Barkov, J.-M. Barnola, I. Basile, M. Bender, J. Chappellaz, M. Davis, G. Delaygue, M. Delmotte, V. M. Kotlyakov, M. Legrand, V. Y. Lipenkov, C. Lorius, L. Pépin, C. Ritz, E. Saltzman, and M. Stievenard, 1999: Climate and atmospheric history of the past 420,000 years from the Vostok ice core, Antarctica. *Nature*, **399**: 429–436.

Pinot, S., G. Ramstein, S. P. Harrison, I. C. Prentice, J. Guiot, M. Stute, and S. Joussaume, 1999: Tropical paleoclimates at the last glacial maximum: Comparison of Paleoclimate Modeling Intercomparison Project (PMIP) simulations and paleodata. *Climate Dynamics*, **15**: 857–874.

Ram, M., and G. Koenig, 1997: Continuous dust concentration profile of pre-Holocene ice from the Greenland Ice Sheet Project 2 ice core: Dust stadial, interstadials, and the Eemian. *Journal of Geophysical Research*, **102**: 26641–26648.

Rind, D., 1998: Latitudinal temperature gradients and climate change. *Journal of Geophysical Research*, **103**: 5943–5971.

Rind, D., 2000: Relating paleoclimate data and past temperature gradients: Some suggestive rules. *Quaternary Science Reviews*, **19**: 381–390.

Rodbell, D. T., G. O. Seltzer, D. M. Anderson, M. B. Abbott, D. B. Enfield, and J. H. Newman, 1999: An ~~15,000-year record of El Niño–driven alluviation in southeastern Ecuador. *Science*, **283**: 516–520.

Sandweiss, D. H., J. B. Richardson, III, E. J. Reitz, H. B. Rollins, and K. A. Maasch, 1996: Geoarchaeological evidence from Peru for a 5000 years B.P. onset of El Niño. *Science*, **273**: 1531–1533.

Seltzer, G. O., 1994: A lacustrine record of late-Pleistocene climatic change in the subtropical Andes. *Boreas*, **23**: 105–111.

Sheets, P., 2000: The effects of explosive volcanism on simple to complex societies in ancient Middle America. *In* Markgraf, V. (ed.), *Interhemispheric Climate Linkages*. San Diego: Academic Press, Chapter 5.

Sowers, T., and M. Bender, 1995: Climate records covering the last deglaciation. *Science*, **269**: 210–214.

Steig, E. J., 1999: Paleoclimate: Mid-Holocene climate change. *Science*, **286**: 1485–1487.

Steig, E. J., E. J. Brook, J. W. C. White, C. M. Sucher, M. L. Bender, S. J. Lehman, D. L. Morse, E. D. Waddington, and G. D. Clow, 1998: Synchronous climate changes in Antarctica and the North Atlantic. *Science*, **282**: 92–95.

Stocker, T. F., 1998: The seesaw effect. *Science*, **282**: 61–62.

Stocker, T. F., 2000: Past and future reorganizations in the climate system. *Quaternary Science Reviews*, **19**: 301–320.

Stute, M., M. Forster, H. Frischkorn, A. Serejo, J. F. Clark, P. Schlosser, W. S. Broecker, and G. Bonani, 1995: Cooling of tropical Brazil (5°C) during the last glacial maximum. *Science*, **269**: 1000–1003.

Thompson, L. G., 1977: Variations in microparticle concentration, size distribution and elemental composition found in Camp Century, Greenland, and Byrd Station, Antarctica, deep ice cores. *In* International Symposium on Isotopes and Impurities in Snow and Ice. *International Association of Hydrological Science*, **118**: 351–364.

Thompson, L. G., E. Mosley-Thompson, and P. A. Thompson, 1992: Reconstructing interannual climate variability from tropical and subtropical ice-core records. *In* Diaz, H. F., and V. Markgraf (eds.), *El Niño: The Historical and Paleoclimatic Record of the Southern Oscillation*. Cambridge, UK: Cambridge University Press, pp. 295–322.

van der Hammen, T., and H. Hooghiemstra, 2000: Neogene and Quaternary history of vegetation, climate and plant diversity in Amazonia. *Quaternary Science Reviews*, **19**: 725–742.

Villalba, R., R. D. D'Arrigo, E. R. Cook, G. Wiles, and G. C. Jacoby, 2000: Decadal-scale climatic variability along the extratropical western coast of the Americas: Evidence from tree-ring records. *In* Markgraf, V. (ed.), *Interhemispheric Climate Linkages*. San Diego: Academic Press, Chapter 10.

Webb, T., III, W. F. Ruddiman, F. A. Street-Perrott, V. Markgraf, J. E. Kutzbach, P. J. Bartlein, H. E. Wright, and W. L. Prell, 1993: Climatic changes during the past 18,000 years: Regional syntheses, mechanisms, and causes. *In* Wright, H. E., J. E. Kutzbach, T. Webb, III, W. F. Ruddiman, F. A. Street-Perrott, and P. J. Bartlein (eds.), *Global Climates Since the Last Glacial Maximum*. Minneapolis: University of Minnesota Press, pp. 514–535.

White, W. B., and R. G. Peterson, 1996: An Antarctic circumpolar wave in surface pressure, wind, temperature and sea-ice extent. *Nature*, **380**: 699–702.

Whitlock, C., P. J. Bartlein, V. Markgraf, and A. C. Ashworth, 2000: The midlatitudes of North and South America during the last glacial maximum and early Holocene: Similar paleoclimatic sequences despite differing large-scale controls. *In* Markgraf, V. (ed.), *Interhemispheric Climate Linkages*. San Diego: Academic Press, Chapter 21.

Yu, Z., J. H. McAndrews, and U. Eicher, 1997: Middle Holocene dry climate caused by change in atmospheric circulation patterns: Evidence from lake levels and stable isotopes. *Geology*, **25**: 251–254.

Zielinski, G. A., P. A. Mayewski, L. D. Meeker, K. Grönvold, M. S. Germani, S. Whitlow, M. S. Twickler, and K. Taylor, 1997: Volcanic aerosol records and tephrochronology of the Summit, Greenland, ice cores. *Journal of Geophysical Research*, **102**: 26625–26640.

Index